HUMIC SUBSTANCES IN SOIL, SEDIMENT, AND WATER

HUMIC SUBSTANCES IN SOIL, SEDIMENT, AND WATER

Geochemistry, Isolation, and Characterization

Edited by
GEORGE R. AIKEN, DIANE M. MCKNIGHT, ROBERT L. WERSHAW
U.S. Geological Survey
Water Resources Division

and

PATRICK MACCARTHY
Department of Chemistry and Geochemistry
Colorado School of Mines

A Wiley-Interscience Publication
JOHN WILEY & SONS
New York Chichester Brisbane Toronto Singapore

Library of Congress Cataloging in Publication Data:

Main entry under title;
Humic substances in soil, sediment, and water.

"A Wiley-Interscience publication."
Bibliography: p.
Includes index.
1. Humic acid. 2. Humus. 3. Soil chemistry.
4. Sediments (Geology) 5. Water chemistry. I. Aiken, George R., 1951-
QE516.5.H86 1985 551.9 84-25696
ISBN 0-471-88274-7

Printed in the United States of America

10 9 8 7 6 5 4 3 2 1

Contributors

George R. Aiken, *U.S. Geological Survey, Federal Center, Denver, Colorado, 80225*

Deborah A. Boran, *National Oceanic and Atmospheric Administration, Atlantic Oceanographic and Meteorological Laboratory, Miami, Florida, 33149*

Irving A. Breger (*deceased*), *U.S. Geological Survey, National Center, Reston, Virginia, 22092*

Y. Debyser, *Institut Francais du Petrole, Department de Geochemie, Rueil Malmaison, France*

R. S. Farnham, *Department of Soil Science, University of Minnesota, St. Paul, Minnesota, 55108*

George R. Harvey, *National Oceanic and Atmospheric Administration, Atlantic Oceanographic and Meteorological Laboratory, Miami, Florida, 33149*

Patrick G. Hatcher, *U.S. Geological Survey, National Center, Reston, Virginia, 22092*

M. H. B. Hayes, *Department of Chemistry, The University of Birmingham, Birmingham, England*

E. W. D. Huffman, Jr., *Huffman Laboratories, Inc., Wheat Ridge, Colorado, 80034*

Ryoshi Ishiwatari, *Tokyo Metropolitan University, Department of Chemistry, Faculty of Science, Tokyo, Japan*

Jerry A. Leenheer, *U.S. Geological Survey, Federal Center, Denver, Colorado, 80225*

PATRICK MACCARTHY, *Department of Chemistry and Geochemistry, Colorado School of Mines, Golden, Colorado, 80401*

DIANE M. MCKNIGHT, *U.S. Geological Survey, Federal Center, Denver, Colorado, 80225*

GARY E. MACIEL, *Colorado State University Department of Chemistry, Fort Collins, Colorado, 80523*

RONALD L. MALCOLM, *U. S. Geological Survey, Federal Center, Denver, Colorado, 80225*

S. P. MATHUR, *Soil Research Institute, Agriculture Canada, Central Experiment Farm, Ottawa, Ontario, Canada, K1A OC6*

LAWRENCE M. MAYER, *University of Maine, Darling Center, Walpole, Maine, 04573*

UWE MUENSTER, *Max-Planck-Institute fur Limnologies, Ploen, Federal Republic of Germany*

R. PELET, *Institut Francais du Petrole, Department de Geochemie, Rueil Malmaison, France*

E. MICHAEL PERDUE, *Chemistry Department, Georgia Institute of Technology, School of Geophysical Sciences, Atlanta, Georgia, 30332*

JAMES A. RICE, *Department of Chemistry and Geochemistry, Colorado School of Mines, Golden, Colorado, 80401*

MORRIS SCHNITZER, *Chemical and Biology Research Institute, Research Branch, Agriculture Canada, Ottawa, Ontario, Canada, K1A OC6*

CORNELIUS STEELINK, *Department of Chemistry, University of Arizona, Tucson, Arizona, 85721*

CHRISTIAN STEINBERG, *Bayer Landsstodt fur Washerwirtschaft, Munchen, Federal Republic of Germany*

F. J. STEVENSON, *Department of Agronomy, University of Illinois, Urbana, Illinois, 61801*

HAROLD A. STUBER, *Huffman Laboratories, Inc., Wheat Ridge, Colorado, 80034*

ROGER S. SWIFT, *Department of Soil Science, Lincoln College, University of Canterbury, Christchurch, New Zealand*

NIKOLAUS M. SZEVERENYI, *Colorado State University, Fort Collins, Colorado, 80523*

E. M. THURMAN, *U.S. Geological Survey, Federal Center, Denver, Colorado, 80225*

M. VANDENBROUCKE, *Institut Francais du Petrole, Department de Geochemie, Rueil Malmaison, France*

ROBERT L. WERSHAW, *U.S. Geological Survey, Federal Center, Denver, Colorado, 80225*

Preface

This book provides a comprehensive and critical overview of the nature and functions of humic substances in diverse environments. Humic substances are a general class of ubiquitous, biogenic, heterogeneous, organic substances. If one climbs the proverbial mountain because it is *there,* then one studies humic substances because they are *everywhere,* and much yet remains to be learned about these substances. The study of humic substances is challenging for several reasons. In chemically characterizing humic substances one is confronted with the "mixture problem" arising from their heterogeneous nature. In geochemical investigations of humic substances major problems are encountered in quantifying processes and in physically separating humic substances from their environment with minimal introduction of artifacts. Furthermore, in the literature on humic substances, there is a surfeit of ambiguous terminology and sometimes a failure to clearly delineate what is fact and what is conjecture. It is the purpose of this book to resolve some of the indefiniteness associated with humic substances, and the theme is embodied in the phrase "what we know, what we don't know, and what we think we know" concerning the nature and functions of humic substances. The chapters are written by experts who are active researchers in the fields of their respective chapters, and who were charged with the task of preparing critical reviews with the above-mentioned theme in mind.

One of the reasons for publishing this book at this time is that many researchers from diverse scientific disciplines have recently become involved in the study of humic substances. The following environments are considered in this book: soil, peat, groundwater, lake, lake sediment, stream, estuary, marine, and marine sediment. The present book should be useful both to the novice and to the experienced researcher in humic substances. While the book should be self-sufficient, extensive documentation and literature references are provided for those who wish to find more detailed information on a given topic. We believe that the present volume will serve as a textbook for a graduate course on humic substances as well as being a useful reference book. The book may be studied as a whole to provide a comprehensive overview of the various aspects of humic substances, or individual chapters may be consulted to provide in-depth treatment of a particular aspect. For the benefit of the reader the individual chapters are extensively cross-referenced. The book is organized into three sections dealing with Geochemistry, Isolation and Fractionation, and Characterization, respectively, of humic substances. A follow-up book to this one will discuss the structural nature of humic substances and their interactions with metal ions and organic chemicals.

Due to the multidisciplinary nature of this book a given reader is likely to encounter unfamiliar terms or definitions that are not completely clear. For this reason we have compiled a glossary of terms which constitutes Appendix A. A Glossary of Chemical Compounds is given in Appendix B.

GEORGE R. AIKEN
DIANE M. MCKNIGHT
ROBERT L. WERSHAW
PATRICK MACCARTHY

Denver, Colorado
Golden, Colorado
February 1985

Acknowledgments and Disclaimers

The U.S. Geological Survey of the United States Department of the Interior has provided critical support to the first meeting of the International Humic Substances Society, and to the other undertakings of the new society, such as collection of the standard aquatic humic substance sample. This publication was made possible, in part, by a grant to the International Humic Substances Society from the U.S. Geological Survey, Department of the Interior, under U.S.G.S. Agreement No. 14-08-0001-G-717.

We are most grateful to all contributing authors for their diligence and cooperation in helping this book become a reality. Dr. I. A. Breger, coauthor on Chapter 10, died during the preparation of this work and we wish to acknowledge his prior contributions to the science of humic substances. Christine Miller provided invaluable help in compiling the glossary; Mark Brenner contributed by compiling the reference section; and Leah Wilson provided critical editorial assistance. Many individuals gave generously of their time and expertise in reviewing individual chapters and we greatly appreciate their contribution. Mary Conway, Life Sciences Editor at Wiley, provided editorial guidance and advice in a most gracious and professional manner. The assistance of Barbara Kemp, Barbara Macklin, Terry Ripsom, and Debbie Sites in typing manuscripts is also appreciated. These few brief lines cannot do justice to our spouses, Ellen Aiken, Larry Espositio, Esther Wershaw, and Helen MacCarthy for the many hours and days borrowed from family time, necessitated in the editing of this book. This work is equally a product of their patience, encouragement, and understanding.

In any area so ill-defined as humic substances there are, of necessity, differences of opinion. While the editors may not agree with all the opinions expressed by the individual authors, it is important that these various viewpoints be provided a forum for expression. The use of trade names or company names in any of the chapters of this book is for identification purposes only and is not intended to imply endorsement by the authors, editors, or their respective employers or sponsoring institutions.

G.R.A.
D.M.M.
R.L.W.
P.MacC.

Contents

1. **An Introduction to Humic Substances in Soil, Sediment, and Water** 1
George R. Aiken, Diane M. McKnight, Robert L. Wershaw, and Patrick MacCarthy

SECTION I. GEOCHEMISTRY OF HUMIC SUBSTANCES

2. **Geochemistry of Soil Humic Substances** 13
F. J. Stevenson

3. **Geochemistry of Humic Substances in Natural and Cultivated Peatlands** 53
S. P. Mathur and R. S. Farnham

4. **Humic Substances in Groundwater** 87
E. M. Thurman

5. **Geochemistry and Ecological Role of Humic Substances in Lakewater** 105
Christian Steinberg and Uwe Muenster

6. **Geochemistry of Humic Substances in Lake Sediments** 147
Ryoshi Ishiwatari

7. **Geochemistry of Stream Fulvic and Humic Substances** 181
Ronald L. Malcolm

8. **Geochemistry of Humic Substances in Estuarine Environments** 211
Lawrence M. Mayer

9. **Geochemistry of Humic Substances in Seawater** 233
George R. Harvey and Deborah A. Boran

10. **Geochemistry of Humic Substances in Marine Sediments** 249
M. Vandenbroucke, R. Pelet, and Y. Debyser

11. **Geochemistry of Humin** 275
Patrick G. Hatcher, Irving A. Breger, Gary E. Maciel and Nikolaus M. Szeverenyi

12. **Nature of Nitrogen in Humic Substances** 303
Morris Schnitzer

SECTION II. ISOLATION AND FRACTIONATION OF HUMIC SUBSTANCES

13. **Extraction of Humic Substances from Soil** 329
M. H. B. Hayes

14. **Isolation and Concentration Techniques for Aquatic Humic Substances** 363
George R. Aiken

15. **Fractionation of Soil Humic Substances** 387
Roger S. Swift

16. **Fractionation Techniques for Aquatic Humic Substances** 409
Jerry A. Leenheer

SECTION III. CHARACTERIZATION OF HUMIC SUBSTANCES

17. **Analytical Methodology for Elemental Analysis of Humic Substances** 433
E. W. D. Huffman, Jr. and Harold A. Stuber

18. **Implications of Elemental Characteristics of Humic Substances** 457
Cornelius Steelink

19. Molecular Size and Weight Measurements of Humic Substances 477
Robert L. Wershaw and George R. Aiken

20. Acidic Functional Groups of Humic Substances 493
E. Michael Perdue

21. Spectroscopic Methods (Other than NMR) for Determining Functionality in Humic Substances 527
Patrick MacCarthy and James A. Rice

22. Application of Nuclear Magnetic Resonance Spectroscopy for Determining Functionality in Humic Substances 561
Robert L. Wershaw

Concluding Remarks 583

Bibliography 585

Appendix A. Glossary of Terms 643

Appendix B. Glossary of Chemical Compounds 661

Appendix C. General References 671

Index 675

HUMIC SUBSTANCES IN SOIL, SEDIMENT, AND WATER

CHAPTER ONE

An Introduction to Humic Substances in Soil, Sediment, and Water

GEORGE R. AIKEN, DIANE M. McKNIGHT, ROBERT L. WERSHAW, and PATRICK MacCARTHY

Humic substances comprise a general class of biogenic, refractory, yellow-black organic substances that are ubiquitous, occurring in all terrestrial and aquatic environments. These humic substances constitute the major organic fraction in soils and have been studied by soil scientists for two centuries (Chapter 2). In fact, the essence of the most commonly employed method for extracting humic substances from soils is found in the works of Achard (1786). These complex materials have occupied the attention of classical chemists such as Liebig, Berzelius, and Wallerius, and a host of other scientists up to the present time. To a large extent, the study of humic substances has been dominated by soil scientists, and most investigations have involved samples that were extracted from soil or peat. Historically, the conclusions reached by soil scientists as to the nature of humic substances have strongly influenced the thinking of other scientists interested in the role of humic substances in other contexts, such as in the formation of kerogen or in the chemistry of trace metals in the aquatic environment.

In recent years, two major trends in the study of humic substances have been evident. First, there has been an escalating interest in humic substances by scientists from a wide diversity of disciplines outside the realm of soil science. Many papers are presently being published by chemists, geochemists, biologists, environmental scientists, and medical researchers on the nature, role, and applications of humic substances. Second, a trend not unrelated to the first is the increased study of humic substances isolated from environments other than soil. In particular, *aquatic* humic substances have been investigated extensively in recent years partially because of the potential health effects arising from the chlorination of municipal drinking water containing low concentrations of humic substances (Bellar et al., 1974; Rook, 1974; U.S. EPA, 1974; Christman et al., 1983).

Compared to the isolation of humic substances from soil or peat, the extraction of reasonable quantities of humic substances from aquatic sources poses severe difficulties due to their very low concentrations in natural waters. However, in recent years significant advances have been made in the technology of isolating humic substances from aquatic environments as discussed by Aiken in Chapter 14.

In addition to their presence and role in diverse environments, humic substances are important because they serve as a major reservoir of organic carbon in soils and the oceans for the global carbon cycle. The global cycling of organic carbon is represented schematically in Figure 1. Table 1, taken from Woodwell et al. (1978) and Woodwell and Houghton (1977), compares the quantity of carbon present in different forms in the biosphere and lithosphere. There are major uncertainties in the quantities of humic carbon in soils and in the rates of degradation of humic substances in soils and in the oceans. For example, not only do estimates for carbon in soil humus range

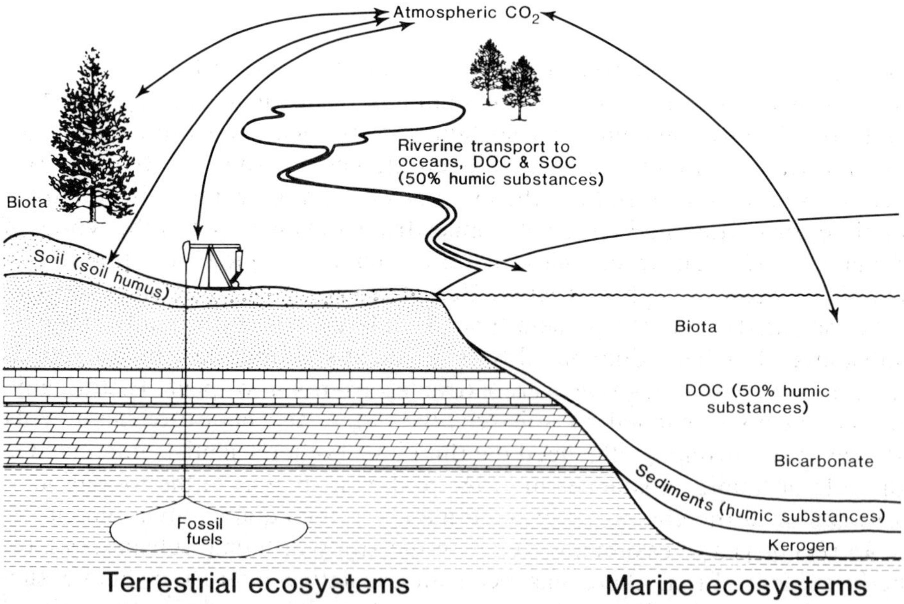

FIGURE 1. Diagram of the global carbon cycle, indicating the importance of humic substances.

TABLE 1. Estimated Size of Major Pools of Carbon in the World Carbon Budget

	$\times 10^{15}$ g C
Atmosphere	638–702
Land	
Biota (plants)	827
Soil humus	700–3000
Oceans	
Biota	2
Organic (50% humic substances)	1700
Inorganic	38,600
Fossil Fuels	10,000

Source. *Woodwell and Houghton (1977) and Woodwell et al. (1978).*

from 700 $\times$ 10^{15} to 3000 $\times$ 10^{15} g C, but the extent to which deforestation and agricultural use of wetlands accelerate the degradation of soil humus is not known. The amount of carbon released to the atmosphere by accelerated degradation of soil humus might be as great as the amount of carbon released by direct harvest of forests (Woodwell et al., 1978). Another example of the uncertainty and the importance of humic substances in the global carbon budget is the question of the origin of the dissolved humic substances in the oceans. Studies of the transport of humic substances in estuaries lead to the conclusion that as much as 50% of the oceanic DOC originated in freshwater and terrestrial environments (Mantoura and Woodward, 1983; and Chapter 8 in this book by Mayer). However, characterization of humic substances isolated from open ocean environments indicates a primarily autochthonous source (Chapter 9 by Harvey and Boran). Clearly, a better understanding of the geochemistry of humic substances will contribute to a better understanding of the world carbon budget.

Although different aspects of humic substances have been carefully studied by many scientists for many years, this work has not led to a fundamental, or even generally accepted, understanding of the nature, origin, and geochemical role of humic substances. In fact, after reviewing the history of this subject, it is difficult to categorize any single discovery or development in humic substance research as constituting a major breakthrough. Information on the nature of humic substances and on their geochemical implications has been steadily accumulating over the years. However, these data yet remain to be integrated within a clear conceptual framework, and, at present, all discussions relating to humic substances are enshrouded, to a considerable degree, in a cloud of vagueness. This, of itself, hinders communication on the subject. Some important questions relating to humic substances are: How exactly are humic substances formed? What is the precise chemical nature of humic substances? What are the detailed mechanisms of their formation in the natural environment? There are many intriguing hypotheses, but no definitive answers to these long-standing fundamental questions. For this reason, the study of humic substances is a bona fide academic exercise per se and does not *necessarily* require practical applications or geochemical significance to lend it justification.

PURPOSE OF THE BOOK

The purpose of this book is to bring together "what we know, what we don't know, and what we think we know" concerning the geochemistry of humic substances in natural environments and the isolation, fractionation, and characterization of these substances. The challenge accepted by the contributors to this book was to present a critical and definitive review of their respective topics, and it is our expectation that the reader of this book will find that this challenge has been met to a significant extent. However, there are many unknowns and uncertainties in the study of humic substances, which lead to differences of opinion among scientists. The editors of this book do not agree with all the opinions expressed in different chapters of this book, and in editing this book an effort has been made to identify opinions and speculations as such.

DEFINITION OF HUMIC SUBSTANCES

One of the first difficulties encountered in the study of humic substances relates to terminology. This is, perhaps, symptomatic of the subject area as a whole in that it is difficult to define precisely that which is known only vaguely. Humic substances do not correspond to a unique chemical entity and, accordingly, cannot be described in unambiguous structural terms. Similarly, humic substances are not biologically predestined to carry out a specific biochemical action and thus cannot be defined in functional terms. As a result, humic substances must be defined operationally. This has led to a plethora of definitions throughout the history of this subject and much yet remains to be done in the standardization of terminology relating the precise usage of the various terms by workers in different areas. With this in mind, we have striven for the adoption of uniform and consistent terminology throughout this book. A number of terms are so important to the present book that we include their definitions here:

Humic Substances. A general category of naturally occurring, biogenic, heterogeneous organic substances that can generally be characterized as being yellow to black in color, of high molecular weight, and refractory.

There are three major fractions of humic substances referred to throughout this book, and these fractions, once isolated from the environment, are operationally defined in terms of their solubilities.

Humin. That fraction of humic substances that is not soluble in water at any pH value.

Humic Acid. That fraction of humic substances that is not soluble in water under acid conditions (below pH 2), but becomes soluble at greater pH.

Fulvic Acid. That fraction of humic substances that is soluble under all pH conditions.

These definitions have survived the test of time because of their practical utility. Water solubility is an effective criterion because of its dependence on important chemical characteristics, such as acidic functional group content, molecular weight, aromaticity, and so on. Other terms that may not be familiar to scientists outside a particular discipline are defined in Glossary of Terms (Appendix A).

Publications on humic substances are dominated by discussions on humic acids and fulvic acids, with relatively little discussion of humin. The former two fractions can be dissolved in aqueous media which facilitates their isolation and study. The geochemistry of humin is discussed by Hatcher et al. in Chapter 11; the presence and nature of humin in various environments are also discussed in a number of other chapters in this book. For example, Stevenson (Chapter 2) discusses humin from soils, Ishiwatari (Chapter 6) provides a rather extensive discussion of humin from lake sediments, and Vandenbroucke et al. (Chapter 10) consider humin in marine sediments.

ORGANIZATION OF THE BOOK

There are three main sections in this book: Geochemistry, Isolation and Fractionation, and Characterization. Each section is organized into a sequence of discrete topics that comprehensively cover the general subject.

Geochemistry of Humic Substances

In the first section of the book, the geochemistry of humic substances in different environments is presented in a sequence going from soils to marine sediments. The environments considered are: soil, peat, groundwater, lake, lake sediment, stream, estuary, marine, and marine sediment. In editing the book, we have been constrained to arrange and discuss the different environments in a linear sequence. Figure 2 presents a schematic diagram of the many possible environmental flowpaths that might be followed by a humic substance. An examination of Figure 2 shows that it actually corresponds to

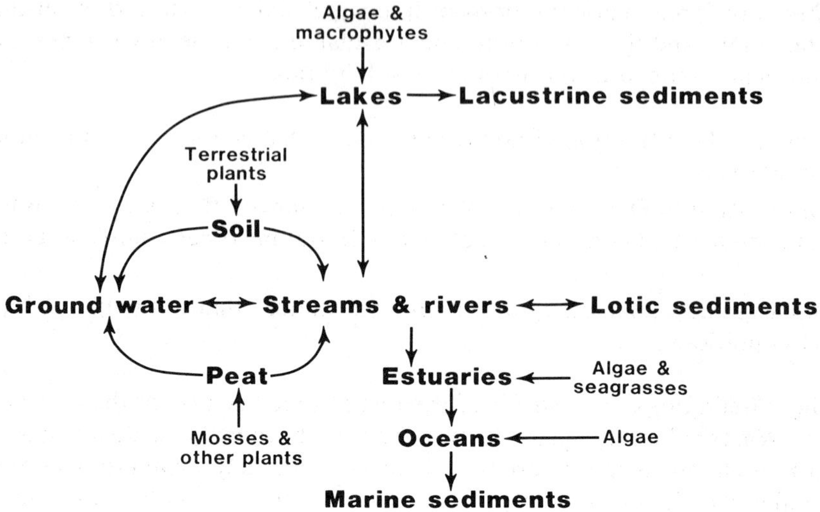

FIGURE 2. Diagram of the many possible environmental flowpaths of humic substances.

the various paths taken by rainwater, fallen on land, in its journey to the ocean. Figure 2 further indicates that all humic substances are interconnected through the medium of water, and that it is primarily by means of water that these substances are transported in the environment.

In comparing these different environments, one also becomes aware of the tremendous range in environmental conditions (particulate versus dissolved, oxic versus anoxic, etc.) and in the length of time humic substances remain in an environment (hundreds of years in soil or deep aquifers compared to only weeks or months in the surface waters of lakes, streams, and estuaries). One must be equally aware of the fact that, in reality, the lithosphere is not compartmentalized into the convenient categories represented by the various chapters in this section of the book. Rather, each environment is part of an overall system where each overlaps and interacts with neighboring environments. This interaction is dramatized in the case of estuaries where the transition from a freshwater environment to a marine environment is pronounced.

The relationship of humin to kerogen, and the role of these substances as precursors to coal and petroleum are discussed by Hatcher et al. in Chapter 11. Schnitzer discusses the nature of nitrogen in humic substances in the last chapter (Chapter 12) of the geochemistry section.

Isolation and Fractionation of Humic Substances

The second section of the book presents ''state-of-the-art'' information on the isolation and fractionation of humic substances from soils and water. The methodologies for isolating and fractionating humic substances from sedimentary samples are not discussed specifically, but are generally similar to those for soil samples, as both classes are of a bulk solid nature. Many advances in the understanding of humic substances have come from improved methodology for isolation and fractionation, and this trend can be anticipated to continue. Although there is no single *best* method to extract humic substances from soil, sediment, or water, there is a need for a *standard* method with which other methods can be compared. Hopefully, the extraction procedures used for the standard collection of humic substances (discussed later in this chapter) will be adopted as a standard method in the future.

Characterization of Humic Substances

The final section of the book describes the application of different methods for chemically characterizing humic substances. The authors present detailed discussions of the strengths and weaknesses in applying these characterization tools, which range from very basic methods, such as elemental analysis, to recently developed methods, such as solid state nuclear magnetic resonance spectroscopy. In discussing these methods, the focus has been on characterization of functionality, which includes acidic functional groups (primarily carboxylic and phenolic), carbonyl groups, hydroxyl groups, and so on. In addition to functional groups, general chemical features, such as aliphatic or aromatic character, are also considered in this section. A more detailed discussion of the structural features of humic substances will be presented in a follow-up book to this, tentatively entitled, *Humic Substances: Structure and Interactions*.

The fact that a humic substance is not a pure compound, but is a heterogenous mixture of many compounds with generally similar chemical properties, places an important constraint on all these characterization methods. Examples of the "multicomponent mixture" problem are presented in the chapters discussing interpretation of elemental analysis, determination of molecular weight, analysis of potentiometric data, and interpretation of infrared and other spectroscopic data.

A conclusion that can be reached from the geochemistry and characterization sections is that it is extremely important to characterize samples from a range of environments. The disparity in the sources of humic substances, and in the geochemical processes controlling their transport and alteration, may be expected to lead to major differences in humic substances from different environments. It is also important to characterize humic substances from many seemingly very similar environments to see if the variation in chemical characteristics in this sample set is indeed less than in a sample set that includes more diverse environments. At this point in the study of humic substances, significant geochemical insight can probably be gained from well-designed descriptive studies.

STANDARD AND REFERENCE COLLECTIONS OF HUMIC SUBSTANCES

As mentioned above, humic substances do not correspond to a unique chemical compound, or even to a unique mixture of compounds, but are defined in operational terms on the basis of their aqueous solubility as a function of pH. Consequently, humic substances isolated from different sources are different, and vary as a function of many factors such as the nature of the soil, sediment, or water source, climatic and botanic environment, and depth of burial. This problem is compounded by the fact that, of necessity, one must adopt somewhat different extractive methodologies for bulk solid samples, such as soils or sediments, compared to extraction from aquatic sources (see Chapter 13 by Hayes and Chapter 14 by Aiken). However, even for a given type of substrate, for example, soil, there is a very wide range of extractive techniques extant in the literature. These techniques differ in terms of the pH values of the extractants, the nature of the chemicals used to prepare the extracting solutions, the presence or absence of complexing agents in the extractant (see discussion by Stevenson in Chapter 2), temperature and duration of extraction, and other parameters. The fractionation, "purification," and other handling procedures applied to the extracted humic substances also vary widely from worker to worker (see Chapter 15 by Swift and Chapter 16 by Leenheer). The variability in humic substances, resulting from differences in source materials and extractive and fractionation techniques, makes the interlaboratory comparison of the experimental results of various researchers difficult. In brief, when inconsistent or conflicting results are reported by various researchers it is not possible, in general, to ascertain whether these discrepancies arise from differences in the humic substance samples, from differences in the experimental approaches of the workers, or from errors on behalf of one or more of the investigators. This undesirable situation is further highlighted by the fact that there is no such thing as *an analytical method for humic substances,* as is evident from reading Section 3 of this book.

These problems were addressed in a paper published by MacCarthy (1976) wherein it was proposed that an international reference collection of humic substances be established. The potential benefits that would result from such a collection were outlined in the paper. That proposal was further promoted by MacCarthy and Malcolm (MacCarthy and Malcolm, 1979; Malcolm and MacCarthy, 1979), and at the Congress of the International Society of Soil Science in 1978 (MacCarthy and Malcolm, 1978) a special working group was formed to examine the possibility of establishing an international standard collection of humic substances. In 1981, an international group of scientists met in Denver, Colorado and formulated a plan for generating a suite of standard humic substances which would be available to researchers worldwide. It was decided to extract and "purify" humic and fulvic acids from the soil, peat, leonardite, and aquatic sources by a carefully controlled extraction procedure. These materials will constitute the standard humic substances. Other samples extracted by a variety of extractive procedures will be referred to as *reference* (as distinct from *standard*) materials. The preparation of these standard and reference materials is presently underway and a detailed report will be published upon its completion. The International Humic Substances Society evolved from these activities as an organization for the promotion and advancement of research on humic substances, and it is this Society which is responsible for preparing standard and reference samples of humic and fulvic acids. It is hoped that these standard and reference samples will contribute to solving some of the numerous problems in humic substances research which are outlined in the following chapters of this book.

SECTION ONE

GEOCHEMISTRY OF HUMIC SUBSTANCES

CHAPTER TWO

Geochemistry of Soil Humic Substances

F. J. STEVENSON

ABSTRACT

Past and present concepts of the nature and origin of humic substances are discussed in terms of the structural chemistry of humic acids and relationships between the various humus fractions. Major consideration is given to the interactions of humic substances with clay minerals and trace elements, to sorption reactions involving pesticides and related organic compounds, and to the role of humic substances in rock weathering and the translocation of mineral matter. Other topics of interest include the contribution of humic substances to the cation exchange capacity of the soil, speciation of trace metals in the soil solution, chemical interactions involving pesticides and humic substances, and radiocarbon dating of soil humus.

Humic substances represent a complex mixture of molecules of various sizes and shapes but no completely satisfactory scheme has been forthcoming for their extraction and purification. Major emphasis needs to be given to the development of procedures leading to the isolation of molecularly homogeneous fractions free of inorganic and organic impurities. Other areas requiring further study are: (1) genetic relationships between humic substances from diverse sources, (2) nature of the building blocks of humic substances, (3) structural arrangement of reactive functional groups, and (4) mechanisms whereby humic substances combine with trace elements and clay minerals.

INTRODUCTION

Humic acids and related pigments, collectively referred to as humic substances, are widely distributed in soils, natural waters, marine and lake sediments, peat, carbonaceous shales, lignites, brown coals, and miscellaneous other deposits. These constituents are best described as a series of acidic, yellow-to-black-colored polyelectrolytes that have properties dissimilar to the biocolloids of living organisms. The current view is that they represent an extremely heterogeneous mixture of molecules, which, in any given soil or sediment, may range in molecular weight from as low as several hundred to perhaps over 300,000 (Dubach and Mehta, 1963; Flaig et al., 1975, Hayes and Swift, 1978; Schnitzer, 1978; Stevenson, 1982).

Humic substances in soil have properties similar to those found in other natural systems. Accordingly, results obtained by the soil scientist on the nature, origin, and reactions of these complex substances are of considerable interest to researchers in several disciplines of science. In this chapter emphasis is given to historical aspects of research on soil humic substances, to the interactions of humic and fulvic acids with soil mineral components, and to the participation of humic substances in geochemical processes.

HISTORICAL BACKGROUND

An exhaustive review of research on soil humus is not possible and only the main points can be discussed. For more detailed information, the reviews of Waksman (1936) and Kononova (1966) are recommended.

Development of Extraction and Fractionation Procedures

The first attempt to isolate humic substances from soil appears to have been made by Achard (1786), who extracted peat with alkali and obtained a dark, amorphous precipitate upon acidification. This alkali-soluble, acid-insoluble material became known by a number of names, of which the term humic acid survived.

Saussure (1804) is usually credited with introducing the term *humus* (Latin equivalent of soil) to describe the dark-colored organic matter in soil. Somewhat later, Döbereiner (1822) designated the dark-colored component of soil organic matter as *Humussäure* or *humus acid*. The origin of the term **humic acid** to identify the alkali-soluble, acid-insoluble fraction is somewhat obscure but it was in common use by the time of Berzelius (1839). Waksman (1936) pointed out that humus acid and humic acid were often used indiscriminately and seldom was any differentiation made between the two; when distinguished, the former referred to the alkali extracted material, whereas the latter referred to the precipitate obtained by acidification. Use of the term *humin* to describe the alkali-insoluble material had a similar development and was often used synonymously with *humus coal*. A summary of names suggested by various investigators for the various humic fractions is given in Table 1.

One of the first comprehensive studies of the chemical nature of humic substances was carried out by Sprengel (1826, 1837). The procedures he developed for the preparation of humic acids became generally adopted, such as pretreatment of the soil with dilute mineral acids to enhance extraction with alkali. Sprengel concluded that, for soils rich in bases, humic acid was in a bound form; consequently, the soil had a neutral reaction (contained "mild humus"). This soil was regarded as highly fertile. On the other hand, for soils poor in bases, the humic acid was believed to be in the free form, with the result that the soil was acid and unproductive (contained "acid humus"). A major contribution of Sprengel to humus chemistry was his extensive studies on the acidic nature of humic acids.

Research on the chemical properties of humic substances was extended by the Swedish investigator Berzelius (1839). One of his main contributions was the isolation of two light-yellow-colored humic substances from mineral waters and a slimy mud rich in iron oxides. They were obtained from the mud by extraction with base (KOH), which was then treated with acetic acid containing copper acetate. A brown precipitate was obtained called "copper apocrenate." When the extract was neutralized, another precipitate was obtained, called "copper crenate." The free acids, *apocrenic* and *crenic acids*, were then brought into solution by decomposition of the copper complexes with alkali. These newly described humic substances were examined in considerable detail, including isolation, elementary composition, and properties of their metal complexes (Al, Fe, Cu, Pb, Mn, etc).

The investigations of Berzelius were developed further by his contemporaries and former students, especially Mulder (1862), who classified humic substances on the bases of solubility and color into the following groups: (1) ulmin or humin—insoluble in alkali; (2) ulmic acid (brown) and humic acid (black)—soluble in alkali; and (3) crenic and apocrenic acids—soluble in water. Several new preparations were also obtained (see Table 1), for which various names were given (glucic acid, apoglucic acid, chlorhumic acid, etc). These terms were later abandoned. Like many of his contemporaries, Mulder was of the opinion that the different humic fractions were chemically individual compounds. He also felt that humic substances were free of nitrogen.

The view of Mulder and others that humic acids were nitrogen-free was strongly criticized by the Russian investigator German (cited by Kononova, 1966), as well as by Hermann (1845). Hermann felt that all plants adsorbed nitrogen from the air and that even artificially prepared humic substances could adsorb nitrogen. A large number of preparations were obtained by Hermann from soil, peat, coal, and artificial preparations, for which various names were given (see Table 1). Hermann concluded that, when humic acids were oxidized to crenic and apocrenic acids, more atmospheric nitrogen was

TABLE 1. Names Suggested for the Common Humus Fractions Obtained on the Basis of Solubility Characteristics[a]

Current Designation	Solubility Characteristics	Sprengel (1826)	Berzelius (1839)	Mulder (1840)	Hoppe-Seyler (1889)	Oden (1919)	Springer (1938)
Humic acid	Soluble in alkali, precipitated by acid	Humus acid	Humic acid	Humic acid	Humic acid	Humus acid	Humic acid
Brown humic acid	Not coagulated from alkali solution in the presence of electrolyte	—	—	—	—	—	Braunhumiñsäure
Gray humic acid	Coagulated in the presence of electrolyte	—	—	—	—	—	Grauhuminsäure
Fulvic acid	Soluble in alkali, not precipitated by acid	—	Crenic acid, apocrenic acid	Glucic acid, apocrenic acid	—	Fulvic acid	—
Hymatomelanic acid	Soluble in alkali, precipitated by acid, soluble in alcohol	—	—	—	Hymatomelanic acid	Hymatomelanic acid	—
Humin	Insoluble in alkali	Humus coal	Humin	Humin, ulmin	Humin	Humus coal	—

[a] *Adapted from Waksman (1936).*

adsorbed. As Waksman (1936) pointed out, the nitrogen in his preparations (as well as those of other investigators) could have originated from the nitric acid or ammonium used for their preparation. The well-known ability of humic substances to combine with ammonia under alkaline conditions was actually demonstrated in several early studies (e.g., Schloesing, 1876). Also, the fact that oxygen is consumed by humic acid in the presence of alkali was demonstrated (e.g., Mulder, 1862).

The later part of the 18th century was characterized by the proliferation of classification schemes for the isolation of new substances from decomposing plant residues, from soil, and from artificial mixtures generated in the laboratory. From this and the earlier work, it became firmly established that humus was a complex mixture of organic substances that were mostly colloidal in nature and that had weakly acidic properties. Information had also been obtained on their interactions with other soil components. As Kononova (1966) pointed out, each investigator repeated the mistakes of his predecessor by regarding the substance isolated as a chemically individual compound of specific composition. However, as the century came to a close, more and more investigators (see review by Waksman, 1936) came to doubt the validity of chemical formulas attached to the various preparation. Also, the concept that synthetic laboratory products were similar to those occurring naturally came under considerable criticism. Of the many names proposed during this period, only the *hymatomelanic acid* of Hoppe-Seyler (1889) survived. This substance is the fraction obtained from humic acid by extraction with alcohol.

Renewed efforts were made early in the 20th century to classify humic substances and to determine their chemical nature and structure. Among the more important contributions are those of Oden (1914, 1919), who classified humic substances into the following groups: humus coal, humic acid, hymatomelanic acid, and fulvic acid. He appears to have been the first to use the term fulvic acid to describe the alkali-soluble, acid-soluble fraction of humus (e.g., see Waksman, 1936). The term humic acid was reserved for the dark-brown to black material that was soluble in alkali but insoluble in acid, and that had a carbon content of about 58%. Hymatomelanic acid, earlier named by Hoppe-Seyler, was believed by Oden to be formed from humic acid during extraction with alkali. This material was lighter in color than humic acid (chocolate brown) and had a higher carbon content (about 62%). The yellow-brown fulvic acids, which were recovered from marsh waters by ultrafiltration, were recognized as being similar to the crenic and apocrenic acids of Berzelius. Oden believed that humus coal and fulvic acids were group designations while humic acid and hymatomelanic acids were chemically individual compounds. Oden confirmed the acidic nature of humic acids by means of potentiometric measurements.

Although workers of earlier periods recognized the occurrence of specific organic compounds in soil (e.g., Berzelius and Mulder), attention had been focused almost exclusively on the so-called humus substances. Starting in

1908, Schreiner and Shorey (1910) initiated studies on identifiable organic chemicals in soil. Over the following three decades, they reported the occurrence in soils of over 40 compounds belonging to the well-known classes of organic chemistry, including organic acids, hydrocarbons, fats, sterols, aldehydes, carbohydrates, and specific nitrogen-containing substances. Their studies on the toxic effects of organic compounds on plants attracted wide attention during this and subsequent periods.

A detailed study of the nature and structure of humic substances was carried out by Shmuk who, in his final work in 1930, reviewed the more important aspects of humus chemistry and the formation of humic substances by soil microorganisms. He regarded humic acid not as a specific compound but as a mixture of closely related substances having similar structural features. By esterification of humic acids with alcohol in the presence of dry HCl, Shmuk (1930) demonstrated the occurrence of esters, thereby indicating the presence of COOH groups. Nitrogen was regarded as a structural component, and in this respect his views differed from those of Oden and others. The source of nitrogen was believed to be the proteins of microorganisms. Shmuk concluded that humic acids contained two major components: an organic nitrogen-containing compound (protein) and the aromatic ring, the latter being derived from lignin. He described humic acid as "a nitrogenous body of an acid nature, the acidity being due to both its power of adsorption, as a result of the colloidal condition of the humic acid, and to the presence of COOH groups."

The origin of humus was also a popular subject for research during this period. Two main theories prevailed, one being that they were derived from the lignified tissues of plant residues and the second that they were formed from simple sugars. With regard to the latter, Maillard's (1916) work deserves special consideration. He conducted extensive studies on the production of dark-colored, humic-like substances (melanins) from mixtures of reducing sugars and amino acids, which he properly attributed to the interaction of the C=O group of the sugar with the NH_2 group of the amino acid, with elimination of water and formation of CO_2 from the amino acid COOH group (through decarboxylation). The reaction between reducing sugars and amino acids or amines to form brown nitrogenous polymers has since been shown to be of great significance in the commercial dehydration of food products and is commonly referred to as the "Maillard reaction."

Maillard's concept that humic substances represented products formed by chemical reactions of simple compounds is in close agreement with some modern views of humus formation, as described later. However, Trusov's concept of humification (reviewed by Kononova, 1966) comes even closer to present-day concepts. In a series of studies on the biochemistry of humus formation, Trusov postulated the following sequence in the humification process: (1) hydrolytic decomposition of plant remains with the synthesis of simple substances of an aromatic nature; (2) oxidation of the latter by microbial enzymes to form hydroxyquinones; and (3) condensation of the quinones into dark-colored products (humic substances).

V. R. Williams (1914), a well-known Russian investigator, also postulated the existence of two stages in the humification process, the first being the decomposition of the original plant residues to simpler compounds, and the second being the synthesis of substances of a more complex nature. In contrast to the views of Maillard, both processes were believed to result from the enzymatic activity of microorganisms.

The view that lignin was the precursor of humic acids was further advanced by Hobson and Page (1932), Waksman (1936), and others. The theory became generally adopted and had a dominating influence on humus chemistry for several decades. Waksman (1936) postulated that proteins combined with modified lignins through the Schiff reaction, as follows:

$$\begin{array}{l} C_{52}H_{46}O_{10}(OCH_3)COOH(OH)_4CO + H_2NRCOOH \rightarrow \\ C_{52}H_{46}O_{10}(OCH_3)COOH(OH)_4C{=}NRCOOH \leftarrow \quad H_2O \end{array} \qquad (1)$$

Concurrent investigations into the chemistry of brown coals and their humic acids, such as those carried out by the German scientist Fuchs (1930, 1931), had a great influence on the study of soil humus. Many of the techniques that were applied to coal humic acids, including the determination of COOH and phenolic OH groups, had general applicability to the study of soil humic substances. Fuch's scheme for the structure of humic acid (Fig. 1) has been widely quoted in the soils literature.

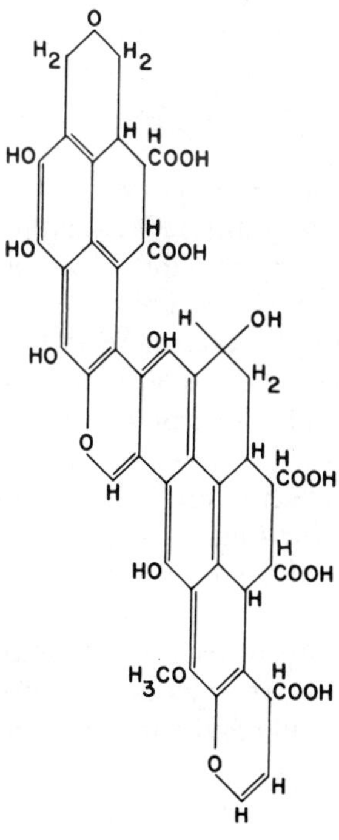

FIGURE 1. Structure of humic acid according to Fuchs.

Finally, toward the end of the premodern era, additional fractions were added to the list of substances that could be separated from humus. On the basis of their behavior toward electrolytes, Springer (1938) subdivided humic acids into Braunhuminsäure (brown humic acid) and Grauhuminsäure (gray humic acid). This was accomplished by redissolving the original humic acid in alkali and adding electrolyte (KCl) to a final concentration of 0.1*N*. The gray humic acids were easily coagulated, had relatively high carbon contents, and exhibited a low degree of dispersion.

At an even earlier date, Waksman (1936) isolated a separate component from the fulvic acid fraction. When the filtrate from the separation of humic acid was adjusted to pH 4.8 another precipitate was formed. This was designated as the *β-fraction* of humus, or the "neutralization fraction" of Hobson and Page (1932). This material was rich in aluminum and was considered to be an "Al-humate." No mention is made of the β-fraction in the book by Kononova (1966).

From time to time, questions arose as to the validity of classification schemes for identifying components of soil organic matter. Page (1930) suggested that the term "humus" be discontinued altogether because of the widely different meanings attached to the word. The term "humic matter" was proposed to describe the dark-colored, high-molecular-weight organic colloids; "nonhumic matter" was suggested for the colorless organic substances resulting from the biological decomposition of plant and animal residues, such as waxes and cellulose. The terms proposed by Page are somewhat analogous to the present-day usage of *humic and nonhumic* substances to designate the two main classes of organic compounds in soil.

Waksman (1936) recommended abandonment of "the whole nomenclature of 'humic acids,' beginning with 'humins' and 'ulmins,' through the whole series of 'humus,' 'hymatomelanic,' 'crenic,' 'apocrenic,' and numerous other acids, and ending with the 'fulvic acid' and 'humal acids'." Notwithstanding, terms such as humic acid, humin, fulvic acids, and others have survived and will undoubtedly continue to be used in the future. Most studies on humus chemistry involve preliminary separations on the basis of solubility characteristics, and abandonment of these terms would cause even greater confusion than their continued use. For example, reference to the alkali-soluble, acid-insoluble material as humic acid is considerably less cumbersome than repeated reference to the "alkali-soluble, acid-insoluble fraction."

With regard to the designation of *humin* as a separate fraction, it is possible that this material consists of portions of other fractions so intimately associated with mineral matter that they cannot be solubilized by extraction with alkali. Also, it is not known whether *hymatomelanic* acid is a distinct chemical entity. This material may be an artifact produced from humic acid during fractionation. The simple process of redissolving the alcohol-insoluble material in alkali followed by reprecipitation with acid results in a further increase in alcohol-soluble material.

The *fulvic acid fraction* has a straw-yellow color at low pH values and turns to wine-red at high pH values, passing through an orange color at a pH near 3.0. There is little doubt that compounds of a nonhumic nature are present. The term *fulvic acid* should be reserved as a class name for the pigmented components of the acid-soluble fraction.

Modern Views of the Origin and Nature of Humic Substances

Biochemistry of Humus Formation

By the mid-20th century, the lignin-protein theory of humus formation, popularized only two decades earlier by Waksman (1936) and others, had been rejected by most investigators. The view gradually became accepted that humic and fulvic acids were formed by a multistage process that included: (1) decomposition of all plant components, including lignin, into simpler monomers, (2) metabolism of the monomers with an accompanying increase in the soil biomass, (3) repeated recycling of the biomass carbon (and nitrogen) with synthesis of new cells, and (4) concurrent polymerization of reactive monomers into high-molecular-weight polymers (Flaig, 1966b; Kononova, 1966). The general consensus was that polyphenols (quinones) synthesized by microorganisms, together with those liberated from lignin, polymerized alone or in the presence of amino compounds (amino acids, etc.) to form brown-colored polymers. A less likely regarded mechanism was by condensation of amino acids and related substances with reducing sugars, according to the Maillard reaction.

Despite the current popularity of the polyphenol theory, a completely satisfactory scheme for the occurrence of humic and fulvic acids in diverse geologic environments has yet to be established. In practice, all pathways may be operative, but not to the same extent in all environments or in the same order of importance. A lignin pathway may predominate in wet sediments, such as peats and swamps. The drastic conditions existing in soils under a harsh continental climate (e.g., some Mollisols) may favor humus synthesis by sugar-amine condensation. The disappearance of amino acids from buried sediments has been attributed to the formation of brown nitrogenous polyelectrolytes by reaction with reducing sugars (Stevenson, 1974).

General Chemical Properties of Humic and Fulvic Acids

With time, it became apparent that humic substances consisted of a heterogeneous mixture of compounds for which no single structural formula could be given. Also, each fraction (humic acid, fulvic acid, etc.) came to be regarded as being made up of a series of molecules of different sizes, few having precisely the same structural configuration or array of reactive functional groups. In contrast to humic acids, the low-molecular-weight fulvic acids contain higher oxygen but lower carbon contents, and they contain considerably more acidic functional groups, particularly COOH. Another important difference is that practically all the oxygen in fulvic acids can be accounted for in known functional groups (COOH, OH, C=O); a high proportion of the oxygen in humic acids occurs as a structural component of the nucleus (e.g., in ether or ester linkages).

The distribution of oxygen-containing functional groups in humic and fulvic acids, as recorded in the recent literature, is summarized in Table 2. For any specific group, a considerable range of values is apparent, even with preparations obtained from the same soil type. The total acidities of the fulvic acids (890–1420 meq/100 g) are unmistakably higher than those of the humic acids (485–870 meq/100 g). Both COOH and acidic OH groups (presumed to be phenolic OH) contribute to the acidic nature of these substances, with COOH being the most important. The concentration of acidic functional groups in fulvic acids would appear to be substantially higher than in any other naturally occurring organic polyelectrolyte.

TABLE 2. Distribution of Oxygen-Containing Functional Groups in Humic and Fulvic Acids Isolated from Soils of Widely Different Climatic Zones (All Numbers in Units of meq/100 g)

	Climatic Zone					
		Cool, Temperate				
Functional Group	Arctic	Acid Soils	Neutral Soils	Subtropical	Tropical	Range
			Humic Acids			
Total acidity	560	570–890	620–660	630–770	620–750	560–890
COOH	320	150–570	390–450	420–520	380–450	150–570
Acidic OH	240	320–570	210–250	210–250	220–300	210–570
Weakly acidic + alcoholic OH	490	270–350	240–320	290	20–160	20–490
Quinone C=O	230	{ 10–180	{450–560	{ 80–150	{ 30–140	{ 10–560
Ketonic C=O	170					
OCH_3	40	40	30	30–50	60–80	30–80
			Fulvic Acids			
Total acidity	1100	890–1420	—	640–1230	820–1030	640–1420
COOH	880	610–850	—	520–960	720–1120	520–1120
Acidic OH	220	280–570	—	120–270	30–250	30–570
Weakly acidic + alcoholic OH	380	340–460	—	690–950	260–520	260–950
Quinone C=O	200	{170–310	—	{120–260	30–150	{120–420
Ketonic C=O	200		—		160–270	
OCH_3	60	30–40		80–90	90–120	30–120

Source. *Stevenson (1982) as adapted from the summary of Schnitzer (1977).*

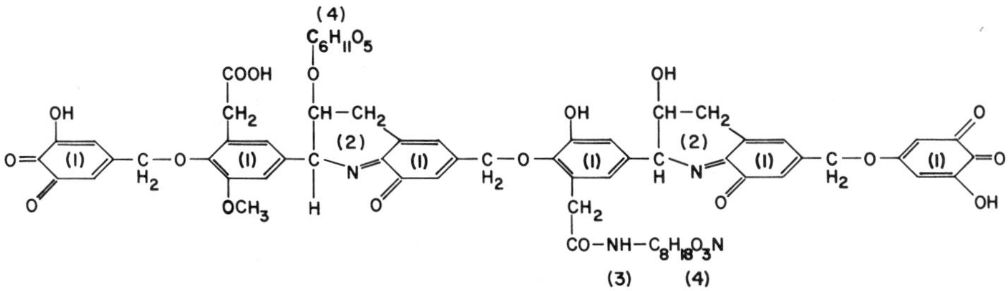

FIGURE 2. Dragunov's structure of humic acid as recorded by Kononova (1966): (1) Aromatic ring of the di- and trihydroxybenzene type, part of which has the double linkage of a quinone group. (2) Nitrogen in cyclic forms. (3) Nitrogen in peripheral chains. (4) Carbohydrate residue.

Structural Characteristics of Humic and Fulvic Acids

According to current concepts, a "type" molecule for humic acid consists of micelles of a polymeric nature, the basic structure of which is an aromatic ring of the di- or trihydroxy-phenol type bridged by —O—, —CH_2—, —NH—, —N=, —S—, and other groups, and containing both free OH groups and the double linkages of quinones. The typical dark color of humic acids, and their ability to be reduced to colorless products with sodium amalgam, is consistent with this concept. In the natural state, the molecule contains attached protein and carbohydrate residues.

"Type" structures, based on the concepts mentioned above, have been proposed for humic acids, none of which can be considered entirely satisfactory. The structures shown in Figures 2–4 all show the presence of aromatic rings of the di- and trihydroxybenzene type, as well as the presence of the quinone group. The data of Table 2 suggest that Dragunov's structure (Fig. 2), as well as Flaig's (Fig. 3), contains an insufficient number of COOH groups (relative to phenolic OH). All three structures show nitrogen as a

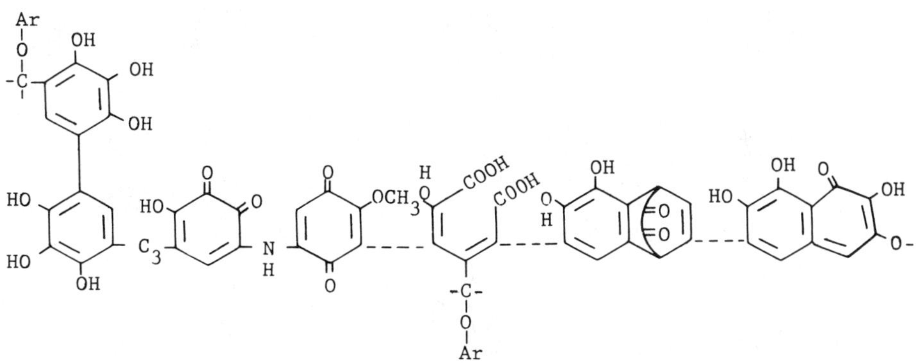

FIGURE 3. Hypothetical structure of humic acid according to Flaig (1960b).

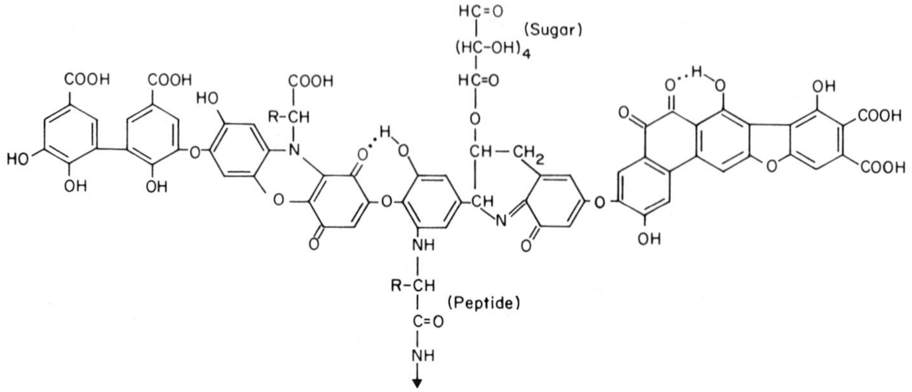

FIGURE 4. Hypothetical structure of humic acid showing free and bound phenolic OH groups, quinone structures, oxygen as bridge units, and carboxyls variously placed on the aromatic ring. From Stevenson (1982).

structural component; Figures 2 and 4 indicate the occurrence of carbohydrate and protein residues.

Schnitzer and Khan (1972) concluded that fulvic acids consist in part of phenolic and benzenecarboxylic acids held together through hydrogen bonds to form a polymeric structure of considerable stability (Fig. 5). Buffle's (1977) model structure of fulvic acid (Fig. 6) contains aromatic and aliphatic components extensively substituted with oxygen-containing functional groups. Both structures show an abundance of COOH groups.

Interrelationships Between Humic Fractions

One useful concept that has evolved over the years, and has been popularized by Kononova (1966), is that the various humic fractions represent a *sys-*

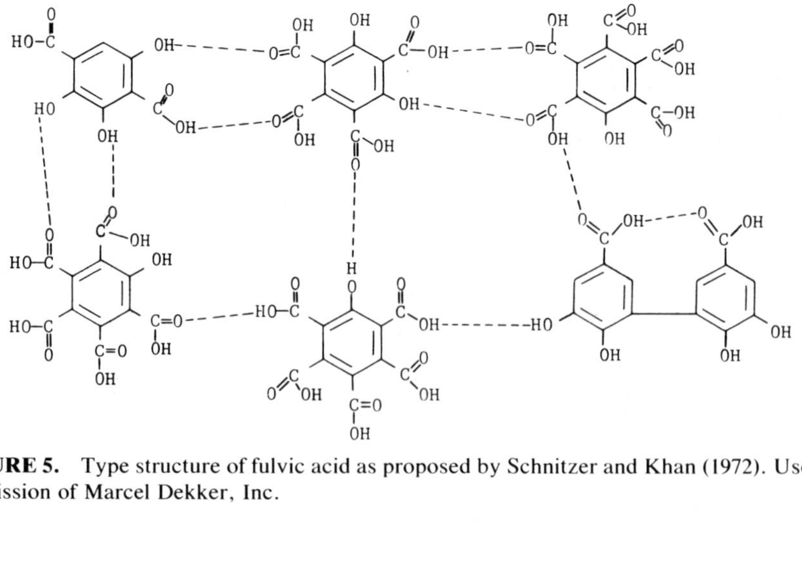

FIGURE 5. Type structure of fulvic acid as proposed by Schnitzer and Khan (1972). Used by permission of Marcel Dekker, Inc.

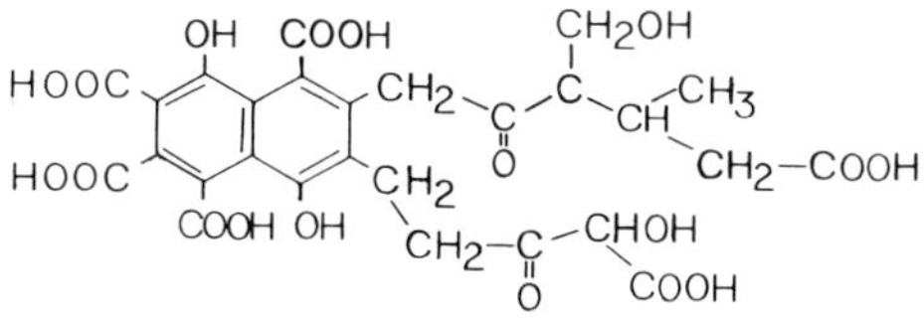

FIGURE 6. Type structure of fulvic acid as proposed by Buffle (1977).

tem of polymers which vary in a systematic way in elemental content, acidity, degree of polymerization, and molecular weight. The proposed interrelationships are shown in Figure 7. No sharp difference exists between the two main fractions (humic and fulvic acids) or their subgroups. The humin fraction (material not extracted with alkali) is not represented but this component may consist of (1) humic acids so intimately bound to mineral matter that the two cannot be separated, and (2) highly condensed humic matter having a high carbon content (>60%) and thereby insoluble in alkali [the ''humus coal'' of Sprengel (1937)].

All soils would be expected to contain a broad spectrum of humic substances, as depicted in Figure 7. However, distribution patterns will vary from soil to soil and with depth in the soil profile. The humus of forest soils

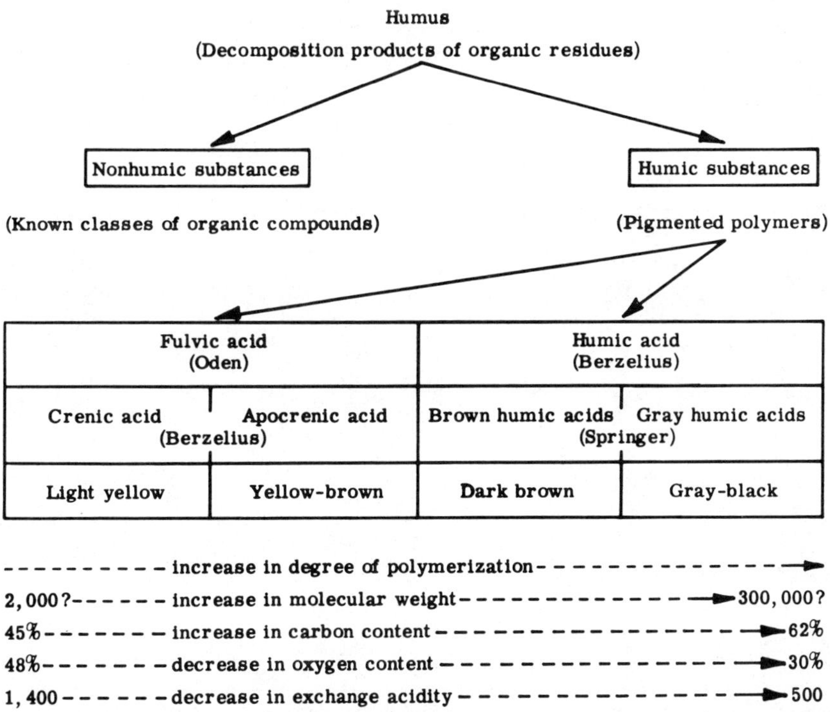

FIGURE 7. Classification and chemical properties of humic substances. See Table 1 for definitions of the various fractions. From Stevenson and Butler (1969) as modified from Scheffer and Ulrich (1960).

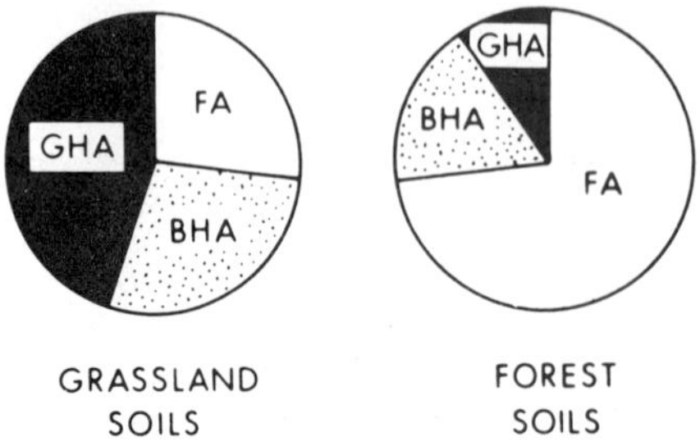

FIGURE 8. Distribution of humus forms in grassland and forest soils. FA = fulvic acid; GHA = gray humic acid; BHA = brown humic acid. Adapted from Stevenson (1982).

(Alfisols, Spodosols, and Ultisols) are characterized by a high content of fulvic acids; that of peat and grassland soils (Mollisols) contains high amounts of humic acid. The humic acids of forest soils are mostly of the brown humic acid type; those of grassland soils are of the gray humic acid type, as illustrated in Figure 8.

The humic acid/fulvic acid ratios of the surface layers from several great soil groups are shown in Table 3. In agreement with the above, soils representative of the Mollisols (Chernozem and Chestnut) have the highest ratios. One difficulty in interpreting published data on humic acid/fulvic acid ratios is that seldom has allowance been made for nonhumic substances present as impurities, particularly in the fulvic acid fraction.

Humic substances in other geologic environments would be expected to have properties similar, but not necessarily identical, to those found in soils. However, direct comparisons cannot yet be made because of lack of stand-

TABLE 3. Humic Acid/Fulvic Acid Ratios of Some Surface Soils as Recorded by Kononova (1966)

Soil[a]	Humic Acid/ Fulvic Acid Ratio	Soil[a]	Humic Acid/ Fulvic Acid Ratio
Chernozem		Gray Forest	1.0
Deep	1.7	Sod Podzolic	0.8
Ordinary	2.0–2.5	Krasnozem	0.6–0.8
Southern	1.5–1.7	Brown desert steppe soil	0.5–0.7
Chestnut			
Dark	1.5–1.7	Tundra	0.3
Light	1.2–1.5		
Serozem			
Typical	0.8–1.0		
Light	0.7		

[a] *Approximate equivalents in the comprehensive soil classification system are: Chernozem and Chestnut, Mollisol; Serozem, Aridisol; Gray Forest, Alfisol.*

ardized extraction, fractionation, and purification procedures. Various names have been used from time to time to describe humus accumulations in wet sediments, including *copropel,* a brown or gray, pulpy, coprogenic substance formed from microscopic plants in the top mud of eutrophic lakes and marshes, *sapropel,* a black mass of humus found in deeper hypolimnetic areas of lakes and bays, *förna,* a pondweed type of sapropel, *dy,* a deposit in dystrophic lakes consisting of an allochthonous precipitate of humic acid and detritus, and *dopplerite,* a deposit of humic substances beneath or within certain peat bogs (Swain, 1963).

ASSOCIATIONS OF ORGANIC MATTER IN SOIL

Most of the humic materials in soils, as well as sediments, occurs in insoluble forms. The ways in which humic substances are bound include the following:

1. As insoluble macromolecular complexes.
2. As macromolecular complexes bound together by di- and trivalent cations, such as Ca^{2+}, Fe^{3+}, and Al^{3+}.
3. In combination with clay minerals, such as through bridging by polyvalent cations (clay–metal–humus), hydrogen bonding, van der Waal's forces, and in other ways as discussed by Greenland (1971) and Theng (1979).

Mechanism (1) is particularly important in peat and other organic-rich sediments, where clay and metal complexes are present in very low amounts in relation to the humus component. A typical example of humic substances bound by polyvalent complexes (item 2) is the Spodosol. These soils have developed under climatic and biologic conditions that have resulted in the mobilization and transport of considerable amounts of iron, aluminum, and organic matter into the B horizon. This illuvial horizon is a rich source of fulvic acids, which are readily separated from the sesquioxides by mild extractants.

When clay minerals are coated with layers of hydrous oxides, the surface reactions are dominated by these oxides rather than the clay and, once again, reaction (2) is of some significance. Allophanic materials, which have the general structure $x\mathrm{SiO_2}\cdot\mathrm{AlO_3}\cdot y\mathrm{H_2O}$, are strong adsorbents of humic substances, which accounts for the exceptionally high levels of organic matter in soils derived from volcanic ash.

Methods for extracting humic substances must take into account the various ways in which organic matter is bound. Free forms of humic and fulvic acids can be recovered by procedures used to displace the soil solution, as well as by extraction with neutral salt solutions or dilute mineral acids. The usual procedure for recovering humic material bound to polyvalent cations is by extraction with a chelating agent, the most popular being sodium pyrophosphate ($Na_4P_2O_7$). Reactions leading to extraction of organic matter by $Na_4P_2O_7$ were postulated by Alexandrova (1960) to be as follows:

$$R(COO)_4Ca_2 + Na_4P_2O_7 \rightarrow R(COONa)_4 + Ca_2P_2O_7 \qquad (2)$$

$$2[RCOOX(OH)_2](COO)_2Ca + Na_4P_2O_7 \rightarrow 2[RCOOX(OH)_2](COONa)_2 + Ca_2P_2O_7 \qquad (3)$$

where X is a trivalent cation.

More drastic extraction procedures are required for extracting organic matter intimately bound to clay minerals, such as with caustic alkali. The subject of organic matter extraction is discussed in considerable detail in Chapter 12.

Clay–Metal–Humus Complexes

In most mineral soils, practically all of the humic material occurs in association with clay (item 3), probably as a clay–metal–humus complex. Clay and organic colloids are negatively charged and the positively charged polyvalent cation (M^{2+}, M^{3+}) serves to neutralize the charges while at the same time linking the two colloids together. A schematic diagram of a clay–humate complex is shown in Figure 9.

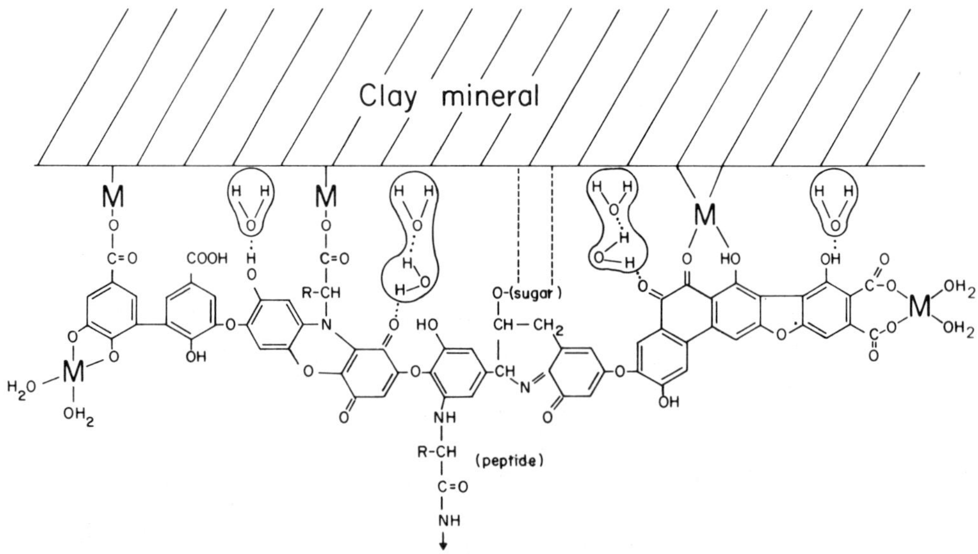

FIGURE 9. Schematic diagram of a clay–humate complex in soil. From Stevenson and Ardakani (1972).

Evidence that most of the soil organic matter occurs in association with clay has come from studies where unbound organic matter, consisting of free humic and nonhumic substances plus undecayed or partially modified plant remains, is removed by flotation in a liquid of density intermediate between the free material and the clay–organic complex (see Greenland, 1965b). Solutions of density between 1.8 and 2.0 have been used, such as a benzene–bromoform mixture. Elutriation and sieving methods have also been applied.

The main polyvalent cations responsible for the binding of humic and fulvic acids to soil clays are Ca^{2+}, Fe^{3+}, and Al^{3+}. The divalent Ca^{2+} ion does not form strong coordination complexes with organic molecules, and humic matter bound in this manner should be rather easily displaced by a monovalent cation, which may account for the small quantities of humic materials that can be displaced when calcium-saturated soils are leached with an NH_4^+ salt or dilute mineral acid. In contrast, Fe^{3+} and Al^{3+} form coordination complexes with organic compounds and strong bonding of humic substances is expected through this mechanism. In this case, displacement of the bound metal is difficult and extraction may require a strong chelating agent or drastic treatment with caustic alkali.

The proportion of the clay surface in any given soil that is coated by organic substances will depend on organic matter content and the kind and amount of clay. In soils containing exceptional amounts of organic matter, such as many prairie grassland soils, practically all of the clay may be covered with a thin layer of organic matter.

Relative Contribution of Humic Substances and Clay to the Cation-Exchange Capacity (CEC) of Soils

Both clay and organic matter contribute to the cation-exchange capacity (CEC) of the soil. The contribution from humic and fulvic acids is due largely to the ionization of COOH groups, although some contribution from phenolic OH and NH groups is expected.

From 25 to 90% of the total CEC of the top layer of mineral soils is believed to be due to organic matter. As one might expect, practically all of the CEC of highly organic soils (peats), as well as the humus layers of forest soils, is due to organic matter. For these special cases, the greater the degree of humification the higher is the CEC. The small amounts of organic matter normally found in sandy soils is extremely important in retaining cations against leaching.

The method used most recently to determine the relative contribution of organic matter and clay to the CEC has been by measurement of the total CEC for a range of soils having variable organic matter and clay contents followed by regression analysis of the accumulated data. Regression equations relating CEC with organic matter and clay contents have been calculated by Hallsworth and Wilkinson (1958), Helling et al. (1964), McLean et al. (1969), and Drake and Motto (1982), among others. The assumption is made that the compositions of the organic matter and clay are identical from one sample to another and that the soils vary only in the amounts of the components present. For this reason, regression equations can only be used for predicting the contribution of organic matter to the CEC within a confined geographical and climatic zone.

The contribution of organic matter to the CEC of the soil depends to some extent upon soil pH. The results of Helling et al. (1964) show that for each unit change in pH, the change in CEC of organic matter is several fold greater than for clay. The CEC of organic matter is much more strongly influenced by soil pH than is the CEC of clay.

Regression equations obtained by Hallsworth and Wilkinson (1958) for five major soil types are given in Table 4. The contribution of carbon to the CEC, indicated by the b_2 coefficients, ranged from 1.12 to 5.12 meq/g of carbon. As expected, the CEC of organic matter increased markedly with increasing pH. On the assumption that organic carbon constituted 58% of the organic matter, a CEC of 297 meq/100 g was estimated for organic matter of the Chernozem and Sierozem soils; for the miscellaneous acid soils from the same area, a CEC of 134 meq/100 g of organic matter was estimated.

The CEC of the soil is determined not only by humus content but by the kind and amount of clay present. The CEC of different clays is of the order of 3–5 meq/100 g for kaolinite, 30–40 meq for illite, and 80–150 meq for montmorillonite. As noted in Table 2, exchange acidities of humic acids usually range from 485 to 870 meq/100 g; for fulvic acids, values up to 1400 meq/100 g have been recorded. These comparisons explain why humus can

TABLE 4. Regression Equations Relating CEC of Some Australian Soils to Organic Carbon (x_2) and Clay (x_1)

Soil Type	Description	Number of Soils	Mean pH	Regression Equation
Chernozemic and Sierozemic	Medium to heavy textured soils of low rainfall area	88	7.20	$Y = 4.12 + 0.82x_1 + 5.12x_2$
Stony Downs	Medium to heavy textured soils of low rainfall area; carbonate concentrations in subsoil and often in surface soil	24	7.00	$Y = 4.37 + 0.44x_1 + 2.25x_2$
Euchrozems	Derived from bauxite Laterites; slight acid and well-supplied with organic matter	19	6.65	$Y = 11.48 + 0.01x_1 + 5.01x_2$
Miscellaneous acid soils	Selected from the same climatic zone as the Chernozemic and Sierozemic group	65	5.58	$Y = 5.13 + 0.23x_1 + 2.27x_2$
Alpine humus	A group of acid alpine soils	15	4.82	$Y = 3.60 + 0.18x_1 + 1.12x_2$

Source. Hallsworth and Wilkinson (1958).

make a significant contribution to the CEC of the soil even though the amounts present may be quite low relative to clay. Organic matter is particularly important as a cation exchanger in soils where kaolinitic-type clays predominate.

In the natural soil, the CEC of organic matter and clay cannot be considered additive because some sites are lost through associations between the two. Also, many of the organic sites may be tied up as complexes with polyvalent cations. Thus, CEC values for organic matter *in situ* will be somewhat less than for isolated soil components. Schnitzer (1965a) suggested that two types of CEC for organic matter should be considered: (1) "measured" CEC as determined by exchange with an appropriate cation and (2) "potential" CEC, the sum of the above and CEC due to blocked sites. The "blocked" sites, which may exceed the "measured" ones, are exposed when organic matter is extracted from soil. Some blocked sites may be released when soils are limed (McLean et al., 1969).

Microaggregates in Soil

Information concerning clay–metal–humus complexes has come from studies using sonic vibration to examine soil aggregates. The concepts derived from these studies have been summarized by Bremner (1968) as follows:

1. Microaggregates in soil consist largely of clay and organic colloids linked together through polyvalent cations. These microaggregates can be presented as

$$[(\text{C—P—OM})_x]_y$$

where C indicates clay, P the polyvalent metal ion (Ca^{2+}, Mg^{2+}, Fe^{3+}, Al^{3+}, etc.), OM the humified organic matter, and C—P—OM the clay-size particles ($<2\ \mu m$), x and y are finite whole numbers dictated by the size of the primary clay particle.
2. The bonds linking the C—P—OM particles into the larger $(\text{C—P—OM})_x$ and $[(\text{C—P—OM})_x]_y$ units can be disrupted by mild shaking if the interparticle bonds are weakened, such as by substitution of Na^+ for the polyvalent metals. This reaction is undoubtedly of considerable importance in the extraction of humic substances with dilute alkalies (discussed below).
3. Microaggregates are formed by a mechanism that is a reversal of what occurs when soil particles are dispersed by water shaking. The reversible processes of dispersion (D) and aggregation (A) can be represented as follows:

$$[(\text{C—P—OM})_x]_y \underset{\text{A}}{\overset{\text{D}}{\rightleftharpoons}} y(\text{C—P—OM})_x \underset{\text{A}}{\overset{\text{D}}{\rightleftharpoons}} xy(\text{C—P—OM})$$

Emerson's (1959) theory of aggregate formation has been widely quoted in the literature. According to his theory, soil crumbs are formed from units of colloidal clay, or domains, and coarser particles of silt and sand (quartz) cemented together by humus. A domain was defined as "a group of clay crystals having suitable exchangeable cations which are oriented and sufficiently close together for the group to behave in water as a single unit." His model of a soil crumb is shown in Figure 10. Four possible types of bonds are shown: A, quartz–organic matter–quartz; B, quartz–organic matter–domain; C, domain–organic matter–domain (organic matter positioned between the faces of two clay domains, between two edges, and between an edge and a face); and D, domain–domain, edge–face.

The "clay" domains of Emerson may in reality exist partly as clay–humus and/or clay–metal–humus domains. In soils that are well supplied with organic matter, and which are often well aggregated, most of the clay will be coated with organic matter.

In concluding this section, it should be noted that a variety of compounds may be responsible for the formation of stable aggregates in soils. Several lines of work indicate that a major role is played by the polysaccharides (see review by Harris et al., 1966).

Soil Wettability and Water Repellancy

Humic substances may be partly responsible for the condition of water repellancy or nonwettability that has been observed for citrus groves, burned-over areas of forest soils, and turf. In most cases repellancy has been associated with coarse-textured sandy soils.

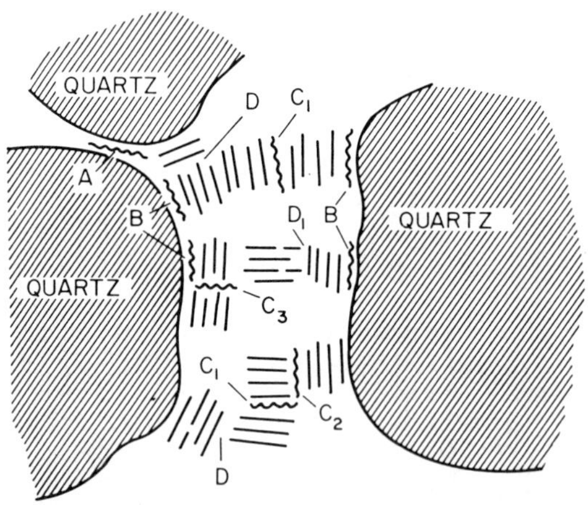

FIGURE 10. Possible arrangements of organic matter, clay domains, and quartz to form a soil crumb: A, quartz–organic matter–quartz; B, quartz–organic matter–clay domain; C, clay domain–organic matter–clay domain (C_1 = face–face, C_2 = edge–face, C_3 = edge–edge); D, clay domain–clay domain, edge–face. From Emerson (1959), used by permission of Oxford University Press, Oxford.

The nonwettable condition of sandy soils has often been attributed to fats and waxes, but evidence in support of this hypothesis is circumstantial (e.g., see Jamison, 1945). Miller and Wilkinson (1977) concluded that fulvic acids synthesized by fungi were responsible for localized water-repellent areas on sand golf greens.

TRACE METAL INTERACTIONS

The ability of humic substances to form stable complexes with polyvalent cations has been well established (Stevenson, 1982). The formation of these complexes facilitates the mobilization, transport, segregation, and deposition of trace metals in soils, sediments, sedimentary rocks, and biogenic deposits of various types. Organic complexing agents play a key role in the chemical weathering of rocks and minerals, and they function as carriers of metal cations in natural waters (Stevenson and Ardakani, 1972; Stevenson, 1983). A portion of the trace metals found in soils and sediments, as well as coal and other biogenic deposits, occurs in organically bound forms. Some of the trace elements, notably boron, copper, iron, manganese, molybdenum, and zinc, are essential for plant growth.

A schematic diagram showing the organic matter reactions involving metal ions in soil is given in Figure 11. The metals present in the solution phase as charged species, and as soluble metal–organic complexes are shown to be influenced by the activities of higher plants and microorganisms, both of which serve as sources of ligands for complex formation; some metals are held in insoluble humate complexes and are nonleachable.

Caution needs to be exercised in attributing all natural phenomena of trace metal cycling to humic substances. Organic matter may be the dominant driving force in some systems but of little significance in others. In

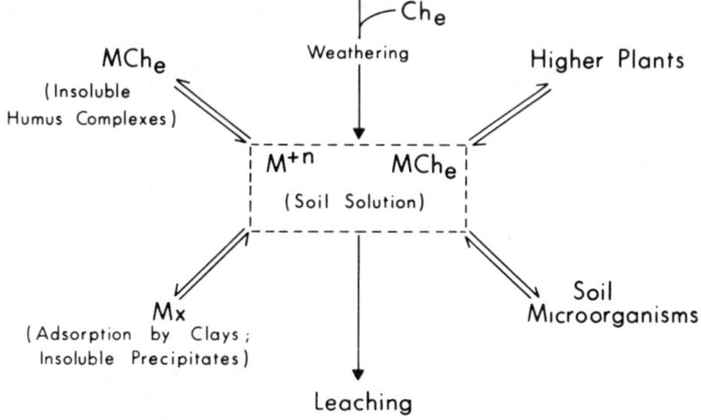

FIGURE 11. Schematic diagram of organic matter reactions involving micronutrients in soil. From Stevenson and Ardakani (1972) as modified from Hodgson (1963).

some cases, such as rock weathering, an effect due to humic substances cannot easily be distinguished from complexation due to low-molecular-weight biochemical compounds (organic acids, phenolic acids, lichen acids, etc.). The impact of trace metal–organic matter interactions in soils and related environments has been discussed by Saxby (1976), Siegel (1971), Reuter and Perdue (1977), Turekian (1977), Jackson et al. (1978), Stevenson and Ardakani (1972), Stevenson and Fitch (1981), and Stevenson (1983).

Nature of Trace Metal Interactions with Humic Substances

The great importance of humic and fulvic acids in modifying the chemical properties of trace metals in the soil environment requires that some consideration be given to the mechanisms whereby they combine with metal ions. Their ability to form complexes with metal ions can be attributed to their high content of oxygen-containing functional groups, including COOH, phenolic-alcoholic, and enolic-OH, and C=O structures of various types. Amino groups may also be involved. Structures commonly considered to be present in humic substances, and that have the potential for binding with metal ions, include the following:

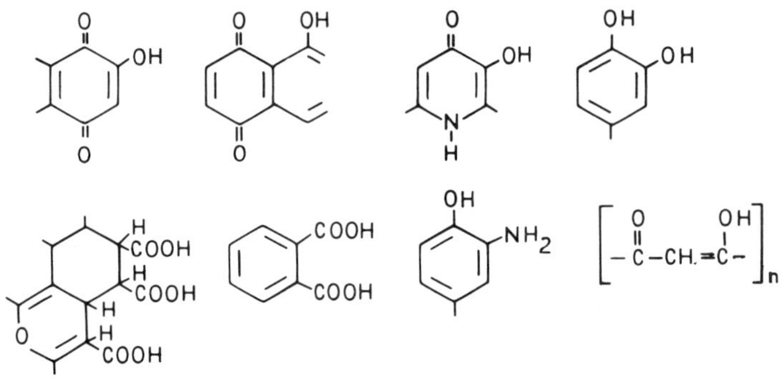

Schnitzer (1969) and Gamble et al. (1970) postulated that two types of reactions are involved in metal–fulvic acid interactions, the most important one involving both phenolic OH and COOH groups. A reaction of lesser importance involved COOH groups only. The two reactions are:

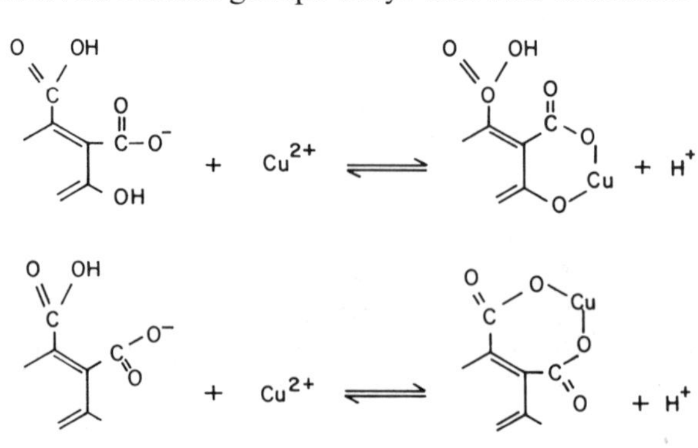

The formation of phthalate-type complexes (bottom reaction) is likely because humic acids have been shown to contain COOH groups that are located on adjacent positions of aromatic rings (Stevenson, 1982). Positive proof for the formation of salicylate-like ring structures (top reaction) has yet to be achieved.

Results of infrared spectroscopy studies have confirmed that COOH groups, or more precisely carboxylate (COO—), play a prominent role in the complexing of metal ions by humic and fulvic acids. Some evidence indicates that OH, C=O, and NH groups may also be involved (Vinkler et al., 1976; Boyd et al., 1979; Piccolo and Stevenson, 1981). The suggestion has been made (Piccolo and Stevenson, 1981) that, in addition to the above, complexes may be formed with conjugated ketonic structures, according to the following reactions:

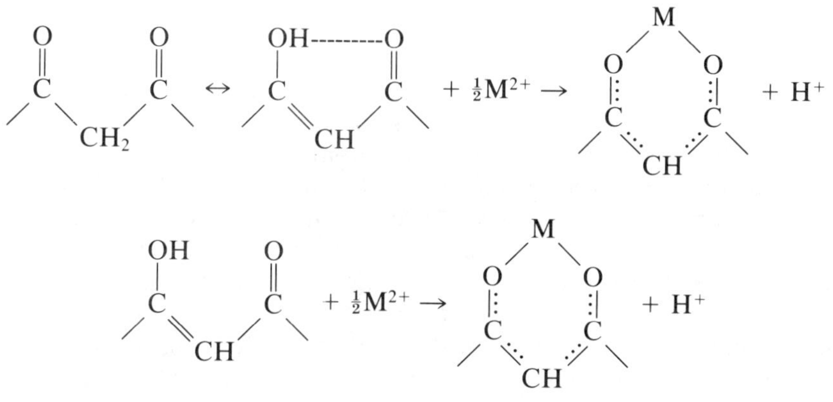

Considerable controversy exists as to the extent to which COO^- linkages are covalent or ionic. The asymmetric stretching vibration of COO^- in ionic bonds occurs in the 1630–1575 cm^{-1} region; when coordinate linkages are formed, the frequency shifts to between 1650 and 1620 cm^{-1}. Frequency shifts with metal–humate complexes have been variable and slight, a result that may be due to the formation of a mixture of complexes. Interpretations in the 1620 cm^{-1} region are further complicated because of interference from covalent bonding with other groups (Piccolo and Stevenson, 1981).

Results of electron spin resonance (ESR) spectroscopy studies have also been inconclusive. Lakatos et al. (1977b) reported that Cu(II) was bound to humic acid by a nitrogen donor atom and two carboxylates. On the other hand, McBride (1978) concluded that only oxygen donors (i.e., COO^-) were involved; furthermore, a single bond was observed. Goodman and Cheshire (1973, 1976) and Cheshire et al. (1977) obtained evidence suggesting that copper retained by a peat humic acid after acid washing was coordinated to porphyrin groups, from which they concluded that a small fraction of the copper in peat was strongly fixed in the form of porphyrin-type complexes. In other work, Bresnahan et al. (1978) observed that the ESR spectrum for a copper–fulvic acid system was influenced by the copper/fulvic acid ratio. At high ratios, a single site for hydrated Cu(II) was indicated while at low ratios two sites were reported. An alternative interpretation of the ESR spectrum obtained by Bresnahan et al. (1978) has been given by Goodman (1980).

Metal Ion Binding Capacity of Humic Substances

Approaches used to determine the binding capacities of humic substances for metal ions include coagulation (Rashid, 1971), proton release (van Dijk, 1971; Stevenson, 1976a,b, 1977), metal ion retention as determined by competition with a cation-exchange resin (Zunino et al., 1972; Crosser and Allen, 1977), dialysis (Zunino and Martin, 1977), anodic stripping voltammetry (Guy and Chakrabarti, 1976; O'Shea and Mancy, 1976), and ion-selective electrode measurements (Buffle et al., 1977; Bresnahan et al., 1978). In general, the maximum amount of any given metal ion that can be bound is approximately equal to the content of acidic functional groups, primarily COOH. Exchange acidities of humic substances vary greatly but they generally fall within the range of 1.5–5.0 meq/g. For copper, this corresponds to retention of from 48 to 160 mg per gram of humic acid. Assuming a carbon content of 56% for humic acids, one Cu atom would be bound per 20 to 60 carbon atoms in the saturated complex. Lees (1950) arrived at a value of one copper atom per 60 carbon atoms for a peat humic acid.

Factors influencing the quantity of metal ions bound by humic substances include pH, ionic strength, molecular weight, and functional group content. For any given pH and ionic strength, trivalent cations are bound in greater amounts than divalent cations; for the latter, those forming strong coordination complexes (e.g., Cu) will be bound to a greater extent than weakly coordinated ones (e.g., Ca and Mg).

Solubility Characteristics

Humic and fulvic acids form both soluble and insoluble complexes with polyvalent cations, depending on degree of saturation. Because of their lower molecular weights and higher contents of acidic functional groups, metal complexes of fulvic acids are more soluble than those of humic acids. Attempts have been made to subdivide humic acids on the basis of molecular weight by fractional precipitation with ammonium sulfate at pH 7 (Theng et al., 1968).

A number of processes affect the solubility characteristic of metal–humate and metal–fulvate complexes in soils, as well as in natural waters. A major factor is the extent to which the complex is saturated with metal ions. Other factors affecting solubility include pH, adsorption of the complex to mineral matter (e.g., clay), and biodegradation. Under proper pH conditions, trivalent cations, and to some extent divalent cations, are effective in precipitating humic substances from very dilute solutions; monovalent cations are generally effective only at relatively high particle concentrations.

Flocculation of humic substances in natural water can result from changes in water chemistry. Thus, cation-induced coagulation of humic colloids is of importance in the removal of bound Fe and other elements from river water during mixing with seawater in estuaries (see Chapter 8). The concentrations of trace metals in the ocean are extremely low, a result that has been attributed by Turekian (1977) to the role of particles (organic and inorganic) in sequestering metals during every step of the transfer from the continent to the ocean floor.

Estimation of Organically Bound Trace Elements

Two main methods have been used to estimate organically bound forms of trace metals in soils and sediments, namely, extraction with a chelating agent (e.g., pyrophosphate) and release by chemical oxidation of organic matter. Procedures for estimating organically bound metals have generally been carried out in conjunction with a more extensive fractionation of trace metals. In the fractionation scheme of McLaren and Crawford (1973), soluble plus exchangeable copper was determined by neutral salt extraction ($CaCl_2$), copper specifically adsorbed on clay by extraction with dilute acetic acid, organically bound copper by extraction with $K_4P_2O_7$, oxide occluded copper by treatment with oxalate under ultraviolet light, and mineral lattice copper by HF digestion of the final soil residue. For 24 contrasting soil types, from one-fifth to one-half of the copper occurred in organically bound forms, with most of the remainder being found as oxide occluded or in association with clay minerals.

A sequential extraction procedure was used by Tessler et al. (1979) for the partition of trace metals in sediments. The following five fractions were obtained: (1) exchangeable—extraction with $1M$ $MgCl_2$ at pH 7.0 or $1M$ NaOAc at pH 8.2; (2) bound to carbonates—leaching of residue with $1M$ NH_4OAc at pH 5.0; (3) bound to Fe–Mn oxides—residue extracted with $0.3M$ $NH_2OH \cdot HCl$ in 25% (v/v) HOAc; (4) bound to organic matter—oxidation of residue with $HNO_3 \cdot H_2O_2$ followed by extraction with NH_4OAc-$HClO_4$. The fraction of the trace elements accounted for in organically bound forms varied from one trace element to another and was of the order of 25% for Cu.

Speciation of Trace Metals in the Soil Solution and Natural Waters

Natural waters from all sources, including soils, lakes, streams, estuaries, and the ocean (see review of Stevenson, 1983) have been found to contain trace metals in organically bound forms. The micronutrient cations in displaced soil solutions have also been shown to occur partly in organically bound forms (Geering et al., 1969). Trace metals that would ordinarily convert to insoluble precipitates (as carbonates, sulfides, or hydroxides) at the pH values found in many soils, sediments, and natural waters are undoubtedly maintained in solution through complexation. The interaction of Al with organic matter is believed to be of considerable importance in controlling soil solution levels of Al in acid soils (e.g., see Bloom et al., 1979).

Several indirect approaches have been used in attempts to estimate organically bound forms of trace metals in the soil solution. In one method, a complexing agent is added that forms a complex which can be removed from the system with an immiscible solvent (Geering et al., 1969). A second technique has been to pass the solution through a cation-exchange resin, in which case cationic forms are adsorbed; complexed forms pass through. In both approaches, the amount of complexed metal is taken as the difference between the amount removed and total concentration in solution (Hodgson et al., 1966). Estimates obtained in this way are undoubtedly high.

The concentration of free metal ions in water can be determined directly by use of an ion-selective electrode (ISE), or by anodic stripping voltammetry (ASV). A major limitation of ISE is its rather low sensitivity; furthermore, only a few commercial electrodes are available (e.g., Cu^{2+}, Pb^{2+}, Cd^{2+}, Ca^{2+}). In both methods, electrode response is affected by pH, ionic strength, and sorption of organics on the electrode surface (Brezonik et al., 1976; Blutstein and Smith, 1978; Greter et al., 1979).

Reduction Properties of Humic Substances

Humic substances have the ability to reduce oxidized forms of certain metal ions, a typical case being the reduction of Fe(III) (Szilagyi, 1971; Goodman and Cheshire, 1972; Lakatos et al., 1977a; Skogerboe and Wilson, 1981). Other examples include reduction of Mo(VI) to Mo(V) and Mo(III), vanadium (V) and V(IV), and Hg(II) to Hg(0) (Goodman and Cheshire, 1975, 1982; Skogerboe and Wilson, 1981). Reduction of ionic species is of considerable importance in soil and water systems because the solubility characteristics of the metal ions (and hence mobilities) are modified. Evidence for reduction of vanadium by humic substances has been provided by electron spin resonance (ESR) spectroscopy (Goodman and Cheshire, 1975; Cheshire et al., 1977; McBride, 1980a,b). The ESR approach has been used also in conjunction with Mössbauer spectroscopy to obtain information on oxidation states and site symmetries of Fe bound by humic and fulvic acids (Senesi et al., 1977c; Griffith et al., 1980).

SORPTION OF ORGANIC CHEMICALS

Adsorption by organic matter is a key factor in the behavior of many compounds introduced into soils and sediments as pesticides or noxious waste organic chemicals. Bioactivity, persistence, biodegradability, leachability, and volatility of the organic chemical are all affected. Numerous studies, reviewed by Hayes (1970), Adams (1973), and Stevenson (1972, 1976b), have shown that the rate at which any given adsorbable pesticide must be applied to achieve adequate pest control can vary as much as 20-fold, depending on the nature of the soil and the amount of organic matter it contains. Soils that are black in color (e.g., most Mollisols) have higher organic matter contents than those that are light in color (e.g., Alfisols), and pesticide application rates must often be adjusted upward on the darker soils in order to achieve the desired result. Results of studies correlating the adsorption of herbicides to organic matter, clay, and other soil properties are tabulated in Table 5.

TABLE 5. Organic Matter, Clay, and Other Soil Properties Correlated with Adsorption Parameters of Herbicides[a]

		Correlation Coefficient[b]			
Compound	Number of Soils	Organic Matter	Clay	CEC	pH
s-Triazines					
Ametryne	34	0.41*	0.14	0.19	−0.37*
Atrazine	25	0.82**	0.65**	0.63**	−0.28
Propazine	25	0.74**	0.71**	0.69**	−0.41*
Prometon	25	0.26	0.60**	0.55**	−0.42*
Prometryn	25	0.40*	0.68**	0.63**	−0.49
Simazine	25	0.83**	0.77**	0.79**	−0.39
Simazine	65	0.72**	0.12	0.52**	0.04
Simazine	32	0.62**	0.27	0.54**	−0.35
Simazine	18	0.82**	0.48**	0.84**	−0.40
Substituted ureas					
Diuron	34	0.73**	0.37*	0.58**	0.10
Diuron	32	0.89**	0.28	0.56**	−0.03
Linuron	11	0.90**	0.06	0.57*	−0.14
Neburon	7	0.76*	−0.37	0.19	0.14
Picloram	6	0.90*	0.55	0.65	—
Phenylcarbamates					
CIPC	32	0.85**	0.16	0.38*	0.48*
Other					
Diphenamid	11	0.91**	0.16	0.60*	0.11

[a] *Adapted from Stevenson (1976) as recorded from literature data.*
[b] **Significant at p = 0.05.*
***Significant at p = 0.01.*

Organic Matter Versus Clay as Adsorbent

Organic matter and clay are the soil components most often implicated in the adsorption of organic chemicals. However, individual effects are not easily ascertained because in most soils organic matter is intimately bound to the clay, probably as a clay–metal–organic complex. Thus, two types of surfaces are normally accessible to the molecule, namely, clay–humus and clay alone. Accordingly, clay and organic matter function more as a *unit* than as separate entities and the relative contribution of organic and inorganic surfaces to adsorption will depend on the extent to which the clay is coated with organic substances. As can be seen from the schematic diagram shown in Figure 9, the interaction of organic matter with clay still provides an *organic surface* for adsorption.

The relative importance of organic matter in adsorbing organic compounds will also be influenced by the chemical properties of the organic compounds. Cationic organic molecules, such as the bipyridylium herbicides (e.g., Diquat and Paraquat), are held primarily by a cation-exchange mechanism and partition between organic matter and clay will depend on the relative contribution of each to the CEC of the soil. Neutral but polar organic chemicals are held by both organic matter and clay, with preference usually being shown for the former. Hydrophobic organic molecules are held almost exclusively by the soil organic matter as noted later.

Potential Chemical Reactions Involving Pesticides and Organic Substances

The organic fraction of the soil has the potential for promoting the nonbiological degradation of many organic chemicals applied to soils as pesticides (see Stevenson, 1976b). Nucleophilic reactive groups of the types believed to occur in humic and fulvic acids (e.g., COOH, phenolic-, enolic-, heterocyclic-, and aliphatic-OH, amino, heterocyclic amino, imino, semiquinones, and others) are known to produce chemical changes in a wide variety of pesticides. Of additional interest is that humic substances have the capability of bringing about a variety of reductions and associated reactions, as discussed by Crosby (1970). The known occurrence of stable free radicals in humic and fulvic acids further implicates organic matter in chemical transformations of pesticides. The heterocyclic ring of amitrole, for example, is known to be highly susceptible to attack by free radicals (Kaufman et al., 1968). Reactions of the type shown in Figure 12 are believed to be involved in hydroxylation of the chloro-*s*-triazines.

Fulvic acids, being more soluble than humic acids, may have a special function with regard to the fate of organic compounds applied to soil as pesticides. Ogner and Schnitzer (1970a) suggested that fulvic acids act as carriers of alkanes and other normally water-insoluble organic substances in aquatic environments, and it is possible that these constituents also function

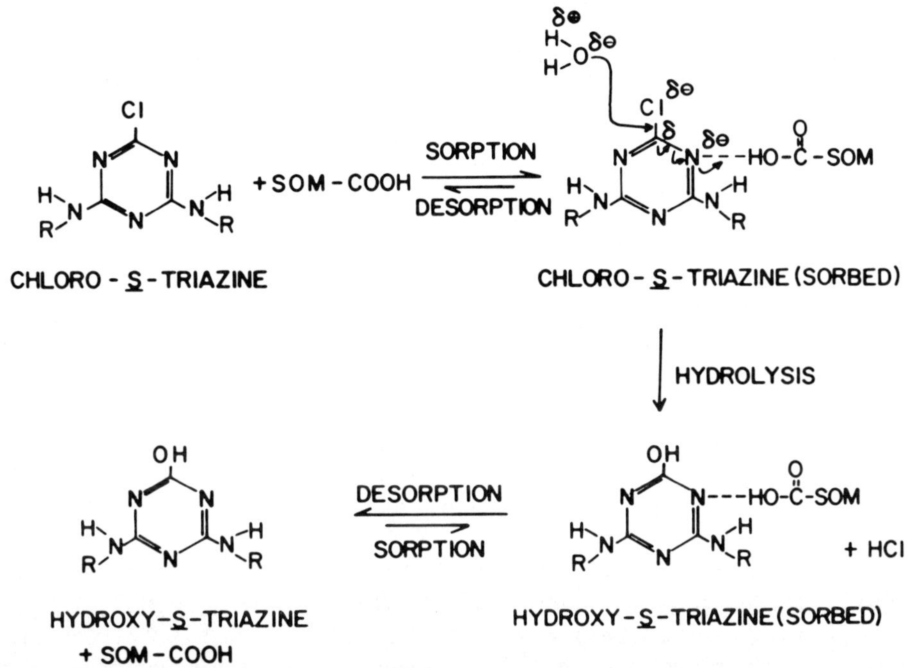

FIGURE 12. Proposed model for the sorption-catalyzed hydrolysis of the chloro-*s*-triazines by soil organic matter. From Armstrong and Konrad (1974), reproduced by permission of the Soil Science Society of America, Madison, WI.

as vehicles for the transport of pesticides. According to Ballard (1971), the downward movement of the insecticide DDT in the organic layers of forest soils is due to water-soluble, humic-like substances.

Adsorption Mechanisms

Bonding mechanisms (Stevenson, 1982) for the retention of organic chemicals by humic substances in soil include *ion exchange, hydrogen bonding, van der Waals forces, and coordination through an attached metal ion* (ligand exchange).

Ion Exchange and Protonation

Adsorption through ion exchange is restricted to those organic chemicals that either exist as cations (e.g., the herbicides Diquat and Paraquat) or become positively charged through protonation. Whether or not protonation occurs will depend upon (1) the nature of the compound in question as reflected by its pK_a and (2) the proton-supplying power of the humic colloids. Reactions leading to adsorption of the *s*-triazines, as postulated by Weber et al. (1969), are shown by the following equations:

$$\mathrm{T + H_2O \rightleftharpoons HT^+ + OH^-} \tag{4}$$

$$\mathrm{RCOOH + H_2O \rightleftharpoons R{-}COO^- + H_3O^+} \tag{5}$$

$$\mathrm{RCOO^- + HT^+ \rightleftharpoons R{-}COO{-}HT} \tag{6}$$

$$\mathrm{RCOOH + T \rightleftharpoons R{-}COO{-}HT} \tag{7}$$

where R is the organic colloid, T the *s*-triazine molecule, HT^+ the protonated molecule, and H_3O^+ the hydronium ion.

Equation (4) represents pH-dependent adsorption through protonation in the *soil solution* while (5) represents ionization of the colloid COOH group. Ionic adsorption of the cationic *s*-triazine molecule, formed by reaction (4), is shown by Equation (6). Adsorption through direct *protonation* on the surface of the organic colloid is shown by reaction (7).

For anionic organic molecules, such as the phenoxyalkanoic acids, repulsion by the predominantly negatively charged surface of organic colloids may occur. Positive adsorption of anionic molecules at pH values below their pK_a values can be attributed to adsorption of the *unionized* form of the compound to organic surfaces, such as by hydrogen bonding between the COOH group and C=O or NH_2 groups of organic matter, as follows:

$$\mathrm{R{-}\overset{\displaystyle O}{\overset{\|}{C}}{-}OH{\cdots}O{=}C{-}O.M.}$$

Hydrogen Bonding, van der Waals Forces, and Coordination

Adsorption mechanisms for retention of nonionic polar organic molecules, such as phenylcarbamates and substituted ureas used as herbicides, are illustrated in Figure 13. The great importance of hydrogen bonding in retention is suggested. Other adsorption mechanisms include van der Waals forces (physical adsorption), ligand exchange ($—M^{z+}$ ..... O=C), and, for organic molecules containing an ionizable COOH group, a salt linkage through a divalent cation on the organic exchange site. For chlorinated phenoxyalkanoic acids, such as the herbicide 2,4-D, physical adsorption to aromatic constituents of organic matter may be involved; hydrogen bonding will be limited to acid conditions where COOH groups are unionized.

Solubility Effect and Partitioning

Partitioning into hydrophobic media has been proposed as a mechanism for retention of nonpolar organic molecules by soil organic matter. Sorption by

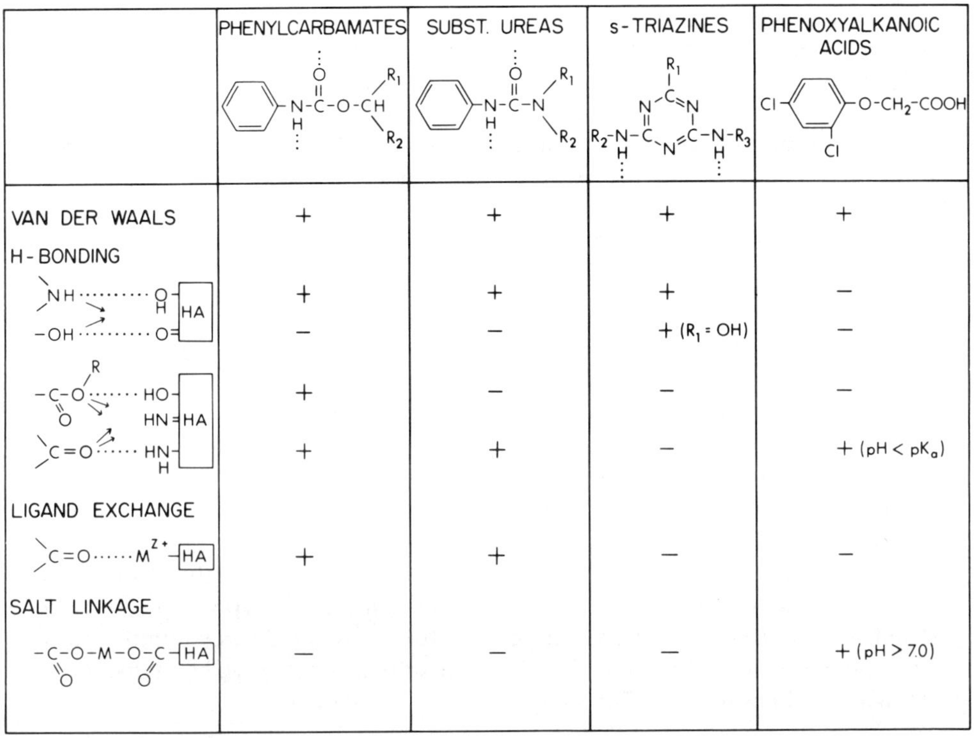

FIGURE 13. Typical bonding mechanisms for some of the common herbicide types by humic acid (HA). From Stevenson (1972).

this means has been described as being due to the "solubility effect" (Chiou et al., 1979). Particular attention has been given to this mechanism for the retention of polynuclear hydrocarbons by soils and sediments. Active surfaces for hydrophobic adsorption include fats, waxes, and resins, as well as possible aliphatic side chains on humic and fulvic acids.

When adsorption of nonpolar organic compounds by soil is expressed as a function of organic carbon content, a constant, K_{oc}, is obtained which is a characteristic property of the compound being adsorbed:

$$K_{oc} = \frac{K_d}{\%\ \text{organic C in soil}}$$

where K_{oc} is the Freundlich K_d constant divided by percent carbon in the soil.

A direct relationship has been found to exist between K_{oc} and the compound's octanol–water partition coefficient, K_{ow} (see Karickhoff et al., 1979 and Khan et al., 1979). This allows for the estimation of K_{oc} from a knowledge of K_{ow}. It should be noted that K_{oc} can also be estimated from the compound's water solubility (Chiou et al., 1979).

The utility of the above relationships is that the adsorption properties of a wide range of nonpolar compounds by soil can be predicted without extensive screening. All that is required is a knowledge of the organic carbon content of the soil and the octanol–water partition coefficients (or water solubilities) for the compounds being tested. Caution must be exercised in using this approach, as exceptions can be expected and specific compounds may deviate from accepted rules (Mingelgrin and Gerstl, 1983).

Chemical Binding of Pesticides and Their Decomposition Products

Substantial evidence exists to indicate that pesticide-derived residues, such as chloroamines produced by partial decomposition of such herbicides as the acylanilides, phenylcarbamates, and phenylureas, can form stable chemical linkages with components of soil organic matter and that such binding greatly increases persistence of the pesticide residues. Two main mechanisms can be envisioned: (1) direct chemical attachment of the residues to reactive sites on colloidal organic surfaces and (2) incorporation into the structures of newly formed humic and fulvic acids during humification. These aspects of the binding of pesticide residues in soil have been discussed elsewhere (Stevenson, 1976b).

GEOCHEMICAL ASPECTS OF SOIL HUMIC SUBSTANCES

Soil Formation

Horizon differentiation in mineral soils is considered to be due to four basic processes: additions, removals, transfers, and transformations in the soil system (Simonson, 1959). In most, if not all, soils, one or more of these processes is influenced by the presence of humic substances.

Additions of humus occur at an early stage in horizon differentiation in soils. The effect is not limited to visible accumulations in the surface (A_1) horizon, but, beginning with advanced stages in soil development, considerable migration occurs into the lower soil horizons, often in association with clay or metal ions. Worm and root channels and ped surfaces (aggregate, crumb, or granule formed by natural processes) become coated with dark-colored mixtures of humus and clay. In some soils, streaks or tongues result from the downward seepage of humus. Illuvial humus also appears as coatings on sand and silt particles. Localized accumulations of sesquioxides (Fe and Al) and humus are common. In many soils, a secondary maximum in humus coincides with the accumulation of clay.

Humic matter is characteristically found in all horizons of the soil profile, even the C horizon, and a variety of processes are thereby affected. For example, organic matter may be involved in a process known as gley formation, a condition that occurs under impeded drainage or a high water table and where a strong reducing condition leads to a soil layer having a light gray color tinged with blue or green, due presumably to Fe^{2+}. Organic constituents may also be involved in clay movement through active dispersion by humic substances. The presence or absence of organic matter, as evidenced by color, is an important criterion in the classification of soils.

It cannot be assumed that humic material in the subsoil has the same chemical composition as that of the surface layer. The conditions under which plant remains undergo decay in the subsoil are different from those in the top layer; furthermore, the pH of the subsoil is usually higher or lower than the surface soil, and decomposition often proceeds under more anaerobic conditions. Another factor to consider is that root material from which part of the organic matter is formed may be chemically different from the leaves and stems which become incorporated with the surface layer. It should also be pointed out that a portion of the organic matter in the subsoil, even in the Mollisol, has migrated from the uppermost part of the soil and has been deposited in the lower horizons. In this respect, the profile can be considered to be a *natural chromatographic column,* in which selective elution of humic compounds occurs, the kinds and amounts being dependent upon soil conditions. In general, the humic acid/fulvic acid ratio decreases with depth, suggesting selective eluviation of fulvic acids.

Weathering of Rocks and Minerals

Both low-molecular-weight biochemical compounds and the higher-molecular-weight humic and fulvic acids have been implicated in the degradation of mineral matter in nature. The ability of many microorganisms, particularly the lichen fungi, to bring about the weathering of rocks or minerals due to the synthesis of biochemical chelating agents is well known. Notwithstanding, the concept that humic substances also play a role has persisted. Evidence both for and against their involvement can be cited. Loughman (1969) has questioned the effectiveness of humic acids as weathering agents whereas Ong et al. (1970) and Baker (1973) have concluded that they act as solubilizing agents for a number of minerals and metals. As can be seen from Table 6, data obtained by Baker (1973) show that humic acids exhibit an activity of the same order as that of several simple organic chelating agents. On this basis, it would appear that humic substances may affect metal ion mobility in environments where appreciable quantities of these substances are present in soluble forms. The ability of humic substances to decompose the common soil minerals (e.g., biotite, muscovite, illite, kaolinite) has been demonstrated by Huang and Keller (1971), Schnitzer and Kodama (1976), and Tan (1980), among others. Because of their low molecular weights, fulvic acids may be particularly effective in dissolving silicate minerals.

TABLE 6. Comparison of the Action of Humic Acid and Several Low-Molecular-Weight Compounds in Solubilizing Metal Ions[a]

Sample[b]	Element Determined	Humic Acid	Salicylic Acid	Oxalic Acid	Alanine
		Micrograms of Metal Extracted in 1 Hour			
Galena	Pb	200	130	95	<5
Sphalerite	Zn	30	30	20	20
Bornite	Cu	190	260	650	15
Chalcocite	Cu	3800	4450	9750	1530
Bismuthinite	Bi	550	180	4820	55
Stibnite	Sb	45	<5	580	<5
Pararammelsbergite	Ni	9800	10,500	7620	1730
Hematite	Fe	470	<3	80	20
Pyrolusite	Mn	1000	4200	15,500	520
Calcite	Ca	10,500	11,900	980	1400

[a] *Adapted from Baker (1973).*
[b] *All extractants 0.1% w/v.*

Whereas positive proof is lacking that humic substances participate directly in weathering processes, it has been definitely established that humic substances play a role in the migration and enrichment of commercially important elements (Ag, Au, Cu, Ni, Zn, Pb, U, and V) in sediments and biogenic deposits. This topic is beyond the scope of the present chapter and the reader is directed to the reviews mentioned earlier for additional information.

A unique concept of the role of humic substances in weathering processes and subsequent translocation of trace elements to biological systems has been advanced by Zunino and Martin (1977). The first stage was described as an attack on insoluble mineral matter by simple organic chelates (e.g., lichen acids, organic acids, phenols) excreted by pioneering microorganisms. These were termed *type-I complexes*. With time and through the action of microorganisms, the trace elements become sequestered by newly synthesized humic substances to form *type-II complexes,* which were believed to be more stable than type-I, thereby preventing loss of essential metal ions by percolation to groundwaters. With an increase in trace element content, a point is reached at which the chelating sites become saturated and less stable complexes are formed, the so-called *type-III complexes*. Simple organic compounds were believed to compete successfully for the metal ions of type-III to form complexes of type-I.

In practice, a whole series of events are involved in the translocation of trace metals from primary rocks and minerals and their subsequent concentration in the subsoil or in bottom sediments. They include release from rocks and silicate minerals from land masses of the earth through biochemical weathering by microorganisms, assimilation by plants and subsequent translocation of trace elements to leaf tissues, incorporation into the humus layer following leaf fall, and movement to sediments in seepage waters as soluble organic complexes. Such a sequence has been described by Fraser (1961) for the accumulation of toxic amounts of copper in a forest peat.

Translocation of Mineral Matter in Terrestrial Soils

Abundant evidence exists for the downward translocation of metal ions in soil as soluble complexes with organic matter (Simonson, 1959). The process has been termed "cheluviation" by Swindale and Jackson (1956). Complexation results in differential movement of metal ions according to their ability to form coordination complexes with organic ligands; iron, aluminum, and other strongly bound elements being eluted to a greater extent than weakly bound ones.

Although cheluviation occurs to some extent in essentially all soils, the process is most pronounced in Spodosols (Podzols). These soils have developed under climatic and biologic conditions that have resulted in the mobilization and transport of considerable quantities of sesquioxides into the subsoil. The organic surface layer of the soil, A_0, consisting largely of acid decomposition products of forest litter, is underlain by a light-colored eluvial horizon, A_2, which has lost substantially more sesquioxides than silica. The A_2 horizon is underlain, in turn, by a dark-colored illuvial horizon, B, in which the major accumulation products are sesquioxides and organic matter.

Stobbe and Wright (1959) reviewed the major processes involved in the formation of Spodosols and reported that the prevailing concept was that polyphenols, discrete organic acids, and possibly other complexing substances in waters percolating from the surface litter bring about the solution of sesquioxides and the formation of soluble metal–organic complexes. Nevertheless, the concept has persisted, with good reason, that humic substances (particularly fulvic acids) also play a role. The classical studies of Deb (1949) show that humus can carry with it from 3 to 10 times its weight of iron oxide and that different fractions of humus vary in their ability to maintain oxides in solution. The humus-protected sol theory assumes that a fulvic acid solution percolating downward through the soil profile forms complexes with Fe and Al until a critical metal/fulvic acid ratio is reached, following which precipitation occurs. Partial decay of organic matter in the B horizon would further increase the metal/fulvic acid ratio. Once started, the accumulation process would be self-perpetuating since the free oxides thus formed would cause further precipitation of sesquioxide-humus sols. On periodic drying, the organic matter complex may harden, thereby restricting movement below the accumulation zone.

The theory that organic substances are responsible for the translocation of sesquioxides in the Spodosol has not been universally accepted. For example, Anderson et al. (1982) concluded that the insoluble organic matter–metal complexes of Spodosol B horizons are formed *in situ* by the interaction between humic substances in drainage waters and previously precipitated inorganic sesquioxides. In other work, Farmer et al. (1983) attributed the formation of the B horizon in hydromorphic humus podzols to coprecipitation arising from the mixing of organic-rich surface waters with aluminum from groundwater.

Diagenetic Transformations in Humic and Fulvic Acids

In a mature soil, the balance of humus is maintained by the continued synthesis of new material as part of the old is mineralized; consequently, the chemical nature of the humic substances remains constant over time. The humus of each soil may have its own characteristic "equilibrium composition," both with regard to chemical nature and composition.

When the source of plant raw material for humus synthesis is cut off, such as when eroded soil is transported to sediments, humus is exposed to successive cycles of biological attack, with concomitant changes in chemical composition. Easily decomposable compounds, such as proteins and carbohydrates, are attacked first, with the result that these constituents are eliminated at the expense of resistant molecules such as humic and fulvic acids.

With time, chemical processes play an increasingly important role in the diagenesis of humic substances. Larger molecules can be formed by condensation of low-molecular-weight biochemical compounds, such as through the reaction of amino acids with reducing sugars and phenols (Stevenson, 1974). Humic substances (humic acid, fulvic acid, etc.) are ultimately altered to kerogen or coal-like products.

Blom et al. (1957) prepared a diagram illustrating the distribution of oxygen-containing functional groups in lignites and coals are related to their carbon and oxygen contents. From published data on the carbon, oxygen, and functional group contents of humic and fulvic acids it has been possible to extend this diagram to pigments having higher oxygen, but lower carbon, contents. The modification is given in Figure 14. Admittedly, further refinements are needed, particularly with respect to changes in OH groups. Nevertheless, the diagram provides evidence for a diagenetic relationship between humic substances and coal. It is apparent that if humic and fulvic acids are involved in the coalification process, COOH groups disappear first followed in order by OCH_3 and OH groups.

Paleohumus

Modified remnants of humic substances are found in buried soils (paleosols) of all ages. This organic matter, or paleohumus, is of interest in geology and pedology because of its importance as a stratigraphic marker and as a key to the environment of the geologic past. The occurrence of dark-colored humus

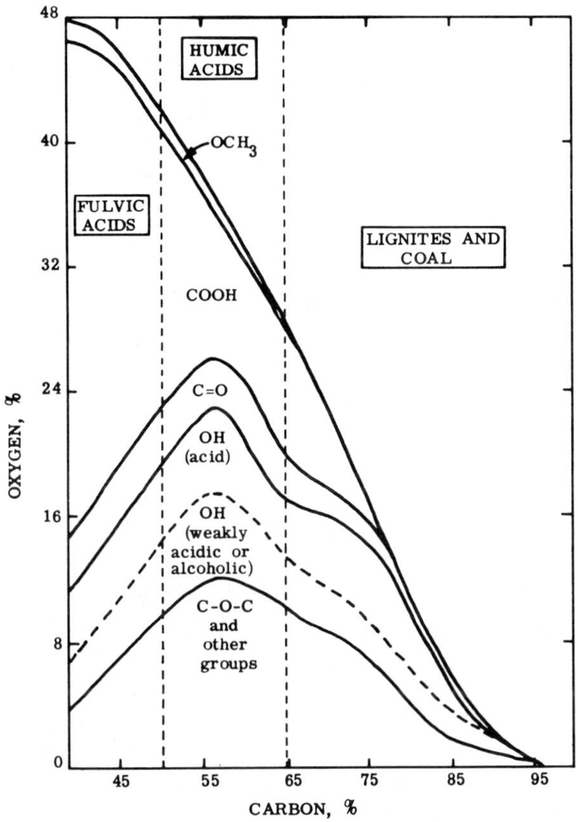

FIGURE 14. Oxygen-containing functional groups in fulvic acids, humic acids, coal, and lignite as related to carbon and oxygen contents. From Stevenson and Butler (1969) as modified from Blom et al. (1957).

zones, when used in conjunction with other pedological observations, has served as a basis for establishing the identification of buried soils, from which it has been possible to draw conclusions relative to climate, vegetative patterns, and the geography of former land surfaces. Very little research has been carried out on the chemical nature of paleohumus but the humic acids contained therein would be expected to have properties intermediate between those of present-day soils and lignites or brown coals.

The humus of buried soils offers the possibility for establishing the ages of sediments through dating of carbon by the ^{14}C method. However, this approach is usually attempted only as a last resort, the preferred method being the ^{14}C dating of fossil plant remains. Contamination of the paleosol by recent organic matter is nearly always a problem, such as through penetration of the buried soil by plant roots and by downward movement of soluble organics in percolating waters. Principles involved in ^{14}C dating of paleosols have been discussed by Ruhe (1969).

Radiocarbon Dating of Soil Humus

Absolute ages cannot be determined for the humus of present-day soils because of continued decomposition of old humus and resynthesis of new humus by microorganisms. The term "mean residence time" (MRT) has been used to express the results of ^{14}C measurements for the *average age* of modern humus. The resistance of humus to biological decomposition has long been known but not until the advent of ^{14}C dating was it possible to express average age on a quantitative basis.

Typical MRTs obtained for carbon in the surface (A_1) horizons of several soil types are recorded in Table 7. Considerable variation in mean ages are evident (210–1900 years). In general, the humus of Mollisols (e.g., Chernozem) appears to be older than for other soil types. As expected, the MRT of organic matter increases with soil depth (see Paul and Veen, 1978).

TABLE 7. Mean Residence Times of Organic Matter in Some Typical Soils

Sample Description	Mean Residence Time (years)	Reference
Cheyenne grassland soil, North Dakota		
(a) Virgin soil	1175 ± 100	Alexander[a]
(b) Clean cultivated area	1900 ± 120	Alexander
Bridgeport loam, Sheridan, Wyoming		
(a) Surface sod layer and corn–fallow-corn plot	3280	Alexander
(b) Continuous wheat plot	1815	Alexander
Black Chernozemic (Mollisol), Saskatchewan, Canada	870 ± 50	Campbell et al. (1967)
Gray-Wooded Podzolic (Alfisol), Saskatchewan, Canada	250 ± 60	Campbell et al. (1967)
Orthic Black Chernozem (Mollisol), Saskatchewan, Canada	1000	Paul et al. (1964)
Black Chernozems (Mollisol), Saskatchewan, Canada Two catenas examined		
(a) Crest of catena to depression	545 to present	Martel and Paul (1974)
(b) B horizons	700 to 4000	Martel and Paul (1974)

[a] *See Paul et al. (1964).*

TABLE 8. Mean Residence Times (MRTs) for Different Organic Matter Fractions of a Chernozemic Black Soil (Melfort Silt Loam) as Reported by Campbell et al. (1967)

Component	Mean Residence Time (years)
Unfractionated soil	870 ± 50
Acid extract of soil	325 ± 60
Fulvic acid	495 ± 60
Humic Acid	
Total sample	1235 ± 60
Acid hydrolysate	25 ± 50
Nonhydrolyzable	1400 ± 60
Humin	
Total sample	1140 ± 50
Acid hydrolysate	465 ± 50
Nonhydrolyzable	1230 ± 60

The MRTs of individual humus fractions also vary, as can be seen from Table 8. The MRT of the original soil (Melfort silt loam) was 870 years whereas the MRTs of individual fractions ranged from 25 to 1410 years. For both humic and fulvic acids, the acid hydrolyzable fractions (6*N* HCl for 18 hours) had lower MRTs than the nonhydrolyzable fractions. This result is in accord with expectations, because the hydrolyzable material would include the carbon of readily decomposable substrates (e.g., carbohydrates and amino acids). Stability of the major humus fractions followed the order: humin = humic acid > fulvic acid (Campbell et al., 1967a).

The fundamental assumption underlying ^{14}C dating is that ^{14}C production by cosmic radiation has been constant for at least 70,000 years and that the rate of carbon exchange between the atmosphere and biosphere has remained constant. These conditions appear to have been fulfilled, although deviations are known to have occurred. Since about 1870, the distribution of ^{14}C has been complicated by combustion of fossil fuel, which has diluted the ^{14}C in the atmosphere by release of ^{12}C as CO_2. More recently the explosion of thermonuclear bombs has added a large amount of ^{14}C to the atmosphere: In 1964 the ^{14}C level of the atmosphere was twice the natural level. Even now, the ^{14}C content of the atmosphere is still much higher than the previous normal level. Enrichment of the atmosphere with "bomb" radiocarbon has been used as the basis for determining rates of movement and turnover of organic carbon in soils (O'Brien and Stout, 1978; Paul and Veen, 1978).

The significance of radiocarbon dates in contributing to our knowledge of the turnover of carbon in soils has been discussed by Paul and Veen (1978). Principles involved in dating pedogenic events by ^{14}C dating of the humus in buried soils have been outlined by Ruhe (1969).

SUMMARY AND CONCLUSIONS

Humic substances constitute the bulk of the organic matter in most terrestrial soils. The functions they perform are multiple and varied and include the weathering of rocks and minerals, mobilization and transport of metal ions, and formation of stable aggregates by combination with clay minerals. Humic substances make a significant contribution to the cation-exchange capacity of the soil, and they are involved in the sorption of organic molecules applied to soils as pesticides.

Although excellent progress has been made in recent years toward the characterization of humic substances, research in this area will have to be expanded if continued progress is to be made. Humic substances represent a complex mixture of molecules having various sizes and shapes, but no completely satisfactory scheme has been forthcoming for their isolation, purification, and fractionation (see Chapters 12 and 14). There is some evidence to indicate that, in any given soil, the various fractions obtained on the basis of solubility characteristics (e.g., humic acid, fulvic acid, and others) represent a system of macromolecules whose chemical properties (elemental composition, functional group content) change systematically with increasing molecular weight. This concept requires verification.

There is an urgent need for the development of procedures leading to the isolation of molecularly homogeneous fractions free of inorganic and organic impurities and suitable for characterization studies. Other problems remaining to be solved are: (1) nature of the building blocks of humic substances from diverse origins; (2) structural arrangement of reactive functional groups and the manner whereby stable complexes are formed with metal ions; and (3) genetic relationships between humic and fulvic acids from various sources and their role in the formation of coal and other naturally occurring carbonaceous substances.

CHAPTER THREE

Geochemistry of Humic Substances in Natural and Cultivated Peatlands

S. P. MATHUR and R. S. FARNHAM

ABSTRACT

Peatlands, or organic soils, are water-logged deposits of partly decomposed plant debris. Ecological variation and diversity in extent of domination by water influence the degree of humification in peatlands. Humification in peatlands does not coincide with decomposition as well as in mineral soils. Measurement and extraction of peatland humus are hampered by the presence of large proportions of unhumified material. Therefore, most studies on the characterization of peatland humus have focused on humic acids, or on pyrophosphate extracts of the organic soils. Pyrophosphate extracts contain less unhumified materials than alkali extracts but are far less effective in extracting the peat humic substances.

Elemental composition, functional group analyses, spectral properties, and characterization of acid hydrolysates have shown that peat humic acids tend to be similar to those from mineral soils. NMR spectroscopy has revealed that peat fulvic acids are largely carbohydrate in nature while the residue of alkali extraction is not all humin.

Drainage, liming, fertilization, and cultivation of peatlands all enhance decomposition and humification. Since humus is more compact than plant debris, cultivated organic soils subside (lose surface elevation typically by about 2.5 cm/yr). Humification and subsidence threaten to destroy valuable cultivated organic soils. A new method is now being used to retard the subsidence of cultivated peatlands by about a factor of 2. It is based on recent observations (1) that the rate of decomposition and humification of cultivated organic soils is determined by activities of soil enzymes, and (2) that of all the feed and fertilizer elements, copper is most effective in inactivating the soil enzymes. The recommended rates of copper application have been found to be effective, safe, and economic. The addition of copper tends to increase plant-availability of manganese, zinc, and iron, perhaps due to displacement of these metals by copper from sites of strong complexation to weaker ligands.

The humus in well-decomposed organic soils also plays a role in the bioavailability and retention of minor elements, pesticides, and pesticide degradation products, but this role is currently not well understood.

INTRODUCTION

Humic substances in most terrestrial and aquatic systems are a distinct quantitatively small fraction of the total mass of the physical environment. This chapter deals with an environment in which humic substances are an integral, characteristic, and substantial constituent of a wide continuum of related organic materials. These organic materials coexist in surficial deposits or organic soils that contain up to 97% organic matter in the solid phase. The moisture content of some of these deposits is often 90% by weight. Worldwide, such soils cover about 500 million hectares (Table 1) equal to nearly half the area of Europe, and have a carbon content close to 0.67 billion (10^{12}) metric tons (Bramryd, 1980). These deposits are called mires and peatlands in Europe and North America. Soil scientists call these organic deposits *Histosols* (Greek word *Histos* meaning tissue) as the term rightly refers to an accumulation of living bodies (mostly plants), their residues, and by-products.

Organic matter accumulates wherever continual or seasonal dominance of water inhibits degradation and humification of plant residues by excluding the air needed for the complete microbial oxidation of the detritus. Even when the net production of phytomass exceeds mineralization only marginally, over thousands of years this condition can result in the development of Histosols. Humic substances occur in both the solid and liquid phases of Histosols, but they have not received the same attention as the corresponding humic substances in soil.

Humic substances in peatlands are often hypothesized to be precursors in the formation of coal. In Table 2 the elemental analyses of several Canadian peats are compared with those of different coals. The data on peats show that the carbon, sulfur, and nitrogen contents increase with increasing degree of decomposition and the oxygen content decreases. The sulfur con-

TABLE 1. Estimated Areas of Peatlands Deeper than 30 cm in Various Countries in 10^6 Hectares[a]

Country	Area
Canada	170
USSR	150
USA/Alaska	30
Indonesia	26
Finland	10.4
United States (minus Alaska)	10.24
Sweden	7.0
China	3.48
Norway	3.0
Malaysia	2.36
Great Britain	1.58
Poland	1.35
Ireland	1.18
West Germany	1.11
Iceland	1.0
East Germany	0.550
Cuba	0.450
The Netherlands	0.250
Japan	0.200
New Zealand	0.150
Denmark	0.120
Italy	0.120
Hungary	0.100
Yugoslavia	0.100
Uruguay	0.100
France	0.090
Switzerland	0.055
Argentina	0.045
Czechoslovakia	0.031
Austria	0.022
Belgium	0.018
Australia (Queensland)	0.015
Romania	0.007
Spain	0.006
Israel	0.005
Greece	0.005
Bulgaria	0.001

[a] *Kivinen and Pakarinen (1980).*

TABLE 2. Elemental Composition of Peat and Coal[a]

Types	Ultimate Analysis (Dry, Ash-Free Basis)				
	Carbon	Hydrogen	Oxygen	Nitrogen	Sulfur
Sphagnum moss peat[b]	48–53	5.0–6.1	40–46	0.5–1.0	0.1–0.20
Decomposed sphagnum peat[c]	56–58	5.5–6.1	34–49	0.8–1.2	0.1–0.30
Highly decomposed peat[d]	59–63	5.1–6.1	31–34	1.0–2.7	0.2–0.50
Lignite coal	70–80	4.7–5.0	20–25	1.0–1.5	0.6–0.8
Bituminous coal	80–85	5.0–6.0	9–12	1.1–1.4	1.0–2.0

[a] *Tibbetts (1981).*
[b] *Equivalent to fibric in U.S. classification system.*
[c] *Equivalent to hemic in U.S. classification system.*
[d] *Equivalent to sapric in U.S. classification system.*

tents of peats are lower than those of coals. The data for coals (lignite and bituminous) show higher carbon contents than for peats but much lower oxygen contents. The hydrogen contents for both peats and coals are quite similar.

TYPES OF PEATLAND

Histosols vary in their botanical origin, extent of humification, present flora, and physical and chemical properties, mainly due to differences in the source and nature of their aqueous environment. The water may originate mainly from direct precipitation (rain, snow), ground seepage, runoff, meandering streams, riverine deltas, or from *littoral* zones of lakes or marine environments. The water may be retained mainly due to (1) absorption by the organic mass, (2) low hydraulic conductivity or water movement especially when the organic mass is well humified, (3) the presence of nearly impervious mineral sublayers (clay or rocks), (4) topography (basins), or (5) high water tables in the region. The impounded water is usually augmented by a marked surplus of precipitation over evapotranspiration. It is therefore natural that peatlands are widely extant in temperate and subarctic zones which had been glaciated, while in the warmer climes they tend to be restricted to high elevations, riverine floodplains, deltas, and estuaries. Table 1 depicts the geographical distribution of peatlands.

Peatlands can be classified from various, often overlapping, perspectives (Farnham and Finney, 1965; Moore and Bellamy, 1974; Clymo, 1983; Gore, 1983). The primary criterion may be the present surface flora (e.g., mangrove, sawgrass); chief botanical origin (e.g., *sphagnum,* sedges); nature of the water, its mineral and flow properties (e.g., high moor, *low moor,* salt marsh, *bog, fen, mire*); degree of decomposition (e.g., *sapric, hemic, fibric, peat, muck*); or intended use (e.g., horticultural, industrial, fuel). Most classifications use water chemistry, botanical composition, and decomposition criteria at one level or another. Similarly, peatlands receive attention from diverse perspectives.

Table 3 compares the most commonly used peat classification systems based primarily on degree of decomposition. The U.S. system (*Soil Taxonomy,* USDA, 1975) and the Canadian system (Day, 1968) are essentially similar except for terminology and are based on the amount of unrubbed or rubbed fiber content.

The three decomposition classes of organic materials of the U.S. system have the following properties:

1. *Fibric Organic Material.* Least decomposed type, with lowest ash content and bulk density, highest saturated water content, the greatest amount of plant fiber, and a low pyrophosphate index. Three well-defined subtypes are the high acid, relatively undecomposed sphagnum moss peats, and the less acid hypnum moss peats, and the reed-sedge type peats.
2. *Hemic Organic Material.* Moderately decomposed with intermediate bulk density, fiber content, saturated water content, and variable acidity.
3. *Sapric Organic Material.* Most decomposed type of organic material, with high bulk density, high ash content, lowest acidity.

The Swedish system of Post and Granlund (1926) is based on a 10-grade scale of humification determined by the turbidity, color, and structure of the wet peat material when squeezed.

TABLE 3. Comparison of Peat Classification Systems Related to Decomposition

	Decomposition Classes		
Systems	Low	Medium	High
United States[a]	Fibric	Hemic	Sapric
Canada[b]	Fibric	Mesic	Humic
Sweden[c]	1–2–3	4–5–6	7–8–9–10
Soviet Union[d]	10–20–30	40–50–60	70–80–90
I.P.S.[e]	Weak	Moderate	Strong

[a] Soil Taxonomy, *USDA (1975).*
[b] *Day (1968).*
[c] *Post and Granlund (1926).*
[d] *Kazakov (1953).*
[e] *Kivinen (1980).*

The Soviet system as described by Kazahov (1953) is based on a microscopic examination of the amount of fibrous and amorphous material contained in a smear of peat on a slide.

The International Peat Society system (I.P.S.) described by Kivinen (1980b) is a three-grade scale based on the relative degree of decomposition. In addition, this system includes the type of plant remains and the trophic levels (oligotrophic, mesotrophic, eutrophic) of the peat material.

IMPORTANCE OF PEAT HUMUS

Organic soils are a storehouse of plant nutrients and water, and provide a desirable medium for producing high quality vegetable crops that grow grit-free in or on the ground. Thus they attract agriculture in spite of the difficulties of management and initial drainage. Peat is also useful as an ameliorant of mineral soils, as a base for slow-release nitrogen and phosphorus-organic fertilizers, and as a medium for hothouse culture. Yeasts and other microbes can be grown on peats or peat derivatives and the resulting biomass can be used as fodder or extracted to yield food supplements such as proteins (single-cell protein). By-products of such microbial cultures (alcohols, vinegar, enzymes) can also be useful (Fuchsman, 1980).

Where the choice of energy sources is restricted, or when imports become expensive, the carbon in peatlands offers an attractive alternative (Farnham, 1978). Peat may be burned as such, converted to liquid or gaseous fuel, or used to grow biomass for energy. Also, organic deposits formed on marine shores during the last ice age, and subsequently buried under sediments and water during deglaciation, offer an opportunity to study coalification (Hatcher et al., 1983a).

Histosols often act as a transition phase between terrestrial and aquatic environments, and between uplands and lowlands. Histosols thus offer an ancillary or primary habitat for many organisms (Clarke-Whistler and Rowsell, 1983; Speigt and Blackith, 1983), and influence the hydrology, productivity, and ecology of adjacent ecosystems. For example, frogs living on a Histosol habitat feed mainly on animals from the nearby shore, forest, and grassland ecosystems, not on coinhabitants of the bog (Mason and Standen, 1983). Thus, perturbation of peatlands can have far-reaching consequences. For example, disturbance of coastal swamps can promote eutrophication of estuarine environments or promote contamination with metals inhibitory to primary producers.

Over time, the world peatlands may have accumulated approximately 150 $\times 10^{15}$ g of carbon and estimates of the net annual accumulation vary from 0.21 to 0.30 $\times 10^{15}$ g of carbon. Since Histosols contain a sizable portion of the total carbon on the earth's surface, the size of the carbon fluxes in or out of these soils can influence CO_2 levels in the atmosphere and thus affect climate. Although cultivation of Histosols is not very well quantified, it is evident that utilized peatlands may play an important role in the global carbon budget (discussed in Chapter 1) and may indeed already be contributing significantly to the recent increments in atmospheric CO_2 (Aramentano, 1979; Bramryd, 1980).

All the above interests in peatlands are influenced by the amount of humus in the organic matrix. Humus retains metals, resists biodegradation, contains plant nutrients and energy-rich carbon bonds but tends to retain and conduct less water than less-humified peat. Also, a humified organic soil (Saprist, muck) is less prone to biodegradation, and is poorer in soluble *phenolic* compounds, than a raw peat (Fibrist). The ability of a bog and its water to preserve materials (e.g., butter), sanitize (as used for litter for animals), and heal skin infections is probably due to soluble phenolic compounds and humic substances (Fuchsman, 1980). Consequently, there is a renewed interest in the content and nature of humus in peatlands, in its formation, and in the effects of humification on the nature of the peats and their potential usages. Earlier interest in peat humus stemmed from a perceived ease of extraction due to lack of complexing mineral matter, and for its importance in classification and stratification of organic deposits. The presence of humus in an organic soil profile represents past environments where aerobic soil microbial activity occurred to a certain extent.

HUMIFICATION IN PEATLANDS

Humification and mineralization in Histosols are complex transformations involving physical, chemical, enzymic, faunal, and microbial processes. These occur in a succession of microenvironments varying in their redox and chemical properties.

As plant tissues senesce and die, three processes may ensue almost simultaneously. First, enzymes within the dead but sterile and physically intact cells cause proteolysis and other autolytic degradations. The released *amino acids,* sugars, *tannins,* phenols, and *quinones* may be oxidized by chemical or enzymatic catalysis to produce humus-like pigments, or proto-humus as discussed by Stevenson in Chapter 2. This was well illustrated by Cohen (Given and Dickinson, 1975) who observed cellular material of partially polymerized leuco-*anthocyanins* in residues of *Rhizophora mangle* deposited in a mangrove swamp in Florida. The autolytic reactions may be prominent in situations where microbial decomposition is slow due to acidity, anaerobiosis, or lack of basic nutrients.

Second, the *epiphytic* microflora may start heterotrophic activity on moist plant material even before it falls to the ground. Soil animals may also initiate their activities at this stage. Soil arthropods and annelids fragment plant litter, thereby increasing the surface area amenable to microbial activity, modify organic substrates, and concentrate mineral nutrients by fecal production.

Soil animals also disseminate microbial decomposers either by carrying them on their body surface or by passage through their alimentary canals in viable state. Some members of the soil fauna also act as predators of yeasts, bacteria, and fungi (Wynn-Williams, 1982; Tadros and Varney, 1982).

At the state of decomposition before the plant material falls on the water or ground surface, humification in peatlands does not differ from humus formation in mineral soils. The aboveground environment of organic as well as mineral soils allows growth of aerobic bacteria, actinomycetes, and fungi. However, the presence of hydrophobic waxes on leaves, and tannins in tree barks, favors organisms capable of utilizing or tolerating waxes and tannins.

In environments where the level of water is above the soil surface, such as in mangrove swamps, plant debris falls through water into a zone where the dissolved oxygen is not sufficient for complete mineralization, but the conditions are not suitable for *obligate anaerobes*. Where the water level is at or below the soil surface the conditions are still not similar to those in most mineral soils. Peatlands generally lack minerals bearing phosphorus, potassium, and other base elements. Thus peat microbes have to depend largely on the debris itself for their mineral nutrition, particularly in bogs where the main source of water is precipitation rather than runoff or groundwater. Indeed, under these conditions, many plants translocate mineral nutrients from their aboveground parts to belowground parts for storage before the leaves die (Mason and Standen, 1983). Such environments also tend to be acidic—a condition that promotes the leaching of some essential minor elements. The acidity, as well as contact with air at the surface, favor fungi over bacteria. Below the surface, where much of the acidity is generated due to incomplete decomposition of plant debris to organic acids rather than to CO_2 and water, the oxygen-deficient conditions become more optimal for certain bacteria. In these environments, organisms that can use carbon and sulfur as electron acceptors to produce CH_4 and H_2S are mostly anaerobic bacteria. In the anaerobic zone decomposition may not be accompanied by humification to the same extent as under aerobic conditions due to a lack of the oxygen that promotes humus formation. The oxygen may help generate the free radicals postulated to be required for humus formation. Indeed when anoxic layers of organic deposits are exposed to air they are seen to become darker in color as humification occurs. Under natural conditions, periodic exposures to air, caused by recession of the water table, may also cause such episodes of humification. The humic substances formed in peatlands tend to be more soluble than those in mineral soils due to the paucity of oxygen, metals, and microbial activity which may induce or cause further aggregation and transformations. Some of the humic substances formed in peatlands may therefore be translocated by water movement.

The more soluble humic substances can coat insect bodies, fecal material, and plant materials in lower horizons of the soils. Complexation of enzyme proteins by humic substances (Burns, 1978) inhibits their catalytic activities. An envelope of humic substances around even readily decomposable materials can bind and reduce the activity of enzyme proteins and therefore retard enzymatic degradation of the inner organic materials. This is important because a large part of the primary production in Histosols is below the surface (Mason and Standen, 1983). Precipitation of humic substances on animal body surfaces (containing proteins or chitins) in the Histosol sublayers also interferes with exchange of respiration gases between the animals and the water or air trapped around some animal bodies (for plastron respiration) (Speigt and Blackith, 1983; Mason and Standen, 1983), thus curtailing the survival and activities of the animals.

The free-standing water in peatlands does not transmit light well enough to support phytoplankton growth due to the presence of dissolved humic substances. Therefore in the aquatic phase of peatlands, light limitation and the acidity of the water favor detrital food chains based on bacteria to the near exclusion of autotrophic food chains based on phytoplanktons (Speigt and Blackith, 1983). Phytoplankton-based food chains are characteristic of many aquatic environments, such as lakes and oceans.

The sequence of the decomposition process can be expected to be similar in the organic and mineral environments because water-soluble carbohydrates, organic acids, and proteins are degraded faster than polyphenolic substances such as lignin. Micromorphological analysis of peat deposits, however, shows that the sequence is in fact not clear-cut. Microscopic examination of thin slices of peat materials in different stages of decomposition reveal humus-like pigments in dead but physically intact cells, in amorphous masses of dark materials that lack anatomical features, in or on fecal pellets of insects, and in proximity of dead or live bacteria and fungi. All these may occur in intimate contact with roots barely colonized by decomposer microbes (Levesque, 1981; Given and Dickinson, 1975).

Many earlier studies were made to assess microbiological activities in various layers of peat deposits by counting viable fungal propagules and bacteria; and the latter were differentiated into various types of aerobes and anaerobes. Anaerobic bacteria, capable of growing with little or no oxygen, and usually producing CH_4 or H_2S, tended to be a far greater proportion of the microbial flora in the deeper anoxic layers than in the surface layer (Given and Dickinson, 1975).

It is known that the presence of large populations of microbes in peat does not necessarily indicate high microbial activity because even alien microbes can survive quiescently in organic soils much longer than in mineral soils (Tate, 1978; Mathur, 1983). Some of the microbes found in peat sublayers may have been carried there by small particles of humified materials during lowering of the water table, and prevented from rising back to the surface with the water due to conglomeration and enmeshing with fine fibers. Evidence of such occurrences was recently obtained by carbon dating and palynological analyses (Dinel, Levesque, and Mathur, unpublished data).

Recent evidence suggests (Mathur, 1982a,b) that enumeration of viable cells or measurement of biomass may not reflect soil biochemical processes fully and truly in the case of organic soils. Many studies indicate that abiotic (not associated with living organisms) enzymes help determine the decomposition (and humification) rate of their organic surroundings. To comprehend this, one must realize that although enzymes are essential to all life, life is not essential for all enzyme activity. Most of the enzymes added to a soil or organic deposit by plants, animals, and microbes are ephemeral due to their inactivation by heat, oxygen, and metals. A proportion of the enzymes, nonetheless, is stabilized in soil by association with humus or clay colloids. These soil enzymes continue to function in time and space far removed from their parent cells (Burns, 1978). In mineral soils, a large part of the potential enzyme activity is not realized mainly due to the lack of organic substrates—or compounds on which the enzymes act. In organic soils this limitation is minimal so that the rate of decomposition is found to be positively related to the levels of stabilized and accumulated enzymes that catalyze degradative reactions (Mathur 1982a,b). A better understanding of the role of enzymes in the humification of organic soils is hindered by a lack of proper criteria and methods for distinguishing between functional enzymes that are ephemeral (transient) in the soils and those that are accumulated in the stabilized (immobilized) form.

MEASUREMENT OF THE EXTENT OF HUMIFICATION

It is desirable to have at least a semiquantitative estimate of the extent of humification, or degree of decomposition of a peat material, since this is a prominent feature of all aspects of natural and economic roles of peatlands. Many avenues are available for achieving this objective because humification involves numerous reactions and results in various products and effects as discussed by Stevenson in Chapter 2.

In search of a suitable method of measuring humification in peats, Levesque et al. (1980a,b), Levesque and Mathur (1979), and Mathur and Levesque (1980) conducted a series of studies with up to 92 peat materials from distinct deposit layers of various peatland areas. Twenty-five physical, chemical, and botanical properties related to humification were measured (Table 4), and simple linear correlation analyses were performed between all possible pairs of properties. The object was to determine which property relates with a maximum number of the other properties in the most consistently logical and widely based manner. Properties that were lacking in significant or consistent correlation with others were excluded from the resultant Table 5. It was concluded that no single method of measuring humification is totally satisfactory mainly due to the heterogeneity and diversity of the organic debris in its physical, chemical, and morphological features resulting from variation in botanical origin (Levesque et al., 1980b). As discussed below all individual methods of quantifying humification in Histosols are beset with limitations.

TABLE 4. Properties of Some Peat Materials[a]

Properties	Mean	Range	Coefficient of Variation (%)
1. Mineral matter (%)	3.70	0.3–8.9	75.4
2. pH value	3.76	3.0–5.2	22.8
3. Calcium (mg/g)	6.92	1.0–15.0	67.1
4. Cation-exchange capacity (meq/100 g)	148.5	121–224	18.6
5. Carbon (%)	50.69	46.7–54.9	4.8
6. $-OCH_3$ (%)	2.04	0.77–3.40	36.5
7. C/N ratio	45.64	15.4–99.2	48.5
8. Bulk density (g/cm^3)	0.11	0.064–0.195	21.8
9. Water-holding capacity (%)	616.1	366–937	22.9
10. Total pentoses (mg/g)	0.63	0.15–1.43	51.7
11. Total hexoses (mg/g)	34.2	13–79	58.9
12. Total sugars (mg/g)	35.5	13.8–81.3	58.3
13. Arabinose/xylose ratio	0.91	0.10–2.33	99.01
14. Pyrophosphate (PP) index	17.8	6.7–63.4	68.0
15. Phenolics in PP extractants (mg/g)	14.65	6.2–34.0	46.6
16. Calorific value (Btu/lb)	9272	8310–10,520	6.7
17. Mosses (%)	28.8	6–77	96.7
18. Sedges (%)	37.1	7–83	76.1
19. Wood (%)	34.1	7–83	75.7
20. Fiber unrubbed (%)	51.59	26.0–93.3	28.2
21. Fiber rubbed (%)	28.92	5.6–90.0	61.8
22. Particles $<75\ \mu m$ (%)	34.01	15.0–64.6	41.0
23. Particles >1 mm (%)	10.30	1.1–30.6	66.7
24. Initial rate of biodegradation ppm carbon lost/day	124.8	59.8–319.2	53.7
25. Steady rate of biodegradation ppm carbon lost/day	55.1	22.3–207.8	75.6

[a] *Levesque and Mathur (1979) and Levesque et al. (1980b).*

As decomposition progresses mineral content and pH increase and percent carbon decreases. Suspended mineral particles and dissolved ions like calcium, however, may also be gained by peat deposits through incoming water and air.

Bulk density of the peats increases as they are transformed to the more compact humus and increase in mineral content. Humus has higher cation-exchange capacity, nitrogen content, and calorific value (energy released on combustion) than fresh plant debris, but it is generally poorer in methoxyl content and in capacity for retaining water. Hydrolyzable sugars represent easily decomposable *hemicellulose* (xylose polymers), *cellulose* (glucose polymers), and microbial polysaccharides (rich in arabinose) that are by-products of decomposition. The use of all these as bases for methods of

TABLE 5. Significant Correlations of Each Individual Property with Others in the List[a]

Properties	Total Number of Correlations	Correlated with the Listed Properties: Positively	Correlated with the Listed Properties: Negatively
1. Mineral matter (%)	12	2,3,4,8,13,19	5,7,9,11,12,17
2. pH value	3	1,4	7
3. Calcium (mg/g)	9	1,4,8,13	7,9,11,12,17
4. Cation-exchange capacity	6	1,2,3,8	9,5
5. Carbon (%)	8	22	1,4,10,13,20,21,23
7. C/N ratio	10	11,12,17,21	1,2,3,8,13,18
8. Bulk density	10	1,3,4,13,19	7,9,11,12,17
9. Water holding capacity	4		1,3,4,8
10. Total pentoses	6	20,21,25	5,14,22
11. Total hexoses	9	7,12,17,21	1,3,8,13,18
12. Total sugars	10	7,11,17,20,21	1,3,8,13,18
13. Arabinose/xylose	8	1,3,8	5,7,11,12,22
14. Pyrophosphate index	5	22	10,17,20,21
17. Mosses (%)	13	7,11,12,20,21,25	1,3,8,14,18,19,22
18. Sedges (%)	6		7,11,12,17,19,21
19. Wood (%)	4	1,8	17,18
20. Fiber, unrubbed	10	10,12,17,21, 23,24,25	5,14,22
21. Fiber, rubbed	13	7,10,11,12,17,20, 23,24,25	5,14,18,22
22. Particles <75 μm	8	5,14	10,13,17,20,21,23
23. Particles <1 mm	5	20,21,25	5,22
24. Initial rate of biodegradation	3	20,21,25	
25. Stable rate of biodegradation	6	10,17,20,21,23,24	

[a] *Levesque and Mathur (1979) and Levesque et al. (1980b).*

determining the extent of humification is limited by the diversity of botanical origin which causes these properties to vary widely.

Visual and microscopic examination of peats are helpful only to a limited extent for determining botanical origin, because the characteristic anatomical features become indiscernable as the material becomes amorphous due to decomposition. Estimation of the percent of amorphous material (degree of decomposition or the von Post scale) (Post and Granlund, 1926) by microscopic examination, as a measure of the extent of humification, is handicapped by the fact that all amorphous material is not humus.

Peat materials are often richer in nonhumified plant residues than in humus. Exhaustive alkaline extraction and estimation of humic substances are therefore hampered by the presence of hydrophobic lipids, and by coextraction of phenolic compounds from undecomposed plant residues, particularly from lignified tissues. Artifacts may occur also during the solvation, separation, or concentration steps of the procedure, as discussed by Hayes in Chapter 13. Some of the problems can be mitigated by prior extraction of the lipids, soluble phenols, and sugars, as in the various schemes of proximate analysis discussed by Walmsley (1973).

Kaila (1956) took a different route to overcome the problems. He extracted 1 g of dry peat for 18 hours with 100 mL of $0.025M$ aqueous sodium pyrophosphate rather than with strong alkali. The filtered extract was diluted fivefold, its absorption of 550 nm wavelength light was measured, and the optical density reading was multiplied by 100. The resultant pyrophosphate index has been used widely although there is little theoretical basis for assuming that the color intensity of a peat extract should be closely related to the extent of humification or that the extraction would be even semiquantitative in the presence of significant amounts of mineral matter. The pyrophosphate may extract nonhumified phenols although they may not form humus-like compounds as readily as during an extraction with strong alkali. In any case, the optical measurement is insensitive to the nature and extent of aggregation of the chromophores. Consequently, Schnitzer (1967) found that the pyrophosphate index did not correlate with the carbon in humic or fulvic acids obtained by exhaustive alkaline ($0.5N$ NaOH) extraction of the peat samples without any pretreatment. When some of the mineral matter was removed from the samples by treatment with HCl-HF solution (0.5% HCl + 4.8% HF in water), the pyrophosphate index of the washed soils correlated positively with the carbon in fulvic acids, but negatively with the carbon in humic acids extracted by NaOH. As Kaila (1956) had pointed out, a NaOH extract of peat contains considerably more nonhumic materials than the pyrophosphate extract.

It is logical that as decomposition and humification progress the proportion of biodegradable materials decreases, and consequently the potential rate of biodegradation, or percent carbon lost per unit time through respiration, decreases. However, this presupposes that there are no other constraints on the decomposition process. That condition is hard to achieve and has its own complications. Nevertheless, by reasoning that the decomposition of natural peat materials is restrained almost exclusively by lack of oxygen and suboptimal temperature, at least in the short term, Levesque and Mathur (1979) measured the rate of respiration in peat materials (Table 4) under optimal conditions of aeration and temperature. The respiration or biodegradation rates (Table 4) were found to correlate with the rubbed fiber content of the peat samples much better than with other properties (Levesque and Mathur, 1979) (Table 5). The rubbed fiber content is the percent of material, by volume, that does not pass through a 100 mesh screen (pore size 0.15 mm) upon rubbing under running water. Levesque and Mathur (1979) suggested that rubbed fiber content offers the single most reliable criterion for determining the amount of nonhumified material, and thus the extent of humification and decomposition. This and other proposals await further examination for applicability to a wide variety of peat materials. In the meanwhile, no single method of measuring extent of humification in Histosols should be used in total disregard of all others.

NATURE OF PEAT HUMUS

Peat humus has been studied for nearly as long as soil humus although much less intensely or effectively. The considerable problems attending extraction of humic substances from mineral soils (Hayes, Chapter 13 of this book) are increased in peats by the presence of nonhumic compounds in proportions often greater than that of humus. The main origin of the nonhumic substances in organic deposits may vary from algae, for some aquatic peats, to trees. The consequent diversity in aromatic and aliphatic compounds of different solubilities can influence the extraction and fractionation of fulvic acids, humic acids, or humins from Histosols. As indicated before, the soluble nature of some of the humus formed in peatlands may cause movement of some humic substances more than of others. Even the apparent advantage for extraction stemming from a lack of humus-binding clays is partly negated by a probable concentration of the largely nonclay mineral matter on the humus. Consequently Schnitzer (1967) found that washing peats with an aqueous HCl-HF solution increased the alkali solubilization of humic substances twofold from peats and fourfold from mucks (Table 6). The extent of the change caused by the acid pretreatment was such that the ratio of humic-C to fulvic-C was completely reversed so that the more humified mucks appeared to be almost exclusively fulvic acids. It is noteworthy that in a group of peats and mucks of similar botanical origin, the more humified mucks contained most of their humus as humic acids (Table 7) (Preston et al., 1981).

It is likely that elements such as copper and iron form intermolecular linkages between fulvic and humic acids, thus assisting aggregation. Therefore the separation of humic and fulvic acids would be influenced by the presence of such metals by determining the size of the humic macromolecules.

Determination of the amounts of humic and fulvic substances in peats and mucks of various botanical origin and mineral content to establish whether advances in humification preferentially favor the increase of one or the other would therefore be achieved only when proper methods of extraction and fractionation are established and followed in full cognizance of the problems discussed in this and other chapters. The fractionation of humin from min-

TABLE 6. Yields of Humic Acids (HA) and Fulvic Acids (FA) from Some Organic Soils Extracted with 0.1 or 0.5*N* NaOH With (B) or Without (A) Prewashing with a Solution of HCl-HF[a]

					Percent of Total Carbon in:					
			Total Carbon (%)		HA		FA		HA + FA	
Soil Type	Pyrophosphate Index	Ash (%)	A	B	A	B	A	B	A	B
Peats	22.80 ± 6.53	11.15	46.31	44.59	2.93	11.98	4.33	3.14	7.26	15.12
		±1.65	±1.08	±0.97	±0.39	±3.09	±0.76	±0.73	±0.97	±3.16
Peaty mucks or	51.2 ± 3.68	9.26	47.40	44.94	4.43	16.78	1.93	2.58	6.36	19.36
mucky peats		±1.09	±1.41	±0.88	±1.16	±2.90	±0.70	±0.52	±0.07	±2.35
Mucks	94.3 ± 5.87	18.79	39.47	37.88	6.04	0.88	4.21	40.71	10.25	41.59
		±2.35	±2.09	±2.15	±1.39	±0.32	±0.61	±7.12	±0.94	±7.26

[a] *Variations given are standard errors of the means (Schnitzer, 1967).*

TABLE 7. Some Properties of and Percent Distribution of Carbon in Peats and Mucks from Sackville, N.B., Canada[a]

	Ash (%)	Pyrophosphate Index	—OCH_3 (%)	Total Carbon (%)	Percent of Total Carbon in: Humic Acids	Fulvic Acids
Peat 1	10.0	12.2	2.27	47.16	14.2	7.9
Peat 2	8.4	14.8	3.05	49.10	11.2	5.5
Muck 1	29.1	51.0	1.62	37.55	31.7	11.9
Muck 2	31.3	100.0	1.32	36.62	44.4	8.3

[a] *Mathur et al. (1980) and Preston et al. (1981).*

eral soils is not hampered by the presence of nonhumic organic substances as much as by mineral matter. In the case of organic soils, however, humin is often taken as the residue after alkali extraction. One may follow the alkali extraction with a HCl-HF treatment to remove most of the clay and nonclay mineral matter. Even then, neither the residue of this treatment nor an alkali extract of the residue may be pure humin due to the presence of unhumified organic matter and the chars that result from occasional fires in many peatlands. In most studies, however, the residue of alkaline extraction is considered to be humin particularly if the samples had been prewashed with a mild acid like 0.01N HCl as in a study by Zelazny and Carlisle (1974). Since all eleven samples from three soil profiles were moderately decomposed (Medisaprist), the variations in proportion of the three humic substances (Table 8) were probably due to the mineral contents of the soils and state of aeration which affects the process of humification in the various depth layers.

As indicated before, the distribution of carbon in humic substances is often determined as part of a proximate analysis in which soluble phenols, carbohydrates, lipids, and lignins are removed by acid hydrolysis or selective solvation to obtain data of the type presented in Table 9. Perhaps a less

TABLE 8. Distribution of Organic Matter in 11 Florida Mucks (with Standard Errors of the Means[a])

Total		Humin		Humic Acids		Fulvic Acids	
Range	Mean	Range	Mean	Range	Mean	Range	Mean
29.1 to 93.7	74.04 ±21.30	35.1 to 86.2	65.12 ±4.12	7.2 to 54.6	20.17 ±4.22	2.9 to 35.8	14.74 ±2.75

[a] *Zelazny and Carlisle (1974).*

TABLE 9. Classification and Chemical Data for Different Peat Types[a]

Type of Peat	Botanical Type	Degree of Decomposition (%)	Ash (%)	Bitumen	Readily Hydrolyzable Substances	Hemi-cellulose	Humic Acids	Fulvic Acids	Nonreadily Hydrolyzable Substances	Nonhydrolyzed Residue
High moor	Scheuchzeria–Sphagnum	5	2.6	2.7	50.6	24.6	8.5	8.8	20.7	9.2
	Sphagnum (hollow)	15–20	5.0	8.5	42.4	18.0	9.6	15.3	16.8	6.7
	Sphagnum–Cotton grass	40	2.6	10.9	26.2	11.4	25.6	16.2	7.3	14.4
Low moor	Carex	30	5.4	3.7	31.7	17.6	29.5	12.6	5.8	23.3
	Carex	35	8.3	5.0	29.8	14.5	36.9	19.6	4.1	4.8
	Reed	30	9.2	3.3	29.6	16.8	40.1	9.7	5.4	11.9
	Reed	35	10.9	2.8	25.5	9.0	41.8	12.7	3.3	13.9
	Wood	50	11.7	3.0	27.9	10.9	37.4	16.0	3.9	12.8
Transitional	Cotton grass	70	4.1	22.3	6.6	1.5	45.4	11.9	4.2	9.6
Tropical (Ceylon)	Grass	25	15.5	2.8	12.5	—	34.7	18.2	1.2	27.0
	Wood and grass	50	31.8	8.1	13.1	—	58.5	8.2	1.2	10.0
Submerged	Hypnum Carex	30	7.8	8.3	10.5	3.6	20.0	6.1	10.6	43.7

[a] *Data from Rakovskii et al. (1963) as percent of organic matter.*

tedious method of obtaining humic substances from organic soils would be to physically separate the decomposed material through flotation, or sieving, or both, and then extract the fine material only. For example, Schnitzer and Desjardins (1965, 1966) and Schnitzer (1967) sieved air-dried organic soils and used only material that was less than 2 mm in size. Since this fraction is obviously not entirely humic substances, its solubility in 0.1*N* NaOH, following an acid pretreatment, varied from 6 to 100% (Schnitzer, 1967). Also, because some humus may be present on surfaces of larger particles, the yields of humic and fulvic acids from peats and mucky peats or peaty mucks were similar although their pyrophosphate indices were not (Table 6).

Elemental composition of humic substances in peats and mucks is considered to be similar to that of corresponding humic substances in mineral soils (Manskaya and Drozdova, 1968; Preston et al., 1981) (Table 10). The distributions of nitrogen and types of amino compounds in peats and mucks, or

TABLE 10. Elemental Composition of Humic Substances from Various Organic Soils with Standard Errors of the Means

Organic Soils	Number of Samples	Carbon (%)	Hydrogen (%)	Nitrogen (%)	Sulfur (%)	Oxygen (%)
				Humins		
Florida mucks[a]	11	57.8 ± 0.6	4.7 ± 0.2	2.9 ± 0.1	0.4 ± 0.1	34.2 ± 0.6
				Humic Acids		
Florida mucks[a]	11	57.0 ± 0.3	4.5 ± 0.2	3.3 ± 0.1	0.4 ± 0.1	34.8 ± 0.3
Florida mucks[b]	11	56.5 ± 0.5	5.9 ± 0.1	3.9 ± 0.2	1.0 ± 0.0	32.7 ± 0.6
Sackville peats[c]	2	53.3 ± 0.8	5.6 ± 0.1	4.0 ± 0.3	1.5 ± 0.2	35.8 ± 0.8
Sackville mucks[c]	2	49.6 ± 0.1	5.2 ± 0.3	3.4 ± 0.5	0.8 ± 0.1	41.1 ± 0.8
				Fulvic Acids		
Florida mucks[a]	11	54.5 ± 1.3	5.3 ± 0.4	1.9 ± 0.2	0.8 ± 0.2	37.6 ± 1.1
Sackville peats[c]	2	44.7 ± 1.1	6.7 ± 0.9	3.4 ± 0.1	1.2 ± 0.2	44.2 ± 2.2
Sackville mucks[c]	2	42.6 ± 0.1	6.8 ± 1.0	3.3 ± 0.3	1.2 ± 0.1	46.3 ± 1.3

[a] *Zelazny and Carlisle (1974).*
[b] *Volk and Schnitzer (1973).*
[c] *Preston et al. (1984).*

their associated humic and fulvic acids, have been found to be similar to those in mineral soils (Preston et al., 1981).

Functional group analyses of Histosols and their humic substances are of special interest because they may reflect the botanical origin of the soil and the extent or rate of humification. For example, Table 11 presents data on 20 organic soils widely varying in their extent of humification. The soils were prewashed with HCl-HF, total hydroxyls were determined by H_2SO_4-catalyzed acetylation with acetic anhydride, and total acidity by baryta absorption under nitrogen. Carboxyl groups were estimated by the calcium acetate method. The difference between total acidity and carboxyl content was taken as phenolic hydroxyls, and the difference between phenolic and total hydroxyls was taken as alcoholic groups. Carbonyl groups were measured by oximation, and by treatment with 2,4-dinitrophenyl hydrazine. Values obtained by the two methods were averaged. The methoxyl (—OCH_3) groups were determined by the Zeisel method. Although the results (Table 11) revealed a logical trend in the carboxyl and hydroxyl groups, the methoxyl groups increased with humification contrary to the expectation that demethoxylation occurs as lignin is humified. The reason for the anomaly could well have been that the peats were originally poor in OCH_3 groups due to low content of lignins as in *sphagnum* peats (Fuchsman, 1980), while the mucks had more herbaceous or woody components.

Volk and Schnitzer (1973) concluded that variations in the functional group components and spectral properties of humic acids from a group of Florida mucks indicated that higher rates of humification were related to (1) greater amounts of carboxyl, phenolic hydroxyl, quinone, and ketonic carbonyl groups; (2) fewer alcoholic hydroxyl groups and aliphatic structures, as per IR evidence; and (3) increments in E_4/E_6 ratios and free-radical contents as revealed by ESR spectroscopy (Table 12).

By using methods analogous to those in the above studies Zelazny and Carlisle (1974) found that oxygen-containing functional group levels in humin and humic and fulvic acids from all series and layers of the Florida mucks were similar (Table 13), irrespective of their extents of humification or rates of subsidence.

The distribution of nitrogen and types of amino compounds in peats and mucks, or their humus fractions, have been found to be similar to those in mineral soils (Preston et al., 1981). The "unknown" nitrogen component of peats has not yet received the attention given to the same in mineral soils (Schnitzer, Chapter 12 of this book).

Attempts at characterizing peat humus through spectroscopy and chemical degradation procedures (Walmsley, 1973; Stevenson, 1974; Fuchsman, 1980) also have shown that peat humic substances are similar to those from mineral soils. For example Levesque et al. (1980b) found that humic substances extracted by sodium pyrophosphate from 10 different peat materials that varied in botanical origin and extent of humification yielded aliphatic and phenolic compounds and benzenecarboxylic acids (Table 14) in amounts

TABLE 11. Major Functional Groups (meq/g) in Organic Soils Varying in Their Extent of Humification[a]

Types of Soil	Pyrophosphate Index	COOH	Phenolic OH	Alcoholic OH	C=O	OCH_3
1. Peats	23 ± 4	1.9 ± 0.21	3.6 ± 0.16	2.4 ± 0.22	2.7 ± 0.19	0.4 ± 0.10
2. Peaty mucks or mucky peats	51 ± 4	2.0 ± 0.21	3.5 ± 0.16	1.2 ± 0.22	2.7 ± 0.19	0.6 ± 0.10
3. Mucks	94 ± 3	2.5 ± 0.15	3.3 ± 0.11	0.9 ± 0.15	3.1 ± 0.13	0.8 ± 0.07

[a] *Schnitzer and Desjardins (1980).*

TABLE 12. Functional Groups (meq/g) and Spectroscopic Properties of Humic Acids from 11 Samples of Florida Mucks[a]

Functional Group or Property	Range	Mean	Standard Error of the Mean
Total acidity	4.0–8.6	5.9	0.4
Carboxyls	2.8–6.8	4.0	0.3
Phenolic OH	1.0–2.9	1.9	0.2
Alcoholic OH	0.6–5.5	3.4	0.5
Quinone C=O	0.6–2.4	1.4	0.2
Ketonic C=O	0.6–7.0	2.4	0.5
E_4/E_6 ratio	4.4–6.3	5.6	0.2
Free radical spin, $g \times 10^{-17}$	1.4–6.7	3.2	0.5

[a] *Volk and Schnitzer (1973).*

and ratios nearly independent of the botanical origin and humus contents of the materials. In addition, the compounds identified were the same as those obtained by oxidation of humic substances from mineral soils (Levesque et al., 1980b).

Preston and Ripmeester (1982) compared both solution and solid state ^{13}C NMR spectra of four organic soils with those of their humins, humic acids, and fulvic acids. The application of NMR to humic substances is discussed by Wershaw in Chapter 22. The humins appeared to be very similar to the whole peats and mucks, as were the humic acids, except that the latter were marginally higher in aliphatic (paraffinic) carbon. The fulvic acids, on the other hand, appeared to be largely carbohydrates, with high carboxyl contents. Further attempts at quantifying the structural parameters of the humic materials from the organic soils were hampered by considerable peak overlap, broadness of peaks, lack of simple line shapes, and low signal-to-noise ratios. It was also realized that relative intensities of CPMAS (*c*ross-*p*olarization *m*agic *a*ngle *s*pinning) signals vary differentially with acquisition con-

TABLE 13. Functional Groups in Humic Substances from 11 Florida Muck Samples, meq/g with Standard Errors of the Means[a]

	Total Acidity	Carboxyls	Phenolic OH	Alcoholic OH	Carbonyls
Humins	5.1 ± 0.2	2.0 ± 0.2	3.1 ± 0.2	3.6 ± 0.3	2.6 ± 0.2
Humic acids	7.2 ± 0.4	3.1 ± 0.2	4.2 ± 0.3	1.3 ± 0.3	1.3 ± 0.1
Fulvic acids	8.6 ± 0.4	4.0 ± 0.2	4.6 ± 0.2	0.8 ± 0.2	4.3 ± 0.1

[a] *Zelazny and Carlisle (1974).*

TABLE 14. Summary of Types of Compound Obtained by Oxidation of 1.0 g of Pyrophosphate-Soluble Peat Humus with CuO-NaOH or $KMnO_4$[a]

Location of the Peat Layer	Pyrophosphate Index	Main Plant Precursor[b]	Type of Compounds: mg from Each Oxidation: Aliphatics		Phenolics		Benzenecarboxylics		Ratio of Benzene Carboxylates to Phenols	
			CuO	$KMnO_4$	CuO	$KMnO_4$	CuO	$KMnO_4$	CuO	$KMnO_4$
Keswick, Ontario	6.7	S	5.9	0.1	32.8	1.0	16.7	3.5	0.51	3.5
St. Clair, Quebec	9.3	M	3.4	4.2	13.9	1.7	6.5	4.6	0.47	2.7
Ormstown, Quebec	10.4	W	14.6	0.2	36.3	0.2	9.8	3.2	0.27	16.0
St. Chrys., Quebec	12.3	S	34.3	0.0	47.5	0.8	23.9	4.9	0.50	6.1
St. Clair, Quebec	12.5	M	21.3	0.0	50.4	0.2	20.7	3.6	0.41	18.0
Keswick, Ontario	15.4	S	12.2	0.3	17.5	0.1	17.9	5.9	1.02	59.0
Farnham, Quebec	26.8	W	8.3	0.0	48.2	1.1	3.5	10.3	0.07	9.4
Ormstown, Quebec	29.2	S	9.6	0.1	19.4	0.3	16.4	5.6	0.84	18.7
Farnham, Quebec	35.4	WS	15.2	0.3	66.7	0.6	14.3	10.4	0.22	17.3
Farnham, Quebec	39.8	S	17.8	0.5	53.6	1.1	16.1	10.6	0.47	9.6

[a] *Levesque et al. (1980b).*
[b] *W = wood; S = sedges; M = Moss.*

ditions used in obtaining the spectra and metal contents of the humic material. For example, Preston et al. (1981) observed, by varying acquisition parameters, that increments in copper content of the organic soils from 400 to 3000 μg/g caused greater loss in the intensity for carbohydrate regions of the spectra of the four organic soils than in other regions. However, it appears that under appropriate acquisition conditions, solid-state CPMAS NMR spectra may be useful in obtaining comparative semiquantitative estimates of the relative abundances of carbon types in organic soils of similar origin but humified to varying extents.

As the origin of Histosols is so variable, and the conditions in which they are formed so diverse, humic substances in these soils offer a great opportunity to explore the role of various compounds and reactions that may contribute to the synthesis of humic substances (Stevenson, Chapter 2 in this book). The limited number of generally descriptive studies made thus far suggests that humic substances in Histosols generally resemble those in mineral soils, with one exception. As a result of NMR studies, Hatcher et al. (1983a) suggested that some marine peats contain humic substances with largely aliphatic carbon atoms, perhaps because the sediment is derived almost exclusively from marine algae. There is a need for investigations in which comparisons are made between similarly extracted humic substances from various peats that differ mainly in origin or conditions of formation.

HUMIFICATION OF UTILIZED PEATLANDS

The water-logged deposits of carbon and plant nutrients in peatlands are mined for fuel, soil ameliorants, and greenhouse substrates; managed for forestry and forage production; and cultivated for agriculture. The agronomic or silvicultural usages often follow mining of the surface peat for fuel or greenhouse substrates. On a global scale, about 4.5 million hectares of peatlands are being mined for fuel and garden peat, 0.5 million hectares are being used for forestry, and a few million hectares are under agricultural use. Consequently about 22% of the global peatlands are in economic use (Kivinen, 1980a; Kivinen and Pakarinen, 1980). The percentage of peatlands in use in Poland and Germany, however, is up to 75% of the total. All economic usages of peatlands require drainage, except for the small area now being mined by hydraulic dredge methods in which the peat is removed as a slurry.

The removal of water by drainage causes aeration of the peat. The resultant surge in decomposition and humification is often further accelerated by the liming, fertilization, and tillage needed for forestry or agriculture. Bramryd (1980) and Aramentano (1979), among others, have assessed the impact of the decomposition and humification of drained peatlands on the global carbon and nitrogen cycles to be considerable. As the humus produced is more compact than the undecomposed plant residues, humification also contributes to the loss in volume of the organic deposit. The loss manifests itself as a slow but continual decline in surface elevation of the cultivated organic soils by 1 mm to 7 cm/yr. This phenomenon is called subsidence. The large diversity in the rates of subsidence is indeed the most tangible evidence of the influence of various environmental and other factors on the rate of humification of Histosols. As indicated before, no single method has been found to be totally satisfactory for measuring or comparing humification among various peatlands. Subsidence thus offers the best available perspective for considering the relative effects of various factors that influence the rate of humification in peatlands (Mathur, 1982c).

Polyphenolic plant constituents resist microbial decomposition more than polysaccharides. Consequently, peats richer in cellulose decompose faster and to a greater extent than woody peats. As climate influences all biological processes, Eggelsmann (1976) found that within a group of low moors the subsidence rate was correlated with the Lang factor (millimeters of annual precipitation/mean temperature in °C). Consequently, Eggelsmann (1976) observed that a maximum rate of subsidence, 7 cm/yr, occurred at the hottest and driest site which was in Greece. Maintaining a high water table in a managed peatland helps to reduce humification and subsidence because the water curtails the amount of air in soil pores. The level at which the water table may be maintained safely and conveniently is partly determined by the types of crop grown. The crop choice also influences the level to which soil pH has to be raised by liming. Liming mobilizes phenolic compounds (Morita, 1975) which may be transported into ground and surface waters. Since some of these phenolic compounds are bacteriostatic or bactericidal, their removal enhances microbial processes. In any case, microbial decomposition and concomitant humification are at their maxima at or near neutral pH. For example, Frercks and Puffe (1959) noted that about 50% of the increase in the decomposition rate due to rising pH occurs between pH 4 and 6. Most agricultural crops require a soil pH above 5 or 5.5. As decomposition and humification progress, and the peat becomes denser, the voids between soil particles tend to become smaller. Such micropores are more apt to be anaerobic than macropores, and therefore the lack of oxygen slows down humification with time. The lack of oxygen can also occur if root density in soil is as great as it is under grass. Peatlands where grass predominates therefore tend to decompose and subside less than those under row crops (Schothorst, 1977; Mathur, 1982c).

EFFECTS OF HUMIFICATION ON PEATLANDS

As discussed in the above section, humification and decomposition of drained and cultivated peatlands cause subsidence. Up to 70% of the continual loss in surface elevation by 1–70 mm/yr is due to biochemical processes (Stephens and Speir, 1970; Mathur, 1982c). Subsidence threatens to phase out of usage vast areas of highly profitable cultivated organic soils in the United States, Canada, and elsewhere (Stephens and Speir, 1970; Mathur, 1982c). As the soils decompose, humify, and subside, the increasing humus content provides sites for strong complexation of some essential micronutrients and retention of various pesticides used for crop protection. In addition, the water-holding capacity and hydraulic conductivity of the soils decrease because humified peats are more compact and retain less water than raw peat.

Subsidence

The humification and resultant subsidence of cultivated organic soils, typically 2.5 cm/yr, not only threaten the existence of these soils but also necessitate periodic intensification of the drainage systems, and protection of the fields against flooding from surrounding higher groundwater or surface water bodies. Therefore, development of methods to control subsidence is an important area of research and requires the application of what is known about the process of humification and chemical properties of humic substances.

Conventional Methods of Reducing Subsidence

Maintaining a high water table in the field is the most obvious method of slowing down humification and subsidence because the water excludes air that promotes oxidative decomposition as well as gives buoyant support to the surface layer (Mathur, 1982c). However, the method is not entirely satisfactory in many cases because it increases chances of (1) periodic flooding of crops if rainfall is heavy, (2) difficulties in tillage, and (3) damage to crops by the H_2S produced during anaerobic decomposition or by the lack of oxygen needed for root respiration. Consequently, water tables are generally not maintained as high as would be necessary to curtail the decomposition substantially.

Most vegetable crops grown on organic soils require a soil pH of about 5.0. However, the aims of achieving uniform crop growth and quality within a few years of cultivation often necessitate liming the soils to a pH of about 5.5. Such surplus applications of lime ensure that those areas of a field which are more acidic than others, due to natural variation, have a pH of at least 5.0 even before uniform mixing and complete dissolution of the lime is achieved in up to 5 years. As the soils humify and gain in mineral content, the buffering capacity of the soils decreases and the pH rises. This happens to a greater extent if the sublayers that are continually mixed into the top layer are not much more acidic than the surface layer, as in Florida peatlands. It is not known whether large-scale application of sulfur to such soils would be practical and beneficial in significantly reducing humification and subsidence. The sulfur will be oxidized by soil bacteria to H_2SO_4 which can decrease or maintain the soil pH at or about 5.0.

Overlaying of sand on peat pastures as practiced in Europe helps reduce subsidence. The water table can be kept at the mineral–organic interface while the sand is able to bear the weight of cattle, thus preventing the animals from sinking into the wet peat. In certain fen peats used for growing cereals, fields are kept flooded from fall to spring to permit sedimentation of mineral matter from river water. This mineral matter is then mixed into the organic layer, thus helping to stabilize the humus in clay–humus complexes.

Retardation of Humification and Subsidence by Copper

Since most of the polymers, such as cellulose and lignins, that form the bulk of the organic matter in soils and forest litter are water-insoluble and of large molecular size, they do not permeate microbial cell walls. The decomposer soil microbes therefore excrete soluble enzymes that degrade the polymers into water-soluble oligomers and monomers that are then absorbed and utilized. As discussed previously, most of the enzyme molecules in soil are ephemeral. Some, however, survive due to immobilization on humus and clay colloids. These are truly soil enzymes. Activity levels of the soil enzymes in forest litters and peats are found to be correlated with the rate of decomposition of the organic components of soil (Tyler, 1976; Ross and Speir, 1979; Freedman and Hutchinson, 1980; Mathur and Levesque, 1980; Mathur and Sanderson, 1978, 1980a,b; Mathur, 1982a,b).

Complexation of metals such as lead, mercury, silver, gold, and copper by the amino, carboxyl, and thiol groups of enzyme proteins, particularly at the active sites, inactivates the biocatalysts. Of all the feed and fertilizer elements essential to life, copper is most effective in such enzyme inactivation. Consequently, the concentration of residual fertilizer copper, in each of four different areas of cultivated organic soils in Canada, was found to be negatively correlated with their enzyme activity levels and overall rates of decomposition (Mathur and Rayment, 1977; Mathur and Sanderson, 1980a,b) (Fig. 1). Most noticeably, the decomposition rate under field conditions declined by 70% as the concentration of total copper in soils increased from 100 to 300 μg/g (Mathur et al., 1979a) (Fig. 2). The reason for such variation in copper levels in cultivated organic soils is that although copper is generally recognized as essential to farming on organic soils the recommended application rates vary from 0.5 kg/ha periodically to 58 kg/ha initially, followed by 14 kg/ha per year. This is due to lack of definite knowledge of the residual effect of fertilizer copper. Extensive research led to the recommendation that the humification and subsidence of cultivated organic soils can be retarded by about 50% by gradual additions of copper (5–15 kg/ha per year) to attain and maintain copper concentrations of 100–400 μg/g in soils with bulk densities of 0.1–0.4 g/cm^3 (Mathur, 1982a,b; Levesque and Mathur, 1983b; Mathur and Levesque, 1983a,b).

The effect of copper on decomposition of organic soils is persistent because (1) the levels of copper required for mitigating subsidence are not

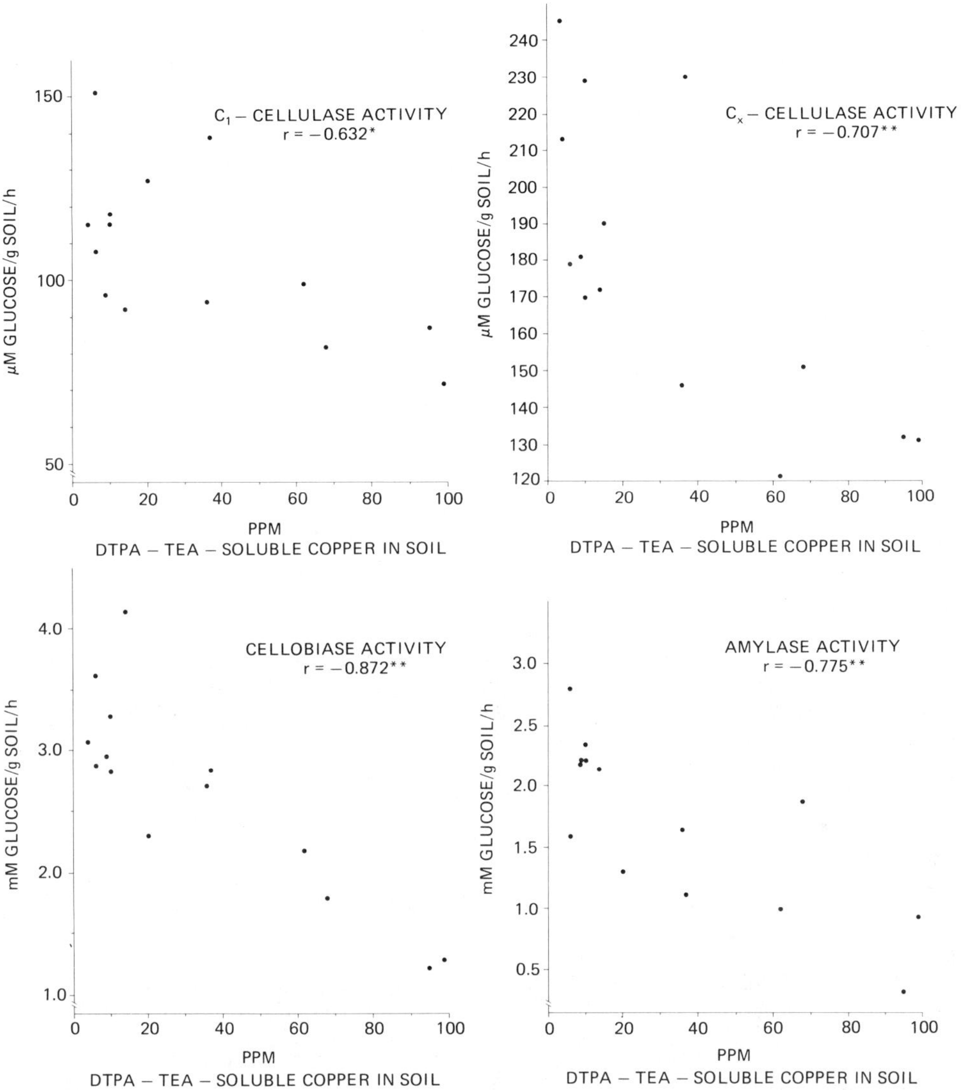

FIGURE 1. The correlations between DPTA–TEA soluble copper contents and enzyme activities of the investigated soils.

directly bactericidal in organic soils (Mathur, 1983); (2) the copper in organic soils remains in the relatively biodynamic and acid-hydrolyzable fraction of humus (Preston et al., 1981); and (3) copper is complexed by proteins more strongly than by humus (Mathur et al., 1980).

This effectiveness of copper in limiting organic soil degradation is in agreement with the supression of decomposition observed in forest litters contaminated with zinc, nickel, and copper (Tyler, 1976; Freedman and

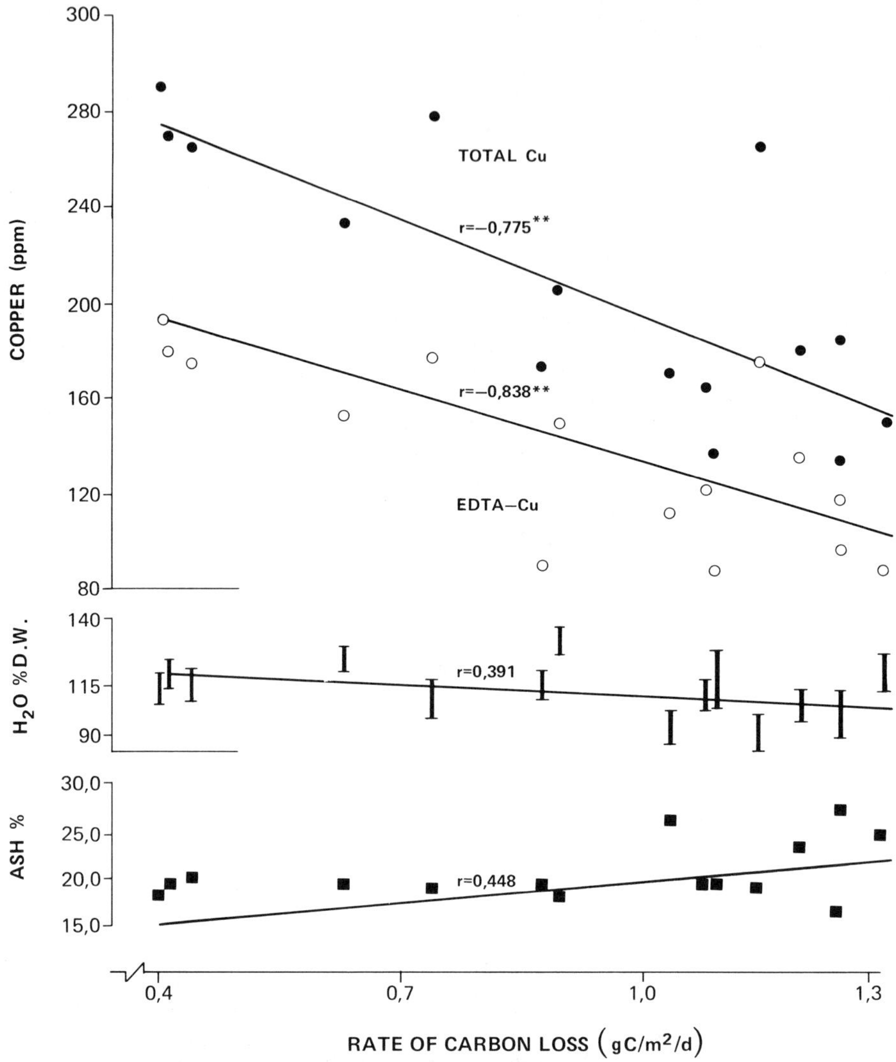

FIGURE 2. Correlations between carbon loss, as carbon dioxide in the field vs. total copper, EDTA–copper, water content, and percent ash. Statistical significance indicated by asterisks for significance at the 1% level. Total copper is expressed as solid circles, and EDTA–copper as empty circles.

Hutchinson, 1980). In these studies copper was found to be more effective than the zinc or nickel in suppressing decomposition.

Another important consideration in the use of copper to reduce subsidence is a possible effect on the availability of plant nutrients. Studies of a natural cupriferous bog that has contained elevated levels of copper for centuries revealed that the distributions of carbon and nitrogen in the soils and humus fractions were affected quantitatively but not qualitatively by the copper. In addition to this evidence, it has also been shown that the ability of organic soil biota to provide ammonia and nitrate for crops is not affected by the soil-copper concentrations used for reducing subsidence (Mathur and Preston, 1981). One reason for this is that the ammonification and nitrification processes that mineralize soil-nitrogen to plant-available forms occur mostly within microbial cells, where the macromolecular complexes of soil-copper do not permeate. Another reason that nitrogen availability is not reduced by copper additions may be that immobilization of ammonia and nitrate in soil by the microbial biomass is reduced for lack of dissolved carbon sources.

Greenhouse and microplot studies have shown that even in poorly limed peats the 100 μg/g concentration of copper would not be phytotoxic while even two to three times the 400 μg/g copper level required in well-humified mucks would not cause any physiological stress on grain, bulb, leaf, root, or cole crops (Mathur et al., 1979b; Levesque and Mathur, 1983a,b; Mathur and Levesque, 1983a,b; Mathur et al., 1983a,b). The normal antagonisms between plant absorptions of copper and iron, manganese or zinc were found mostly offset because the added copper displaced the other metals from stronger to weaker ligands that are more accessible to plants (Mathur and Levesque, 1983a). Recent evidence confirmed an earlier suggestion that soil copper begins to be phytotoxic only when total copper in a soil is equivalent to 5% of the soil's cation-exchange capacity (CEC). Since the CEC of organic soils usually falls between 100 and 150 meq/100 g soil, the safe loading limit for copper in these soils should be 1600–2400 μg/g soil. Only a few percent of the copper in organic soils is in plant-available forms (Mathur and Levesque, 1983a).

Copper is not held as strongly in mineral as in organic soils due to a relative lack of humus in the former. Shallow organic soils are often mixed with their mineral sublayers for agronomic convenience. However, the weight-basis copper concentrations in the mineral–organic soil mixtures would be lower than in the organic layer because the mineral sublayers are severalfold denser than organic soils. Indeed, greenhouse studies have shown that the recommended levels of copper would have no adverse effects even on crops with underground edible portions grown in mixtures of the organic soil with various mineral sublayers (Levesque and Mathur, 1983a, 1984; Mathur and Levesque, 1983a). The copper concentrations of crops grown would be much below maximum safe levels of copper in food. Copper becomes toxic to plants before reaching levels unsafe for animals feeding on the plants, except for sheep on a molybdenum-deficient diet.

Humus can hold copper in concentrations of up to 1% of its weight. Indeed some bogs through which copper-enriched artesian waters flow naturally contain up to 100,000 μg/g copper probably in their humus as Cu^{2+}, as $CuOH^{+}$, and as occlusions in iron or manganese oxides, and as mixed oxides. It was therefore found that copper applied to the surface soil did not move downward into groundwater except when a single application of 1500

$\mu g/g$ was made to an acidic peat that contained little humus (Mathur et al., 1984). Any copper that may be carried to surface waters through soil erosion is unlikely to affect any organisms as it would be in nonbiocidal forms. Indeed, because the copper is not in inorganic forms it also fails to hydrolyze organophosphorus pesticides (Belanger and Mathur, 1983). Nonetheless, it would be advisable not to apply organophosphorus compounds simultaneously with inorganic copper for reducing the subsidence rate.

Humification and Minor Element Fertilization of Organic Soils

Most peatlands are deficient in iron, manganese, zinc, copper, boron, and molybdenum due to lack of clay minerals and the leaching of some of these elements under wet acidic conditions. When virgin organic soils are drained, limed, and fertilized for agronomic use, the humus content increases and the paucity of the minor elements is accentuated due to increased phytomass production and decreased return of plant residue to the soils. With continued cultivation and resultant biological oxidation the mineral content increases. This tends to correct minor element deficiencies. However, the change in mineral content is often accompanied by an increase in soil pH beyond the range of 5.0–5.8 that is most suitable for plant-availability of the nutrient metals. Also, the continued humification and slow mineralization results in a lower level of saturation of the humus with nutrient metals, and an increase in molecular size of humic complexes of di- and trivalent metals. Humic complexes of copper, zinc, and manganese, unlike sodium and potassium humates, tend to be water-insoluble and are thus less available to plants unless the molecular weights of the humates, or their metal contents, are low. Recent studies have shown that metals such as copper are associated with humic rather than fulvic acids in Histosols (Petruzzelli et al., 1975; Petruzzelli and Guidi, 1976; Preston et al., 1981). It is logical that as the humus content increases more than the metal content in soil, more and more of the metal nutrients would tend to be bound at sites of greater complexing strength where they would be less accessible to plants. These points were well illustrated by Ennis and Brogan (1961).

Petruzzelli et al. (1975) showed that while only 11% of the copper added to an alkaline extract of soil was resistant to removal by the Amberlite IR 120 resin, the same resin could not remove 66% of the copper naturally present in the humic substances. Consequently, it was surmised that when copper is added to organic soils a part of it is bound to low-molecular-weight humic or fulvic acids from which it can be extracted by the chelates in root exudates. Time, humification, and lower intensity of site coverage favor the retention of the copper at stronger, and perhaps sterically hindered, complexation sites (Petruzzelli et al., 1975; Petruzzelli and Guidi, 1976). Therefore, it is advisable to periodically apply small dosages of copper to organic soils, once every 1–3 years, particularly if the soil contains large amounts of humus, even if the total amount of copper in these soils is many times the requirements of crop nutrition. Although much less work has been done with zinc and manganese, this reasoning should generally apply to these metals as well.

Retention of Pesticides in Organic Soils

Many pesticides or their degradation products tend to be adsorbed by or incorporated into soil organic matter, particularly humus. For example, Mathur and Morley (1975, 1978) showed that even the relatively unreactive insecticide methoxychlor was unextractably retained in model chemical and fungal humic acids to levels of 263 and 50 ppm, respectively.

The complexes formed are considered to be analogous to "clathrate compounds" (Khan, 1982) wherein the pesticide or its products are believed to be entrapped in the internal voids of sieve-like humus molecules.

Organic soils generally receive about twice the amount of a pesticide required for the same effect in a mineral soil. The reasons are (1) the poor bioavailability of the pesticide or herbicide for the target organism (due to pesticide retention by humus) and (2) on a volume basis, organic soils contain more water than mineral soils so that more solute is required in the former for equal and effective concentrations of the solute. Even moderately decomposed organic soils are comparatively rich in humus of which hundreds of kg/ha are synthesized annually. Since humus can be biodegraded, albeit slowly, some of the xenobiotic compounds may be released from the humus by soil microbes, as has been amply demonstrated (Hsu and Bartha, 1974; Mathur and Morley, 1975; Khan, 1982). The presence of these slowly released pesticides may lead to undesirable uptake of the chemicals by maturing crops and damage to crops susceptible to herbicides applied for a different crop in the previous year (Morris and Penny, 1971; Khan et al., 1976a,b; Belanger and Hamilton, 1979). Such problems, of course, do not arise when the herbicides are applied as foliar sprays. Also, the behavior of soil-applied pesticides is influenced by various physical and chemical properties of the soils (see, e.g., Belanger et al., 1982).

The protracted presence of pesticides or their degradation products (Mathur and Saha, 1975, 1977) in organic soils may evoke the evolution of resistant target organisms, as well as increase chances of perturbation of desirable microbial activities in these soils.

SUMMARY AND CONCLUSIONS

World peatlands have been estimated to cover about 500 million hectares. Although the resource data are incomplete, these estimates indicate that approximately 90% of the peatlands occur in glaciated areas of the northern hemisphere and only 10% in equatorial regions and the southern hemisphere.

The formation of humic substances in a peatland environment is a complex humification process which is principally due to certain enzymatic and microbial activities. These organic matter transformation processes are influenced by the nature of the peat-forming plants and certain physical and chemical properties within a particular peatland. In very acidic or low nutrient peatlands a very different microflora may exist than in a more eutrophic and less wet situation. In the former situation humification may be retarded and the peatland plants will be preserved and thus accumulate. On the other hand, in less acid environments with moderate amounts of nutrients and periodic water-table fluctuations humification proceeds relatively rapidly and leads to decomposed organic soils: such is the case in drained and cultivated organic soils.

Oxidative degradation appears to be the main mechanism governing humification. In determining the degree of decomposition or humification of peat materials the estimation of total fiber content correlates well with relative biodegradability and is a more reliable method than either the pyrophosphate index or bulk density values.

The processes and agents of decomposition and humification are similar to those in mineral soils although anaerobic microbes and soil enzymes play a greater role in degradation of organic matter in peatlands than in mineral soils. It is not clear whether the enzymes affect humification in the organic soils directly. It is generally recognized that the lack of oxygen in peatlands alters the relationship between the rates of humification and decomposition so that it is different from that in mineral soils, although this has not been established experimentally. It has been indicated that sodium hydroxide extracts of organic soils, particularly their anaerobic zones, contain considerable amounts of nonhumic materials. Extraction with pyrophosphate yields less nonhumified material but does not dissolve all the humus in the peat. If the sodium hydroxide extraction is preceded by an acid treatment, the humus removed tends to be richer in fulvic acids than in humic acids.

The presence of large proportions of nonhumified material of diverse origins and properties hampers the dissolution, fractionation, and estimation of humic substances in peatlands. No satisfactory methods exist for these purposes. Most studies of peat humus therefore focus on pyrophosphate extracts or humic acids obtained by sodium hydroxide extraction.

Elemental composition, functional group analyses, spectral properties, and characterization of acid hydrolyzates have shown that peatland humic acids are similar to those in mineral soils. NMR spectroscopy has revealed that peat fulvic acids are largely carbohydrate in nature while the residue of alkaline extraction is not all humin. It is suspected but not proven that humic substances in peatlands are more soluble, due to lack of clays and mineral elements, than those in mineral soils.

Drainage, liming, fertilization, and cultivation of peatlands all enhance decomposition and humification. Since humus is more compact than plant debris, cultivated organic soils subside (lose surface elevation typically by 2.5 cm/yr). Humification and subsidence threaten to phase out of use vast areas of valuable cultivated organic soils. A new method is now being used to retard the subsidence by about one-half. It is based on recent observation (1) that the rate of decomposition and humification of cultivated organic soils is determined by activities of soil enzymes, and (2) that of all the feed and fertilizer elements, copper is most effective in inactivating the soil enzymes. The recommended rates of copper application have been found to be effective, safe, and economic. The addition of copper tends to increase plant-availability of manganese, zinc, and iron, perhaps due to displacement of these metals by copper from sites of tight complexation to weaker ligands.

The humus in well-decomposed organic soils also plays a role in the bioavailabilty and retention of minor elements, pesticides, and pesticide degradation products.

There are several areas of research on humic substances in peatlands that should be pursued. Data on the chemistry of tropical and southern hemisphere peats are very scanty. We need to study the chemistry of the peat-forming plants of these areas, as well as the more highly decomposed peats of tropical regions such as Malaysia and Indonesia.

It is suggested that benchmark sites (type locations) of the various peat types of the world be selected. The criteria for classification could be that recently proposed by the International Peat Society (1980), which include three important properties—botanical composition, degree of decomposition, and the trophic status (base saturation) of peat.

Large samples of each type could be collected and analyzed in several different laboratories, utilizing the latest techniques for the various humic substances. The classification and analytical methods would be standardized. In this way the extensive reserves of organic matter contained in the world's peatlands could be adequately characterized as to chemical substances.

We also need to study in more detail the accumulation rate of organic matter in natural peatlands occurring in widely different geographic locations and developing under different local environmental conditions.

Also, the effect of increasing carbon dioxide levels in the atmosphere on the growth of peatlands needs further investigation, as does the release of carbon dioxide due to drainage and cultivation.

CHAPTER FOUR

Humic Substances in Groundwater

E. M. THURMAN

ABSTRACT

This chapter discusses and reviews the amount of organic carbon and humic substances in groundwater and the characterization and origin of humic substances in groundwater. Because the concentration of organic carbon in groundwater is commonly less than 1 mg carbon per liter, it is difficult to isolate and characterize humic substances, and there are relatively few studies of these substances in groundwater.

The major study that has been done (Thurman, 1979) is reviewed in detail. In this study humic substances from the water in five aquifers were isolated by sorption onto macroporous resins. The samples came from the following aquifers: Biscayne (Florida), Laramie–Fox Hills (Colorado), St. Peters (Minnesota), and Madison and Red River (South Dakota). The concentration of humic substances in groundwater increased with the concentration of dissolved organic carbon (DOC) from 0.04 mg C/L (21% of DOC of St. Peters) to 8.6 mg C/L (66% of DOC of Biscayne). Humic substances from groundwater have less oxygen and less color than humic substances from surface water. They are similar to humic substances in surface waters in carboxyl content (5–6 meq/g), in molecular weights (1000–5000), and in binding constants for copper (log K *= 5.6 at pH 6.3). Based on these characterizations, humic substances in groundwater may originate either from humic substances in soil interstitial waters, or from* kerogen, *the organic matter deposited with the sediments of the aquifer.*

INTRODUCTION

There is growing public concern with contamination of groundwater by hazardous wastes, and increased interest in the amount, nature, and interaction of humic substances in groundwater. Humic substances may play a role in the complexation of trace metals and the solubilization and transport of relatively insoluble organic compounds. There is also interest in the biogeochemistry of organic carbon and its movement through canopy drip, interstitial waters of soil, shallow groundwater, and surface water (Telang et al., 1981; Meyer and Tate, 1983). Humic substances are transformed and removed in various parts of this cycle, and their study in groundwater provides more information concerning geochemical processes.

There are relatively few studies on humic substances in groundwater (Thurman, 1979; Telang et al., 1981). These substances are difficult to isolate and characterize because the concentration of organic carbon in groundwater is commonly less than 1 mg carbon per liter (mg C/L) (Leenheer et al., 1974). This has been a major deterrent to studying humic substances in groundwater. However, with resin isolation procedures developed over the last 10 years (Aiken et al., 1979; Thurman, 1979; Chapter 14 this book), it is now possible to isolate humic substances from groundwater, even though concentrations are micrograms per liter.

This chapter discusses four aspects of humic substances in groundwater: (1) the concentration of organic carbon and humic substances in groundwater; (2) isolation procedures for humic substances; (3) the elemental composition, molecular weight, and other characteristics of humic substances from five specific groundwaters; and (4) the geochemistry and origin of humic substances in groundwater.

Concentration of Organic Carbon in Groundwater

As water penetrates the surface of soil it enters the unsaturated zone, also called the *vadose* zone. The unsaturated zone varies in thickness depending on the amount of precipitation, *evapotranspiration,* and underlying geology (Fig. 1). The unsaturated zone contains soil moisture and acts as an important barrier to organic matter from the surface. In this zone both biological decay and adsorption processes decrease the concentration of organic carbon in the shallow interstitial waters of soil (Wallis, 1979; Telang et al., 1981; Dawson et al., 1981; Meyer and Tate, 1983). The preceding studies also show that dissolved organic carbon decreases with depth, especially during periods of dryness when the soil is damp but not saturated. When soil is wet during the spring, greater concentrations of dissolved organic carbon are present. As shown in Figure 1, dissolved organic carbon decreases from a median concentration of 20 mg C/L for interstitial water in the A *horizon* to 2 mg C/L in the C horizon. These data are summarized from studies listed previously.

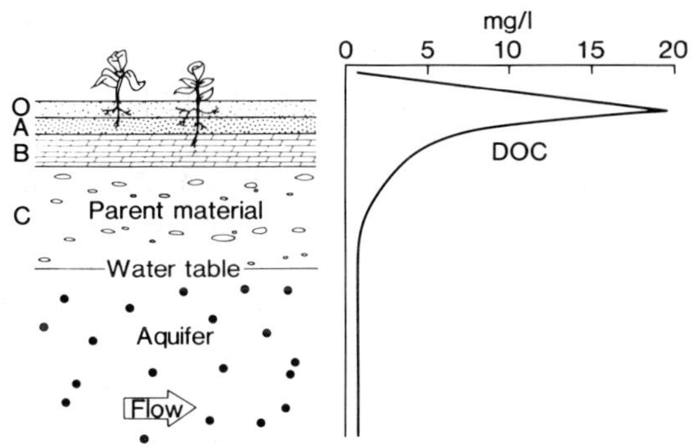

FIGURE 1. Changes in concentration of organic carbon from precipitation to interstitial waters of soil (unsaturated zone) and in saturated zone of groundwater based on Leenheer et al. (1974), Wallis (1979), Dawson et al. (1981), Antweiler and Drever (1983), Meyer and Tate (1983), and Thurman (1985).

In the north temperate climates, organic matter is flushed from soil and plant litter in the unsaturated zone during the spring runoff (Antweiler and Drever, 1983; Thurman, 1985) and during storm events in summer and fall (Meyer and Tate, 1983). This flush carries organic carbon and humic substances into the saturated zone of groundwater or into streams and rivers by surface runoff, and is a major contributor of organic carbon to groundwater and surface water. This process is an important process in biogeochemistry and needs further study.

Table 1 shows the median concentrations of organic carbon from seven different types of aquifers. The median concentration of dissolved organic carbon is 0.7 mg C/L for sand and gravel, limestone, and sandstone aquifers. Only igneous aquifers with 0.5 mg C/L had a lower median concentration of organic carbon. In a study based on 50 samples from various types of aqui-

TABLE 1. Median Concentration of Organic Carbon in Various Types of Aquifers[a]

Aquifer	DOC[b] (mg C/L)
Sand and gravel	0.7
Limestone	0.7
Sandstone	0.7
Igneous	0.5
Oil shales	3.0
Humic colored	10.0
Petroleum associated	100.0

[a] *After Leenheer et al. (1974), Thurman (1979), and Feder and Lee (1981).*
[b] *DOC is dissolved organic carbon.*

fers, Leenheer et al. (1974) showed that the median concentration of organic carbon in the saturated zone is less than 1 mg C/L. This study also found that the concentration of dissolved organic carbon did not correlate significantly with depth of the sample or inorganic chemistry of the sample.

There are groundwaters containing concentrations of organic carbon greater than 1 mg C/L. They generally originate from aquifers receiving recharge from organically rich waters, such as the Biscayne groundwater in Florida (Thurman, 1979; Feder and Lee, 1981). In this aquifer, where recharge from the Everglades is a rich source of dissolved organic carbon, the concentration of humic substances is 10 mg C/L, similar to that found in marshes and swamps.

Other groundwaters that contain greater concentrations of dissolved organic carbon include those in contact with sediments rich in kerogen (the organic matter deposited with sediments). For example, groundwater in oil-shale regions commonly has concentrations of dissolved organic carbon of 2–4 mg C/L (Leenheer and Noyes, in press). *Trona* water (groundwater containing sodium carbonate) is an extreme example; it has concentrations of dissolved organic carbon of 40,000 mg C/L (Thurman, 1985). Groundwaters associated with petroleum and oil-field brines contain large amounts of organic acids and natural gas. For example, dissolved organic carbon may be as much as 1000 mg C/L in oil-field brines (Willey et al., 1975), and volatile organic carbon from natural gas may be hundreds of milligrams per liter.

Concentration of Humic Substances in Groundwater

The concentration of humic substances in groundwater increases with the concentration of dissolved organic carbon. Thurman (1979) measured the concentration of humic substances in five groundwaters using the isolation method of Thurman and Malcolm (1981). In these deep groundwaters (greater than 150 meters) humic substances account for 12–33% of the dissolved organic carbon. In colored groundwater humic substances account for 65% or more of the dissolved organic carbon. Table 2 shows the amount of organic carbon and humic substances in the five aquifers studied. The aquifers consisted of three lithologies: dolomite, sandstone, and limestone; and the inorganic chemistry of the water was of two types, calcium bicarbonate and calcium sulfate. Neither the inorganic chemistry of the water nor the geology of the aquifer correlated with the concentrations of organic carbon and humic material in the groundwater. It was, rather, the origin of the recharge water that controlled the concentration of dissolved organic carbon and humic substances.

Wallis (1979) and Telang et al. (1981) measured dissolved organic carbon and humic substances in groundwater of the Marmot Creek system in Canada and found that humic substances were the dominant fraction of the dissolved organic carbon. Humic substances contributed approximately 900 μg C/L and 90% of the dissolved organic carbon. No characterization of the humic material was done.

TABLE 2. Concentration of Organic Carbon and Humic Substances in Selected Groundwaters[a]

Aquifer	DOC[b] (μg C/L)	Humic Substances (μg C/L)	Geology of Aquifer	Chemistry of Aquifer
St. Peters (Minnesota)	200	40	Sandstone	Calcium sulfate
Madison (So. Dakota)	300	100	Limestone	Calcium bicarbonate and sulfate
Red River (So. Dakota)	500	100	Dolomite	Calcium bicarbonate and sulfate
Laramie–Fox Hills (Colorado)	700	80	Sandstone	Calcium bicarbonate
Biscayne (Florida)	13,000	8600	Oolitic limestone	Calcium bicarbonate

[a] *Thurman (1979).*
[b] *DOC is dissolved organic carbon.*

Hydrophobic Organic Substances

Leenheer and Huffman (1976) designed a hydrophobic classification of dissolved organic carbon using resin adsorption as a means of fractionation. The procedure is based on adsorption chromatography onto XAD resins. Those organic substances that adsorb onto the resins with pH adjustment (low pH for acids and neutral pH for bases) are termed "hydrophobic;" those organic substances that do not absorb are "hydrophilic." Fulvic and humic acids are classified as hydrophobic substances. This classification procedure makes it possible to measure indirectly the amount of humic substances in water. A more detailed explanation of this procedure is presented by Leenheer (1981).

The hydrophobic/hydrophilic fractionation of organic carbon in groundwater is different than the fractionation of organic carbon in surface water. Table 3 shows that the amount of hydrophobic material in the groundwaters studied was less than 35%, and in surface waters the amount of hydrophobic

TABLE 3. Hydrophobic/Hydrophilic Split on Dissolved Organic Carbon from Groundwater

Aquifer	Hydrophobic (%)	Hydrophilic (%)
Red River	58	42
Laramie–Fox Hills	21	79
Madison	33	67
St. Peter	35	50
Surface Water	50	50

material is 50–60% (Malcolm et al., 1977; Stuber and Leenheer, 1978). Results of other studies of groundwater using the hydrophobic/hydrophilic fractionation procedure (Malcolm et al., 1981) reveal that dissolved organic carbon in groundwater is more hydrophilic than that in surface water. This is an important difference between organic matter in groundwater and surface water: the longer residence time of organic matter in groundwater results in hydrophobic substances being either adsorbed onto the aquifer solids or degraded into simpler organic acids by bacteria in the aquifer.

ISOLATION OF HUMIC SUBSTANCES FROM GROUNDWATER

A useful procedure for the isolation of humic substances from groundwater is that of Thurman and Malcolm (1981) with the following modifications. Hydrochloric acid should be added immediately to the water sample to prevent the precipitation of iron hydroxide, and the sample should be evacuated with a vacuum pump to remove hydrogen sulfide. If the sulfide is not removed, it can react to form both elemental sulfur and polysulfides that adsorb onto and clog the XAD resin (Leenheer and Noyes, in press; Thurman, 1979).

Samples may be collected in 45 L glass bottles after pumping the wells for approximately an hour to remove water from the casing and to obtain representative samples of the aquifer. Because sediment is rarely found in groundwater samples, filtration is usually unnecessary.

After concentrating the humic substances using the resin procedure, eluate from the XAD resin is passed through an Enzacryl-gel column to remove low-molecular-weight acids, which are a resin contaminant (Thurman and Malcolm, 1981). Finally, the hydrogen-saturated eluate is freeze dried. This step further purifies the sample of low-molecular-weight acids, and a hydrogen-saturated product of humic material remains. More detailed procedures for the resin methodologies are given in Thurman and Malcolm (1981) and in Chapter 14 of this book. Leenheer (1981) describes an alternative method of isolating humic substances from groundwater.

NATURE OF HUMIC SUBSTANCES IN GROUNDWATER

Elemental Composition

Humic substances from groundwater contain more carbon and less oxygen than humic substances from surface water. As shown in Table 4, humic substances from groundwater commonly contain greater than 60% carbon and less than 30% oxygen, while humic substances from surface water contain an average of 52% carbon and 42% oxygen (Thurman and Malcolm, 1981). There are at least two hypotheses to explain this difference in elemen-

TABLE 4. Elemental Composition of Humic Substances from Groundwater on an Ash- and Moisture-Free Basis

Aquifer	C	H	O	N	Ash
	Fulvic Acid				
St. Peters	62.3	6.3	30.2	0.5	2.2
Laramie–Fox Hills	62.7	6.6	29.1	0.4	1.1
Biscayne	55.4	4.2	35.4	1.8	0.4
Madison	56.5	5.8	—	—	—
Red River	58.5	5.7	—	—	—
	Humic Acid				
Laramie–Fox Hills	62.1	4.9	23.5	3.2	5.1
Biscayne	58.3	3.4	30.1	5.8	10.4

tal composition. One hypothesis is that kerogen, organic matter deposited with the sediments in an aquifer, is a source of humic material in groundwater (Thurman, 1979). Kerogen is enriched in *aliphatic* carbon and humic substances originating from kerogen would have a greater carbon content that humic substances in surface water. Another possibility is that, in the anaerobic environment of an aquifer, microbes use humic oxygen as an electron acceptor (Thurman, 1979), lowering the content of oxygen in the humic substances.

Oxygen is a major constituent of the functional groups of humic substances; however, with the exception of carboxyl groups, the structural position of oxygen in humic substances from groundwater has not been determined. The carboxyl content of humic substances from groundwater is nearly identical to the carboxyl content of humic substances from surface water, 5–6 meq/g (milliequivalents per gram); therefore, oxygen depletion must occur in other functional groups, such as in carbonyl, hydroxyl, or ether functional groups. More studies on the nature of oxygen in humic substances and especially on the nature of oxygen in humic substances from groundwater would be valuable.

The hydrogen content of humic substances from groundwater is greater than that from surface waters, and the atomic ratio of hydrogen to carbon is greater for humic substances from groundwaters (1.2) than for humic substances from surface waters (1.0–1.1 from Thurman, 1985) (see data in Table 5). The slightly greater H/C ratio of humic substances from groundwater indicates that aliphatic carbon may be more abundant in humic substances from groundwater.

The nitrogen content of humic substances from groundwater is similar to that of humic substances from surface water. Data from the St. Peters and Laramie–Fox Hills aquifers suggest that the nitrogen content of humic substances from groundwater may be somewhat lower than the average nitrogen

TABLE 5. Atomic Ratio H/C for Humic Substances from Groundwater

Aquifer	H/C Ratio
Fulvic Acid	
Laramie–Fox Hills	1.27
St. Peter	1.22
Madison	1.24
Red River	1.14
Biscayne	0.90
Humic Acid	
Laramie–Fox Hills	0.91
Biscayne	0.70

content in surface water humic substances (Table 4), but this is based on a limited set of samples.

Color and Absorbance

Humic substances from groundwater are considerably less colored per unit of carbon than humic substances from surface water; their absorbance at 465 nm (a wavelength commonly used for color in Standard Methods, 1971) is 3–10 times less than the absorbance of humic substances from surface water. Table 6 compares absorbances of humic substances from groundwater with absorbances of humic substances from surface water. Absorbances of samples from the Red River, St. Peters, and Laramie–Fox Hills aquifers are considerably less than absorbances of humic substances from an average surface water. Only the Madison and the Biscayne samples are similar in color to humic substances from surface water. The Biscayne is

TABLE 6. Absorbance at 465 nm for Humic Substances from Groundwater and Surface Water

Aquifer	Fulvic Acid (L/mg C/cm)	Humic Acid (L/mg C/cm)
St. Peters	16	118
Laramie–Fox Hills	35	42
Madison	122	522
Biscayne	145	104
Red River	36	74
Surface Water	120	240

a shallow groundwater (15 m) with a recharge from surface water and humic substances in this aquifer originate in terrestrial marshes, causing them to be more colored. There is no apparent explanation for the greater degree of color in samples from the Madison aquifer.

The lack of color in humic substances from groundwater indicates that they contain fewer *chromophores,* such as conjugated double bonds, *aromatic* rings, and *phenolic* functional groups. It has been suggested that these groups serve as color centers in humic substances (Schnitzer and Khan, 1972; Oliver and Thurman, 1982). The increased aliphatic content (thus lower aromatic content) of humic substances from groundwater is also consistent with decreased color.

Molecular Weight

Thurman et al. (1982) measured the molecular weight of aquatic humic substances from different environments and compared them with published results of molecular weights measured by chromatography, ultrafiltration, *colligative properties,* and X-ray scattering. They found that fulvic acids from groundwaters had radii of gyration (an approximation of molecular diameter) of 4.7 Å (angstroms) to 14 Å for monodisperse samples (Table 7).

The samples from both the St. Peter and Biscayne aquifers had a molecular weight range of 500–750. This suggests that these humic substances are of low molecular weight, about as low as have been found in natural waters (Thurman et al., 1982). The humic substances from the Madison and the Laramie–Fox Hills aquifers had larger radii of gyration and greater molecular weight. Thurman et al. (1982) concluded than an average molecular weight for humic substances from surface water is 1000–2000; most humic substances from the groundwaters in this study were within that range.

TABLE 7. Radii of Gyration of Humic Substances from Groundwater[a]

Aquifer	Radii of Gyration	Molecular Weight
	Fulvic Acid	
St. Peters	4.7	500
Biscayne	5.3	500–750
Madison	9.8	1500–2500
Laramie–Fox Hills	14.0	5000–10,000
Red River	—	—
	Humic Acid	
Biscayne	8.8	1000–2000
Trona water	12.0	2500–5000

[a] *After Thurman et al. (1982).*

Infrared Absorbance

Figure 2 shows the infrared scan of the fulvic acid from the Madison (A), St. Peters (B), and Red River (C) aquifers, and that of humic and fulvic acids from the Laramie–Fox Hills aquifer. The first major absorption is 3400 cm^{-1} and is related to hydroxyl groups in the samples. This is typical of all humic substances (Schnitzer and Khan, 1978). The second absorption, at 2960 cm^{-1}, indicates that aliphatic C—H is present in the humic material. This is consistent with the increased H/C atomic ratio of humic material from groundwater, and suggests that there is more aliphatic carbon in humic substances from groundwater than in humic substances from surface water. The third absorption at 1725 cm^{-1} represents carboxyl groups present in the humic material. Absorptions seen at 1385 and 1465 cm^{-1} are the result of C—H deformation (Bellamy, 1960). In conclusion, the major difference in the infrared spectra of humic substances from groundwater and humic substances from surface water, which have been analyzed, is the increased absorption at 2960 cm^{-1}, probably caused by greater aliphatic carbon.

Carbohydrate Content

The carbohydrate content of humic substances from groundwater was determined by the Mollisch test (Clapp, 1969) on all samples. In this assay, *polysaccharides* within the humic material are hydrolyzed with sulfuric acid into monomers that are then determined by a colorimetric test.

All samples were at the detection limit of the method, approximately 0.1% by weight. In comparison, humic substances in soil commonly vary from 5 to 10% carbohydrates (Stevenson, 1982). The isolation method with XAD resin does not isolate polysaccharides unless they are part of the structure of the humic material (see Chapter 14). In a previous study (see Chapter 7) carbo-

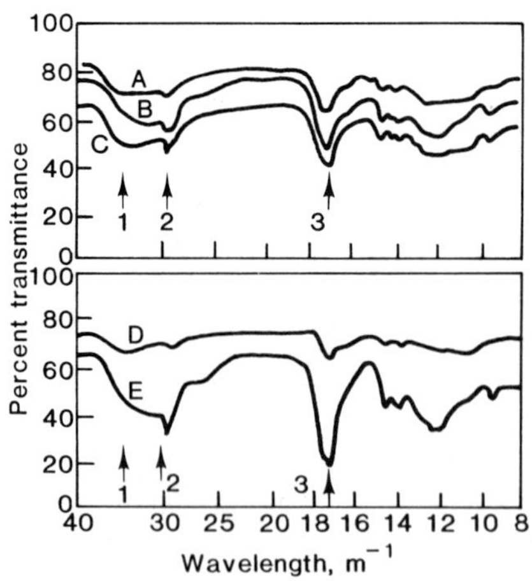

FIGURE 2. Infrared spectra of fulvic acid from the Madison (A), St. Peters (B), and Red River (C) aquifers, and of humic and fulvic acids from the Laramie–Fox Hills aquifer (D and E, respectively) (Thurman, 1979).

hydrates were found to account for 1–4% of the organic carbon. Apparently, humic substances from groundwater have lost carbohydrate components through microbial decay. Microbial degradation is more important for groundwater humic substances because of the longer residence time of humic substances in groundwater compared to humic substances in surface water. The longer residence time in groundwater is also important in understanding the carbon isotope data discussed below.

Carboxyl Content

Unfortunately, only two analyses were performed for carboxyl content of humic substances from groundwater, the Biscayne and Laramie–Fox Hills. The Biscayne has a carboxyl content of 6.3 meq/g and the Laramie–Fox Hills has a carboxyl content of 3.8 meq/g. These values span the range that have been found for humic substances from all natural waters, about 3–7 meq/g (see Chapter 7 on humic substances from rivers).

This carboxyl content, based on a titration of the humic material in its hydrogen saturated form to pH 8, is a measure of strong acidity, presumably from carboxyl groups. Other functional groups that may titrate in this range are strong phenolic groups and *enolic* hydrogens. More data are needed on the functional group content of humic substances in groundwater.

Carbon Isotopes

The stable carbon isotope ($^{13}C/^{12}C$) fractionation on several humic substances from groundwater was measured; the carbon isotope fractionation was −25.6 for the Laramie–Fox Hills sample and −26.4 for that of the Biscayne (see Table 8). These results are similar to fractionation values in humic substances from surface water, and to those reported in the soil literature (Nissenbaum, 1973; Stuermer et al., 1978).

TABLE 8. Stable Carbon Isotopic Fractionation for Fulvic Acid from Groundwater

Aquifer	$^{13}C/^{12}C$ Fractionation
Biscayne	−26.4
Laramie–Fox Hills	−25.6
Soil	
Mollisol	−22.5
Podzol	−24
Nissenbaum (1973)	−30
Algal Organic Carbon	−18

Therefore, the values for the groundwater humic substances indicate a terrestrial rather than an aquatic origin. Based on the fractionation of stable carbon isotopes by algae and higher plants, Nissenbaum (1973) concluded that ratios of -18 indicate algal origin and ratios of -25 to -30 indicate terrestrial origin.

The ^{14}C age of humic material from one groundwater, the Biscayne aquifer, was measured and found to be 660 (± 50) years before present. The analysis required 2 g of hydrogen-saturated humic material and so is unique for humic substances in groundwater. The measurement was done by M. Stuiver on 2 g of sample, and the age has an error of ± 50 years. Since the carbon may not be from a single source, this age is an average of all carbon sources. There is no way of knowing how much carbon was from recent organic matter and older organic matter, or how much was "dead" organic matter (containing no ^{14}C). Therefore, only a simple interpretation can be made.

For example, the fulvic acid from the Biscayne is not "young," that is, recent organic carbon. If this were the case the age would be much younger because of *"bomb"* ^{14}C in the sample. Fulvic acid from the Suwannee River in northern Florida had a younger average age than the 1953 standard, which indicates that humic material in the Suwannee is recent organic matter (Thurman and Malcolm, 1983). The Biscayne aquifer may contain some recent carbon, but it must be a small amount for the combined age to be 660 years before present. Likewise, the fulvic acid cannot contain much kerogen from the aquifer; if it did, the age would be much older because "dead" carbon from a kerogen source would contain no ^{14}C and would greatly increase the age of the sample. Recharge to the Biscayne aquifer is from the Everglades and represents organic carbon from a swamp source. Most of the carbon in the fulvic acid is neither recent nor dead; rather it ranges from 50 to 1000 years old, with an average age of 660 years. This means that there has been degradation of fulvic from the Biscayne aquifer compared to fulvic acid from the Suwannee River. Yet the humic substances from the Biscayne are the most similar to humic substances in surface water, suggesting that alteration of humic substances in groundwater takes considerably longer than does alteration of humic substances in soil.

E_4/E_6 Ratio

The fulvic acids from each aquifer were scanned in both visible and ultraviolet wavelengths. The scans were featureless; absorbance increased with decreasing wavelength, and there were no peaks. For this reason the E_4/E_6 ratio (absorbance of the sample at 465 and 665 nm) was also determined on each of the samples. Kononova (1966) used this ratio as a measure of humification. Campbell et al. (1967b) found an inverse relationship between E_4/E_6 ratios and mean residence time of humic materials in soil, and Chen et al. (1977) present a convincing case for a relationship between E_4/E_6 and molecular weight, with higher ratios meaning lower molecular weight.

TABLE 9. E_4/E_6 Ratios of Humic Substances from Groundwater

Aquifer	E_4/E_6 Ratio Fulvic Acid	E_4/E_6 Ratio Humic Acid
Laramie–Fox Hills	4.8	4.4
Biscayne	14.5	8.3
St. Peter	7.2	2.4
Red River	15.0	2.5
Madison	17.6	7.8
Mean	11.8	5.1

Values of E_4/E_6 in Table 9 indicate that humic acid in groundwater has a consistently lower E_4/E_6 ratio than fulvic acid. In previous work, Thurman et al. (1982) found that humic acid in surface water was larger in molecular size and weight than fulvic acid; this may also be true for humic acid in groundwater. Humic acid in soil is generally thought to be more humified than fulvic acid (Schnitzer and Khan, 1978), which may also be the case for humic and fulvic acids in groundwater. It appears, therefore, that in groundwater humic acid is older than fulvic acid, larger in molecular weight, and more *humified*.

Conclusions on Characterization

The concentration of humic substances ranges from 30 to 100 μg C/L in most groundwater; colored groundwaters may have concentrations of 1000–10,000 μg C/L. The percentage of hydrophobic organic substances in groundwater is less than in surface water, reflecting that probable absorption and degradation have occurred. The E_4/E_6 ratio and elemental analysis of humic substances from groundwater indicate that they are less humified and contain less aromatic character than humic substances from surface water. The infrared spectra show that humic substances from groundwater are more aliphatic than those in surface water; this is also indicated by the H/C ratio of 1.1–1.2. Finally, the elemental analysis of humic substances from groundwater indicates they contain more carbon and less oxygen than humic substances from surface water. Most of the oxygen is in carboxyl groups, which are present at 3.8–6.3 meq/g, constituting 50–60% of the oxygen in the humic material.

GEOCHEMISTRY OF HUMIC SUBSTANCES

Metal Complexation: Copper

As explained in the first section of this chapter, low concentrations of humic substances occur in groundwater from the saturated zone. Thus, geochemical reactions involving humic substances may play a less important role in the saturated zone than in the unsaturated zone or in surface water. Metal complexation by humic substances has been studied in detail for copper in surface water, and Mantoura (1981) reviews many of the articles and the relative magnitude of the binding constants. In general, the copper binding constant for humic substances is between 10^5 and 10^6 at pH 6. McKnight et al. (1985a) found that many humic substances from several surface waters and one groundwater (Biscayne) have copper binding constants in that range and a concentration of binding sites approximately equal to 1 μeq/mg C in humic substances.

If this number of sites is also found in humic substances in groundwater (the one sample that has been analyzed suggests that this is the case), then a groundwater containing 100 μg C/L of humic substances will have 0.1 μeq of metal binding sites per liter. This would bind approximately 6 μg/L of copper for a humic concentration of 100 μg C/L. Other divalent metal ions, such as calcium, would compete for this copper binding site. The binding strength for calcium would be 100–1000 times less (Mantoura, 1981), but the concentration of calcium in groundwaters is commonly 40 mg/L (2 meq/L). Thus, the calcium would probably tie up most of the binding sites for copper.

It is still possible that some copper is bound by humic substances, even when present at trace concentrations. There are stronger binding sites that are present at 0.1 μeq/mg C and have binding constants of about 10^9 (McKnight et al., 1983a); these sites might bind copper despite high concentrations of calcium. However, they would bind only about 0.01 μeq/L in a sample containing 100 μg C/L as humic material, corresponding to less than 1 μg/L of copper. It seems that the low concentration of humic material in the saturated zone in groundwater reduces the role that humic material plays in geochemical reactions such as metal binding.

In groundwater such as the Biscayne, with 8.6 mg C/L, there are 8.6 μeq/L of copper binding sites. Calcium plus magnesium in this water is 8 meq/L. If copper is bound 1000 times more strongly than calcium and magnesium, a reasonable assumption, based on the review by Mantoura (1981) and earlier work by Mantoura et al. (1978), then, at least 50% of the sites are available to bind metal ions. This means that only in colored groundwater are humic substances important in binding metal ions. In groundwater in the unsaturated zone, with DOC concentrations of 10–20 mg C/L, humic substances would also be able to complex and transport metals from the soil horizon.

Organic acids called hydrophilic acids are also present in groundwater. Although the chemical nature of these acids is unknown, they are thought to be similar to humic material, with more carboxyl and hydroxyl character and lower molecular weights (Thurman and Malcolm, 1981; McKnight et al., 1985b; Leenheer, 1981). These hydrophilic acids may have as many binding sites (1 μeq/mg C) as humic substances; if so, an additional number of binding sites for metal ions may be present. At this time the binding strength and number of sites in this material are unknown.

Other binding sites to consider are those present on the aquifer solids. These sites consist of silicic acid binding sites on clays, silts, and sands of the aquifer. Generally, the concentration of organic matter in the aquifer is less than 0.1% as organic carbon and is probably not significant. It is not in the scope of this chapter to discuss the binding of metal ions by inorganic solids.

Humic Substances and Pollutants

Humic substances in groundwater may have the ability to interact with less soluble organic pollutants and transport them in the aquifer, an area of study that has not received much attention. In measurements of the binding of *DDT* by aquatic humic substances, Carter and Suffett (1982) found that binding constants are weak for aquatic fulvic acids but increase for humic acids. Aquatic humic substances may increase the solubility of relatively insoluble compounds such as DDT by as much as 2–4 times, so the potential for transport of insoluble compounds by humic substances in the aquifer exists. However, as the water solubility of the pollutant increases, the ability of humic substances to move them probably decreases. Given that most groundwater pollutants (tetrachloroethylene, benzene, toluene) are quite water soluble compared to DDT and that aquatic humic substances are present in low concentrations in groundwater, the effect of transport by aquatic humic substances is probably insignificant. Further studies are needed in this area.

Origin of Humic Substances in Groundwater

From the data and conclusions presented in the characterization section, it is obvious that humic substances in groundwater are different from humic substances in surface water. At least two hypotheses may be proposed for the origin of humic substances in groundwater.

One is that humic material originates in overlying soils. Soil interstitial waters leach organic matter from the unsaturated zone and transport it to the aquifer. The humic material is transported from the oxidizing environment of the soil to the reducing conditions of the aquifer (not all groundwaters are reducing, but those of this study were), where it undergoes chemical alteration.

Another hypothesis is that humic substances are leached from kerogen in the sediment of the aquifer. Both hypotheses will be examined in light of the characterization data in order to see which makes the most chemical and geological "sense."

Hypothesis 1

Humic substances are from terrestrial sources in overlying soils. Humic substances in the Biscayne aquifer are quite similar to those in surface water in elemental analysis, carbohydrate content, color, molecular weight, $^{13}C/$

^{12}C fractionation, and infrared spectra. The Biscayne is a shallow aquifer in a carbonate sandstone (Thurman, 1979), with recharge from the Everglades which contain humic-rich surface water. Humic substances in this aquifer may originate in surface water. The humic material in the Biscayne sample, with a radio carbon age of 660 ± 50 years before present, is "older" than humic material from the Suwannee River, which drains swamps in northern Florida. This suggests that, in spite of residence time in the aquifer, no major alteration of humic material occurs in the reducing conditions of the aquifer.

It appears that humic substances come into groundwater from recharge waters and are altered slowly, if at all, in the reducing conditions of the aquifer. Because there is no evidence of humic material in the Biscayne originating from kerogen in the carbonate sands, the kerogen hypothesis seems inapplicable to the Biscayne aquifer. However, there is some alteration of humic substances in the Biscayne compared to those in surface water. The oxygen content in the Biscayne is 30% for humic acid and 35% for fulvic acid, while the oxygen content for both humic and fulvic acids in the Suwannee River is 39%. In soil, humic acid contains 36% oxygen and fulvic acid contains 45% oxygen (Schnitzer and Khan, 1978). The oxygen depletion in humic substances in groundwater may be caused by decarboxylation reactions or reduction of oxygen functional groups to hydrocarbons, reactions that may be both chemical and biological. At this time, neither process has been examined.

Sorption of humic substances from soil on aquifer solids should also be considered. Selective sorption may occur on clays present in the aquifer. Because the alumina sites on the clays are weakly basic, they may be good binding sites for humic substances, which are weak acids. However, if the aquifer consists mainly of sands and gravels, the sorption process may be minor. These ideas are speculative; no studies are reported in the literature.

Hypothesis 2

Humic substances are leached from kerogen in the sediment of the aquifer. The elemental composition, infrared spectra, and low color per unit carbon in the humic material all suggest that kerogen is a source of humic material for the Laramie–Fox Hills, St. Peter, Madison, and Red River aquifers. The elemental composition of the material indicates it is more aliphatic than humic substances from surface water, such as the Suwannee River (Thurman and Malcolm, 1983). The infrared spectra show stronger absorptions at 2940–2980 cm^{-1} than for humic substances from surface water, suggesting that aliphatic carbon is present in the humic material. Little is known of the chemical structure of kerogen from aquifers in this study, but the environments of deposition indicate what types of organic matter might have been deposited. The Madison aquifer is a dolomitic limestone deposited in shallow seas containing productive algal activity. The St. Peter aquifer is a beach sand, also of marine origin. The Laramie–Fox Hills aquifer is a marine sand,

and the Red River aquifer is a dolomite. These aquifers all could have accumulated organic matter from marine and algal sources. Humic substances from marine sources have been shown to be rich in aliphatic carbon by ^{13}C NMR (Hatcher et al., 1980b). It seems reasonable that, if humic substances originate from kerogen of these rocks, they should contain aliphatic carbon and have the low amount of color per unit carbon characteristic of humic substances of algal origin (Thurman, 1985).

Other evidence supporting the kerogen hypothesis is that kerogen is enriched in carbon and hydrogen and depleted in oxygen, as is the case for humic substances from the Red River, Madison, St. Peter, and Laramie–Fox Hills aquifers. The only conflicting evidence is the $^{13}C/^{12}C$ fractionation, which suggests a terrestrial source of humic material.

FUTURE STUDIES

One study yet to be done on humic substances in groundwater is isolation of the hydrophilic acid fraction by weak-base ion exchange (Leenheer, 1981). This method isolates 80–90% of the organic acids found in groundwater and reduces the amount of water needed, making it easier and faster to study the nature of hydrophilic humic substances. More detailed characterizations could then be made using ^{13}C NMR, molecular weight, metal-binding constants, and derivatization studies, as has been done with humic substances from surface waters. The degradation of humic substances by microorganisms and their role in the transformation of humic material in groundwater should also be examined. Both areas of study are presently being pursued as part of a new research thrust on hazardous wastes in groundwater by the U.S. Geological Survey.

ACKNOWLEDGMENTS

I thank Ron Malcolm for his help and guidance in studying humic substances in groundwater. He planned much of the work presented in this chapter and helped in isolation and characterization of the samples. He also was my thesis advisor during this research. Others who helped in sample collection include Pat Avery and George Aiken. I thank Diane McKnight for work on determination of binding constants for copper.

CHAPTER FIVE

Geochemistry and Ecological Role of Humic Substances in Lakewater

CHRISTIAN STEINBERG and UWE MUENSTER

ABSTRACT

In lakes, the pool of dissolved organic carbon (DOC) is dominated by dissolved humic substances (up to 80% of the DOC). Lake humic substances are similar to soil humic substances in that carboxyl, hydroxyl, phenol, and probably methoxyl groups are of major significance. Fluorescence spectra of DOC may be interpreted in terms of the different geochemical origins of DOC (e.g., allochthonous versus algal derived). One or more moieties of dissolved humic substances are produced autochthonously; mechanisms may include polymerization of phenols (promoted by transition metals), Maillard condensations, or oxidation via phenolase systems. Aliphatic structural units in dissolved humic substances provide a flexible conformation to the humic substance "molecule."

Losses of dissolved humic substances from the water column occur via adsorption onto surfaces of minerals and by cleavage upon exposure to UV radiation or ozone. Cleavage of humic "molecules" seems to be an important step in the decomposition of humic substances by microbes. Easily degradable substrates in the DOC pool (glucose, lactate, etc.) appear to stimulate microbial degradation of humic substances, either by a priming effect of an easily degradable substrate or by bacterial cometabolism. These organic nutrients also enhance uptake of metals from metal–organic complexes. Dissolved humic substances complex or sorb major cations, trace metals, trace anions, and hydrophobic pollutants (e.g., pesticides), and thereby change both bioavailability and geochemical cycling of these substances. Furthermore, humic substances inhibit precipitation of calcium carbonate and can catalyze certain photochemical and redox reactions.

Dissolved humic substances can bind microbially significant substrates such as carbohydrates and proteins and it may be that this interaction reduces the concentration of proteins and carbohydrates usable by microorganisms to below threshold levels.

Comparison of humic substances from different lakes indicates a high variation in concentration, composition, and molecular weight. The extent to which differences in methodology contribute to this variation has not been evaluated. Temporal and spatial distributions of dissolved humic substances and humic-associated organic substances are presented for five representative lakes. General parameters (UV absorbance, DOC measurements with or without fractionation on the basis of molecular size) do not adequately reflect the dynamic nature of various humic substances in lake ecosystems.

INTRODUCTION: AQUATIC HUMIC SUBSTANCES AND CARBON CYCLING IN LAKES

Studies of detrital organic materials form a relatively young branch of limnology, and were pioneered by Birge and Juday (1926, 1934) and Ohle (1934, 1935, 1937). As indicated by Wetzel (1983), the data of Birge and Juday on a large number of Wisconsin lakes provide an introduction to the chemical characterization of dissolved organic matter in lakes. The total organic carbon content of natural waters ranged from 1 to 30 mg C/L. Average values from over 500 Wisconsin lakes were: dissolved organic matter (DOM), 15.2 mg C/L, and particulate (living and dead) organic matter (POM), 1.4 mg C/L.

Ohle's work on "organic colloids" was influenced strongly by soil humus science. Even in his early papers (1934–1937), Ohle went beyond pure chemical analyses, and discussed his results from a limnological perspective. For example, Ohle (1935) described the adsorption of phosphorus and iron by organic colloids and pointed out the significance of such phosphorus sinks in nutrient cycling and primary production. Ohle's papers were instrumental in changing the role of limnochemistry as an "illustrative tapestry" (Schindler et al., 1975) into a fundamental aspect of limnology essential for understanding living processes in aquatic systems.

The complexity of the geochemistry and ecological role of humic substances in lakes is apparent when one examines the position of aquatic humic substances in the carbon cycle (Fig. 1; Melzer and Steinberg, 1983).

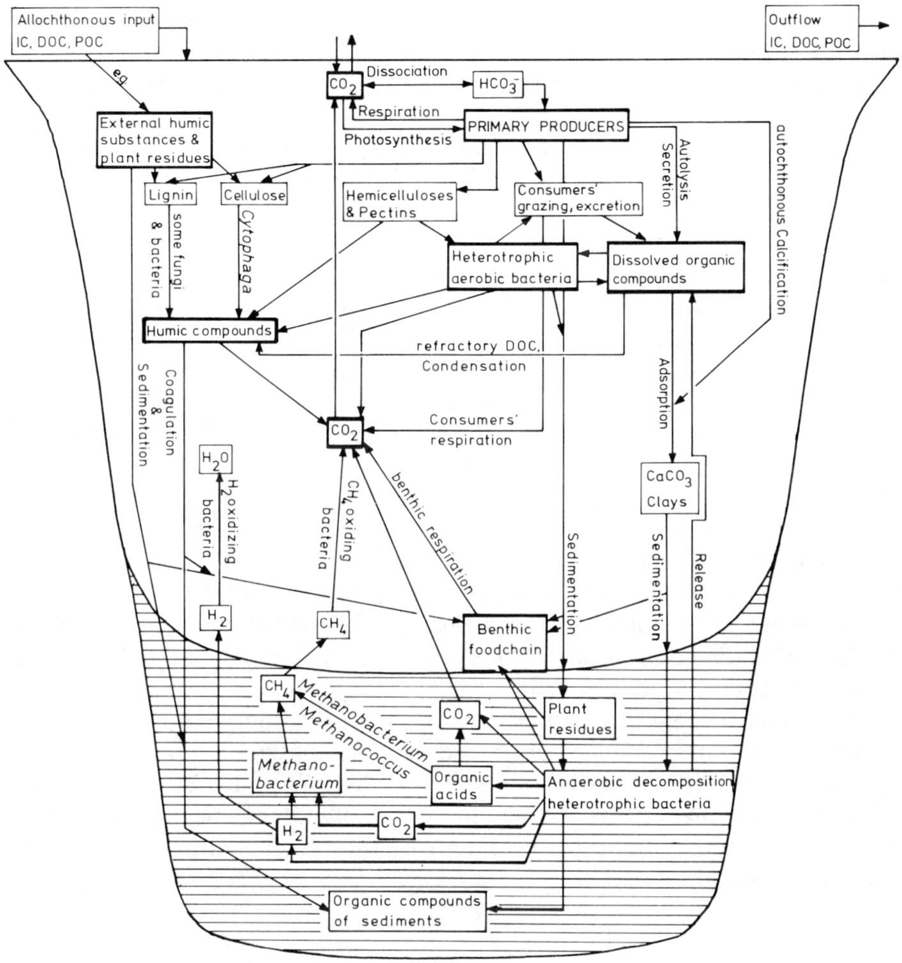

FIGURE 1. Carbon cycling in a lake ecosystem (from Melzer and Steinberg, 1983).

In principle, this figure is valid for both lakes and running water. In clear lakes, the input of carbon via autochthonous production (and the humification process per se) is of greater significance than in rivers and bogs where allochthonous humic material predominates.

Detritus represents the total pool of nonliving organic carbon that is available to the ecosystem as dissolved and particulate matter. In most lake ecosystems the organic carbon in the detritus compartment is much greater than the organic carbon in living organisms. The detrital compartment is very important to the functioning of the ecosystem, as important as the chemical and physical environments. Furthermore, transformations brought about by utilization of dissolved and particulate detritus by organisms as an energy source are identical; only the rates of transformation differ (Wetzel,

1983). Similarly, a distinction between dissolved humic substances and other organic compounds is justified only when a different origin or a varying degree of bacterial/chemical transformation is known. For example, in the energy flux in a lake ecosystem from primary production to permanent sediments, it makes no difference if the organic carbon is incorporated into a humic or fulvic acid molecule or if it is bound into chitin in a *Daphnia's* exoskeleton. Both forms are detritus, and both represent nonpredatory losses of organic carbon. Humic substances, at least quantitatively, are the most significant component of detritus in lake ecosystems.

We believe that an ecosystem approach as outlined above will provide more detailed insight into the function of aquatic humic substances in lakes than can studies based on chemical analyses alone.

Several authors (e.g., Tipping and Cooke, 1982) point out that the terminology used to discuss organic matter in natural waters is ambiguous. It is therefore appropriate to note that by "humic substances" we mean simply the soluble brown organic material which can be extracted from natural water by adsorption onto various resins, such as Amberlite XAD or polyvinyl pyrolidone (PVP). A further discussion of the isolation of humic substances from water is found in Chapter 14 of this book. Unless stated otherwise, the word "humic" does not imply a distinction between humic and fulvic acids. In many studies, however, there is no clear differentiation between humic substances and dissolved organic matter in general. The latter approach is adequate for studying ecosystem energy flux. When referring to chemical species of dissolved organic material, however, a more sophisticated terminology is obviously needed. We must stress, however, that the functional interrelationships of dissolved organic matter, aquatic humic substances, and aquatic organisms that occur within an ecosystem should not be overlooked.

CHEMICAL AND PHYSICOCHEMICAL CHARACTERIZATION OF DISSOLVED HUMIC SUBSTANCES

General Chemical Characteristics

Elemental Analysis

The elemental compositions of several fulvic acids from different waters are presented in Table 1. This table also includes data from two lakes within the blast zone of Mt. St. Helens, Washington, that received large amounts of dissolved organic material from the pyroclastic flows of the eruption on May 18, 1980. Concentrations of carbon, hydrogen, oxygen, nitrogen, and phosphorus in fulvic acids from these two lakes were very similar to those of a nearby unaffected lake and well within the range of concentrations commonly found for aquatic fulvic acids. Although elemental analysis is impor-

TABLE 1. Chemical Characteristics of Aquatic Humic Substances

	Elemental Analysis (% by weight)							
	Carbon	Hydrogen	Oxygen	Nitrogen	Phosphorus	Sulfur	Chlorine	Ash
Yellow organic acids[a]	54.6	5.6	39.1	1.2	—	—	—	1.4
Water humic substances[b]	43	5.5	—	1.1	—	—	—	—
Soil humic substances[b]	45–63	3–6	—	0.5–5	—	—	—	—
Aquatic fulvic acid several sources[c]	47–53	4–5	35–40	0.5–1.5	1	1	—	—
Lake Bansee[d]	45	5	47	2	—	1	—	—
Lake Hohlohsee[e]	46.9	3.3	—	0.9	—	1.5	—	2.5
Lakes of Mt. St. Helens Region[f]								
South Fork Castle Lake[g]	51.7	5.0	37.7	0.7	0.2	2.1	1.3	0.1
Spirit Lake[g]	51.6	4.9	37.5	1.0	0.2	4.3	0.4	1.1
Merril Lake[h]	51.3	5.0	39.5	0.7	—	0.5	0.4	1.4
Lake Celyn[i]								
Humic acid	50.2	3.1	44.8	1.9	—	—	—	—
Fulvic acid	43.5	2.7	51.6	2.2	—	—	—	—

[a] *Shapiro (1957).*
[b] *Gjessing (1976).*
[c] *McKnight et al. (1982).*
[d] *Frimmel et al. (1980).*
[e] *Eberle and Feuerstein (1979).*
[f] *McKnight et al. (1985a).*
[g] *Influenced by pyroclastic flow after Mt. St. Helens eruption.*
[h] *Not influenced.*
[i] *Wilson et al. (1981a).*

tant in characterization of humic substances (see Chapter 18), these data show that elemental analysis alone may not provide much information about the origin and function of humic substances.

Functional Group Analysis

Analysis of functional groups such as carboxyl, phenol, carbonyl, or methoxyl (Table 2) increases our understanding of the chemical structure of humic substances and can be used to explain the behavior of humic substances in various humification processes (Gjessing, 1976). Carboxyl and phenolic hydroxyl groups clearly predominate, although in some cases methoxyl groups are quantitatively important as well (Muenster, 1982).

Relative to soil humic substances, humic substances from Lake Celyn, Wales, and fulvic acids from lakes near Mt. St. Helens contain larger amounts of reactive acidic functional groups (especially carboxyl groups). The reason for this is not known. In Lake Celyn, 24% of the humic acid carbon is carboxyl and 40% is aromatic, suggesting that the Lake Celyn humic acids are largely of terrestrial origin (M. A. Wilson et al., 1981a).

Muenster (1982) found for Lake Plussee DOC that, as in other studies, carboxyl, hydroxyl, and methoxyl groups were most abundant, and carbonyl and methyl groups were of minor importance. Wilson and Kinney (1977) found that the OH/COOH ratio of fulvic acid from Smith Lake, Alaska, was 0.71, which is similar to other humic materials listed in Table 2. In contrast,

TABLE 2. Functional Groups in Humic Substances

	meq/g			
	Total Acidity	Carboxyl	Phenolic Hydroxyl	Methoxyl
Lake Celyn humic acid[a]	8.9	5.9	3.0	—
Lake Celyn fulvic acid[a]	11.0	8.9	2.1	—
Merril Lake fulvic acid[b]	—	4.3	0.8	—
South Fork fulvic acid				
Castle Lake fulvic acid[c]	—	5.5	2.8	—
Spirit Lake fulvic acid[c]	—	5.2	1.7	—
Soil humic acid[d]	—	3.9	1.9	—
Peat[e]	8.3	2.4	5.9	1.9
Aquatic fulvic acid[c]	—	4–6	1–3	—

[a] *Wilson et al. (1981).*
[b] *McKnight et al. (1985a).*
[c] *McKnight et al. (1982).*
[d] *Wagner and Stevenson (1965) (from Gjessing, 1976).*
[e] *Riffaldi and Schnitzer (1973) (from Gjessing, 1976).*

the OH/COOH ratio for Lake Plussee DOM lies significantly above 1.0. Visser (1982) compared functional group contents of humic materials of both terrestrial and aquatic origins. He observed that aquatic humic acids contained more COOH and fewer phenolic OH groups than their terrestrial counterparts. Similar to what has been observed on humus from terrestrial environments, aquatic fulvic acids are richer in carboxylic and phenolic groups than their humic acid counterparts. Furthermore, Visser found that with progressive humification the COOH content of fulvic acids of microbial origin increased, whereas the number of phenolic OH groups diminished in the case of the humic acids.

Acidity Including Isoelectric Focusing Studies

The acidic character of DOM in lakes has been characterized by two different methods: DOC fractionation which utilizes various resins (Leenheer and Huffman, 1976) and isoelectric focusing (Gjessing and Gjerdahl, 1975). Results from both methods lead to the same conclusion that organic acids predominate. The DOC fractionation results for several mountain lakes (McKnight et al., 1983) show that hydrophobic acids, of which fulvic acid is the major component, and hydrophilic acids are the two major fractions, accounting for more than 80% of the DOC. In the only isoelectric focusing study of nondystrophic lakes, Muenster (1982) found that 70–80% of the DOC in Lake Plussee focused at pH 1.5–2.5. Under normal pH conditions in hardwater lakes such as Lake Plussee acidic functional groups are apparently totally dissociated and demonstrate a high ionic potential. This result is similar to that of Gjessing and Gjerdahl (1975), who demonstrated that about 80% of the DOM from several dystrophic Norwegian lakes had an isoelectric point lower than pH 2.0.

Fluorescence Studies

Dissolved organic substances possess fluorescence properties (Shapiro, 1957; Black and Christman, 1963a; Povoledo and Gerletti, 1963; Hall and Lee, 1974; Smart et al., 1976; Larson and Rockwell, 1980; Stewart and Wetzel, 1980). Fluorescence intensities and fluorescence spectra are commonly measured following excitation at wavelengths between 325 and 427 nm. The fluorescence spectra of aquatic humic substances of diverse origin (as tabulated by Larson and Rockwell, 1980) are quite similar, which is a major limitation of using fluorescence spectra in characterization. Fluorescence intensity is also affected by pH and the presence of metals. Furthermore, fluorescence intensity alone may be a poor predictor of DOC concentration. Stewart and Wetzel (1981a) found little correspondence between DOC and fluorescence, which could be explained by greater levels of internal quenching and shielding in compounds of larger apparent molecular weight. A lake-to-lake comparison implicated a calcium-related selective loss of high-molecular-weight humic substances, which could invalidate the use of fluorescence as a predictor of DOC concentrations in hardwater systems even after correcting for seasonal changes in pH (Stewart and Wetzel, 1981a).

Despite these limitations, there are advantages of fluorescence measurements of dissolved humic substances. For example, Ghosh and Schnitzer (1980a) were able to differentiate between soil fulvic acids and humic acids, since fulvic acids exhibited additional excitation bands at 360 nm. Further studies are needed to reveal if this phenomenon also occurs in freshwater humic substances.

As mentioned above, weak UV irradiation (325–427 nm) is frequently used for excitation. When far-UV radiation is used, however, increased fluorometric response and more information on structural properties of dissolved organic compounds can be obtained, since organic molecules found in natural waters have a greater absorbance at shorter wavelengths. Stabel (personal communication) attempted to differentiate between probable sources of DOM in lakes using fluorescence spectra of DOM from various inland waters, such as softwater, hardwater, and saline systems. With exci-

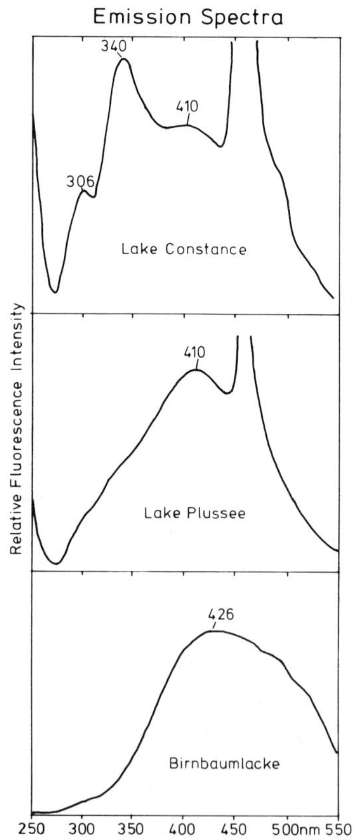

FIGURE 2. Fluorescence spectra of DOC of different origins. Excitation at 230 nm, both slits 5 nm.

tation at 230 nm, three types of emission spectra could be distinguished: predominantly autochthonous DOM exhibited three peaks (306, 340 [main peak], and 410 nm). DOM of terrestrial origin had a single peak of 410 nm, while in saline lakes the fluorescence spectra had one peak at 426 nm (Fig. 2). The fluorescence spectra of DOM from various algal cultures suggested that the peak at 340 nm in the first spectrum type could probably be attributed to extracellular material released by phytoplankton.

Resin Adsorption Studies

Dissolved humic and fulvic acids can be isolated from the total DOC pool by adsorption onto resins as discussed by Aiken in Chapter 14 of this book. Muenster (1982) found that 77–86% of the DOC (measured by UV absorbance and organic carbon) adsorbed onto XAD resin at pH values of 2.0–3.0. Adsorption onto PVP resin was slightly less effective (70–76% at pH values of 2.0–3.0). These differences in adsorption efficiency may be explained by organic material other than humic substances being retained on these resins.

Adsorption can be used to differentiate between free dissolved monomeric phenols and monomeric carbohydrates. Muenster (1982) showed that the sum of both monomeric substance classes in Lake Plussee water comprised between 3 and 10% of the total DOC concentration depending on the resin used. Perdue and co-workers (Lytle and Perdue, 1981; Sweet and Perdue, 1982) utilized Amerlite XAD-7 resin to adsorb humic substances and determined free dissolved amino acids and humic-acid-associated amino acids, as well as monosaccharides, polysaccharides, and humic-bound saccharides in the Williamson river system (Oregon). As a mean, less than 4% of the dissolved amino acids occurred in free dissolved form, and only 2.6% of the total carbohydrates were monosaccharides. This observation as well as some ecophysiological implications are described more extensively in a later section.

Gel Permeation Chromatography Studies

Gel permeation chromatography (GPC) is a commonly used method in the characterization of DOM isolated from aquatic habitats as discussed in Chapter 16 by Leenheer. Humic acids are predominately of high molecular weight (up to about 300,000 daltons), while fulvic acids are commonly believed to have molecular weight values of less than about 1000 daltons (Dawson et al., 1981). However, when interpreting GPC fractionations, one should bear in mind that this method does fractionate DOM, but not always according to molecular weight or size. There are several possible artifacts (Gjessing, 1976). For example, Muenster (1982) compared the GPC fractionation of DOM from Lake Plussee with and without prior concentration by evaporation. He found an oligomer fraction in the concentrated material which was not found in the unconcentrated water, and showed that in the concentrated material DOM of high apparent molecular weight was retarded, and DOM of low apparent molecular weight was eluted too early. These artifacts are produced by high concentrations of electrolytes in the concentrated sample (Gelotte, 1960).

The chemical and biochemical behaviors of humic substances can also be changed by GPC. Frimmel and Sattler (1982) studied the complexation/adsorption of trace metals by dissolved humic substances and discovered that the affinity of humic substances for metals markedly increased following GPC. Similarly, Stewart and Wetzel (1982) observed that all Sephadex G-100 fractions of dissolved humic material obtained from the aquatic macrophyte *Typha* were more stimulatory to ^{14}C assimilation by algae than were the same humic substances that had not been fractionated. The observations indicated that the gel, eluent, or processing procedure (e.g., lyophilization, reconstitution, cleavage during separation) either reduced the toxicity of the humic substances or enhanced its stimulatory nature or affinity toward trace substances.

Polyacrylamide Gel Electrophoresis Studies

Polyacrylamide gel electrophoresis (PAGE) provides a versatile, gentle, high-resolution method for fractionation and physicochemical characterization of polyelectrolytes, for example, proteins or humic substances on the basis of molecular size, conformation, and net charge (see Chapters 15 and 16). From mobility (relative to arbitrary ion or moving boundary) measurements at several gel concentrations, PAGE allows for calculations of molecular volume, surface area, radius, free mobility, and valency (Chrambach and Rodbard, 1971). High-resolution PAGE results in very small sample volumes (1–3 mL), and preconcentration is not required.

The results of Muenster (1982) using PAGE to fractionate DOC from Lake Plussee water provide additional information on the physicochemical properties of DOM. PAGE separation of Lake Plussee DOC resulted in three different fractions (Fig. 3). F1 exhibited high electrophoretic mobility and F2 low mobility. F3, the third PAGE fraction, did not penetrate into the gel at all. Since the gels allowed the migration even of substances having a molecular weight of 1,000,000 daltons, F3 was believed to have a very small molecular charge rather than a molecular weight exceeding 1,000,000 daltons. Muenster (1982) also found that the pattern of apparent molecular weight obtained by GPC could not be confirmed by PAGE.

According to Chrambach and Rodbard (1971), spherical molecules (e.g., some proteins) show decreasing electrophoretic mobility with increasing gel concentrations (and thereby, decreasing gel pore sizes). Plotting mobility versus increasing gel concentrations (Ferguson plot), proteins, for example, yield straight lines with negative slope. The opposite, however, is true for humic substances from Lake Plussee (Fig. 4). Both fractions F1 and F2 yield

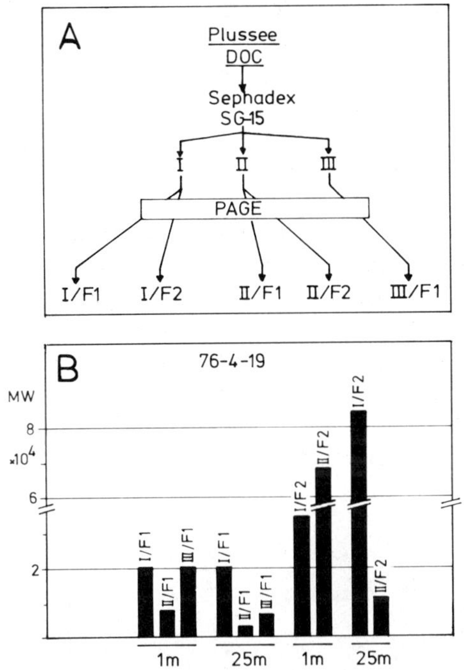

FIGURE 3. PAGE separation of Lake Plussee DOC after GPC (from Muenster, 1982): (A) Separation scheme. I = apparently high-molecular-weight fraction, II = apparently oligomeric fraction, III = apparently low-molecular-weight fraction, F1 and F2-PAGE fractions of high and low electrophoretic mobility, respectively. (B) Apparent molecular weights derived by PAGE.

straight lines with positive inclinations in all samples (collected from 1, 4, 7, 15, and 25 m). Apparently, these two fractions increase in free electrophoretic mobility with decreasing gel pore size. In this experiment it appears that humic substances do not have a spherical molecular configuration (see also Ghosh and Schnitzer, 1980b). In electric fields, mobility is affected more by the molecular net charge than by molecular shape or conformation. This phenomenon cannot be explained in terms of a rigid molecular conformation. Thus, we believe that dissolved humic substances probably occur more or less in a certain, perhaps globular, molecular conformation under "normal" conditions; they may, however, unfold when moving within an electric field, as for instance within a PAGE system, or, in a more ecophysiological sense, within an electrochemical double layer at a cell surface. If this is true, we suggest that aquatic humic substances must have a high content of

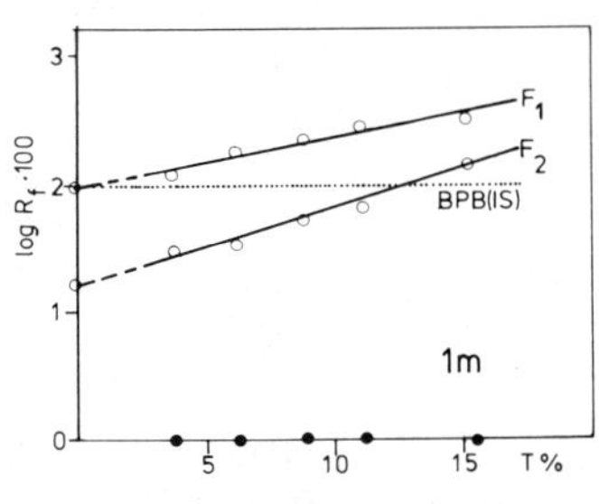

FIGURE 4. Ferguson plot of Lake Plussee DOC from 1 m depth, June 30, 1979 (from Muenster, 1982). F1 and F2 are PAGE fractions of high and low electrophoretic mobility. BPS(IS) = Bromphenol Blue as the internal standard.

aliphatic structural units with single bonds which allow free rotation within the molecule rather than being highly condensed and thereby of a more or less rigid molecular conformation. Chemical analyses of dissolved humic substance degradation products provide additional evidence for this belief.

Characterization of Chemical Structure

Based on the assumption that lignin is essential in humification processes, many researchers have suggested that humus "molecules" have a high degree of aromaticity. An example is given in Figure 5*a*, as proposed by Gamble and Schnitzer (1974). The failure to find aliphatic compounds in early degradation studies, especially dicarboxylic acids, supported the highly aromatic model for aquatic humic substances (Gjessing, 1976).

In more recent studies, several authors have found a relatively high proportion of aliphatic chains in aquatic humic substances; this is in significant

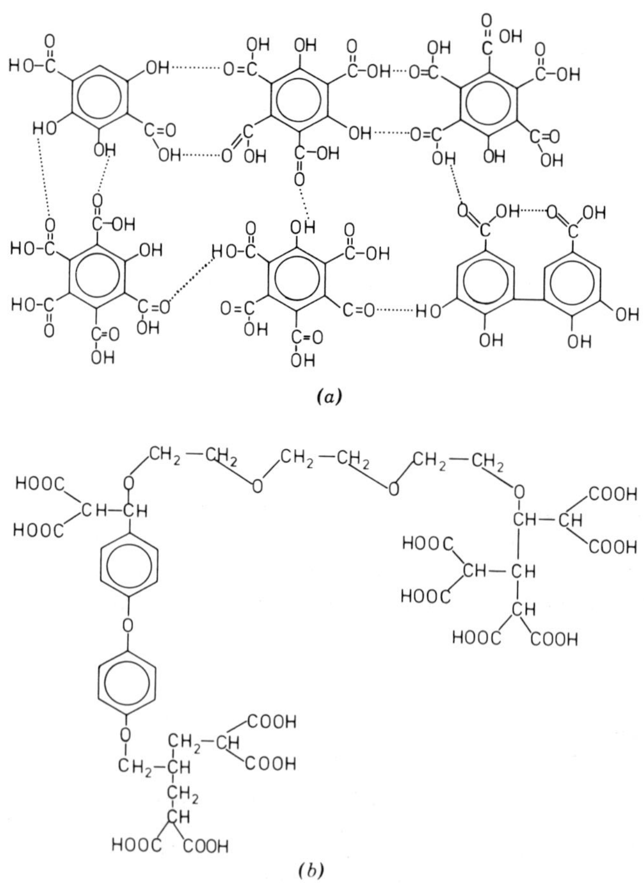

FIGURE 5. Hypothetical structures of humic substances from freshwater and from activated sludge systems: (*a*) from Gamble and Schnitzer (1974); (*b*) from Bergmann (1978).

contrast to soil humic substances (Wilson et al., 1978; De Haan et al., 1979). De Haan et al. (1979), for example, studied freshwater fulvic acids from Tjeukemeer (The Netherlands) using Curie point pyrolysis–mass spectrometry (unfortunately without gas chromatography before mass spectrometry). It should be noted that a major problem with this method is that the spectra are complex and not necessarily representative of all the original humic material. Pyrolysis of products of freshwater fulvic acids resembled those of soil fulvic acids, although freshwater fulvic acids were richer in aliphatic compounds than fulvic acids from soils. If this is generally true, it would indicate that lignin degradation is less important in lakes than in soils.

In a detailed study on methylated permanganate-degradation products of aquatic humic and fulvic acids (originating from Black Lake, North Carolina, and Lake Drummont, Virginia) by means of gas chromatography/mass spectrometry, Liao et al. (1982) found methyl esters of benzenecarboxylic acids, furancarboxylic acids, aliphatic mono, di, and tribasic acids, and (carboxyphenyl) glyoxylic acids (Table 3). The degradation products of fulvic and humic fractions from the two lakes were qualitatively similar, but dis-

TABLE 3.

type of degradation products	possible sources in humic macromolecule(s)

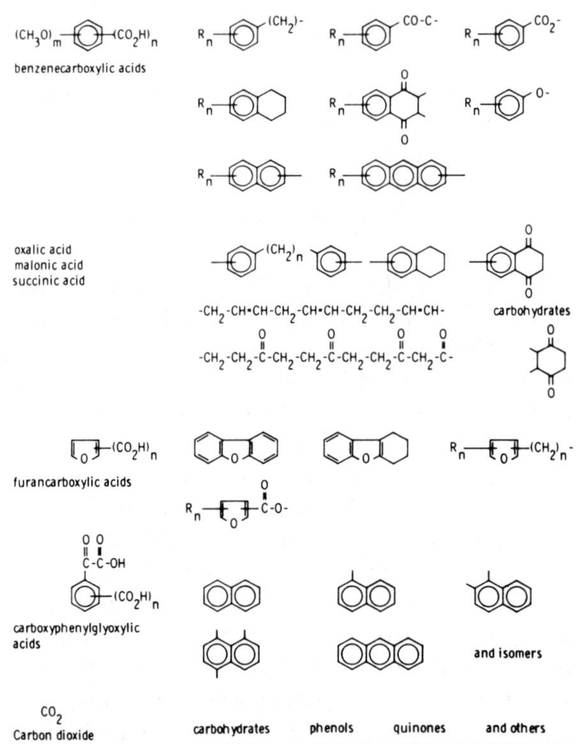

[a] R = H, OH, CO_2H, or alkyl substituents.

tinct quantitative differences were found. The authors stress that aquatic humic substances contain both aromatic and aliphatic components. The aromatic rings contain mainly three to six alkyl substituents, and polynuclear aromatic and fused-ring structures may also be present. The data of Liao et al. (1982) suggest that the principal aliphatic components of the original natural material are composed of relatively short saturated chains (two to four methylene units).

Except for the fused rings, the structural units of fulvic and humic acids published by Liao et al. (1982) are similar to those proposed by Bergmann (1978) who studied humic substances originating in activated sludge systems. Bergmann's proposed humic unit (Fig. 5*b*) also contains relatively small amounts of aromatic rings as lignin degradation products. If the ratio of autochthonous to allochthonous humic substances further increases, as in the sea (see Chapter 9), aromaticity will also decrease.

ECOPHYSIOLOGICAL INTERACTIONS

Removal of Aquatic Humic Substances by Chemical Processes, Photolysis, and Adsorption

Humic substances can be removed from the water column by two physicochemical mechanisms:

1. Adsorption onto surfaces such as suspended particulates, and carbonate or hydrous metal oxide precipitation.
2. Cleavage by UV irradiation.

Adsorption onto Surfaces

Gloor et al. (1981) investigated adsorption of DOM in Lake Greifensee by colloidal alumina and found that after 10 hours the DOC was reduced by 50% at pH 5.9 and by 40% at pH 8.3 (Fig. 6). Furthermore, gel permeation chromatography showed that molecules with an apparent molecular weight of >500 daltons were absorbed, and the degree of adsorption increased with an increase in molecular size. Gloor et al. (1981) conclude that adsorption may regulate the removal of apparently high-molecular-weight organic compounds from natural aquatic ecosystems, especially in systems such as alpine lakes with a high input of inorganic particulate matter.

David and Gloor (1981) also studied the adsorption of DOM fractionated by gel permeation chromatography on colloidal alumina. Organic compounds with an apparent molecular weight of 1000 daltons formed strong complexes with the alumina surface, while lower-molecular-weight compounds were weakly adsorbed. Electrophoretic mobility measurements indicated that alumina particles suspended in the original lakewater were very

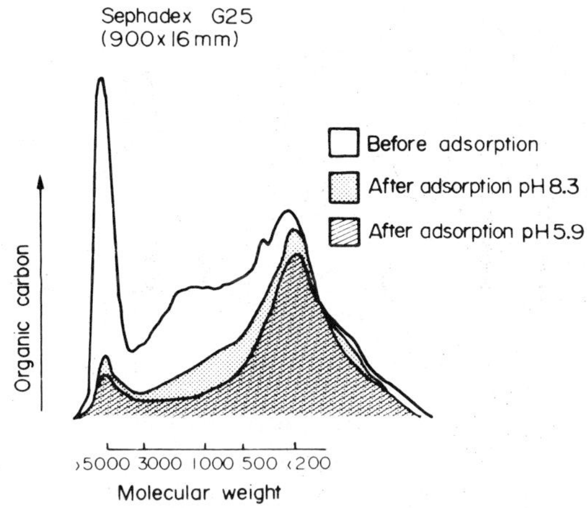

FIGURE 6. Changes in apparent molecular weight distribution of Lake Greifensee DOC after adsorption onto γ-Al_2O_3 (from Gloor et al., 1981).

negatively charged due to adsorbed organic matter. Tipping (1981) found iron oxide to behave similarly.

The significance of the removal of DOM by absorption was also shown by Stabel and Steinberg (1976a) in a study of Lake Walchensee, an alpine lake in Bavaria. They compared the DOC concentration and the fractionation of DOM by gel permeation chromatography in samples from a deep, microbially inactive layer, before and after snowmelt and heavy rainfall had brought a large amount of suspended material into the lake. About 65% of the DOC was adsorbed by the suspended material, and the fractionation results (Fig. 7) correspond to the laboratory results of Gloor et al. (1981) (Fig. 6), except that apparently smaller molecules were also adsorbed in Lake Walchensee water.

Gjessing (1976) found a markedly increased adsorption of aquatic humic substances onto montmorillonite with decreasing pH, which could be explained by charge neutralization of the humic substances. For soil humic substances, adsorption is assumed to be partly interlamellar, which requires

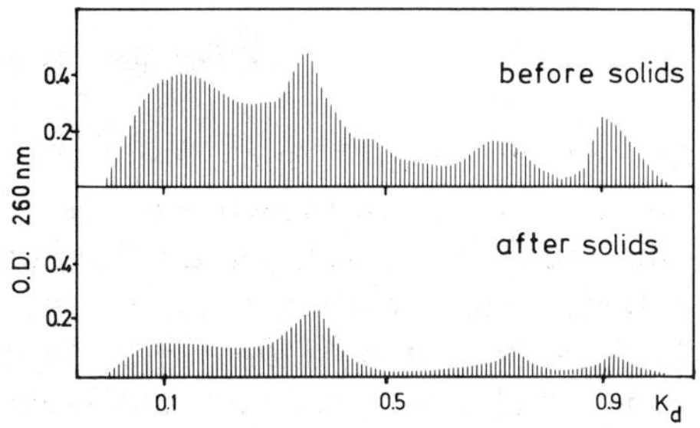

FIGURE 7. Changes in apparent molecular weight distribution (Sephadex G-15) of Lake Walchensee DOC after natural inputs of suspended solids (after Stabel and Steinberg, 1976a).

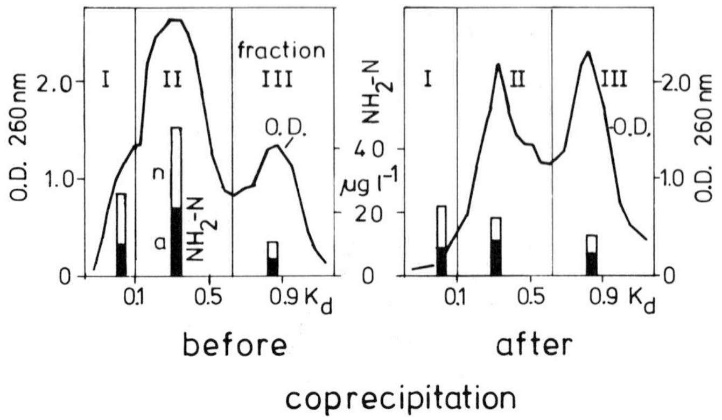

FIGURE 8. Changes in apparent molecular weight distribution (Sephadex G-15) of Lake Schöhsee DOC optical density (O.D.) at 260 nm and proteinaceous matter after coprecipitation by $FeCl_3$ at pH = 4.3 (after Steinberg, 1976). NH_2—N = proteinaceous nitrogen, n = neutral, and a = acidic amino acids.

uncharged humic molecules. Studies have shown that most aquatic humic substances become electrically neutral between pH 2 and 1.

Besides surface adsorption, adsorption of humic substances may occur via ionic linkages involving polyvalent cations or oxides as intermediates between humic materials and clay particles (Greenland, 1965a; Kodama and Schnitzer, 1968; Gjessing, 1976). Steinberg (1976) investigated the effects of ferric iron precipitation at pH 4.3 on humic substances and proteinaceous material from Lake Schöhsee in northern Germany using gel permeation chromatography (Fig. 8). More dissolved UV-absorbing substances were precipitated, in the fraction of high apparent molecular weight (≥1500 daltons), than dissolved proteinaceous material (85% vs. 18%). In the lower apparent molecular weight fraction II, 34% of the UV-absorbing, and 62% of the associated peptides, were precipitated. The concentrations of free dissolved amino acids increased after ferric iron flocculation, suggesting release of proteinaceous material originally associated with humic substances. Within the proteinaceous material, neutral amino acids were affected more than acidic amino acids, indicating that adsorption of uncharged molecules may predominate in precipitating proteinaceous compounds.

Cleavage by UV Irradiation

Since 1899, when Whipple (quoted in Strome and Miller, 1978) recorded the photo-bleaching of brown water humic substances, many studies have been published on the role of UV radiation and oxidizing agents in degrading natural organic matter (Gjessing and Gjerdahl, 1970; Allen, 1976; Gjessing, 1976; Strome and Miller, 1978; Francko and Heath, 1979; Gilbert, 1980; Gloor et al., 1981; Stewart and Wetzel, 1981b).

Strome and Miller (1978) and Gilbert (1980) found that photolysis caused a gradual decrease in average molecular size. Allen (1976) noted a greater UV reactivity of the larger-size fractions of dissolved organic material, but found that intermediate-size fractions were almost resistant to photolysis. In general, cleavage by UV radiation seems to be a prerequisite to microbial decomposition of aquatic humic substances (Strome and Miller, 1978). Bacteria are capable of utilizing humic photolytic products for growth and are able to reduce humic absorbance to some extent. Light as a "priming agent" is generally very effective in enhancing ability of bacteria to degrade humic substances.

Stewart and Wetzel (1981b) examined the following humic substances with respect to their photolytic behavior: *Typha* leachate, Contech fulvic acid, and hypolimnetic water from Lawrence Lake (Fig. 9). UV radiation was sufficient to cause substantial losses in sample absorbance and fluorescence in most humic substances within a few minutes, depending upon the material and the nature of the UV radiation. In water samples collected from the hypolimnion of Lawrence Lake, loss of absorbance was much slower than loss of fluorescence, suggesting that, in that environment, absorbing materials were more resistant to photolysis. Using fluorescence/absorbance ratios as an index of relative molecular weight, Stewart and Wetzel (1981b) suggested that, in their experiments, photolysis of lower-molecular-weight humic substances proceeded more rapidly than photolysis of high-molecular-weight humic substances. These conclusions contradict the statement of Allen (1976), but this may be due to differences in humic substances or in methodology.

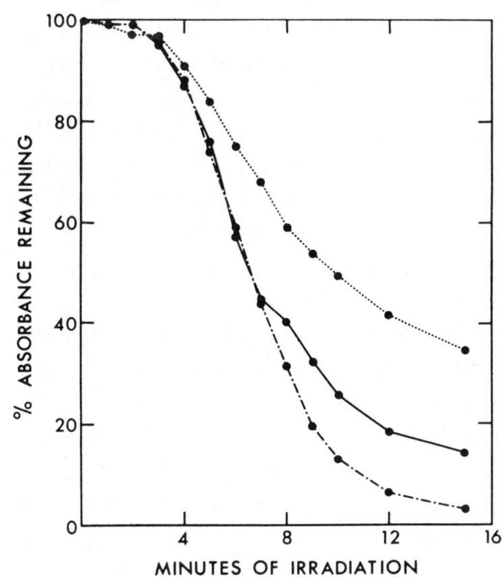

FIGURE 9. Percent of absorbance (250 nm, 1 cm path length cells) remaining during exposure to high-intensity UV radiation (from Stewart and Wetzel, 1981b): ——— = *Typha* leachate; —·—·— = Contech fulvic acid; ··· = Lawrence Lake, hypolimnetic water sample.

Francko and Heath (1979) studied the influence of UV radiation on the phosphate complexation properties of dissolved humic substances from Crazy Eddie Bog in the central United States. In this bog, filterable phosphorus compounds were largely associated with humic substances of high apparent molecular weight. These fulvic acid–phosphorus associations resisted hydrolysis by alkaline phosphatase, but released orthophosphate upon irradiation with low doses of UV radiation. The turnover time of the fulvic acid–phosphorus compounds was calculated to be less than 1 hour at the surface of a lake on a cloudless day.

Complexation of Metals and Trace Anions by Aquatic Humic Substances

Metal Complexation

Ohle (1935, 1937) noted the specific uptake of iron by humic "colloids" and discussed the ecological significance of this finding with respect to regulation of nutrient availability for primary producers. Since then, the binding of trace metals by aquatic humic substances has been investigated by many researchers (see Gjessing, 1976). Complexation of trace metals by humic substances may affect aquatic biota in two ways: (1) by decreasing the toxicity of certain metals (such as copper) and (2) by increasing the availability of some metals (such as iron) that are otherwise very insoluble at the pH and Eh of most lakes. At this time, the role of humic substances in reducing the toxicity of trace metals is more clearly understood than the role in enhancing trace metal availability. For example, Sunda and Lewis (1978) showed in a laboratory study that complexation of copper by dissolved humic substances in river water decreased the toxicity of copper to an estuarine diatom by decreasing the free copper ion concentration. Podolski (1979) observed a decreased uptake of cadmium by *Daphnia magna* in the presence of humic acids. Hongve et al. (1980) working with natural phytoplankton assemblages found that detoxification of metals by humic substances in bog water was not significant with Cd(II), but was significant with Zn(II), Pb(II), Hg(II), Cu(II), and a mixture of these five metals. In a field study of a copper sulfate treatment of a reservoir, McKnight (1981) found that most of the added copper was complexed by dissolved humic substances, and showed that one green algal species that became dominant in the reservoir after the $CuSO_4$ treatment would not have grown if the added copper had not been complexed by dissolved humic substances. Dissolved organic material other than humic substances, such as material exuded by algae, may also be important in complexing and detoxifying trace metals (McKnight and Morel, 1979, 1980). De Haan et al. (1981a) observed in Tjeukemeer that the copper-binding capacity was influenced by allocthonous humic substances from surrounding polders and by autochthonous organic material produced by the productive phytoplankton populations in the lake.

The enhancement of iron uptake by the presence of humic substances has been demonstrated in laboratory studies (Provasoli, 1963; Prakash et al., 1973), but the biochemical mechanisms have not been elucidated. For iron, the active excretion by microorganisms of iron-binding compounds, such as siderophores by blue-green algae, is also very important (Murphy et al., 1976). Recent studies of iron uptake by algae (Anderson and Morel, 1982) may lead to a better understanding of the role of humic substances in iron uptake. On the other hand, humic substance-metal complexes are also thought to inhibit biological activity. For instance, Jackson and Hecky (1980) describe the depression of primary productivity caused by making iron unavailable to phytoplankton by humic substances rather than light attenuation or inhibition of enzymes.

Complexation/Adsorption of Trace Anions, with Special Reference to Phosphate

Because phosphorus is commonly a limiting nutrient for algal growth in lakes, the possible chemical interaction between phosphate and dissolved humic substances may be even more important than iron complexation in regulating phytoplankton growth. Under various conditions of low pH and low redox potential, dissolved humic substances may associate with orthophosphate in the presence of iron (cf. Ohle, 1935, 1937; Francko and Heath, 1979, 1982; Stevens and Stewart, 1982) and even in the presence of manganese (Steinberg and Baltes, 1984) and probably render it inaccessible to phytoplankton. Since concentrations of dissolved humic substances can often be 2–3 orders of magnitude greater than concentrations of orthophosphate, even low binding affinities between these two materials may place substantial constraints upon available phosphate (Stewart and Wetzel, 1981b). These authors attempted to quantify interactions of ^{32}P-labeled orthophosphate and dissolved humic substances using gel permeation chromatography. They were unable to demonstrate binding of ^{32}P-orthophosphate to dissolved humic matter under conditions similar to those in the epilimnion of their study site, Lawrence Lake. This may be attributed to high concentrations of calcium, which competes with iron for binding sites in the humic "molecule." However, their negative results may also reflect an inadequate experimental design, for they actually tested only short-term uptake of phosphate by humic substances rather than previously sorbed phosphate.

Steinberg and Baltes (1984) studied the influence of iron, manganese, and cadmium on the association of phosphate with dissolved humic substances and reached two conclusions:

1. The addition of low quantities of iron and manganese causes humic substances to sorb phosphate in significant quantities.

2. High concentrations of manganese ions and successive additions of cadmium ions lead to increases in low-molecular weight phosphates and decreases in high-molecular weight phosphates, which are most likely due to catalytic cleavage of high-molecular weight phosphorus compounds.

The first results can be interpreted based on the work of Tipping and Higgins (1982) and Francko and Heath (1979, 1982). Under the experimental conditions, iron and manganese (but not cadmium) may form a hydroxide colloidal phase that incorporates phosphate, as well as humic substances, but does not coagulate. Dissolution of the hydroxides upon acidification would release the adsorbed phosphate. If the high-molecular-weight humic–phosphorus complexes described by Francko and Heath are similar to these postulated colloids, UV radiation probably disrupts the phenolic groups of the humic substances in the colloids, thereby reducing the affinity of iron or manganese for phosphate, which is then released. These complexes are refractory to enzymatic hydrolysis. There may be steric inhibition of the enzyme by the phosphorus–metal hydroxide–humic complexes or inactivation of enzymes by dissolved humic substances (Baxter and Carey, 1982). Francko and Heath (1982) suggest that orthophosphate sorbed to ferric iron–dissolved humic substance complexes may be released by a mechanism involving UV-induced photoreduction of ferric iron to the ferrous state.

A study of the influence of dissolved humic substances on carbon assimilation and alkaline phosphatase activity by Stewart and Wetzel (1982) illustrates the nutritional significance of the interaction between humic substances and phosphorus. They found that mixed natural assemblages of algae and bacteria exhibited low rates of ^{14}C assimilation and high rates of dissimilation of recent photosynthate when amended with low concentrations of unfractionated humic substances. The extent of the inhibition of ^{14}C assimilation was greatest in the smaller microorganisms (1–5 μm). In different algal–bacterial assemblages, additions of dissolved humic substances markedly enhanced community alkaline phosphatase activity, particularly under low-light regimes. Humic substances of low apparent molecular weight were much more stimulatory to both ^{14}C assimilation and alkaline phosphatase activity than humic substances of high apparent molecular weight, supporting the belief that molecular weight is an important determinant of interactive capacity. Stewart and Wetzel (1982) invoked two hypotheses to explain increases in alkaline phosphatase activity in response to humic substances:

1. Humic substances might sequester organic phosphorus-containing molecules and render phosphate available only through enzymatic hydrolysis. If so, production and release of organophosphorus compounds by the microflora would gradually result in decreased phosphate availability. Biotic equilibrium would be established after increases in alkaline

phosphatase activity allowed more phosphorus to become available to the microflora.

2. Humic substances stimulated the growth of either bacteria or algae, and increases in competition between members of these two groups for phosphate, caused one or both groups to increase its alkaline phosphatase activity.

Inhibition of Calcium Carbonate Precipitation and Similar Processes by Aquatic Humic Substances

In hardwater lakes, epilimnetic decalcification proceeds vigorously during the summer months, for the precipitation of $CaCO_3$ is enhanced by increases both in water temperature and in photosynthetic activities of phytoplankton (Wetzel, 1983). As epilimnetic decalcification proceeds, phosphate and dissolved organic matter are lost from the epilimnion via co-precipitation (Otsuki and Wetzel, 1973; Wetzel and Otsuki, 1974; Rossknecht, 1980). In a detailed study, Reynolds (1978) suggests mechanisms for calcite crystal growth, and the inhibition of calcite growth by DOM. In the absence of inhibiting species, calcite grows by a spiral dislocation mechanism. Polyphenolic substances such as tannins or humic acids selectively sorb to the spiral dislocations and force the crystal to grow by a slower surface nucleation process. Stewart and Wetzel (1981b) studied the inhibition of calcite precipitation by metal-free fulvic acid and found that fulvic acid concentrations of greater than 2.0 mg C/L inhibited all calcite precipitation (see Fig. 10). Reddy (1978), however, found that a concentration of 10 mg/L of a different humic acid caused only 75% inhibition within 11 days of incubation. Inhibition of calcite precipitation by the Contech fulvic acid used by Stewart and Wetzel was much more complete (100%) at lower concentrations over shorter intervals of time, suggesting that either the higher-molecular-weight humic acid used in Reddy's experiments was less inhibitory to the calcite precipitation process, or that in the experiment of Stewart and Wetzel, the more natural conditions (e.g., naturally occurring nonuniform nuclei, photosynthetic removal of dissolved carbon dioxide) favored the inhibiting effects of fulvic acid on decalcification (Stewart and Wetzel, 1981b).

From an ecological point of view, Stewart and Wetzel (1981b) discussed some interrelationships of the various geochemical properties of humic substances in a lake with respect to calcite precipitation. During exposure to sunlight of an intensity sufficient to cause photolysis of aquatic humic substances, water temperature also increases and photodegradation products, for example, carbon dioxide and carbonates, accumulate. The increase in water temperature substantially decreases the solubility of calcite, and photodegradation products alter both water pH and buffering capacity. In addition, losses of humic substances through photolysis may allow calcite precipitation to proceed by removal of threshold quantities of dissolved humic substances which would normally inhibit the formation of calcite crystals.

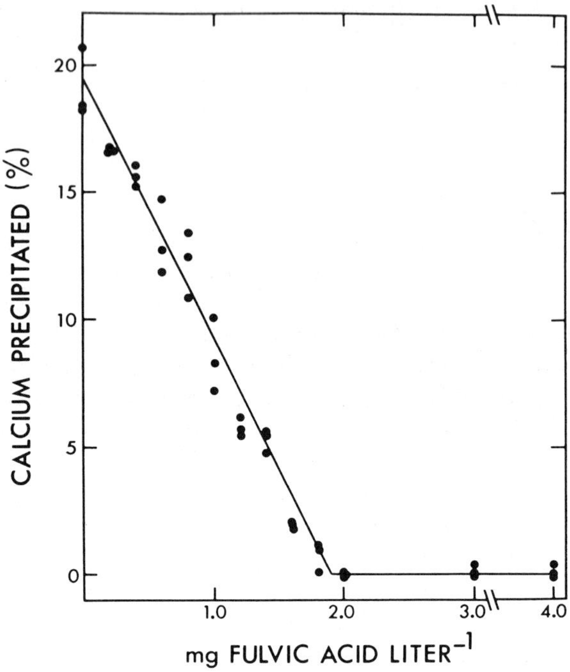

FIGURE 10. Inhibition of calcium carbonate precipitation by fulvic acid (from Stewart and Wetzel, 1981b).

The interrelationship between the many variables, such as the concentration and type of humic substances, the amount of UV radiation, pH, water temperature, and the carbonate–bicarbonate equilibrium, consequently lead to a nearly intractable matrix of potential outcomes. There is evidence that aquatic humic substances affect not only calcium carbonate precipitation but also many other precipitation processes (e.g., Davis and Gloor, 1981). For instance, colloids added to surface water samples become negatively charged due to the adsorption of humic substances and change coagulation properties accordingly (Tipping and Cooke, 1982). Tipping and Higgins (1982) measured the colloid stability of hematite (Fe_2O_3) particles in the presence of different amounts of humic substances. Their results show that adsorbed humic substances enhance the colloid stability of the hematite particles. Presumably an adsorbed layer prevents initiation of aggregation.

Further studies are required to understand the dependence of these processes on varying environmental conditions, especially in the trophogenic zones of lakes. These studies should be designed to answer questions such as: (1) Are such colloids stable under photolytic conditions? (2) To what extent do seasonal fluctuations in pH alter colloidal stability? (3) How do seasonal changes in di- and monovalent cations change the reactivity of aquatic humic substances with respect to colloidal stabilization?

Adsorption of DOM onto surfaces of particulate and colloidal matter is not only a contribution to the organic carbon cycle, but is also a process that significantly modifies the surface properties of suspended matter. For example, adsorbed organic matter will have a significant effect on the sorption of trace substances (e.g., metals). The particulate material covered by humic substances in lakes may be very effective in removing trace metals from solution by the formation of metal–organic complexes and may significantly change the bioavailability of metals (Davis and Gloor, 1981).

Degradation of Aquatic Humic Substances and Associated Substrates by Microbes

Studies of the biodegradation of aquatic humic substances (Shapiro, 1957; Ryhaenen, 1968; De Haan, 1974, 1977; Stabel et al., 1979; Steinberg and Herrmann, 1981) or humic substance–organic trace substance associations (Stabel, 1977; Steinberg, 1977a) are relatively few. In pioneering work on biodegradation, Shapiro (1957) stated that the yellow organic acids from Lower Linsley Pond were reasonably stable to biological decomposition over a period of 4.5 months, while Ryhaenen (1968) observed an increase in biodegradation of humic substances from Finnish lakes after amendments of phosphorus and nitrogen.

De Haan (1974, 1977) studied the effect of fulvic acids from Tjeukemeer on the growth of *Pseudomonas* and an *Arthrobacter* species. He found that fulvic acids in the culture medium had a stimulatory effect on a species of *Pseudomonas* which was explained in terms of the co-metabolism of fulvic acids. De Haan suggested that the same enzyme system is involved in both the lactic acid and the fulvic acid oxidation. This may be valid if the fulvic acid "molecules" are largely aliphatic, which has recently been shown for Tjeukemeer fulvic acids (De Haan et al., 1979).

In another study on biodegradation, De Haan (1977) demonstrated that the addition of benzoic acid to water from Tjeukemeer stimulated bacterial growth and decreased both the fluorescence and the color of the water, indicating disappearance of fulvic acids. The stimulation of bacterial growth probably resulted from a priming effect of benzoic acid on the slow microbial decomposition of fulvic acids rather than an effect of cometabolism, for the cultured bacterium, a strain of *Arthrobacter,* was also able to directly utilize fulvic acids without the amendment of benzoic or lactic acid. De Haan showed further that, during the growth of the *Arthrobacter* strain from Tjeukemeer on benzoic acid in the presence of fulvic acids, fulvic acids of relatively small molecular size were formed from fulvic acids of larger molecular size, suggesting that microbial activity may play a substantial role in the seasonal fluctuations in the molecular size distribution of fulvic acids in Tjeukemeer. De Haan et al. (1981b) indirectly demonstrated the influence of the biota on aquatic humic substances by analyzing seasonal variations of its chemical composition. The proportion of benzoate-metabolizing bacteria in

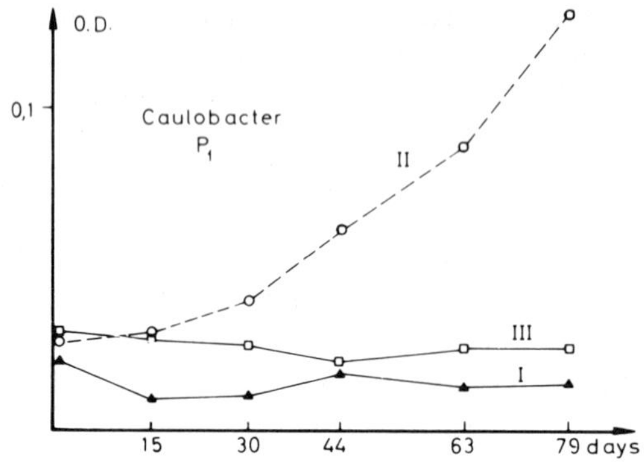

FIGURE 11. Growth of an oligocarbophilic bacterium, *Caulobacter* sp., on GPC fractionated DOM from Lake Plussee (from Stabel et al., 1979). I = apparent macromolecules (initial carbon concentration = 50 mg/L) II, = apparent oligomers (60 mg/L), III = low-molecular-weight materials (45 mg/L). Growth was measured as turbidity at 546 nm.

the total heterotrophic flora seems to reflect the rate of fulvic acid biodegradation in lakes enriched in humic substances (De Haan, 1983). It can be concluded that, in addition to UV irradiation (Strome and Miller, 1978), microbial degradation may also be responsible for the seasonal fluctuation in the molecular size distribution of fulvic acids.

Stabel et al. (1979) tested the ability of several oligocarbophilic bacterial strains to utilize DOM from lakes. Oligocarbophilic bacteria are able to grow only on low concentrations of DOM, especially on refractory organic matter. DOM from Lake Plussee was fractionated by gel permeation chromatography and the growth of eight oligocarbophilic and one saprophytic strain was recorded for each molecular weight fraction of the DOM. After long lag periods, oligocarbophilic bacteria were able to degrade apparently smaller molecules (as shown with a *Caulobacter* strain in Fig. 11), while the saprophytic strain showed no growth. The apparently larger molecules were much more resistant to microbial attack. Additions of glucose seemed to support cometabolism of the labile and the refractory molecules, especially with high-molecular-weight DOM fractions. Glucose amendments caused little difference in rate of degradation of the apparently smaller fraction.

In experiments of similar design, Steinberg and Herrmann (1981) tested the effect of metal–organic compounds on nonaxenic cultures of *Chlorella fusca* (a green alga) and microorganisms adapted to aquatic humic substances. Three fractions of DOM were obtained. The intermediate fraction supported growth better than either of the other fractions, especially with the algae–bacteria system. The DOC in all cultures declined, but the best correlation to algal growth occurred in the presence of apparently oligomeric dissolved organic matter. Microorganisms adapted to aquatic humic substances were more efficient in "utilizing" organic iron compounds than were algae–bacteria systems. Contrary to the findings of Stabel et al. (1979) and Steinberg (1977a) and De Haan (1974, 1977), Steinberg and Herrmann (1981) found that amendments of organic nutrients (glucose) did not increase biodegradation of aquatic humic substances, but organic amendments did accelerate losses of iron and other metals.

Autochthonous Production of Dissolved Humic Substances

Aquatic humic substances in freshwater ecosystems are often believed to be largely of terrestrial origin (e.g., Gjessing, 1976). However, metabolic pathways in decomposition of organic debris occurring in soil environments are, in all likelihood, similar to those in aquatic ecosystems. In terrestrial systems, the decomposition of lignin assumes quantitative significance because of the dominance of lignin in terrestrial plants, whereas in large, deep lakes the decay of lignin-free algae biomass increases in significance.

There is indirect evidence for autochthonous production of humic substances in aquatic systems. Larson and Hufnal (1980) found that transition metal oxides (MnO_2, ZnO, CuO) and cations (Mn^{2+}, Fe^{3+}) promoted the polymerization of dissolved catechol (1,2-dihydroxybenzene) and other catechol and pyrogallol derivatives. Colored catechol polymers formed much more rapidly in stream water than in deionized water buffered to the same pH. Sediment or clay increased polymerization, a phenomenon known from soil science (cf. Wang et al., 1978). Autochthonous production of aquatic humic substances may also occur via enzyme-mediated oxidations, many of which incorporate phenolase (cf. Martin et al., 1975; Steinberg, 1977a; and De Haan et al., 1981b; see Fig. 12). A third autochthonous pathway of humic substance production which may occur in lakes is the "Browning reaction" between sugars and amino acids, known as Maillard condensations (Stuermer, 1975). These reactions involve the formation of a Schiff base between the amino nitrogen in the amino acid and the aldehyde or ketone groups of sugars, with subsequent rearrangements, cyclizations, and decarboxylations to form complex, brown-colored mixtures referred to as melanoidins. However, the condensation of amino acids and sugars alone is insufficient to account for the abundance of aliphatic structures in humic substances.

All three suggested polymerization routes proceed more efficiently at higher reactant concentrations; hence, microscale heterogeneity may be important in the formation of aquatic humic substances. There are many possible examples:

1. Decaying organisms contain the precursors in high concentrations.
2. Organic matter in particles may be in dynamic equilibrium with that in solution.
3. Lipids, fatty acids, and triglycerides mainly occur as particles or colloids.

4. Filter-feeding organisms concentrate organic matter in their guts and in fecal pellets.
5. Surface active molecules may exist as aggregates from a few molecules to the much larger micelles and colloids.

In all these instances, organic molecules are brought within bonding distance and intermolecular reactions become possible (after Stuermer, 1975).

In conclusion, autochthonous humification processes in lakes must occur. The ratio of autochthonous to allochthonous input varies, of course, from lake to lake and depends on factors such as the specific ratio of watershed size to lake area, watershed structure, hydrologic input into the lake, productivity within a lake, and the relative sizes of pelagic and littoral zones of a lake.

Control of Proteinaceous Matter and Carbohydrates

Humic substances can play an important role in aquatic ecology by several mechanisms which have been discussed. Humic substances may also control the bioavailability of dissolved proteinaceous material and carbohydrates. In field and laboratory studies, Stabel (1977) and Steinberg (1977a) showed that carbohydrates associated with humic substances had a greater microbial availability than proteinaceous matter associated with humic substances,

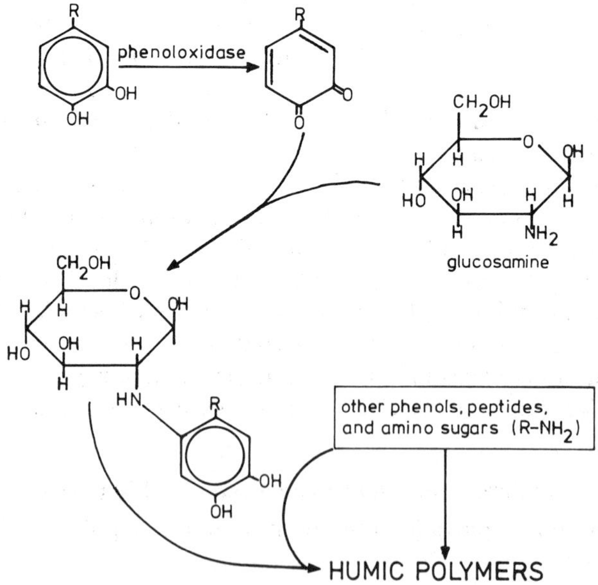

FIGURE 12. Oxidative polymerization of phenol derivatives involving amino sugar units (modified after Martin et al., 1975).

and postulated that this was due to different kinds of binding to humic substances. Sweet and Perdue (1982) showed that in Upper Klamath Lake, in Oregon, polysaccharides accounted for nearly all dissolved sugars. The humic-bound carbohydrate fraction was rather small. However, they analyzed only a few samples and did not test the biological availability of the different saccharide fractions. Lytle and Perdue (1981) fractionated amino acids using resin adsorption (XAD-7) and found that, on the average, more than 96% were humic-bound in Williams River, and to a lesser degree in Upper Klamath Lake. Comparing these studies, it appears that humic substances in Upper Klamath Lake have more bound amino acids than carbohydrates, which supports the microbiological findings of Stabel (1977) and Steinberg (1977a).

Using PAGE prior to specific staining of humic substances, carbohydrates, and proteinaceous substances, Muenster (1982) showed that relatively strong bonds existed between humic substances and proteinaceous matter. This strong association is already known in plant physiology, as tannins from cell walls form strong complexes with proteins (Mason, 1955). With carbohydrates, but not with proteinaceous matter, Muenster observed considerable release of monomeric substances during the adsorption process onto resins (XAD-2, PVP). Thus, only weak hydrogen bonding and interionic forces between those carbohydrates found in the monomeric fraction and humic substances could have been established.

In gel permeation studies, Stabel and Steinberg (1976b) and De Haan and De Boer (1978) found that carbohydrates and proteinaceous matter eluted simultaneously. De Haan and De Boer (1978) concluded that these substances could only be associated by weak, acid-labile bonds, such as hydrogen bonds. The findings of Stabel and Steinberg (1976b) and Steinberg (1976) revealed that this was true only for one fraction, but not for all proteins and peptides. Some of this material seemed to be bound more tightly to the humic molecule. Witthauer and Kloecking (1981) concluded that there most likely exist acid-labile, reversible, hydrogen bonds between carbohydrates and dissolved humic substances. They could not find any evidence for the existence of phenolic glycoside bonds that have been postulated.

These papers support to some extent the hypothesis that there is different microbial availability of carbohydrates and proteins caused by differences in bond strengths between these compound classes and dissolved humic substances.

Our understanding of control of microbially significant substrates, however, is not consistent with other studies that conclude that carbohydrate moieties connected with dissolved humic substances are resistant to microbial attack (Steelink, 1977; Gocke et al., 1981). These contradictions illustrate again the urgent need of studies on the ecophysiological significance of humic substances in aquatic ecosystems and reflect our limited knowledge of interactions between humic substances and microorganisms.

DISTRIBUTION OF DISSOLVED HUMIC SUBSTANCES IN LAKE ECOSYSTEMS

Variations Among Lakes

Since there exist no lakes with identical water chemistry, basin morphology, watershed structure, and climate, one can expect large lake-to-lake differences with respect to types and concentrations of DOM. Selected examples are given in Table 4.

There have been relatively few studies of the DOC pool of lakes over long periods of time. Because there are even fewer papers which discuss exclusively dissolved humic substances over long periods of time, we have included studies describing other fractions of the DOC pool (e.g., amino acids, peptides, proteins, or carbohydrates). The examples presented in Table 4 unfortunately differ in methodology and seasonal variation in the humic substances. Thus, lake-to-lake comparisons are nearly impossible.

A high lake-to-lake variation in DOM content is found even in restricted geographic regions. For instance, at the end of winter stratification in 1969, Pennanen (1975) collected water samples from seven lakes in southern Finland and fractionated the dissolved humic substances using gel permeation chromatography. She observed the relative abundances shown in Figure 13. The data indicate local differences with respect to depth within a given lake, and between closely adjacent lakes in the fractionation of DOM. Pennanen found distinct differences between intensely colored and relatively uncolored waters. If the water was intensely colored, there was also an appreciable colloidal fraction (c.f. Lakes Hakojärvi and Lappajärvi in the water sedi-

TABLE 4. Concentrations of Dissolved Organic Materials in Various Lakes[a]

Lake	DOC/ COD (mg/L C)	Humic Substances (mg/L)	Phenols (μg/L)	Carbohydrates (γg/L C)	Amino Acids (μg/L C)
Esthwaite Water, U.K.[b]	—	1.5–3.0	—	—	—
Lake Plussee, FRG upper epilimnion[c]	10.5	—	2800	944	—
Lake Schöhsee, FRG upper epilimnion[d]	3.5	—	—	62	470
Tjeukemeer, The Netherlands[e]	48	—	—	—	1160

[a] *Data refer to mean values of about one annual cycle each.*
[b] *Tipping and Woof (1983).*
[c] *Muenster (unpublished data).*
[d] *Stabel (1977) and Steinberg (1977a).*
[e] *De Haan et al. (1979).*

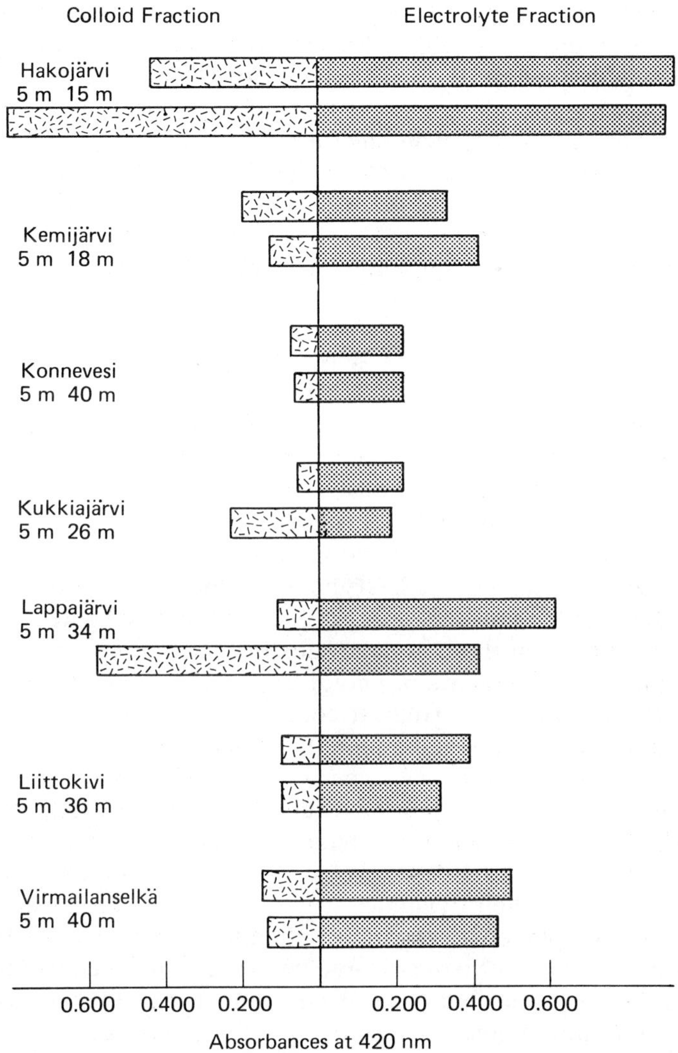

FIGURE 13. Distribution of two different GPC fractions of humic material in seven Finnish lakes at two different depths (5 m below surface and 1 m above sediment). *Left:* colloidal; *right:* dissolved fraction, respectively (from Pennanen, 1975).

ment interface). If the water was relatively uncolored, the colloidal fraction showed a tendency toward a more uniform distribution within the water column (e.g., Lake Konnevesi).

In a similar study, Hama and Handa (1980) fractionated DOM from three Japanese lakes of different trophic status using gel permeation chromatography. Distribution patterns of DOC, carbohydrates, amino acids, proteins, and organic acids were determined. Bimodal distribution patterns of DOC were observed in all samples. Compounds of low apparent molecular weight were relatively abundant. Carbohydrates were found exclusively as polysaccharides and eluted within the void volume. Free amino acids and peptides were distributed within the apparent molecular weight ranges of 200–1400 daltons. Five organic acids, including formic, acetic, propionic, butyric, and lactic acids, appeared in the lowest molecular weight fraction. However, these acids accounted for only minor portions of the total DOM. Carbohydrates, amino acids, proteins, and organic acids accounted for only 16.5–20.6% of the total DOC, so that about 80% of the DOC remained uncharacterized.

Temporal and Spatial Distributions

A conspicuous feature that has been studied for many lakes is the relatively small amount of change in concentration of DOC with depth or season within a lake. The constancy of the DOC pool persists regardless of whether the sampling at each meter of depth is done at weekly or biweekly intervals (Wetzel et al., 1972; and summarized in Wetzel, 1983).

This finding suggests that the DOC pool is comprised largely of refractory organic carbon compounds that are relatively resistant to bacterial decomposition and that inputs of these organic substrates are approximately equal to their slow microbial degradation, photolysis, or sedimentation after adsorption to particulate surfaces. Highest concentrations typically occur during summer stratification in the epilimnion (8–9 mg C/L DOC), and consistently fluctuate to a great extent in the upper strata. The deeper-water layers exhibit a temporally more constant DOC content (4–6 mg C/L). Secretion of DOC by planktonic algae and littoral flora contributes to DOC in the epilimnion. Decomposition of labile, secreted organic compounds often is very rapid (48 hours), and their dynamics would not be observed using the sampling frequency employed for the more generalized picture of the DOC pool.

Among different lakes there are large differences in the concentration and spatial and temporal distributions of DOM. Summarized below are our field studies of DOM and humic substances in different lake ecosystems.

Esthwaite Water, England

Esthwaite Water (Tipping and Woof, 1983, a and b) is a softwater lake in the English Lake District. In this lake, humic substances extracted by butan-1-ol were present almost exclusively in the dissolved and colloidal size classes. The small fraction of humic substances in the particulate form should not be discounted, since it may explain a hypolimnetic accumulation of these substances. Humic carbon comprised 60–70% of the DOC in winter and early spring, but only 30–40% in summer. The higher proportion of nonhumic DOC during summer months presumably resulted from increased biological production of DOC.

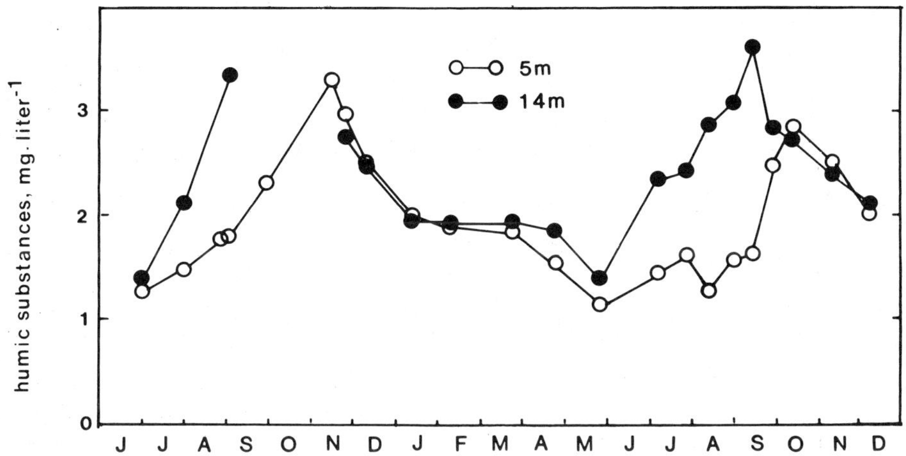

FIGURE 14. Concentrations of humic substances at 5 m and 14 m in Esthwaite Water, 1980–1981 (after Tipping and Woof, 1983).

During winter months, when the lake was well mixed, humic substances were uniformly distributed throughout the water column (Fig. 14). As the season of thermal stratification proceeded, however, the concentrations became different. The concentrations at a 5 m depth declined to about 1.5 mg/L and remained at this level throughout the summer; the concentrations at 14 m increased to reach maxima of 3 mg/L. The maximum values could be due to iron oxide precipitation or to anaerobic decomposition of settled particulate detritus.

In a more recent paper, Tipping and Woof (1983b) present evidence for a mechanism whereby humic substances accumulate by co-sedimentation with iron oxide particles. They also follow the movements of iron during its redox cycling in the hypolimnion and sediment. Their attempts to distinguish epilimnetic and hypolimnetic humic substances by spectroscopy and gel chromatography, however, were unsuccessful, and so provide little evidence for actual mechanisms responsible for the hypolimnetic accumulation of humic substances.

Tjeukemeer, The Netherlands

Tjeukemeer is an alkaline, humic-substance-rich, polder lake in the northern Netherlands (De Haan et al., 1979, 1981b). In winter the lake receives humic-substance-rich water from the surrounding peaty polders. In summer, evaporation losses are compensated for by input of humic-substance-poor water from the Ijsselmeer. De Haan (1972a) demonstrated that the fulvic acids from Tjeukemeer could be separated into three different molecular weight fractions. The relative proportion of each fraction appeared to depend on the type of water present. In winter, about 70% of the fulvic acid

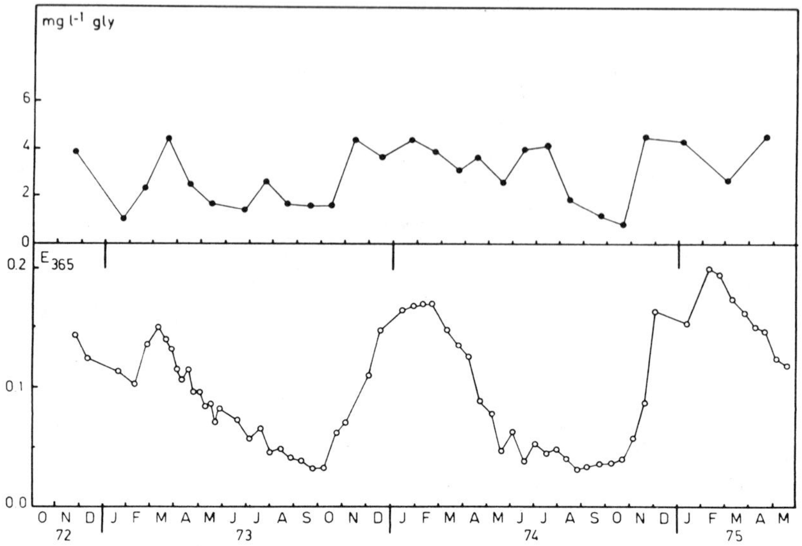

FIGURE 15. Content of dissolved amino acids (glycine equivalents) (upper curve) and absorbance at 365 nm (lower curve) as measures of fulvic acid content in Tjeukemeer, October 1972 to May 1975 (from De Haan et al., 1979).

was found to have an apparent molecular weight exceeding 5000 daltons, whereas in summer, this value fell to 50%.

The annual pattern of the dissolved amino acids (DAA) in Tjeukemeer (expressed in mg/L as glycine equivalents) is shown in Figure 15. During the period of study, the average amount was 2.9 mg/L. Since primary production is relatively low in winter, the high amount of fulvic and amino acids in the Tjeukemeer during winter suggest that the amino acids were transported into the lake together with the fulvic acids from the surrounding polders. In summer, the concentration of DAA varied from 1 to 4 mg/L. The ratio of amino acid carbon to total carbon reflects the dynamics of the dissolved amino acids. Distinct maxima not exceeding 13% of the total carbon could be observed during spring, summer, and autumn. De Haan and De Boer (1978) suggested that large amounts of amino acids were related to populations of senescing algae. The concentration of dissolved amino acids per unit of fulvic acid was greater during summer than in winter. The greatest amounts of amino acids were found following blooms of green and blue-green algae in May and June.

In a later paper, De Haan et al. (1981b) report on the seasonal variation in the composition of fulvic acids determined by Curie point pyrolysis–mass spectrometry. Water samples were collected monthly between January and September, 1978, and fulvic acid fractions were isolated by gel permeation chromatography. Differences in the chemical composition (particularly striking in the pyrograms of fractions of high apparent molecular weight) were explained in terms of fragment molecules attributable to polysaccharides, proteins, and/or phenolic polymers. In winter, the fulvic acids contained a relatively high proportion of aromatic groups, and in summer, they were relatively rich in carbohydrates. These data indicate that during summer the composition of the fulvic component was partly determined by production processes. For instance, pyrograms for June and July samples were more similar to pyrograms of the winter samples than to those of the other summer samples. Aromatic groups derived from proteins appeared to be responsible for this phenomenon. These findings were related to the June–July decline of the cyanophyte bloom, which resulted in relatively high concentrations of extracellular proteins.

Pyrograms of high- and low-molecular-weight fulvic acids showed that the seasonal composition of the low-molecular-weight fulvic acids was more variable than that of the apparently high-molecular-weight fulvic acids. Assuming that microbial activity can change the composition of fulvic acids, this observation agrees with the common opinion that high-molecular-weight fulvic acids are relatively refractory with respect to degradation by microbiota.

The seasonal variations observed in the composition of the fulvic acids occurred simultaneously with changes in their concentrations and molecular weight distribution. This suggests that the same processes (mainly hydrology and microbial activity) are involved both in the seasonal changes of the composition concentration and the molecular weight distribution of fulvic acids.

As stated above, De Haan and co-workers did not consider photolysis of fulvic acids by UV irradiation as an important determinant of fulvic acid molecular weight distribution pattern or chemical composition. Since cleavage of fulvic acid "molecules" and subsequent degradation to CO_2 and H_2O by UV irradiation can occur especially in summer, the observed fluctuations in Tjeukemeer may be attributed to phytolysis as well as degradation by bacteria. This is supported by findings of Strome and Miller (1978) and Gilbert (1980), who demonstrated an enhanced biodegradability of fulvic acids after UV irradiation.

Lake Schöhsee, Northern Germany

Lake Schöhsee is an oligotrophic hardwater lake of the Plön lake district (Stabel, 1977; Steinberg, 1977a). Results of DOM fractionations for the summer months are shown in Figure 16 (Steinberg and Stabel, unpublished data). DOC concentrations in the upper epilimnion and of the sediment–water interface vary little, fluctuating between 2.5 and 4.5 mg C/L. This agrees with the finding of Wetzel et al. (1972) for Lawrence Lake. The constancy, however, also seems to be valid for distribution patterns of molecular weight. Except for June 1973, in the upper epilimnion, 70–90% of the DOC resided in the fraction with apparent molecular weight of <10,000

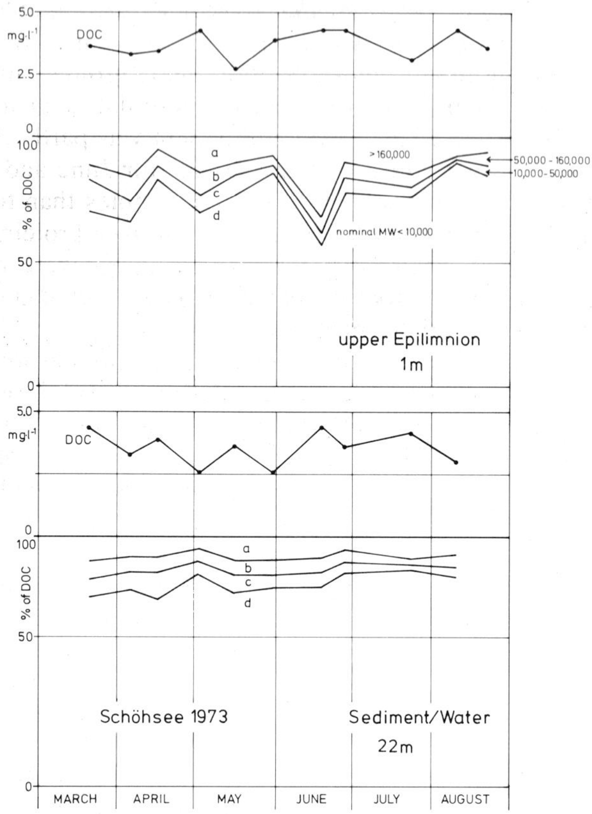

FIGURE 16. Fractionation of Lake Schöhsee DOC (glucose equivalents) by ultrafiltration (Stabel and Steinberg, unpublished data).

daltons. The other three molecular weight fractions usually did not exceed 10% of the DOC. The relatively high concentrations of DOC in the fraction of highest apparent molecular weight in June were probably caused by large populations of phytoplankton.

Additional investigations on extracellular dissolved proteinaceous matter in Lake Schöhsee were completed by Steinberg (1977a), and on carbohydrates by Stabel (1977). An annual study of different water layers (upper epilimnion, epilimnion–metalimnion interface, metalimnion, and sediment–water interface) showed that maximum concentrations of proteinaceous matter in the epilimnion were related primarily to phytoplankton fluctuations during stratification. During circulation periods, synchronous maxima of proteins, peptides, and amino acids occurred throughout the water column. The concentration maxima in the two lower strata were apparently correlated more to sedimentation and resuspension processes than to phytoplankton dynamics. A lower threshold concentration of proteinaceous matter (about 65–80 μg/L amino nitrogen) was found for the lake; this concentration proved to be lake specific (Steinberg, 1977b) and may correspond to the fraction of microbially resistant humic-bound proteins, peptides, and amino acids.

The four fractions derived by ultrafiltration made up on the average:

1. Nominal MW $\leq$ 10,000 daltons	58–60% of the total proteinaceous nitrogen
2. Nominal MW 10,000–50,000 daltons	10–12% of the total proteinaceous nitrogen
3. Nominal MW 50,000–160,000 daltons	12–15% of the total proteinaceous nitrogen
4. Nominal MW $\geq$ 160,000 daltons	11–17% of the total proteinaceous nitrogen

Relative to the DOC content, the fractions of higher molecular weight contained more proteinaceous matter than fractions with lower apparent molecular weights (i.e., 10,000 daltons). It is interesting to note that the distribution patterns of the proteinaceous matter showed a reduction of the mean concentrations of the two smaller fractions which contained most of the colored substances within the thermocline. This is probably due to the specific productivity and respiration rates in the thermocline. The major fractions of DOC seem to be more refractory.

In Lake Schöhsee, the annual average contribution to the DOC pool by carbohydrates was 1–2% (Stabel, 1977). In the higher molecular weight fractions, this percentage occasionally increased to 50–60%. The yearly average of dissolved carbohydrate–carbon was 69 μg C/L in the upper epilimnion, 58 μg C/L in the epilimnion–metalimnion interface, 43 μg C/L in the metalimnion, and 32 μg C/L above the sediment. High concentrations (up to 250 μg C/L) were measured in some samples while others did not contain any sugars at all. Unlike proteinaceous matter, no lower threshold concentrations for carbohydrates could be observed. Thus, carbohydrates appear to be more available to microbiota than proteinaceous matter, suggesting again different kinds of bonding between proteinaceous material and fulvic acids and carbohydrates and fulvic acids.

Ultrafiltration fractionations revealed that there rarely were traces of mono- and oligosaccharides. Most of the carbohydrates were associated with macromolecules (>160,000 daltons).

Summarizing parameters, DOC and molecular weight fractionation studies reveal a relatively high degree of constancy in time and space within a given limnetic ecosystem. High variations in the proteinaceous and carbohydrate pools of Lake Schöhsee, over short time periods however, indicate that the apparent constancy could be seriously misleading. But carbohydrates and proteins are not the only chemical constituents associated with aquatic humic substances.

Lake Plussee, Northern Germany

Lake Plussee (Muenster, in preparation) is a eutrophic hardwater lake of the Plön lake district. During the summer, concentrations of dissolved combined phenolic compounds oscillate drastically over short periods of time (as shown in Fig. 17 for epilimnetic waters), although DOC concentrations (measured as COD in glucose–carbon equivalents) were much more stable. Fluctuations in combined phenolic compounds correlate poorly with phytoplankton standing crop (Fig. 17, lowermost panel). Thus, distribution patterns of free phenols and phenolic compounds may result primarily from abiotic factors (e.g., photolysis, allochthonous inputs by rainstorms, or adsorption onto autochthonous calcite) or biotic ones other than release by phytoplankton (e.g., biodegradation after photolysis). These processes, which have not yet been quantified, obviously influence the upper water layers most, since absolute concentrations of phenolic compounds (as well as oscillations within the concentrations) are significantly lower in the deeper-water layers. Perhaps many of the phenols in the deeper strata occur in a particulate state, adsorbed onto sedimentary matter. Alternatively, total phenolic concentrations are really lower in the deeper strata; if true, the reasons remain obscure.

Additional evidence for the dynamic fluctuations of humic substances within the DOC pool is provided by Muenster (1982) using PAGE techniques with water from Lake Plussee. Depth profiles from June 19 and June 30, 1979, are given in Figure 18. Great fluctuations occur with depth on a given sampling date, and from one sampling date to the next. For instance, the apparent molecular weight of the fraction which moved rapidly, Fl, was between 2.0×10^4 and 3.5×10^4 daltons in the shallow strata on June 19, 1979. Twelve days later, the apparent molecular weight of this fraction at the lake's surface increased to about 4.5×10^4 daltons. With the slowly moving fraction F2, the corresponding changes are even more pronounced, suggesting condensation reactions in the DOC pool (allochthonous inputs were negligible during that time). Similar changes were observed with the molecular net charges (equivalent to molecular valency). Both fractions F1 and F2 increased in molecular net charge in epilimnetic DOM from June 19 to June 30, 1979, whereas in the hypolimnion, a slight reduction in the parameters for fraction F2 was observed. In the metalimnion (thermocline) most of the parameters of each fraction differed from epi- and hypolimnetic DOM. Thus, the significance of metalimnetic metabolism is indicated again.

None of the measured biological and microbiological parameters, for example, phytoplankton biomass (chlorophyll-a and phaeopigments), primary production, or uptake kinetics for glucose and carbon dioxide in the dark, coincided with any changes in the PAGE fractions (Fig. 18). Consequently, future studies are required to reveal causes and consequences of the highly dynamic nature of DOC and dissolved humic substance pools in freshwater systems. It must be emphasized that measurements of summarizing parame-

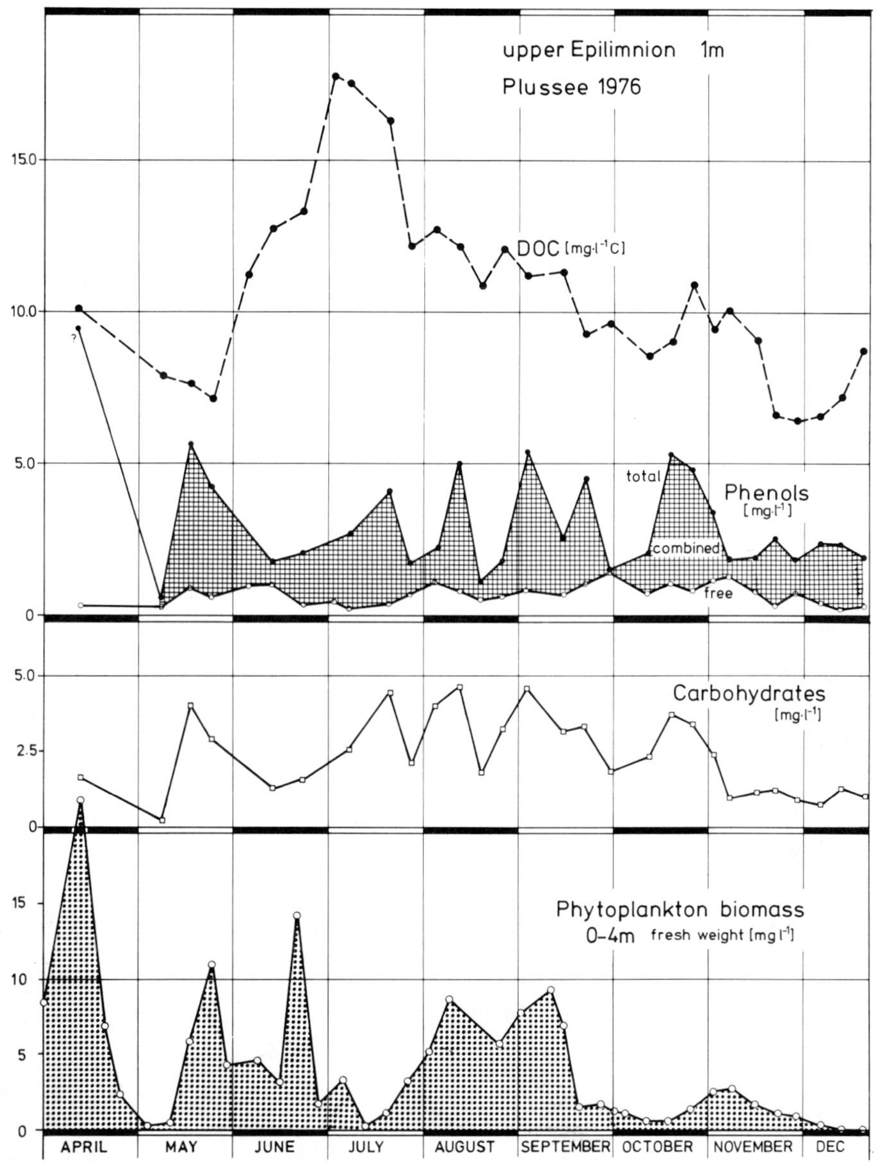

FIGURE 17. Distribution of DOC, phenols, and carbohydrates in Lake Plussee epilimnion, 1976 (from Muenster, in preparation). Algae biomass after Hickel (1978).

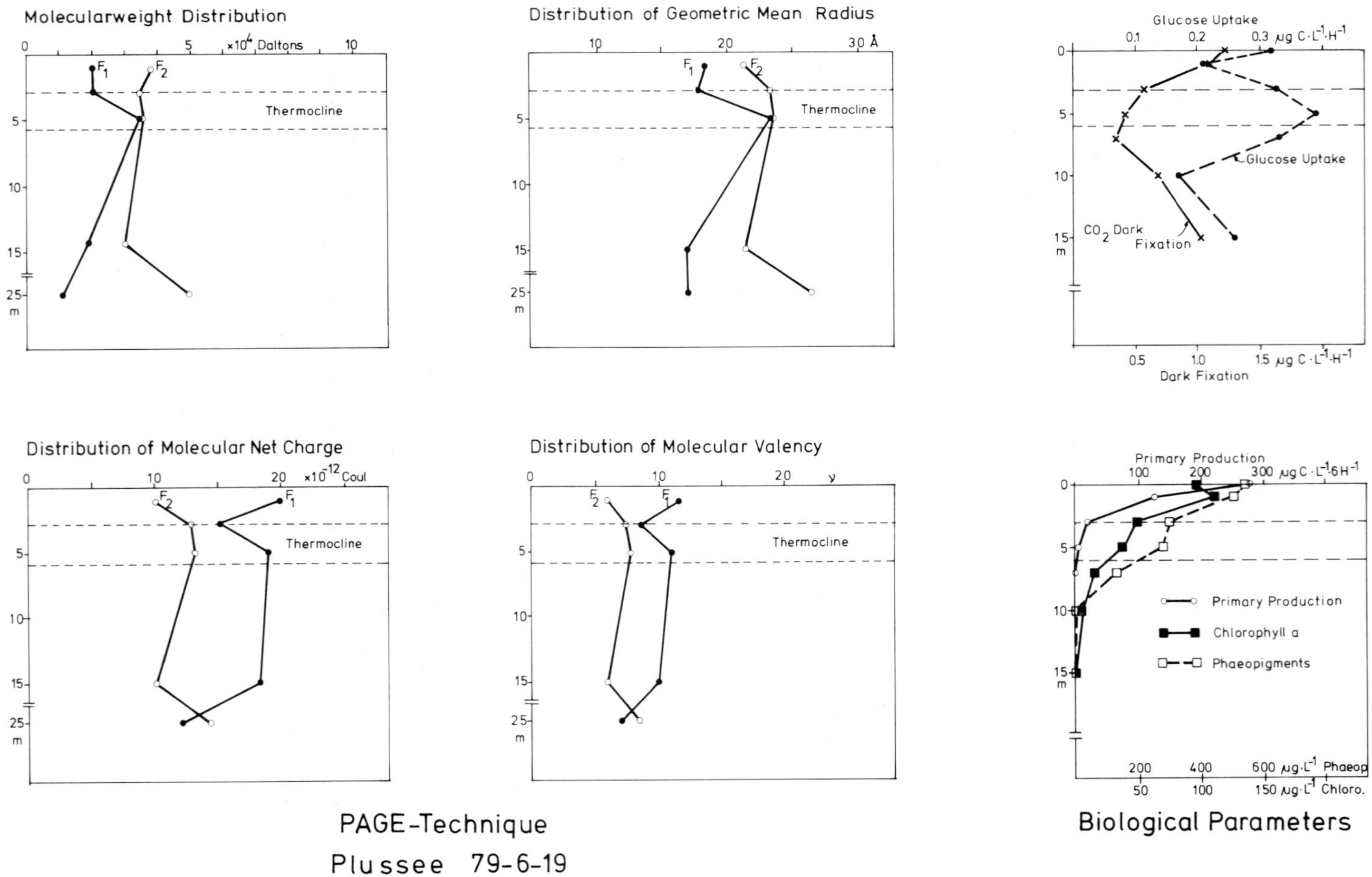

FIGURE 18. Distribution of PAGE parameters in space and time in Lake Plussee in relation to biotic parameters (from Muenster, 1982).

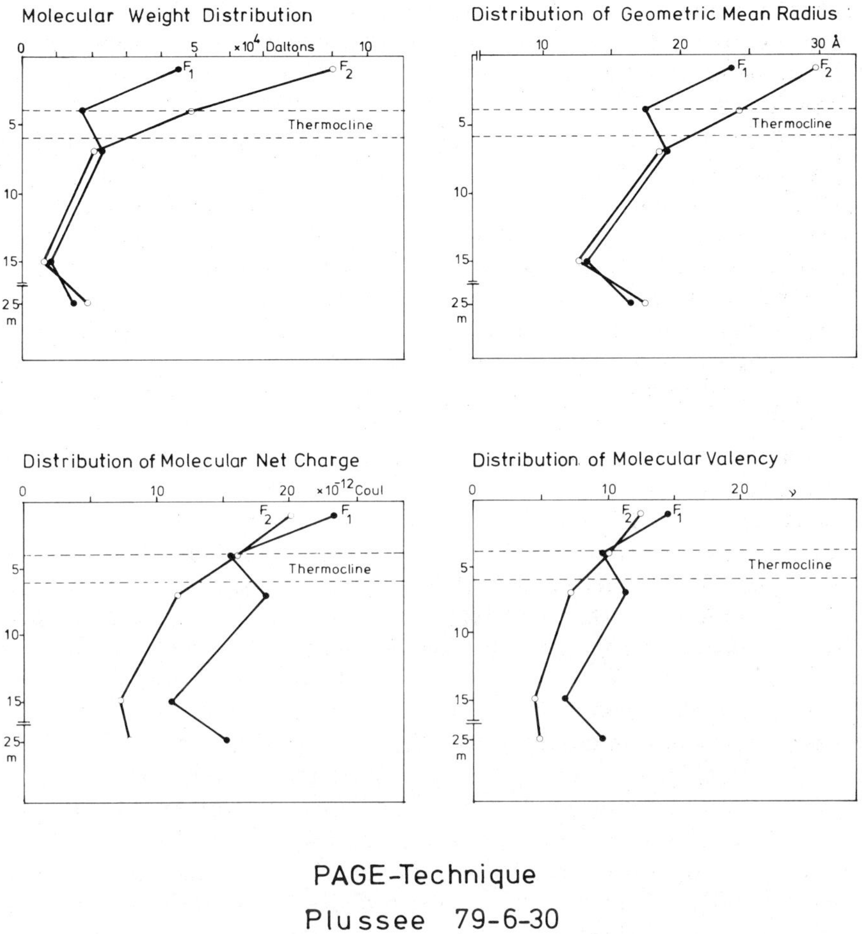

FIGURE 18. Continued.

ters alone (i.e., absorbance, DOC, fluorescence, etc.) seriously underestimate the dynamic nature (and significance) of DOM and aquatic humic substances in the majority of cases.

CONCLUSIONS

Several times in this chapter we pointed out what little is known about the geochemistry of humic substances and their ecological significance in lakes. Some contradictory findings in different reports have also been discussed. In the following, we want to summarize briefly what we should know and what we do not know.

Additional studies should be carried out applying, for instance, fluorometric and spectrometric methods to reveal differences of humic substances of different chemical nature or geochemical origin. Degradation product analyses should confirm chemical characteristics and elucidate further geochemical pathways in production and degradation of humic substances in lakes. For example, future studies will need to specifically address mechanisms of autochthonous humic substance production from algal detritus. Compared to soil humus, dissolved humic substances in lakes contain higher contents of aliphatic structural units. Aliphatic content can probably act as a measure of autochthonous production.

Adsorption of aquatic humic substances onto surfaces changes their molecular weight distribution pattern. However, there is apparently little knowledge of how the chemical composition and the microbial availability is altered by this process under environmental conditions. Photolysis by UV radiation also results in changes of the molecular weight distribution, but, there is some contradictory evidence concerning which molecular weight fractions are affected most.

Complexation of metals and trace anions (e.g., phosphate) by humic substances leads to a decrease of toxicity of certain metals toward microorganisms and increases the availability of some metals, but decreases phosphorus resources. At this time, the role of humic substances in reducing the toxicity of trace metals is more clearly understood than the other roles. Adsorption of micropollutants by aquatic humic substances may enhance their toxicity toward microorganisms in many cases. The specific modes of action, however, are mostly unknown. On the other hand, dissolved humic substances may assist in the degradation of organic pollutants. Under environmental conditions, the predominating process is not obvious.

There is some, at least partly contradictory, evidence that different binding/adsorption mechanisms exist for microbially significant substrates (e.g., carbohydrates and proteinaceous matter), resulting in different degrees of microbial availability of these substance classes. But more unequivocal data seem to be necessary. From soil science, it is known that humic substances bind/adsorb enzymes, especially hydrolases, and reduce their activity (Burns, 1982; Ladd and Butler, 1975). Except for a single paper by Baxter and Carey (1982), equivalent reports on the interactions of dissolved humic substances in lakes are lacking.

There is also little knowledge about the mechanisms and pathways involved in microbial uptake and utilization of humic compounds and/or associated substrates. We do not know if they are really taken up completely or if the cleavage is done extracellularly. As mentioned above, one bacterial species (*Arthobacter*) is reported to utilize fulvic acids directly. Nothing is known on the existence of further microbial species able to utilize humic substances directly.

Based on a great lack of studies on spatial and temporal distributions of dissolved humic substances, these substances are still believed to be the refractory component of the DOM of lakes. Such studies are made difficult by the unsuitable methods to measure humic substance dynamics in nature. However, there is evidence from studies on Lake Plussee DOM that the phenolic component of the DOM pool showed temporal and spatial variations. This observation needs further confirmation by studies in other lakes.

ACKNOWLEDGMENT

Thanks are due to Drs. H.-H. Stabel, E. Tipping, and A. J. Stewart for providing valuable preprints to the authors. We are also grateful to A. J. Stewart and to the book's editors, especially to Dr. D. McKnight, for improving the style of our manuscript. This paper is dedicated to Prof. Dr. W. Ohle, Plön.

CHAPTER SIX

Geochemistry of Humic Substances in Lake Sediments

RYOSHI ISHIWATARI

ABSTRACT

Humic substances are the most abundant organic constituents in lake sediments and play a central role in the geochemical cycle of carbon in lakes, lake sediments, and sedimentary rocks. Humic substances in lakes and lake sediments originate both from aquatic organisms living in the lake (autochthonous) and from organic matter that is washed into the lake from surrounding soils and streams (allochthonous). The chemical characteristics of the lake humic substances indicate that they are mainly autochthonous in normal freshwater lakes. The humic substances in the uppermost layer of a freshwater lake sediment form very rapidly from dead phytoplankton cells. In the deeper layers of the sediment the formation of humic substances continues to take place but at a slower rate. Humic substances undergo diagenetic changes with burial of the sediment. These changes include a gradual decrease with depth of burial of humic and fulvic acids and a concomitant increase of humin.

INTRODUCTION

Organic matter in lake sediments is both autochthonous (produced by aquatic organisms such as phytoplankton, etc.) and allochthonous (derived from surrounding soils and higher plants). Therefore, the organic matter content and molecular composition in lake sediments are dependent on various factors such as trophic level, climate, drainage into the lake, and chemical, biological, and geological characteristics of the surrounding environments. The discussion in this chapter is limited to humic substances from normal freshwater lakes since there are insufficient data from other lake types.

Geochemical studies of humic substances in lake sediments are aimed at (1) recognizing the molecular nature, formation, behavior, and fate of humic substances in lake sediments, (2) determining their role in material cycles in lakes and lake sediments, and (3) understanding a general picture of organic processes with respect to humic substances in sedimentary environments (hydrosphere in a broad sense). Organic matter in lake sediments is situated at the initial stage of the long-term carbon cycle in the earth's crust. Since in sedimentary environments the nature of chemical reactions on organic matter in both lakes and oceans is considered similar, understanding organic matter in lake sediments will contribute greatly to a better understanding of the geochemical cycle of carbon in the earth's crust.

Other than the author and his collaborators, the following scientists have carried out studies on humic substances in lake sediments. Karavaev and co-workers (Karavaev and Budyak, 1960; Karavaev et al., 1964) studied humic and fulvic acids in some Russian lakes using chemical oxidation, infrared spectroscopy, and other methods to point out their aliphatic character. Povoledo and collaborators (Povoledo and Gerletti, 1963, 1968; Povoledo et al., 1975; Povoledo and Pitze, 1979) extracted fulvic and humic acids from Italian and Canadian lake sediments and characterized their molecular weight distribution and lipid components. Kemp and co-workers (Kemp and Mudrochova, 1973, 1975; Kemp and Wong, 1974; Kemp and Johnston, 1979) extracted humic and fulvic acids from the Great Lakes and studied their molecular weight distribution and nitrogen-containing components. Bourbonniere (1979) and Bourbonniere and Meyers (1978, and unpublished) extracted humic and fulvic acids from Lakes Huron and Michigan and characterized them by spectroscopy, NaOH hydrolysis, and other methods. Otsuki and Hanya (1967) examined the infrared spectrum of humic acid from a Japanese lake (Lake Haruna) and discussed its precursors.

Humic substances are clearly the major constituent of organic matter in lake sediments and therefore should play the central role in the geochemical cycle of carbon compounds in lakes, lake sediments, and sedimentary rocks. The characteristics and geochemistry of humic substances in lake sediments is described in this work by citing research done by the author and his co-workers, some of which is unpublished.

In this chapter, humic substances, humic acid, fulvic acid, and humin refer to material extracted from lake sediments that were initially extracted with an organic solvent. Where a different extraction method was used, it is described in the text.

Limnological data of Lake Haruna (Japan) are given next (Otsuki and Hanya, 1967, Ishiwatari et al., 1980a) because the results on humic substances from this lake will be described in some detail in this chapter. Lake Haruna is situated on the top of Mt. Haruna in Gumma Prefecture and was formed as a caldera lake about 40,000 years ago. The lake has no river inflow and the supply of water to the lake is from groundwater and precipitation. The lakewater flows out of the lake through two streams. This lake is a representative mesotrophic lake in Japan. The phytoplankton is dominated by diatoms, *Asterionella* sp. being the most common. The lake sediment is composed mostly of diatomaceous-gyttja, and the average sedimentation rate is estimated to be 0.65–0.63 mm/y over the past 1400 years. Three species of annual aquatic rooted plants, *Myriophyllum spicatum* L., *Hydrilla verticillatz* Casp., and *Potamogeton crispus* L. are growing at several places along the lake shore. *Hypomesus olidus* is the sole species of fish in the lake.

ABUNDANCE OF HUMIC SUBSTANCES IN LAKE SEDIMENTS

Biochemical compounds such as carbohydrates, proteins (amino acids), and lipids present in humic substance fractions (humic acid, fulvic acid, and humin) pose a problem in determining and characterizing humic substances in various environments, especially in freshly deposited lake sediments where relatively large amounts of those biochemicals are present. Common separation methods cannot separate "true" humic substances from nonhumic substances. According to Riffaldi and Schnitzer (1972b), 6*N* HCl hydrolysis efficiently removes nonhumic substances from soil humic acid. However, the following questions remain unresolved:

1. Are these biochemicals only a mixture with humic substances?
2. Can we completely remove the biochemicals by this procedure?
3. How should we consider the humic substances released into solution by this procedure?

Since most chemical studies on humic substances in lake sediments have been reported on samples without HCl hydrolysis, it should be kept in mind that these humic substances contain a relatively large amount of biochemicals.

Approximate abundance of humic substances was reported by the author (Ishiwatari, 1970; Ishiwatari et al., 1966). Air-dried sediment samples from Japanese lakes (Lakes Haruna, Shoji, Nishinoumi, Yamanaka, Nakatsuna, and Kizaki) were extracted by organic solvent (ethanol–benzene or methanol–acetone–benzene). Humic substances were then extracted from the preextracted sediment with 0.1*N* NaOH solution for two different extraction durations (6 hours and 1 month). The summarized results given in Table 1 indicate that humic substances extracted over a 1-month period amounted to

TABLE 1. Abundance of Humic Substances in Lake Sediments[a]

	Total Organic Matter		Composition (% of Total Organic Carbon in Sediment)						
				6 Hour Extraction			1 Month Extraction		
	C (mg/g)	N (mg/g)	Lipids[b]	Humic Acid	Fulvic Acid	Humin	Humic Acid	Fulvic Acid	Humin
Range	41.9–52.6	3.19–4.76	4.3–7.2	8–24	6–23	59–76	16–31	18–31	43–51
Average	44.2 ± 8.0	4.00 ± 0.66	5.8 ± 1.0	17 ± 6	11 ± 6	67 ± 6	22 ± 6	25 ± 5	47 ± 3

[a] *Ishiwatari (1970). Sediments taken from Lakes Haruna, Shoji, Nishinoumi, Yamanaka, Nakatsuna, and Kizaki.*
[b] *The organic substances soluble in organic solvent.*

TABLE 2. Biochemicals and Nonbiochemicals in Lake Sediments

Name of Lake	Total organic Carbon (mg/g)	Percentage of Total Organic Carbon				Reference
		Lipids	Protein	Carbohydrates[a]	Nonbiochemicals[b]	
Haruna (0–8 cm)	51.5	7.7[c]	29.0[d]	20	43.3	Ishiwatari (1975b)
Biwa (10 m)	10.1	6.5[e]	17.8[f]	18.2	57.5	Handa (1972, 1973)
Huron (0–3 cm)	—	5.0 ± 1.3[c]	19.8 ± 1.0[g]	4.8 ± 0.5	70.4 ± 2.7	Kemp and Johnston (1979)
Erie (0–3 cm)	—	5.0 ± 2.1[c]	12.8 ± 1.8[g]	2.9 ± 0.7	79.3 ± 2.8	Kemp and Johnston (1979)
Ontario (0–3 cm)	—	3.8 ± 1.6[c]	18.5 ± 3.6[g]	3.3 ± 1.3	73.5 ± 3.9	Kemp and Johnston (1979)

[a] *Determined by Anthrone method.*

[b] *Total organic carbon minus biochemicals (lipids, protein, and carbohydrates) carbon.*

[c] *Extracted by methanol/acetone/benzene.*

[d] *Organic N × 6.25.*

[e] *Extracted by chloroform/methanol (2 : 1).*

[f] *Determined by a ninhydrin method.*

[g] *Including aminosugars: determined by ion-exchange chromatography after HCl hydrolysis (Kemp and Mudrochova 1973).*

41–51% of the total organic matter in these sediments (on a carbon basis); the 6-hour treatment with alkali extracted only 58% of the humic substances extracted by the 1-month treatment. In addition, the ratio of humic acid to fulvic acid was different between the 1-month extraction (0.95 average for six lakes) and the 6-hour extraction (1.93), which suggests that the relative abundance of humic acid, fulvic acid, and humin is dependent upon the extraction procedure used. A general discussion of the importance of extraction procedures is presented in Chapter 13. Based on these observations it is imperative that the extraction procedures used be described clearly when data are presented. The amount of humin given in Table 1 was calculated by subtracting the amount of extractable humic substances (humic acid + fulvic acid) from that of the total organic matter.

Humin can be isolated by dissolution of the mineral matrix (after extracting humic acid and fulvic acid) with a mixture of HF and HCl [e.g., 46% HF/6*N* HCl (1 : 1)]. The amount of humin actually isolated by the above procedure is expected to be lower than the amount of "calculated" humin, because a significant amount of organic matter may be released into solution by degradation or dissolution during the isolation procedure. According to Morinaga et al. (unpublished), isolated humin accounted for 24–44% (average: 32.2%) of the total organic carbon of freshwater lake sediments (Lakes Haruna, Yunoko, Suwa, Biwa, Shoji, Motosu, and Nakanuma). These values correspond to approximately 68% of the calculated humin obtained after the 1-month extraction previously described.

In addition to direct extraction of humic substances, the amount of humic substances has been estimated by subtracting the amount of biochemicals (sum of lipids, amino acids or proteins, and carbohydrates) from the total organic matter in the sediments (Kemp and Johnston, 1979). In this chapter, this difference is called "nonbiochemicals," although no doubt there is much overlap between nonbiochemicals and extracted humic substances. As shown in Table 2, nonbiochemicals amount to 42–58% of the total organic matter in two Japanese lake sediments, but in the Great Lake (North America) sediments nonbiochemicals amount to 70–79% of the total organic matter on average. The latter values are close to those observed for marine sediments (Ishiwatari, 1979).

CHEMICAL CHARACTERISTICS OF HUMIC SUBSTANCES FROM LAKE SEDIMENTS

Elemental Composition

Elemental composition is one of the most essential characteristics of humic substances. Average elemental composition of humic substances was calculated and results are presented in Table 3. As shown, data for fulvic acid and humin are rare compared to those for humic acid. Interestingly, the average

TABLE 3. Average Values of Elemental Composition of Humic Substances from Lake Sediments[a]

Number of Samples:	Humic Acid (22)	Fulvic Acid (5)	Humin (2)
Element (%)			
Carbon	52.05 ± 3.61[b]	44.98 ± 3.90	53.82 ± 4.50
Hydrogen	5.67 ± 0.65	5.12 ± 1.24	4.88 ± 1.00
Nitrogen	5.63 ± 1.08	7.63 ± 0.56	4.17 ± 0.68
Oxygen	36.55 ± 4.27	42.27 ± 4.69	36.78 ± 4.62
Atomic Ratio			
H/C	1.30 ± 0.13	1.34 ± 0.24	1.08 ± 0.13
N/C	0.093 ± 0.018	0.147 ± 0.020	0.068 ± 0.016
O/C	0.533 ± 0.094	0.716 ± 0.146	0.518 ± 0.108

[a] *Data from Ishiwatari (1967b), Ishiwatari and Machihara (1983), Ishiwatari et al. (1980b), Karavaev and Budyak (1960), Karavaev et al. (1964), Kemp and Mudrochova (1975), Kemp and Wong (1974), Povoledo et al. (1975), and Stuermer et al. (1978).*
[b] *Standard deviation.*

value for humic acid is essentially the same as reported previously by the author (Ishiwatari, 1967b), and compares to humic acid derived from soil and marine sediments in the following way:

Carbon content (%)	: Soil(57.94)	> Marine(52.31)	~ Lake(52.05)
Atomic H/C ratio	: Marine(1.42)	> Lake(1.30)	> Soil(0.98)
Atomic N/C ratio	: Lake(0.093)	> Marine(0.058)	~ Soil(0.055)

These features for lake sediment humic acids suggest that they are closely related to their precursory materials (e.g., phytoplankton) and show a relatively low degree of humification.

As shown in Table 3, the carbon content of fulvic acid is lower than that of humic acid while nitrogen displays the opposite trend. Humin appears to show slightly higher carbon and lower nitrogen contents than humic acid.

Molecular Weight Distribution

Molecular weight distributions of humic and fulvic acids from lake sediments were determined by gel filtration using Sephadex gels (Ishiwatari, 1971; Kemp and Wong, 1974). Humic and fulvic acids ranged from molecular weights of less than 700 to over 200,000. Table 4 demonstrates that fulvic acid contains greater amounts of low-molecular-weight fractions than humic acid. According to Ishiwatari (1971), the apparent molecular weights of humic acid decreased significantly when hydrolyzed by acid or alkali. A humic acid fraction with molecular weight larger than 100,000 was collected by gel

TABLE 4. Apparent Molecular Weight Distribution of Humic Acids and Fulvic Acids from Lake Sediments as Determined by Sephadex Gel Permeation Chromatography

	Molecular Weight Range (% of Total Organic Matter)[a]			
Sample	<5000	5000–100,000	>100,000	Reference
Humic Acid				
Lake Haruna	10	20	70	Ishiwatari (1971)
Lake Kizaki	32	25	43	Ishiwatari (1971)
Lake Ontario	5–6	42–43	52	Kemp and Wong (1974)
Lake Erie	8	58	35	Kemp and Wong (1974)
Fulvic Acid				
Lake Ontario	28–30	43	27–29	Kemp and Wong (1974)
Lake Erie	34	41	25	Kemp and Wong (1974)

[a] *Using absorbance at 254 nm as a measure of organic matter concentration.*

chromatography and hydrolyzed with 6*N* HCl by refluxing. Only 6% of the humic acid fraction remained in the molecular range of over 100,000, the majority (60%) having changed into fractions with molecular weights ranging from 5000 to 10,000.

Ultraviolet and Visible Spectroscopy

Visible spectroscopy is a simple but important characterization method for humic substances (Kumada, 1977). Absorption spectra of humic and fulvic acids extracted from lake sediments were measured by Ishiwatari (1967a, 1970) and Ishiwatari et al. (1966). Table 5 summarizes optical properties of lake humic substances. Crude humic substances from lake sediments did not show any maxima or minima in the ultraviolet and visible spectra, similar to most soil humic substances. Upon purification by acid precipitation–base dissolution cycles, humic acid spectra revealed several features (Table 5). A group of lake humic acids (Lakes Aoki, Kizaki, and Nakatsuna in Table 5) showed relatively high E_{600} values (absorbance at 600 nm at a sample concentration of 1 mg/mL). These lakes are connected by a river. The high E_{600} values were explained by significant contribution of soil humic acid from the surrounding area. Another noteworthy feature was that some humic acids gave a very weak shoulder near 410 nm. This slight shoulder was due to chlorophyll-derived pigment (Ishiwatari, 1973) and was not removed completely by organic solvent extraction and gel filtration. The amount of this pigment was estimated to be 0.2% (as pheophytin a) for Lake Haruna humic acid.

TABLE 5. Optical Properties of Humic Substances from Lake Sediments[a]

	Humic Acid[b]		Purified Humic Acid[b]		Fulvic Acid[c]	
	E_{600}[d]	E_{400}/E_{600}[e]	E_{600}[f]	E_{400}/E_{600}	E_{600}[d]	E_{400}/E_{600}
Range	1.1–6.2	4.1–6.9	1.0–3.3	4.1–5.1	0.01–0.23	9.1–151
Average	2.8 ± 2.0	5.1 ± 1.2	1.8 ± 0.8	4.7 ± 1.1	0.11 ± 0.08	24.0 ± 2.8

[a] *Ishiwatari (1967b) and Ishiwatari et al. (1966). Sediments from Lakes Haruna, Shoji, Nishinoumi, Yamanaka, Nakatsuna, Kizaki, Aoki, Kawaguchi, and Motosu.*
[b] *Measured in 0.1N NaOH solution (light path = 10 mm).*
[c] *Measured in 0.05N H_2SO_4 solution (light path = 10 mm).*
[d] *Absorbance at 600 nm and sample concentration of 1 mg carbon/mL.*
[e] *Ratio of absorbance at 400 nm to that of 600 nm.*
[f] *Absorbance at 600 nm and sample concentration of 1 mg organic matter/mL.*

Povoledo et al. (1975) also observed absorption bands at 410 and 670 nm for lake humic acids extracted without prior organic solvent extraction. They concluded that the pigments responsible for these bands were chlorophyll derivatives, notably pheophytin a, and reported the approximate content in humic acid from a Canadian lake sediment was 0.1%. They reported that other pigments (e.g., carotenoids) were also present in lake humic acids without prior organic solvent extraction. Bourbonniere and Meyers (1978) extracted humic substances from Lake Huron surficial sediment without prior organic solvent extraction and measured visible spectra in 0.05*N* $NaHCO_3$ solution (pH 7.3–8.7) which gave E_{465}/E_{665} values of 3.92 for humic acid and 10.66 for fulvic acids.

Infrared Spectroscopy

Infrared spectra of humic substances from lake sediments were studied by several authors (Ishiwatari and Hanya, 1965; Ishiwatari et al., 1966; Otsuki and Hanya, 1967; Ishiwatari, 1970; Kemp and Mudrochova, 1973, 1975; Bourbonniere and Meyers, 1978). These authors report surprisingly similar infrared spectra of humic acids.

Figure 1 gives a typical infrared spectrum of lake humic acid. The interpretation of infrared spectra of humic substances is discussed in depth by MacCarthy and Rice in Chapter 21. The similarity of infrared spectra of humic acids from different lakes suggests a similarity of the aspects of chemical structure that are related to their infrared absorptions. However, infrared spectroscopy is not sensitive enough to uncover minor structural differences among humic acids. In fact, humic acids were separated by organic solvents (chloroform, methylethylketone, methanol, dimethylformamide) into various fractions (Ishiwatari, 1969b, 1973). Infrared spectra of two of these fractions, the chloroform-soluble fraction and the methylethylketone-

FIGURE 1. Infrared absorption spectrum of humic acid from Lake Haruna sediment (Ishiwatari, 1967a).

soluble fraction, showed large amounts of methylene and carboxyl bands. Since these fractions account for only 5–8% of the total humic acids, these infrared spectra do not shed much light on the complete spectrum of humic acid.

A hydrolysis study (Ishiwatari, 1967a) showed the chief feature of infrared spectra of lake humic acid to be the presence of an absorption band at 1540 cm^{-1}, probably arising from peptide bonds. No infrared evidence to show the aromatic structure of humic acid has been obtained.

Several authors (Ishiwatari et al., 1966; Ishiwatari, 1970a; Kemp and Mudrochova, 1973; Bourbonniere and Meyers, 1978) have published infrared spectra of fulvic acids. These studies show that absorption spectra of fulvic acids extracted from the same sediment sample are not necessarily the same. This may be due to the difficulty of removing inorganic materials and to the existence of many kinds of organic materials in the fraction. However, absorption bands for most fulvic acids appear essentially the same as those for humic acids except for a carboxyl band at 1740 cm^{-1} for fulvic acids and at 1720 cm^{-1} for humic acids. To date no reliable data on humin have been obtained by infrared spectroscopy.

NMR and ESR Spectroscopy

Humic acids from two lakes (Lakes Haruna and Kizaki) were separated by organic solvents into various fractions and characterized by ^{1}H NMR spectroscopy (Ishiwatari, 1973). The results clearly showed the lake humic acids to be aliphatic in character (presence of a large peak in 1.0–1.4 ppm (δ value) range characteristic of acyclic methylenes) with no aromatic protons (absence of a peak in 6.0–8.0 ppm range). The lack of aromatic protons could be due to heavily substituted benzene rings or to the strong relaxation effect of the spins of unpaired electrons as proposed for soil humic acids (Atherton et al., 1967), but the most probable explanation is a low concentration of aromatic rings in the humic acids from lake sediments. Using ^{1}H NMR data and applying the Williams method (R. B. Williams, 1958), it was estimated that 40–50% of the total carbon in the lake humic acids forms cyclic structures that are alicyclic rather than aromatic.

No further ^{1}H NMR studies have been conducted for freshwater lake humic substances. However, recent ^{1}H NMR spectra of marine humic acid appear similar to those of lake humic acid. Hatcher et al. (1980b) estimated ^{1}H aromaticity of marine humic acid to be 2–3% and the ^{13}C aromaticity to be 9–14%. They also estimated ^{13}C aromaticity of marine humim to be 7% (Hatcher et al., 1980c).

A few researchers have done research on ESR spectra of lake humic substances. Atherton et al. (1967) studied free radicals in humic acids from lake sediments and found that ESR spectra are divided into two classes: Class I having four-line spectra with a breadth of 1.75–1.9 gauss and Class II giving ill-defined, structureless spectra without clear peaks and with a breadth of 1.75–1.9 gauss. They considered that the humic acids contained semiquinone free radicals. However, these results do not seem representative of ESR characteristics of humic acids from normal freshwater lake sediments, because the lake sediments studied are largely soil derived (Atherton et al., 1967). According to Atherton et al. (1967), the Class I spectrum is characteristic of acidic sphagnum peat and the Class II spectrum is observed for mull humus. ESR spectra of freshwater lake humic substances were measured by Ishiwatari (1974) and Stuermer et al. (1978), generally giving a single symmetrical line devoid of any fine structure. The following data for both humic acids and humin were obtained: spin concentration $(4.5–5.7) \times 10^{17}$ spins/g; linewidth 3.8–6.5 gauss; g-value, 2.0022–2.0032. From the g-values the author concluded that free radicals in sedimentary humic acids are probably (1) semiquinone radicals in a condensed ring system or (2) radicals of carbon and nitrogen in the molecule (Yen et al., 1962).

Alkaline Permanganate Oxidation

Alkaline permanganate oxidation is a method for characterizing humic substances which has been used extensively for structural elucidation. Alkaline permanganate oxidation of humic substances produces various organic compounds, which are determined by gas chromatography–mass spectrometry (GS–MS). The yield of degradation products is usually low, and this severely limits the utility of the data.

Ishiwatari (1975a, unpublished) methylated a lake humic acid with BF_3/methanol, resulting in two fractions: a benzene-soluble fraction from which *n*-hexane-soluble materials were removed (64% of the initial humic acid) and

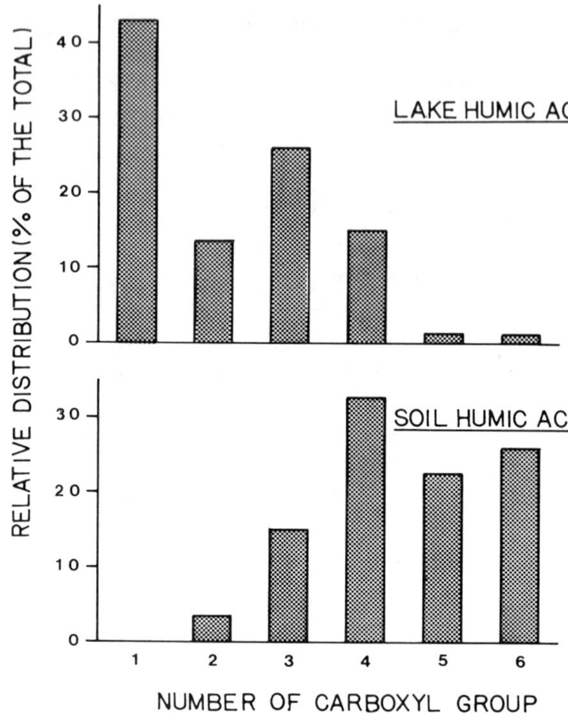

FIGURE 2. Distribution of benzenecarboxylic acids in alkaline permanganate degradation products of lake sediment (Ishiwatari, 1975a) and soil humic acids (Hansen and Schnitzer, 1966).

a benzene-insoluble fraction (36%). Both fractions were oxidized by alkaline $KMnO_4$ at 60°C for 6 hours. GC–MS analysis revealed that the degradation products consisted of (1) normal C_8-C_{30} monocarboxylic acids, (2) branched C_8-C_{16} monocarboxylic acids, (3) isoprenoid C_{14} and C_{15} acids, (4) normal C_6-C_{24} α,ω-dicarboxylic acids, and (5) benzene mono-to-hexacarboxylic acids. There was no essential difference in the degradation products between benzene-soluble and benzene-insoluble fractions. The most striking features were (1) the highly aliphatic character of the degradation products and (2) the difference of soil humic acid (Hansen and Schnitzer, 1966) from lake sediment humic acid in the relative abundance of benzenecarboxylic acids. As shown in Figure 2, benzenecarboxylic acids with fewer substitutions are more abundant in lake humic acid than in soil humic acid.

Machihara and Ishiwatari (1981) oxidized humin isolated from lake sediments (Lakes Haruna, Biwa, Motosu, and Shoji) in alkaline permanganate solution at 60°C for 1 hour. This oxidation condition proved to be suitable for degradation of aliphatic lake humic substances (Machihara and Ishiwatari, 1983). The amounts of degradation products identified by gas chromatography ranged from 3.5 to 5.5% of the initial weight of humin. Figure 3 gives

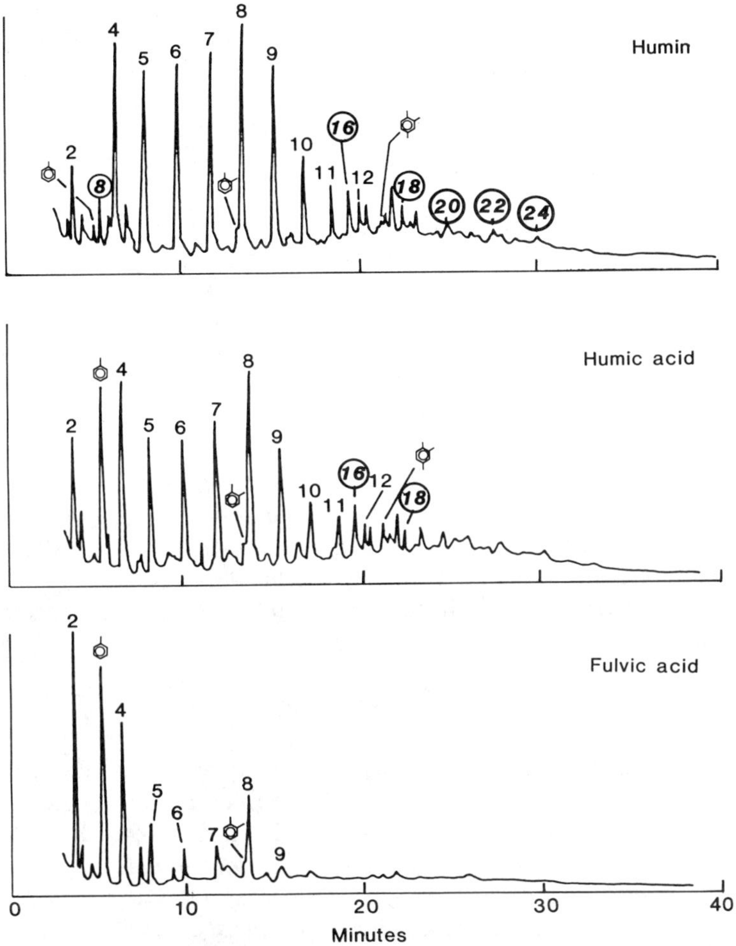

FIGURE 3. Gas chromatograms of alkaline permanganate degradation products of humic substances from a lake sediment (Lake Haruna) (Ishiwatari and Machihara, 1983). The carbon numbers of the dicarboxylic acids are indicated by the arabic numbers and monocarboxylic acids by the primed arabic numbers. The structure ⌬ indicates benzenecarboxylic acids.

typical gas chromatograms of alkaline permanganate degradation products of humic substances from a Lake Haruna sediment (Ishiwatari and Machihara, 1983). Aliphatic acids (C_4-C_{10} α,ω-dicarboxylic acids + C_{16}-C_{24} monocarboxylic acids) accounted for more than 90% of the degradation products identified. Benzene mono-to-tricarboxylic acids were minor products. Degradation products for lake humin are quite different from those for soil humin. Upon oxidation of soil humin obtained near Lake Haruna, benzenecarboxylic acids accounted for approximately 30% of the degradation products. Moreover, according to Khan and Schnitzer (1972), aromatic acids accounted for 76% of the oxidation products of a Canadian soil humin.

In order to obtain information on low-molecular-weight degradation products, Morinaga et al. (unpublished) oxidized humin isolated from lake sediments (Lakes Suwa, Nakanuma, Yunoko, Haruna, Shoji, Motosu, and Biwa) in alkaline permanganate solution at 60°C for 1 hour. Major oxidation products were carbon dioxide [amounting to 26–42% (average: 31 ± 7%) of the initial humin carbon] and oxalic acid [amounting to 15–26% (average: 20 ± 4%)].

As minor oxidation products, C_1-C_3 monocarboxylic acids and C_4 α,ω-dicarboxylic acid were obtained (1–2% of the initial humin carbon). The humin was then hydrolyzed by 6*N* HCl at 110°C for 24 hours, and the unhydrolyzable part of humin was oxidized. Interestingly, degradation products of the unhydrolyzable part of humin were almost the same as those of the original humin although 30–49% (by weight) of the original humin was released into solution on HCl hydrolysis.

Oxidative degradation was conducted for fulvic acid obtained after dialysis (Ishiwatari and Machihara, 1983). As shown in Figure 3, oxalic acid, *n*-C_4 and *n*-C_8 α,ω-dicarboxylic acids, and benzoic acid were major degradation products for fulvic acid. Oxalic acid accounted for 44% of α,ω-dicarboxylic acid in the degradation products and was considered to have been derived predominantly from carbohydrates and amino compounds present in the fulvic acid.

Other Analysis

Oxygen-Containing Functional Groups. Few quantitative determinations of oxygen-containing functional groups have been done for lake sediment humic substances (Karavaev et al., 1964; Ishiwatari, 1969a). Potentiometric methods are most commonly used for these determinations and are discussed in Chapter 20 by Perdue. Karavaev et al. (1964) reported the total acidity of humic acid and fulvic acid to be 4.3 meq/g and 6.9 meq/g, respectively. Ishiwatari (1969a) determined carboxyl groups by reaction with calcium acetate (Blom et al., 1957) and phenolic hydroxyls with barium hydroxide accompanied by corrections for carboxyls according to Kononova (1961). Humic acid gave 2.3 meq/g for carboxyls and 2.3 meq/g for hydroxyls, which were smaller than those reported for soil humic acid. Consequently, the carboxyls and hydroxyls accounted for 21 and 10% of the total oxygen in the humic acid, respectively.

Aromaticity Estimation by Air Oxidation. Aromaticity of a lake humic acid was estimated by the method established by Mazumdar et al. (1959) for coal (Ishiwatari, 1969a). The method consists of heating powdered humic acid at 170°C for a long period (e.g., 600 hours) under a current of air. End-groups of humic acid are oxidized to carbon dioxide by this treatment, leading to the formation of hydroxyl, carbonyl, and carboxyl groups while the aromatic skeleton remains unaffected. Thus, carbon aromaticity (a ratio of aromatic carbon to the total carbon) and hydrogen aromaticity (a ratio of aromatic hydrogen to the total hydrogen) of a lake humic acid were estimated to be 36 and 14%, respectively. The carbon aromaticity was calculated to be 14% after correction for contributions by carbohydrates and proteinaceous materials in the humic acid to aromaticity (these materials were proved to cause a positive error).

Vacuum Pyrolysis. Fukushima (1982) applied vacuum pyrolysis at 500°C to characterize lake humic acid and humin and analyzed organic-solvent-soluble pyrolysates by GC–MS. The results showed that normal alkanes (C_{14}-C_{32}) and normal alkenes (C_{14}-C_{28}) were present in the pyrolysates although their amounts were extremely small (0.001% of the initial weight for humic acid and 0.003% for humin).

Isotopic Analysis. Isotopic study of humic substances is important for obtaining information on precursors and processes involving formation and diagenesis (Nissenbaum and Kaplan, 1972; Stuermer et al., 1978). However, little work has been carried out for humic substances in lake sediments. Table 6 lists isotopic data of lake sediment humic substances.

Characteristics of the Hydrolyzable Part of Humic Substances

Hydrolysis is a mild degradation method for characterizing humic substances. It is believed that hydrolysis products may be closely related to the starting materials of humic substances, as mentioned later. Hydrolysis studies of lake humic substances were conducted by several authors (Ishiwatari, 1967a, 1970; Kemp and Mudrochova, 1973; Bourbonniere, 1979; Bourbonniere and Meyers, unpublished) and many compounds were detected in their hydrolysates.

Amount and Approximate Composition of the Hydrolyzable Part

Ishiwatari (1967a, 1970) hydrolyzed humic acid from Lake Haruna with HCl, H_2SO_4, or NaOH under reflux for 1–38 hours. The hydrolyzable portion of the humic acid reached approximately 50% of its initial weight (Table 7).

Approximately 70–90% of the nitrogen in the initial humic acid passed into solution on acid hydrolysis. Measurement of the infrared absorption band of humic acid at 1540 cm^{-1} implied that nitrogen in the hydrolysate was derived from protein-like material. Thus, 60–75% of the hydrolyzable portion of the humic acid was accounted for by protein-like materials. In addition, 4–10% of the hydrolyzable portion was accounted for by carbohydrates, which were determined by an Anthrone method and expressed as "glucose equivalent."

Humin samples from seven freshwater lakes (Lakes Suwa, Nakanuma, Yunoko, Haruna, Shoji, Motosu, and Biwa) were hydrolyzed with 6*N* HCl

TABLE 6. Isotopic Data of Lake Sediment Humic Substances

Name of Lake	$\delta^{13}C$ (‰)	$\delta^{15}N$ (‰)	δD (‰)	Reference
Humic Acid				
A Florida lake	−24.3	+2.32	−78.1	Stuermer et al. (1978)
Lake Haruna	−21.0(1); −26.5(2) −25.47(3)	—	—	(1) Nissenbaum and Kaplan (1972); (2) Ishiwatari (unpublished); (3) Ishiwatari et al. (unpublished)
Fulvic Acid				
Lake Haruna	−24.04	—	—	Ishiwatari et al. (unpublished)
Humin				
A Florida lake	−27.74	+2.38	−95	Stuermer et al. (1978)
Lake Haruna	−25.30	—	—	Ishiwatari et al. (unpublished)
Lake Biwa (sediment depth 11 m)	−24.6	—	—	Ishiwatari (1977)
(45 m)	−27.5	—	—	Ishiwatari (1977)
(56 m)	−25.2	—	—	Ishiwatari (1977)
(130 m)	−25.5	—	—	Ishiwatari (1977)

TABLE 7. Composition of the Hydrolyzable Part of Lake Sediment Humic Acid[a]

Condition of Hydrolysis		Hydrolyzed Organic Matter (% of Humic Acid)	Composition of Hydrolysate (% of Humic Acid)					
				Nitrogen Compounds				
Medium	Duration (hours)		Carbohydrates	Total N		Amino N	Ammonium N	Unaccounted N
0.1N H_2SO_4	17	34.1	3.61	3.48	21.8[b]	1.30	0.79	1.39
0.1N H_2SO_4	38	36.7	3.17	3.76	23.5[b]	1.91	0.82	1.03
3N H_2SO_4	25	49.7	2.04	4.50	28.2[b]	—	0.80	—

[a] *Ishiwatari (1970).*

[b] *Values calculated as protein.*

at 110°C for 24 hours (Morinaga et al., unpublished). Weight loss (carbon) after hydrolysis amounted to 30–49% (average: 36%) of the initial humin. Nitrogen loss ranged from 42 to 87% (average: 72%) of the initial humin.

Detailed Composition of Hydrolyzable Materials

Amino Acids. Kemp and Mudrochova (1973) determined amino acids and amino sugars by ion-exchange chromatography in 6*N* HCl hydrolysates of humic and fulvic acids from Lake Ontario sediments. They obtained total amino acids of 21.5% for humic acid and 12.6% for fulvic acid. Total amino sugars accounted for only 1.9 and 1.3% for humic acid and fulvic acid, respectively. They found the amino acid distribution in the humic acid resembled that of zooplankton and suspended sediment samples, with the exception of glycine which was higher in the sediments. This lends support for the assumed autochthonous nature of lake sediment organic matter. On the other hand, basic amino acid concentrations were low in the fulvic acid and its amino acid distribution resembled the combined form in the interstitial waters.

Gas chromatography was used to analyze amino acids in 6*N* HCl hydrolysates of fulvic acid, humic acid, and humin from lake sediments (Lakes Suwa, Nakanuma, Yunoko, Haruna, Shoji, Motosu, and Biwa) (Yamamoto, 1983). Table 8 gives an example of analytical results of amino acids (Lake Haruna). The total amino acids for the seven-lake sediments accounted for 3–16% of humin, 11–21% of humic acid, and 4–24% of fulvic acid. The percentage of amino nitrogen in the total nitrogen in each fraction was 20–44% for humin, 21–36% for humic acid, and 4–30% for fulvic acid.

In the seven lakes studied by Yamamoto (1983), amino acid distribution of fulvic acid, humic acid, and humin resembled each other. However, after detailed examination of amino acid distribution, the following regularities were found to exist in almost all humic substances studied:

1. The relative abundance of basic amino acids and neutral hydrophobic amino acids increased in the order of fulvic acid $<$ humic acid $<$ humin.
2. The relative abundance of acidic amino acids and neutral hydrophilic amino acids decreased in the order of fulvic acid $>$ humic acid $>$ humin.

These regularities hold for the amino acid distribution in humic acid and fulvic acid reported by Kemp and Mudrochova (1973).

Carbohydrates. Carbohydrate (neutral sugar) content and relative distribution were determined by gas chromatography for humic substances from Lake Haruna sediments (Uzaki and Ishiwatari, 1983; Uzaki, unpublished). The results indicated that the concentration of total carbohydrates in fulvic acid (16.8% of fulvic acid) was higher than those in the other humic fractions (2.4–4.0%). Carbohydrate distribution in three humic fractions resembled

TABLE 8. Distribution of Amino Acids in Sediment and Sedimentary Humic Substances in Lake Haruna[a]

Amino Acid	Sediment	Humic Acid	Fulvic Acid	Humin
Acidic				
Aspartic acid	10.9	12.9	16.4	11.5
Glutamic acid	10.0	9.0	16.7	9.4
Total acidic	20.9	21.9	33.1	20.9
Basic				
Arginine	0.5	0.3	0.3	0.2
Lysine	5.1	3.1	1.9	4.1
Ornithine	0.3	0.3	1.2	0.3
Total basic	5.9	3.7	3.4	4.6
Neutral Hydrophobic				
Phenylalanine	4.5	2.8	2.4	3.6
Tyrosine	2.5	1.1	0.4	1.5
Valine	5.8	7.0	5.4	7.3
Leucine	7.1	6.3	3.5	6.3
Isoleucine	3.5	4.4	2.8	4.7
Proline	5.6	5.3	3.5	6.1
Total neutral hydrophobic	29.0	26.9	18.0	29.5
Neutral Hydrophilic				
Glycine	14.1	21.5	18.6	15.6
Alanine	11.5	13.1	13.3	13.0
Serine	8.2	6.4	7.1	8.1
Threonine	6.2	5.2	5.4	5.6
Hydroxyproline	1.2	0.8	0.8	1.3
Total neutral hydrophilic	41.2	47.0	45.2	43.6
Sulfur-Containing				
Methionine	1.5	0.1	0.1	0.2
Cystine	0.3	0.0	0.0	0.0
Total sulfur-containing	1.8	0.1	0.1	0.2
Others				
γ-Aminobutyric acid	1.2	0.5	0.4	1.1
Total amino acids (mg/g)	26.5	197	177	160

[a] *Yamamoto (1983). Values in relative molar %.*

each other—glucose, galactose, and mannose being relatively abundant. The average relative composition of carbohydrates (% of the total carbohydrates) for humic acid, fulvic acid, and humin (in this case, a sediment residue of humic and fulvic acids extraction) was: glucose, 31.9%; galactose, 16.7%; mannose, 15.4%; xylose, 9.5%; arabinose, 7.8%; ribose, 1.7%; fucose, 6.1%; and rhamnose, 9.2%. The carbohydrate distribution appeared similar to that in the microbial decomposition residue of planktonic materials in lakewaters (Ochiai, unpublished).

Fatty Acids and Other Organic Acids. Fatty acids were extracted from Lake Haruna sediment humic acid by refluxing with BF_3/methanol (Ishiwatari, 1975a). The fatty acids consisted of a series of long-chain saturated (C_{11}-C_{34}: maximum at C_{16}), unsaturated (C_{16}, C_{18}, and C_{24}), and branched (C_{15} and C_{17}) monocarboxylic acids. The fatty acids amounted to 0.2–0.3% of the humic acid.

Fatty acids in humin from Lake Haruna sediments were analyzed for the fraction obtained by solvent (benzene/methanol 6 : 4) extraction followed by saponification (2*N* KOH aqueous solution at 200°C for 3 hours) extraction (Yamamoto and Ishiwatari, 1981). The fatty acids were composed of normal C_{12}-C_{30} saturated monocarboxylic acids (maximum at C_{16}), unsaturated (C_{16} and C_{18}), and branched (C_{13}, C_{15}, and C_{17}) monocarboxylic acids. The fatty acid distribution in humin resembled that in humic acid. Total fatty acids accounted for 1.0% of the humin and probably originated from algae, bacteria, and higher plants.

Bourbonniere and Meyers (unpublished) hydrolyzed humic substances from Lake Huron sediments with 5*N* NaOH at 170°C for 12 hours under a nitrogen atmosphere and found the following organic acids: n-C_{16} and n-C_{18} monocarboxylic acids; lactic acid, 2-hydroxybutanoic acid, 3,4-dihydroxybutanoic acid, oxalic acid, and succinic acid. It was proposed that the smaller organic acids were derived from cellulose-related materials. 2-Hydroxybenzoic acid, 4-hydroxybenzoic acid, 2,5-dihydroxy-3-pentenoic acid, and vanillic acid were also observed. It was believed that 4-hydroxybenzoic acid and vanillic acid originated from lignin and that the ratio of 3,4-dihydroxybutanoic acid to vanillic acid indicates the proportion of cellulose to lignin. The proportion was in the order of fulvic acid > humic acid > humin.

Concept of Chemical Structure of Humic Substances

The analytical results obtained clearly indicate that lake humic substances are aliphatic in nature. It is also clear that a significant amount of biochemical compounds (amino acids, carbohydrates, fatty acids, etc.) are released from humic substances by hydrolysis or solvent and saponification extractions.

A portion of these biochemical compounds may be associated with the extracted humic substances. However, as already described, the humin which had been hydrolyzed by 6*N* HCl at 110°C for 24 hours gave essentially the same oxidative ($KMnO_4$) degradation products (aliphatic C_4-C_{14} α,ω-dicarboxylic acids as major products) as untreated humin. Moreover, stepwise (eight steps) oxidative ($KMnO_4$) degradation of humin produced similar degradation products (aliphatic C_8-C_{18} monocarboxylic and C_5-C_{16} α,ω-dicarboxylic acids and small amounts of benzenecarboxylic acid; Machihara and Ishiwatari, 1980). These facts indicate that the major part of humin forms aliphatic structures with biochemical compounds distributed uniformly in the humin matrix. These compounds are firmly linked within the humin matrix by unknown bonds.

Figure 4 gives a primitive structural model of a lake humin. This model is a slight modification of the previous one (Machihara and Ishiwatari, 1980), which was deduced from the results of alkaline permanganate degradation experiments of Lake Haruna humin. The lake humin has a large percentage of oxygen-containing components which are probably melanoidin-like material (see the next section). The oxygen-containing components connect with other clusters and are readily decomposed to carbon dioxide and oxalic acid on oxidation or hydrolyzed into soluble materials. The clusters of melanoidin-like material consist of oxygen-containing components, polymethylene chains, and small amounts of aromatic rings. They produce carbon dioxide, oxalic acid, aliphatic mono- and dicarboxylic acids, benzenecarboxylic acids, and so on upon oxidation. This model is considered valid for lake sediment humic acid as well as for humin, although the content of the functional groups is different from that in humin.

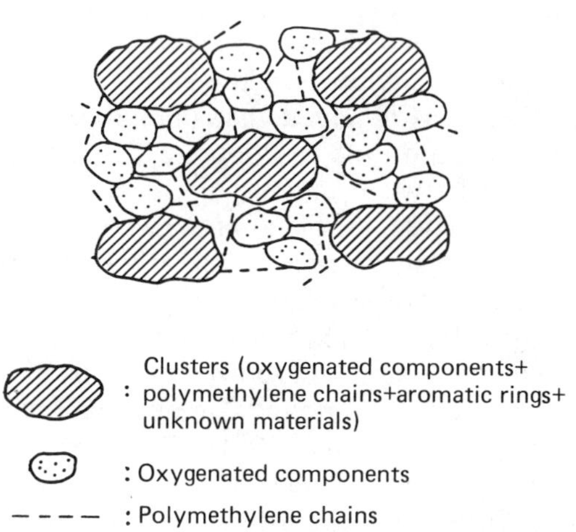

FIGURE 4. A structural model of lake sediment humin (a modification of Machihara and Ishiwatari, 1980).

GEOCHEMISTRY OF HUMIC SUBSTANCES IN LAKE SEDIMENTS

Formation of Humic Substances

Several ideas have been presented in the literature on the formation process of humic substances in marine sediments. Abelson (1967) claimed that polymerization of unsaturated fatty acids in phytoplankters after their death accounts for the formation of kerogen in marine sediments. Abelson and Hare (1971), Hoering (1973), and Hedges (1976) studied reactions between carbohydrates and amino acids under laboratory conditions as a possible formation reaction of humic acid and humin in sediments. They prepared a number of artificial humic acids by reacting glucose with amino acids. The synthetic products resembled natural humic acid and humin. A comprehensive review was published by Abelson (1978).

Nissenbaum and Kaplan (1972), on the basis of their isotopic study, presented the idea that humic acid formation and transformation in marine sediments proceed by the following pathway: (1) degraded cellular material → (2) water-soluble complex containing amino acids and carbohydrates → (3) fulvic acids → (4) humic acid → (5) kerogen.

Chemical reactions of formation of humic substances in lake sediments may be essentially the same as those in marine environments. Lake sediment humic substances are expected to represent an earlier stage of humification than those from marine sediments.

Outline of Humic Substances Formation Processes

The aliphatic nature and other chemical characteristics of lake humic substances undoubtedly indicate that the predominant part of humic substances in normal productive freshwater lake sediments originates from aquatic organisms, primarily phytoplankton. On the basis of oxidative degradation studies and other data, it is believed that humic substances in lake sediments are formed by Maillard-type reactions, which are nonenzymic browning reactions occurring between amines (e.g., amino acids) and reducing compounds (e.g., carbohydrates), or, amino-carbonyl reactions (Reynolds, 1963, 1965). In addition to amino acids and carbohydrates, phytoplankton-derived lipids take part in these amino-carbonyl reactions, since lipid-derived compounds with polymethylene chains constitute a part of the building block of humic substances. Amino-carbonyl reactions of amino compounds with lipids (aliphatic carbonyl compounds) are known to occur under certain conditions (Reynolds, 1965; Suyama, 1981). Therefore, all biochemical materials which constitute aquatic living matter participate in the formation reaction of humic substances. According to this idea, biochemical compounds (amino acids, carbohydrates, fatty acids, etc.) found in humic substances should be regarded as intermediate forms which are finally changed into humic substances.

The formation reaction of humic substances in lakes proceeds rapidly and is almost completed in decaying phytoplankton and/or in surface sediments. The major characteristics of humic acid and humin in sediments are determined by this formation stage, reflecting primarily organic constituents of phytoplankton.

Maillard Reaction Between Carbohydrates and Amino Acids or Proteins

As suggested by the works of Abelson and Hare (1971), Hoering (1973), and Hedges (1976) on synthetic and natural humic acids and humin, the Maillard reaction involving carbohydrates and amino acids is a probable mechanism of formation of lake humic substances. However, the apparently rapid formation of lake humic substances makes us consider that reactions between carbohydrates and proteins in dead phytoplankton or in surface sediments are also probable.

Yamamoto (1983) studied the reaction of glucose with casein (milk). The proportions employed were 3 : 1, 1 : 1, and 1 : 3 (weight ratio). They were reacted at temperatures ranging from 50°C to boiling temperature for 17 hours to 7 days. Amino acid distribution in the resulting synthetic fulvic acid, humic acid, or humin was analyzed. A significant amount of humic acid or humin was rapidly formed by this reaction. For example, in the reaction with proportions of 3 g casein and 1 g glucose in 16 mL water at 80°C, 74% of the initial material was changed into humin after 7 days. This is important because humin is a dominant fraction of humic substances in lake sediments and a very long reflux time seems to be necessary to obtain a sufficient amount of humin for the reaction of carbohydrates with amino acids (Abelson, 1978). It was also found that, other than for the basic amino acids, the distribution of amino acids (free + combined) in the synthetic fulvic acid, humic acid, and humin resembled that in natural humic fractions. Yamamoto proposed from his experimental results that synthetic fulvic acid, humic acid, and humin are formed almost simultaneously in the reaction of casein with glucose. This mechanism of formation of humic substances is also considered to take place in the natural environment.

Participation of Lipids in the Formation of Humic Substances

In order to understand how lipids participate in the formation of humic substances, Ishiwatari and Machihara (1983) isolated lipids (organic solvent extracts), fulvic acid, humic acid, and humin from a surface sediment of Lake Haruna and examined aliphatic acids with long methylene chains in alkaline permanganate degradation products. The results for humic substances have already been described. Figure 5 gives a gas chromatogram of degradation products of lipids, the major degradation products being C_2 (1.4 mg/g of lipids) and C_4-C_{12} α,ω-dicarboxylic acids (45.7 mg/g), n-C_8 to n-C_{10}

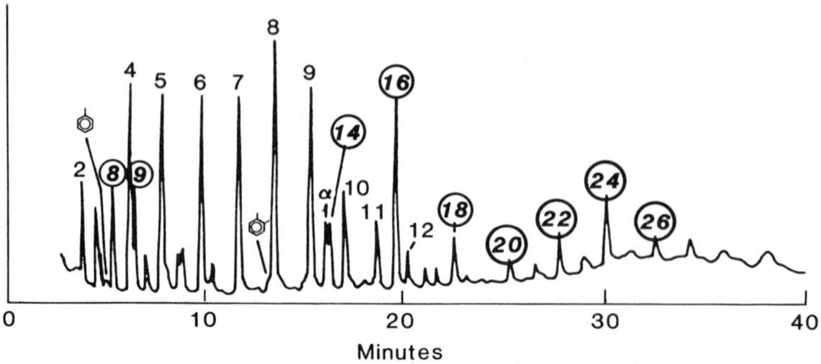

FIGURE 5. A gas chromatogram of alkaline permanganate degradation products of lipids from Lake Haruna sediment (Ishiwatari and Machihara, 1983). Abbreviations are the same as in Figure 3; α indicates 6,10,14-trimethylpentadecan-2-one.

(6.8 mg/g) and n-C_{14} to n-C_{26} (23.1 mg/g) monocarboxylic acids, and C_{18}-isoprenoidal ketone (3.0 mg/g).

The distribution pattern of α,ω-dicarboxylic acids for lipids resembled those for humic acid and humin (Fig. 3). This fact clearly indicates the common origin for the polymethylene chains in lipids, humic acid, and humin, which means that phytoplankton-derived lipids actively took part in the formation of humic acid and humin. The relative abundance of polymethylene chains in lipids and humic substances was estimated on the assumption that the yield of production of aliphatic acids from polymethylene chains by alkaline permanganate oxidation was the same for these organic fractions. The following estimations resulted: 42% (% of the total amount of polymethylene chains in the sediment) for humin, 38% for lipids, 19% for humic acid, and 1% for fulvic acid.

The abundance of polymethylene chains in humic acid and humin cannot be explained by the formation of these humic substances by polymerization of fulvic acid which has an extremely small amount of polymethylene chains. It is therefore concluded that, as far as polymethylene chains are concerned, the mechanism of sequential formation of humic substances proposed by Nissenbaum and Kaplan (1972) is not valid for humic acid and humin.

Ishiwatari and Machihara (1983) estimated roughly the degree of contribution of lipids to humic acid and humin by assuming that all polymethylene chains (C_4-C_{14}) in these fractions were derived from sedimentary lipids. Surprisingly, by this calculation 43% of the humic acid carbon and 74% of the humin carbon were derived from lipids. These extremely high values are in conflict with $\delta^{13}C$ calculations. Using $\delta^{13}C$ data, the degree of contribution of lipid to humic acid and humin was estimated on the assumption that (1) humic acid and humin were formed from lipids and nonlipid materials and (2) $\delta^{13}C$ of humic acid and humin were the simple sum of $\delta^{13}C$ of lipid (-30.56

ppt) and nonlipid material whose $\delta^{13}C$ were represented by that of fulvic acid (−24.04 ppt). Calculations indicate the degree of lipid contribution to be 21% for humic acid and 19% for humin.

Diagenetic Changes of Humic Substances

In this section, diagenetic changes of humic substances are tentatively divided into two stages and their changes in abundances and chemical characteristics are discussed. First is the earlier stage of diagenesis (approximate ages of 1–10 years, at surface sediment, to 10^6 years), where lake sediments are still soft and do not suffer from intensive geothermal effects; second is the later stage of diagenesis (ages older than 10^7 years), where chemical reactions of humic substances take place by geothermal action.

The Earlier Stage of Diagenesis

In the preceding section it was stated that humic substances are formed very rapidly in dead phytoplankton bodies (cells) and at the uppermost layers of sediments. In the surficial and deeper sediments, the formation reaction of humic substances continues to take place, but more slowly compared to the initial stage of its formation. This is due to the relatively low temperatures and reducing conditions commonly found in the sediments of productive lakes. Moreover, some qualitative changes of humic substances are expected in sediments.

Transformation of Biochemicals to Nonbiochemicals: Formation of Humic Substances. Evidence of humic substances formation is not clear in surficial sediments. Ishiwatari (1975b) determined the amount of biochemicals (lipids, amino acids, carbohydrates) in the surface sediments (24 cm approximate age 400 years) of Lake Haruna. The organic matter fraction not determined as biochemicals accounted for 43–36% (average: 42%) of the total organic matter and showed no trend with depth. Transformation of biochemicals to nonbiochemicals appeared to take place very slowly. According to Handa (1972, 1973), the percentage of biochemicals in the total organic matter in a 200m-long sediment core sample from Lake Biwa decreased gradually from 40 to 20% in the period of approximately 500,000 years, as shown in Figure 6. No clear trend of transformation of biochemicals to nonbiochemicals up to 10,000 years was observed in the sediments of the Great Lakes (Ontario, Erie, Huron) (Kemp and Johnston, 1979).

A possibility of diagenetic conversion of lipids into humin by polymerization in sediments was claimed by Shioya and Ishiwatari (1983) on the basis of a laboratory heating experiment. Isolated lipids from Lake Haruna sediments were heated in a nitrogen atmosphere at 125–370°C for 1–7 days. A significant amount of lipids (maximum 50% of the initial weight) were poly-

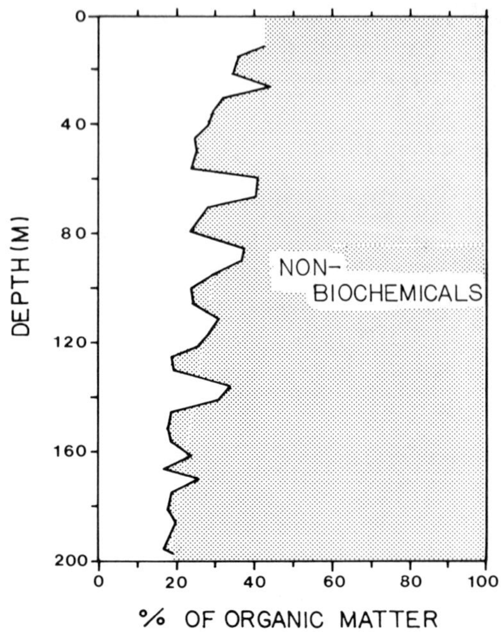

FIGURE 6. Vertical change of biochemicals and nonbiochemicals in Lake Biwa sediments (Handa, 1972, 1973).

merized into humin-like matter on heating at 175°C. It was concluded from this kinetic study that half of the lipids in a young sediment are converted into humin in 10^4–10^5 years at 0–30°C.

Transformation of Humic Substances: Relative Abundance. The humic substances which were extractable with NaOH (humic acid and fulvic acid) decreased slightly with depth in Lake Haruna sediments (Ishiwatari, 1975b), as shown in Figure 7A. This decreasing trend in Lake Haruna was confirmed by later studies. According to Ishiwatari et al. (1980a), alkali-extractable humic substances were determined by colorimetry (measurement of absorbance at 400 nm) and their amounts accounted for 22% of the total organic matter at 0–10 cm in depth, 16% at 30–40 cm, 12% at 60–68 cm, and 13% at 105–115 cm (about 1400 years). Bourbonniere (1979) observed similar trends with depth in sediments extracted without prior organic solvent treatment from Lakes Huron and Michigan. In Lake Huron sediment (65 cm, ~560 years), the relative abundance of fulvic acid slightly decreased while that of humin increased with depth. A Lake Michigan sediment (42°20′N 86°50′W; 98 cm, ~5400 years) showed clear vertical trends where fulvic acid exhibited a continuous decrease (35 → 13% of the total organic matter on a carbon basis) and humin showed a corresponding increase (64 → 86%) with depth. Humic acid was a minor component (below 4%). However, Nriagu and Coker (1980) did not recognize any trend with depth in the amount of humic

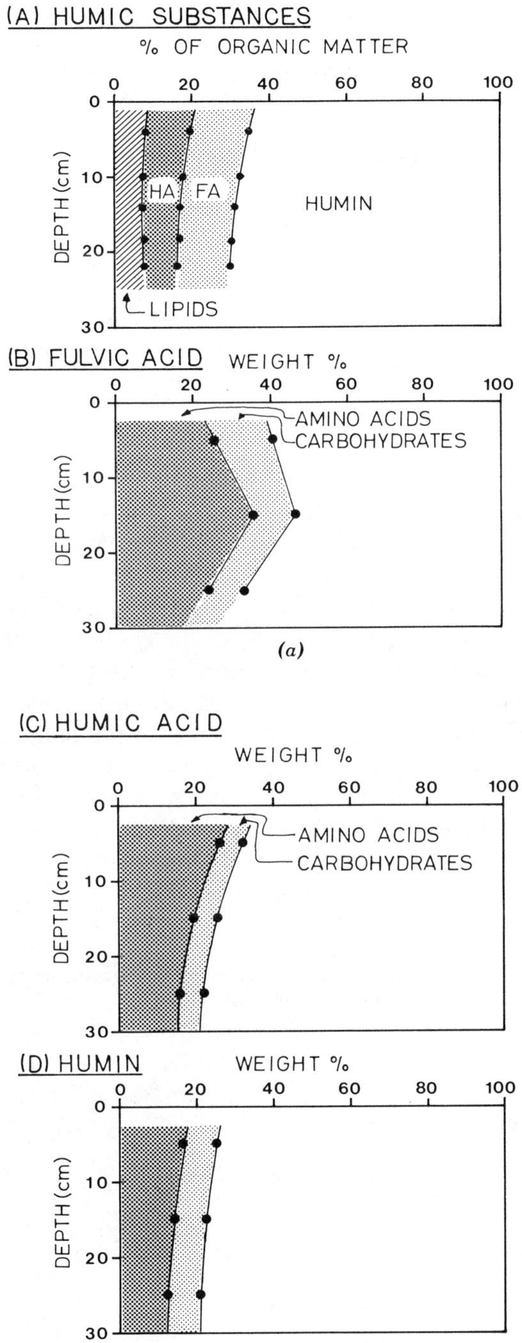

FIGURE 7. Vertical changes in humic substances and their carbohydrate and amino acid contents in Lake Haruna sediments (Ishiwatari, 1975; Yamamoto, 1983; Uzaki, unpublished).

TABLE 9. Humic Acids and Humin Isolated from Lake Biwa Sediments[a]

Depth (m)	Total Organic Matter Concentration[b] (mg/g)	Abundance (% of Total Organic Matter)		Elemental Analysis of Humin		
		Humic Acid[c]	Humin	Carbon (%)	Hydrogen (%)	Atomic H/C Ratio
11	16.2	4.3	6.2	61.54	6.57	1.27
45	10.2	0.0	5.8	64.06	6.92	1.29
56	11.0	0.0	11.9	64.60	6.56	1.21
130	6.6	0.0	64.0	—	—	—

[a] *Ishiwatari (1977).*
[b] *Total organic matter = total organic carbon × 1.6.*
[c] *Ash content was not determined.*

acid and fulvic acid in Lake Ontario sediments (0–40 cm), ranging from 9 to 15% of the total organic matter for humic acid and 2 to 4% for fulvic acid.

Ishiwatari (1977) isolated humic acids and humin from samples at various depths (11–130 m) of Lake Biwa sediment. As shown in Table 9, a small amount of humic acid was extracted from sediments of 11 m depth, but no humic acid was obtained from sediments in deeper layers (45–130 m) although alkali extracts were yellow-colored. Humin isolated from sediment samples increased with depth from 6.2% of the total organic matter to 64%, and at 130 m in depth accounted for 80% of the nonbiochemicals. Ishiwatari and Kawamura (1981) again measured approximate amounts of alkali-extractable humic substances in the long sediment core sample of Lake Biwa by colorimetry (at 400 nm). The ratios of alkali-extractable humic substances to the total organic matter decreased gradually with depth, as shown in Table 10.

TABLE 10. Vertical Changes of Extractable Humic Substances in Lake Biwa Sediments[a]

Depth (m)	Number of Samples	Total Carbon (mg/g)	Extractable Humic Substances[b] (mg/g)	Extractable Humic Substances / Total Organic Matter[c]
0–10	18	16.7 ± 4.7[d]	6.15 ± 2.30	0.204 ± 0.096
10–50	23	13.3 ± 3.6	4.87 ± 2.10	0.203 ± 0.103
50–100	10	12.8 ± 2.3	4.01 ± 0.79	0.174 ± 0.046
100–200	7	10.1 ± 1.1	2.96 ± 0.95	0.163 ± 0.055

[a] *Ishiwatari and Kawamura (1981).*
[b] *Extractable with 0.5N NaOH.*
[c] *Total carbon × 1.8.*
[d] *Standard deviation (σ).*

In summary, fulvic acid and humic acid decrease gradually with depth and the relative abundance of humin increases in this stage of diagenesis.

Transformation of Humic Substances: Chemical Characteristics. Ishiwatari (1975b) studied chemical characteristics of humic acid in Lake Haruna sediment (0–24 cm) and recognized the following trends with depth:

1. Increase of carbon content.
2. Increase of C/H and C/N ratios.
3. Increase of color (E_{400}) intensity.
4. Decrease of proteinaceous portion, which was associated with decrease of infrared absorptions at 1640 (amide I + C=O, etc.) and 1540 cm^{-1} (amide II).
5. Decrease of carbohydrate-like portion, which was associated with decrease of 1000–1020 cm^{-1} band (C—O—C).

A study by Bourbonniere and Meyers (1978) on humic substances extracted without prior organic solvent treatment from a Lake Huron sediment (0–65 cm, about 560 years) did not exhibit the same trends. The H/C ratios of humic acid and fulvic acid showed no trend with depth: the N/C ratio appeared constant for humin, that for humic acid increased slightly with depth, and that for fulvic acid showed a slightly decreasing trend with depth including several significant fluctuations. Although no clear trend with depth in the E_4/E_6 (E_{465}/E_{665}) values was observed for fulvic acid, values for humic acid showed a progressive increase with depth. This latter tendency is similar to that for Lake Haruna sediment. The ratio of the infrared absorption band at 1655 cm^{-1} to that at 1730 cm^{-1} for fulvic acid showed a slightly decreasing trend with depth. Bourbonniere and Meyers (1978) explained some of these trends in N/C, E_4/E_6, and 1655 cm^{-1}/1730 cm^{-1} ratios by transformation of fulvic acid to humic acid in sediments. However, further evidence to support their explanations is needed.

A gradual decrease of carbohydrates with depth in humic substances in Lake Haruna sediments was shown by Uzaki (unpublished). Uzaki analyzed carbohydrates (neutral sugars) in humic acid, fulvic acid, and humin [Fig. 7(B)–(D)].

Vertical changes of amino acids in humic substances from Lake Haruna sediments (0–30 cm) were studied in detail by Yamamoto (1983). As shown in Figure 7(B)–(D), amino acid content and the percentage of amino acid nitrogen in the total nitrogen in humic acid decreased gradually with depth. Amino acids in humin showed a vertical trend similar to those in humic acid. However, amino acids in fulvic acid appeared to show no clear trend with depth. Yamamoto recognized the following vertical trend in relative amino acid composition in humic substances:

1. The relative concentration of basic amino acids (lysine and arginine) in every humic fraction decreased with depth.
2. For humin, relative concentrations of acidic amino acids (aspartic acid and glutamic acid) decreased with depth while neutral amino acids increased with depth.
3. Acidic amino acids in fulvic acid fraction increased slightly with depth.
4. The relative concentration of proline and hydroxyproline which are, or are expected to be, low in reactivity in the Maillard reaction, increased with depth in every humic fraction.

In conclusion, in the early stage of diagenesis, biochemical compounds and the hydrolyzable part (including biochemical components) of humic substances decrease gradually with depth primarily by degradation and, in part, by Maillard-type reaction (i.e., humification).

The Later Stage of Diagenesis

The long-term fate of humic substances may be understood by comparing humic substances in recent sediments to those in ancient sediments. Huc and Durand (1977) studied humic substances in a Green River shale (United States, Eocene) and a Messel shale (Germany, Eocene). They found organic carbon content to be 28.0% for a Green River shale and 29.7% for a Messel shale. The Green River formation was deposited in large shallow lakes under a subtropical climate. The organic matter in this shale is mainly composed of microscopic algae and other organisms, perhaps accompanied by nonlacustrine organic components such as wind-blown or water-borne pollens and waxy spores. The Messel shale was deposited in a series of shallow swampy lakes linked by slow-moving fluvial systems. Analysis of fossil plants and pollens indicates that a hot, damp tropical climate existed at the time of deposition (Huc and Durand, 1977). After Soxhlet extraction with $CHCl_3$ and subsequent removal of carbonates (2*N* HCl), fulvic acid and humic acid were extracted by 0.1*N* NaOH + 1% sodium pyrophosphate solution. Humic acid was separated from fulvic acid by centrifugation of the acidified extract (pH 2). Humin (kerogen) was obtained by dissolution of a shale which had been extracted by $CHCl_3$ with 4*N* HCl at 70°C and then with 4*N* HCl/40% HF (1 : 3 to 2 : 3). The major findings were as follows:

1. The content of humic acid and fulvic acid in these shales was extremely low (0.4% humic carbon/total carbon for the Green River shale and 6.1% for the Messel shale), and the ratio of fulvic acid to humic acid was 0.28 for the Messel sample.
2. Infrared absorption bands related to amide I and II (1640 and 1540 cm^{-1}) were lacking in humic acid.

Huc and Durand (1977) inferred from these findings that humic acid lost solubility in alkalis by losing oxygen-containing functional groups (e.g., C=O) via diagenesis. These findings correspond well to the observations of humic substances in the earlier stages of diagenesis.

Ishiwatari et al. (1977, unpublished) studied thermal alteration of humic substances from marine and lake sediments by laboratory heating experiments. Laboratory heating experiments are a powerful tool providing insight into the natural evolution process of these substances which proceeds predominantly under geothermal influence. Humic substances (humic acid and humin) isolated from marine and lake sediments were heated in a nitrogen atmosphere at 150–410°C for various times (5–120 hours). Table 11 shows results of heating experiments conducted for humic acid from Lake Haruna. There were no essential differences between lake humic substances and marine humic substances in their behavior in the heating experiments. In an earlier stage, humic acid and humin lost their oxygen-containing functional groups when heated, generating predominantly CO_2 and H_2O (13–38% of the initial humic acid), and thus their H/C and O/C ratios decreased (Table 11). On heating, humic acids gradually lost their solubility in alkali (NaOH) owing to the loss of oxygen-containing functional groups, thus changing into

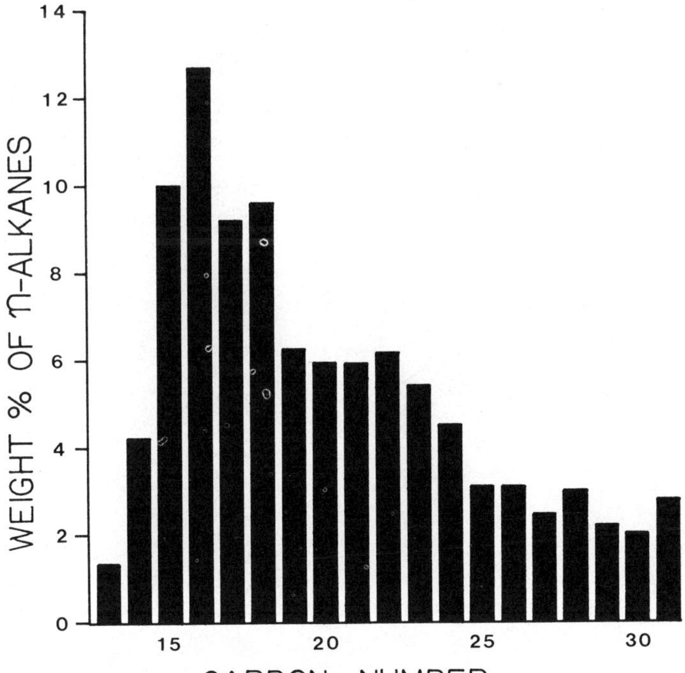

FIGURE 8. Distribution of normal alkanes generated from Lake Haruna sediment humic acid upon heating at 330°C for 24.2 hours.

TABLE 11. Variation in Various Properties of Lake Haruna Sediment Humic Acid Upon Heating

Sample No.	Heating Condition: Temperature (°C)	Heating Condition: Time (hours)	Residual Organic Matter (% of Initial Sample)	Ash (%)	Elemental Analysis (on Ash-Free Basis): C (%)	H (%)	N (%)	O (%)[a]	Atomic H/C Ratio	ESR Properties: Spin Concentration × 10^{17} (spins/g)	ESR Properties: g-Value	Solubility in NaOH[b]	Evolved n-Alkanes[c] (μg/g Initial Sample)
Unheated			100.0	0.0	55.81	5.82	5.15	33.22	1.24	4.5	2.0032	++	65
H-9	154	24.0	77.8	1.6	62.82	5.50	5.89	25.79	1.04	15.9	2.0032	++	79
H-8	200	24.0	67.4	1.6	70.85	5.66	6.41	17.08	0.95	22.0	2.0029	+	105
H-7	255	24.0	62.1	5.1	76.70	5.81	6.68	10.81	0.90	65.9	2.0029	±	260
H-10	300	24.0	58.5	3.0	77.55	5.59	6.39	10.47	0.86	72.8	2.0028	–	828
H-13	330	24.2	51.9	—	75.45	4.93	ND[d]	ND	0.78	ND	ND	ND	976
H-11	370	24.3	54.1	4.2	82.70	4.31	6.37	6.62	0.62	153.6	2,0028	–	660

[a] *Calculated by difference.*
[b] *Qualitatively: ++ soluble, + slightly soluble, ± almost insoluble, – insoluble.*
[c] C_{13}-C_{31} *alkanes.*
[d] *Not determined.*

humin (kerogen). In the next stage of heating, humic-acid-derived humin as well as original humin cleaved their carbon–carbon bonds and released aliphatic hydrocarbons (Table 11; Fig. 8). The remaining portion of humin became more aromatic as heating continued, approaching a graphite-like structure (Ishiwatari et al., 1977; Ishiwatari and Machihara, 1982). In accordance with these changes, free radical concentration in humic acid and humin increased upon heating probably due to the elimination of functional groups and alkyl groups.

Comprehensive reviews and monographs on insoluble organic matter (kerogen) in sedimentary rocks in relation to petroleum genesis have been published (e.g., Durand, 1980; Tissot and Welte, 1978).

CONCLUDING REMARKS AND THE ROLE OF FUTURE RESEARCH

The chemical characteristics of humic substances in productive lake sediments are determined primarily by organic composition of aquatic organisms (e.g., phytoplankton) as precursors. Humic substances (humic acid and humin) remain in sediments because they are relatively hydrophobic, causing them to be more rapidly deposited than other hydrophilic organic materials and to be less easily attacked by microorganisms (Vallentyne, 1962).

The contribution of allochthonous organic matter to lake sediment humic substances is variable. In this chapter, we did not discuss in depth the contribution of allochthonous materials, such as soil organic matter, to sedimentary humic substances. However, it is important in the study of lake humic substances to determine how much authochthonous or allochthonous organic matter contributes to the formation of lake humic substances. Therefore, a method must be established for estimating the degree of contribution of authochthonous or allochthonous organic matter to lake humic substances.

Another important area of research not covered here concerns the geochemical role of lake sediment humic substances. Some general aspects regarding the role of humic substances in aquatic environments as materials having the ability to form complexes with metal ions, minerals, and nonhumic organic compounds have been reviewed (e.g., Schnitzer and Khan, 1972; Stevenson, 1982). A review on the ecological and geochemical importance of water and sedimentary humic substances was published by Jackson (1975). The geochemical role of lake humic substances is similar to that of humic substances in other environments. Research is needed on qualitative and quantitative relationships between water humic substances and sedimentary humic substances. Otsuki and Wetzel (1973) claimed that precipitated $CaCO_3$ transports water humic substances to bottom sediments in certain lakes. A possible role of humin as a carrier and an agent of transformation for nonhumic compounds in contemporary aquatic environments was studied by Ishiwatari et al. (1980b) and Yamamoto et al. (1981). Humin isolated from a lake sediment was mixed with nonhumic compounds (fatty acid methyl esters, alkylbenzene sulfonates, carbohydrates, or amino acids) and examined for interaction with humin. The lake sediment humin interacted with these compounds. This phenomenon is similar to that reported for soil humic acids (Schnitzer and Khan, 1972) and soil and marine sediment humin (Abelson and Hare, 1970, 1971). This suggests that these types of interactions of sedimentary humic substances with nonhumic organic compounds in water is possible in the lake environment.

In the field of organic geochemistry in lake sediments, extensive work has been done on lipid (organic-solvent-soluble) materials (Barnes and Barnes, 1978; Cranwell, 1982). Chlorophyll pigments, important constituents in phytoplankton, are extractable by organic solvents. However, when solvent extraction is not done before isolation of humic substances, these chlorophyll pigments are delivered into humic fractions. In some researches (e.g., Povoledo et al., 1975; Bourbonniere and Meyers, 1978), the organic solvent extraction procedure was omitted before extraction of humic substances. Results are similar for other biochemical compounds such as fatty acids, carbohydrates, and amino acids. Therefore, careful and detailed investigation on the standardization of isolation and purification procedures of humic substances and on the forms of existence of these biochemicals in sediments and humic substances is needed. In addition, investigation of the relationship between humic substances and nonhumic substances should be conducted in order to gain an understanding of the formation and diagenesis of humic substances in lake sediments.

CHAPTER SEVEN

Geochemistry of Stream Fulvic and Humic Substances

RONALD L. MALCOLM

ABSTRACT

A brief history of the study of stream humic substances is divided into four periods: "Previous to 1950," the "Awakening Period" (1950–1964), the "Sephadex Period" (1964–1973), and the "Resurgence Period" (1973–present). The major researchers, their philosophy and research approaches, and significant research findings for each period are discussed. A state-of-the-art characterization of representative stream fulvic and humic acids from the Ogeechee river in Georgia is presented. A critical review of the geochemistry of stream humic substances relative to occurrence, origin, theories of formation, structure, acid properties, metal associations, organic solute associations, and biological aspects is summarized. The study of humic substances in water has been strongly influenced by soil science in the past, but the individuality of the science of stream humic substances is emerging. The new science is founded on recent breakthroughs in stream humic chemistry which have resulted from a large influx of interdisciplinary talented researchers, innovative experimental designs, and the application of advanced theories and mathematical modeling. The science of stream humic substances is advancing and changing so rapidly that approaches and theories of only a decade ago are now outdated.

INTRODUCTION

What we know today about stream humic substances is essentially the result of the past 30 years of growth pains in this emerging science. Thirty years may seem a very short period, but for scientists who have worked and experienced most of their scientific career in this time period it has been both exciting and tumultuous. Rapid growth in any science is characterized by resistance to discarding outdated, unsupported assumptions that prejudice the foundation of the science, by slowness to accept new data and theories that refute older assumptions and theories, and by the polarization of thought and philosophy between various researchers and research groups. For these and other reasons, it is virtually impossible to gain a consensus on what are stream humic substances and what we definitely know about them, but an attempt is made here to present the "present-day majority" opinion, recognizing that there are opinions outside this realm on certain subjects. It is recognized, and it is highly probable, that the "majority opinion of today" may be the "minority opinion of tomorrow" on certain aspects of stream humic substances, because research in this area is now extensive. The objectives of this chapter are to present a brief, critical review of past research on stream humic substances, to discuss the amounts and distribution of these substances in rivers, to present a state-of-the-art characterization of typical stream humic and fulvic acids, and to discuss some of the more relevant aspects of their reactivity and geochemistry in aquatic systems. These objectives will be discussed in view of what we know, what we think we know, what we don't know, and what we should know about stream humic substances.

The focus of this chapter is on the geochemistry of stream humic substances and should provide the reader with an appreciation of the dynamics, importance, and uniqueness of streams within the hydrologic system. Streams should not be envisioned only as arteries connecting lake, ground water, and soil environments (which are considered in previous chapters) with estuaries and oceans (which are presented in following chapters), nor as integrators of humic substances from upgradient, but rather they should be viewed as a different and unique aquatic environment where stream humic substances also have a different and unique character.

HISTORY OF RESEARCH ON STREAM HUMIC SUBSTANCES

Previous to 1950

Previous to the 1950s, there was not much awareness of organic water quality and the importance of humic substances in water. Although the phenomenon of water coloration and staining, taste, and odor problems in drinking water, and the ability of streams to assimilate sewage wastes were vaguely associated with humic acids, stream humic substances were studied by few scientists. The study of stream humic substances was only of passing academic interest because uncolored drinking waters could readily be found to replace colored sources, colored waters were not known to cause any environmental or health-related problems, and the rivers seemed to be a nonexhaustable repository for human waste. Also, because there was general consensus among scientists that coloration in water was synonymous with the presence of humic substances, there was no need to investigate the issue. It was also the general consensus that, without investigation, organic coloration was the result of leaching soil humic substances into streams during rainfall.

Ossian Aschan (from Finland) is acclaimed as the first scientist to devote most of his scientific career (1905–1935) to extensive research on aquatic humic substances (Aschan, 1908, 1932). He established stream humic substances, precipitated with $FeCl_3$ from colored waters, to be organic acids composed of carbon, hydrogen, oxygen, nitrogen, phosphorus, and sulfur. Birge and Juday (1922) studied organic lake constituents which could lead to the formation of organic color. One of their findings (1934) was that organic color-producing compounds in lakes, and presumably streams, were low in, and perhaps devoid of, nitrogen.

Selman A. Waksman, the noted microbiologist and soil humic chemist of the early part of this century, devoted an entire chapter in his book on soil humus (1938) to one of the first reviews and discussions of stream humic substances. A later review by Black and Christman (1963a) summarized the studies on stream humic substances by several other researchers previous to 1950. Some of the significant findings in these reviews were that organic color in water was present as negatively charged colloidal material (Saville, 1917), that absorption spectral analysis could be used for determining the concentration of organic substances (Demmering, 1936), and that certain natural organic substances produced intense color which followed Beer's Law (Datsko, 1949).

In this chapter, the three decades from the early 1950s to the present are arbitrarily divided into three periods based upon the philosophy of studying stream humic substances, the principal methods of their study, and instrumental capabilities. The three periods are the "Awakening Period" from 1950–1964, the "Sephadex Period" from 1964–1973, and the "Resurgence Period" from 1973 to the present.

Awakening Period (1950–1964)

During the "Awakening Period" many scientists first became aware of the extensive existence of colored waters throughout the world and found the study of stream humic substances to be a relatively virgin area of research. These scientists were influenced by soil humic studies (Flaig and Beutelspacher, 1954; Kumada and Aizawa, 1958; Kononova, 1961; Schnitzer and

Gupta, 1964; Bartholomew and Clark, 1965) and became familiar with the previous pioneering soil organic research (Mulder, 1862; Schreiner and Shorey, 1914; Oden, 1919; Waksman, 1938). Therefore, this period was dominated by the application of methods, approaches, instrumentation, and philosophies used by soil scientists in studying soil humic substances to characterize stream humic substances. Stream humic substances were assumed to be the same as soil humic substances (Ponomareva and Ettinger, 1954; Shapiro, 1957; Wilson, 1959; Black and Christman, 1963a).

Stream humic substances were isolated from water by vacuum concentration (Shapiro, 1957; Barth and Acheson, 1962; Black and Christman, 1963a; Midwood and Felbeck, 1968), freeze concentration (Black and Christman, 1963a; Shapiro, 1967), and solvent extraction (Shapiro, 1957; Barth and Acheson, 1962; Black and Christman, 1963a). Almost all humic substances studied were obtained from dilute organically colored waters, because sufficient quantities of stream humic substances for intensive studies could be isolated with the least effort, and because many uncolored waters were not believed to contain humic substances. Because the isolation of stream humic substances was so time consuming it was considered acceptable to use soil humic substances obtained by extraction of soils as representative of stream humic substances in water quality studies.

Spectrophotometers, fluorometers, and infrared analyzers were the principal instruments used to define the color characteristic of stream isolates (Shapiro, 1957; Skopintsev and Krylova, 1955; Delhez, 1960; Black and Christman, 1963a; Packham, 1964). Although these methods were state-of-the-art for that period, the resulting characterizations were of limited value, because they were conducted on gross and complex mixtures of many different dissolved organic constituents in addition to the humic substances present in the stream waters.

Paper chromatography was another popular method of studying organic water constituents (Shapiro, 1957; Packham, 1964; Midwood and Felbeck, 1968) and also was of limited value in characterizing humic substances and colored organic acids, because the complex mixtures often produced broad diffuse bands and streaks. Some of the spot-tests and chromatograms of Shapiro (1957) demonstrated the presence of colorless acids and suggested the presence of phenols and enols.

The works of Packham (in England), Black, Christman, and Shapiro (in the United States) are frequently associated with this period. The extensive investigations on stream humic substances by Russians, such as Goryunova (1952, 1954), Ponomareva and Ettinger (1954), Fotiyev (1970), and numerous others are generally unappreciated outside the USSR, because their works were not commonly translated into English.

Sephadex Period (1964–1973)

During this period, the influence of soil science on the study of stream humic substances began to wane. Many believed that new and innovative techniques and approaches were necessary to advance the knowledge of stream humic substances. Findings to that date had supported the hypothesis that humic substances were a complex mixture of similar polymeric materials. Further progress could only be made if the stream humic substances (which were a mixture themselves) were separated and characterized apart from the complex mixtures of other compounds in water. These sentiments were bolstered by the application of exclusion chromatography (Sephadex gel) and ultramembrane filter (such as Dionex) technology to the study of stream humic substances. By these techniques, humic substances of supposedly similar size fractions could be isolated and characterized because they were relatively free of other types of dissolved constituents in water.

Exclusion chromatography (Sephadex) was first tauted as an excellent method free from interferences for the determination of molecular weight and to accomplish desalting of humic substances with essentially no loss of organic matter (Gjessing, 1975; Gjessing and Lee, 1967, Holty and Heilman, 1971). Later, the same author and others (Gjessing, 1973, 1976) retracted these claims stating that excluded fractions were not "clean" and the usefulness of Sephadex in estimating molecular size distribution was in doubt. Due to uncontrollable interactions between the gel and aquatic humic substances such as specific sorption, reversible sorption, and the effect of ionic-strength gradients (Eaker and Porath, 1967; Lindquist, 1967; Tryland and Gjessing, 1975), Sephadex chromatography should be used with extreme caution. A discussion of these and other problems associated with Sephadex gel chromatography is presented in Chapter 19 of this book.

The use of ultrafiltration membranes, as a technique to fractionate dissolved organic matter, has suffered from the same general limitations as Sephadex (Gjessing, 1973). With washing, the retained fraction can be made relatively free of inorganic ions and lower-molecular-weight organic compounds if there is no interaction between the compounds of interest (Gjessing, 1973). A coarse separation of humic substances into several general molecular size or molecular weight ranges may result from continuous washing of a series of ultrafiltration membranes, but exact interpretations are limited and should be based upon numerous factors of the individual water sample. Many of these factors are discussed in detail by Buffle et al. (1978) and in Chapter 19 of this book.

The methods of isolation and physical characterization for stream humic substances during the "Sephadex Period" were essentially the same as during the "Awakening Period." One new isolation technique that became popular was freeze-drying. Large volumes (in excess of 1000 L) of colored waters were often freeze-dried and the resulting residues were characterized (Leenheer and Malcolm, 1973). Inorganic particulates and dissolved salts were removed from colored water samples by filtering the water through 0.45 μm filters and hydrogen saturating the filtrate by cation exchange prior to freeze-drying (Leenheer and Malcolm, 1973). This modification was successful in removing most of the clay colloids, but was unsuccessful in removing dissolved solids, such as sulphates which caused severe interferences.

Sulfates were converted to sulfuric acid during hydrogen saturation, which oxidized the organic matter during freeze-drying.

Because nondegradative techniques were unsuccessful in elucidating the structure of stream humic substances, several oxidative degradative techniques were attempted in an effort to discern the chemical structures of these substances (Felbeck, 1965b; Christman, 1970; Ogner and Gjessing, 1975). An array of oxidation products was usually obtained and identified by chromatographic techniques. The summation of identified oxidation products rarely accounted for more than 10% of the carbon in the original material. Most of the identified oxidation products were usually aromatic acids, which supported the hypothesis that humic substances were composed primarily of aromatic structures.

In an excellent work by Reuter et al. (1983), serious shortcomings in the interpretation of previous oxidation data were demonstrated. They established that mild oxidation of stream humic substances produced large quantities of oxalic acid indicating stream humic substances to be predominantly aliphatic in nature, whereas strong or severe oxidation produced structures stabilized by a $4n + 2$ overlapping π-electron system, attributed mainly to a series of benzene carboxylic acids. Structural interpretation based on the products of severe oxidation would lead one to falsely conclude a greater degree of aromaticity in stream humic substances than actually exists in the unoxidized material.

The Resurgence Period (1973–Present)

The resurgence in aquatic humic studies has been attributed to the flurry of health-related research after Rook (1974) discovered natural humic substances to be the major source and precursor of trihalomethanes (THMs) during chlorination of drinking water. There is no doubt that these health-related studies contributed to the revival, but the stage was already set for this resurgence by the timely integration of several factors. Some of these factors included the development of quantitative aqueous carbon analyzers, the rapid advances in instrumental organic analysis (especially NMR), the development of XAD-resin technology for the isolation of humic substances from water, and the host of capable environmental scientists entering the field of humic studies.

Previous to the 1970s there were no accurate and direct methods for determining the low concentrations of organic carbon in water. The four most common methods of determining organic constituents, including humic substances, in water were COD (chemical oxygen demand), permanganate oxidation, cobalt color comparison, and spectrometry. All these methods had serious limitations; therefore, literature values obtained by these methods should be interpreted with caution and may be only approximations and rough indices of organic water quality. Some of the most serious limitations include incomplete oxidation of organic compounds; oxygen consumption by inorganic compounds; interferences of ions, compounds, and particulate material on color and oxygen demand; and the interferences which gross mixtures have on spectrometric measurements.

During the early 1970s, a number of commercial instruments were manufactured for the direct analysis of carbon in water. The improvement and development of these instruments made possible the rapid and accurate determination of dissolved organic carbon (DOC) and suspended organic carbon (SOC) in water by 1974—the beginning of the resurgence.

Aqueous organic carbon analyzers helped revolutionize humic studies during the last decade. Not only could organic carbon measurements be made quickly and accurately on any water sample, but organic carbon analysis, used as a monitoring detector like conductivity and fluorescence, opened a new dimension in experimental design and approaches. In conjunction with other methods, the amount and distribution of humic substances in water in relation to other carbon-containing compounds could be determined.

The usefulness of microporous, synthetic resins as a second contribution to the resurgence of humic studies was much slower in development than the aqueous carbon analyzer, because the chemistry of the sorption and desorption processes on the resins was not well understood. Shapiro (1957) was one of the first to consider anion-exchange resins as a method of concentration, but he dismissed the idea because of low elution recoveries of humic substances from the resin. Packham (1964) was the first to use anion-exchange resins as a preparative method of isolating humic substances from water. He reported overall recoveries of less than 50%. Riley and Taylor (1969) first reported the possible utility of XAD-1 (microporous, uncharged, styrene-divinylbenzene resin) for concentrating humic substances from seawater. Their discovery was essentially unnoticed until Stuermer and Harvey (1974) and Mantoura and Riley (1975) used the method for preparative isolation. After that, modifications of the method began to flourish and uncharged XAD resins became the most popular sorbents for isolating humic substances from water. As a result of extensive investigations by Leenheer and Huffman (1976), Malcolm et al. (1977), Thurman et al. (1978), Aiken et al. (1979), and Leenheer (1981), XAD-8 has been demonstrated to be the best resin for isolating humic substances from natural waters. The use of XAD resins in isolating humic substances is discussed by Aiken in Chapter 14.

A third contributing factor to the resurgence has been the enormous advances in scientific instrumentation which enable the scientists to probe details that were unknown a decade ago. Developments in nuclear magnetic resonance spectroscopy have added a new dimension to structural studies of humic substances (Vila et al., 1976; Miknis et al., 1979a,b; Hatcher et al., 1980b). High-resolution gas chromatography–mass spectroscopy has been indispensible in identifying natural organic substances and degradation products of humic substances (Havlicek et al., 1979; Christman et al., 1980; Reuter et al., 1983). Fast atom bombardment has been useful in structural studies of polymeric substances (Barber et al., 1981; Taylor, 1981) and is a potentially valuable tool in humic substances research (Saleh et al., 1983). Development of new liquid chromatographic packings, detectors, and techniques shows great promise in separation of humic substances into a family of closely related macromolecules. Natural humic substances have been fractionated by pH gradient elution (MacCarthy et al., 1979) and other solvent gradient elutions (Thurman and Malcolm, 1979; Personal Communication, F. Y. Saleh, North Texas State University). Present research on derivatization should produce humic substances with more hydrophobic character which are more amenable to liquid chromatographic separation.

OCCURRENCE AND DISTRIBUTION OF STREAM HUMIC SUBSTANCES

Stream humic substances are defined in this chapter as comprising that portion of the organic substances in stream waters which passes through a 0.45-μm membrane filter and, upon acidification to pH 2 with HCl, has a column distribution coefficient (k') of greater than 50 on XAD-8 resin at 50% breakthrough of the column for the visually dark-colored, nonspecific amorphous carbonaceous material. Stream humic substances are composed of both stream fulvic acids and stream humic acids. Aquatic humic acids are those stream humic substances which are insoluble and form precipitates at pH 1. Stream fulvic acids are those stream humic substances which are soluble at pH 1. In most streams, approximately 90% of the stream humic substances are stream fulvic acids. This definition excludes any colored or uncolored specific organic compounds which may be physically admixed with the nonspecific humic substances during isolation or concentration from water. After analysis of numerous stream waters of the United States, the author has found that more than 95% of the colored organic substances are sorbed onto XAD-8 at a k' of 50, which results in a concentration factor of 10 or greater. Other resins, such as XAD-1, XAD-2, and XAD-4, are unacceptable because of nonquantitative sorption, size exclusion, and irreversible sorption of a portion of the humic substances (Aiken et al., 1979).

Humic substances occur in every natural water sample which has been analyzed for their presence. The amount and composition of humic substances vary considerably from soils, surface waters, and groundwaters, but their "ubiquity" in water is without question. Humic substances in water occur as a size continuum ranging from dissolved through colloidal, to particulate phases. The dissolved phase is a predominant phase in most streams and is the phase emphasized in this chapter. The geochemical activity and reactivity of dissolved and particulate organic phases are thought to be of sufficient difference in magnitude to merit the separation based on size at 0.45 μm.

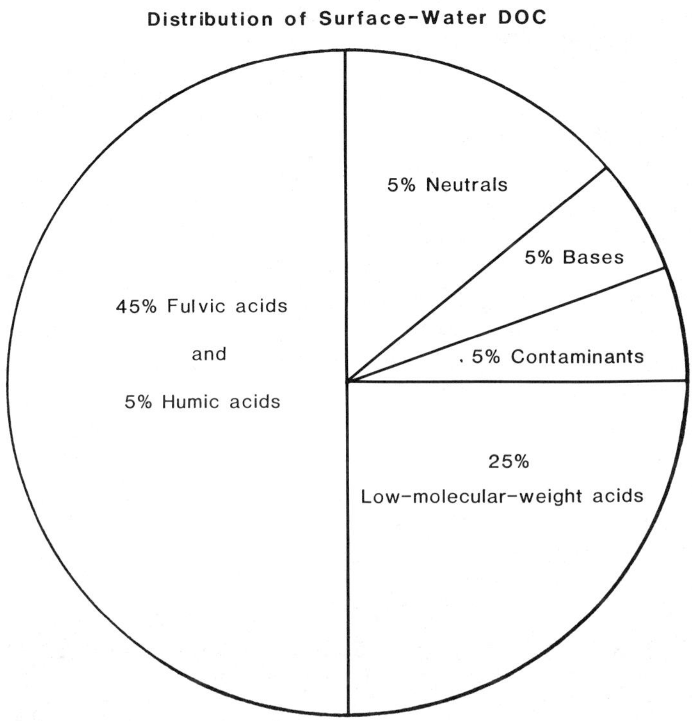

FIGURE 1. Distribution of surface water DOC in rivers of the United States.

The average concentration of dissolved humic substances in visually uncolored surface waters of the United States is 2.2 mg C/L or 4.4 mg/L expressed as humic substances. The value of 4.4 mg/L is based on elemental data presented later in this chapter showing that stream humic substances are approximately 50% organic carbon. The average dissolved organic carbon (DOC) for the same data of several hundred data points throughout the United States is 5 mg C/L with a range between 1.5 and 10 mg C/L. Figure 1 shows how the dissolved stream organic carbon is distributed into various classes. From this figure, it can be seen that approximately 50% of the DOC in uncolored United States streams occurs as humic substances. Therefore, once the DOC of a surface water is determined, the concentration of humic substances can be estimated on an approximate basis. It is not clear whether this approximation can be applied to other streams of the world, but this warrants investigation.

The DOC of organically colored stream waters is extremely variable, ranging from approximately 5 mg C/L to greater than 50 mg C/L. Visible color is rarely observed in natural waters of DOC <5 mg C/L. Unlike uncolored natural water with a relatively constant proportion of the DOC as humic substances, the organically colored natural waters vary considerably in the proportion of DOC as humic substances. For example, humic substances represent 80% of the 40 mg C/L DOC Suwannee River water in Georgia as compared to 48% humic substances of the 33 mg C/L DOC Pleasant River water in Maine.

Approximately 90% of the humic substances in uncolored waters occur as fulvic acid and only 10% or less as humic acid. The humic–fulvic acid separation is accomplished at pH 1.0 on humic substance concentrations in excess of 500 mg C/L. At humic concentrations less than 500 mg C/L humic–fulvic separations may be inadequate due to the kinetics of slow precipitate formation. There is a general trend in which the proportion of humic acid increases above 10% with increasing DOC concentrations in organically colored surface waters.

The humic–fulvic acid separation is believed to be a useful and meaningful characterization and separation procedure for stream humic substances. The clear-cut differences between stream humic and fulvic acids as discussed in the next section of this chapter strongly support the conclusion that it is a valid and meaningful separation.

CHEMICAL AND PHYSICAL CHARACTERIZATION OF STREAM HUMIC SUBSTANCES

Because an exact structure or family of structures cannot be written for humic substances, we must continue to characterize these substances on the basis of several different chemical and physical properties. No one or group of these characteristics rigidly define stream humic substances, but the greater the number of general characteristics which fit an "idealized" conceptual model of the majority of researchers is the present "nonexacting" standard. A minimally recommended characterization would include the DOC of the water sample from which the humic substances were isolated; the percentage of the DOC as humic substances; the fulvic/humic acid ratio; the method of isolation; elemental analysis for carbon, hydrogen, oxygen, nitrogen; ash content of isolate; functional group analysis; and color properties. Molecular size, ^{13}C NMR, and proton NMR are also rapidly becoming important methods of characterization. Molecular-size measurements are discussed by Wershaw and Aiken in Chapter 19, and ^{13}C NMR and proton NMR spectroscopy are discussed by Wershaw in Chapter 22. A more satisfactory elemental analysis should also include phosphorus, sulfur, and halogens with the summation of elemental constituents approximating 100%. Methods of elemental analysis are discussed by Huffman and Stuber in Chapter 17, and the interpretation of elemental data is discussed by Steelink in Chapter 18 of this book. Analysis of acidic functional groups are discussed in by Perdue in Chapter 20. Other useful characterizations include ^{14}C age,

TABLE 1. Characterization of Ogeechee Stream Fulvic Acid (Collected December 1981)

1. Elemental analysis (in percent on a moisture-free and ash-free basis).

C = 54.56	N = 0.87	% ash = 0.86
H = 4.97	S = 0.74	
O = 38.20	P = 0.62	

2. Carbon distribution by solid-state CPMAS ^{13}C NMR (in percent).

0– 55 ppm (C—C)	= 36	110–145 ppm (C=C)	= 12
55– 65 ppm (C—O)	= 8	145–160 ppm (ϕ—O)	= 5
65– 95 ppm (C′—O)	= 16	160–195 ppm (COOH)	= 16
95–110 ppm (anomeric)	= 3	195–225 ppm (C=O)	= 4

3. Aliphatic carbon percentage = (0–110 ppm) = 63%;
   Aromatic carbon percentage = (110–160 ppm) = 17%.
4. Functional groups:

COOH (titration)	= 6.4	Alcoholic OH (^{13}C NMR)	= 5.1
COOH (^{13}C NMR)	= 6.8	Phenolic OH (titration)	= 1.6
Carboxyl (^{13}C NMR)	= 1.7	Phenolic OH (^{13}C NMR)	= 2.1
Methoxyl (^{13}C NMR)	= 3.4		

5. Molecular weight = 650–950; radius of gyration = 6 Å.
6. Percent carbohydrate $\sim$5%.
7. Percent of total nitrogen as amino acids $\sim$20%.

TABLE 2. Characterization of Ogeechee Stream Humic Acid (Collected December 1981)

1. Elemental analysis (in percent on a moisture-free and ash-free basis).

C = 55.94	N = 1.27	% ash = 1.13
H = 4.13	S = 0.93	
O = 36.52	P = 0.25	

2. Carbon distribution by solid-state CPMAS ^{13}C NMR (in percent).

0– 55 ppm (C—C)	= 23	110–145 ppm (C=C)	= 21
55– 65 ppm (C—O)	= 8	145–160 ppm (ϕ—O)	= 9
65– 95 ppm (C′—O)	= 12	160–195 ppm (COOH)	= 16
95–110 ppm (anomeric)	= 4	195–225 ppm (C=O)	= 7

3. Aliphatic carbon percentage = (0–110 ppm) = 47%;
   Aromatic carbon percentage = (110–160 ppm) = 30%.
4. Functional groups:

COOH (titration)	= 4.7	Alcoholic OH (^{13}C NMR)	= 4.3
COOH (^{13}C NMR)	= 6.8	Phenolic OH (titration)	= 1.9
Carboxyl (^{13}C NMR)	= 3.0	Phenolic OH (^{13}C NMR)	= 3.9
Methoxyl (^{13}C NMR)	= 3.4		

5. Molecular weight = 2000–3000; radius of gyration = $\sim$10 Å.
6. Percent carbohydrate $\sim$10%.
7. Percent of total nitrogen as amino acids $\sim$25%.

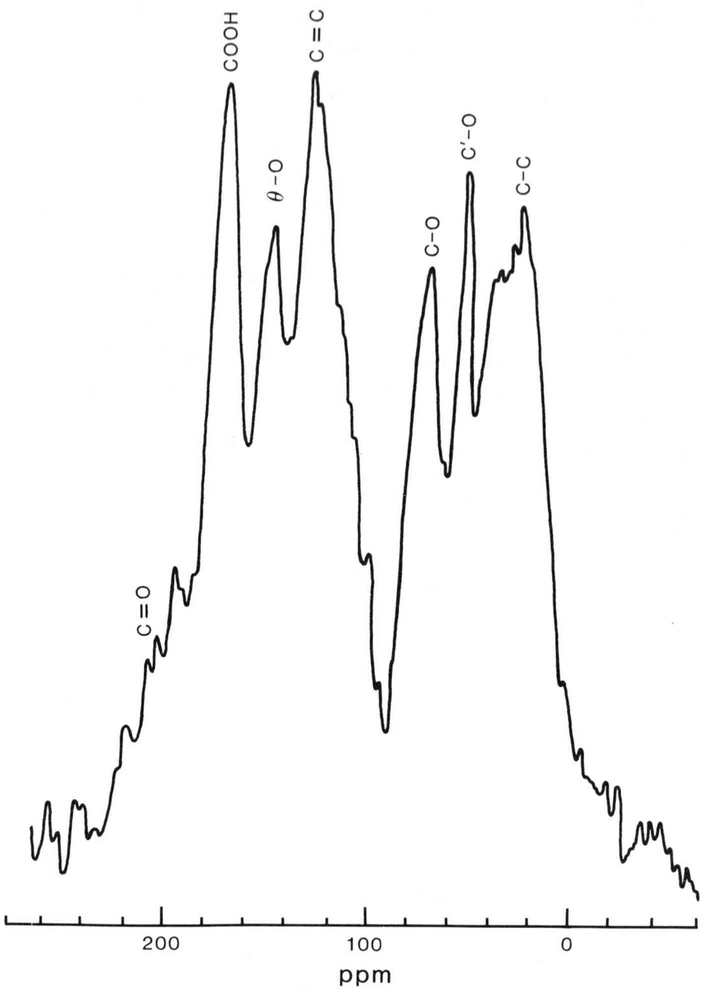

FIGURE 2. The cross-polarization, magic-angle-spinning, solid state ^{13}C NMR spectrum of stream humic acid from the Ogeechee River near Louisville, Georgia during December 1981.

$^{16}O/^{18}O$ ratios, proton NMR, total organic halide (TOX), amino acid composition, pentose and hexose sugar composition, infrared spectra, and trihalomethane potential. The characterization data of typical stream humic and fulvic acids isolated from uncolored streams are given in Table 1 and 2 and in Figures 2, 3, 4. A similar characterization of stream fulvic acids in colored waters is given by Thurman and Malcolm (1983). The stream fulvic acids in colored or uncolored waters are found to be very similar.

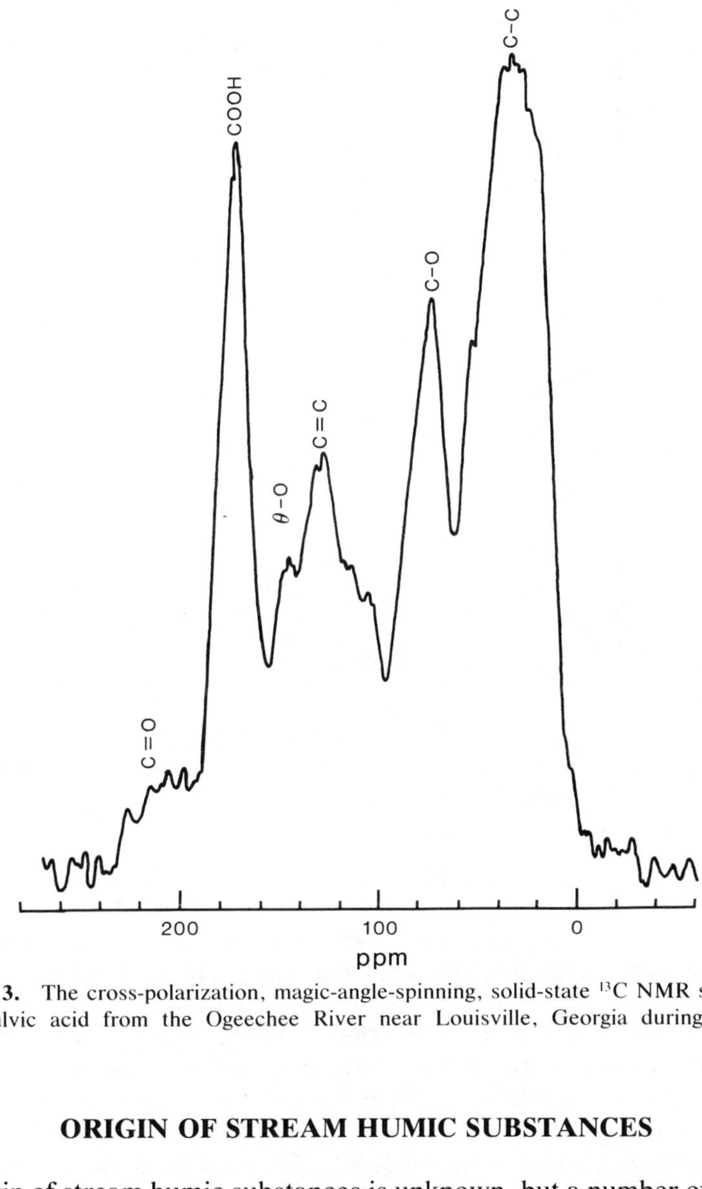

FIGURE 3. The cross-polarization, magic-angle-spinning, solid-state ^{13}C NMR spectrum of stream fulvic acid from the Ogeechee River near Louisville, Georgia during December 1981.

ORIGIN OF STREAM HUMIC SUBSTANCES

The origin of stream humic substances is unknown, but a number of theories and assumptions abound. The common sweeping conclusion, found so often in the literature, that stream humic substances are the same as soil humic substances and, therefore, must have their origin in soil is unsupported. If the same "logical" reasoning were applied to biology, any animal with two eyes, four legs, and a tail, which runs about the landscape would be an elephant! It would appear that this aspect of the science has retrogressed

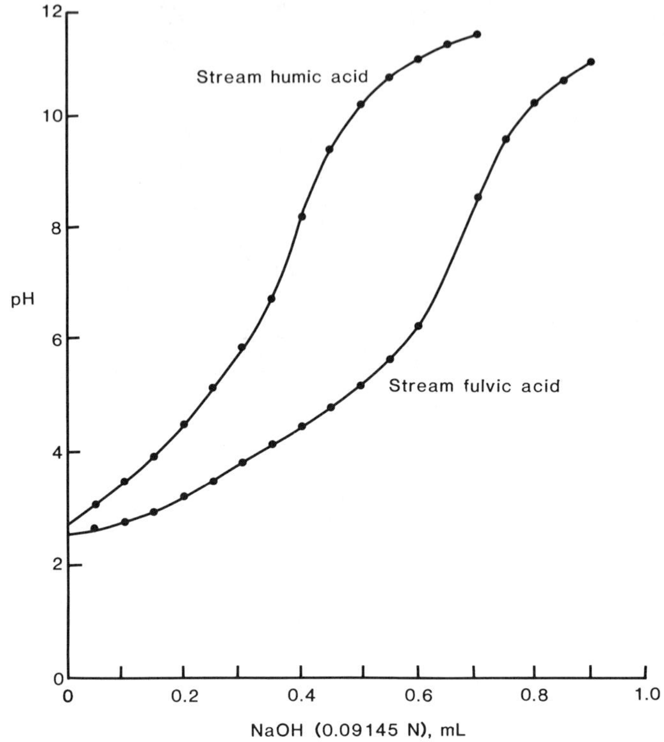

FIGURE 4. Titration curves for stream humic and fulvic acids isolated from the Ogeechee River near Louisville, Georgia during December 1981.

rather than progressed over the past 50 years since Waksman's (1938) review of the subject. His conclusions were that aquatic humic substances differed from soil humus in source and composition. Only a part of stream humic substances is derived by wind and water currents from the soil, litter, and decaying terrestrial vegetation (the allochthonous source); the other part of stream humic substances is produced during decomposition of aquatic plants and animal residues (the autochthonous source). Humic substances in mineral soils originate largely from higher plants composed primarily of cellulose and lignin; however, substances in water have been contributed by diatoms and algae which consist largely of hemicellulose and proteins and by aquatic animals rich in proteins and chitinous substances. Waksman also states that sewage effluents may be an important source of aquatic humus. Secondary sewage effluents have been shown to contain large amounts of humic substances (Rebhun and Manka, 1971).

Weber and Wilson (1975) faced this issue when they stated that it is merely an assumption that stream humic substances are the same as soil humic substances and that such has never been proven. There is no doubt, as Waksman (1938) and others (Malcolm and Durum, 1976; Reuter and

Perdue, 1981) state, that soil is one source of stream humic substances or humic substance precursors, but to assume soil as "the source" or "only source" is exaggerated. To do so is to totally ignore other known sources such as groundwater, decaying vegetation and litter, canopy drip, sewage, and autochthonous material.

The issues at the heart of the controversy are: (1) the differences in perspectives of soil as a true source of stream humic substances or merely as a transient transport medium between the true source and the stream; and (2) the relative contribution of soil as compared to other sources. The litter on top of the mineral soil is the principal area of decomposition and humus formation. During rainfall, the organic substances that leach from the vegetative canopy and litter layer along with the humic substances formed in the litter layer are flushed into the soil. According to Kennedy (personal communication, U.S. Geological Survey, Menlo Park, California), over 70% of the water entering the stream during rainfall events has had intimate contact with the soil as overland flow and interflow. The incident rainfall has a residence time of minutes, hours, or days in the soil before entering the stream. The soil may be considered as primarily a short-term or transient storage medium of laterally percolating interflow waters containing humic substances and humic precursors leached from the vegetative canopy and litter layer.

A third issue is that even though the soil is one source of stream humic substances, it is not necessary that soil and stream humic substances have the same composition. If they were of the same composition, then stream humic substances would be primarily humic acids, because the humic acid to fulvic acid ratio in soil is approximately 3 : 1. However, as previously discussed in this chapter, stream humic substances are approximately 90% fulvic acids. One may say that fulvic acids are leached from soils in preference to humic acids. This may be true, but no one has shown water leachates of soil to contain fulvic acid of the same composition as in the bulk soil. Beck et al. (1974) state that meteoric waters percolating through soil will selectively mobilize nonrepresentative fractions of the soil organic matter. It should be emphasized that even if stream humic substances are the same as soil humic substances, one can not infer that one is the source of the other, but that the same precursors and humification process are probably operable in both soil and stream environments.

If we do not know with certainty the potential roles of soil as different sources of stream humic substances, it is speculative to postulate the relative contribution of soil as compared to other sources of stream humic substances. Needless to say, there is an urgent need for experiments to determine the quantitative contribution of all the postulated sources of stream humic substances.

Two additional sources of stream humic substances which are generally neglected but could be seasonally important are humic substances in groundwater and erosion or scouring of the stream bed during high discharges. As described by Thurman in Chapter 4 of this book, groundwater contains humic substances in low concentrations. The source of humic substances in deep-seated groundwaters is believed to have been present in the formation before burial in the form of humic substances per se or as organic detritus which degraded to humic substances. The addition of humic substances in recharge waters is postulated to be insignificant, because the organic content of percolating waters decreases with depth due to sorption or organic solutes on mineral surfaces and decomposition and assimilation of organic solutes by microbes.

Shallow groundwaters are the major contributor to the base flow of many streams. Under such low-flow conditions, the humic substances in groundwater may also be a major factor controlling stream humic substances in terms of both quantity and quality. Groundwater humic substances have been shown to be different from most stream humic substances; therefore, possible seasonal differences in stream humic substances, especially at low flow, may reflect the influence of groundwater contributions. Even though groundwater may be the sole contributor to base stream flow, the influence of groundwater humic substances may not be manifested, because the extensive contact of the humic substances with mineral surfaces and microbial processes may not have been sufficient in shallow groundwater flow for the humic substances to have fully developed a unique composition different from lateral soil interflow into streams.

Soluble humic substances are known to form and accumulate in bottom sediments of streams. The concentration of humic substances in such sediments may be an order of magnitude to several orders of magnitude greater than in the overlying stream water. There would be a natural gradient for these substances to diffuse into the stream water. The release of the source of stream humic substances during rainfall and high flow by scouring of the bed could be one contribution to the observation that dissolved organic substances increase in concentration with increasing stream discharge.

THEORIES FOR THE FORMATION OF STREAM HUMIC SUBSTANCES

The mechanism(s) for the formation of stream humic substances is not known. Several theories of formation of these substances have been formulated, but none have been supported with adequate systematic data. Most of the theories of formation are common to soil humic substances and few are unique to stream humic substances. Five of the general or overall theories are the following:

1. The first needs no theoretical discussion; it merely assumes that stream humic substances consist of soil fulvic acid which has been leached or eroded from soils.

2. A second theory, closely aligned with the first, assumes stream humic substances to be formed within the stream by the same processes as soil humic substances, whatever they may be.
3. A third theory speculates that stream humic substances are soil fulvic acids leached from soil in the initial stages of humification and then modified, transformed, or aged by stream humification processes which result in humic substances unique to this aquatic environment.
4. A fourth theory postulates that stream humic substances are formed by a unique stream humification process, whereby simple reactive moieties are polymerized and condensed into humic substances unique to the stream environment.
5. A fifth prevalent theory is that stream humic acids are formed by continuation of the polymerization process to form larger molecular units of fulvic acid which are called humic acid (Beck et al. 1974).

There are numerous variations to these general theories.

A few of the specific theories of formation of stream humic substances merit additional discussion. A commonly accepted theory today is that stream humic substances represent heterocondensation polyphenolic polymers of reactive simple monomers (Schnitzer and Khan, 1972; Wetzel, 1975). The reactive monomers are polyphenols, amino acids, simple sugars, and quinones, which are present in all waters as a result of cell lysis, cell exudates, and as intermediates of microbial degradation of plant and animal detritus. A variation of this theory postulates that stream humic substances are composed of an infinite number of random heteropolymers, the composition of which is determined by the relative amounts of monomers available for incorporation at any given time. Such a theory precludes the elucidation of a definite structure for stream humic substances. Another variation of this theory, which is more plausible to the author, is that the polymerized and condensed monomers have a limited or fixed arrangement in the macromolecular product. In this theory, the polymerization process could be stopped or limited by a deficiency or low concentration of any one of the reactive monomers. This theory would result in stream humic substances being more homogeneous from various environments and would not preclude the elucidation of a definite or limited number of structures for stream humic substances. The contribution of microbiological or abiotic factor(s) to the polymerization and condensation reactions is unknown. The general aspects of the theory are supported by the incorporation of ammonia, polyphenols, and certain xenobiotic compounds into humic-like polymers (Bollag et al., 1980).

Another common theory of formation for stream and soil humic substances is a modified remnant of the old ligno-protein theory of Waksman (1938). This theory states that soil and stream humic substances are degradation products of lignin from terrestrial plants which are composed primarily of lignin and celluloses. This theory has been especially promoted by many oceanographers, who advocate an exclusive aromatic lignin signature for stream humic substances, which they believe to be the same as soil humic substances. They also postulate that stream humic substances have a degraded condensed ring structure or are highly aromatic in nature. They advocate marine humic substances to be formed by an entirely different mechanism from different precursors from stream and soil humic substances (Rashid and King, 1970; Nissenbaum and Kaplan, 1972; Stuermer and Harvey, 1974); thus, the marine humic substances can readily be differentiated from stream or soil (terrestrial) humic substances. Many aspects of this theory and associated assumptions appear to be questionable because (1) stream fulvic acids are highly aliphatic in nature; (2) soil humic substances are a questionable major source of stream humic substances; and (3) stream humic substances are conservative upon entering saline ocean waters or estuaries and do not precipitate near shore, but may be mixed and distributed throughout the ocean (refer to Chapter 8 by Mayer in this book).

Other possible theories of stream humic substance formation are the browning reactions (the black substances produced by the action of acids on sugars) (Haworth, 1971) and the melanoidin reaction (the dark pigments produced when glucose, alanine or ammonia, and phenolic substances react with one another) (Hoering, 1973; Filip et al., 1974; Ertel and Hedges, 1983). These theories are discussed in detail by Stevenson in Chapter 2 of this book.

It is probable that new insights into the formation of stream humic substances will come from more detailed characterization of hydrophilic acids (Leenheer and Huffman, 1976), which account for 20–30% of the DOC in stream waters. These are highly branched and substituted organic acids, some of which are slightly colored, but are excluded from the definition of stream humic substances because they have a k' of less than 50 on XAD-8 resin. These hydrophilic acids probably have a lower molecular weight and have a greater number of acidic functional groups per carbon atom than stream fulvic acids. The hydrophilic acids may also include organic polyacids which are precursors of stream humic substances, as well as degradational products of stream humic substances.

An intensive research effort should be focused on the mechanisms of formation of humic substances to evaluate the merits of the various theories. Unfortunately, these types of studies are very time consuming and require highly trained biochemists and process biologists.

Stream ecologists are the major proponents of two recent theories, the river continuum concept (Vannote et al., 1980) and the nutrient spiraling concept (Elwood et al., 1983), which they believe may be important frameworks for studying stream humic substances. These theories emphasize the diverse and dynamic aspects of different reaches within the same stream, the differences among streams, and the differences in seasonal patterns among streams in various geographical regions. The theories predict enormous variations in stream humic substances with time, season of the year, stream discharge, vegetation of stream basin, and processing of the organic matter within the stream, with changes in carbon sources and biological populations within the stream.

STRUCTURAL FEATURES OF STREAM HUMIC SUBSTANCES

As stated previously, the exact structure of none of the humic substances has been elucidated. The accumulated data on stream humic substances, the many recent advances in stream humic studies, and the better resolution of ^{13}C NMR spectrometers enable a few comments to be made about the structure of stream humic substances.

We know that the core or dominant structures of stream humic substances are primarily aliphatic and not aromatic, contrary to what has been theorized for decades. ^{13}C NMR data indicate that only 16–20% of the carbon in fulvic acid is aromatic carbon. Even the humic acid is only comprised of approximately 30% aromatic carbon which also indicates that its nature is primarily aliphatic and not aromatic.

All indications are that stream fulvic acids have an approximate molecular weight of 1000, are completely dissolved in water, and have a radius of gyration smaller than 10 Å. These data suggest that fulvic acids are much smaller solutes than originally conceived in years past. Stream humic acids are slightly larger with a molecular weight of approximately 3000, have radii of gyration of 10–20 Å, and are completely dissolved in water. The size and shape of humic and fulvic acids will vary with pH depending on the net negative charge, inter- and intramolecular hydrogen bonding, the concentration of humic substances, the ionic strength, and the hydration of the molecules.

The carbon skeleton of the molecule of stream fulvic acid is highly substituted with oxygen-containing functional groups for every three to four carbon atoms (Thurman and Malcolm, 1983). The major oxygen-containing functional groups are carboxyl, with minor amounts of phenolic and methoxyl. The relative contribution of ether groups to hydroxyl groups is presently unknown. Stream humic acid is also highly substituted with oxygen-containing phenolic, methoxyl, and carboxyl functional groups. The molecule has two to three unsubstituted aliphatic side chains of 6 to 10 carbon atoms in length.

Nitrogen is a definite structural component of both stream humic and fulvic acids. There is approximately one nitrogen atom per molecule in fulvic acid, but all fulvic acid molecules may not contain nitrogen. Stream humic acids contain two to three nitrogen atoms per molecule. The percentage of the nitrogen as α-amino nitrogen is unknown, but is expected to be approximately one-third of the nitrogen.

GEOCHEMICAL REACTIVITY OF HUMIC SUBSTANCES IN STREAMS

Acid Characteristics

One of the first and foremost aspects to remember about stream humic substances is that they are polyprotic acids and many of their properties are determined by carboxyl and phenolic functional groups. As shown in Table 1, the carboxyl acidity is dominant in both fulvic and humic acids, but the relative proportion of the phenolic acidity is higher in humic acids. Typical titration curves for hydrogen-saturated stream humic and fulvic acids are shown in Figure 4. The S-shaped character of the titration curve is somewhat similar to that of a monoprotic acid, but it is believed to be that of a mixture of polyprotic acids with a continuum of pK_a values (acid dissociation constants). Gamble (1970) has referred to the pK_a groups in a soil humic acid between 2.5 and 3.5 as highly acidic carboxyl (Type I, Acidity), between 4 and 5 as moderately acidic carboxyl (Type II, Acidity), and between 9 and 10 as phenolic hydroxyl groups (Type III, Acidity). The ionized and unionized acidic functional groups of humic substances make the moieties polar and/or charged, respectively, which imparts water solubility to these substances.

Fulvic acids, humic acids, and lower-molecular-weight acids can be major factors in determining the pH of natural stream waters. The active acidity and potential acidity of these compounds may also exert a considerable buffering capacity in streams. In most streams with a pH between 5 and 8, fulvic and humic acids are present as organic polyanions, with the magnitude of this charge being pH dependent (the higher the pH, the higher the negative charge due to the continued ionization of carboxyl and phenolic functional groups).

Humic substances, as polyanions, have several implications on water quality which include anion–cation balance, alkalinity titration, cation-exchange reactions, metal complexation, and many more. First of all, organic anions have an effect on the anion–cation charge balance among dissolved solutes. The relative importance of organic acid anions in accounting for the anionic charge in water is variable depending on the concentration of organic acids, the pH, and the amount and nature of the dissolved inorganic solids. The maximum expression of anionic charge of organic acids occurs in organically colored waters of low specific conductance and moderate to low pH. In such waters neglecting the contribution of organic anions to the total anionic charge can lead to serious errors in calculating the anion–cation charge balance. These waters, where the organic constituents dominate practically all aspects of water quality, are found throughout the world, especially in the colder regions of both northern and southern latitudes. Large regions of this type are found across northern Europe and Russia (Gjessing, 1976). The geochemistry of these waters in the southeastern coastal plain of the United States is presented by Beck et al. (1974) and in the Amazon region by Leenheer (1980).

The protonation of organic acid anions during alkalinity titration can cause serious errors in alkalinity determination and the use of the data for carbonate speciation and equilibrium chemical modeling. Errors in alkalinity depend on the same factors as for anion–cation balance and may vary from negligible values to as much as 1270% error as documented by Beck et al. (1974). Accurate values for alkalinity and carbonate speciation can be determined by direct measurement of CO_2 evolved from acidified water samples.

Metal–Humic Associations in Water

The geochemistry of colored waters has always been associated with metal chemistry, especially with iron. The association of humic substances with metals varies enormously, is affected by many variables, and often requires rather rigorous mathematical and theoretical chemical expressions (Gamble, 1970; MacCarthy and Smith, 1979; Perdue and Lytle, 1983a) for an exacting understanding. The importance of humic substances in complexing metal ions in normal uncolored river waters was seriously underestimated by Stumm and Morgan (1970), but their erroneous assumptions about the low concentrations of stream humic substances and their low complexing ability were subsequently arighted in Reuter and Perdue's (1977) paper which demonstrated aquatic humic substances to be potentially significant complexers of trace metals in all freshwater streams even in the presence of thousand-fold excesses of major cations.

The association of positively charged metal ions with negatively charged organic anions goes beyond the concept of cation exchange to specific exchange, complexation, and possibly chelation. The fact that protons are released into solution is evidence of metal complexation with acidic functional groups on humic substances. Other functional groups on stream humic substances containing nitrogen, phosphorus, and sulfur may also be potential complexing sites, but they are believed to be minor compared to the abundance of carboxylic and phenolic functional groups. The relatively high stability of humic–metal complexes and the ability of humic substances to complex polyvalent metal ions such as iron, aluminum, and copper have led scientists to postulate that chelates (a specific type of complexation where a ring structure is formed between the metal ion with two or more electron donor groups on the organic ligand) may be formed. Although chelation of metal ions with humic substances has never been proven experimentally, the general and incorrect usage of chelation as synonomous with complexation has lowered the scientific credibility of metal–humic research in the scientific community.

An incomplete list of some of the factors that influence metal–humic complexation includes concentration of humic substances and metal ions, competing ligands and metal ions, source of humic substances, type and speciation (charge) of the metal ion, pH, ionic strength, and temperature. In a recent review of metal–humic complexation, Saar and Weber (1982) stated that the effect of pH is important, because "the hydrogen-ion concentration determines which forms of the fulvic acid and metal ions are prevalent; different forms of these species have different tendencies to enter a complex. Another way to look at the effect of pH on complexation is to consider that H^+ ions compete with metal ions for anionic binding sites on fulvic acid, and OH^- ions compete with fulvic acid for the cationic metal ion. As pH is raised, fulvic acid becomes more available for complexation and the metal ion becomes less available. Some intermediate pH most favors complexation between fulvic acid and metal ions." Conditional stability constants tend to increase with pH up to a certain point and then decrease. A good example of how complexation is affected by type of metal ion, concentration of fulvic acid, and concentration of metal ion is shown in a diagram (Fig. 5) from Reuter and Perdue (1977). Only 8% of Cu(II) is complexed at 1 mg/L of fulvic acid; whereas 50% was complexed at the same pH at 10 mg/L of fulvic acid. The concentration of fulvic acid may also affect metal speciation for weakly bound metal ions like Cd(II) (Saar and Weber, 1979). In general,

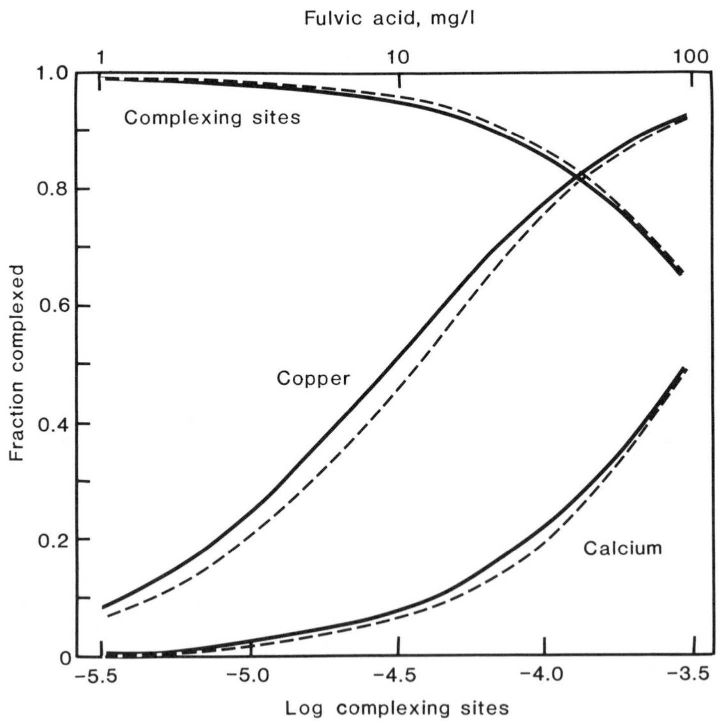

FIGURE 5. Degree of complexation of Cu(II) and Ca^{2+} as a function of fulvic acid concentration. Total Cu(II) concentrations 1×10^{-8} *M* (solid lines) and 1×10^{-5} *M* (dashed lines).

hydrogen, aluminum, iron, lead, and copper have larger stability constants than nickel, chromium, cobalt, zinc, or manganese. Stability constants for calcium, magnesium, and cadmium are the lowest. Theoretically, monovalent ions, like potassium and sodium, can compete for complexing sites on stream humic substances only if their concentration is very high (Gamble, 1973).

Humic–metal associations in water occur both in true solution and colloidal forms; it is often impossible to differentiate between the two forms. This has been well documented in the coagulation literature where efforts have been made to remove both iron and humic substances from drinking waters. The products of humic–metal association, at moderate to low pH values and low concentrations of the metal ion, tend to be dissolved complexes (Shapiro, 1957; Ghassemi and Christman, 1968; Perdue et al., 1976). As either the pH increases to near neutrality or the metal to humic ratio increases, the adducts may be either in solution or in a fine colloidal state. If the concentration of humic substances is sufficiently high and the concentration of the metal ion low, the humic ligand will compete well with OH^- to keep iron and other trace metals complexed in solution or as a finely divided stable colloid. At high metal-ion concentrations (as with high doses of alum in coagulation studies) slightly above or below neutrality (Sridharan and Lee, 1972), the metal ions hydrolyze, resulting in the formation of copious metal–humic flocs. For a more detailed discussion of the precipitation process refer to Ong et al. (1970).

In no other aspect of humic studies have so many approaches and techniques been attempted to establish quantitative chemical reactions and associations than between metals and humic substances. It would appear that the older and tried complexation methods of titration (Gamble, 1970; van Dijk, 1971; Stevenson et al., 1973) and the ion-exchange method of Schubert (1948), as presented in Martell and Calvin (1952), (Schnitzer and Hansen, 1970; Ardakani and Stevenson, 1972; MacCarthy and Mark, 1977), lead to as meaningful and comparable values as newer methods. The newer methods such as equilibrium dialysis (Truitt and Weber, 1981; Saar and Weber, 1983), chelating ion-exchange resin (Figura and McDuffle, 1980), volumetric methods (Wilson et al., 1980; Weber, 1982) fluorescence quenching (Saar and Weber, 1980), ultrafiltration (Sposito et al., 1979), competing-ligand differential spectroscopy (Tuschall and Brezonik, 1983), gel chromatography (Ghassemi and Christman, 1968), and electron paramagnetic resonance spectroscopy (Senesi and Schnitzer, 1977) all have been especially promising and may be very useful under certain environmental conditions, but all seem to be limited in various aspects (Saar and Weber, 1982).

There is no doubt that metal complexation by humic substances is significant in the natural environment, but for any given stream the relative importance of organic to inorganic metal binding can be determined only after a thorough evaluation of stream chemistry. Mantoura et al. (1978) included complexation by humic substances in a calculation of metal speciation in an average river water. Of all the metals considered, they concluded that only copper would be significantly complexed by fulvic acid.

The competitive metal complexation of stream humic substances does not follow the theoretical Irving–Williams stability series; therefore, the relative order of complexation and the magnitude of the stability constant are variable. All the methods of stability-constant determination are limited in some way and often yield inconsistent values between and among methods (Saar and Weber, 1982).

Association with Other Organic Solutes

The size and shape of stream and soil humic substances vary with the intra- and intermolecular associations of humic substances themselves (Wershaw and Pinckney, 1971; Thurman et al., 1982). Because humic substances engage in self-association, it is not surprising that stream humic substances can interact with other organic solutes. This association may occur by a variety of mechanisms including hydrogen bonding, cation and anion exchange, hydrophobic adsorption, and partitioning within the humic molecules; all are considered to be "reversible equilibrium binding" mechanisms. It is possible that one or more of these types of associations may exist at any one time between humic substances and organic solutes. The association of stream humic substances with DDT (Porrier et al., 1972; Carter and Suffet, 1982), with polycyclic aromatic hydrocarbons (PAHs) (Gjessing and Berglind, 1981), and with petroleum hydrocarbons (Van Vleet and Quinn, 1977) markedly increased the apparent water solubility of these compounds. In none of these experiments have the stream humic substances or associations been well characterized, but the associations have been assumed to occur by hydrophobic interactions. Other investigations have shown a strong association between soil humic substances and DDT (Wershaw et al., 1969), dialkyl phthalates (Ogner and Schnitzer, 1970b; Matsuda and Schnitzer, 1973), and cholesterol (Hassett and Anderson, 1979). These high "binding capacities" are not believed to be representative of stream humic substances nor may the associations occur by the same mechanisms.

The "binding" or association between stream humic substances and other organic solutes depends upon the mechanism of binding, the nature of the organic humic substances, pH, concentration of humic substances, ionic strength, metal complexation, and other undetermined factors. All these factors can be readily understood to influence the polarity, charge, and hydrophobicity of humic substances and solutes as well as the formation of and the competition for binding sites on humic substances. Most of these aspects of "solute binding" by humic substances, although poorly defined, can be readily understood to apply to a reversible equilibrium process. Stream humic substances are postulated to have lower "binding capacities" for hydrophobic pollutants and other solutes than soil humic substances, because they are smaller in size, lower in molecular weight, and more polar, which tends to give them more of a true solution than micellar character in water. Such characteristics should hinder hydrophobic interactions and limit them to surface sorption. The larger, higher-molecular-weight, and less-charged soil humic and fulvic acids tend to form more micelle-like, colloidal moieties (especially humic acids) than true solution. Such micelles would be expected to have a much greater "binding capacity" for hydrophobic pollutants. The binding mechanism has also been suggested to be partitioning rather than surface sorption (Chiou et al., 1979). The supporting data for this conclusion (Chiou et al., 1983) are linearity of the sorption isotherm to high solute concentrations in water, the low-heat effect for solute sorption, and the lack of solute competition in soil sorption. The pH, ionic strength, and extent of metal complexation should have direct effects upon size, shape, and charge of the humic molecules, thus also having a marked effect on organic solute "binding capacity."

The reactivity, fate, and distribution of bound solutes are certainly changed by association with stream humic substances. The rate of photolysis of certain organic compounds (Zepp et al., 1981a,b), the rate of volatilization of polychlorinated biphenyls (Griffin and Chian, 1980), the bioaccumulation of polynuclear aromatic hydrocarbons in fish (Leversee, 1981), the rate of humic acid induced acid–base catalysis (Perdue, 1983), and the rate of microbiological decomposition are some specific examples. The octyl ester of 2,4-D (2,4 DOE) was predicted by theoretical and mathematical models and found by experimentation to be resistant to base hydrolysis when bound to humic substances (Perdue, 1983). The same model predicted the humic acid catalyzed hydrolysis of atrazine as demonstrated by Li and Felbeck (1972).

The factors for solute binding by humic substances are altogether different if the binding mechanism is covalent bonding which may be considered "ultimate" binding, an essentially irreversible process. Bollag (1983) has shown that such a binding process is operable in the incorporation of amines, phenols, and xenobiotic compounds into humic substances and other polymers by abiotic and enzymatic cross-coupling reactions. If compounds are incorporated by this mechanism, they lose their identity as a specific compound, and their fate is entirely associated with the breakdown, decomposition, or degradation of stream humic substances.

Biological Aspects

Stream humic substances are well known for their general resistance to both microbiological and chemical breakdown (Waksman, 1938; Shapiro, 1957; Midwood and Felbeck, 1968). The humic substances content of natural samples containing bacteria and fungi in excess of 10^8 organisms per milliliter of water has been observed to be essentially constant for months. BOD values are typically near zero (Black and Christman, 1963a; Gjessing, 1976; Beck et al., 1974). Minor changes within water samples may occur due to degradation of nonhumic constituents, sedimentation of colloidal matter, or flocculation and aggregation of humic substances upon standing in the presence of dissolved metals. Additional evidence of chemical stability and of the dominance of strong covalent bonding in aquatic humic substances rather than weak hydrogen and hydropholic bonding is that 2 hours of oxidation by refluxing with acid permanganate does not completely destroy humic materials (Shapiro, 1957).

There is evidence that, after long periods of time or with acclimated cultures of organisms where humic substances are the sole sources of carbon, a small fraction of humic substances are slowly degraded (Gjessing, 1976). It has also been observed that groundwater humic substances, after long periods of burial, are gradually depleted of carbohydrate and nitrogen moieties (Lytle and Perdue, 1981; Sweet and Perdue, 1982).

From a practical standpoint, stream fulvic acids which comprise over 90% of the stream humic substances are not an important food source for aquatic organisms, but all stream humic and fulvic acids are positive influences on biological growth in respect to phosphorus and nitrogen nutrient cycling, trace metal availability, and limiting potential metal toxicity.

Phosphorus and nitrogen occur in stream humic substances as structural components and as functional groups. The nitrogen and phosphorus contents of stream sediment humic and fulvic acids are much higher than their instream counterparts. Therefore, any process or condition that would favor the release of sediment humic substances into the dissolved phase would potentially favor biological processes. Such sediment release and cycling processes have been demonstrated (Koenings and Hooper, 1976; Stewart and Wetzel, 1981b). Because nitrogen and phosphorus are usually two of the most limiting nutrients to microbial and algal growth, even the smallest amount of nitrogen or phosphorus mobilization would contribute to the nutrient pool of aquatic plants and animals. The nitrogen in stream humic substances may be released as soluble amino acids (De Haan and De Boer, 1978; De Haan et al., 1981a) or as mineralized nitrogen. The phosphorus may be released as colloidally associated phosphorus (Stewart and Wetzel, 1981b), as mineralized or hydrolyzed phosphorus from organic phosphate groups, as inosital phosphate (Hong and Yamane, 1981), or as some low-molecular-weight organic constituent that is readily assimilated into the natural soluble organic phosphorus pool (Christman and Minear, 1971; Gjessing, 1976).

SUMMARY AND CONCLUSIONS

Prior to the last decade, the majority of research on humic substances was conducted in the area of soil science. Research on aquatic humic substances developed slowly until the 1970s. The recent rapid development of environmental and health-related sciences has catapulted humic substances into the forefront for a number of reasons: (1) humic substances are now known to be ubiquitous to all environments including soils, groundwaters, streams, estuaries, and oceans; and (2) humic substances are very reactive and important participants in many geochemical reactions and processes. Stream humic substances are intimately related to all aspects of water quality, environmental pollution, and the well-being of biota, including humans.

The major emphasis of the recent resurgence in humic studies has been in a broad spectrum of geochemistry and applied environmental sciences. Research on stream humic substances is now conducted by biologists, engineers, geologists, medical scientists, chemists, ecologists, mineralogists, and many other professionals in addition to soil scientists and hydrologists. Their research has demonstrated that stream humic substances can complex metal ions, associate with other organic solutes, dissolve clay minerals, influence the cation–anion balance, form trihalomethanes upon chlorination, and impart acidity and coloration to stream waters. Thus, stream humic substances markedly affect not only the concentration, fate, distribution, and reactivity of many organic and inorganic stream constituents, but the biology and microbiology of these constituents as well.

Several factors that have contributed to the resurgence in basic research on stream humic substances are advances in XAD-8 resin technology for humic isolation, innovations and new instrumental capabilities in organic chemistry (especially proton and ^{13}C NMR), and the obvious need for basic research to better interpret various aspects of applied research. XAD-8 resin technology has enabled the preparative quantitative concentration and isolation of stream humic substances which are free from admixture with other stream organic and inorganic constituents. Characterizations of stream humic substances can now be made without the limitations and interferences associated with past characterizations of variable and undetermined inclusions or mixtures of nonhumic–organic stream constituents.

The new findings and instrumental characterizations have dramatically changed the old concepts of the formation and structure of stream humic substances. ^{13}C NMR data have established these substances to be primarily aliphatic, not aromatic in nature. Stream fulvic acids contain only 16% aromatic carbons. Low-angle X-ray scattering has shown stream fulvic acids and humic acids to have radii of gyration of 5–7 Å (molecular weight of approximately 1000) and 9 and larger Å (molecular weight of approximately 2000), respectively. This is almost an order of magnitude lower in molecular weight than previously assumed. The ^{14}C age of stream humic substances is 20 years or less; this is in extreme contrast with the previously assumed age of several hundreds of years. The new findings and characterizations are very dramatic; however, the exact sources, the mechanism(s) of formation, and the exact structure(s) of stream humic substances are unknown.

The new findings also indicate that stream humic substances are very different from soil, groundwater, estuarine, or marine humic substances.

Because stream humic substances are much younger, less aromatic, lower in molecular weight, and less intense in color per carbon atom than soil humic substances, it is likely that soils are but a minor source of stream humic substances. The available data suggest a gradual sequential change or trend in humic substances to lower molecular weight, less intensity in color per carbon atom, and less aromaticity with the progression from soils, to streams, to groundwaters. Although humic substances from all environments have many characteristics in common, the relative abundance of various features such as aromaticity, molecular weight, color per carbon atom, functional group ratios, elemental contents, and fulvic/humic acid ratios differ from one environment (e.g., stream water) to another (e.g., ocean water). It would also appear that stream humic substances are distinct in many ways from humic substances in other environments and it may be suggested that the sources and/or humification process may also be unique in stream waters.

Stream humic substances are similar to other aquatic habitats in that fulvic acids predominate by a factor of approximately 9 : 1 over humic acids. The average concentration of stream humic substances of 2.2 mg C/L is higher than for most other aquatic environments, but this concentration is sometimes attained in the other aquatic environments.

The fulvic/humic acid separation appears to be a meaningful, valid, and useful separation of stream humic substances. Stream humic acids are always higher in nitrogen content, aromatic carbons, methoxyl and phenolic reactive groups, higher in molecular weight, and more intense in color per carbon atom than stream fulvic acids.

It is expected that future research will continue to define the uniqueness of humic substances in the stream environment. Although enormous strides have been made in stream humic substances research in the past decade, the solution of and formulation of exact structure and molecular shapes remain in the distant future. Some of the eminent research needs for the future are the lability of phenolic groups, the effect of free-radical quenchings on the resolution of phenolic groups by ^{13}C NMR spectroscopy, the relative contribution of carbohydrates to ethers in the 60–95 ppm chemical shift in ^{13}C NMR spectra, the relative contribution of biotic and abiotic processes in the formation of humic substances, the number of monomer precursors of humic substances, the exact mechanisms of the humification process in different environments, the effect of temperature and climate on the humification process, the extent of photon (sunlight) induced free-radical decomposition of humic substances, and so many others. Although the list is lengthy, the rapid understanding and progress made in the recent past give hope and encouragement that with the continued focus on research activity, with the present analytical and instrumental excellence, and with the intense desire of talented researchers, there is great promise in the continued resolution of the mysteries of stream humic substances.

ACKNOWLEDGMENTS

The research presented in this chapter was partially supported by Interagency Agreement No. AD-14-F-1-559 between the U.S. Environmental Protection Agency (EPA) and the U.S. Geological Survey. The research cooperation of Dr. Fred Kopfler of the EPA Health Effects Research Laboratory in Cincinnati, Ohio is appreciated.

The solid-state ^{13}C NMR spectra presented in this paper were obtained on equipment at the Colorado State University Regional NMR Center, funded by National Science Foundation Grant No. CHE-8208821. The special assistance of Dr. Gary Maciel, Professor of Chemistry at Colorado State University, is appreciated.

CHAPTER EIGHT

Geochemistry of Humic Substances in Estuarine Environments

LAWRENCE M. MAYER

ABSTRACT

The concentrations and nature of humic substances in estuarine water and sediments are similar to humic substances of the terrestrial and marine environments, and often appear to consist of simple mixtures of these two endmembers. The bulk of aquatic humic substances derives from input by rivers. Mixing of river water with seawater causes several chemical changes in riverine humic substances, including increasing saturation of ion-exchange sites with alkaline earth ions, contraction of the molecules, and aggregation of high-molecular-weight materials, including most of the humic acid fraction. Aggregation may induce retention of some humic substances in estuarine sediments, but affects only a small proportion of total humic substance pool. The bulk of the riverine humic substances passes through the estuary to the sea, undergoing little aggregation or metabolism during passage. Humification processes should be intense in estuaries compared to the open ocean, but have been observed in only a few studies. Humic substance complexation of trace metals should decrease with increasing salinity if riverine humic substances were the only ligands of importance; however, sparse field data suggest that estuarine- or marine-derived ligands are more important than those of riverine origin. It is likely that humic substance complexation of organic pollutants is important in pollutant cycling, but the estuarine chemistry of this interaction has received little study.

INTRODUCTION

The estuarine environment is a transitional zone between the terrestrial and marine environments—not only in a geographic sense (Fig. 1) but also with respect to many biological and chemical characteristics. The water and sediments found in the estuarine zone derive primarily from inputs from surrounding environments, and it is thus not surprising that many of the characteristics of humic substances found within estuaries should also reflect these various sources. Humic substances serve in a similar role in estuaries as in other environments—with respect to features such as a repository for biologically refractory organic matter, pollutant scavenging, and so on. However, the estuarine zone is unique in several respects. First, it is a zone in which materials contained within any introduced water mass will soon encounter a change in salinity, which may have a strong effect on the chemical reactivity of humic substances. Second, it is a zone of higher primary productivity than most of the marine environment, thereby provid-

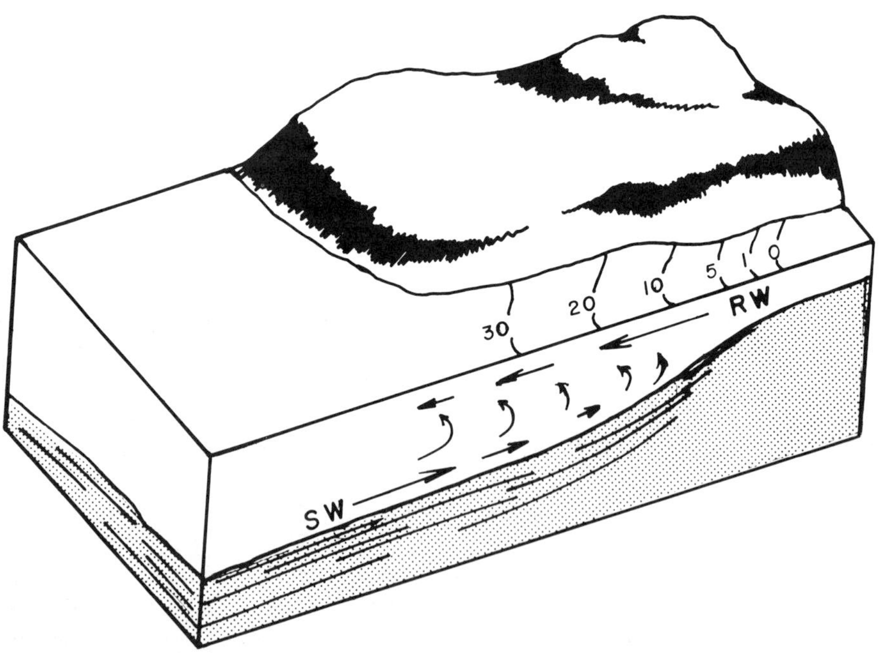

FIGURE 1. Diagrammatic view of an estuary, showing typical, nontidal water circulation and resultant salinity distribution (parts per thousand).

ing abundant precursor organic materials for the humification process. Third, it is a zone of shallow water depths relative to the rest of the ocean, which has important implications for the types of indigenous plant life and resultant humic precursors, as well as for the fraction of organic matter produced in the water column which reaches the sediments.

Often the study of humic substances in estuaries has been undertaken because estuaries are more easily sampled than the open ocean, and not because of the unique aspects of estuaries per se. It is only in the past few years that systematic studies of humic materials, traversing the salinity gradient, have been carried out. Techniques used in most studies have been those borrowed from the classical fields of soil humus studies; it seems likely that in time the techniques will evolve in response to the unique chemical processes that humic materials undergo in the estuarine zone. Quantitatively considered, the literature on humic materials in estuaries lacks the extent of the geographical or topical coverage of the soils literature. This paucity of data causes many conclusions drawn so far to be quite tentative and in need of corroboration.

CONCENTRATIONS AND NATURE OF ESTUARINE HUMIC SUBSTANCES

It is difficult to define humic substances in an unambiguous manner. Operational definitions commonly in use (reviewed in Chapter 1) derive from chemical properties; these properties can in turn be associated with a wide variety of organic structures. It is my opinion that the essential characteristic of humic substances derives from the fact that they originate from normal biochemical compounds which are then altered by processes other than normal biochemical reactions to result in a product that is biologically refractory. If these alterations consist of condensation reactions the result is a heterogeneous set of macromolecules (Kononova, 1966) or "heterocondensates." Some humic substances may consist instead of partial degradation products of biogenic molecules. That most such molecules tend to be of relatively high molecular weight, are acidic, and are generally colored is of geochemical interest, but these properties are not necessarily indicative of a refractory character. Care must be taken to prevent definitions of humic substances based on these properties from obscuring the study of the nature and organic geochemical pathways of refractory humic substances. Certainly a major problem in the discussion of refractory humic substances in estuaries is that there is presently no analytical criterion for determining their existence, much less quantifying their abundance.

Techniques used to separate humic substances from either a dissolved or sedimentary matrix rely on their acid–base solubility properties. In some of the separation techniques, it is likely that some nonhumic material accompanies the humic substances during the purification processes—to an extent depending on the sample and the technique. Relatively little effort has been made to eliminate nonhumic substances from isolates. Likewise, it seems unlikely that most purification steps used to date are completely efficient in isolating refractory humic substances. The isolation of aquatic humic substances is discussed by Aiken in Chapter 14.

The yield of humic substances obtained in the various isolation techniques varies considerably. The two most common methods for quantifying yields have been elemental (carbon and nitrogen) and spectral (usually absorbance in the 250–520 nm range) analyses. While each method has its advantages and disadvantages, clearly a serious problem with the latter is that absorptivities of humic substances are quite variable. In this chapter, I have deemphasized quantitative conclusions regarding humic substance concentrations (expressed as weight carbon or weight organic matter) resulting from the spectral techniques such as that of Martin and Pierce (1971).

Aquatic Humic Substances

The isolation of humic substances from estuarine waters has usually been performed by acidification of the water sample followed by either filtration, to yield humic acids only, or adsorption onto a resin such as XAD-2, which appears to allow recovery of both the humic and fulvic acid fractions (Stuermer and Harvey, 1977a). Table 1 provides a listing of determinations of humic and fulvic acids in estuarine waters by a variety of investigators, along with the techniques employed. The concentrations of dissolved humic and fulvic acids found typically range from undetectable to less than 2 mg C/L, with most values in the tens of μg C/L for the humic acid fraction and hundreds of μg C/L for the fulvic fraction. Higher values are usually found at lower salinities. However, it must be stressed that few analyses are available; most of these are from estuaries of the northeastern United States and a different isolation technique has been used in virtually every study cited.

Nevertheless, it seems reasonable to infer that aquatic humic substance concentrations in estuaries are intermediate between those of rivers and those of the open ocean, as would be expected from a mixture of river water and seawater. The concentrations of humic acids at intermediate salinities do not correspond, in a linear fashion, to the relative proportions of river water and seawater (Fig. 2), primarily because of a removal from solution of a portion of the riverine contribution. This behavior is in contrast to that of total dissolved organic carbon (DOC) which generally does show linear mixing lines when plotted versus salinity (Sholkovitz et al., 1978; Moore et al., 1979; Laane, 1980; Fox, 1983a; Mantoura and Woodward, 1983).

The chemical character of dissolved humic substances in estuaries also reflects its mixed origin, with a number of parameters exhibiting values intermediate between the riverine and oceanic endmembers. These properties include UV–visible absorptivities, C/N ratios, total acidity, molecular weight (Preston, 1979), amino acid content (Fox, 1981), and perhaps $^{13}C/^{12}C$

TABLE 1. Concentrations of Aquatic Humic Substances in Estuaries[a]

Estuary	Isolation Method	Humic Substance Concentrations (mg C/L)	DOC (%)	Reference
Elemental Analysis				
Various middle Atlantic estuaries—United States	Acidification to pH 2, Whatman GFC filtered	HA: 0–1.8	0–20	Fox (1983a)
Delaware Bay	Acidification to pH 2, glass fiber filtered	HA: 0.05–0.29	7–20	Sharp et al. (1982)
Vineyard Sound, Massachusetts	Acidification to pH 2, adsorbed onto XAD-2	HA: 0.027–0.033 FA: 0.15–0.17	0.6–0.7 2.9–3.8	Stuermer and Harvey (1977a)
Quisset Harbor, Massachusetts	Acidification to pH 2, adsorbed onto XAD-2	HA: 0.027–0.108 FA: 0.26–0.358	— —	Stuermer and Harvey (1977a)
Spectral Analysis				
Amazon, Brazil	Acidification to pH 1–1.5, 0.4-μm filtered (river water HA as spectral standard)	HA: 0.017–0.169	3–5(?)	Sholkovitz et al. (1978)
Connetquot, New York	Precipitation with acetic acid–isoamyl alcohol, Whatman GFC filtered (soil HA as spectral standard)	HA: 0–1.2	—	Hair and Bassett (1973)

[a] *HA = humic acid and FA = fulvic acid.*

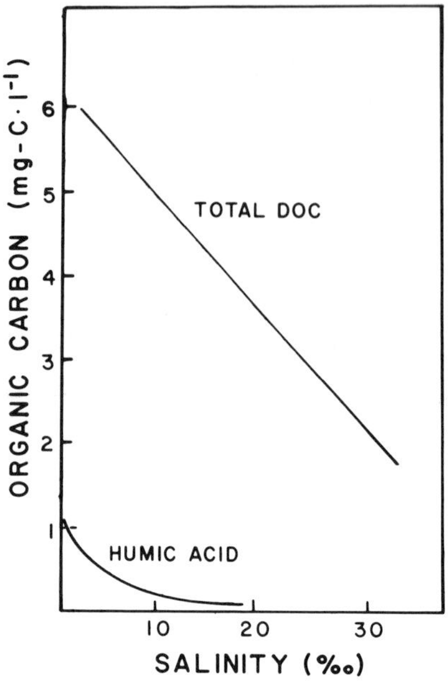

FIGURE 2. Schematic plot of typical distributions of total dissolved organic carbon (DOC) and humic acid carbon versus salinity.

isotope ratios (Stuermer and Harvey, 1974). Plots of these properties against salinity sometimes show nonlinearities related to selective removal or modification of certain components of the humic pool during estuarine mixing.

Sedimentary Humic Substances

Problems of intercomparability of data are similar for sedimentary and aquatic samples. Cronin and Morris (1982) discussed the large changes in the amounts and apparent nature of sedimentary humic substances resulting from variations in extraction technique. Factors showing the greatest variability among investigators include the pretreatment of samples for carbonate removal or lipid extraction, concentration and duration of the base extraction step, and the pH of the fulvic–humic acid separation step.

The concentration of humic and fulvic acids in unpolluted estuarine sediments tends to fall in the range of 10–68% of the total sedimentary organic carbon (Palacas et al., 1968; Brown et al., 1972; Huc and Durand, 1973; MacFarlane, 1978; Jones and Jordan, 1979), although values in calcium carbonate-rich sediments can be considerably lower (Palacas et al., 1968). This range of values is typical of the marine environment in general (Rashid and King, 1969; Huc and Durand, 1973; Nissenbaum, 1973) and is also similar to that found for lake sediments (Ishiwatari, 1966) and soils (Kononova, 1975). Sediments in shallow waters such as estuaries usually have higher organic carbon concentrations than those from deeper areas (Mayer et al., 1981;

Premuzic et al., 1982), so it may be presumed that humic substance concentrations are higher as well.

The ratios of humic to fulvic acids in estuarine and coastal sediments range from 0.4 to 3.4, the higher values being associated with areas or sediments having a terrestrial influence (Palacas et al., 1968; Brown et al., 1972; Huc and Durand, 1973; Pelet and Debyser, 1977; MacFarlane, 1978). These values are also consistent with those from other marine and terrestrial environments (Ishiwatari, 1966; Kononova, 1975; Stuermer et al., 1978; Cronin and Morris, 1982). Other parameters measured on coastal humic substances, such as elemental composition, spectral properties, organic components, stable isotope ratios, or ^{14}C ages (Pelet and Debyser, 1977; Stuermer et al., 1978; Benoit et al., 1979; Nissenbaum, 1979) are consistent with terrestrial or marine humic compounds, or a mixture of these two endmembers.

PASSAGE OF RIVERINE HUMIC SUBSTANCES THROUGH ESTUARIES

Allusion has been made above to changes that occur in the humic substances introduced to estuaries by the riverine source. This section reviews the chemistry of these changes, and considers their effect on the delivery of riverine humic substances to the oceans. Because riverine humic substances derive from zones of low ionic strength, the rapid increase in both the types and concentration of dissolved salts upon estuarine mixing should have important effects on their ion-exchange properties, their conformations in solution, and their solubility.

Aquatic humic substances have a considerable ion-exchange capacity—typically 4–14 meq/gC—resulting primarily from ionized carboxyl and phenolic hydroxyl groups (Rashid and Prakash, 1972; Huizenga and Kester, 1979; Preston, 1979; Chapter 20 in this volume). The composition of exchangeable cations associated with these humic substances depends on the composition of the freshwater with which they are associated. Mantoura et al. (1978) demonstrated that in world average river water (Livingston, 1963), humic substances can be expected to have approximately 20% of their binding sites unassociated with a cation, while the remainder will be bound primarily to calcium, and to a lesser extent magnesium, ions. In a freshwater endmember of similar total dissolved solids content, but much lower concentrations of calcium and magnesium, they found 75% of the humic sites to be free, with 15% bound to calcium and 6% bound to magnesium. These calculations are consistent with the relatively high electrophoretic mobilities found for organic matter adsorbed to iron oxides in waters relatively low in alkaline earths as compared to waters with higher alkaline earth concentrations (Tipping and Cooke, 1982). Upon mixing with seawater, the increasing alkaline earth concentrations bind to a major fraction of the carboxylate sites, with magnesium becoming more abundant than calcium because of its higher concentration in seawater (Mantoura et al., 1978; Mantoura and Woodward, 1983). Whether or not the organic material becomes completely saturated with alkaline earth ions upon mixing with seawater is unclear. Using one set of stability constants, Mantoura et al. (1978) predicted virtually complete saturation, while use of another set of constants (Mantoura and Woodward, 1983) yielded a prediction of about one-third of the acidic sites dissociated. Electrophoretic work on particles with and without organic coatings is consistent with a retention of a significant amount of negative charge by natural organics, upon mixing with seawater (e.g., Loder and Liss, 1982). However, it is not clear whether the residual negative charge is due to carboxylate sites or some other type of site.

The conformation of dissolved humic substances has received little attention (Varney et al., 1983). The abundance of ionized groups at the pH and alkaline earth concentration of freshwaters will tend to cause macromolecules to stretch out in response to mutual charge repulsion of ionized groups on the same molecule. Reuter (1977) used a combination of viscosimetric and gel permeation chromatographic measurements to demonstrate that the size of aquatic humic substances is reduced upon entering estuarine waters. This reduction in size presumably results from reduction of charge repulsion due to some combination of complexation of carboxylate groups by alkaline earth ions and the reduction in electrical double-layer field strength by the high ionic strength of the estuarine waters. These conformational changes should be most important for the higher-molecular-weight fractions; riverine fulvic acids of molecular weight 500–1000 may not be amenable to large conformational changes.

Dissolved organic matter in rivers also undergoes a change in average size due to aggregation of some portion of its high-molecular-weight component. This aggregation was first recognized as a result of two sets of observations. First, laboratory experiments in which water containing terrigenous aquatic humic substances was mixed with seawater resulted in slow precipitation of brownish colloidal material (Moore and Maynard, 1929; Sieburth and Jensen, 1968). Second, studies of the UV and visible absorbances of coastal waters of northern Europe demonstrated a loss of river-derived colored organic matter (e.g., see Jerlov, 1955). Brown (1975) used a combination of ultrafiltration and absorbance measurements to show that the high-molecular-weight component of river organic matter was preferentially lost. The involvement of operationally defined humic substances in this aggregation process was suggested by the estuarine surveys of Hair and Bassett (1973), who found dissolved humic acids present at low salinities to be apparently replaced by particulate humic acids at higher salinities. In a series of papers, Sholkovitz ad co-workers (Sholkovitz, 1976; Eckert and Sholkovitz, 1976; Sholkovitz et al., 1978) used a combination of laboratory mixing experiments and field measurements of dissolved humic acids to demonstrate that riverine humic acids were indeed lost from solution during mixing with seawater.

More recently, Fox (1983a) and Sharp et al. (1982) have quantified the extent of humic acid removal using carbon measurements of the isolated humic acid fractions.

Two aspects of the aggregating organic matter in estuaries, then, are that it is at least partially humic and is at least partially of high molecular weight. Fox (1983b) examined more closely the degree of overlap of fractions defined as (1) precipitable by seawater (the SFR fraction), (2) the humic acid fraction (c.f. Table 1), and (3) the high-molecular-weight fraction, defined as that material filterable by a nominal 100,000 MW ultrafilter (the UFR fraction). These extracts were examined for elemental composition, hydrolyzable amino acid distribution, carbohydrate content, and NMR spectra. Some of these data are summarized in Table 2. In column A, it is seen that each of the extractions removed less than one-third of the riverine dissolved organic carbon. There were considerable similarities among the various extracts for the Mullica River but not the Broadkill. However, the potential overlap from the data of column A is reduced by the results of column B, in which generally less than one-half of the SFR and UFR fractions were shown to be composed of operationally defined humic acid. The proximate composition of the different extracts is seen in columns C and D. The discrepancy between hydrolyzable carbohydrate content and the carbohydrate content indicated by the NMR data led Fox (1981) to suggest that a significant portion of the carbohydrates, and indeed in the total acidity of these fractions, derives from acidic mucopolysaccharides.

It is clear from these data that although there is some overlap among the various fractions there are considerable differences as well. These differences may result in part from the particular operational definitions used to separate the fractions in Fox's study. For example, Sholkovitz (1976) and Sholkovitz and Copland (1981) have shown that the extent of humic acid precipitation from river water varies with the salt content, in estuarine mixing experiments, and with the pH, during acidification. It also seems reasonable to expect that the nature of the precipitated material would vary with changing salt or acidity. The degree of overlap to be expected among fractions precipitated in different manners should then depend considerably on the operational parameters used in their separation. Extension of the analytical approach used by Fox (1981) would be most valuable in determining the nature of the aggregation process.

Determinations of average molecular size of humic substances in estuaries are consistent with a preferential loss from the high-molecular-weight fraction. In all cases reported (Preston, 1979; Gillam and Riley, 1981), the high-molecular-weight fraction has been found to decrease with increasing salinity.

There is also some evidence to suggest that the colloidal material in estuaries is dominated by marine-derived organic matter rather than that of terrigenous origin (Sigleo et al. 1982; Zsolnay, 1979). However, it is unclear whether this trend is due to loss of terrigenous material because of aggrega-

TABLE 2. Proximate Analysis of Humic Acid (HA), Salt-Precipitable Fraction (SFR), and Ultrafiltered Fraction (UFR) of Extracts from the Mullica and Broadkill Rivers[a]

		A Extract Carbon	B HA Carbon in Extract	C H-CHO-C	D Proton Assignments from Amino Acid Analysis and NMR			
River	Extract	River DOC	Extract Total Carbon	Extract Total Carbon	Amino Acid	Carbohydrate	Aromatic	Unidentified
Mullica	HA	19	—	8	16	33	17	34
	SFR	16	26	2	7	46	21	26
	UFR	33	33	2	7	30	17	44
Broadkill	HA	18	—	10	19	57	18	8
	SFR	3	7	0	0	70	8	20
	UFR	25	59	9	7	62	17	8

[a] *Data from Fox (1983b). All proportions expressed as percentages. Column A gives HA, SFR, and UFR as proportion of dissolved organic carbon (DOC). Column B gives proportion of humic acid carbon in SFR and UFR fractions. Column C gives hydrolyzable carbohydrate carbon (H-CHO-C) as a proportion of extract carbon. Column D gives the proportions of compound classes in each fraction as determined from amino acid analysis and NMR spectra.*

tion or simply dilution of terrigenous colloidal material by marine organics produced *in situ*.

Fox (1983a) has shown that riverine humic acids, as defined by pH 2 acidification followed by filtration, can behave conservatively in mixing experiments combining certain river waters with seawater; that is, they show no evident aggregation. This surprising result was found for rivers for which the field data indicated a loss of dissolved humic acid with salinity. The Broadkill was one of these rivers; it is notable that there was very little humic acid in the SFR fraction (Table 2). An important implication of this result is that salt-induced aggregation may not be the mechanism by which all riverine humic acids are lost from estuarine waters, and that other chemical or perhaps biological reactions are responsible.

A considerable stimulus to the study of humic substance aggregation has derived from interest in the simultaneous aggregation of iron colloids and other trace metals. Organic material has been established as a peptizing agent for the stabilization of iron colloids in river waters (Boyle et al., 1977; Moore et al., 1979). Although humic substances from diverse environments have been shown to be capable of this type of peptization (Ong and Bisque, 1968; Tipping and Cooke, 1982), they have not been shown conclusively to be responsible. Upon mixing with seawater, iron colloids aggregate with a rapidity much greater than can be accounted for by their concentration. Mayer (1982b) has suggested that the organic matter aggregating with the iron colloids enhances the latter's aggregation kinetics. It seems likely that there is an incomplete overlap between the organic fraction associated with iron colloids in river water and the humic substances which aggregate upon mixing with seawater. Evidence for this nonoverlap comes from the differing extents of iron colloid and humic acid aggregation in response to salinity, as well as the different kinetics of aggregation of the two substances (Eckert and Sholkovitz, 1976; Fox, 1981; Mayer, 1982b). The precipitation of iron colloids by acidification to pH 2 may be a reaction in which riverine iron colloids peptized by humic or some other organic substances behave like humic acids, or it may represent a co-flocculation of previously separated humic acids and iron colloids. For example, Ghassemi and Christman (1968) showed iron and colored organic material to be separated in Sephadex gel filtration profiles at pH 7.5 but coincident in the same water acidified to pH 5.5.

The mechanism(s) of the aggregation process induced by seawater is not well understood. The vastly greater enhancement of humate aggregation by alkaline earth ions relative to sodium ions (Eckert and Sholkovitz, 1976; Preston, 1979) indicates that the reaction is not simply one of electrolyte-induced reduction of the electrical double layer thickness followed by van der Waals coagulation (Boyle et al., 1977). Rather, the high affinity of humic carboxylate groups for divalent ions (e.g., Mantoura et al., 1978) suggests a strong role of charge neutralization, converting relatively hydrophilic colloids or molecules into relatively hydrophobic ones (Ong and Bisque, 1968;

Eckert and Sholkovitz, 1976). Subsequent to this charge neutralization, van der Waals coagulation may be the driving force responsible for precipitation. Ion–dipole interactions (Theng, 1979), in which calcium or magnesium ions bridge and connect functional groups such as carboxylates, may also play a role. Simple precipitation of insoluble calcium and magnesium humates (Boyle et al., 1977; Preston, 1979) is perhaps an equivalent process. Hydrophobic interactions between the organic materials do not appear to be important (Mayer, 1982b; Mantoura and Woodward, 1983).

The kinetics of the aggregation reaction have been found to be rapid for humic acids (Fox, 1981) and organic carbon (Mayer, 1982b). Aggregates form to a size retained by 0.5–1.2 μm filtration within 1 hour. Over time spans of hours to days, the aggregation reaction can continue to the point where visible aggregates form and begin to settle from suspension (Sieburth and Jensen, 1968; Höpner and Orliczek, 1978). In the case of iron, the kinetics of this slow, continuing reaction depend on the turbulence of the suspension (Mayer, 1982b).

Aggregation of dissolved humic substances can also occur with particulate materials in the estuarine water column. Preston and Riley (1982) showed that the adsorption of riverine humic substances onto kaolinite, montmorillonite, and illite increased with increasing salinity and dissolved humic substance concentration. Adsorption increased in the order kaolinite $<$ illite $<$ montmorillonite, which they ascribed to increasing cation-exchange capacity of the clays. They found considerable quantitative differences between the extent of adsorption of riverine versus extracted sedimentary humic substances, indicating the importance of using materials of proper origin in experiments of this type.

Studies of the quantitative extent of the aggregation of humic substances during estuarine mixing have shown that only a minor portion of the total DOC is so affected (Sholkovitz et al., 1978; Fox, 1983a), resulting in linear plots of DOC versus salinity (Fig. 2). Even estuaries with high suspended particulate loads show little or no loss of DOC from the water column (Mantoura and Woodward, 1983). If riverine DOC is composed of one-third to one-half humic substances (Thurman and Malcolm, 1981; Chapter 7 in this book), then only a small portion of even this component would be expected to be lost from solution. Systematic surveys of the concentration of total dissolved humic substances—humic acids plus fulvic acids—as a function of salinity, using, for example, an extraction method such as that of Mantoura and Riley (1975), have not been reported. However, if fluorescence or UV absorbance measurements can be used as an index of humic substances, profiles of these parameters generally show close to conservative mixing profiles with salinity (Fig. 3) in estuaries (Zimmerman and Rommets, 1974; Postma et al., 1976; Dorsch and Bidleman, 1982; Willey and Atkinson, 1982; Carlson and Mayer, 1983), although not always in larger mixing basins such as the Baltic Sea (Brown, 1977; Kalle, 1966).

Another form of removal of riverine humic substances from solution is indicated by enrichment of phenolic materials in the sea surface microlayer

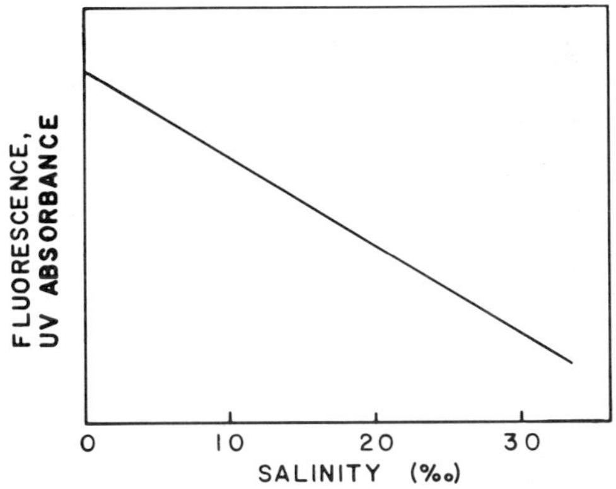

FIGURE 3. Schematic plot of fluorescence and ultraviolet absorbance intensities commonly observed in estuaries.

of estuaries (Carlson and Mayer, 1980). This enrichment probably results from the well-known surface activity of humic substances. Because of the miniscule volume contained within the surface microlayer, however, this exsolution will likely have only an insignificant effect on humic substance flux through an estuary. Evidence for this contention is the linear mixing behavior, versus salinity, of dissolved surface-active materials (Hunter, 1983).

RETENTION OF RIVERINE HUMIC SUBSTANCES IN ESTUARINE SEDIMENTS

It has not been established that the material which aggregates during estuarine mixing is lost from the water column. Mayer (1982a) and Wilke and Dayal (1982) showed that little of the iron aggregated during estuarine mixing is removed by gravity or suspended sediment scavenging; this conclusion may also apply to organic aggregates formed, assuming they are either associated with the iron colloids or are also of low density and therefore not amenable to gravitational settling (Prakash, 1971).

Considerable organic matter of terrigenous origin is found in estuarine sediments (Sackett and Thompson, 1963; Shultz and Calder, 1976; Pocklington and Leonard, 1979). Mayer (1982a) showed that an iron enrichment in estuarine sediments could accompany a riverine organic matter enrichment, and that the C/Fe ratio of the excess iron and organic matter was similar to that found in aggregates formed by mixing river water and seawater. These findings are consistent with input of iron–humic aggregates into estuarine sediments. However, the terrigenous organic material in estuarine sediments may also result from nonhumic particulate organic matter, which makes up a large fraction of the total organic load of many rivers (Meybeck, 1982).

Another mechanism for retention of riverine humic material in estuarine sediments is biodeposition (Prakash, 1971) by either pelagic or benthic animals. Incze et al. (1982) and Stephenson and Lyon (1982) have shown upper estuarine filter-feeding bivalves to incorporate terrigenous organic matter into their biomass; whether this material is humic in nature is not known, but it seems reasonable that flocculated humic substances may be caught up during filter-feeding and delivered to the sediments as part of the animals' pseudofeces. Cloern (1982) has shown that benthic filter-feeding can process a significant portion of the water column in an estuary such as San Francisco Bay.

HUMIFICATION

Estuaries seem a probable zone to observe humification in the marine environment, for the following reasons. First, the rate and extent of humification are likely related, in some positive manner, to the concentration of precursor compounds. This relationship should hold regardless of the specific pathway(s) of humus formation (e.g., see Gagosian and Stuermer, 1977; Chapter 9 in this book). Any area of high organic production would thus qualify, and estuaries are demonstrably more productive than most areas of the ocean. Second, analytical and experimental work have shown that phenolic compounds are particularly conducive to humification reactions (Flaig et al., 1975). Estuaries represent an area of relatively high phenolic input in that they (1) receive terrestrial, aromatic compounds in river runoff, and (2) often contain more productive local sources of marine-derived aromatic compounds than most marine environments. These local sources include vascular plants that use lignin as a structural component, such as seagrasses or mangroves, and macroalgae, such as the *Phaeophyta,* that make a variety of phenolic compounds. Organic production in noncoastal oceanic areas, on the other hand, is dominated by planktonic organisms that produce relatively little aromatic material.

Plant detritus provide sites of obviously high local concentrations of organic matter. Humification of macrophyte detritus has been observed, in laboratory experiments, by Rashid and Prakash (1972) and Rice (1982). Accompanying the humification, the nonprotein nitrogen of the detritus also increased (Rice, 1982), suggesting humification rather than microbial growth as the explanation for commonly observed nitrogen enrichment in decaying detritus (Newell, 1965; Harrison and Mann, 1975). This process is apparently similar to nitrogen increases observed during decay of terrestrial detritus (Suberkropp et al., 1976). Macrophytic debris may serve, then, as important sites of humification in estuaries.

Exudates of *Phaeophyte* macroalgae, which are rich in phenolic compounds, have been found to humify quite readily in solution (Sieburth and Jensen, 1969; Rashid and Prakash, 1972). The humic substances thus formed condense to the extent that they become susceptible to eventual precipitation (Sieburth and Jensen, 1968, 1969). That such humic material is present in a high-latitude estuary was indicated by Sieburth and Jensen (1968); however, there has been no quantification of its importance to date. Exudation of phenolic materials by the *Phaeophyta* is likely to be of importance primarily during summer months (Carlson and Mayer, 1983), indicating a seasonality of precursor availability for this type of reaction.

Slow incorporation of a variety of amino acids and sugars into natural high-molecular-weight material, in sterile incubations, has been observed by Carlson and co-workers (unpublished data) in estuarine waters. These incubations were carried out in prefiltered waters using 10–100 nanomolar (nM) spikes of ^{14}C-labeled monomers. Incorporation of as much as 15% over a period of several weeks was observed, using gel filtration chromatography to differentiate free and complexed pools. Accompanying this uptake by preformed high-molecular-weight material was a tendency toward loss of label onto container walls and filters, implying a high particle reactivity for the "humic" substances thus formed. Such a high particle reactivity could explain the relatively low concentrations of dissolved humic substances in the oceans.

Humification in estuarine sediments has been postulated to occur in a similar fashion, with small organic precursors condensing to form fulvic acids which then further condense to humic acids and thence to kerogen (Nissenbaum et al., 1971). There is some evidence consistent with this pathway. Stable carbon isotope data of extracted humic compounds from estuarine sediments indicate local planktonic source material rather than allochthonous terrigenous humic inputs (Nissenbaum and Kaplan, 1971). Christensen and Blackburn (1982) demonstrated an incorporation of radiolabeled acetate into dissolved high-molecular-weight compounds in sedimentary pore waters, which may represent an abiotic condensation. In addition, the proportion of high-molecular-weight dissolved organic carbon relative to total dissolved organic carbon in pore waters has been found to increase downcore (Nissenbaum et al., 1971; Krom and Sholkovitz, 1977). Evidence from the relative ratios of total humic and fulvic acids downcore is sparse, and equivocal in supporting this reaction sequence. Palacas et al. (1968) found either no change or an increase in the fulvic to humic acid ratio with depth in a Florida estuary, while Brown et al. (1972) found a decrease in the fulvic to humic acid ratio in Saanich Inlet. The relative importance of humification reactions in the pore water as compared to the solid phase has not been investigated.

INTERACTIONS WITH TRACE METAL IONS AND ORGANIC POLLUTANTS

The ability of humic substances to interact with trace metals and organic pollutants is well known; for details the reader is referred to reviews by Schnitzer and Khan (1972), Jackson et al. (1978), and Mantoura (1981). In this section only those aspects relevant to estuarine processes are discussed. Again regarding estuaries as mixing zones between the riverine and marine environments, it seems reasonable to discuss pollutant–humic substance interactions as they are affected by the transition between a low ionic strength medium with high concentrations of humic substances to one of high ionic stength and low concentrations of humic substances.

Metal Ions

Metal complexation by humic substances in estuaries has received little systematic attention. Only a few metals have been studied, with most of the emphasis on copper because of its obvious importance to plant production.

A number of studies have shown that increases in ionic strength cause a decrease in binding of trace metals by humic substances (Schnitzer and Hansen, 1970; Gamble et al., 1977; Stevenson, 1977). These studies have demonstrated the ionic strength effect with a variety of electrolytes, but it is clear that alkaline earth metals are especially effective in competing with trace metals (c.f. Stumm and Morgan, 1981, p. 376). This competition results from the strong attraction between the alkaline earths and carboxylate groups, which are likely the most important sites for metal complexation (Schnitzer and Khan, 1972). In addition to competition by seawater cations for the humic ligand, the increases in alkalinity and salinity in the transition from river water to seawater are accompanied by an increase in inorganic ligands capable of competing for many trace metals, such as chloride for softer metals (more polarizable, e.g., Cd) and hydroxide and carbonate for the harder metals (less polarizable, e.g., Mn) (Mantoura, 1981). Another potential cause of reduced binding of metals with increased salinity is the change in conformation of humic molecules. Bresnahan et al. (1978) found a dramatic decrease in copper binding sites on fulvic acid between pH 6 and pH 5, which they ascribed to conformational changes in the fulvic acid molecules resulting in decreased exposure of binding sites at lower pH. If the same types of conformational changes occur to riverine humic molecules during mixing with seawater, similar losses in binding capability might be expected.

Speciation models for metals in an estuarine mixing zone have been calculated for a number of metals assuming constant humic ligand concentration and selectivity coefficients (Mantoura et al., 1978). Examples of two typical plots are shown in Figure 4, in which it is seen that the influence of humic substances is most important in the low salinity zone. Incorporation of the observation that humic ligands are likely present in lower concentration at the higher salinity end of typical estuaries can be expected to magnify the decreasing importance of humic complexation with increasing salinity. Experimental evidence for this trend was found in electrophoresis work by Musani et al. (1980), who found dilution of seawater to increase the proportion of metals bound to added sedimentary humic acids. In addition, Willey

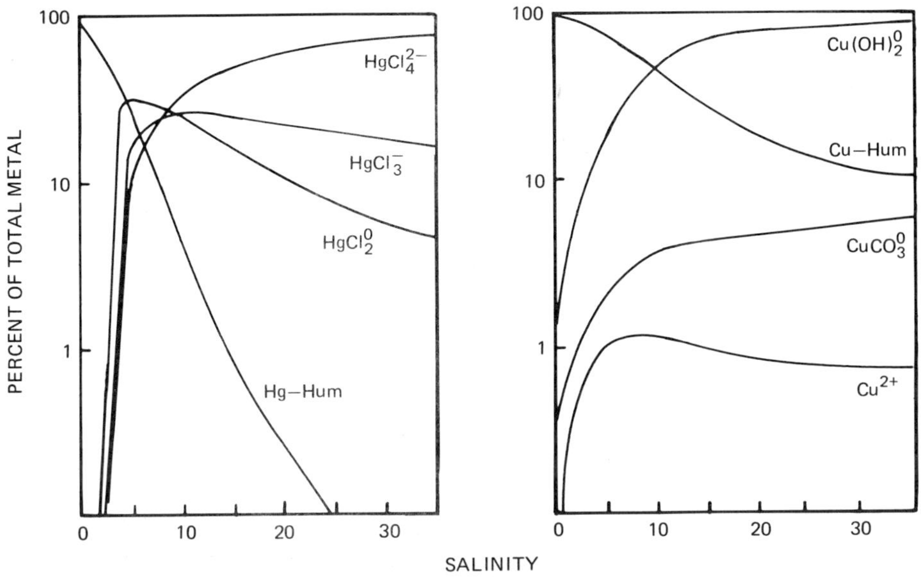

FIGURE 4. Calculated equilibrium speciation of mercury and copper during estuarine mixing of hypothetical river water with seawater. Hum = humic substance. Adapted from Mantoura et al. (1978).

(1984) observed, in both laboratory and field studies, an occasional slight increase in the fluorescence of riverine material during mixing with seawater. She hypothesized that one mechanism to explain this increase is the displacement of fluorescence-quenching metals by magnesium or calcium of seawater origin.

There are few field data with which to test these predictions. Smith (1976) examined the complexation capacity of various molecular weight fractions for copper along an estuarine transect. He found an increase in complexation capacity with increasing salinity for the low-molecular-weight fraction, accompanied by a decrease in capacity for the high-molecular-weight fraction. Kramer (1984) found a decrease in copper complexation capacity by 0.45-μm-filtered water samples of increasing salinity in the Scheldt estuary. This decrease was beyond that predicted by dilution of the river water end-member. Ultrafiltration experiments on waters of different salinities indicated that the loss of complexation capacity was due to loss from the filterable fraction of high-molecular-weight riverine ligands. Mayer et al. (1983) have also shown that the colloidal fraction of riverine ligands is particularly important in chromium complexation, resulting in an incorporation of complexed chromium into flocs upon mixing with seawater. Montgomery and Santiago (1978) found higher proportions of zinc and copper to be organically bound in coastal waters of Puerto Rico than in adjacent river waters. Srna et al. (1980) found maximum binding capacity for copper at intermediate salinities in a lagoonal system. In general, the field results are consistent with a loss of complexation ability due to dilution or flocculation of the colloidal fraction, but are also frequently indicative of production of organic ligands in the estuary or introduction from the shelf waters. Potential sources of these ligands include phytoplankton (Foster and Morris, 1971; Morris, 1974; McKnight and Morel, 1979; Wallace, 1982), macrophytes (Ragan et al., 1979; Sueur et al., 1982), and sedimentary pore waters (Sugai and Healy, 1978; Wallace, 1982). It is by no means established that the compounds capable of enhanced metal binding in estuaries are humic in nature. Instead, they may represent low-molecular-weight compounds exuded by algae to detoxify metals, as suggested by the importance of the low-molecular-weight fraction (Smith, 1976) and the association of copper–organic binding with times of relatively high phytoplankton productivity (Foster and Morris, 1971).

Prakash (1971) reviewed the potential importance of riverine humic materials in stimulation of coastal phytoplankton productivity, which he ascribed to complexation of toxic trace metals. However, the actual responsibility of the riverine material for such complexation has yet to be demonstrated.

Little work has been done to compare the nature of ligands in riverine, estuarine, and coastal waters. Preston (1979) found similar selectivity coefficients for copper with humic compounds isolated from different salinity regimes of the Tamar estuary. His results are made uncertain by lack of knowledge of the molecular weights of the compounds, but it appeared that the selectivity for copper decreased with increasing salinity. The stability constant data of Mantoura et al. (1978) also show similar selectivities for copper by aquatic humic substances from river, lake, and marine waters, which would imply that little variation in selectivities should be found along an estuarine salinity gradient.

An important consequence of humic complexation of metals in the riverine or estuarine environment is the subsequent protection of the riverborne metals from scavenging by resuspended sediment during passage of river water through an estuary (Yamazaki et al., 1980; Sholkovitz and Copland, 1981). The apparently high efficiency of passage of humic materials through the estuary should lead to correspondingly high efficiencies of passage for associated metals. The likely importance of the colloidal fraction of riverine ligands in metal complexation makes it important to determine the effect of flocculation processes on their transport through the estuary.

A further interaction between humic substances and metals, which has received increasing attention recently, is the organically catalyzed photoreduction of higher-valence metals to a lower-valence form (Anderson and Morel, 1982; Francko and Heath, 1982; Sunda et al., 1982; Collienne, 1983). Photoreduction effectively solubilizes the metals, rendering them more available to uptake by phytoplankton. Both iron and manganese have been found to be susceptible to this type of reaction. Because of the importance of iron and manganese oxides to the cycling of several other elements, such as some trace metals and phosphate, the implications of photoreduction extend to a variety of important geochemical cycles. The high concentrations of humic substances and colloidal iron in the estuarine zone, relative to the open ocean, make it important to investigate further this reaction.

Organic Pollutants

The interactions between humic substances and organic pollutants can be expected to lead to ecological effects analogous to those resulting from humic–metal interactions. Specifically, the associations formed will affect both the transport of organic pollutants through the estuarine zone and the availability of the organics to biota. In addition, organic pollutants are susceptible to biological and nonbiological degradation, the extent and nature of which will be affected by speciation.

The types of chemical interactions between humic substances and pollutant organics have been reviewed by Wershaw and Goldberg (1972) and Schnitzer and Khan (1972). A variety of chemical bonding mechanisms may lead to solubilization of otherwise insoluble organics, including complexation and adsorption of pollutants on humic macromolecules and peptization by humic substances of micellar aggregates of organics such as hydrocarbons. The type of interaction will depend on the nature of both the organic pollutant and the humic material, the concentration of each, and likely other environmental factors such as temperature, salinity, and so on.

The salinity dependence of solubilization will depend on the chemistry of the interaction. Carter and Suffet (1982) found that adding small amounts of electrolytes (calcium and sodium chloride) to humic solutions slightly increased their complexation of DDT, as did lowering the pH. They suggested that neutralization of the humic charge, perhaps in combination with salting out of the DDT, was responsible for enhancing the hydrophobic effect and consequently increasing association. Means and Wijayaratne (1982) found the opposite salinity dependence, over a range more closely duplicating estuarine salinities, for binding of atrazine and linuron to colloidal organic material from the Chesapeake Bay. However, it is not clear whether the salinity dependence they tested was determined using the same colloidal material in experiments at different salinities or colloidal material collected at different salinities. Different behaviors might be expected in these two cases, because possible loss of higher-molecular-weight colloids from the filterable fraction during estuarine mixing might lead to an apparent lower binding capacity. For example, Khan (1973) found humic acids to bind greater amounts of bipyridylium herbicides than did fulvic acids; Hassett and Anderson (1979) found cholesterol to complex preferentially to the highest-molecular-weight dissolved organic matter in river water; and Poirrier et al. (1972) found DDT to bind to colloidal material in river water which was readily removable by centrifugation. Boehm and Quinn (1973) observed complex relationships between ionic strength and the solubilization of several hydrocarbons, but at high ionic strength all compounds tested showed decreasing ability to pass a filter with increasing salinity. In their experiments, humic substances served as peptizing agents for micellar aggregations of hydrocarbons; the salinity dependence of peptization would not necessarily exhibit the same type of response to salinity as would complexation of organic monomers to humic substances. Therefore, the data are not yet available to generalize about the effect of salinity on organic complexation reactions in estuaries.

Humic substances will likely influence the passage of organic pollutants through estuaries. In the absence of humic substances, the increasing salinity encountered by riverborne organic pollutants should drive the dissolved–particulate partitioning of hydrophobic pollutants toward the particulate phase, as salt decreases the solubility and therefore enhances adsorption reactions. Complexation with humic substances should protect the pollutants from suspended sediment scavenging (Meyers and Quinn, 1973) in the same manner as for trace metals. To the extent that the pollutants associate with the fraction of humic substances that flocculate and settle out in the estuary, they will still be incorporated into the sediments, albeit via a different pathway. It should be reiterated, however, that these predictions are based on laboratory studies and need field testing.

METABOLISM OF HUMIC SUBSTANCES

The metabolism of humic substances is a subject that has received little attention, particularly so for estuarine systems. However, the mass balance calculations of Berner (1982) suggest that a major portion of the terrigenous organic carbon entering the marine environment must be metabolized in some manner; the possibility that some fraction of this metabolism occurs in estuaries must therefore be considered.

While humic substances are quite refractory of course with respect to metabolism, they have been shown to be amenable to some usage by bacteria, particularly if a labile "priming" compound or a cometabolite is available (De Haan, 1972b, 1977; Steinberg, 1977a; Rifai and Bertru, 1980). Estuaries may represent a zone of relatively high concentrations of such labile compounds as a result of their high productivity. However, there is no evidence at present to support this hypothesis. Carlson and Mayer (1983) found a loss of UV absorbance during estuarine mixing which was equal to that found in short-term mixing experiments, indicating no losses of UV absorbing material—and so, by inference, humic substances—in the estuary beyond that predicted from colloid aggregation. Fluorescence versus salinity plots, a similarly potential indicator of humic substance passage through estuaries, typically show similar conservative mixing curves (Zimmerman and Rommets, 1974; Postma et al., 1976a; Willey and Atkinson, 1982). The linear mixing curves for total DOC found in most estuaries are also consistent with little, if any, metabolism of any significant portion of this fraction.

On the other hand, the loss of the humic acid fraction during estuarine mixing found by Fox (1983a), in estuaries in which sea-salt was apparently unable to induce aggregation, led him to suggest that biological removal of some sort might be operating. If some portion of the humic acid fraction found by Fox (1981) is indeed composed of acidic mucopolysaccharides, these compounds would likely be quite amenable to bacterial attack (Fenchel and Blackburn, 1979, p. 67). However, the lack of any evidence for appreciable metabolism of terrestrial humic substances in estuaries suggests that such metabolism, if indeed required for mass balance considerations (Berner, 1982), must be sought in sedimentary, shelf, or deep-sea environments.

SUMMARY AND CONCLUSIONS

On the basis of few data from a limited range of geographical areas, it appears that humic substances in estuarine zones exhibit many attributes of the transitional nature of the environment. Aquatic humic substances show concentrations and chemical characteristics intermediate between those found in river and ocean waters, indicating relatively little *in situ* production, consumption, or chemical change. Sedimentary humic substance concentrations are somewhat higher than are usually found in the ocean, reflecting the high primary productivity and shallow water depths, but are chemically similar to either riverine or oceanic endmembers. The actual nature of estuarine humic substances is poorly known, but this problem is no worse than for humic substances from most environments.

The aquatic humic substances in estuaries are dominated by inputs from rivers. The introduction of riverine humic substances into a medium of seawater salinity leads to virtual saturation of their carboxylate sites, which in turn leads to contraction or coiling of the macromolecules. The reduced charge also allows aggregation of the high-molecular-weight fraction of the humic substances. This aggregation occurs with other, nonhumic organic fractions as well as with iron colloids, leading to particulates of micrometer size. The fulvic acid fraction appears to be largely unaffected by this aggregation process, while most of the humic acid fraction may aggregate at intermediate salinities. The relationships among the humic and nonhumic fractions before and during the aggregation process are poorly understood; our present confusion is likely compounded by our inability to distinguish, or even chemically define, the humic substance fraction.

While the bulk of riverine humic substances pass through estuaries with little retention, it is clear that some terrigenous organic matter is retained in the sediments. However, neither the nature of the retained organic material nor its mechanism of delivery to the sediments is understood. Little, if any, metabolism of riverine humic substances appears to occur during the estuarine passage.

High precursor concentrations, coupled with an abundance of phenolic materials, suggest that humification should be readily observable in estuaries. There are indications that humification may occur in the dissolved phase, particularly if algal exudates are abundant. Macrophytic debris may be an important site for humification. Sediment humification processes have been suggested on the basis of downcore increases in high-molecular-weight DOC, along with stable isotope data. However, the actual or relative importance of all these sites of humification in estuaries has yet to be demonstrated.

Stability-constant data coupled with speciation calculations imply that humic complexation of trace metals should decrease with increasing salinity, leading to negligible influence of humic substances on trace metal speciation at high salinities. However, field data suggest that relatively high proportions of copper can be complexed by organic ligands at intermediate or high salinities in estuarine zones, implicating either a marine or an *in situ* source of ligands. The nature, source, and humic character of the ligands in estuaries have not been determined. The wide variety of potential humic interactions with organic pollutants, coupled with a paucity of studies of their salinity dependence, render impossible any generalizations regarding this subject. Humic interactions with both metals and organic pollutants should have an important influence on transport and immobilization of these materials in estuaries. However, very few field data are available to test the implications of the largely experimental nature of studies carried out to date.

ACKNOWLEDGMENTS

I thank L. Schick for her assistance with the preparation of this chapter, and E. Boyle and L. Fox for their useful comments. Partial support for this work came from NSF grants OCE 7920244 and ISP 8011448, and NOAA Sea Grant R/LRF-45.

CHAPTER NINE

Geochemistry of Humic Substances in Seawater

GEORGE R. HARVEY and DEBORAH A. BORAN

ABSTRACT

Marine humic substances have received very little attention because of their low concentration in seawater and the difficulties associated with obtaining enough sample for study. Recent developments have allowed gram quantities to be isolated conveniently. Seawater fulvic and humic acids have now been isolated from several diverse environments and in different seasons of the year. Spectral, chemical, and physical studies indicate that seawater humic substances are very similar wherever they are formed. Based on these observations a hypothesis is proposed: seawater humic substances are derived from the free radical cross-linking of unsaturated lipids released into the water column. The hypothesis has been tested by laboratory synthesis by the proposed pathway and by correctly predicting the results of several chemical and biological reactions. A few problems with the structure still remain. The percent hydrogen is too low, the functionality of the nitrogen is not known, and there is more ^{13}C in the humic substances than in the starting lipids.

INTRODUCTION

Gelbstoff, the yellow acidic organic matter of seawater, was first recognized over 40 years ago by Kalle (1938, 1966). If he had named this class of organic matter something different, for example, Kalleic acid, much confusion would have been avoided during the past half century. Gelbstoff became known as humus (humic and fulvic acids), as defined by soil scientists. The terminology has hindered investigations and perceptions of marine humic substances ever since. Given the history and the state of the art, in this chapter we accept the classical definitions of fulvic and humic acids described elsewhere in this book with the additional requirement that the final product of isolation should be insoluble in ordinary organic solvents, for example, ether, methylene chloride, or acetone. This latter criterion thus excludes fatty acids, benzoic and salicylic acids, and other small molecules as part of marine humic substances. It cannot be overemphasized that it is important to dissociate marine humic substances from terrestrial humic substances.

Even in his early work, Kalle surmised that Gelbstoff was a relatively stable product of phytoplankton metabolism which was forming in seawater remote from terrestrial influence. He also observed that the intensity of the blue fluorescence emitted from UV-irradiated seawater increased with increasing salinity. He concluded that in addition to the Gelbstoff of coastal waters there was also a Gelbstoff unique to the open sea.

Further work on differentiating marine and coastal runoff humic substances was hindered by the lack of suitable isolation techniques for the marine material since concentrations in the open sea rarely exceed 0.25 mg/L. Sorption of marine humic substances from seawater onto solid phases is now a standard technique and can be used to extract gram quantities of marine humic substances for chemical and physical studies (see Aiken, Chapter 14). Sieburth and Jensen (1968) first used rolled nylon stockings as an adsorbant but the method suffered from contamination. Kerr and Quinn (1975) used a specially treated charcoal and obtained quantitative recovery of the dissolved colored substances in seawater. Riley and Taylor (1969) introduced the use of cross-linked polystyrene resins, specifically Amberlite XAD-2. This polymer is now the most widely used for open-ocean work (Stuermer and Harvey, 1974; Bada et al., 1982; Harvey et al., 1983) and in estuaries (Mantoura and Riley, 1975). These isolation methods have made available sufficient quantities of seawater humic substances for detailed chemical studies.

Based on comparison of data from UV, fluorescence, and NMR spectroscopy, and from carbon isotope determination for humic substances isolated from coastal and open ocean environments, the authors have concluded the following: (1) other than its metal complexation and redox functions, the only resemblance between humic substances from open ocean (marine) and terrestrial environments is that they are both colored organic acids soluble in water, and (2) marine humic substances are formed *in situ* and only in the coastal zone is there an admixture of terrestrially derived humic substances from rivers. However, this second conclusion has not yet been reconciled with the observations discussed by Mayer in Chapter 8 that riverine humic substances, generally behave conservatively in the passage through estuaries to the oceans. In this chapter, the discussion will be confined to autochthonous dissolved marine humic substances which are assumed to be represented by oceanic dissolved humic substances collected far from the nearest terrestrial influence.

A good review and discussion of the hypothetical and presumed functions ascribed to marine humic substances has been given by Pocklington (1977). His challenge to marine chemists was to either define Gelbstoff structurally and what it does, once and for all, or stop using its existence in scientific arguments to explain uninterpretable results. This challenge was the first attack on the common reference to marine humic substances—as in soil humic chemistry—as the "trash barrel" of biogenic substances. Prevailing views of marine geochemists were, and to a large extent still are, that marine humic substances are a macromolecule composed of amino acid–carbohydrate condensation products with some fatty acids attached through ester linkages along with a little bit of everything else in the sea.

It is most unlikely that the dissolved organic matter in seawater can randomly polymerize or polyaggregate at the micro- or nanomolar concentrations that prevail. Such processes might become important in estuaries or at the sediment–water interface, but not in the open sea. In a previous study of the nature of seawater humic substances we worked with a single composite sample isolated from the Sargasso Sea (Stuermer and Harvey, 1974). The elemental composition of the fulvic acid isolated from the Sargasso Sea was compared with a fulvic acid isolated from the B_h horizon of a podzol (Stuermer and Payne, 1976). The seawater sample had a lower oxygen content, higher nitrogen content, and a higher H/C ratio than the soil sample (Table 1). At that time it was unknown if the Sargasso Sea sample was representative of seawater humic substances in general or merely typical of the Sargasso Sea, although a similar acidic material isolated by Kerr and Quinn (1975) had similar properties.

In 1979 our laboratory began a seasonal survey of dissolved humic substances in the Gulf of Mexico. High and low productivity sites both near shore and far from land were chosen. The goal of this study was to obtain gram quantities of marine humic substances from each site and season of the

TABLE 1. Ash-Free Elemental Composition of a Marine Fulvic Acid and a Soil Fulvic Acid[a]

Sample	% C	% H	% N	% O (by difference)	% Ash	H/C ratio
Sargasso Sea fulvic acid	50.0	6.8	6.4	36.9	3.37	1.61
Soil fulvic acid[b]	46.7	4.5	0.5	44.3	2.35	1.15

[a] *Stuermer and Payne (1976).*

[b] *Isolated from the B_h horizon of a podzol in Falmouth, Massachusetts.*

TABLE 2. Concentrations of Dissolved Organic Carbon (DOC), Marine Humic Acid, and Marine Fulvic Acid in the Gulf of Mexico[a]

Station	Type sample	Season (Fall/Spring)	Depth (m)	DOC (mg/L)	Humic (μg/L)	Fulvic (μg/L)
Mississippi Outflow	Coastal waters	F	3	1.32	114	550
		F	3	1.50	126	1270
		Sp	4	1.10	17.6	904
Gulf Loop	Open ocean	F	20	—	26	724
		F	20	0.96	3	190
		Sp	20	0.50	14	234
		Sp	55	0.54	9	58
Yucatan	Intermediate coastal/open ocean	Sp	10	0.63	29	223
Campeche	Intermediate coastal/open ocean	Sp	5	0.98	105	754
Cape San Blas	Intermediate coastal/open ocean	Sp	4	3.19	336	577
		Sp	55	2.20	10	13.9

[a] *Harvey et al. (1983).*

year so that detailed chemical and biological studies could be conducted to determine the structure of marine humic substances and their role in controlling the biological quality of seawaters. A small pilot plant was constructed in a portable van equipped with six stainless steel drums with a capacity of 1500 L (Tokar et al., 1983). Water was brought into the drums by means of a gas lift system (Tokar et al., 1981) at a rate of 20 L/min. The filled drums were pressurized with nitrogen to force the water through 800 mL columns of XAD-2 resin at 500 mL/min. With these methods we were able to obtain gram quantities at each station on each cruise within the strict time constraints of a ship schedule. In all seasons the open-ocean waters contained 150–250 μg/L of marine fulvic acid and ten times less marine humic acid (Table 2). Productive coastal waters contained 400–800 μg/L of marine humic substances, 90% of which are marine fulvic acid. The highest concentrations were found during the spring bloom on the continental shelf. These concentrations were comparable to concentrations in the Sargasso Sea (Stuermer and Harvey, 1974). Recently, the same range of marine humic substance concentrations has been found in the equatorial Pacific.

Upon spectral and chemical analyses of the marine fulvic acids and marine humic acids from this collection it became obvious that, unlike soil humus, marine humic substances are not a random potpourri of marine organic matter, but exhibit regular and consistent structure in all marine environments behaving like a "reagent" in their properties. Differences occur only in the relative proportions of protons on paraffinic, aromatic, and oxygenated carbons and in their degree of interaction with transition metals (Piotrowicz et al., 1983a). Thus, we were led to the conclusion that marine humic substances are formed by a narrow suite of reactions that are possible in all marine environments and that these reactions lead to a class of compounds with consistent properties. This chapter will describe the structures of marine fulvic acid and humic acid, evidence for the proposed pathway for the formation of marine humic substances, and the functions of humic substances in seawater.

FORMATION AND STRUCTURE

The mechanism of formation and class structure of marine humic substances presented here was first proposed by Harvey et al. (1983). Since that time the proposed pathway and structures of marine fulvic acid and marine humic acid have been tested and the predicted chemical and physical behaviors of marine humic substances have been supported in several experiments.

The equivalent weight and molecular size data indicate that marine humic substances are approximately 900–1200 daltons and are composed of two to four fatty acid chains (Harvey and Boran, 1982b). The key to the pathway (Fig. 1) is in the NMR spectra in that all the marine fulvic acids and marine humic acids had 79–94% methyl and methylene protons and 2–24% protons on oxygenated carbon. The question then posed was: What hydrocarbon-like class of compounds, present in all marine environments, is capable of

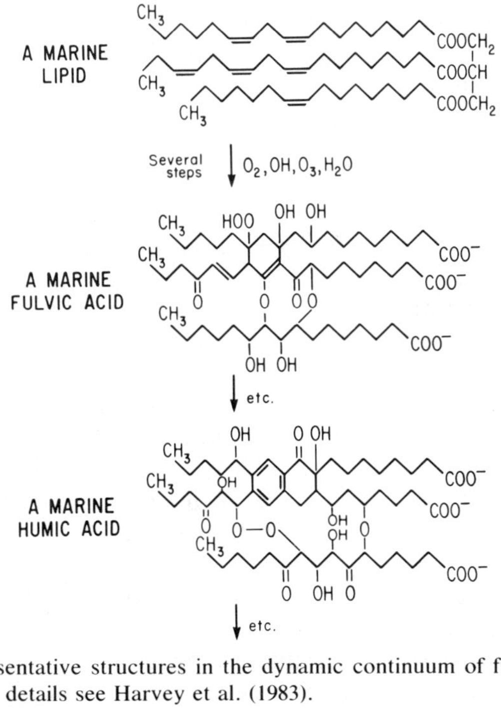

FIGURE 1. Representative structures in the dynamic continuum of fulvic and humic acid formation. For more details see Harvey et al. (1983).

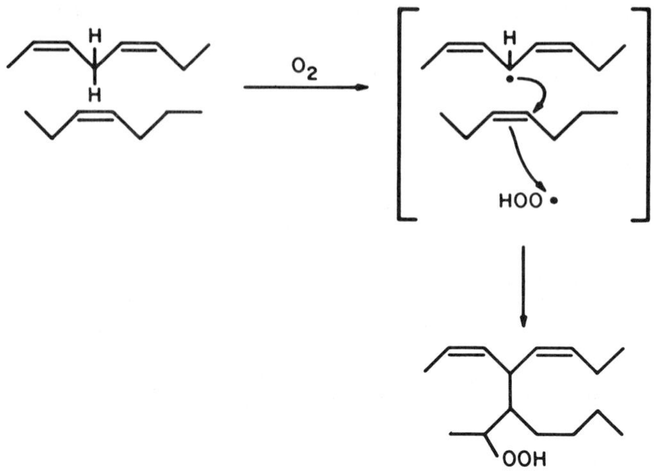

FIGURE 2. Initiation of autoxidation of a polyunsaturated lipid.

reacting to form water-soluble organic acids mainly composed of saturated carbon–hydrogen bonds? Since marine plants and animals generally have a much greater unsaturated lipid content than terrestrial organisms, and these lipids are known to react spontaneously with oxygen, polyunsaturated lipids were chosen as a testable class of precursors.

Our hypothesis is that marine humic substances are formed from the free radical autoxidative cross-linking of unsaturated lipids released into the water by plankton. This hypothesis was arrived at for several reasons which are discussed in more detail in Harvey et al. (1983). Briefly, autoxidations are accelerated by light and catalyzed by transition metals, both of which are abundant in surface seawater.

The precise mechanism and kinetics of autoxidation of unsaturated lipids have recently been elucidated (Porter et al., 1981). Initiation is by abstraction of an allylic hydrogen atom by triplet oxygen ($^3O_2$) to produce an allylic radical (Fig. 2). The radical may react with oxygen to give a hydroperoxide, or with another site of unsaturation, to give a new C—C bond. At some point in the autoxidation pathway (Fig. 1) the fatty acid precursor becomes soluble in water and has the properties of a fulvic acid. A small amount of the cross-linking at this point may lead to substituted cyclohexane rings which are ultimately the source of the aromaticity present only in the marine humic acid. At the marine fulvic acid stage of this dynamic continuum the structures are still quite flexible and can be solvated by water in a number of configurations, soluble at all pH values. As intramolecular cross-linking continues the structures become less flexible, especially if the rigid aromatic rings form. The structures eventually become insoluble at low pH (marine humic acid). Apparently, the path leading to marine humic acids is not favored; the marine fulvic acid/marine humic acid ratio is generally greater than 10 (Harvey et al., 1983). The dominant pathway is probably oxidative cleavage of the marine fulvic acids, mainly by singlet oxygen ($^1O_2$), the photoexcited state present only in daytime. The marine fulvic acids are cleaved into smaller and smaller fragments, for example, low-molecular-weight aldehydes, and acids until the material no longer fits the definition of a humic substance. This hypothesis was tested by allowing a seawater solution of a marine fulvic acid, isolated from the open ocean, to stir for 2 days in Miami sunlight fully exposed to air (Harvey, unpublished). Less than 1% was recovered as marine humic acid and only half of the original marine fulvic acid was recovered. The "new" marine humic acid was insoluble in acid and did have aromatic protons.

The unsaturation in marine fatty acids is generally on the middle-carbons of the 16 to 22 carbon chains, that is, on C-9, 12, 15, and so on. Cross-linking leaves the carboxyl end relatively unhindered and free. Three predictable consequences of this are:

1. Esterification should be as facile as with a free fatty acid, unlike terrestrial humic substances (Schnitzer and Khan, 1972).
2. Oxidation should give a series of dicarboxylic acids up to about C-9.
3. At alkaline pH, the carboxylate groups will electrostatically repel each other (Fig. 3). The repulsion increases the molecular size of the marine humic substances and imparts an apparent higher molecular weight (Fuoss and Strauss, 1948).

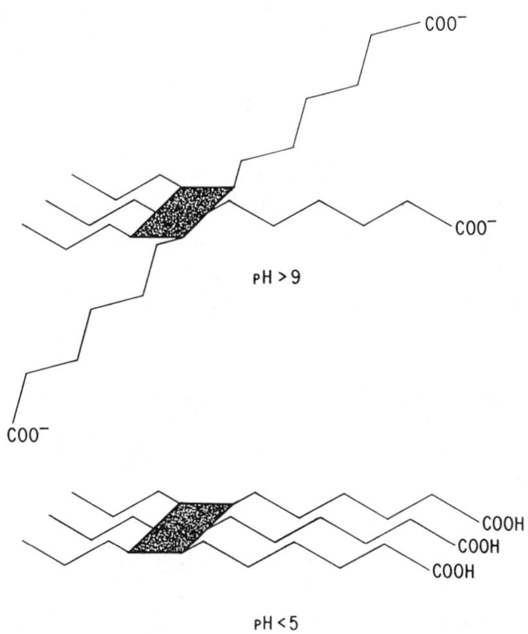

FIGURE 3. Dependence of molecular size on pH due to carboxylate repulsion (upper) and intramolecular hydrophobic bonding (lower).

Our laboratory has confirmed prediction 1 and 2 (Harvey et al., 1983). Another laboratory has confirmed the third prediction with an estuarine fulvic acid (Varney et al., 1983). Gel permeation high-pressure liquid chromatography in acid solution by Weisel and Zika (unpublished) and Varney et al. (1983) showed apparent molecular weights of 700–900. At high pH values the apparent molecular weights were 10,000–70,000. Adsorption and ionic strength effects in gel permeation chromatography vary significantly between model compounds and other families of compounds of unknown molecular size. Comparisons should be used with caution until the determined sizes are confirmed by alternative methods as was done by Thurman et al. (1982). These problems are discussed further in Chapter 19 by Wershaw and Aiken. The most important observation from the Wiesel and Zika work is that the retention times of all the marine fulvic acids and marine humic acids examined differed by less than 10%. Thus, the difference between marine fulvic acid and marine humic acid is not molecular size but internal structure. Other chemical, spectroscopic, and physical data supporting the proposed structure are given in Harvey et al. (1983).

SYNTHESIS

The final test of any proposed structure or family of structures is synthesis from known starting materials and known reactions (Woodward, 1956). Autoxidation of oils, though it does lead to mixtures, is not a purely random process. Oxygen will react mainly at activated sites such as allylic (Fig. 2), benzylic, unsaturated, and heterosubstituted carbon. Only enzyme-regulated autoxidations, such as the transformation of arachidonic acid into the prostaglandin family (Nelson et al., 1982) lead to the same discrete compounds consistently. But we can infer that marine fulvic acid and marine humic acid are mixtures of similar functional groups from their spectra and titration curves. So, unregulated interaction of unsaturated lipids in seawater with ambient air and light were the conditions chosen (Harvey and Boran, 1982a). The structures of the compounds used as starting materials are shown in Figure 4. They range in reactivity with oxygen from trilinolein, with 21 allylic or unsaturated sites, to tripalmitin, with no sites and which was used as a procedural control.

The natural marine fulvic acid and marine humic acid were removed from 10 L of seawater with XAD-2 resin (Harvey et al., 1983). After readjusting the pH to 8.2, 5 mg of one of the particular lipids was added to the column effluent every half hour, with stirring, up to a total of 5 mg/L. The solution was allowed to stir for 1, 2, or 7 days, depending on the starting material. The solution was then acidified to pH 2 and extracted as above with XAD-2. The yield of product was 50–100 mg.

The isolated substances were similar to marine fulvic acids. Tripalmitin gave no product and in fact never dissolved. The products were pale yellow powders soluble in acid and base but insoluble in ether or methylene chlo-

Trilinolein (1):

C_5H_{11}-CH=CH-CH_2-CH=CH-(CH_2)-COOCH$_2$
C_5H_{11}-CH=CH-CH_2-CH=CH-(CH_2)-COOCH
C_5H_{11}-CH=CH-CH_2-CH=CH-(CH_2)-COOCH$_2$

Glycerol-1,2-dioleate-3-palmitate (2):

C_{10}-H_{21}-CH=CH-(CH_2)-COOCH$_2$
C_{10}-H_{21}-CH=CH-(CH_2)-COOCH
$C_{15}H_{31}$-COOCH$_2$

Glycerol-1,3-dioleate-2-palmitate (3):

$C_{10}H_{21}$-CH=CH-(CH_2)-COOCH$_2$
$C_{15}H_{31}$-COOCH
$C_{10}H_{21}$-CH=CH-(CH_2)-COOCH$_2$

Tripalmitin (4):

$C_{15}H_{31}$COOCH$_2$
$C_{15}H_{31}$COOCH
$C_{15}H_{31}$COOCH$_2$

FIGURE 4. Glycerides used as starting materials for the autoxidative synthesis of marine fulvic and humic acids. Compound (4) was used as a procedural blank.

ride. Each synthetic product had IR (Fig. 5) and NMR (Fig. 6) spectra that closely matched one or more spectra from our library of natural marine fulvic acids. Furthermore, the synthetic fulvic acid protected marine phytoplankton from copper toxicity as well as the natural marine fulvic acid (Ortner et al., 1983). Thus, evidence to date indicates that the natural processes leading to marine fulvic acid were simulated.

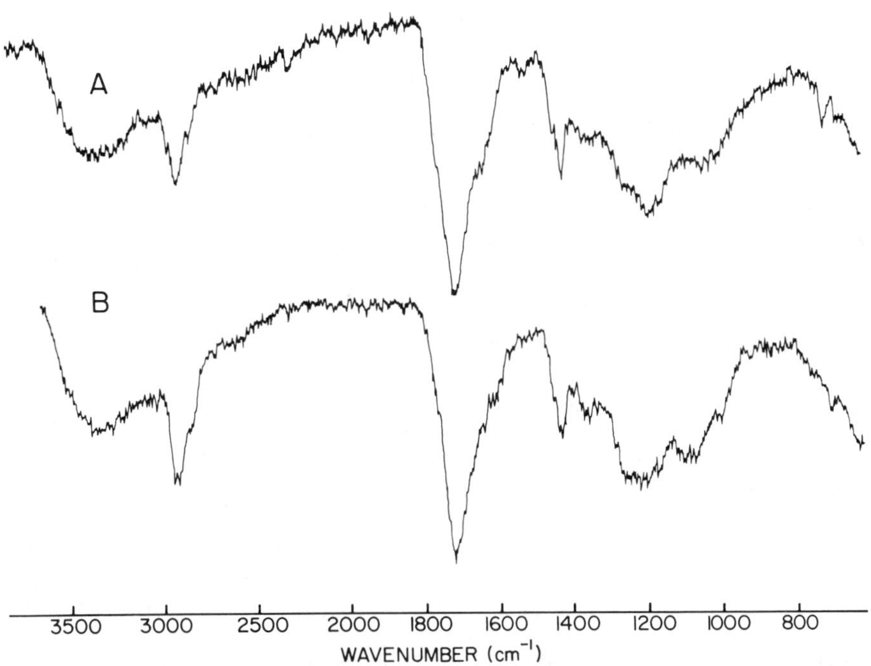

FIGURE 5. Infrared spectra of (A) a natural marine fulvic acid isolated from the Gulf of Mexico and (B) a synthetic fulvic prepared by the autoxidation of trilinolein.

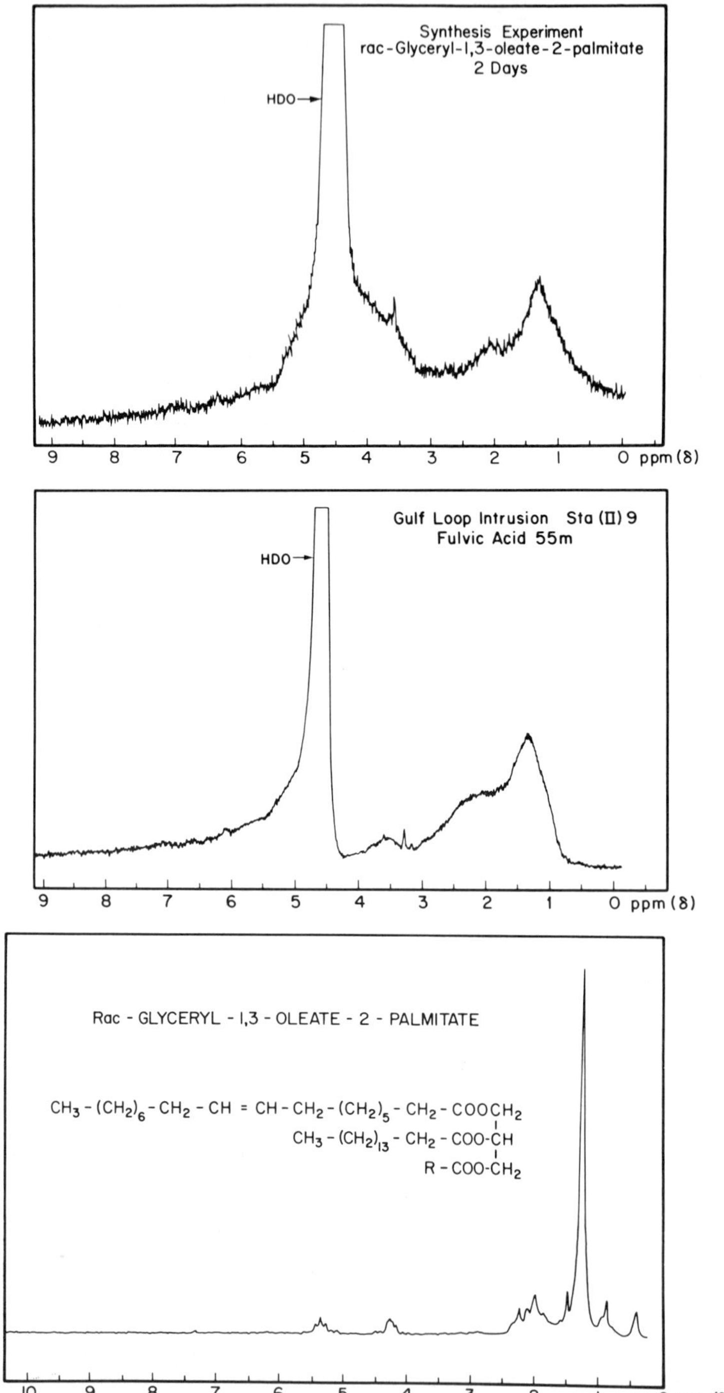

FIGURE 6. Proton NMR spectra of a synthetic fulvic acid (*top*), a natural fulvic acid isolated from open-ocean seawater (*middle*), and the starting glyceride (*bottom*).

TABLE 3. Carbon Isotope Values for Natural and Synthetic Fulvic Acids

Substance	Depth of Isolation (m)	$\delta^{13}C$
MFA (R03R04)	4	−21.5
MFA (R03R01)	25	−22.0
MFA (R01A04)	1100	−20.6
From trilinolein	—	−19.7
From glycerol-1,2-dioleate-3-palmitate	—	−17.6
Trilinolein	—	−28.1

Marine lipids are more depleted in ^{13}C ($\delta^{13}C = -25$ to -28) than the total dissolved organic carbon pool ($\delta^{13}C = -21$) or marine fulvic acid ($\delta^{13}C = -19$ to -21) (Stuermer and Harvey, 1974). It was thought that lipids could not be the source of marine fulvic acid because an isotope effect or fractionation of six to eight parts per mil was too great. The success of the synthesis experiments allowed us to test the validity of this criticism of the proposed pathway. The trilinolein used as starting material (Supelco) is derived from safflower oil and has a $\delta^{13}C$ of −28‰. The synthetic fulvic acid synthesized from that trilinolein had a $\delta^{13}C$ of −19‰. The $\delta^{13}C$ results of two synthetic and three natural marine fulvic acids are summarized in Table 3. The carbon fractionation is very large for the types of reaction involved but the results are in the same direction and magnitude as the natural process. The unprecedented fractionation certainly deserves further investigation.

FUNCTIONS OF SEAWATER HUMIC SUBSTANCES

Metal Complexation

The functions of seawater humic substances can be correlated with its family or class structure (Fig. 1). The relative amounts of carboxylate and other oxygen functional groups in marine humic substances allow for strong to weak transition-metal complexation (Batley and Florence, 1974; Piotrowicz et al., 1983a, 1983b). Metal complexation can be detected by anodic stripping voltammetry (Piotrowicz et al., 1982), selective ion electrodes (Sunda and Guillard, 1976), and bioassay (Ortner et al., 1983). These methods have confirmed that marine humic substances play a major role in the chemistry and biology of some toxic metals in seawater by various degrees of metal–organic association. Laboratory experiments in which previously isolated marine fulvic acid or marine humic acid was added to seawater have

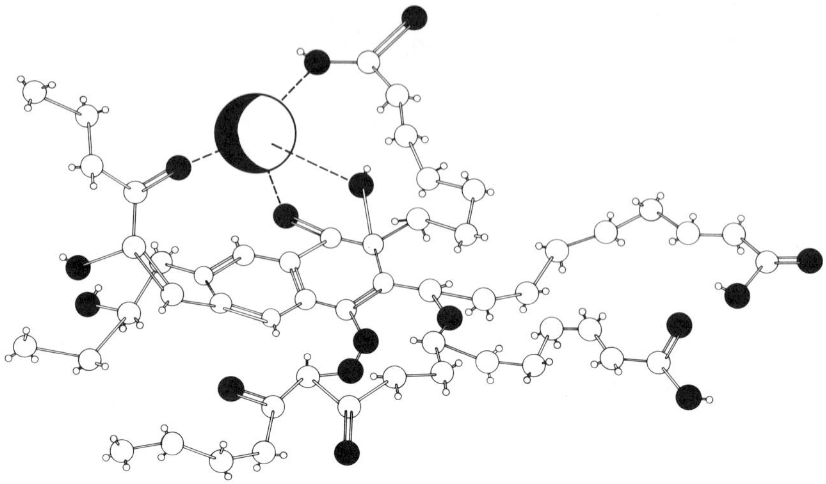

FIGURE 7. One of several possible interactions between a marine humic acid and a metal ion. ○ = carbon, ∘ = hydrogen, and ● = oxygen.

shown that the copper and zinc complexes formed are as strong as those found naturally (Piotrowicz et al., 1983b). Metal complexation by marine humic substances decreases the mobility of the metal and inhibits biological uptake (Ortner et al., 1983). Interestingly, marine humic substances caused a slight increase in the uptake of plutonium and americium into marine diatoms (Fisher et al., 1983). The marine fulvic acid and marine humic acid may preferentially complex Cu(II), Cd(II), and Zn(II) away from binding sites on the diatom cells, affording access to those sites by the radionuclides. In these same experiments the synthetic organic complexing agents ethylenediaminetetraacetic acid (EDTA) and nitrilotriacetic acid (NTA) totally prevented uptake of plutonium and americium. This illustrates that synthetic complexing agents should not be used as model compounds for natural systems.

A copper organic complex has been isolated from seawater (Mills et al., 1982). The ESR spectrum of the complex revealed two kinds of copper–oxygen bonds. Copper binding to nitrogen and sulfur ligands was not seen in their sample, but this possibility was not excluded. An illustration of how copper could be bound by two kinds of oxygen-containing ligands is shown in Figure 7 for a marine humic acid. The copper is associated with an oxygen-rich, π-electron-rich cavity in the mid-portion of the molecule. One of the flexible carboxylate chains is folded back to bond with the copper, completing the cavity. Many other configurations are possible, of course, but the illustration is consistent with the anodic stripping voltammetry data and the ESR spectrum.

Redox

The continuum of reactions leading to marine fulvic acid and marine humic acid is an oxidative process (Fig. 1). Therefore, reduction must be simultaneously occurring in the seawater. Very likely candidates for functioning as the electron acceptors in these redox reactions would be the higher oxidation states of the transition metals. Two possible processes are illustrated in Figure 8. Such a redox system has been demonstrated recently (Sunda et al., 1983). Insoluble manganese IV oxide (actually $MnO_{1.8}$ or MnOx) was slowly reduced in seawater containing a previously isolated marine humic acid, to soluble Mn^{2+}. The reaction was greatly accelerated in sunlight. The rate of dissolution of the MnOx was linearly related to the quantity of marine humic acid. After 6 hours a suspension of MnOx and 2 mg/L of dissolved marine humic acid was converted to 22% soluble manganese in the light while 9% dissolved in the dark. A marine fulvic acid behaved similarly in its reducing power (Stone, 1983). Manganese is an essential nutrient for marine plankton. The photoreduction of insoluble forms to the available, soluble species is of considerable biological importance. Since the ultimate source of the reducing agent (marine humic substances) is the biological community itself, the production of such compounds could be viewed as an ecological feedback mechanism. It is not known at this time if marine humic acid will reduce Fe(III) to Fe(II) in seawater as dissolved terrestrial humics do (Miles et al., 1981). Also, photoreduction of the highly toxic Cr(VI) needs to be investigated.

Skopintsev (1982) has pointed out the enormous potential of marine humic substances as an organic nutrient resource (15×10^{18} kcal). We are unaware of any published work demonstrating that dissolved humic substances can serve as a heterotrophic food source. Microbial enzymes should be capable of using the free carboxyl end of the marine humic substances as a simple fatty acid. However, this is conjecture and should be investigated by microbiologists and planktologists.

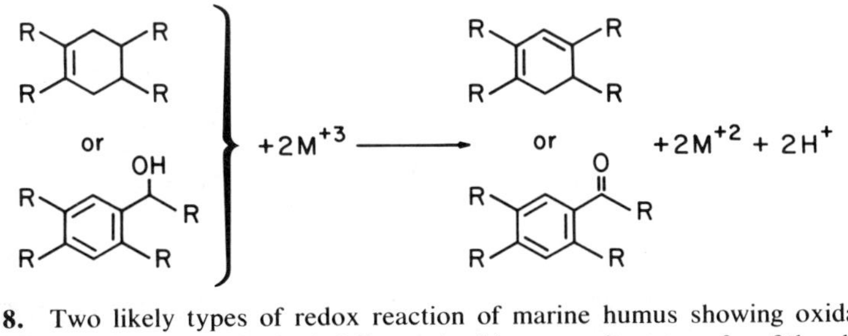

FIGURE 8. Two likely types of redox reaction of marine humus showing oxidation of a cyclohexene (*top*) and a benzylic alcohol (*bottom*) with concomitant transfer of the electrons to a metal, M.

OTHER RESEARCH PROBLEMS

The structures for marine humic substances presented are inconsistent with elemental compositions of about 48% carbon, 6% hydrogen, 3% nitrogen, and 40–50% oxygen usually found. As the proton NMR spectra (e.g., Fig. 9) have been interpreted, there are too many methyl and methylene protons (60–94%; Harvey et al., 1983) and too few protons on nitrogen-, sulfur-, and oxygen-bearing carbons to have only 6% hydrogen. This has not been seen as a problem in the past because without the aid of NMR one merely added up the carbon, hydrogen, nitrogen, and sulfur values, subtracted the result from 100, and assumed that the difference was oxygen. The proton NMR spectra preclude such high carbon-bound oxygen percentages. Hydrogen compositions of 6% are only consistent with structures such as glucose (6.6%), or naphthalene (6.2%), that is, highly aromatic or saccharide-like structures, which are clearly excluded by the NMR. One possible explanation for this discrepancy may be the incorporation of small amounts of heavy elements into the marine humic substances during the free radical cross-linking. In support of this possibility one marine humic acid sample was analyzed by neutron activation. In addition to traces of several heavy metals the sample contained 0.7% chlorine, 0.07% bromine, and 0.05% iodine (Harvey et al., 1983). The halogens would add to the molecular weight but would not be detected by spectral analysis. Further work is required on this problem.

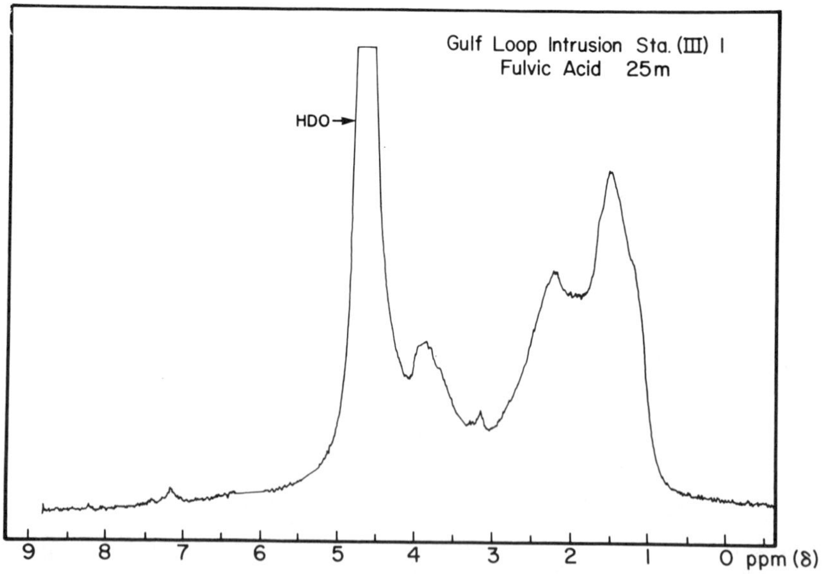

FIGURE 9. Proton NMR of a marine fulvic acid showing the high amounts of methyl and methylene protons in the 0.9–2.5 ppm range. Note the small amount of aromatic protons at 7.1 ppm.

The nature of the 3–5% nitrogen in the marine fulvic acid and marine humic acid is not known. This is a general problem with humic substances which is discussed by Schnitzer in Chapter 12. It is possible that some of this nitrogen content is an artifact of the extraction procedure in which NH_4OH is used to elute the XAD-2 resin. If the nitrogen is present in amine functional groups, then it may play a substantial role in metal complexation. We do know that it is not present as hydrolyzable amino acid moieties. Bada et al. (1982) isolated marine humic substances from the open Pacific Ocean using our extraction method. The purified marine humic acid was subjected to the standard protein hydrolysis procedure, that is, 6*N* HCl at reflux for 18 hours. No amino acids were found. This evidence, however, does not exclude the presence of amino acids bonded into the main structure through carbon–carbon bonds which would not be hydrolyzed. The solution to the nitrogen problem may lie in more extensive ESR studies (e.g., Mills et al., 1982).

In addition to the problems of biological utilization and the ^{13}C fractionation mentioned before, the nature of the abyssal humic substances needs to be investigated. It is widely accepted that the humic substances in deep ocean, that is >3000 m is biologically refractory (Barber, 1968) and very old (Williams, 1968; Skopintsev, 1972). Recent evidence of the ^{14}C age of the food of benthic fish (>5000 m) indicates that 80% is of very recent origin (Williams, 1983). We have recently isolated and studied marine fulvic acid from 2000 m in the Gulf of Mexico. The NMR spectrum indicates that this deep marine fulvic acid is similar to surface fulvic acid. These conflicting observations must be rectified because they are very germane to any understanding of the cycle of organic carbon in the seas.

ACKNOWLEDGMENTS

We thank C. P. Weisel (NRC postdoctoral fellow, NOAA) and R. Zika (University of Miami) for permission to cite their unpublished gel permeation chromatography work, and B. Wrenn for technical assistance. The ^{13}C analyses were kindly provided by J. and P. Gearing of the Graduate School of Oceanography, University of Rhode Island. This work was supported by the Office of Marine Pollution Assessment, U.S. National Oceanic and Atmospheric Administration.

CHAPTER TEN

Geochemistry of Humic Substances in Marine Sediments

M. VANDENBROUCKE, R. PELET, and Y. DEBYSER

ABSTRACT

Humic substances occur in recent marine sediments and their amounts and compositions depend both on (1) the sedimentation environment, which influences the nature of the organic input and importance of alteration processes, and (2) the evolutionary stage of the organic matter. A brief review points out some chemical features of sedimentary humic substances revealed by a number of analytical methods: elemental analysis, infrared spectroscopy, nuclear magnetic resonance, degradative methods, and isotopic analysis. Based on samples obtained during the ORGON cruises, the main factors controlling the composition of marine humic substances present in the water–sediment interface are the origin of organic input and the redox potential of the environment during transportation of the organic matter to the sediment. This chapter discusses possible origins of humic substances in marine sediments and suggests that they can result either from polycondensation reactions of the degradation products of organic matter, or from oxidative degradation reactions. Formation of primary humic substances (polycondensation) leads mainly to insoluble humin—humic and fulvic acids are mainly secondary humic substances formed by oxidation of humin. Humic substances in marine sediments can be derived from organic material produced either in terrestrial or marine environments. Evolution of humic substances during burial of the sediment and their relation with the kerogen of ancient sediments is tentatively proposed to depend on their molecular size and functionality.

INTRODUCTION

A major emphasis of this chapter is on the geochemistry of humic substances in recent marine sediments, with only passing reference to ancient sediments. The reason for this is twofold. First, the organic matter of ancient sediments yields only small quantities of products soluble in aqueous alkaline solvents which makes their study uninteresting; the result is that a survey of the literature shows few reports on humic substances in ancient sediments. Second, this laboratory was strongly involved in the ORGON group for the study of organic geochemistry of recent sediments.

The term "humic" comes from the latin *humus,* meaning soil of vegetation, and indeed the discovery in soils of organic compounds which differed from living matter brought about the founding of pedology. But it is clear that forget-me-nots in a raygrass meadow under oak trees cannot be found at the bottom of the oceans. Thus, the origin of marine humic substances is *a priori* different from the humification processes described on land by pedologists. To begin with, the term "humic substances" here only means "substances extractable by the same processes used by pedologists to extract humic substances from soils." In the last section, we will go further and show that humic substances from soils and sediments have many common features.

HUMIC SUBSTANCES IN RECENT AND LESS RECENT SEDIMENTS

Extraction of Humic Substances from Marine Sediments

Before the complex mixtures of organic compounds in sediment can be analyzed, they must first be extracted and separated into fractions with similar chemical features. Organic solvents which do not alter chemical structures (such as hydrocarbons or chloroform) can extract only a small to moderate amount of the organic matter from sediments. If the initial organic matter has numerous hydrophilic functional groups, as is the case in recent sediments, it is more soluble in aqueous solutions, particularly when the pH and ionic strength are modified by addition of alkaline or acidic substances.

Unfortunately, no generally accepted, common procedure for extraction of humic substances from sediments exists. The results reported here were obtained using the procedure described by Debyser and Gadel (1981), which

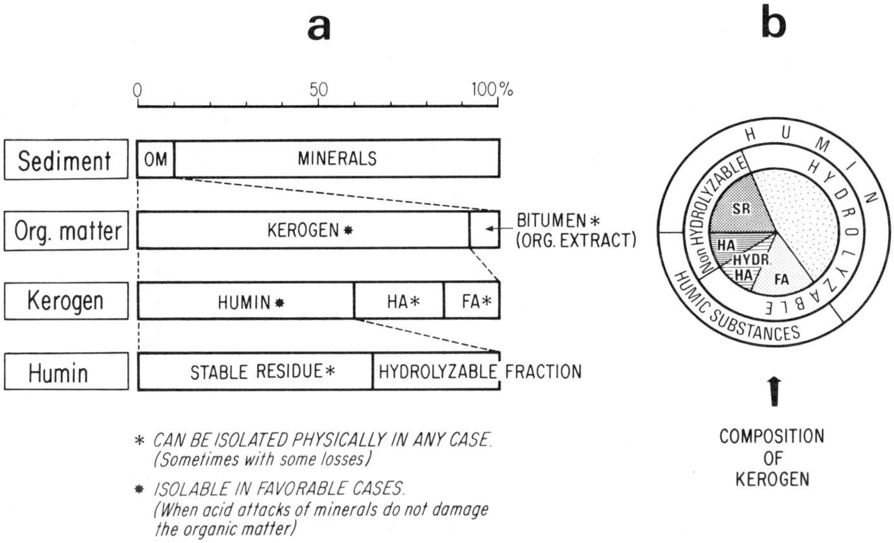

FIGURE 1. Composition of mineral and organic fractions in sediments. The stable residue is representative of kerogen in mature ancient sediments.

involves the use of $0.10N$ NaOH and $1.0N$ $Na_4P_2O_7$ solutions. Recent results show that the amount of humic substances extracted is very sensitive to the procedure, whereas the elemental composition and infrared spectra of the extracts are relatively unaffected. The residual organic matter in the sediment (humin, see Fig. 1 for terminology) can be isolated by nonoxidative acidic attacks on the mineral fraction. Carbonates are generally destroyed with $6N$ HCl, and silicates with a 1 : 2 mixture of $6N$ HCl and 40% HF. This procedure was originally devised for the isolation of kerogen in ancient sediments (Durand and Nicaise, 1980) and is less applicable with recent sediments where humin is readily hydrolyzable. The insoluble residue that remains after chloroform extraction, decarbonation by cold $2N$ HCl, alkaline extraction, and acid attack on the mineral fraction is defined as the stable residue. Figure 1 shows the relationships between the various organic and mineral fractions in a recent sediment.

Factors Influencing Amount and Composition of Humic Substances in Marine Sediments

Evolutionary Stage

The amount of humic substances present in sediment depends on the evolutionary stage of the sediment. Geological evolution is a kinetic concept, depending on time and temperature. Geochemists divide sedimentary evolution into three main stages: *diagenesis, catagenesis,* and *metagenesis*. Metagenesis is the most advanced stage, where organic matter is almost reduced to a carbonaceous residue. During the preceding stage, catagenesis, thermal degradation of the organic matter in sediment generates hydrocarbons (Tissot and Welte, 1978; Durand, 1980). The first evolutionary stage, diagenesis, is subdivided into early diagenesis, when organic matter loses mainly nitrogen, and diagenesis *sensu stricto,* when organic matter loses mainly oxygen (Pelet, 1980). Humic substances are still present during the latter stage of diagenesis, but the amount that can be extracted from the sediment has decreased. All these transformations occur during burial of the sediment, under the influence of time and temperature. During early diagenesis, the influence of biota can also be important.

Sedimentation Environment

The amount and composition of the humic substances in sediments also depend on the *sedimentation environment* (geography and climate), and on the relative productivity of the continental and marine environments. These factors influence both the nature of organic input and the extent of presedimentary alteration. The chemical composition of the organic input depends on the climate, the proximity of the shore, and the relative contribution of marine and continental biomass to the sediment. The extent of alteration is related to the distance between where the biomass is produced and where sedimentation occurs, along with the depth and oxygen content of the water. Particulate organic matter may be protected from degradation by a low surface to volume ratio (large particle size) or by association with minerals. These factors limiting degradation can also cause variations in amounts and composition of humic substances according to grain-size fractions, mainly in marine organic matter (Jocteur-Monrozier et al., 1983). Finally, the rate of accumulation and the extent of bioturbation affect the oxygen content of the surficial sediment.

PRINCIPAL CHEMICAL FEATURES OF SEDIMENTARY HUMIC SUBSTANCES

Major Elements

The major organic elements in sedimentary humic substances are carbon, hydrogen, oxygen, nitrogen, and sulfur. Samples are frequently compared according to their atomic H/C, O/C, N/C, and S/C ratios. Minerals, in the form of ash, are sometimes abundant and can affect some determinations of elemental composition. Silica and alumino-silicates interfere with the determination of organic oxygen. Pyrite, which is not destroyed by acid during preparation of stable residues, must be taken into account in determination of organic sulfur (Durand and Monin, 1980).

Atomic H/C ratios in marine-sediment humic substances may vary between 0.9 and 1.7, atomic O/C ratios between 0.2 and 1.0, and atomic N/C ratios between 0.02 and 0.10. The elemental compositions of representative fulvic acids, humic acids, and stable residues of two reference marine sediments are compared in Figure 2, where the atomic H/C ratio is plotted as the ordinate, with the N/C ratio on the left and the O/C ratio on the right of the abscissa. One of the reference marine sediments is a sample from the Kerguelen Islands that contains marine autochthonous organic matter, and the other is a sample from the Mahakam Delta, which contains terrestrial organic matter (Debyser and Gadel, 1979). In general, for most sediments, fulvic acids have higher atomic H/C and O/C ratios than humic acids and stable residues. This is due, in the case of H/C ratios, to a lower carbon content, hydrogen being similar for each species, and in the case of O/C ratios, both to higher oxygen and lower carbon contents in the fulvic acids. Comparisons of humic acids with stable residues reveal higher O/C ratios in humic acids, but similar H/C ratios in the two fractions. However, it should be pointed out that the acidic treatments used in preparing stable residues sometimes lead to hydrolysis of organic matter, and that fulvic acid cannot be recovered without losses. As a consequence, the sum of these three fractions does not represent the total kerogen.

Nitrogen to carbon ratios in humic acids are generally slightly higher than N/C ratios in fulvic acids. In stable residues, most of the nitrogen occurs in the hydrolyzable fraction and is lost during sample preparation. Therefore, the N/C ratios of stable residues are always considerably less than the N/C ratios of humic acids.

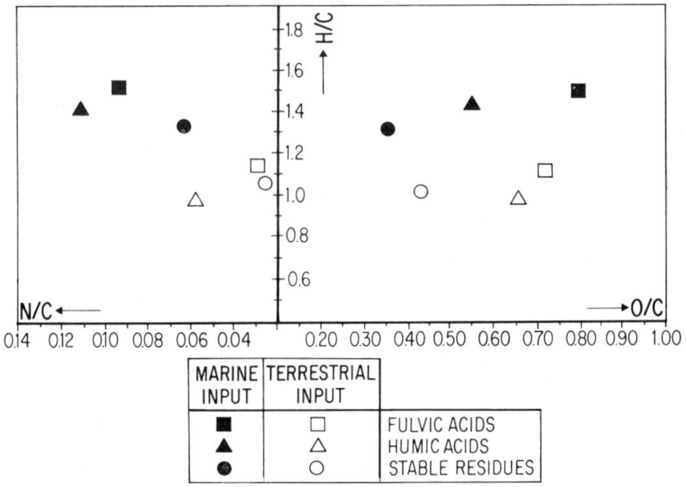

FIGURE 2. Elemental composition of reference humic substances. Marine input: sediment from Kerguelen Islands; terrestrial input: sediment from Mahakam Delta (Indonesia). After Debyser and Gadel (1981).

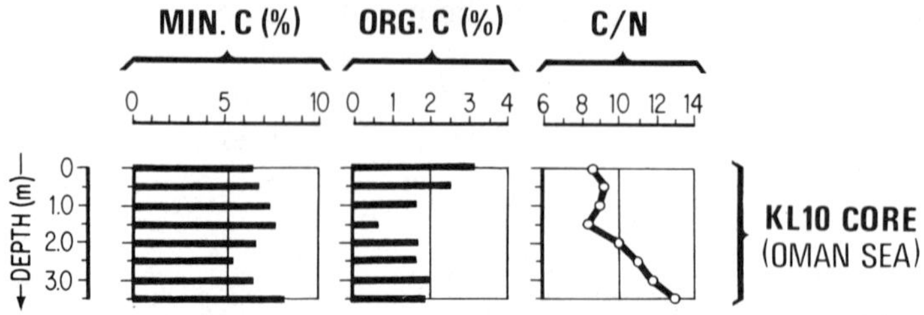

FIGURE 3. Evolution of C/N ratio of a homogeneous core from Oman Sea, showing the rapid loss of nitrogen during burial of the sediment.

Elemental composition of humic substances changes as a function of geological evolution. Nitrogen content decreases in the first few meters of the sediment. Figure 3 shows the N/C ratio in a core sample from a deep marine sediment of the Oman Sea. This sediment has a very homogeneous organic and mineral carbon content with depth (Debyser and Gadel, 1981). Atomic O/C ratios show much less variation in humic acids and stable residues than in the fulvic acids. The examples in Figure 4 correspond to peat and lignite samples deposited in a deltaic environment (Mahakam Delta; Boudou, 1981). The variation of the atomic O/C ratio is presented as a function of the organic carbon content in the samples. The carbon content of the sedimentary organic matter increases with increasing depth of burial from 0 to 1500 m, which corresponds to the entire diagenesis stage of evolution. In these samples, humic acids have higher O/C ratios than the corresponding stable residues.

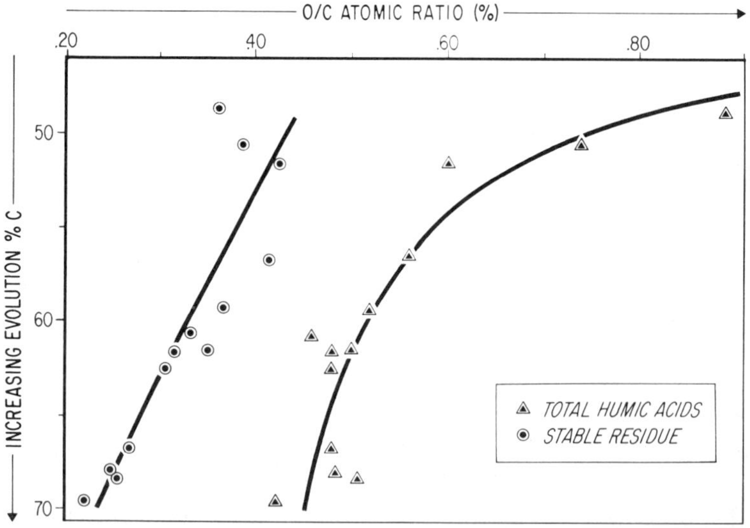

FIGURE 4. Evolution of O/C ratio in humic acids and stable residues in peats from the Mahadam Delta.
After Boudou (1981).

By plotting H/C ratios versus O/C ratios on a van Krevelen diagram, Tissot et al. (1974) devised a classification scheme which characterizes kerogens into Type I, II, or III, depending on the elemental composition of the kerogen and its evolutionary path. Type I kerogens are typical of lacustrine deposits with intense bacterial reworking, such as the Green River shales, Type II kerogens are typical of open marine environments such as the Toarcian Shales of the Paris Basin, and Type III kerogens are typical of continental input such as the Logbaba cores of the Douala Basin. Using a plot of atomic H/C ratios versus O/C ratios for humic acids and stable residues from recent and immature ancient sediments (Fig. 5) Huc and Durand (1977) demonstrated that the overall difference between the humic acids and stable residues for Type III kerogens is due to more CO_2 and H_2O in the humic acids than in the stable residues, and for Types I and II kerogen to more CO_2 in the humic acids than in the stable residues. The CO_2 and H_2O referred to are related to oxygenated functional groups of the humic acid, which are lost during diagenesis with the generation of CO_2 and H_2O.

Variations in the atomic S/C ratios are more difficult to observe because of experimental uncertainties linked to the determination of organic sulfur. In surface sediments, S/C ratios are within the range 0.01–0.03 and generally decrease from fulvic acids to humic acids to stable residues. In stable residues, sulfur content increases rapidly in the few first meters of burial (see Table 1; Debyser and Gadel, 1981). The presence of sulfur in humic substances from marine sediments was previously noted by Nissenbaum and Kaplan (1972).

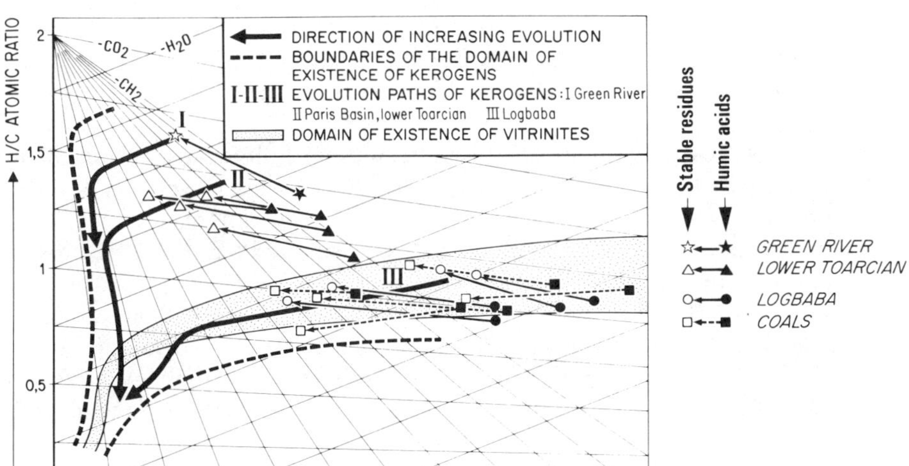

FIGURE 5. Elemental analyses of humic acids and related kerogens in immature ancient sediments.
After Huc and Durand (1977).

TABLE 1. Variation of S/C Atomic Ratio of Humic Substances in Core KL9 from the Oman Sea[a]

	S/C Atomic Ratio		
Depth (m)	Fulvic Acid	Humic Acid	Stable Residue[b]
0.0	0.032	0.011	0.006
0.5	0.022	0.015	—
1.0	0.025	0.016	0.013
1.5	—	0.014	—
2.0	0.025	0.017	—
2.5	0.025	0.018	—
3.0	0.026	0.024	—
3.5	0.026	0.026	—
4.0	0.028	0.022	—
4.5	0.030	0.025	0.036
5.0	0.026	0.026	—

[a] *Debyser and Gadel (1981).*
[b] *Nonhydrolyzable part of humin.*

Functional Groups as Determined by Infrared Spectroscopy

When infrared spectroscopy is applied to humic substances, a number of functional groups can be determined (see MacCarthy and Rice, Chapter 21 in this book). The procedures for the preparation of KBr pellets and recording of the spectra have been described elsewhere (Robin et al., 1977; Debyser and Gadel, 1981). The spectra can be compared in a quantitative way (for related samples or compounds) by means of planimetry of the absorption bands.

Spectra of fulvic acids must be interpreted carefully; purification of this fraction leads to losses that can reach 50%. Comparison of spectra of different humic fractions of the same sample, such as a sediment from the Oman Sea (Fig. 6) illustrates several differences. Oxygenated functional groups are more important in fulvic and humic acids than in stable residues. Particularly important are the absorption bands at 3400 cm^{-1} (OH from alcohols, acids, etc.), 1710 cm^{-1} (C═O from quinones, ketones, carboxylic acids), 1250 cm^{-1} (C—O from alcohols, esters, ethers) and 1050 cm^{-1} (C—O from carbohydrates). Absorption at 1050 cm^{-1} is nearly absent in stable residues. Aliphatic content increases from fulvic acids to humic acids and stable residues (bands between 2870 and 2960 cm^{-1}) and the shape of the aliphatic bands (2900–2950; 1450; 1375) indicates that fulvic acids contain mainly CH groups.

Table 2 presents the variation of absorption coefficients with burial depth in a core from the Oman Sea. The only variation at 5 m burial in humic and

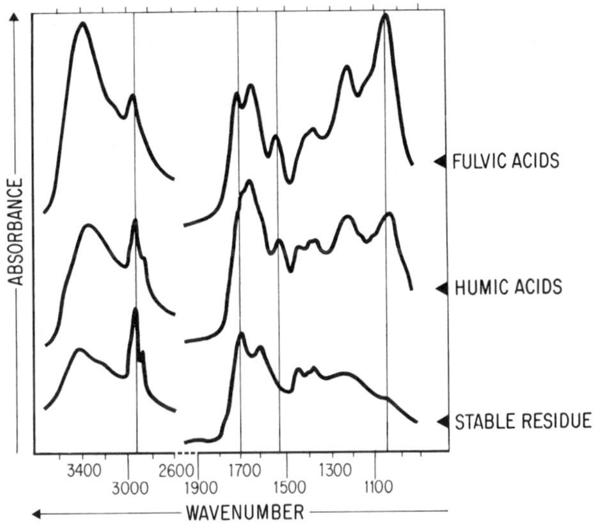

FIGURE 6. Comparison of infrared spectra of fulvic acids, humic acids, and stable residues, from a recent sediment from Oman Sea.

fulvic acids is a decrease of the amide band at 1540 cm^{-1}, due to peptidic linkages. This observation corresponds to the downcore increase of C/N ratio in sediments (Fig. 3).

Structural Information from ^{13}C and ^{1}H NMR Spectroscopy

Infrared spectroscopy is most effective for functional groups containing heteroatoms such as oxygen or nitrogen. Infrared spectroscopy cannot effectively determine aromaticity, because CH bands do not respond well, partic-

TABLE 2. Evolution of Absorption Coefficients (K) for Some IR Bands of Humic Substances in a Core from the Oman Sea[a]

	Fulvic Acid			Humic Acid			Stable Residue	
Depth (m)	K 2920 (Aliphatic Bands)	K 1710 (C=O)	K 1540 (Amide II)	K 2920	K 1710	K 1540	K 2920	K 1710
0.00	34.8	73.2	33.1	39.1	53.4	19.5	60.7	60.7
0.50	35.3	75.9	31.8	41.4	55.6	19.4	—	—
1.00	32.8	78.4	34.7	42.0	55.2	15.8	—	—
1.50	—	—	—	40.5	54.5	13.9	—	—
2.00	41.1	84.0	24.3	40.6	54.5	11.4	—	—
2.50	34.1	77.6	26.5	44.5	54.4	12.4	—	—
5.00	—	—	—	37.8	55.3	12.1	—	—

[a] *Debyser and Gadel (1981). Units are arbitrary.*

ularly in the case of aromatic structures. Elemental analysis can provide general information on aromaticity, but a method that distinguishes carbonaceous structures according to their chemical environment would be more valuable. Technical improvements allowing the application of nuclear magnetic resonance spectroscopy to humic substances have been developed (Wilson, 1981). This promising technique and its general application to humic substances are treated by others in this book (Hatcher, Breger, Maciel, and Szeverenyi, Chapter 11; Wershaw, Chapter 22). Hatcher et al. (1980b) have compared ^{1}H and ^{13}C NMR spectra from marine-sediment humic acids (Mangrove Lake, Bermuda) to those obtained from a peat (Minnesota peat). The aliphatic carbon region (0–2.5 and 0–50 ppm) of their spectra is not easy to explain, as the chemical shifts vary significantly. However, it seems that long aliphatic chains are nearly absent, and that aliphatic substituents are highly branched. A major peak at 75 ppm in the ^{13}C NMR spectra of the Mangrove Lake sample is attributed to carbohydrate carbon. The aromatic carbon region is much more apparent in the humic acids from peat. Many peaks appear in that region, and they certainly correspond to specific chemical environments, but the carbon structures associated with these peaks have not been ascertained at the present time.

^{13}C NMR spectra were obtained by Hatcher et al. (1980c) for "demineralized humins" (i.e., stable residues), by the CPMAS technique on the Mangrove Lake sample and an Everglades peat. There is a greater similarity between the humic acids and stable residue from the Minnesota peat and the Everglades peat than between the humic acids and stable residue from the Mangrove Lake sediment sample. In particular, the humic acids from Mangrove Lake sediments are much less aliphatic than the corresponding stable residue and contain mainly carbohydrate chains, while carbonyl groups are more abundant in the stable residue.

Functional Group Analysis and Degradative Techniques

Since these techniques are specifically addressed elsewhere in this book, some typical applications to humic substances in sediment will be commented upon only briefly here.

Functional Group Analysis

The main types of functional groups analyzed in humic compounds are oxygenated groups (carboxyl, phenolic hydroxyl or total hydroxyl, and carbonyl) and nitrogen-containing groups (amine). The various analytical techniques used in the analysis of those groups can be found in Rashid and King (1970) and in Jocteur-Monrozier and Jeanson (1981); those authors, as well as Huc et al. (1974), have applied these techniques to humic substances from different marine sediments. In general, they have found that total acidity and particularly, phenolic acidity is much less in marine-sediment humic

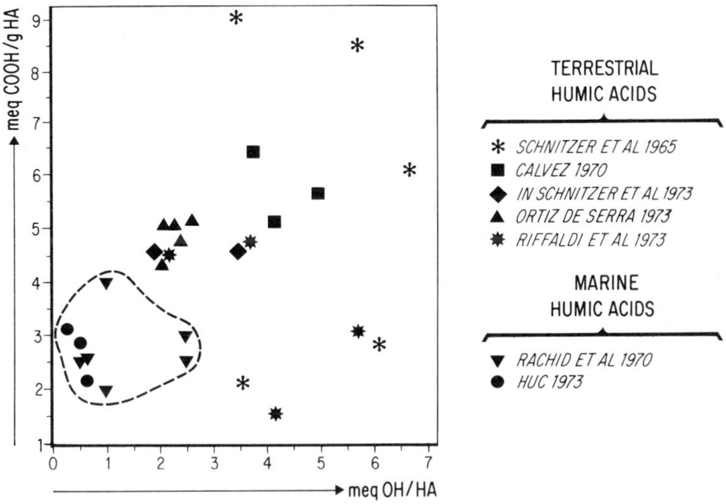

FIGURE 7. Acidic functional groups in marine and terrestrial humic acids. After Huc et al. (1974).

acids deriving from marine biomass than in humic acids deriving from continental biomass (Fig. 7).

Mineralization of the organic matter begins by transformation of the α-amino groups into ammonia. This conversion is observed in the acid-soluble fraction obtained when carbonates are removed by acid treatment of the sediment prior to alkaline extraction. In this fraction, the ratio of ammonium nitrogen to α-amino nitrogen increases steadily with burial in the first few meters of the sediment (Jocteur-Monrozier and Jeanson, 1981). However, these data are not always easy to interpret due to the dissolution, diffusion, and adsorption of ammonium nitrogen on clay minerals.

Degradative Techniques

There are two types of degradative methods: thermal (pyrolysis) and chemical (oxidation, reduction, and hydrolysis). Flaig et al. (1975) and Schnitzer (1978) review these methods. In general, degradative methods must be applied carefully; identical functional groups in different environments can be affected to varying degrees by mild but selective reagents. Strong reagents can separate the sample into small molecular fragments bearing relatively little information, and recombination of degradation products can further obscure the interpretation of the data.

Acid hydrolysis is often used to release amino acids and carbohydrates from the total sediment or from isolated humic acids. This method was first applied to marine sediments by Degens et al. (1964). However, since the apparent distribution of amino acids and sugars seems to depend strongly on the hydrolysis procedure, these products are not easily related to the organic matter examined (Morris, 1975; Pelet and Debyser, 1977), although the ratio of pentose to hexose sugars may be diagnostic (Morris, 1975). For this reason, it seems advisable to compare hydrolyzates only by bulk analytical methods. Elemental analysis is a most useful method for this purpose.

Two problems occur in using acid hydrolysis for quantitative determinations: the occurrence of secondary reactions between some of the compounds released by the hydrolysis (phenols, indoles, furans, etc.) and the dissolution of metals (calcium, aluminum, iron), which subsequently precipitate when the solution pH is neutralized. The latter problem could perhaps be overcome by an initial hydrolysis with water in an autoclave (Stefan and Jocteur-Monrozier, 1983), which would dissolve most of the polysaccharides without dissolving the minerals.

Degradative methods based on pyrolysis are the subject of renewed interest due to the identification power offered by gas chromatography–mass spectrometric systems (GC–MS) (Wershaw and Bohner, 1969; Martin et al., 1977; Meuzelaar et al., 1977; Bracewell and Robertson, 1976). There are two main pyrolysis techniques: (1) controlling the decomposition kinetics by temperature programming and (2) the use of quasi-instantaneous heating (e.g., Curie point pyrolysis). The later technique avoids most recombination reactions, but does not allow kinetic control. The pyrolysis effluent can be detected directly (Rock–Eval method) or after chromatographic fractionation.

Rock–Eval pyrolysis (Espitalie et al., 1977) was applied to immature ancient marine sediments by Herbin and Deroo (1979) and to recent marine sediments by Debyser and Gadel (1981). This technique results in information similar to elemental analysis, but the procedure is much faster, cheaper, and easier. However, it seems necessary to modify this method for use in recent sediments, where organic matter is thermally labile and rich in oxygen. This problem is currently under investigation at the Institut Francais du Petrole.

Pyrolysis–GC is being used increasingly in the field of petroleum geochemistry for rapid comparison of samples by fingerprinting. The method used by the Institut Francais du Petrole (Saint-Paul et al., 1980; Durand and Paratte, 1983) has been applied to humic substances in sediments. It is a low-temperature (475°C) pyrolysis with intermediate trapping of the effluent with liquid nitrogen (with the exception of CH_4) followed by GC analysis (Dexsil 300 packed column).

Figure 8 shows pyrograms from fulvic acids, humic acids, and kerogens of surficial marine sediments from the Mahakam Delta (Indonesia) and from the Black Sea. Peaks are due primarily to saturated, unsaturated, and aromatic hydrocarbons. Benzene and toluene peaks have been tentatively identified, as well as peaks due to *n*-alkanes and *n*-alkenes. Fulvic acids and humic acids behave very differently upon pyrolysis: fulvic acids produce no methane and very few hydrocarbons, except benzene and toluene. Humic

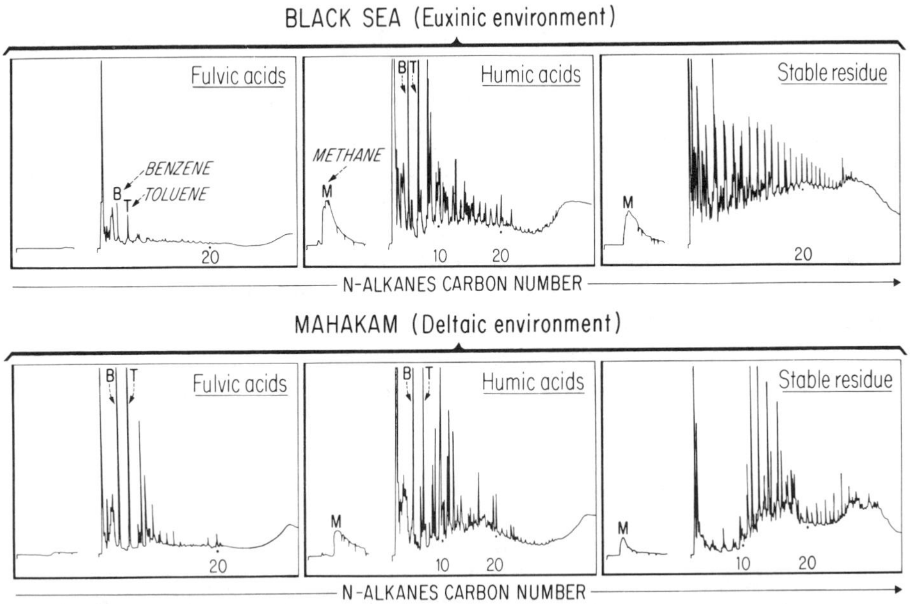

FIGURE 8. Pyrolysis–gas chromatography of fulvic acids, humic acids, and stable residues from marine sediments containing terrestrial organic input (Mahakam Delta) and planktonic organic input (Black Sea).

acids react similarly to stable residues, but generate almost no hydrocarbons beyond C_{20}. These limited results, however, are consistent with those reported by Poutanen and Morris (1983), who extracted humic and fulvic acids from recent sediments accumulating on the continental shelf of Peru. These humic and fulvic acids were then extracted with a chloroform/methanol mixture. The extract from the humic acid contained a proportion of lipids and pigments. In contrast, only very small amounts of lipids were found in fulvic acids (20 times less than in humic acids). This implies that the lipids are associated with the humic acids.

Ishiwatari et al. (1977) conducted pyrolysis experiments using controlled kinetics on humic acids and stable residues in recent marine sediment from offshore California. Their results correlate with those in Figure 8: kerogens and stable residues generate liquid compounds, including *n*-alkanes; humic acids, which contain many acid groups, generate mainly CO_2 and H_2O, with few hydrocarbons and liquid compounds.

Isotopic Composition

Nissenbaum and Kaplan (1972) suggest an autochthonous origin for some marine humic substances based on differences in the isotopic composition of marine and terrestrial organic matter. These two environments generally do not have the same source of carbon, resulting in the incorporation of different percentages of ^{13}C into the organic matter of each environment. Humic acids from temperate marine sediments, where organic input is mainly planktonic, have $\delta\ ^{13}C$ values between -17 and $-22‰$. The mean isotopic composition of carbon in the associated marine plankton is $-19‰$. Humic acids from coastal and littoral sediments, where organic input is mainly continental, generally have $\delta\ ^{13}C$ values between -25 and $-27‰$, similar to that of common higher plants. These data represent only general trends, as isotopic composition depends also on climate and metabolism of the organisms. Fulvic acids are isotopically heavier (generally 1–2‰) than humic acids, due to differences in chemical structure and to the abundance of oxygen-containing functional groups in fulvic acids. As was suggested (Galimov, 1980), these functional groups are isotopically heavier.

INFLUENCE OF SEDIMENTARY ENVIRONMENT

Origin of Organic Input

Terrestrial organic matter is derived mainly from higher plants: lignin and cuticular waxes are its most stable constituents. In contrast, organic matter

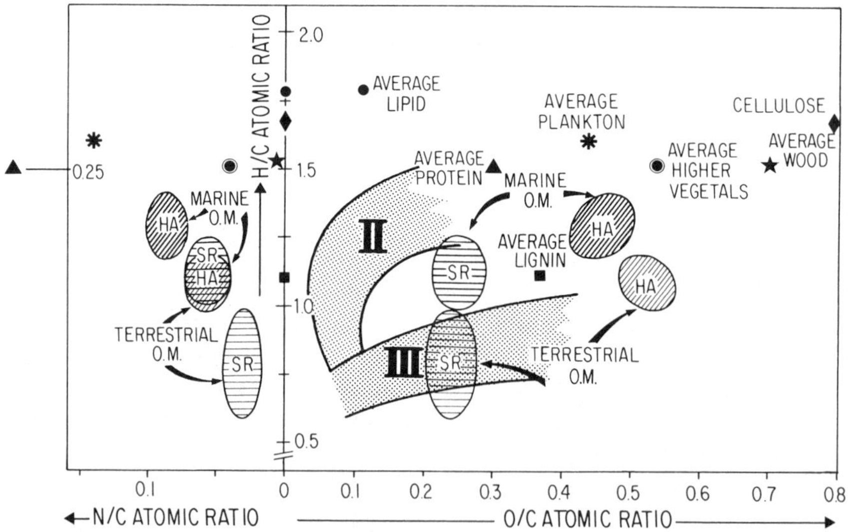

FIGURE 9. Average elemental composition of humic acids and stable residues from marine and terrestrial organic matter, compared to average elemental composition of some biopolymers.
After Pelet (1981).

produced in the oceans is mainly planktonic and is essentially devoid of lignin and waxes. Terrestrial organic matter and marine organic matter can both be transported to a sedimentary basin in varying amounts depending on the productivity of each of the environments and the geographical conditions.

Figure 9 shows the elemental compositions of several humic acids and stable residues isolated from marine sediments with either terrestrial (Amazon deep sea fan, Debyser et al., 1978) or marine (Aden Gulf and Oman Sea, Debyser and Gadel, 1981) organic input. The mean compositions of some constituents of the living organic matter, calculated by Pelet (1983), are also represented. In the humic acids and stable residues, the sedimentary organic materials are clearly differentiated according to their origin. Terrestrial organic matter is depleted in hydrogen and nitrogen relative to marine organic matter. Thus, humic compounds can be roughly related, according to their origin, to Type II or Type III kerogen (Tissot et al., 1974). The differences in composition between marine and terrestrial humic substances are less than those observed in the kerogen series. This can be attributed to the greater variability in elemental composition of humic acids, due to climate (Schnitzer, 1978), different types of plankton (Pelet, 1983), or differential alteration during transportation of amorphous and particulate continental organic matter (Table 3; Pelet, 1978). Attributing continental or marine origin to sedimentary organic matter on the basis of elemental analysis alone is highly controversial (see Table 4).

TABLE 3. H/C Atomic Ratios in Humic Acid from Continental Organic Matter

Climate		Transportation Form		Alteration		
Hot	Cold	Particulate	Pseudo-solution[a]	Subaerial[b]	Subaquatic[b]	H/C
	+	+		+		1.3[c]
+		+		+		0.9–1.1[d]
	+		+	+	+	1.35[e]
+			+	+	+	1.15[f]

[a] *True solution, colloidal solution, or organomineral complex; in all cases size of particles is much smaller than in particulate form.*

[b] *Subaerial means in the atmosphere, subaquatic means in water. In soils, alteration of land plants is mainly subaerial, but when these soils are eroded by water and brought to lakes or sea, where sedimentation occurs, subaerial alteration is followed by subaquatic alteration.*

[c] *Schnitzer (1977).*

[d] *Boudou (1981).*

[e] *Debyser et al. (1977).*

[f] *Debyser et al. (1978).*

TABLE 4. Elemental Analysis of Humic Acid from Soils Under Various Climates[a] and from Various Marine Planktonic Constituents[b]

	Atomic H/C	Atomic O/C	Atomic N/C
Humic Acid from Soils			
Arctic	1.324	0.438	0.066
Temperate	1.068	0.450	0.072
Subtropical	1.039	0.490	0.063
Tropical	1.143	0.475	0.075
Humic Acid from Marine Plankton			
Dinoflagellates	1.52	0.50	0.080
Diatomaceae	1.37	0.42	0.065
Coccoliths	1.29	0.46	0.097

[a] *After Schnitzer (1978).*
[b] *After Pelet (1983).*

Influence of Alteration in the Sedimentary Environments

There are two stages in the alteration process that lead to the final incorporation of biological material into sedimentary organic matter. The first stage, *presedimentary alteration,* takes place on the continent and/or in the water column. It begins with the autolysis of cells and use by other organisms of most of the assimilable compounds. The resulting mineralization is very important; direct measurements on sediment traps in various oceanic sites show that below 2000 m at least 98% of the organic matter derived from plankton at the surface has disappeared (Honjo, 1980; Suess, 1980; Weser et al., 1982).

The second stage, *synsedimentary alteration,* is related to the consumption of organic matter by benthic organisms and bacteria within the sediment. Its mechanism and influence on the composition of the residual organic matter are largely unknown, but it is apparent from the decrease in amounts of organic carbon observed along a core that this synsedimentary alteration is quantitatively much less important than presedimentary alteration. In some cases, the decrease in the amount of organic matter along a core is not as great as expected and is on the same order as variations caused by climatic differences or minor changes in the amounts of organic input. When a decrease in organic carbon content is observed, generally the amount of carbon deposited in the surficial sediment is decreased by a factor not greater than 2 (Pelet, 1983). The overall effect of presedimentary alteration on the composition of organic matter deposited in marine sediments is seen by comparing mean elemental compositions of plankton with marine-sediment humic compounds and mean elemental composition of higher

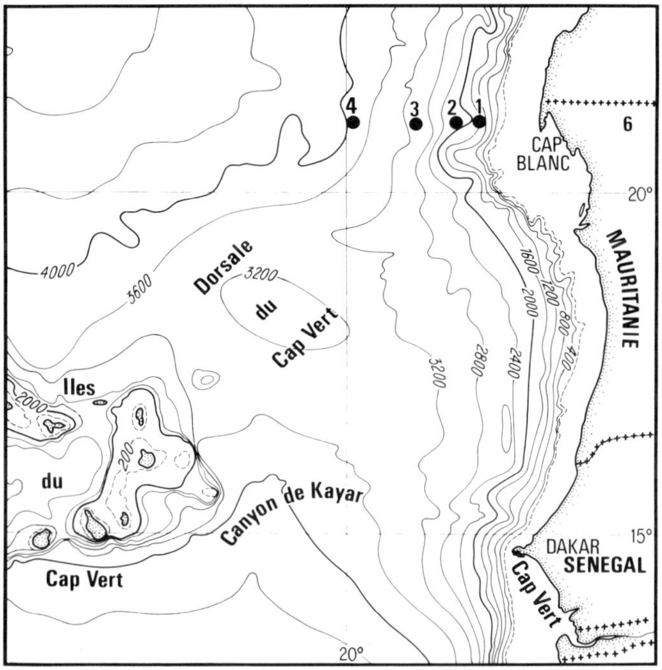

FIGURE 10. Geographic setting of ORGON III—Cape Blanc sampling. After Pelet (1981).

plants with terrestrial humic compounds (Fig. 9). In going from precursor organisms to humic substances, decreases in atomic H/C and O/C ratios are observed for both. In marine samples, a decrease in the N/C ratio is also observed.

The specific influence of transport distance and duration of the marine transportation stage has been demonstrated. The ORGON III cruise (Mauritania) produced one example of this stage of presedimentary alteration by sampling surficial sediments at four stations along the Cape Blanc transect (Fig. 10). Planktonic organic matter is produced by the Mauritania upwelling in a restricted area onshore from core 1 (Huntsman and Barber, 1977). There is no organic matter input from the land, which is desert. Presedimentary alteration increases with water depth and distance from shore. Synsedimentary alteration in each sample is similar (except for sample site 2, where it is lower) as shown by sedimentation rates and *meiofauna* amounts in Table 5. The amount of organic matter lost as the transport distance increases can be estimated by the decrease in organic carbon concentration of the surficial sediments at each of the sampling sites (Table 5). In going from sample site 1 to sample site 4, the decrease is about 5 : 1. This is not high compared to total presedimentary alteration (Suess, 1980). Table 6 shows the corresponding variation in composition of sedimented organic matter. With increased

TABLE 5. Environmental Parameters of Surficial Sediments of the Cape Blanc Area

Sample Number	Water Depth (m)	O_2 Content of Bottom Water (ppm)	Distance to the 200 m Isobath (km)	Mean Age of Sediment (yr)	Meiofauna Amount (No./cm^2)	TOC in Sediment (%)
1	1900	7.3	35	5500	40	2.28
2	2500	7.4	70	2500	2	2.08
3	3250	7.5	135	5500	143	0.91
4	3750	7.4	215	9000	60	0.43

transport, fulvic acid and the hydrolyzable fraction increase, the proportion of stable residue decreases (Debyser and Gadel, 1979), and the proportion of ammonia in the hydrolyzable nitrogen increases (Jocteur-Monrozier and Jeanson, 1979). The elemental analyses of humic acids and stable residues (Fig. 11) show a decrease in the H/C ratio with transport distance, and also a slight decrease in O/C and N/C ratios, mainly in the stable residues. If the H/C ratio versus O/C ratio diagram alone is used, these latter points appear shifted to Type III kerogens which are derived from higher plant material. The N/C ratios of the stable residues (0.06), however, are still characteristic of planktonic organic matter ($\sim$0.06) rather than terrestrial organic matter ($\sim$0.03).

The Cape Blanc samples suggest two possible conclusions: (1) increasing alteration increases the amounts of fulvic acid and of the hydrolyzable fraction at the expense of the stable residue; and (2) alteration as well as origin and evolutionary stage of the organic matter influence the position of the representative points in the H/C, O/C diagram. Presedimentary alteration depends primarily on the nature of the organic matter. On land, terrestrial organic matter decays and is transported in a highly oxygenated and reactive environment. Only the most stable compounds, because of their chemical structure or because they are protected by association with minerals, will reach the marine environment; subsequent alteration of this material will be low. In contrast, degradation of marine organic matter depends largely on the depth of the water column and redox conditions.

TABLE 6. Composition (%) of Kerogen in Surficial Sediments of the Cape Blanc Area

Sample Number	2*N* Hydrolysate	Fulvic Acid	Humic Acid	Hydrolyzable Part of Humin	Stable Residue	Total Hydrolysate	Total Humic Acid
1	14	8	18	25	35	39	26
2	11	10	39	20	20	31	49
3	16	16	23	29	15	45	40
4	28	19	23	23	7	51	42

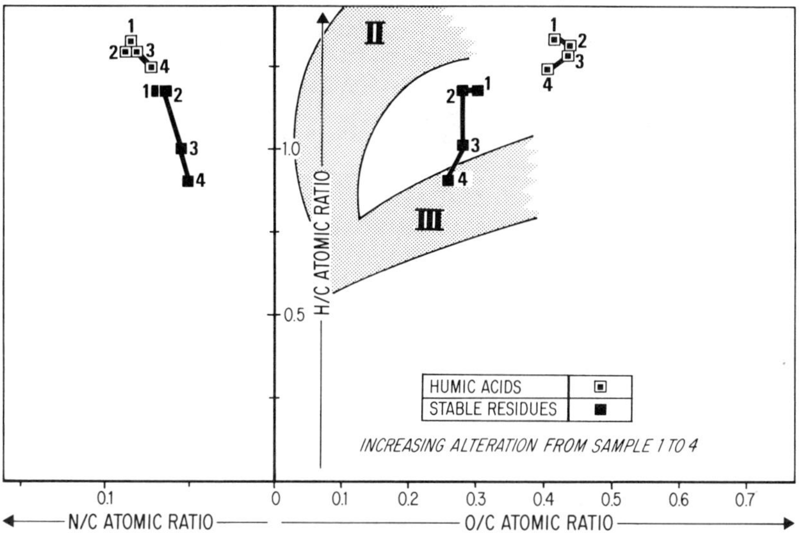

FIGURE 11. Elemental analyses of humic acids and stable residues from the Cape Blanc samples.
After Pelet (1981).

An estimation of the influence of transport distance on the composition of terrestrial and marine organic matter sedimented in the marine environment was performed by Pelet (1979), Boudou (1981), and Debyser and Gadel (1983). These data (see Table 7) are only indicative, due to their small number (essentially all samples come from ORGON cruises); however, some comparisons of the effects of alteration on terrestrial and marine organic matter during transport can be made. A long transportation period increases the amount of the hydrolyzable fraction and fulvic acid in both terrestrial and marine organic matter, and the amount of total humic substances for marine organic matter. These increases can be related to fragmentation of the organic macromolecules by oxidative cleavage. Oxygen fixation can also occur to varying extents, according to the origin of the organic matter. For marine organic material, oxygen is fixed in both humic acids and stable residues; the O/C ratio is constant or increases, and the amount of fulvic acid, created by oxygenation of the other fractions, increases. Nitrogen is eliminated by mineralization of proteins, yielding ammonia, as indicated by the decrease of N/C ratios in humic acids. For terrestrial organic matter (Table 7), the increase of the H/C and N/C ratios in humic acid is related to their decarboxylation (loss of CO_2). The O/C ratio does not vary. The amount of stable residue decreases in kerogen, and stable residues lose aliphatic side chains, becoming condensed and more aromatic, as indicated by the significant decrease of the H/C ratio. This interpretation assumes that chemical differences between the various fractions of terrestrial organic matter (particulate versus amorphous) can be disregarded, which may not be true.

TABLE 7. Influence of Presedimentary Alteration on Composition of Marine and Terrestrial Organic Matter (OM) in Surficial Sediments under 3–4000 m of Water

	Characteristics of Transportation			% Carbon of OM in			Elemental Analysis—Atomic Ratios					
							Humic Acid			Stable Residue		
Origin of OM	Size/Shape of the OM	Redox Status of Environment	Alteration Intensity	2*N* Hydrolysate	Humic Acids	Fulvic Acids	H/C	O/C	N/C	H/C	O/C	N/C
Marine	Amorphous	Reducing	Low	15	15	5	1.35	0.40	0.090	1.20	0.30	0.060
Marine	Amorphous	Oxidizing	High	70	20	20	1.25	0.45	0.075	1.10	0.30	0.055
Terrestrial	Particulate	Oxidizing[a]	Low	40	25	15	1.00	0.55	0.040	1.00	0.35	0.020
Terrestrial	Amorphous	Oxidizing[a]	High	70	15	25	1.20	0.50	0.075	0.80	0.25	0.030

[a] *In beginning.*

POSSIBLE ORIGINS OF HUMIC SUBSTANCES FROM MARINE SEDIMENTS

Incorporation of Terrestrially Derived Humic Substances into Marine Sediments

The weathering of soil by wind and water contributes detritus to the marine sedimentary environment. The organic fraction of this detritus input is composed primarily of the debris and degradation products of higher plants; the mineral fraction consists of clay and quartz. The characteristics of these soil humic substances have been determined by soil scientists. Organic substances, which survive the severe biological and chemical oxidation conditions in soils, are quite stable and are transported without significant modification during weathering. Soil humic substances are found in marine sediments nearly in their original form.

Formation of Humic Substances by Polycondensation of Molecules from the Degradation of the Organic Matter

Most humic substances in soil result from the biological and chemical degradation of dead organisms. The conditions under which humic substances are formed are not clear; these conditions, however, have been simulated in reactions between model substances, and analyses on the resulting compounds using the same methods as used on naturally occurring humic compounds reveal several similarities. The formation of humic substances may result from oxidative alteration of organic fragments, microbial synthesis, or chemical condensation after biological degradation or autolysis of living biomass (Felbeck, 1971). In addition to condensation products, oxygen-containing molecules such as carbohydrates and uronic acids are present, mainly in fulvic acids (Ishiwatari, 1975a; Hatcher, 1980).

Condensation reactions are probably the most important in the mass balance, and the mechanisms governing these reactions are readily extended from soil humic to sedimentary substances. The principal constituents of vascular land plants are carbohydrates in the form of cellulose and hemicelluloses, and phenols, which are present as tannins, flavonoids, and other plant pigments, and as building units in lignin (Given, 1972). Proteins are also present, but generally in low amounts. Condensation reactions in these compounds are mainly related to the oxidative condensation of polyphenols through quinoid derivatives (Flaig et al., 1975). Polycondensates formed in this way are more stable when α-amino compounds are present (Andreux et al., 1979). Nitrogen does not participate directly in the condensation bonds and its concentration is extremely variable.

In the marine environment, organic matter is composed largely of carbohydrates and proteins (Gagosian and Lee, 1981; Harvey and Boran, Chapter 9 in this book). These compounds condense to form brown compounds called melanoidins (Maillard, 1913). In this case, nitrogen participates in the condensation bonds, and the macromolecules formed contain more nitrogen than the products of the oxidative condensation of polyphenols discussed in the previous paragraph. This nitrogen is generally more resistant to hydrolysis (Jocteur-Monrozier and Jeanson, 1981).

Oxidative Degradation of Organic Material

As shown previously, presedimentary alteration of planktonic organic matter during transportation causes an increase in the percentage of fulvic acid in relation to both the total humic extract and total organic matter. It has been demonstrated (Flaig et al., 1975) that if soil humic acid is stored for long periods in solution, fulvic acid is formed. Thus, it is possible that in the marine environment some sedimentary fulvic acid is formed by oxidation of other fractions of the organic matter. This hypothesis explains several structural features of fulvic acid: its low molecular weight, high amount of oxygenated functional groups, and low content of aliphatic chains. There is no reason to think that oxidative degradation reactions similar to those leading to a part of the fulvic fraction could not also form humic acid from humin, although humic acid content is more difficult to monitor for analytical reasons (nature and amount of extract, extraction procedure, and others). Biochemical and chemical oxidation reactions take place in soils as well as in the marine environment, and this could explain, in part, why humic and fulvic acids are often more abundant in the highly oxygenated soil environments than in some oxygen-depleted marine environments. Oxygen is present in marine sediments in the surface layers of the sediment, in the bioturbation zone, and where bottom water can percolate through the sediments. Wherever oxygen is present in the sediment, oxidative alteration can take place, transforming humin to humic acid and fulvic acid, and finally to CO_2, NH_3, and H_2O. Once buried in the sediment in an oxygen-poor environment, the rate of oxidation of organic matter is greatly decreased and diagenesis begins.

DIAGENETIC TRANSFORMATION OF HUMIC SUBSTANCES IN MARINE SEDIMENTS

The amounts of hydrolyzable organic matter and humic substances extracted by alkaline aqueous solutions decrease with the burial depth of a sediment. This is demonstrated by Huc et al. (1980) on Black Sea sediments where the organic matter is mainly autochthonous (Fig. 12). Boudou (1981), who studied diagenesis of terrestrial organic matter deposited in marine deltaic sediments, also noted this decrease (Table 8). The amount of hydrolyzable organic matter decreases very rapidly as a function of depth. Humic substances disappear more slowly, and their decrease can be followed dur-

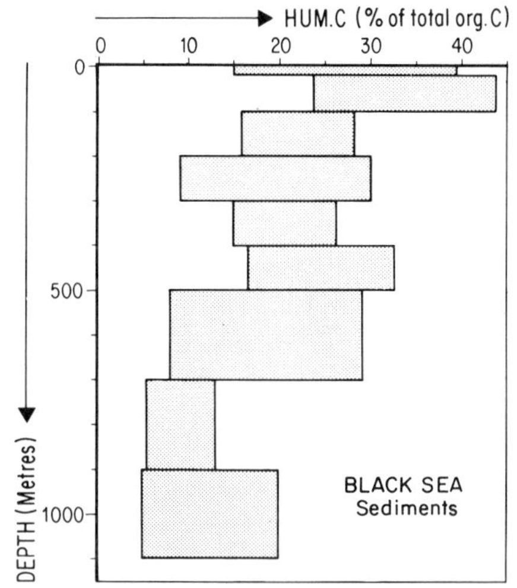

FIGURE 12. Variations in amounts of humic compounds during diagenesis in Black sea sediments.
From Huc (1980).

ing the whole diagenetic stage; here the kerogen, in which the proportion of stable residue becomes increasingly important, loses mainly oxygen. A discussion of the transformation and incorporation of humic substances into the insoluble organic matter of ancient sediments follows.

Fulvic Acid

The ratio of fulvic acid to humic acid decreases with the burial depth of a sediment (Brown et al., 1972; Ishiwatari, 1975; Huc and Durand, 1974). Some authors (Nissenbaum and Kaplan, 1972) believe this diagenetic decrease is a result of progressive condensation into humic acid and then humin. Others think, however, that the progression from fulvic acid to humic acid and humin is not the only possible mechanism to explain the decrease in fulvic acid content (Ishiwatari, 1975b; Jocteur-Monrozier, 1981; Pelet, 1983; Poutanen and Morris, 1983).

Based on the data and speculations presented earlier, it is probable that fulvic acid does not undergo reactions decreasing its solubility. Fulvic acid seems to result principally from the oxidation of organic matter; the formation of fulvic acid decreases with burial depth, because there is no more oxygen in the sediment. The fraction already present in the sediment is eliminated from the sediment by solubilization in the pore water or mineralization by bacteria.

Humic Acid

Elemental analyses of humic acid from samples at increasing burial depths show a general decrease in the O/C and N/C ratios (Ishiwatari, 1975a; Huc and Durand, 1977). As Table 8 shows, this decrease is also seen in terrestrial

TABLE 8. Evolution of Humic Substances in Peats, Mahakam Delta, Indonesia[a]

Atomic O/C (Peat)[b]	Percentage of the Kerogen: FA + HA[c]	Percentage of the Kerogen: Hydrolyzable Part of Humin	Percentage of the Kerogen: Stable Residue	Atomic O/C (FA + HA)
0.67	33.0	8.5	58.5	0.89
0.65	24.1	12.3	63.6	0.74
0.60	36.2	0.7	63.1	0.60
0.49	40.6	0.2	59.2	0.56
0.42	35.4	0.2	64.4	0.52
0.40	41.6	0.1	58.3	0.46
0.39	29.8	0.2	70.0	0.50
0.37	26.9	0.1	72.0	0.48
0.37	25.4	0.3	74.3	0.48
0.28	16.3	0.2	83.5	0.48
0.27	11.9	0.2	87.9	0.51
0.25	11.9	0.1	88.0	0.48
0.23	8.3	0.1	91.6	0.42
0.14	2.1	0	97.9	—

[a] *After Boudou (1981).*
[b] *This parameter is a measurement of increasing evolution with burial.*
[c] *FA = fulvic acid; HA = humic acid.*

organic matter (Boudou, 1981). Infrared analyses demonstrate that defunctionalization occurs with increasing burial depth. In the case of nitrogen-containing groups (amide bands at 1680 and 1540 cm^{-1}), this defunctionalization occurs very quickly. Oxygenated groups disappear more slowly than nitrogen-containing functional groups. The defunctionalization process may promote the insolubilization of humic acid and its incorporation into kerogen. In a given sample of sedimentary organic matter, the stable residue is always poorer than humic acid in oxygen, but their elemental compositions follow parallel trends (Fig. 5).

Very little is known about if and when humic acid molecules are incorporated into stable residues. No information is available concerning aliphatic chains, aromatic structures, the distribution of the residual functional groups, or molecular size of the incorporated molecules.

SUMMARY AND CONCLUSIONS

Humic substances in marine sediments originate from both marine and terrestrial sources of organic matter, depending on the nature of sedimentary input. In some cases, a set of criteria based on chemical properties makes it possible to determine their origin. However, these criteria are less clear-cut than those established for kerogens.

The formation and evolution of humic substances can tentatively be separated into three stages. The first stage is formation of "primary humic substances" from breakdown products of cellular constituents of dead organisms. These substances can be either fairly small stable molecules, such as carbohydrates and amino acids, or macromolecules resulting from condensation reactions between more reactive breakdown products. The distribution of "primary humic substances" between fulvic acid, humic acid, and humin is not known, but a strong tendency toward highly condensed structures is expected. Condensation reactions between functionalized molecules like these breakdown products are known to occur very easily; molecules escaping condensation are metabolized quickly and do not become part of the preserved organic fraction.

The second stage is formation of "secondary humic substances" through chemical and biological oxidative degradation. Oxidative degradation produces smaller and smaller molecules and results in a decrease of the humin fraction and an increase of the fulvic acid fraction. Humic acid might increase or decrease, according to the rates of different degradation reactions. Biological oxidative degradation is probably more efficient than chemical degradation at this stage.

The third stage is incorporation into the sediment. Once these substances are incorporated into sediment and buried, oxygen and biota are no longer present. Fulvic acid, the smallest and most soluble fraction, decreases due to diffusion and mineralization. Humic acid undergoes insolubilization by defunctionalization and evolves with humin into the kerogen of ancient sediments.

The formation and evolution of humic substances, in our opinion, is the key to understanding the mechanisms by which kerogen forms. This knowledge is important in that it can be used to estimate the petroleum potential of a sedimentary series, and further research in this area is of both economic and scientific interest.

CHAPTER ELEVEN

Geochemistry of Humin

PATRICK G. HATCHER, IRVING A. BREGER,*
GARY E. MACIEL, and NIKOLAUS M. SZEVERENYI

ABSTRACT

Humin is the insoluble fraction of sedimentary humic substances. Little is known of its chemical composition and of its individuality as a class of humic substances, primarily because its macromolecular nature and complexity have precluded detailed analyses by conventional methods used for organic structural analyses. The advent of solid-state ^{13}C NMR has allowed new structural information to be obtained in studies of humin. This has provided us with some new perspectives on its composition and on the geochemical processes responsible for its origin. This chapter discusses our understanding of the geochemistry of humin obtained by use of the new technique while providing a review of the existing knowledge. We focus on the formation of humin, its composition, and the processes that result in its transformation to coal and kerogen in ancient sediments.

NMR spectra of humin from three major types of depositional environments, aerobic soils, peats, and marine sediments, show significant variations that delineate structural compositions. In aerobic soils, the spectra of humin show the presence of polysaccharides and aromatic structures most likely derived from the lignin of vascular plants. However, another major component of humin is one that contains paraffinic carbons and is thought to be derived from algal or microbial sources. Hydrolysis of the humin effectively removes polysaccharides, but the paraffinic structures survive, indicating that they are not proteinaceous in nature. The spectra of humin differ dramatically from that of their respective humic acids, suggesting that humin is not a clay–humic acid complex.

* Deceased.

In peat, humin is composed of lignin-derived structures, polysaccharides, and a large concentration of paraffinic structures as determined from the NMR spectra. Examination of vertical profiles of peat shows that the polysaccharides are degraded very rapidly with depth, whereas the lignin and paraffinic structures are selectively preserved. As in the case of aerobic soils, hydrolysis has little effect on the paraffinic structures. Treatment of the humin with sodium paraperiodate to remove the lignin also has little effect on the paraffinic structures. These structures, thought to be from algal or microbial components of the peat environment, are major components of the peat and survive diagenetic reactions with time and burial to become important constituents of coal and kerogen.

Humin from a marine algal deposit, a sapropel, is composed almost entirely of polysaccharides and paraffinic structures. With depth in the sapropel, the polysaccharides are degraded and lost and the paraffinic structures are selectively preserved. These structures are eventually transformed over time to kerogen in oil-producing shales. The paraffinic structures, which undoubtedly originate from algae or other microorganisms in this sapropel, are highly branched and have carboxyl, ether, and amide functional groups associated with them.

The major conclusions that can be drawn from these studies are that humin differs in many respects from associated humic acids, suggestive of the fact that humin is not a clay–humic acid complex, and that humin is composed of a significant fraction of paraffinic carbons derived most likely from algal or microbial sources.

INTRODUCTION

Humin is commonly defined as the class of sedimentary humic matter that remains insoluble when sediments are treated with dilute alkali to extract the soluble humic and fulvic acids. Because of its insolubility and macromolecular nature, humin has been the least studied of all humic fractions. The classification of humin as a separate class of humic substances was initially proposed at the turn of the century by Oden (1919), and this classification has been in use since then. Because of the many similar analytical characteristics (e.g., elemental compositions, functional group compositions, and infrared spectra) between humin and humic acids, and because of the known association of humin with inorganic clays, Khan (1945) and later Kononova (1966) regarded humin as being no more than a clay–humic acid complex. Consequently, Stevenson (1982) has recently questioned whether humin should be considered a separate class of humic substances. Treatment of humin with HF to destroy clays in many instances renders humin soluble in alkali (Stevenson, 1982).

Notwithstanding this possibility that humin is nothing more than humic acid, few comprehensive studies have been made of its origin and composition. Most chemical analyses of humin are reported in studies that have primarily focused on the more soluble components and the major thrust of the work was on humic and fulvic acids. In part, this is related to the fact that humin's insolubility has limited its chemical characterization. Humin is only amenable to study by methods that do not require solubilization, and, in applying even these methods, the high mineral content usually interferes. Recent developments in ^{13}C nuclear magnetic resonance (NMR) spectroscopy of solids have made a significant impact in our ability to examine, nondestructively, the structure of humin with minimal interference from mineral matter. The major topic of this chapter is to report on the recent studies of humin and related substances by solid-state NMR methods while summarizing much of the existing knowledge of its origin, composition, and transformation. It is important to mention here that the limited number of conventional studies of humin has not yielded, individually, as much structural information as that which has been obtained recently from NMR studies. Thus, we believe that most of the new structural information about humin is primarily derived from NMR, and, in keeping with the goals of this book, we emphasize the new NMR data which bear specifically on our understanding of the geochemistry of humin.

Humin has been shown to have nearly identical functional group and elemental compositions (Table 1) as humic acids (Schnitzer and Khan, 1972). Wright and Schnitzer (1961) and Schnitzer and Khan (1972) note that humin is slightly less aromatic than humic acids. This could presumably be related to aliphatic polysaccharides which, operationally, can contribute more to humin than to humic acids because of their general insolubility in dilute alkali. Acton et al. (1963) and, more recently, Saiz-Jimenez et al. (1979) indicate that complex polysaccharides account for a significant fraction of humin. Thus, soil humins are presently considered similar to humic acids but having a greater polysaccharide content. If one defines humin strictly on the basis of an extraction scheme whereby it simply represents the organic residue after extraction, then we can expect polysaccharides to be major components of humin in surface horizons of soils and other sediments. However, Stevenson (1982) does not consider polysaccharides and even lignin from fresh plant remains to be components of humic substances even though, operationally, they contribute to this insoluble isolate of soils. No doubt the use of the term humin must be examined carefully in light of these considerations. The scope of this chapter is not to address the use of the terminology, but to provide geochemical structural information about a component of sediments which has been operationally defined as humin. Thus, we shall consider humin as that fraction of sedimentary organic matter which is insoluble in organic solvents, dilute acid, and dilute alkali.

Although some soil scientists have long considered it to be a humic acid–clay complex, those scientists dealing with humic substances in peats consider humin to be a macromolecular condensate of humic acids (Flaig, 1972; Spackman et al., 1974; Casagrande et al., 1980). Evidence for this relation-

TABLE 1. Elemental and Functional Group Compositions of Humin and Humic Acids from a Variety of Sediments

Samples	C[c]	H[c]	N[c]	O[c,d]	Total Acidity	Carboxyl (meq/g)	Phenolic OH (meq/g)	Alcoholic OH (meq/g)	Carbonyl (meq/g)	Methoxyl (meq/g)
Aerobic soils[a]										
Humic acids	56.2	4.7	3.2	35.5	6.7	3.6	3.9	2.6	2.9	0.6
Humin no. 1	55.4	5.5	4.6	33.8	5.9	3.8	2.1	ND	4.8	0.4
Humin no. 2	56.3	6.0	5.1	31.8	5.0	2.6	2.4	ND	5.7	0.3
Peat[b]										
Everglades, Florida										
Humic acids	52.4	5.6	3.2	38.8						
Humin	56.8	5.5	3.8	33.9						
Mangrove Lake, Bermuda										
Humic acids	58.8	3.6	1.7	35.9						
Humin	55.8	4.7	2.4	37.1						
Marine Sediments										
Walvis Bay, Namibia										
Humic acids	51.9	6.7	5.1	36.3						
Humin	58.2	7.0	6.0	28.8						
Mangrove Lake, Bermuda										
Humic acids	45.3	6.2	4.3	44.2						
Humin	54.2	6.9	4.4	34.5						

[a] *Schnitzer and Khan (1978).*
[b] *Hatcher (1980).*
[c] *Moisture-and-ash-free*
[d] *Oxygen by difference*

ship is based on the fact that humin from peat has similar bulk chemical properties as humic acids (e.g., elemental compositions, infrared spectra, functional group compositions, see Table 1). This hypothesis has been widely accepted.

Little has been done to relate the structure of humin from peat to that of humic acids until recently. Using solid-state ^{13}C NMR techniques, Hatcher et al. (1980c) show that humin in peat is composed of aromatic structures derived from lignin but also contains large amounts of aliphatic structures that are paraffinic in nature. Further studies with ^{13}C NMR (Preston and Ripmeester, 1982; Hatcher et al., 1983a) show that the relative proportions of paraffinic and lignin-derived structures vary considerably among peats. The spectra also show the presence of polysaccharides as major components. Hydrolysis of the polysaccharides effectively removes the peaks for carbohydrates while enhancing the signals for lignin-derived and paraffinic carbons (Preston and Ripmeester, 1982). The ^{13}C NMR spectra of humin are significantly different from those of humic acids from the same sources. First, spectra of humic acids show significantly less carbohydrate carbon, related, undoubtedly, to the fact that polysaccharides in peat are generally insoluble in dilute alkali but contribute mostly to the insoluble residue (humin). Second, the spectra of humic acids show greater concentrations of aromatic carbons relative to paraffinic carbons.

The insoluble macromolecular humin of modern aquatic sediments, called "proto-kerogen" by Philp and Calvin (1976) and Stuermer et al. (1978), is thought to be the primary precursor of kerogen in ancient carbonaceous shales. Attempts have been made to identify the origin and structure of such material, but its insolubility has presented a major barrier to its structural delineation. As a result, attempts have been made to degrade chemically the macromolecular structure to smaller fragments, that are more easily analyzed, in hopes of obtaining useful structural information on the original material (Ishiwatari et al., 1980b). Infrared techniques have been utilized to determine some of the functional groups associated with the humin of aquatic sediments (Huc and Durand, 1977) and solid-state ^{13}C NMR spectroscopy has been utilized as a means of obtaining structural information (Hatcher et al., 1980c, 1983a).

Elemental analyses have indicated that humin of aquatic sediments is similar to humic acids of aquatic sediments (Table 1) and significantly more aliphatic than terrestrial humin owing to its high atomic H/C ratio of approximately 1.5 (Huc and Durand, 1973; Stuermer et al., 1978). This conclusion is confirmed by infrared analyses (Debyser and Gadel, 1977), chemical degradation experiments (Machihara and Ishiwatari, 1980), and NMR (Hatcher et al., 1983a). Humin of aquatic sediments also contains relatively more nitrogen than terrestrial humin, presumably because of the higher protein content of aquatic organisms (Stuermer et al., 1978).

Some attempts have been made to define more explicitly the chemical structure of humin of aquatic sediments. Philip et al. (1978) suggest that it probably consists of a cross-linked aliphatic nucleus with additional components attached to it as esters. Such a model was developed from results of chemical or thermal degradation techniques. Machihara and Ishiwatari (1980) suggest, on the basis of permanganate oxidation studies, that lacustrine sediment humin is composed of polycondensates of carbohydrates and amino acids (''melanoidins'' of Nissenbaum and Kaplan, 1972) which connect clusters of polymethylene chains, aromatic rings, and other components. On the basis of NMR studies, Hatcher et al. (1983a) suggest that humin of aquatic sediments is a predominantly paraffinic structure that is highly cross-linked and contains significant quantities of methyl, ether, carboxyl, and amide functional groups.

DIAGENESIS OF HUMIC SUBSTANCES

Humic substances in subaqueous sediments, such as peats and aquatic sediments, are generally considered to be precursors of coal (Flaig et al., 1975) and kerogen (Huc and Durand, 1977), respectively. Consequently, interest has been focused on the diagenetic changes that humic substances undergo in the course of their transformation to such bioliths.

Coal is formed from peat and the vascular plant remains that accumulate in peat bogs. Anaerobic conditions are considered mandatory for the accumulation and preservation of peat and the formation of coal. Two major types of coals are known: humic coal and sapropelic coal (see Breger, 1963, 1976). The former are formed from peat accumulations rich in humic substances derived predominantly from vascular plant remains. The latter represent coal formed from algal (boghead coal) or spore (cannel coal) accumulations. In many respects, sapropelic coal can be considered to have an aquatic origin similar to that of humin of aquatic sediments which forms from the accumulation of aquatic nonvascular plant debris in clastic sediments. Conversely, kerogen can also have the properties of humic coals (Breger and Brown, 1962) is the source materials to the sediment at the time of deposition are predominantly derived from vascular plants.

Transformation of Peat and Other Terrestrial Sediments

In peat, humic substances are thought to be formed by microbiological degradation of vascular plants (Manskaya and Drozdova, 1968). Of the few recent studies of humic substances in peat (Spackman et al., 1974; Breger et al., 1978; Casagrande et al., 1980), no systematic trends were observed with depth in profiles that indicate diagenetic changes toward lignite or the early stage of coal formation. However, Manskaya and Drozdova (1968) pointed out that humic acids from the lower horizons in peat tended to have higher carbon and lower oxygen contents in accordance with the trend expected if humic substances were participating in the coalification process. Fischer and

Schrader (1921) first proposed that the lignin components of vascular plants in peat are transformed to humic substances that undergo gradual changes to form coal. Later, Waksman (1936) reinforced this view but argued that humic substances were ligno-protein complexes. Kreulen and Kreulen-van Selus (1956) extended this theory to suggest that humic acids were intermediates between lignin and coal and that condensation of humic acids resulted in increasingly complex coal structures. Although at this time cellulose was generally not considered to take part in the coalification process, as was thought earlier at the turn of the century, Kreulen (1962) sought to revive its importance by pointing out that humic acids could be produced by heating cellulose. Later, Davis and Spackman (1964) supported this view as a result of their heating studies. Consequently, a view that had been discredited in the early 1950s (Varossieau and Breger, 1952) has suddenly received more credence (Teichmuller, 1975). Unfortunately, no strong evidence has been provided to substantiate the fact that cellulose is important in the coalification process.

The general views put forth by van Krevelen (1963) and Stach (1975) on the formation of coal from humic substances are as follows:

1. Plant debris is first converted to humic acids at the peat stage by the action of aerobic microorganisms that break down the lignin and cellulose to simple monomers which condense to form humic acids.
2. The humic acids, when buried to greater depths, condense to form humin and result in chemical changes that involve loss of oxygen functional groups with concomitant coalification.
3. As further condensation occurs, the humic acids are all converted to insoluble macromolecules typical of lignite or subbituminous coal.
4. Finally, humic substances no longer play a role in coalification as the insoluble coal undergoes further diagenesis leading to bituminous coal, anthracite, and eventually graphite.

The latter processes are thought to occur under the influence of reactions of higher energy than those necessary for the formation of coals of lower rank (van Krevelen, 1963).

Changes in the chemical composition of humic substances on their path toward coal formation have been documented by Ihnatowice (1952) and Blom et al. (1957). The most labile functional groups thought to disappear early are methoxyl groups. These are followed by carboxyl, carbonyl, phenolic, alcoholic, and finally etheric groups, which are the most resistant. As oxygen functional groups are gradually lost, the carbon contents of humic substances and coal gradually increase. Van Krevelen (1963) proposed a diagram, that has since taken on his name—the van Krevelen Diagram—which records the changes in elemental composition of humic substances as they are gradually transformed into coal of increasing rank and permits an understanding of mechanisms involved as diagenesis proceeds. Such a diagram has been used extensively to determine the degree of evolution of kerogen (Tissot and Welte, 1978).

Suffice it to say that little evidence has been unveiled to suggest that humic acids condense to form humin in peats. Studies by Flaig (1972) on aerobic soils suggest that humic acids can be produced by condensation of phenolic degradation products of lignin and of microbial degradation products. Many have assumed that such processes are active in peat as well and, as a result, it is thought that the lignin of peat is first converted oxidatively in the uppermost layers to simple monomers that then condense to form humic acids. We can hardly expect the level of oxidation that occurs in soil to be present in peat.

The next stage of the condensation theory suggests that humic acids, formed by oxidative degradation of plant remains, condense to form humin. The evidence for this mechanism is again sparse but evolves from the observation that, with increasing rank, coal becomes less extractable by dilute alkali. Thus, fewer humic acids can be extracted from coal of subbituminous or bituminous rank than can be extracted from lignite or brown coal (Teichmuller, 1975) while peat has much greater concentrations of humic acids. Unfortunately, these observations *do not* constitute evidence for the condensation theory as originally suggested. The same information could be interpreted to indicate that humic acids are products of coalification and that their abundance in young deposits merely reflects the rapid rate at which diagenetic changes proceed (Hatcher et al., 1982b). For example, the major changes in the structures of humic substances occur at the peat stage where biological and chemical processes are both active. In coal of higher rank, the major changes have already occurred at an earlier stage and further changes are only minor structural ones.

Formation of Sapropelic Coal

Boghead coal forms from accumulations of algal sapropels (detrital algal muds) that have been buried and diagenetically transformed (see Moore, 1968; Breger, 1976). These coal types are typically rich in aliphatic constituents (high hydrogen content) as the algae lack the aromatic lignin precursors typical of humic coal. Well-known modern analogues of algal coal include *Balkashite* from the USSR and *Coorongite* from Australia. Very little is known about the transformation of algae into humic substances and then into algal coal. Stach (1975) suggested that the lipid- and protein-rich algae remain preserved in anoxic sapropelic muds in contrast to their fate in peats where they are oxidized. Under the same anaerobic conditions though, he proposed that the proteins and lipids are preserved and these later form bituminous aliphatic substances while cellulose and other plant materials are broken down by microbial activity.

Breger (1976) and Cane (1976) suggested that lipids of algae condense to form an insoluble macromolecular precursor that is rich in aliphatic structures. This precursor eventually is transformed to the algal coal directly. Interestingly, these views were proposed as early as 1930 by Stadnikov when he first made the connection between lipid-rich algal residues such as *Balkashite* and *Coorongite* and algal coal (Stadnikov, 1930).

Transformation of Humic Substances in Aquatic Sediments to Kerogen

Studies on the formation of kerogen in aquatic sediments, the insoluble organic matter in ancient shales, from modern analogues have been more extensive than those dealing with the formation of sapropelic coal because of the greater interest in defining the diagenetic processes leading to the formation of petroleum from kerogen. Most organic geochemists accept the fact that humin, formed predominantly from aquatic plant debris, will produce petroleum as it undergoes increasing metamorphism caused by time and temperature. The geochemical phenomena governing such processes have been reviewed by Breger (1963, 1976), Forsman (1963) and Tissot and Welte (1978).

By examination of the recent literature as reviewed by Tissot and Welte (1978), one gets the impression that the diagenesis of humic substances to form kerogen is well understood. In effect, many consider kerogen to be formed via condensation mechanisms in which aquatic plant substances are microbially degraded to form soluble monomers that condense to form humic "polymers" that eventually condense to form kerogen. The melanoidin pathway (sugar–amino acid condensation products) has been invoked by some to explain the structures formed (Nissenbaum and Kaplan, 1972; Huc and Durand, 1973, 1977; Welte, 1973; Nissenbaum, 1974; Stuermer et al. 1978; Tissot and Welte, 1978).

Evidence for such processes was presented by Nissenbaum et al. (1972) who observed that, with increasing depth in a core of sediments from a reducing marine fjord, the interstitial waters contained increasing amounts of "polymeric" material. They equated this material to melanoidin because it contained significant amounts of carbohydrates and amino acids, and its stable isotopic composition was similar to that of algae that are rich in carbohydrates and amino acids, the raw materials for the formation of melanoidin. The evidence for such processes is, at best, meager and certainly ambiguous. An alternative explanation involves the degradation of carbohydrates and proteins as a function of depth such that the dissolved "polymeric" organic materials, that may be present throughout, gradually become a more dominant fraction of the interstital waters. This "polymer" may in fact be nothing more than polyuronic acids that have a reasonably high solubility and, as such, can be incorporated in interstitial waters of sediments. Ernst (1966) and Hatcher et al. (1980a) have noted that the presence of polyuronic acids in fulvic acids of marine sediments and it is entirely possible that similar substances were extracted by Nissenbaum et al. (1972).

Work of recent years has tended to favor the condensation hypothesis (Huc and Durand, 1973; Welte, 1973; Stuermer et al., 1978) but with little direct supporting evidence. Observations that, with increasing depth in marine sediments, fulvic acids and humic acids decrease in concentration relative to humin have also been interpreted to indicate that fulvic acids and humic acids condense to form the humin that eventually becomes kerogen. Huc and Durand (1977) extended their study of humic substances into ancient sediments and kerogen by examination of deep cores (up to 100 m). Similar observations of the loss of soluble humic acids and fulvic acids as a function of depth and age have led them to invoke a condensation mechanism. However, a decomposition mechanism can be forwarded to explain the same observations: the data do not unambiguously support the condensation mechanism. Additional observations by Huc and Durand (1977) tend to support previous suggestions that diagenesis of kerogen involves loss of oxygen functional groups from humin to the point where the kerogen becomes more hydrocarbon-like (Stadnikov, 1930; Forsman, 1963; Tissot et al., 1971). In a fashion analogous to coal, the more labile oxygenated species (carbonyl and carboxyl groups) disappear first and the more resistant groups (ethers, alcohols) persist much longer. Eventually the structure of kerogen begins to approach that of a predominantly aliphatic hydrocarbon-like macromolecule.

METHODS USED TO ISOLATE AND STUDY HUMIN

Isolation Methods

Although humin is operationally defined as the insoluble organic residue from extraction of sediments with dilute alkali (usually 0.5*N* NaOH), it should, in principle, be a macromolecular substance not belonging to a well-known or well-characterized class of biomolecules such as polysaccharides, proteins, or lignin. Unfortunately, these substances are also insoluble in dilute alkali and are defined as humin operationally. Attempts to isolate chemically or remove these substances by hydrolytic or delignification procedures are likely to alter the organic matter which most nearly represents humin. We are left with no alternative but to use the operational definition for humin, recognizing, at the same time, that polysaccharides, lignin, and proteins can contribute. Refluxing this humin with 6*N* HCl or treating it with HF/HCl will partially remove polysaccharides and proteins with minimal or small changes in the residue (Passer, 1957). Alternatively, samples can be collected from sedimentary zones where microbial activity has effectively removed or decomposed the labile polysaccharides and proteins.

In this chapter, we consider humin to be the residue after successive extraction of sediments by benzene/methanol to remove lipids, dilute acid (1*N* HCl), and 0.5*N* NaOH. In marine sediments, further treatment with concentrated HF/HCl (1 : 1 v/v) is required to concentrate the organic matter by removal of mineral matter. This treatment will partially or totally hydrolyze polysaccharides and proteins while probably having little effect on the humic material (Hatcher et al., 1983a).

Characterization Methods

Insofar as humin is an insoluble macromolecular residue, it has mostly been examined by techniques amenable to solid materials (i.e., elemental analysis, infrared, solid-state NMR, and ESR spectroscopy). Degradative techniques such as oxidation, reduction, and pyrolysis have also been employed. All these methods have been used for the study of humic substances and excellent reviews of the various methods are provided by Schnitzer and Khan (1972, 1978) as well as by Stevenson (1982).

With the advent of solid-state ^{13}C NMR, coals (Bartuska et al., 1977; Maciel et al., 1979; Zilm et al., 1979), kerogens (Garroway et al., 1976; Resing et al., 1978; Vitorovic et al., 1978; Miknis et al., 1979b; Vucelic et al., 1979; Breger et al., 1983), and humic substances (Hatcher et al., 1980c, 1981b, 1983a; Wilson et al., 1981b; Preston and Ripmeester, 1982) have been examined by this technique. The spectra have all demonstrated a wealth of structural information especially for humic substances.

Details of the methods used in this chapter for obtaining ^{13}C NMR spectra of humin with the technique of cross-polarization and magic-angle spinning (CPMAS) have been given by Bartuska et al. (1977). The carbon spectra were obtained at a frequency of 15.1 MHz using a 1 msec contact time and 1 sec cycle time. The samples were placed into a bullet-type rotor and spun at a frequency of about 2.5 kHz at the magic angle (54.7° to the magnetic field).

RECENT NMR STUDIES ON THE ORIGIN AND COMPOSITION OF HUMIN IN SOILS

In this chapter we recognize essentially two soil classes: aerobic soils that are well drained and subjected to continued availability of oxygen, and peat that represents subaqueous deposits of vegetal matter subjected to an early oxidative stage of decomposition in upper layers but to an anaerobic stage of decomposition in the lower layers. Differences have been noted in the composition of humic substances in both of these soil classes (Kononova, 1966; Manskaya and Drozdova, 1968). Because of the differences in depositional environments, it is likely that different processes are operative in the formation of humin in these two classes of sediment.

Humin in Aerobic Soils

At the present, we are unaware of NMR data in the literature on humin of aerobic soils although such data have been presented for whole soils (Wilson et al., 1981b) and for histosols (Preston and Ripmeester, 1982). We recently collected samples of an aerobic soil from southern Georgia. The sandy soil was essentially forming in a fallow agricultural field and the primary contributors to the soil were various grasses. Samples were collected from the upper 5 cm and from 5–10 cm and were later extracted to isolate humic acids and humin.

Solid-state ^{13}C NMR spectra of the humin samples and the humic acids from the upper interval are shown in Figure 1. The humic acids spectrum is similar to those of humic acids from a variety of aerobic soils (Hatcher et al., 1981b). Dominant signals are observed for aromatic (100–160 ppm) and carboxyl carbons (160–190 ppm) with minor signals for aliphatic C—O (50–110 ppm) and paraffinic carbons (0–50 ppm). Like most humic acids from aerobic soils, small or negligible signals are observed for methoxyl (55 ppm) and oxygen-substituted aromatic carbons (150 ppm) as are typically observed in NMR spectra of lignin. The peak at 150 ppm is characteristic of phenolic carbons and its absence in the spectrum of humic acids is conspicuous (Hatcher et al., 1983a), given the commonly accepted belief that phenolic carbons are major constituents of humic acids. The paraffinic carbons are minor components of the humic acids as is typical of humic acids from most aerobic soils.

The ^{13}C NMR spectra of humin from the Georgia soil are notably different from the spectrum of humic acids (Fig. 1). In the surface interval, polysaccharides are dominant with NMR peaks at 72 and 106 ppm. However, at depth in the soil, the peaks for polysaccharides diminish in relative intensity as can be expected. Polysaccharides and carbohydrates, in general, are known to be degraded rapidly in soils (Lowe, 1978).

NMR signals are also observed for aromatic carbons in both humin samples at 100–160 ppm. Although the more intense aromatic peak is at 130 ppm (ring carbons not substituted by electron withdrawing groups) a smaller peak is barely discernible at 150 ppm. This peak provides an indication of the presence of phenolic or methoxyl-substituted aromatic carbons as are typical of lignin. The humin in this soil appears to have small amounts of lignin-derived residues whereas the humic acids either are devoid of these residues or the residues have been altered so much that phenolic or oxygen-substituted aromatic carbons are low in abundance, or the NMR spectra do not differentiate the presence of these carbons in humic acids.

A major peak in the humin spectra is that of paraffinic carbons at 30 ppm. This component is present in greater abundance than the aromatic carbons in both humin spectra in contrast to its relatively minor abundance in the humic acids. Also, the intensity of this paraffinic peak, relative to the aromatic

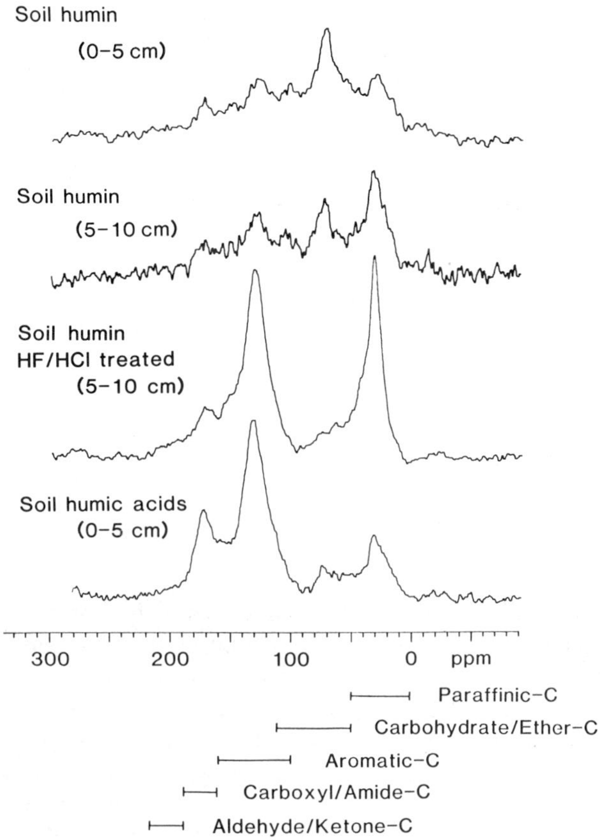

FIGURE 1. Solid-state ^{13}C NMR spectra of humin and humic acids from an aerobic grassland soil from southern Georgia.

peak at 130 ppm, does not decrease with depth, indicating that this component of humin is relatively stable to decomposition compared to the aromatic carbons. It is certainly more stable than the polysaccharides whose peaks show a decrease in relative abundance with depth.

Humin from the lower interval was treated with a 1 : 1 mixture of concentrated HCl and HF to remove the mineral matter which dilutes the organic matter, yielding poor signal-to-noise ratios as are observed in the spectra of humin in Figure 1. The ^{13}C NMR spectrum of this HCl/HF-treated humin (Fig. 1) clearly indicates that paraffinic structures are major components. No significant peaks are seen for carbohydrates (72 and 105 ppm) indicating that the hydrolysis with HCl and HF was effective in removing hydrolyzable carbohydrates. The aromatic and carboxyl-carbon region of the spectrum (100–180 ppm) shows that aromatic structures are significant components of humin, but only a shoulder can be seen at 150 ppm for phenolic carbons.

Comparison of this region of the spectrum with that of the respective humic acids shows that the humin contains significantly less carboxyl (peak at 175 ppm). The peaks at 250 and 0 ppm are spinning sidebands of the aromatic carbon peak. In summary, the primary difference between the spectra of humic acids and humin is the presence of a paraffinic carbon peak at 30 ppm in the spectrum of humin. There is little indication that lignin-like structures exist in either spectrum due to the lack of significant peaks at 55 and 150 ppm for methoxyl and phenolic carbons, respectively.

The presence of paraffinic carbons has been noted in ^{13}C NMR spectra of humic acids and humins from peat (Hatcher et al., 1980c, 1983a) and in spectra of whole soil (Wilson et al., 1981b). Hatcher et al. (1982a) pointed to algal or microbial organisms as the source of these structural components. To our knowledge, this is the first demonstration of the abundance of paraffinic structures in humin from aerobic soils other than histosols. The relatively high intensities of these paraffinic carbons in the humin strongly suggest that paraffinic structures are major components of the organic matter.

Proteinaceous substances could be responsible for signals in the 0–50 ppm region of ^{13}C NMR spectra, but usually the aliphatic signals of proteins center more closely toward 20, 40, and 50 ppm than 30 ppm (Gierasch et al., 1982). Based on studies of humic substances from algal or microbial deposits we have demonstrated that the peak at 30 ppm probably represents a complex series of highly cross-linked paraffinic macromolecules (Hatcher et al., 1980c) which contain carboxyl, amide, and ether linkages but do not appear to be proteinaceous. It is also unlikely that the peak at 30 ppm in the humin spectra (Fig. 1) is proteinaceous, because we would expect the relative abundance of proteins to decrease at a rate equivalent to or greater than polysaccharides as a function of depth in the aerobic soil. The relative intensity of the peak at 30 ppm increased slightly with depth, an indication that the substances responsible for this peak are relatively resistant and probably not proteinaceous.

The comparison between NMR spectra of humin and humic acids in this aerobic soil bears out the fact that some basic structural differences exist between these two soil fractions. The lower polysaccharide content of humic acids compared to humin is expected. The greater relative proportion of carboxyl (or amide) carbons (peak at 175 ppm) in humic acids is another minor difference that was noted. The most important difference is the relative concentration of paraffinic carbons with humin having a much greater concentration than humic acids. Excluding the presence of polysaccharides, it is difficult to imagine that humin, in this instance, is a clay complex of humic acids. If this were the case, the spectra would be nearly identical except for the presence of carbohydrates in humin. It is also difficult to imagine that humin is a condensation product of humic acids. Rather, the comparisons show that either humic acids are decomposition products of the humin (where decomposition selectively alters the structure of individual precursors in humin), or humic acids are genetically unrelated to humin.

Humin in Peat

Many ^{13}C NMR spectra have been published for humin in peat (Hatcher et al., 1980c, 1983a; Preston and Ripmeester, 1982; Dereppe et al., 1983) and all appear to contain peaks for carbohydrates and aromatic, carboxyl, and paraffinic carbons, the proportions of which vary considerably. Hatcher et al. (1983a) pointed out that, while the presence of carbohydrates (polysaccharides) and lignin was expected, the discovery of significant quantities of paraffinic carbons by ^{13}C NMR has made a major contribution to our knowledge of the components of peat humin.

Figure 2 shows solid-state ^{13}C NMR spectra of humin in two peat cores, one from The Everglades, Florida, the other from Mangrove Lake, Bermuda. The spectra of humins from both peats show peaks in similar regions but those from the Mangrove Lake peat are more aromatic. The peat from Mangrove Lake is buried beneath a lacustrine algal sapropel and a sharp contact is observed between the peat and the sapropel at 14.0 m (Hatcher, 1978). In both peat cores, the spectra of humin show large quantities of polysaccharides (peaks at 72 and 106 ppm) in the near-surface layers but their concentrations decrease significantly with depth. Decomposition of

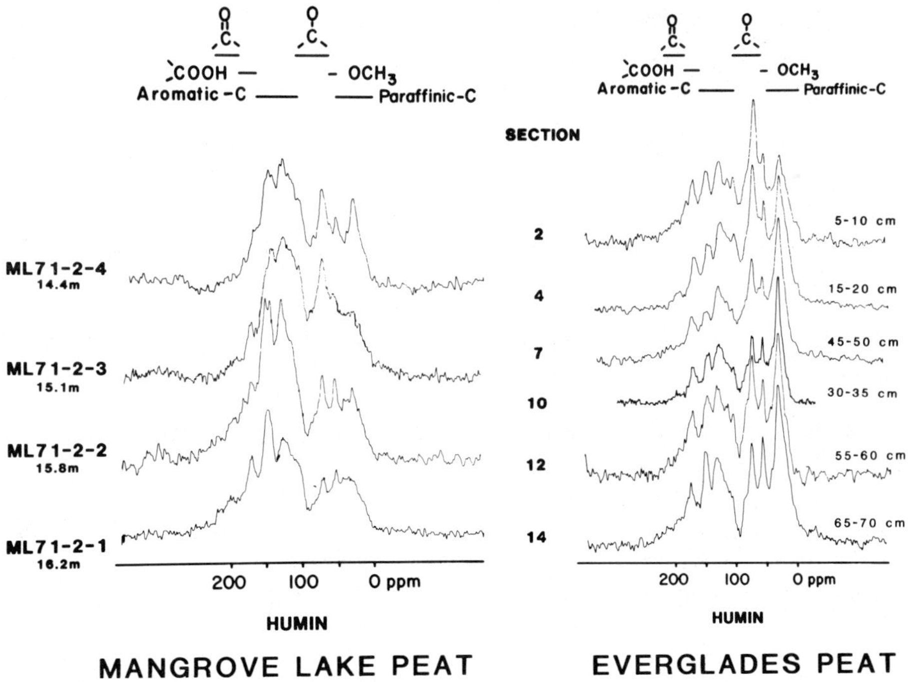

FIGURE 2. NMR spectra of humin isolated from various depth intervals in cores from two peats. The Everglades peat was collected from a sawgrass area in Conservation District 1A west of Fort Lauderdale, Florida. The peat core from Mangrove Lake, Bermuda, was collected and described by Hatcher (1978).

polysaccharides in peat is well known (Exarchos, 1976; Given and Dickinson, 1975) and the spectra show this decomposition well. Aside from the loss of polysaccharides the spectra do not show major structural changes with depth, especially in the Everglades peat where the primary difference between spectra of humin at depth and those at near-surface intervals is the intensity of carbohydrate peaks. All peaks for humin spectra at depth appear to have approximately the same relative intensities when compared to other noncarbohydrate peaks.

Other major peaks in the spectra of humin are those derived from lignin (150, 130, and 55 ppm). In contrast to the humin from soil shown earlier, these peaks are much better resolved, especially the peak at 55 ppm for methoxyl carbon. No doubt, lignin is a major component of humin in peats, and it is likely that it exists in a relatively unaltered state in the peat, even at depth.

Studies of woody tissues buried in anaerobic sediments have shown that lignin can be selectively preserved for thousands and even millions of years, even though cellulose is degraded (Hatcher et al., 1981a, 1982b). Hatcher et al. (1983b) examined a piece of wood, buried in Mangrove Lake sapropel, by NMR and found that the cellulose had been totally eliminated with the spectrum essentially resembling lignin. Figure 3 shows a ^{13}C NMR spectrum of a Douglas fir that had fallen in the rain forest on Mount Rainier, Washington, and had been decomposing for a considerable amount of time in a humid environment. Also shown are NMR spectra of (1) a relatively well preserved cypress log buried several feet in the Okefenokee Swamp, Georgia, and (2) lignin isolated from modern spruce wood by the sodium paraperiodate treatment (Ritchie and Purves, 1947). Clearly, the wood from the Douglas fir has essentially lost its cellulose and other polysaccharides as evidenced by the lack of significant signal intensity at 72 and 106 ppm in the spectrum. The spectrum is very similar to that of periodate lignin. Both the fir and the spruce should have a similar kind of lignin because both are softwood gymnosperms. The buried cypress, however, shows little evidence of having lost its polysaccharides as its spectrum is dominated by peaks assigned to polysaccharides (72 and 106 ppm).

On the basis of these studies on woody tissues, it seems that lignin from vascular plants can be selectively preserved compared to biologically degradable polysaccharides when buried. The same can be expected for the lignin in humin from peat; the spectra shown in Figure 2 consistently demonstrate this selective preservation with increasing depth.

The other major components of peat are paraffinic structures (peak at 30 ppm) observed in NMR spectra and described by Hatcher et al. (1980c). These structures, which probably have the major fraction of the carboxyl or amide groups (peak at 175 ppm) associated with them, also appear to be selectively preserved with depth in the peat (see Fig. 2). Earlier studies have suggested that this paraffinic component of peat is derived from algal or microbial sources (Hatcher, 1980). This may explain why the Mangrove

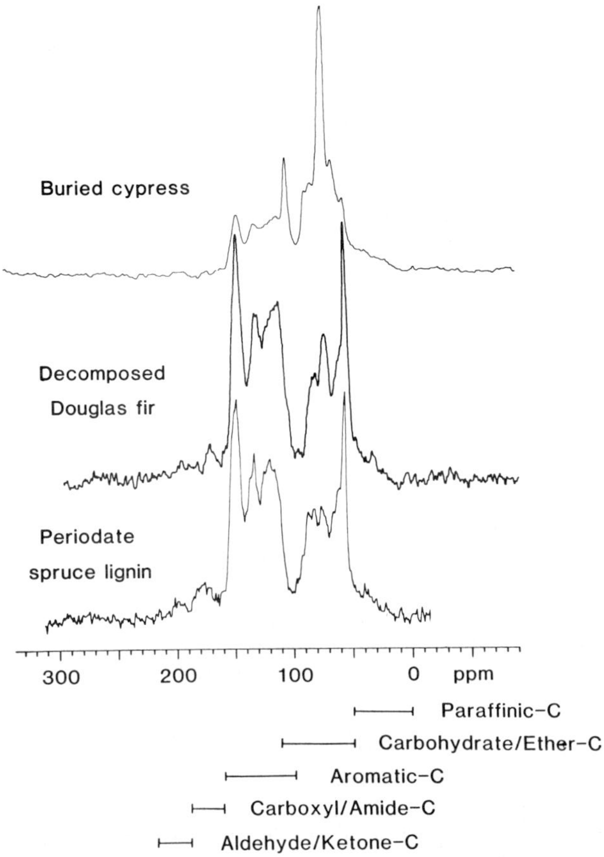

FIGURE 3. ^{13}C NMR spectra of a cypress log buried in the Okefenokee Swamp, Georgia, a Douglas fir tree that had fallen in the rain forest surrounding Mount Ranier, Washington, and lignin isolated from modern spruce wood by the sodium paraperiodate treatment.

Lake peat contains fewer of such components than the Everglades peat; there are apparently fewer algal or microbial contributors in the former.

Several factors lead us to believe that this paraffinic component of peat is macromolecular and nonproteinaceous. First, the peat was treated with a benzene/methanol mixture prior to isolation of humin. Thus, it is unlikely that the paraffinic structures have a significant contribution from lipids. Second, when hydrolyzed in refluxing 6*N* HCl, the humin lost some paraffinic carbons, but mostly its polysaccharides as demonstrated in Figure 4 which shows ^{13}C NMR spectra of humin and its hydrolyzed residue. The paraffinic carbons survive the hydrolysis, demonstrating their resistance. It is unlikely that proteinaceous material would survive such a treatment as an insoluble residue.

Third, the humin samples were subjected to delignification by treating with sodium chlorite according to the method of Passer (1957). Figure 5

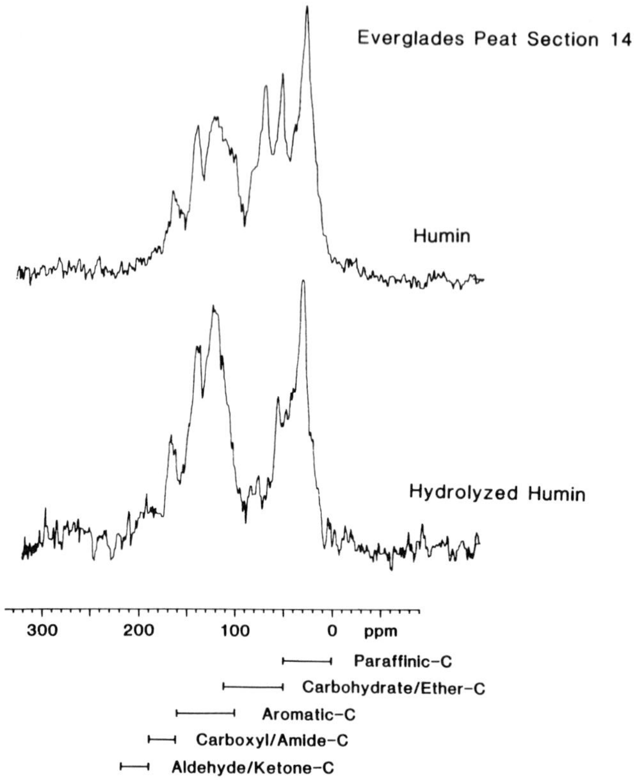

FIGURE 4. ^{13}C NMR spectra of humin isolated from section 14 (65–70 cm) of the peat core from Conservation District 1A in the Everglades, Florida, and of the same humin hydrolyzed by refluxing in 6*N* HCl for 2 hours.

shows the ^{13}C NMR spectra of delignified humin as a function of depth in sawgrass peat from the Everglades. Also shown is a spectrum of humin in the upper layers of the peat. The sodium chlorite oxidation is designed to isolate, as a residue, holocellulose, a polysaccharide fraction containing celluloses and hemicelluloses. That this is, in fact, achieved is shown by the selective loss of peaks related to lignin in the spectra of Figure 5. However, other components, namely, the paraffinic carbons, also survive the treatment and are present in the spectra. Note also that a large fraction of the carboxyl carbons is retained. These are probably associated with paraffinic structures in the lower layers of peat.

By examination of the spectra in Figure 5, it is clear that polysaccharides (holocellulose, peaks at 72 and 106 ppm) are dominant in the delignified humin in the upper layers of peat but diminish in relative concentration with depth. This trend was also observed in the spectra of humin in Figure 2. At depth, the polysaccharides are minor compared to the paraffinic carbons (peak at 30 ppm). Thus, the paraffinic structures in humin are resistant to sodium chlorite oxidation, and their relative increase in concentration with

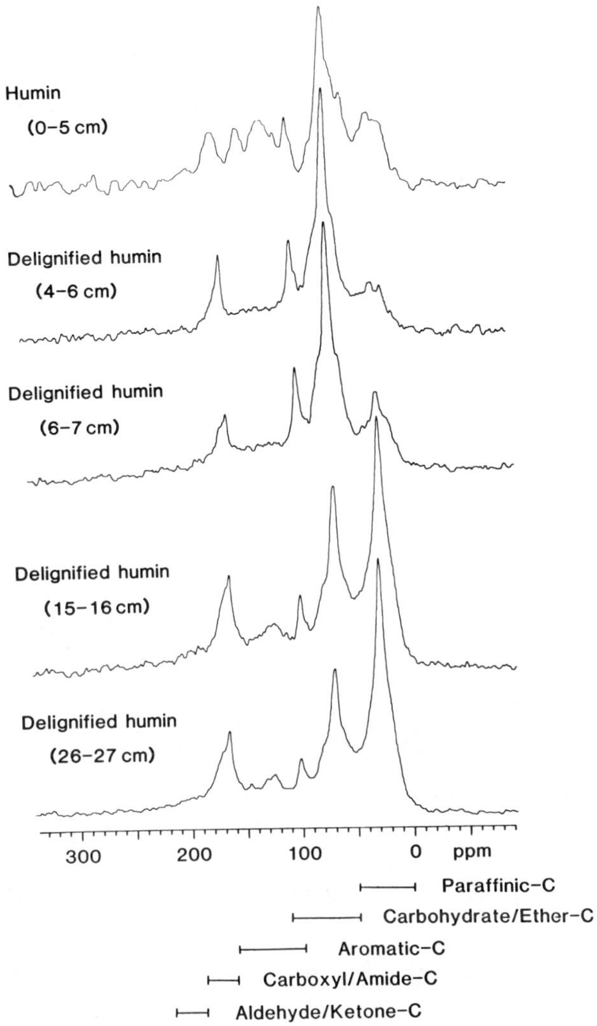

FIGURE 5. ^{13}C NMR spectra of humin and humin that was delignified by treatment with sodium chlorite according to the method of Passer (1957). The peat samples were obtained from a core of sawgrass peat collected near the junction of Alligator Alley and Highway 27, west of Miami, Florida.

depth in the core suggests that they are differentially preserved during diagenesis in comparison to the biologically labile polysaccharides.

RECENT NMR STUDIES ON THE ORIGIN AND COMPOSITION OF HUMIN IN AQUATIC (MARINE) SEDIMENTS

Aquatic sediments include marine and freshwater sediments in which the organic matter is derived primarily from algal and microbial detritus. Several solid-state ^{13}C NMR spectra of humin isolated from aquatic sediments have been reported in the literature (Hatcher et al., 1980c; 1983a). These humins were treated with HF/HCl to concentrate the organic matter from the matrix that is mostly inorganic. As shown in Figure 6, the spectrum of a representative humin from marine sediment, from an algal sapropel from Mangrove Lake, Bermuda, is dominated by paraffinic carbons (0–50 ppm) with a peak at 30 ppm. Minor peaks at 72, 130, and 175 ppm denote the presence of C—O carbons of ethers or carbohydrates, aromatic carbons, and carboxyl

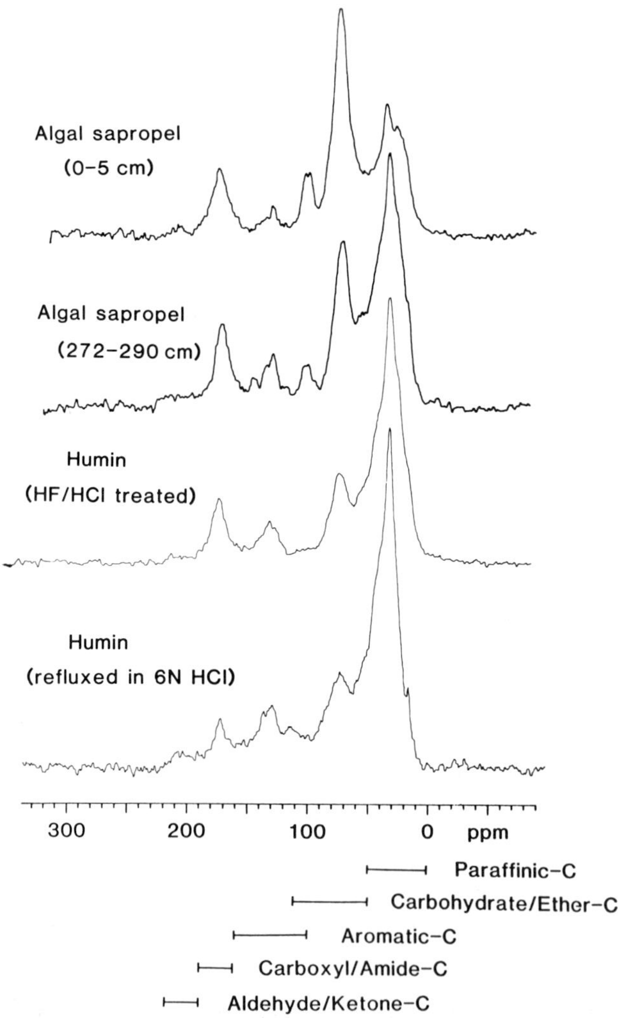

FIGURE 6. ^{13}C NMR spectra of samples of whole sapropel, humin, and hydrolyzed humin from a core of Mangrove Lake, Bermuda. The whole sapropels are from a core collected in 1982. The humin was from a core collected in 1971 (Hatcher, 1978) and is characteristic of humin from all depths in the sapropel.

or amide carbons, respectively. Although this humin was collected from one depth interval, it is representative of humin from all depths (Hatcher et al., 1983b). We have previously suggested that these paraffinic peaks represented complex macromolecular and highly cross-linked paraffinic structures (Hatcher et al., 1983a). Further hydrolysis of this humin from Mangrove Lake had little effect on the NMR spectrum except for a diminution of the peak at 175 ppm. Loss of the amide I and II bands (1640 and 1540 cm^{-1}, respectively) in the infrared spectrum (Fig. 7) and a loss of nitrogen in its elemental composition indicate that some proteinaceous materials were removed by the hydrolysis. The atomic H/C ratios, however, remained unchanged (Hatcher et al., 1983a). This, and the lack of significant change in the NMR spectrum, indicate that the humin in this aquatic system is highly resistant to chemical alteration by hydrolysis.

A similar resistance is noted with regard to diagenetic or microbial alteration. Recent studies of the whole sapropel and humin as a function of depth in Mangrove Lake have conclusively shown that the humin is selectively preserved during diagenesis and that labile substances such as lipids, carbohydrates, and proteins are decomposed and lost (Hatcher et al., 1983b).

An example of this selective preservation is shown in Figure 6 with the NMR spectra of the algal sapropel from a near-surface and a 2.8 m interval. In the surface layers, the organic-rich sapropel is dominated by polysaccharides denoted by peaks at 72 and 106 ppm in the spectrum, and humin accounts for less than 20% of the organic matter. At depth, the polysaccharides and other labile substances are decomposed, and the sapropel contains mostly paraffinic macromolecular humin (approximately 70% of the organic matter). Hatcher et al. (1983b) have shown that the loss of labile

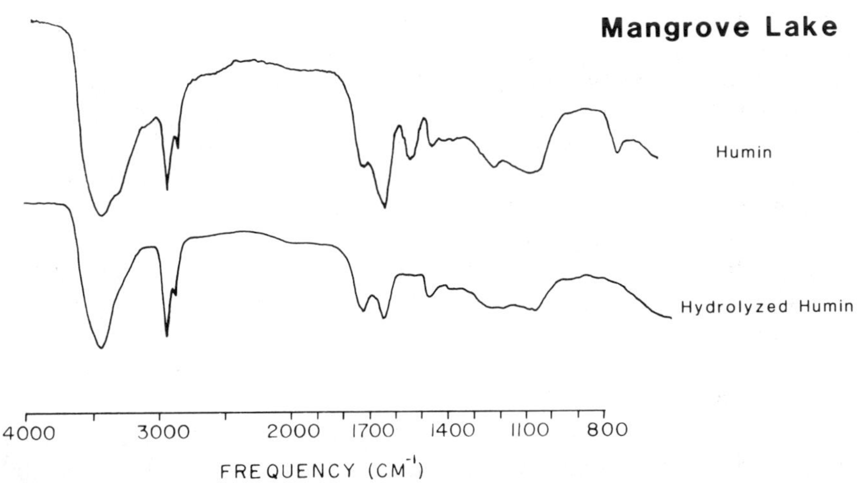

FIGURE 7. Infrared spectra of humin and hydrolyzed humin whose NMR spectra are shown in Figure 6.

organic substances approximately corresponds to the increase in the amount of humin. Spiker and Hatcher (1984) showed that a mass balance of stable carbon isotopes is also consistent with a mechanism of selective preservation.

It is important to note that under the anaerobic conditions existing in the sapropel negligible amounts of humic acids are present. The humic acids that do exist account for less than 5% of the organic matter and are structurally unrelated to the humin. They are essentially polyuronic acids and are only called humic acids because they are operationally extracted as such (Hatcher et al., 1983b). Thus, it is unlikely that humic acids are intermediates in the formation of humin in this environment. Selective preservation is the simplest explanation for the existence of humin.

A similar selective preservation was observed in peat as discussed earlier where an additional component, lignin, was also preserved selectively. However, the major component of humin from Everglades peat was the paraffinic component that also appeared to be selectively preserved relative to the polysaccharides. It is interesting to note the similarity between the spectra of delignified humin at the 15–16 cm interval in peat (Fig. 5) and that of the algal sapropel from Mangrove Lake at the 272–290 cm interval. The similarity between these two spectra infers that similar structural entities are present in these two depositional environments, and it is probable that the two similar structural components are from a common source, namely, algal and microbial remains.

To demonstrate that peat from the Everglades, especially the humin, can be considered as a two-component system, we digitally summed two NMR spectra, one for humin from Mangrove Lake sediments and the other for periodate lignin, varying the relative contribution of each in the summation. The resultant spectra for each summation are shown in Figure 8 along with the spectrum of humin from the Everglades peat. A digital summation containing 60% of the paraffinic humin and 40% lignin yielded a spectrum that is nearly identical to the humin in peat. The major contributor to peat appears to be the paraffinic humin, and it is likely that algal or microbial residues are responsible for this component. Note also that the presence of carboxyl or amide carbon in the spectrum of peat can be accounted for by assuming that it is entirely associated with the paraffinic structures, because the lignin is essentially free of carboxyl carbon.

The occurrence of a predominantly aliphatic humin in marine sediments has been known for some time (Stadnikov, 1930; Breger, 1960; Cane, 1976; Stuermer et al., 1978). This humin was believed to be the precursor of aquatic kerogen in ancient shales and as such has been extensively studied. However, the discovery that a structurally similar aliphatic component exists in humin from peats and even soils is a major new finding that has demonstrated the usefulness of solid-state ^{13}C NMR. The analytical visibility of this component in soil humic substances has been masked by the over-

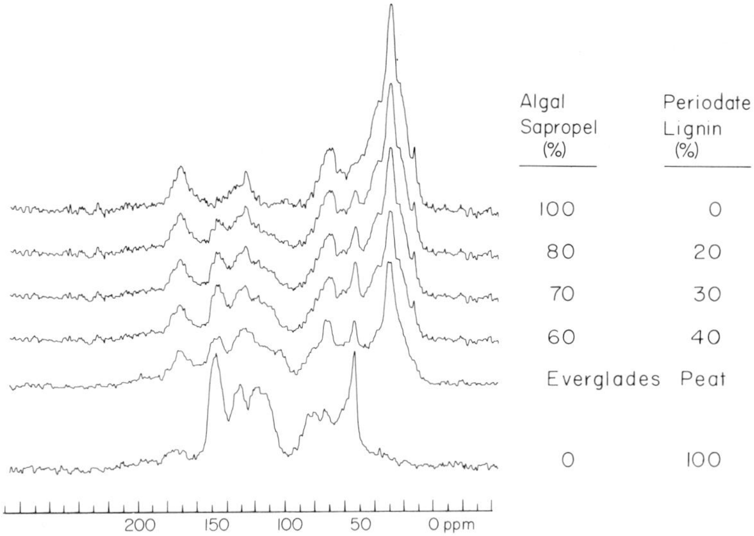

FIGURE 8. ^{13}C NMR spectra of humin from Mangrove Lake (algal sapropel), periodate lignin, and humin from the Everglades peat (Conservation District 1A, section 10, 30–35 cm depth). Also shown are NMR spectra artificially produced by summing spectra of algal sapropel and lignin in the relative proportions indicated.

whelming presence of aromatic structures. Accordingly, the techniques utilized for identification of aliphatic structures were not sufficiently selective for their detection until the advent of NMR.

The NMR data for both the peat and the algal sapropel show that paraffinic components of humin are present in the near-surface intervals. Although they become dominant components at depth due to selective preservation, their concentration in surface intervals is usually small. The question remains: Are these components originally present in algal or microbial residues as suggested by Philp and Calvin (1976) or are they formed rapidly by condensation of soluble substances such as lipids (Stadnikov, 1930; Breger, 1950; Cane, 1976; Harvey et al., 1983) or sugar–amine condensation products (melanoidins, Nissenbaum and Kaplan, 1972)? On the basis of NMR spectroscopy, stable isotope analyses, and individual compound analyses, Hatcher et al. (1983b) have recently suggested that it is more likely that the paraffinic structures exist as macromolecules in algal or microbial cells and do not form by condensation reactions.

GEOLOGIC FATE OF HUMIN

Inasmuch as humin is a possible precursor of fossil fuels, our understanding of its structure and reactivity is of fundamental interest in its origin and formation. Many believe that humin is an intermediate polycondensate in the transformation of low-molecular-weight products of microbial degradation of plant matter to coal and kerogen.

Geologic Fate of Vascular Plant Residues

Recent NMR studies of coalified woods, representing vascular plant remains, show that a multitude of reactions are involved in the evolution of humic coal (Hatcher et al., 1981a, 1982b). The initial diagenetic processes are thought primarily to involve loss of cellulosic or polysaccharide components of the wood with selective preservation of lignin as a macromolecular entity. Eventually the lignin is transformed to lignite with preservation of cellular structure. The lignite is depleted of hydrogen-rich groups in this process, probably by loss of methoxyl groups and the C-3 side chains. We believe that this transformation occurs in a macromolecular state without a soluble humic acid intermediate as is commonly suggested. Thus, the lignite is essentially humin that is further modified by coalification. Eventually, the lignite is transformed to higher rank coal via loss of oxygen functionality. Hatcher et al. (1982b) suggest that loss of oxygen can be effected by chemical degradative processes in which soluble oxygen-rich ''humic acids'' are formed and subsequently lost during compaction and dewatering that occur with coalification. In this model, ''humic acids'' are products of the degradation rather than reactants for condensation. The evidence suggesting such a mechanism as being responsible for coalification of woody tissue is that (1) humic acids are not major components of coalified logs or coal at any rank, (2) carboxyl groups, which are major functional components of humic acids, do not appear to be present in the logs as deduced from NMR spectra, and (3) transformations involving soluble intermediates would destroy the cellular morphology that exists in all coalified logs examined.

Geologic Fate of Humin in Aquatic Sediments

Earlier, we presented some NMR data demonstrating that humin in aquatic sediments is formed in anaerobic environments as a residual fraction from the decomposition of aquatic plants. This humin is characterized by a complex, highly cross-linked paraffinic structure that persists in sediments and is probably transformed to kerogen in older sediments. To examine the transformation of this humin to kerogen or algal coal, ^{13}C NMR spectra were obtained of kerogen samples known to have been derived from algal sources. These kerogens, or algal coals, range from Miocene to Permian and Carboniferous age and their spectra are shown in Figure 9.

It is strikingly apparent that all spectra of aquatic kerogen are predominantly aliphatic with a major peak centered at 30 ppm. This peak is also the most intense peak in the spectrum of humin from Mangrove Lake also shown in Figure 9. Paraffinic structures of the type shown to be present in

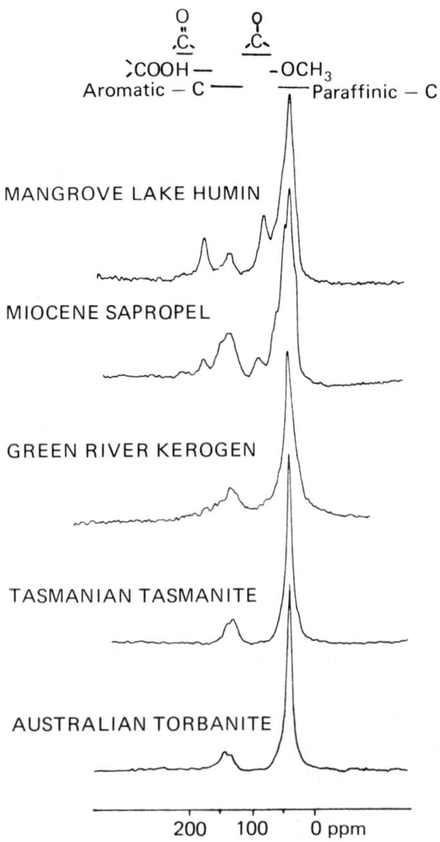

FIGURE 9. ^{13}C NMR spectra of algal kerogens and boghead coals from locations described by Hatcher (1980). The spectrum of humin from Mangrove Lake, Bermuda, is also shown.

humin of Holocene sediments are apparently preserved through time. Elemental data for these samples (Hatcher, 1980) suggest a highly aliphatic structure, and atomic H/C ratios of approximately 1.5 for nearly all samples suggest that these paraffinic structures are highly cross-linked and similar to those of the humin from Mangrove Lake. Published reports dealing with kerogen from oil shale of the Green River Formation have all pointed to the fact that the structure is that of a highly cross-linked paraffin macromolecule (Djuricic et al., 1971; Robinson, 1976; Young and Yen, 1977). Young and Yen (1977) propose that the kerogen is composed of fused alicyclic rings grouped in clusters that are joined by long-chain polymethylene bridges. Such a structure exhibits an atomic H/C ratio of approximately 1.5 which is consistent with the observations made by Hatcher (1980).

The similarity between the various spectra in Figure 9 should be underscored because such spectral similarities reinforce arguments concerning the origin and formation of algal kerogen. In many respects, the spectrum of the Miocene sapropelic coal is nearly identical to that of humin from Mangrove Lake, Bermuda. The only differences appear to be in the relative amounts of carbohydrate/ether carbons having a peak at 72 ppm and of carboxyl carbons at 175 ppm. Both of these functional groups are thought to diminish in concentration during diagenesis (see Tissot and Welte, 1978). Continued loss of such peaks lead to spectra that are similar to those of kerogen from shale of the Green River Formation, Tasmanian tasmanite, and Australian torbanite. These observations are consistent with the elemental data (Hatcher, 1980) that show a significant reduction in the amount of oxygen that is probably associated with functional groups. The Green River kerogen, Tasmanian tasmanite, and Australian torbanite contain much less oxygen than the Mangrove Lake and Miocene sapropelic coal samples and their NMR spectra reflect this by loss of peaks at 72, 175, and 200 ppm attributable to carbohydrate/ether, carboxyl, and carbonyl resonances, respectively.

A mechanism similar to that proposed for loss of oxygen from vascular plant remains during coalification can be invoked to explain the loss of oxygen functional groups, namely, the degradation of humin to oxygen-rich humic acids that are mobilized and lost during compaction of the sediment. This model is in direct contrast to that proposed by Huc and Durand (1977) in which humic acids condense to humin, involving decarboxylation to explain the loss of carboxyl groups.

The condensation model was rejected earlier on the bases that humin is observed to form early in the depositional history of anaerobic sediments without the presence of humic acid intermediates and that selective preservation of an algal or microbial component is the simplest explanation for the observations made with regard to the sapropel from Mangrove Lake, Bermuda. Given the fact that humin, having a general structural resemblance to kerogen in ancient shales, originates early in the depositional record and that the primary difference between this humin and kerogen is the degree of oxygen functionality, it is logical to expect a degradative mechanism in the transformation of humin to kerogen.

CONCLUSIONS

Over the years only a small quantity of data has been accumulated on the chemical composition of humin operationally defined as the insoluble residue on extraction of sedimentary organic matter with dilute alkali. The information has been limited in that we only know of the elemental and functional group compositions of humin. Recently, degradative studies involving pyrolysis–gas chromatography or chemical oxidations followed by gas chromatography and mass spectrometry have indicated that humin is, in a general sense, related to humic acids (Saiz-Jimenez et al., 1979; Schnitzer and Khan, 1978). The advent of solid-state ^{13}C NMR has opened the door to direct, nondestructive structural studies of humin, and the information gained from these studies has shown that some major structural differences exist between humic acids and operationally defined humin (Hatcher et al., 1983a; Preston and Ripmeester, 1982).

Further NMR studies of humin presented here have increased our knowledge of the structural composition of humin from different depositional environments. The following observations and conclusions can be made as a result of these studies.

1. In aerobic soils, humin, operationally defined, differs dramatically from humic acids, being more aliphatic as a result of the presence of polysaccharide and a paraffinic component thought to be derived from algal or microbial sources. If one argues that these components are primary biological macromolecules and should not be considered to be humic substances, then the above comparison is a moot point. Nonetheless, the operationally defined humin in soil does not appear to be a clay–humic acid complex even if one eliminates from consideration the presence of polysaccharide. We are still left with the paraffinic carbons that are major components. Of course one could also consider these to be biomolecules because the NMR studies suggest that these components are present in algae. Clearly, more work is needed to pinpoint the source of these structures in aerobic soils.

2. In peat, humin is also composed of the three structural entities mentioned above. The anaerobic nature of peat precludes the extensive decomposition that occurs in aerobic soils, and biomolecules are likely to be better preserved. Carbohydrates are major components of humin in near-surface intervals but are decomposed and lost with depth in the peat. Lignin and the paraffinic structures are selectively preserved with depth. When the humin of peat is delignified, the paraffinic structures remain. These components are likely to be derived from nonvascular plant contributors to the peat, namely, algae.

3. In an algal sapropel and also in marine sediments, humin and the HF/HCl-treated humin are predominantly composed of the paraffinic structures that evolve as a result of selective preservation with depth in the sediments. In this case as well, the paraffinic structures are most likely derived from algal or microbial sources and could very well be original biomolecular components of the algae.

4. In all aerobic and anaerobic sediments examined, selective preservation of biomolecules appears to be the mechanism by which humin evolves. It is likely that humic acids are products of the degradation of humin rather than precursors.

5. In ancient deposits, the humin is transformed under anaerobic conditions to either coal or kerogen directly without necessarily involving an intermediate "humic acid." The "humic acids" present in ancient sediments may be nothing more than degradation products of humin that are solubilized and lost during compaction.

Although many of the aforementioned conclusions are of a speculative nature and will require further confirmation in order to be widely accepted, NMR has provided us with a powerful means of delineating gross structural changes in organic matter during humification, diagenesis, and coalification. Further studies dealing with these processes should view NMR spectroscopy, especially solid-state NMR, as a major analytical tool that can be combined with other analytical methods available to geochemists.

The problem of foremost interest in future studies should be to ascertain whether operationally defined humin contains significant amounts of clay–humic acid complexes and to determine the genetic relationship between humic acids and humin. Another problem would be to define structurally the paraffinic components of humin and to ascertain whether they are truly biomolecules or are rapidly formed products of decomposition of vegetal matter. Finally, the hypothesis of selective preservation should be tested in other environments.

ACKNOWLEDGMENTS

We thank Elliott C. Spiker and Joseph L. Zelibor, Jr. (U.S. Geological Survey) for assistance in the collection of samples, Lisa A. Romankiw and Marta Krasnow (U.S. Geological Survey) for assistance in preparation of samples, and the G. K. Gilbert Fellowship Program at the U.S. Geological Survey for providing the necessary funding for our studies.

CHAPTER TWELVE

Nature of Nitrogen in Humic Substances

MORRIS SCHNITZER

ABSTRACT

Humic acids and humin contain between 2 and 6% nitrogen, whereas the range for fulvic acids varies from <1 to 3%. From the agricultural point of view, nitrogen is a key nutrient element required by most crops in relatively large amounts. It is also the only element not released by the weathering of minerals; its source is the atmosphere where N_2 is the predominant constituent. A few microorganisms have the ability to fix molecular N_2 and to reduce it to ammonia. In the past, most of the research on nitrogen in humic substances has been concerned with proteinaceous components, amino sugars, and ammonia. This leaves one-third to one-half of the total nitrogen unidentified.

The purpose of this chapter is to highlight our current state of knowledge on the distribution of different types of nitrogen compounds in humic substances and to project future research needs, especially with regard to the identification of the unknown nitrogen in humic substances and the relationship between carbon and nitrogen.

INTRODUCTION

Humic acids and humin contain between 2 and 6% nitrogen, while the nitrogen content of fulvic acids ranges from <1 to 3% (Schnitzer, 1976). Nitrogen is not formed from the weathering of minerals. Its source is the atmosphere where N_2 is the predominant constituent (79%). Nitrogen enters the soil system through fixation of molecular N_2 by microorganisms, and the return of ammonia and nitrate in rainwater. From the agricultural point of view, nitrogen is an essential plant nutrient required in relatively large amounts by most agricultural crops. Humic substances act as suppliers and storehouses of nitrogen for plant roots and microorganisms. The significance of nitrogen is indicated by the fact that it is an important constituent of proteins, nucleic acids, porphyrins, and alkaloids. While a considerable amount of research has been done on nitrogen in humic substances, most of this is limited to qualitative and quantitative determinations of proteinaceous compounds, amino sugars, and ammonia. This leaves about one-third to one-half of the humic nitrogen unidentified. Thus, there is a need for further research in this area.

The biogeochemistry of nitrogen, the distribution of nitrogen in humic substances, the unidentified nitrogen, and efforts to characterize and identify it will be presented in this chapter. In addition, the following question is addressed: Are nitrogen-containing compounds integral components of humic substances?

BIOGEOCHEMISTRY OF NITROGEN

The nitrogen pools on our planet are comprised of atmospheric nitrogen, lithospheric nitrogen (rocks and minerals), and biospheric nitrogen (soil organic matter, plants, and animals). An examination of the global nitrogen distribution shows (Table 1) that 98% of the nitrogen mass of the earth is contained in rocks and minerals where it exists as nitrides of iron, titanium, and other metals and as NH_4^+ ions held in the lattice structure of primary

TABLE 1. Geochemical Distribution of Nitrogen on the Earth[a]

Nitrogen Pool	Nitrogen Mass ($\times 10^{20}$ g)	Percentage of Total Nitrogen Mass
Atmosphere	38.0	1.92
Lithosphere		
Igneous rocks	1930.0	97.86
Sedimentary rocks	4.0	0.20
Biosphere		
Living organisms	0.00038	
Terrestrial organic nitrogen	0.0082	0.02
Ocean-bottom organic nitrogen	0.0054	
Total nitrogen mass	1972.0	100.00

[a] *Porter (1975).*

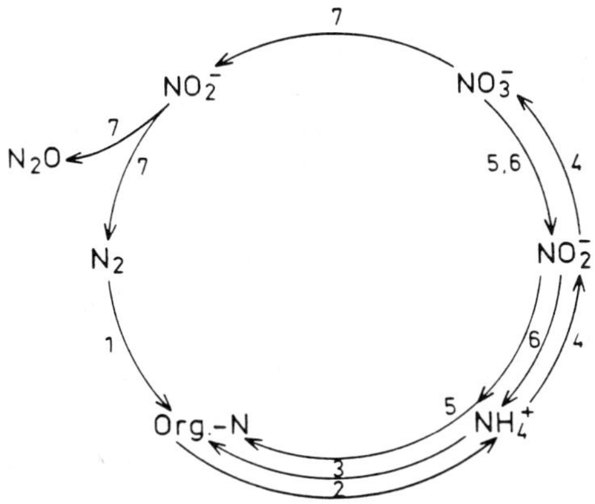

FIGURE 1. Schematic view of biogeochemical nitrogen cycle: 1, nitrogen fixation; 2, mineralization; 3, immobilization; 4, nitrification; 5, nitrate assimilation; 6, dissimilatory nitrogen reduction; 7, denitrification (Rosswall, 1982).

silicate minerals. The amount of nitrogen in the biosphere constitutes only 0.02% of the global nitrogen mass, with most of the biospheric nitrogen found in the humus fraction of soil organic matter. Nitrogen in the atmosphere accounts for 1.96% of the earth's nitrogen, most of which exists as stable molecular N_2 (Porter, 1975).

A schematic view of the biogeochemical nitrogen cycle (Rosswall, 1982) is presented in Figure 1. Of special interest in the context of this discussion are reactions 1, 2, 3, and 4 which involve nitrogen associated with organic matter. These reactions are:

1. Nitrogen → organic nitrogen (nitrogen fixation).
2. Organic nitrogen → ammonium (mineralization or ammonification).
3. Ammonium → organic nitrogen (immobilization or assimilation).
4. Nitrate → organic nitrogen (nitrate assimilation or immobilization).

Nitrogen fixation (reaction 1) may be characterized as the reduction of elemental N_2 to the -3 oxidation state in ammonia. This biological process is catalyzed by nitrogenase, a large anaerobic metalloenzyme. The NH_3 produced is retained by the nitrogen-fixing cells and reacts with glutamate to form glutamine. Newly fixed NH_3 is only rarely released by healthy nitrogen-fixing cells but must pass through an organic form before entering the nitrogen cycle (Smith, 1982).

The mineralization of nitrogen (reaction 2) is carried out mainly by microorganisms. Through this process organically bound nitrogen, the major nitrogen form in soils, is liberated as NH_3. Whether nitrogen is mineralized or immobilized by microorganisms depends on the C/N ratio of the substrate compared to that of the decomposer organisms. If the substrate has a low C/N ratio, nitrogen will be in excess and NH_3 will be liberated.

Immobilization (reaction 3) of nitrogen can occur through both biotic and abiotic processes. For example, NH_4^+ is efficiently immobilized by clay minerals in exchangeable and fixed forms. Exchangeable NH_4^+ is available for biological immobilization. While it has been assumed in the past that nonexchangeable NH_4^+, which is fixed in lattices of clays, had a low biological availability, recent data (Rosswall, 1982) show that in most soils 30–60% of the fixed NH_4^+ is also available for biological uptake. Addition to soils of a substrate with high carbon/nitrogen ratio will bring about rapid microbial immobilization of NH_4^+.

Nitrification is a process through which nitrogen in the form of NH_4^+ is oxidized to NO_2^- and/or NO_3^-, mainly by autotrophic nitrifying bacteria of the genera *Nitrosomonas* and *Nitrobacter*. Recent work (Rosswall, 1982) shows that *Nitrosolobus* is also active in soils in the oxidation of NH_4^+. Nitrification is a key process for determining the fate of nitrogen in soils. Nitrite and nitrate are more mobile than NH_4^+, and therefore are more readily lost through leaching. Nitrite can be reduced to nitrous oxide (N_2O) and dinitrogen (N_2) by denitrifying bacteria. The main factor influencing the nitrification rate is the concentration of available NH_4^+.

The reduction of nitrate via nitrite to ammonia occurs in soils at low rates (Rosswall, 1982). If it were possible to stimulate the reduction of nitrate to ammonia and then its incorporation into the organic matter (reaction 4), large nitrogen losses resulting from denitrification could be prevented so that a high fertility status could be maintained.

DISTRIBUTION OF NITROGEN IN HUMIC SUBSTANCES

What are the major nitrogen-containing compounds that have so far been identified in humic substances?

We first deal with humic substances extracted from soils formed under different climates because climate, as will be shown later, has a profound effect on the types and concentrations of nitrogen compounds present in humic substances.

Amino Acids, Amino Sugars, and Ammonia

As indicated in Table 2, the distribution of nitrogen in the three soils formed under cool temperate climates is similar except that the Solod humic acid contains more amino acid nitrogen and the Chernozem humic acid more NH_4^+ than do the other humic acids. In terms of proportions of nitrogen identified, the order is: Solod humic acid > Chernozem humic acid > Solonetz humic acid. Between 46 and 53% of the nitrogen in the humic acids is identified.

Fulvic acids extracted from soils formed under cool temperate climates (Table 3) contain similar proportions of amino acid nitrogen and amino sugar

TABLE 2. Distribution of Humic Acid Nitrogen in Soils of Cool Temperate Climates[a]

Soil	Percentage of Humic Acid Nitrogen in			Percentage of Humic Acid Nitrogen Identified
	Amino Acid Nitrogen	Amino Sugar Nitrogen	Ammonia Nitrogen	
Solonetz	28.7	1.7	15.1	45.5
Solod	35.7	1.8	15.0	52.5
Chernozem	28.3	1.3	19.8	49.4

[a] *Khan and Sowden (1971).*

nitrogen to humic acids extracted from the same soils, but are richer in amino sugar nitrogen. Between 45 and 59% of the nitrogen in the fulvic acids is identified.

Table 4 shows mean values for the relative distribution of amino acids in acid hydrolyzates of humic acids and fulvic acids extracted from the same soils. These data are expressed as α-amino nitrogen of each amino acid $\times$ 100/total amino acid nitrogen. An inspection of the data in Table 4 indicates, with few minor exceptions, similarities in the amino acid composition of humic acids and fulvic acids. Acid hydrolysis appears to destroy about one-half of the amino sugars and there are losses of threonine and serine (Sowden, 1959, 1969). No corrections are made for their decomposition because the ammonia-nitrogen would then require correction, and a valid correction for it is not possible. The ammonia nitrogen increases with length of time of hydrolysis (Khan and Sowden, 1971).

Table 5 presents data on the distribution of nitrogen in humic acids, fulvic acids, and humin extracted from tropical soils. Especially noteworthy are the relatively high proportions of amino acid nitrogen in all humic fractions, and the unusually high percentages of total nitrogen identified. Humin is especially rich in amino sugars.

TABLE 3. Distribution of Fulvic Acid Nitrogen in Soils of Cool Temperate Climates[a]

Soil	Percentage of Fulvic Acid Nitrogen in			Percentage of Fulvic Acid Nitrogen Identified
	Amino Acid Nitrogen	Amino Sugar Nitrogen	Ammonia Nitrogen	
Solonetz	34.2	5.2	19.3	58.7
Solod	29.4	4.6	18.5	52.5
Chernozem	26.4	3.6	15.1	45.1

[a] *Khan and Sowden (1972).*

TABLE 4. Mean Relative Molar Distribution of Amino Acids in Acid Hydrolyzates of Humic Acids and Fulvic Acids Extracted from Soils Formed Under Cool Temperature Climates[a]

Amino Acid	Humic Acid	Fulvic Acid
Acidic		
Aspartic acid	13.6	12.5
Glutamic acid	9.3	7.3
Basic		
Arginine	1.1	2.6
Histidine	0.9	1.8
Lysine	2.4	3.5
Ornithine	0.8	1.1
Neutral		
Glycine	14.6	11.5
Alanine	10.9	7.6
Valine	6.3	5.6
Leucine	3.0	4.9
Isoleucine	2.9	3.0
Phenylalanine	1.7	2.5
Tyrosine	0.7	1.2
Serine	5.7	5.6
Threonine	6.0	5.6
Proline	4.2	4.0
Methionine	0.3	0.3
Cystine	0.2	0.3
Other amino acids[b]	3.0	2.2
Unidentified	3.3	0.7
Amino sugar ratio	1.1	0.9

[a] *Calculated from data of Khan and Sowden (1971, 1972).*
[b] *Cysteic acid, methionine sulfoxide, OH-proline, alloisoleucine, α-NH_2-butyric acid, OH-lysine, 2,4-diaminobutyric acid, diaminopimelic acid, β-alanine, α-NH_2-isobutyric acid, β-NH_2-isobutyric acid.*

The most striking features of the amino acid composition of the three humic preparations (Table 6) are: (1) the high concentrations of acidic amino acids in the fulvic acids; (2) the relatively low concentrations of basic and some neutral amino acids (phenylalanine, tyrosine, leucine, and isoleucine) in the fulvic acids; and (3) the accumulation of cysteic acid in the fulvic acids. The distributions of amino acids in the humic acids and humin are quite similar.

Differences in the nitrogen distribution in humic substances extracted from soils developed under cool temperature and tropical climates are sum-

TABLE 5. Nitrogen Distribution in Humic Fractions from Tropical Soils[a]

Fraction and Sample No.	Hydrolyzable Nitrogen as a Percentage of Total Nitrogen						Unhydrolyzed Nitrogen	Percentage of Total Nitrogen Identified
	Total Nitrogen[b]	Total	NH_3 N	Amino Sugar N	Amino Acid N	Unidentified N		
Humic Acid								
2	3.96	81.3	14.6	4.8	52.9	9.0	18.8	72.2
3	4.32	81.7	12.9	3.1	47.8	17.9	18.3	63.8
5	5.40	87.5	14.6	3.0	48.1	21.8	12.5	65.7
Fulvic Acid								
2	2.77	93.5	17.0	2.8	39.2	34.5	6.5	50.9
5	1.78	97.7	23.6	1.7	46.2	26.2	2.3	71.5
Humin								
2	1.01	88.3	15.3	9.0	48.0	16.0	11.7	72.3
5	0.43	79.1	17.4	8.1	48.7	4.9	20.9	74.2

[a] *Sowden et al. (1976).*
[b] *On oven-dry basis.*

TABLE 6. Relative Molar Distribution of Amino Acids in Humic Substances (α-Amino Acid Nitrogen of Each Amino Acid × 100/Total Amino Acid Nitrogen) in Tropical soils[a]

	Sample Number						
	Humic Acid			Fulvic Acid		Humin	
Amino Acid	2	3	5	2	5	2	5
Acidic							
Aspartic acid	13.0	11.7	11.8	26.1	23.1	11.8	23.2
Glutamic acid	8.5	8.8	8.6	15.0	20.9	8.2	10.1
Basic							
Arginine	2.0	2.2	2.3	0.5	0.3	1.9	1.9
Histidine	2.3	1.4	1.5	0.9	0.2	2.1	1.4
Lysine	3.3	3.1	3.5	1.9	1.4	2.7	2.6
Ornithine	0.7	0.8	0.7	0.9	0.7	0.6	1.1
Neutral							
Phenylalanine	3.2	3.3	2.9	1.3	0.9	2.9	1.7
Tyrosine	1.5	1.6	1.4	1.2	0.2	1.5	1.2
Glycine	11.2	10.9	11.1	12.6	13.5	13.1	11.2
Alanine	7.6	8.5	8.3	7.4	9.1	8.9	7.9
Valine	5.4	6.2	5.9	4.1	3.6	6.0	3.8
Leucine	5.1	5.8	5.1	3.0	1.7	5.3	4.4
Isoleucine	3.3	3.5	3.5	1.8	1.7	3.0	2.2
Serine	5.0	4.8	4.9	5.2	4.3	5.7	4.2
Threonine	4.9	4.7	5.2	4.4	3.9	5.3	3.8
Proline	4.2	6.2	4.9	4.4	3.3	4.2	2.7
Hydroxyproline	0.7	0.7	0.7	tr[b]	tr	0.8	0.3
Sulphur Containing							
Methionine	0.5	1.0	0.8	0.4	0.3	0.7	0.3
Cystine	0.3	0.1	0.2	0.1	0.2	0.1	0.1
Cysteic acid	0.4	0.5	0.6	1.7	3.4	0.3	0.8
Methionine sulphoxide	0.6	0.2	0.3	0.1	1.0	0.2	0.3
Miscellaneous[c]	1.7	0.9	1.6	0.9	2.9	1.6	2.1
Total amino acid nitrogen (μM/g)	1496.0	1476.0	1856.0	775.0	587.0	346.4	149.5
Amino sugar ratio[d]	1.3	1.3	1.3	1.2	1.4	3.5	2.5

[a] *Sowden et al. (1976).*
[b] *tr = trace.*
[c] *Includes allo-isoleucine, α-NH_2-butyric acid; 2-4-diaminobutyric acid, diaminopimelic acid, β-alanine, ethanolamine, and unidentified compounds.*
[d] *Ratio of glucosamine/galactosamine.*

TABLE 7. Distribution of Hot Acid Hydrolyzable Nitrogen in Humic Substances from Different Climates (Mean Values)

	Amino Acid Nitrogen (%)			Amino Sugar Nitrogen (%)			Ammonia Nitrogen (%)			Percentage of Nitrogen Identified		
Climatic Zone	HA[a]	FA[b]	Humin	HA	FA	Humin	HA	FA	Humin	HA	FA	Humin
Cool temperate	30.9	30.0	—	1.6	4.5	—	16.6	17.6	—	49.1	52.1	—
Tropical	49.6	42.7	48.4	3.6	2.3	8.6	14.0	20.3	16.4	67.2	65.3	63.4

[a] *HA = humic acid.*
[b] *FA = fulvic acid.*

marized in Table 7. These data show that in tropical soils, because of higher microbial activity, greater proportions of the total nitrogen in humic substances occur as amino acid nitrogen and amino sugar nitrogen than in humic substances extracted from soils under cool temperate climates. Thus, climate appears to significantly affect the distribution of nitrogen in humic substances.

The compounds so far identified in humic acids, fulvic acids, and humin are amino acids, which probably are present mainly as polypeptides and proteins, amino sugars (mainly glucosamine and galactosamine), and ammonia which originates from several sources as mentioned above.

Effect of Molecular Weights of Humic Substances on Nitrogen Content

Schnitzer and Skinner (1968) demonstrated that fulvic acid could be separated on Sephadex gels into different molecular weight fractions which varied in total nitrogen content. Their data, summarized in Table 8, show that the total nitrogen content decreases as the molecular weight of the fulvic acid fraction decreases, so that high-molecular-weight fulvic acid fractions contain more total nitrogen than do low-molecular-weight fractions.

Nucleic Acid Bases

Purines and pyrimidines are components of nucleic acids and are known to occur in soils. According to Bremner (1967), the nitrogen in purines and pyrimidines accounts for less than 1% of the total soil nitrogen. Much of the earlier information on the presence of these compounds comes from G. Anderson (1957, 1958, 1961) who identified guanine, adenine, cytosine, thymine, and traces of uracil in acid hydrolyzates of humic acids extracted from three Scottish soils. Anderson used paper and ion-exchange chromatography to detect and estimate the concentrations of nucleic acid bases. More

TABLE 8. Analytical Characteristics of Fulvic Acid and of Fractions Separated from It[a]

Fraction	Yield (g)	Number Average Molecular Weight (MWn)	Percentage Carbon	Percentage Hydrogen	Percentage Oxygen	Percentage Nitrogen
Original fulvic acid	1.000	688	49.50	4.52	45.90	0.75
A	0.164	3570	49.97	5.95	42.73	1.17
B	0.232	1179	50.67	5.51	42.91	0.74
C	0.141	754	51.04	4.66	43.46	0.55
D_1	0.265	337	51.71	5.65	41.90	0.34
D_2	0.056	175	52.16	7.71	40.13	—

[a] *From Schnitzer and Skinner (1968).*

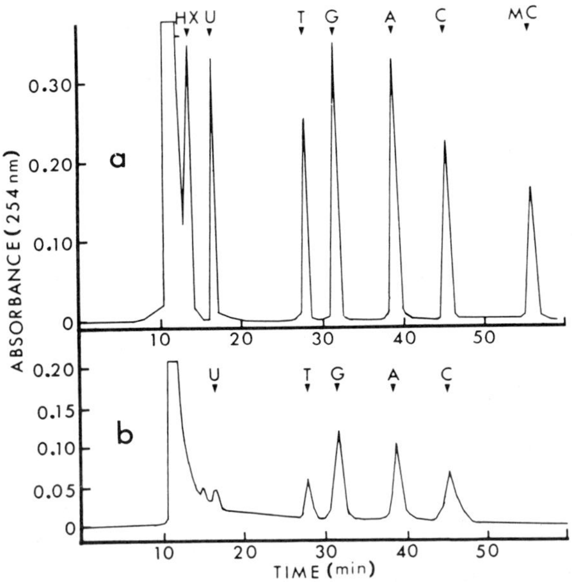

FIGURE 2. Ion-exclusion chromatography of (a) known purines and pyrimidines; (b) purines and pyrimidines extracted from the B_{hf} horizon of a Spodosol. HX = uracil; U = unknown; T = thymine; G = guanine; A = adenine; C = cytosine; MC = 5 methyl cytosine (Cortez and Schnitzer, 1979b).

recently, Cortez and Schnitzer (1979a,b) employed ion-exclusion chromatography for determining the distribution of purines and pyrimidines in a number of humic acids, fulvic acids, and humin extracted from 15 soils of widely differing origins. Figure 2a illustrates the ion-exclusion chromatogram of seven known purines and pyrimidines. A similar chromatogram for purines, and pyrimidines isolated from the B_{hf} horizon of a Spodosol, is shown in Figure 2b. In soils, purines and pyrimidines tend to be associated with humic substances (Cortez and Schnitzer, 1979a). Table 9 lists the concentrations of nucleic acid bases in the three humic fractions. Soils numbered 1–12 are inorganic soils, whereas soils numbered 13–15 are organic soils (Cortez and Schnitzer, 1979b). It is difficult to draw general conclusions from these data. In some cases the nucleic acid bases are evenly distributed among the three humic fractions, in others, humin or fulvic acids are richer in these compounds than are humic acids. In the organic soils, most of the nucleic acid bases are present in the humin. Proportions of total nitrogen in each humic fraction which occur in purines and pyrimidines are presented in Table 10. In the humic acid extracted from soil No. 11 (a Spodosol B_{hf} horizon), 10.7% of the total humic acid nitrogen is present in nucleic acid bases, whereas in the fulvic acid extracted from soil No. 13 (a Mesisol), 18.6% of the total fulvic acid nitrogen is so distributed. The molar distributions of purines and pyrimidines in five humic acids and five fulvic acids are shown in Table 11. The chemical structures of the nucleic acids identified are

TABLE 9. Concentrations in Micrograms per Liter of Purines and Pyrimidines in Humic Fractions of Soil[a]

Soil Number	Humic Acid	Fulvic Acid	Humin
1	19.2(21)[b]	27.7(30)	46.1(49)
2	7.7(30)	8.8(34)	9.5(36)
3	17.6(28)	20.3(32)	25.1(40)
4	18.1(18)	46.1(46)	35.8(36)
5	8.8(27)	21.5(65)	2.7(8)
6	9.4(24)	11.2(28)	19.4(48)
7	6.4(15)	12.9(29)	24.7(56)
8	83.8(20)	77.7(19)	257.5(61)
9	72.4(23)	123.8(39)	117.9(38)
10	128.4(35)	117.3(32)	124.3(33)
11	80.0(27)	147.7(50)	65.3(23)
12	16.9(26)	24.6(37)	24.5(37)
13	200.7(15)	171.0(13)	994.3(72)
14	58.3(11)	5.7(1)	491.0(88)
15	21.9(15)	7.1(5)	121.0(80)

[a] *Cortez and Schnitzer (1979b).*
[b] *Percentage of purines + pyrimidines identified in the three humic fractions.*

TABLE 10. Proportions (%) of Total Nitrogen of Each Humic Fraction in Purines and Pyrimidines[a]

Soil Number	Humic Acid	Fulvic Acid	Humin
1	2.6	3.1	1.7
2	1.7	2.2	0.4
3	1.8	2.8	0.8
4	2.0	6.9	1.3
5	2.1	1.2	10.4
6	2.1	2.2	7.7
7	0.4	0.8	0.8
8	1.3	2.9	2.3
9	2.3	6.0	1.6
10	4.9	13.3	2.1
11	10.7	6.9	6.2
12	2.0	3.8	4.1
13	2.6	18.6	2.8
14	3.7	2.6	1.1
15	0.7	0.8	0.2

[a] *Cortez and Schnitzer (1979b).*

TABLE 11. Distribution of Purines and Pyrimidines in Humic Acids and Fulvic Acids[a]

HA and FA Number[b]	[Purines(g) + Pyrimidines(g)]/ Ash-Free HA or FA	Uracil	Thymine	Guanine	Adenine	Cytosine	Unknown	Guanine + Cytosine	Adenine + Thymine
		Mole %							
HA-1	210.8	0	2	71	11	7	9	78	13
HA-2	352.9	0	0	63	28	9	0	72	28
HA-3	683.4	3	0	58	27	3	9	61	27
HA-4	596.7	2	0	54	21	12	10	67	21
HA-5	810.0	3	0	54	20	17	6	71	20
FA-1	294.3	14	20	39	17	11	0	50	37
FA-2	861.1	3	11	25	13	43	5	68	18
FA-3	648.6	8	28	27	7	27	3	54	35
FA-4	723.5	8	32	18	5	35	2	53	37
FA-5	1086.6	5	24	22	6	37	6	59	30

[a] *Cortez and Schnitzer (1979a).*
[b] *HA = humic acid.*

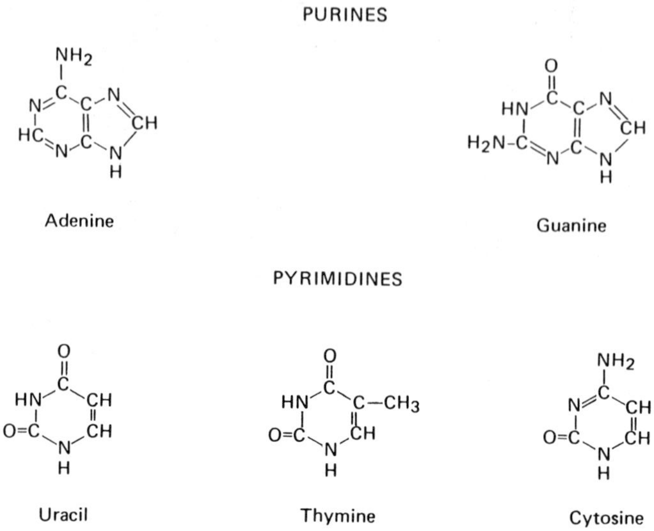

FIGURE 3. Chemical structures of purines and pyrimidines.

presented in Figure 3. The data in Table 11 show that the humic acids are richer in guanine and adenine but poorer in uracil, thymine, and especially cytosine than are the fulvic acids. Average guanine plus cytosine over adenine plus thymine ratios for the humic acids and fulvic acids are >2. The magnitudes of these ratios plus the absence of 5-methyl cytosine suggest a microbial DNA origin for the nucleic acid bases in the humic substances. From the relatively high concentrations of purines and pyrimidines in the humic fractions, from which they are not separated during the extraction and fractionation procedures, one can conclude that they are somehow associated with the humic substances, possibly by hydrogen bonding. It is unlikely that nucleic acid bases are integral components of humic acids and fulvic acids since they can be extracted by reflux with 1*M* HCl, conditions which presumably cause minimal alteration of the chemical structure of humic substances (Riffaldi and Schnitzer, 1972b).

From the data presented here one can conclude that in some humic fractions up to 18.6% of the total nitrogen, or close to 40% of the unidentified nitrogen, may occur as nucleic acid bases. Thus, the nitrogen in purines and pyrimidines appears to constitute a greater proportion of the unidentified nitrogen in humic substances than has been assumed in the past.

UNIDENTIFIED NITROGEN IN HUMIC SUBSTANCES

As has been shown earlier, at least 50% of the nitrogen in humic substances still remains to be identified, and this after more than 100 years of research

$$\underset{\text{protein}}{R\text{-}NH_2} + \underset{\text{lignin}}{R_2C{=}O} \longrightarrow \underset{\text{ligno-protein}}{R_2C{=}N\text{-}R} + H_2O$$

FIGURE 4. Ligno–protein complex (Waksman and Iyer, 1932).

on these materials. During that time numerous proposals on the formation and identity of the unidentified nitrogen have been made. The more prominent of these proposals will be discussed next.

Ligno–Protein

Waksman and Iyer (1932) proposed the formation of ligno–protein complexes in soils. A C═O group of lignin and an NH_2-group of protein, peptide, or amino acid condense to form a Schiff base (Fig. 4). About 75% of the organic matter in soils was considered to be present in such a complex which was thought to be stabilized through the Schiff base. One of the main weaknesses of this proposal is that so far numerous attempts to isolate this type of complex from soils have had little success. Also, organic matter does not contain much material which can be described as lignin or even modified lignin. The current view is that the major part of soil organic matter is not a simple complex of residual plant lignin and microbial protein (Parsons and Tinsley, 1975).

Phenol– and Quinone–Protein

Phenols and quinones can interact with proteins irreversibly by covalent bonding and reversibly through hydrogen bonding. Theis (1945) suggested that ε-lysylamino groups of proteins and quinones interact through covalent bonds as illustrated in Figure 5.

On the other hand, the formation of a hydrogen bond between a phenolic hydroxyl group and oxygen of a peptide bond constitutes a very realistic possibility (Fig. 6). It is likely that in these types of interaction both covalent and hydrogen bonds are formed.

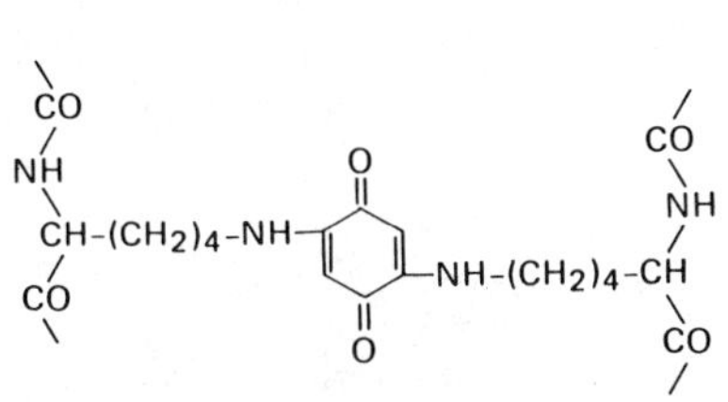

FIGURE 5. Quinone–protein complex (Theis, 1945).

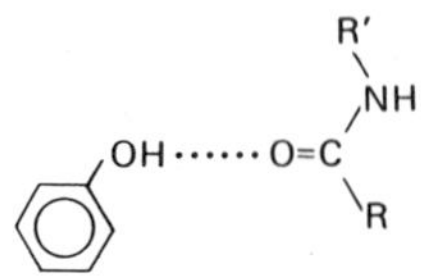

FIGURE 6. Hydrogen bonding between phenolic hydroxyl and oxygen of peptide bond (Ladd and Butler, 1975).

Phenol– and Quinone–Amino Acid

A product resulting from the interaction of glycine with a phenol described by Piper and Posner (1972) is shown in Figure 7. The amino acid cannot be hydrolyzed by hot acid from *N*-(*p*-hydroxyphenyl)glycine. But after oxidation to the nitrogen–(methylcarboxy)quinonimine, the amino acids can be split off by alkaline hydrolysis. These workers believe that acid hydrolysis will remove all amino acids bound by peptide bonds and also those linked to quinone rings, but amino acids bonded to phenolic rings would be released only after subsequent alkaline hydrolysis. Amino acids bound into structures as these may thus account for some of the unidentified nitrogen.

Phenols are readily oxidized chemically or enzymatically to quinones which can interact with amino acids to form aminohydroquinones. The latter can then be oxidized to aminoquinones (Fig. 8). The resulting products are dark colored and release only low yields of amino acids when hydrolyzed with hot acid.

Phenol– and Quinone–Ammonia

It has been suggested that soil organic matter contains nitrogenous complexes that are formed by reactions of phenolic compounds and/or quinones with amino acids and ammonia (Flaig et al., 1975). The required reactants appear to be produced during the decomposition of plant material. In the absence of air and in the presence of ammonia, a hydroxyl group in hydroxyhydroquinone (Fig. 9) can be replaced by an NH_2 group. The aminoresorcinol thus formed reacts in the presence of oxygen with additional hydroxyhydroquinone to produce 7-hydroxyphenoxazone, containing a heterocyclic ring.

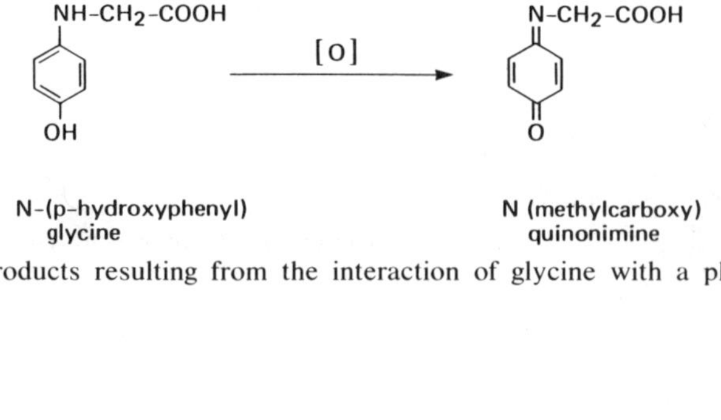

FIGURE 7. Products resulting from the interaction of glycine with a phenol (Piper and Posner, 1972).

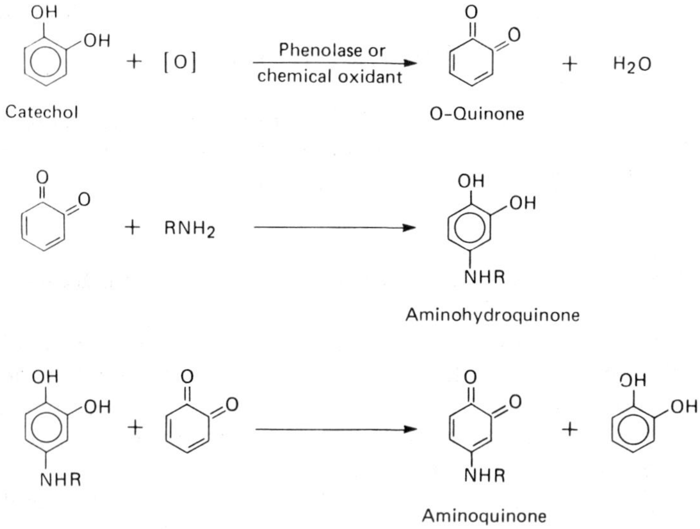

FIGURE 8. Oxidation of catechol in the presence of an amine (Mason, 1955).

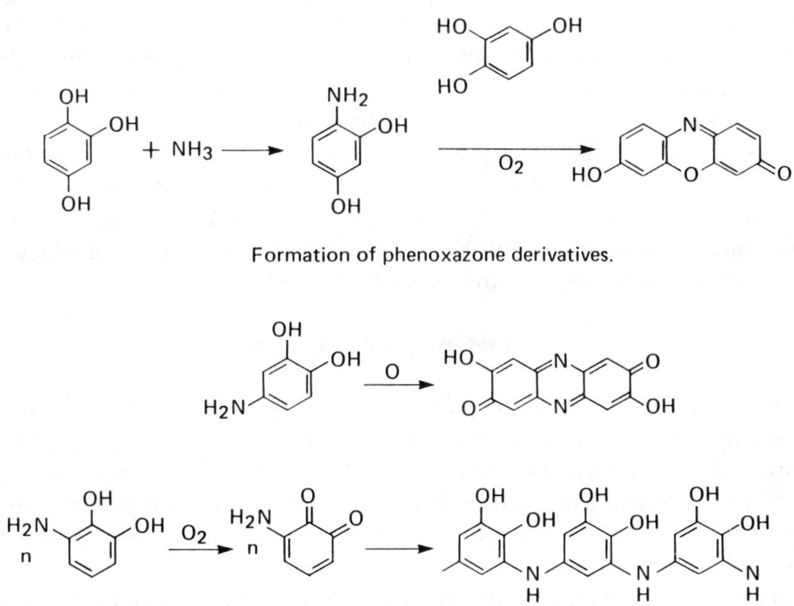

FIGURE 9. Reactions of phenols with ammonia, and autopolymerization reactions (Flaig et al., 1975).

Autopolymerization

According to Horner and Sturm (1957) 1 mole of 4-amino catechol reacts with another mole of the same compound to form a phenazine derivative (Fig. 9). Similarly, 3-amino catechol condenses under oxidative conditions to form a polymer (Fig. 9).

Carbohydrate–Amino Acid (Maillard or Browning Reaction)

The production of brown nitrogenous polymers by condensation of reducing sugars and amino compounds occurs extensively in the dehydration of food products at moderate temperatures and the reaction has been postulated to be of importance in the formation of humic substances in soils (Maillard, 1912; Stevenson, 1982). The major objection is that the reaction proceeds rather slowly at temperatures found under normal soil conditions. However, drastic and frequent changes in soils (freezing and thawing, wetting and drying) together with the intermixing of reactants with mineral matter having catalytic properties may facilitate condensation. An attractive feature of the theory is that the reactants (sugars, amino acids, etc.) are produced in abundance through the activities of microorganisms.

The initial reaction (Fig. 10) involves addition of the amine to the aldehyde group of the sugar to form a Schiff base and the nitrogen-substituted glycosylamine. The latter subsequently undergoes the Amadori rearrangement to form a nitrogen substituted 1-amino deoxy 2-ketose which fragments with the liberation of amine and the formation of 3-carbon chain aldehydes and ketones, such as acetol, glyceraldehyde, and dihydroxyacetone; three water molecules are lost to form hydroxymethyl furfural. All these compounds are highly reactive and polymerize in the presence of amino compounds to form dark-colored products of unknown composition. Maillard thought that humic substances resulted from purely chemical reactions in which microorganisms did not play a direct role except to produce sugars from carbohydrates and amino acids from proteins.

Heterocyclic Nitrogen

It is possible that some of the ''unknown'' soil nitrogen is present in heterocyclic rings. As has been shown above, heterocyclic nitrogen-containing ring compounds can be synthesized in the laboratory in a variety of ways. It is, therefore, somewhat surprising that, except for purine and pyrimidine bases, very few heterocyclic nitrogen compounds have ever been isolated from soils.

The numerous theories advanced on the origin and structure of complex organic nitrogen components of soils fall into two classes: (1) complexes formed by reactions of phenol compounds with proteins, amino acids, and ammonia; and (2) complexes formed by reactions of carbohydrates with

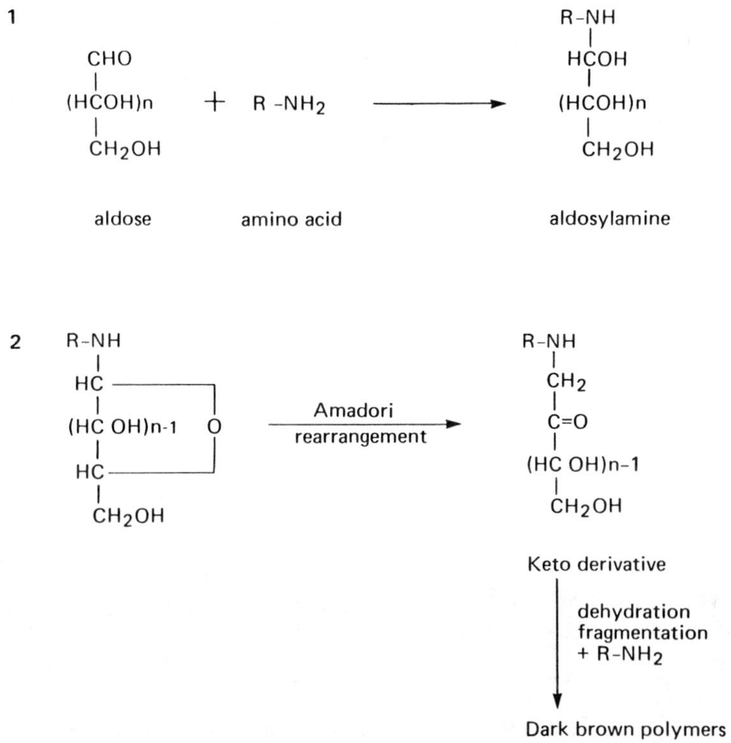

FIGURE 10. Maillard or browning reaction (Stevenson, 1982).

amino compounds and ammonia. Practically all the information available on these nitrogen-containing materials comes from laboratory studies. None of these products has ever been detected in humic substances. The time has now come to either demonstrate the existence of such compounds in humic substances or abandon some of the less realistic theories.

"Pseudo-Amide"

Kickuth and Scheffer (1976) noted that only one-half of the NH_3 released by hot acid hydrolysis originated from amides such as asparagine and glutamine. The remaining NH_3, they proposed, came from "pseudo-amides" which consisted primarily of imino groups attached to quinonoid ring systems, Schiff bases, and enamine structures (see Fig. 11). On the other hand, structures that would not release NH_3 on hot acid hydrolysis were those containing nitrogen as central atoms in heterocyclic rings and nitrogen on the periphery of aromatic or alicyclic rings (Fig. 11). The authors claim that "pseudo-amide" nitrogen was readily available to plants.

Aside from the chemical structures that might produce NH_3 on hot acid hydrolysis suggested by Kickuth and Scheffer, it is well known that NH_3 can

"Pseudo" amide

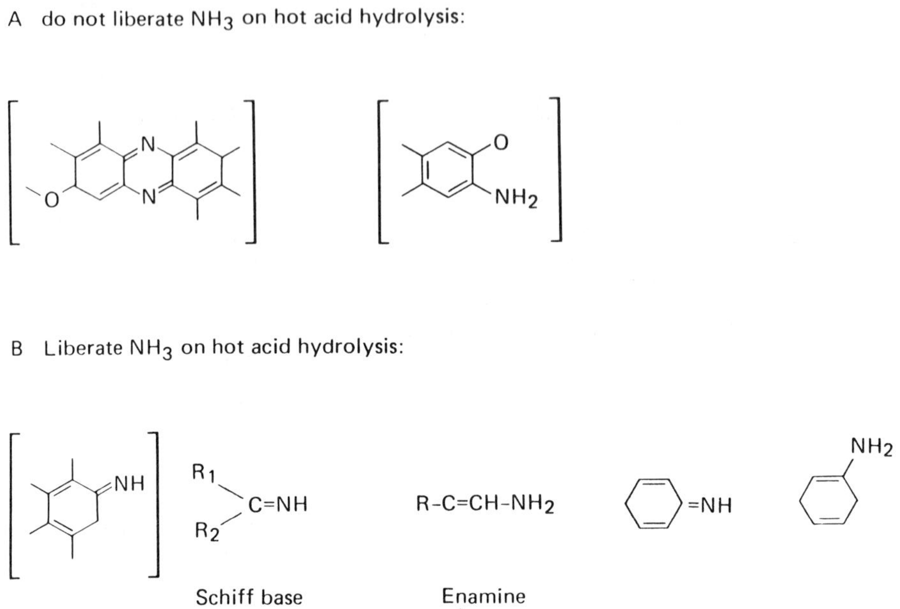

FIGURE 11. Chemical structures that liberate and that do not liberate NH_3 on hot acid hydrolysis (Kickuth and Scheffer, 1976).

also be formed under these conditions from the deamination of hydroxy amino acids, amino alcohols, amino sugars, purines and pyrimidines, and exchangeable and clay-fixed NH_4.

ATTEMPTED CHARACTERIZATION OF THE UNIDENTIFIED NITROGEN IN HUMIC ACIDS AND FULVIC ACIDS

The basic philosophy behind our approach to this problem is to develop a procedure that will allow us to isolate from humic acids and fulvic acids fractions which are rich in unidentified nitrogen but which contain only small amounts of known nitrogen compounds. We believe that the availability of such fractions will make possible the isolation and identification of some of the major unidentified nitrogen components without interference from the many known nitrogen-containing constituents.

Our approach (Schnitzer et al., 1983) is as follows (see also Fig. 12):

1. The humic acid or fulvic acid is hydrolyzed with hot 6*M* HCl for 24 hours.

2. The hydrolyzate is neutralized and the soluble material separated on Sephadex G-25 gel.
3. The highest-molecular-weight fraction is further separated on Sephadex G-50 gel and the second-highest-molecular-weight fraction is separated on Sephadex G-15 gel.

Following this procedure, several fractions have been prepared from both humic acid and fulvic acid which contain between 97.5 and 98.6% of the total nitrogen as unidentified nitrogen, but only 0.84% as amino acid nitrogen, 0% as amino sugar nitrogen, and 0.53% as ammonia nitrogen.

In more recent work, some of the fractions rich in unidentified nitrogen were hydrolyzed with 2*M* H_2SO_4. Acetate derivatives of the products were separated by capillary gas chromatography. A number of peaks were tentatively identified by comparing the mass spectra and analytical characteristics of the components with those of reference compounds of known structures. A number of hydroxy- and ketoindoles, benzylamines, and nitriles were identified in this manner.

The formation of indole rings during the polymerization of quinones and amino compounds has been proposed by Rinderknecht and Jurd (1958) who report that the reaction of phloroglucinol with glycine results in the formation of 3-hydroxyindole (Fig. 13).

Hackmann and Todd (1953) have suggested the formation of indole rings as a result of a rearrangement following the condensation of orthoquinone and a terminal amino group of a protein (Fig. 13).

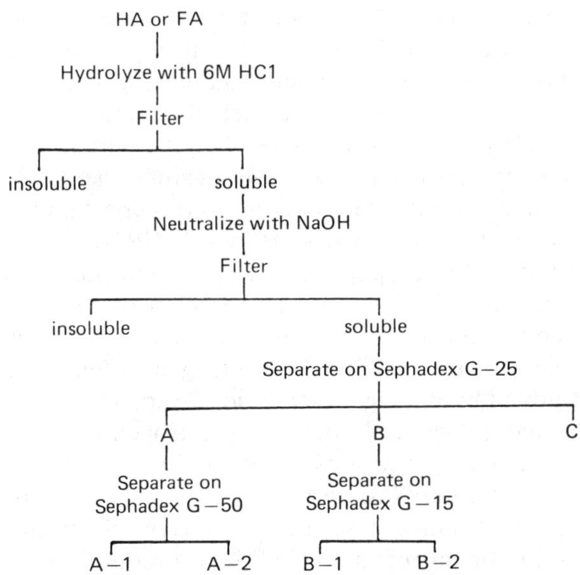

FIGURE 12. Scheme for separating fractions rich in unidentified nitrogen (Schnitzer et al., 1983).

Rinderknecht and Jurd (1958)

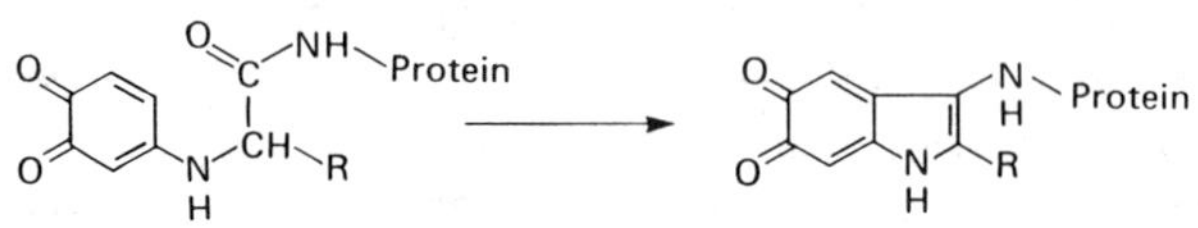

Hackmann and Todd (1953)

FIGURE 13. Rearrangement of the product of the reaction of phloroglucinol with glycine (Rinderknecht and Jurd, 1958). Rearrangement of the product of the reaction of orthoquinone and a terminal amino group of a protein (Hackmann and Todd, 1953).

It is noteworthy that Flaig and Breyhan (1956) have identified indoles in a fusion mixture of a humic acid residue that had resisted 6*M* HCl hydrolysis.

NITROGEN AS INTEGRAL HUMIC CONSTITUENT

The question is often asked whether nitrogen is an integral component of humic acids, fulvic acids, or humin or whether it is a "contaminant" adsorbed on or loosely held by humic substances but not structurally associated with them. A search of the literature indicates support for both sides. Haworth (1971) believes that humic acid contains, or readily gives rise to, a complex aromatic core to which are attached chemically or physically (1) polysaccharides, (2) proteins, (3) simple phenols, and (4) metals. The protein-core attachment appears to be stable against chemical and biological attack. This stability is similar to that conferred on protein by the tanning of leather or the sclerotization process. Haworth (1971) suggests that humic substances are linked to polypeptides by hydrogen bonds, which explains the removal of the latter by hot water. A number of workers (Roulet et al. 1963, Sowden and Schnitzer, 1967, and Khan and Sowden, 1972) report along similar lines that up to 75% of the nitrogen of humic acids and fulvic acids can be removed by passage over cation-exchange resins. These observations suggest that substantial portions of nitrogen components of humic substances are either attached loosely to the humic substances or not attached at all. Flaig et al. (1975), on the other hand, hold that nitrogen is an integral constituent of humic substances. In view of the dominant role of microorganisms in the nitrogen cycle, their production of amino acids, amino sugars, purines, pyrimidines, and other nitrogen compounds, and their likely participation in the synthesis of humic substances, one cannot exclude the possibility that some of the nitrogen is built into the humic structure. It seems to the author that a distinction could be made between nitrogen components which, when isolated, have distinct chemical identities such as proteins, peptides, amino acids, amino sugars, purines, pyrimidines, and so on (group 1) and nitrogen compounds that have become integral constituents of humic substances (group 2) and no longer have the distinctive characteristics of the compounds included in group 1. Probably most of the components of the unidentified nitrogen would fall into group 2. In view of the relative ease with which group 1 compounds can be separated from humic substances, largely in unaltered forms, I would like to suggest that these compounds are only loosely held or adsorbed by humic substances by low-energy bonds such as hydrogen bonds and van der Waal forces. Group 1 compounds, in my opinion, may contain up to 75% of the total nitrogen. Hopefully it will be possible to identify group 2 components (the remaining 25%) in the near future so that more accurate determinations of the nitrogen distribution in humic substances can be made.

FUTURE RESEARCH NEEDS

From the information presented in this chapter it becomes apparent that there exists a voluminous literature on nitrogen in humic substances. But this literature deals mainly with types and concentrations of amino acids and amino sugars and the amounts of ammonia that are found in acid hydrolyzates of humic substances. About one-third to one-half of the total nitrogen still remains to be identified (Sowden et al., 1977). There are indications that this unidentified nitrogen can be transformed by microorganisms (Ivarson and Schnitzer, 1979) and chemically (Schnitzer and Hindle, 1980) to NH_3. However, little is known about the chemical structure of the major components of this very substantial nitrogen fraction. Over the years a number of proposals have been made as to the identities of these constituents. These proposals were based on results of laboratory studies, philosophical considerations, and wishful thinking. Unfortunately, none of the proposed compounds has so far been detected in humic substances.

Taking into consideration current advances in modern chemistry, the availability of sophisticated and powerful instruments, and a greater interest in humic substances, it seems to me that the time has come to assign a higher priority to the chemistry of nitrogen in humic substances than has been the case in the past. There is an urgent need to identify the compounds that constitute about one-half of the total nitrogen so that the relationship between carbon and nitrogen in humic substances can be better understood. The chemistry of nitrogen in humic substances should be of special concern to those working in the agricultural sciences because of the important role of humic nitrogen in soil biochemistry and soil fertility.

SECTION TWO

ISOLATION AND FRACTIONATION OF HUMIC SUBSTANCES

CHAPTER THIRTEEN

Extraction of Humic Substances from Soil

M. H. B. HAYES

ABSTRACT

Consideration of the dissolution of humic substances in a variety of solvents takes account of structures and some physicochemical properties of solutes and solvents. Special emphasis is given to the polyelectrolyte properties of humic substances, and to the secondary forces that must be overcome in the solvent and macromolecular systems before solution takes place.

The classical extraction procedures use aqueous alkaline solvents and these give rise to partially oxidized and slightly degraded humic artifacts. The use of complexing agents, especially certain neutral salt solutions, frees insolubilizing divalent and polyvalent metals and allows the more polar humic components to dissolve in water. However, most humic acids are not dissolved in this way. There is a need for a mild, neutral, or even acidic solvent system that can match the dissolution performance of aqueous sodium hydroxide without giving rise to degradation or artifacts.

Application of solubility parameter data is hindered by the multicomponent nature of humic substances, and homogeneous fractions are needed to obtain the necessary data for the macromolecules. Nevertheless, the available information shows that the best solvents for H^+ humic acids have polar, hydrogen bonding, and proton acceptor solubility parameters greater than 5.

However, for solution to take place self-association of the solvent molecules through hydrogen bonding must be less than that for the solute–solvent systems. Some dipolar aprotic solvents satisfy the necessary criteria, and the best results have been obtained for slightly acidified dimethylsulfoxide.

INTRODUCTION

In highly organic soils humic substances are predominantly associated with each other, with charge-neutralizing cations, and with other organic components of humus. The same associations occur in mineral soils, but in these the humic substances are also held by the inorganic colloidal constituents.

The ''conglomerate'' soil colloid is considered to be composed of associations of clays, various hydroxides, especially those of iron and aluminum, and humus material composed for the most part of humic substances and polysaccharides (Hayes and Swift, 1981). There is evidence to show that divalent and polyvalent cations form bridges between polyanionic humic substances and the negatively charged inorganic colloids (Theng and Scharpenseel, 1975; Theng, 1976, 1979), and it is plausible to consider that these substances are held by coulombic attraction to the pH-dependent positive charges on the colloidal hydroxides of iron and aluminum, and by ligand-exchange reactions forming covalent-type linkages between the humic macromolecules and the colloidal hydroxides (Greenland and Mott, 1978; Mott, 1981; Bolt, 1982). Ideal solvents or extraction systems would isolate homogeneous humic components from the conglomerates, leaving the other components behind.

In Chapter 2 of this book, Stevenson has outlined the modern concepts of the origins and stabilities of humic substances (see also Flaig et al., 1975; Hayes and Swift, 1978; Stevenson, 1982). The available evidence suggests that condensation reactions, many of which are enzymatically catalyzed, are important for the genesis of these substances. There is no evidence for genetic control of the synthesis process, and condensation reactions involving several reactants give rise to numerous macromolecular products. Although the compositions of humic products in a batch from the same environment might largely be similar, it is likely that few molecules in that batch would be exactly the same. Thus, because there are no clear-cut distinctions between the physicochemical properties of the molecules within the humic acid, fulvic acid, and humin gross fractions it would be pointless to try to design solvent systems to isolate relatively homogeneous humic substances from the soil. For the most part the general approach has been to use aqueous alkaline or neutral salt solutions to isolate the maximum amounts of humic substances, and to rely on fractionation techniques, as discussed by Swift in Chapter 15 of this book, to separate the different components of the extracts.

Any attempt to design an effective procedure for the isolation of humic substances from soil should take account of the properties of the solvents or extractants to be used, of the solutes or material to be extracted, and of the types of associations that can exist between these solutes and other soil colloidal constituents. This chapter outlines some aspects of the composition, properties, and associations of soil humic substances which are relevant to their extraction, and it considers some of the properties of solvents that are used or might be considered for use as extractants of humic substances, or of molecules associated with these substances. It then compares the effectiveness and outlines some of the limitations of a selected number of solvents and procedures for dissolving and extracting humic substances.

SOIL HUMIC SUBSTANCES

Some properties of humic substances which are relevant to their solvation and extraction from soils are outlined in this section. The reader is referred to reviews by Schnitzer and Khan (1972), Flaig et al. (1975), Hayes and Swift (1978), Schnitzer (1978), and Stevenson (1982) for more detailed information about the composition, structures, and general properties of these substances.

Composition of Humic Substances

The development of cross-polarization magic angle spinning ^{13}C nuclear magnetic resonance spectroscopy (CPMAS ^{13}C NMR), which allows meaningful spectra of solid samples to be obtained, has greatly increased interest in the composition and structures of humic substances during the past 4 years. This topic is discussed by Wershaw in Chapter 22. Data from the NMR spectra indicate that humic and fulvic acids are composed of aromatic and aliphatic carbon structures, contain carbonyl, carboxyl, and phenolic groups, as well as C—O linkages that could represent alcoholic, ether, or ester groups. Hydrolysis products suggest that carbohydrates and peptides could be associated with these acids.

Interpretations of the published spectra suggest that there are considerable differences in the amounts of aromatic structures in humic substances of different origins. To some extent these differences can depend on the methods used for the calculations. As much as 68% of the components of fulvic acid fractions have been estimated to be aromatic (Schnitzer, 1982), although it would appear that aromatic structures generally do not amount to more than 35% of the mass of the structural components of soil humic acids. This is not surprising because up to 40% of the mass may be lost as volatile and as hydrolyzable material (such as oligo- and polysaccharide, and peptide structures) when humic acids are boiled in 6*M* hydrochloric acid (Atherton et al., 1967). The residual or "core" components left after hydrolysis might well have a predominance of aromatic hydrocarbon structures. Cheshire et al. (1967) conclude from zinc dust distillation studies that these components have mainly fused aromatic structures, but the bulk of the data available indicates that the aromatic compounds in humic substances are single ring and not fused ring structures (Hayes and Swift, 1978).

It would appear that most of the benzene ring structures in humic substances have two or more substituents, such as carbonyl, carboxyl, hydrocarbon, hydroxyl, and methoxyl and other ether functional groups. The evidence from identification of the products of degradation reactions, and more recently from NMR data, suggests that aliphatic hydrocarbons and aliphatic and aromatic ether groups link the core components in the macromolecules, and that carbonyl, carboxyl, and hydroxyl substituents are likely to be attached to some of the aliphatic hydrocarbons (Hayes and Swift, 1978).

Extensive washings and a variety of other purification procedures fail to separate all the carbohydrate- and amino-acid-containing substances from most soil humic acids. It is possible, of course, that polypeptide, proteinaceous, oligosaccharide, and polysaccharide material could merely be sorbed to, or physically entrapped in, these humic structures. However, because of the energy requirement for the removal of the hydrolyzable components, it is likely that they are covalently linked to the core compounds in the humic acid macromolecules. Sugars, oligosaccharides, and polysaccharides could be attached through phenolic and glycoside linkages to the core structures. Peptides and proteinaceous materials might be held through peptide bonds, or through carbon to nitrogen bonds arising from interactions between the carbons α- to the keto groups in quinone structures and free or terminal amino groups in amino acids and peptides (see Hayes and Swift, 1978, p. 186, 241).

The emerging concept is of humic substances that are macromolecular and complex, and composed of substituted aromatic and aliphatic hydrocarbon core materials. It is probable that some aliphatic and aromatic ring compounds are heterocyclic with nitrogen, oxygen, or sulfur as the heteroatoms. It is highly unlikely that the different core components are linked in any regular order, but their behavior on wetting and drying suggests that segments or some side chain components in the structures are substantially hydrophobic. For the most part, however, the primary structures, or the single molecule components, or the monomer units contributing to the macromolecular structures contain significant amounts of polar substituents, incorporating oxygen in a variety of functional groups. Peptide and saccharide materials which are biologically synthesized and have regular sequences of amino acids and sugars, can be linked covalently to the core, especially in the case of the higher-molecular-weight humic acids.

Physicochemical Properties of Humic Substances

Humic substances are polydisperse polyanions, and when fully ionized have cation-exchange capacity (CEC) values ranging from 3 to 6 meq/g for humic acids, and up to 10 meq/g for fulvic acids. When the exchangeable cations are hydrogen, or divalent and polyvalent cations, humic acids are insoluble in water. Fulvic acids are more highly oxidized than humic acids, have lower average molecular weights, and are soluble in water in the hydrogen-ion-exchanged form. The negative charges arise from carboxyl groups and from the phenolic and possibly enolic hydroxyls.

Humic and fulvic acids are soluble in water when the counterions are monovalent because these humate salts readily dissociate. Cameron et al. (1972b) have deduced from molecular weight and frictional ratio data (see also Hayes and Swift, 1978, p. 278) that the molecules adopt random coil conformations in solution. Such coils may be visualized as strands, with polar groups and negative charges distributed along their lengths, which coil randomly with respect to both time and space. The overall shape might be considered spherical, with the mass density of the spheres greatest at the center and decreasing to zero at the outer limits, and with the coils within the boundaries of the spheres perfused with solvent. Dimensions of the spheres would depend on the lengths of the strands, on the charge densities, on the extents of solvation and dissociation of the acid groups, on the nature of the counterions, on the concentrations of salts in the media, and on the extents of the branching and cross-linking of the strands.

A plot of frictional ratio values versus molecular weights for carefully fractionated humic acids soluble in tris buffer [2-amino-2-(hydroxymethyl)-propan 1,3-diol] at pH 9.1 deviated from linearity only for samples with molecular weight values greater than about 3×10^5. This deviation is interpreted as indication of increased branching, or possibly of the occurrence of some cross-linking in the higher-molecular-weight fractions (Cameron et al., 1972b).

Humic Substances in the Soil

The predominant counterions neutralizing the charges on humic polyanions in the soil environment are Ca^{2+}, Mg^{2+}, and H^+. Under certain circumstances K^+, Na^+, Al^{3+}, and Fe^{2+}/Fe^{3+} ions can also contribute significantly to the contents of exchangeable cations.

Strongly hydrated monovalent cations such as sodium are readily dissociated from the conjugate bases of the acidic groups in humic structures. Consequently, the solvated anions repel each other to give the random-coil-type conformations described in the previous section. Soils where sodium ions or other strongly hydrated monovalent cations predominate as the exchangeable species lose their humic substances in drainage waters. When

H^+ ions associate with the conjugate bases, however, the resulting undissociated acids become involved in inter- and intramolecular hydrogen bonding processes. In such circumstances the molecules shrink, water (or solvent) is excluded from the matrix, and precipitation takes place.

Charge-neutralizing divalent and polyvalent cations are only weakly dissociated in humate complexes, and these bring about a contraction or shrinkage of the volume or space occupied by the monovalent ion-exchanged humates (Hayes and Swift, 1978, p. 281). In addition, the higher valence cations form bridges between anionic groups along the strands, and between adjacent strands, pulling them together. The exclusion of solvent and the absence of charge repulsion effects lead to precipitation and give rise to gels or solids. As drying takes place further shrinking occurs, bringing the macromolecules into more intimate contact and thereby enhancing their associations through the involvement of secondary attraction forces. This theory is supported by the fact that dry humic substances are difficult to rewet, and the hydrophobic nature of the dry material could suggest that the less polar components orientate toward the outside of the structures during drying.

The divalent and polyvalent cations which give rise to inter- and intramolecular bridging of the humic molecules can also form bridges between these molecules and negatively charged soil inorganic colloids. Thus, any solvent system useful for the extraction of soil humic substances should either solvate the macromolecules and their counterions, or dissolve the substances with or without solvating the counterions within their structures or involved in their binding to the soil inorganic colloidal constituents. In cases where ligand exchange provides the mechanism of binding of organic colloids to inorganic colloids, the successful solvent system should be capable of breaking the bonds between the two types of colloids. Alternatively, the inorganic components might be disintegrated or dissolved and the organic colloids then separately dissolved in conventional solvents for humic substances.

SOLVENTS AND SOLVENT SYSTEMS

For any solute to dissolve in a solvent, cavities or holes must form in the solvent in order to accommodate the solute molecules. The ease with which solubilization can take place depends on how readily these cavities are formed and on the ability of the solvent to solvate the solute molecules.

Solvent molecules, as well as solute molecules, have attractive and repulsive forces for each other. The overall attractive force between solvent molecules, which enables these to associate in the liquid phase, must be overcome so that cavities can be formed for the solute molecules. When the solute–solute attractive forces are of the same order as those between the solvent molecules, the solute dissolves in the solvent. However, when attraction between solvent molecules is significantly greater than that between the solute molecules, or vice versa, dissolution will not take place.

Secondary forces exist between all molecules. These forces are magnified for macromolecules resulting from associations, through covalent or polar bonds, or coulombic interactions between several smaller molecules. Solution of neutral macromolecules can only take place when the secondary interactions are overcome. A qualitative discussion of the nature of the secondary forces between molecules follows, and an outline is given of some of the solvent properties relevant to the solvation of solute material.

Secondary Forces Between Solvent and Solute Molecules

An appropriate account of secondary interaction forces between atoms and molecules is given by Richards (1980).

Secondary forces are primarily electrostatic in nature. They include the *multipole interactions* of which there are, in principle, an infinite number. However, in practice, only the monopoles, dipoles, and quadrupoles are important.

When two atoms involved in a bond have the same electronegativity, each of the atoms will have an equal share of the electrons in the bonding orbital. The bonds in such cases are *covalent*. When the electronegativities of the atoms are different the bonding electrons are held closer to the more electronegative atom, and the bond is *polar covalent*. Should the difference in electronegativity be great, the electronic charge migration to the highly electronegative atom is complete and *ionic species* are formed.

The *monopole* interaction force is equal to the total net charge on the molecule, and the interaction energy of ionic species is dominated by the monopole interactions.

When two charges q and $-q$ are separated by a distance r the magnitude of the dipole is qr. The dipole has magnitude and direction, and the *dipole moment* is expressed in *debye* (D) units where

$$\mathrm{D} = 3.338 \times 10^{-30}\ \text{coulomb} \cdot \text{meter}$$

In the absence of monopoles, molecules with large dipole moments dominate the multipole interaction energies, and such molecules are polar. The term *polarity* does not refer directly to dipole moment, but polar molecules have the ability to solvate species that have regions of charge deficiency or excess. Water is a highly polar compound, yet it has a relatively small dipole moment compared with the dipolar aprotic solvents acetonitrile, formamide, *N*,*N*-dimethylformamide, and dimethylsulfoxide.

Whereas dipoles, as vector quantities, require three components for specification, the more complex quadrupole requires six components. A neutral, nonpolar, symmetrical molecule can have a quadrupole moment, but the interactions involving these are small.

Induction forces arise when nonpolar molecules become dipolar as they are exposed to an electric field. This field distorts the electron distribution and displaces the center of positive and negative charge. The magnitude of this *induced dipole* moment is proportional to the strength of the field (E) and is expressed as

$$\mu_{\text{induced}} = \alpha\varepsilon_0 E \tag{1}$$

where α is the coefficient of proportionality and is called the *polarizability* of the molecule, and ε_0 is the vacuum permittivity equal to $\mu_0{}^{-1}c^{-2}$, where μ_0 is the vacuum permeability and c is the speed of light.

In practice any polar molecule in the vicinity of another which is less polar will have the effect of polarizing the second molecule. The induced dipole can interact with the dipolar moment of the first molecule, and the two molecules are attracted together in *dipole–induced-dipole interactions*.

Dispersion forces arise because of the time dependence of the charge distribution in molecules. Their origins can be explained by considering the fluctuating electrons in nonpolar molecules. These molecules do not have permanent moments but their fluctuating electron clouds give rise to what might be considered as instantaneous dipoles. The fluctuating dipole in molecule A propagates a fluctuating electric field which travels outward with the velocity of light to induce a similar dipole in B. This second fluctuating dipole in B radiates a fluctuating field back to A, where it interacts with the original fluctuation. The second then behaves in the same way with respect to the first. The two dipoles give rise to an attractive interaction between the two molecules in what is known as an *induced-dipole–induced-dipole* interaction. Dispersion forces are known as the *London* or *van der Waals* forces, although the latter are often considered to include multipole interactions and the induction forces as well.

Hydrogen bonding (both intermolecular and intramolecular) is a very important type of secondary interaction where humic substances, in particular H^+-exchanged humic substances, are concerned. It is responsible for the lack of solubility in water of H^+-humic acids, and it plays a very important role in stabilizing the shapes of biological polymers such as nucleic acids, polysaccharides, and proteins. For two molecules to form an intermolecular hydrogen bond it is necessary to have at least one electronegative atom, such as oxygen or nitrogen, on each molecule and one of these atoms must be bonded to hydrogen to constitute the donor group. The bond to the hydrogen atom in the donor group is strongly polarized, and it is thus associated with a dipole moment directed along the bond pointing toward the hydrogen. The acceptor also has a dipole moment which arises from the separation of the nonbonded electrons and points toward the nucleus of the electronegative atom. The two dipoles enter into a strong dipole–dipole interaction which is the primary source of the energy of formation of the hydrogen bond.

Dipole–dipole interactions are strongest when the dipoles are aligned head to tail. Thus the strongest hydrogen bonds are formed when the two electronegative atoms and the hydrogen are colinear. Since the hydrogen atom has a very small van der Waals radius the two dipoles can approach each other very closely to give relatively strong bonds, with energies that can amount to 20 kJ/mol when added to the much weaker contributions from the multipole, induction, and dispersion forces.

Properties of Solvents

Any study of extraction will benefit from considerations of the solvents that might be used in the extraction process. The acidity or basicity, the ability to hydrogen bond, the relative permittivity, and the dipole moment are all important properties for determining the ability of a solvent to solvate a particular solute. Although such peripheral properties as boiling point, viscosity, and density do not necessarily affect separations, they do provide very valuable information about the suitability of a solvent for a particular extraction.

Consideration of the relative permittivity (K_r) or dielectric constant values can be important when selecting a solvent for a particular solute. The term refers to the ability of the solvent to decrease the coulombic field of an ion. For example, the value for water is of the order of 78.5 which relates to the extent the field is reduced in the liquid, compared with a vacuum, at a distance r from the solute species. Thus water is a very powerful solvent for salts because it greatly decreases the coulombic interactions between the oppositely charged species and hence these do not readily interact to form solid or crystalline salts, as outlined below.

In general, less polar solutes dissolve best in solvents with low K_r values, and the higher values favor dissolution of polar molecules. However, the involvements of specific interactions, especially hydrogen bonding, do not allow definite trends to be established which relate the extent to which a solute is dissolved to the polarity of the solute and the relative permittivity of the solvent (Snyder, 1978).

Consideration of dipole moment values is important for predicting the interactions of solvents with polar or charged solutes. The ability of a solvent to disrupt the ion or molecule associations in a compound depends on the extent to which it can solvate the component molecules or ions and decrease the interactions holding them together. One end of the dipole is attracted electrostatically to the ion of opposite charge, or to the region of the molecule having the appropriate charge excess or deficit. Where specific solute–solvent interactions are not important the dipole moment of the solvent largely determines the orientation of the solvent around the solute molecule and this orientation is essential for the electrostatic solvation process.

The formation of solvent shells around molecules is essential to prevent self-association of the solute species and to allow solution to take place. Solvents other than water which have high dielectric constants, and including some of the dipolar aprotic solvents (defined below), dissolve ionic species by separating and solvating the ions.

Parker (1962) has defined as *dipolar aprotic solvent* materials having dielectric constant (relative permittivity) values greater than 15 and incapable of donating hydrogen atoms to form strong hydrogen bonds. Of the compounds listed in Table 1, acetone, acetonitrile (methylcyanide), *N*-methyl-2-pyrrolidone, *N*,*N*-dimethylformamide, and dimethylsulfoxide are dipolar aprotic solvents. With the exception of acetonitrile the listed solvents have exposed electronegative oxygen atoms, as have the dipolar but protic formamide and *N*-methylformamide, which provide readily available sites for solute to solvent hydrogen bonding. Because they are dipolar, such solvents solvate polar organic molecules, and in so doing form tight solvent shells around the solute species.

Anions are less solvated in dipolar aprotic solvents than in water. In water and protic solvents, the anions solvate by ion–dipole interactions on which strong hydrogen bonding is superimposed. In the cases of the dipolar aprotic solvents, the anions solvate also by ion–dipole interactions, but without the influence of hydrogen bonding. Instead, solvation is aided by less energetic interactions arising from the mutual polarizability of the anions and the solvent molecules (Parker, 1962).

Dack (1976, p. 98) has listed a classification of solvents on the basis of electrostatic factor (EF) values. These values separate solvents into four classes, as outlined in Table 2. EF values represent the product of the relative permittivity and the dipolar moment, and thus they take account of the influence of both of these properties on the electrostatic solvation of solutes. It can be predicted that solvents in classes I, II, and III will have little influence in solvating humic macromolecules. Dack placed water in Class III on the basis of its structure rather than of its EF value, but on the basis of its EF value it is regarded as a Class IV solvent for the purposes of the present discussion.

Hydrogen bonding is a very important property in considering solute–solvent interactions. Pimentel and McClellan (1960) have classified hydrogen bonding solvents into the proton donors, proton acceptors (e.g., keto compounds, ethers, dipolar aprotics), proton donors and acceptors (e.g., alcohols, carboxylic acids, primary and secondary amines, and water), and the *non*hydrogen bonding compounds (e.g., carbon disulfide and paraffins).

Taft et al. (1969) have defined a base parameter, pK_{HB}, which measures the relative strength of the acceptor when a hydrogen-bonded complex is formed using any suitable hydroxyl reference acid. Values of pK_{HB} are not applicable to reference acids involved in intramolecular hydrogen bonding. The reader is referred to the original article for details.

Some pK_{HB} values, from Taft et al. (1969), are given in Table 1: the higher

TABLE 1. Boiling Point, Molar Volume, Refractive Index, Viscosity, Density, Relative Permittivity, Dipole Moment, Electrostatic Factor (EF), and Base Parameter (pK_{HB}) Values for Selected Solvents[a]

Solvent	Boiling Point (°C)	Molar Volume (V, cm^3/mol)	Refractive Index n	Viscosity (η, cP 25°C)	Density (ρ)	Relative Permittivity (K_r)	Dipole Moment (μ)	EF	pK_{HB}
n-Pentane	36	116.2	1.355	0.22	0.61	1.84	0	0	—
Diethylether	35	104.8	1.350	0.24	0.71	4.34	1.36	5.90	0.98
Formic acid	101	37.8	1.371	—	1.220	58.0	—	—	—
Ethanoic acid	118	57.1	1.370	1.1	1.04	6.13	0.83	5.09	—
Pyridine	115	80.9	1.507	0.88	0.98	12.4	—	—	1.88
Acetonitrile	82	52.6	1.341	0.34	0.78	37.5	3.84	144.0	1.05
Acetone	56	74.0	1.356	0.30	0.78	20.7	2.88	59.62	1.18
N-Methyl-2-pyrrolidone	202	96.5	1.468	1.67	1.03	32.0	—	—	2.37
Formamide	210	39.8	1.447	3.3	1.13	109.5	3.37	369.0	—
N,*N*-Dimethylformamide	153	77.0	1.428	0.80	—	36.7	3.82	140.2	2.06
Dimethylsulfoxide	189	71.3	1.477	2.00	1.10	46.6	4.49	209.2	2.53
Ethanol	78	58.5	1.359	1.08	0.79	24.3	1.68	40.82	—
Water	100	18.0	1.333	0.89	1.00	78.5	1.84	144.44	—

[a] *From Taft et al. (1969), Barton (1975), Dack (1976), and Snyder (1978).*

TABLE 2. Classification of Solvents on the Basis of Electrostatic Factor (EF) Values

Class	Solvent Type	EF Value
I	Hydrocarbon	0–2
II	Electron donor	2–20
III	Hydroxide	15–50
IV	Dipolar aprotic	≧50

the value the better the compound as an acceptor in hydrogen bonding. The data show that the dipolar aprotic solvents and pyridine are good acceptors.

The overall tendency of compounds to interact through dispersion forces is related to the refractive index values of the compounds (see Karger et al., 1973): the greater the refractive index the stronger the dispersion interactions. Thus, the dipolar aprotic solvents and pyridine have the strongest influences in dispersion interactions of the compounds listed in Table 1. Where refractive index is used to measure the concentration of solute it is, of course, important to maximize the differences in these values between solute and solvent.

Low viscosity is important for ease of handling and mixing. Table 1 shows that, with the possible exception of *N,N*-dimethylformamide, there is a degree of matching of boiling points and viscosities. When mixed solvents are used, the viscosities of the mixtures are intermediate between the values for the pure components in the mixture. Thus, for a mixture of solvents A and B, the viscosity (η) of the mixture is given by

$$\eta = (\eta_A)x_A(\eta_B)x_B \tag{2}$$

where η_A and η_B are the values for pure A and pure B, and x_A and x_B are the mole fractions of each in the mixture (Snyder, 1978).

Consideration of solvent densities is important for separations using gravity. Again, use can be made of mixtures to regulate density, because the density value for a mixture is close to the arithmetic average of the densities of the pure components in it. For example,

$$\rho \approx \rho_A\phi_A + \rho_B\phi_B \tag{3}$$

where ρ, ρ_A and ρ_B refer to the density values for the mixture, and for solvents A and B, respectively, and ϕ_A and ϕ_B are the volume fractions of A and B in the mixture.

Use of Solubility Parameters for Predicting Solubilities

It was mentioned in the Introduction that solution takes place when the self-attraction forces in solute and solvent molecules are of the same order of magnitude. A good solvent for a (nonelectrolyte) polymer solute will have, for example, a solubility parameter (δ) value close to that of the solute. It is often found that a mixture of two solvents, one having a δ value above and the other a δ value below that of the solute, can provide a better solvent for that solute than either solvent alone.

Hildebrand and co-workers gave rise to the concept of the one-component solubility parameter (δ) more than 50 years ago (see Hildebrand and Scott, 1951; 1962; Hildebrand et al., 1970). It is defined as the square root of the cohesive energy density (CED) and may be expressed as

$$\delta = (-E/V_l)^{0.5} = (\text{CED})^2 \tag{4}$$

where V_l is the molar volume of the liquid and $-E$ is the molar cohesive energy, or the molar energy of vaporization. The unit of solubility is the Hildebrand, expressed as $(\text{cal/cm}^3)^{0.5}$, or 2.046 $(\text{J/cm}^3)^{0.5}$.

The interactions between molecules which produce the cohesive energy characteristic of the liquid phase are described in the section entitled Secondary Forces Between Solvent and Solute Molecules. These involve the dispersion forces, dipole–dipole and dipole–induced dipole interactions, and specific interactions, especially hydrogen bonding. If it is assumed that the intermolecular forces are the same in the vapor and liquid states, then $-E$ is the energy of a liquid relative to its ideal vapor at the same temperature. It can be described as the energy required to vaporize 1 mole of liquid to the saturated vapor phase ($\Delta_l^g U$) plus the energy required for the isothermal expansion of the saturated vapor to infinite volume. Detailed discussion of the theory and derivations is given in the publications by Hildebrand and associates cited above.

The Hildebrand one-component solubility parameter, δ, is appropriate for solutions lacking in polarity and in specific interactions. This parameter, to some extent, is now replaced by the multicomponent solubility parameters which give values for each of the different interaction forces. The total solubility parameter may be expressed as

$$\delta_0 = \left(\sum_i \delta_i^2\right)^{0.5} \tag{5}$$

where δ_i are the empirical estimates of the contributions from dispersion forces (δ_d), polar forces (δ_p), and hydrogen bonding (δ_h). The two quantities δ and δ_0, from Equations (4) and (5), should not necessarily be equal.

There are several excellent reviews of multicomponent solubility parameters, including those by Hansen (1967), Karger et al. (1973), Barton (1975), and Snyder (1978). Barton's review provides all the information necessary to familiarize the reader with the subject, and it contains a comprehensive reference list to the original work that developed the theory and techniques on which the multicomponent approach is founded.

TABLE 3. Data for the Hildebrand (δ), Total (δ_0), Dispersive (δ_d), Polar (δ_p), and Hydrogen-Bonding (δ_h) Parameters[a], and for Proton Donor (δ_a) and Proton Acceptor (δ_b) Parameters[b]

Solvent	δ	δ_0	δ_d	δ_p	δ_h	δ_a	δ_b
n-Pentane	7.0	7.1	7.1	0.0	0.0	—	—
Diethylether	7.4	7.7	7.1	1.4	2.5	—	3.0
Formic acid	12.1	12.2	7.0	5.8	8.1	—	—
Ethanoic acid	10.1	10.5	7.1	3.9	6.6	—	—
Pyridine	10.7	10.7	9.3	4.3	2.9	—	4.9
Acetonitrile	11.9	12.0	7.5	8.8	3.0	—	3.8
Acetone	9.9	9.8	7.6	5.1	3.4	—	3.0
N-Methyl-2-pyrrolidone	11.3	11.2	8.8	6.0	3.5	—	—
Formamide	19.2	17.9	8.4	12.8	9.3	*l*	*l*
N,N-Dimethylformamide	12.1	12.1	8.5	6.7	5.5	—	4.6
Dimethylsulfoxide	12.0	13.0	9.0	8.0	5.0	—	5.2
Ethanol	12.7	13.0	7.7	4.3	9.5	6.9	6.9
Water	23.4	23.4	7.6	7.8	20.7	*l*	*l*

[a] *From Barton (1975).* *l = large*
[b] *From Snyder (1978).*

Table 3 lists the Hildebrand solubility parameter δ, the total solubility parameter δ_0, and the multicomponent parameters for dispersion δ_d, polar δ_p, and hydrogen bonding δ_h forces for a number of solvents. These data are taken from the compilation by Barton (1975). He has pointed out that the data become empirical when multicomponent parameters are used, and thus it is important to use a set of data that are self-consistent. Keller et al. (1971) and Karger et al. (1976) have further subdivided the hydrogen bonding parameter into the acid or proton donor (δ_a) parameter, and the base or proton acceptor (δ_b) parameter. Values for these are listed for some of the compounds in Table 3 from data provided by Snyder (1978). These data are not from the same source as those compiled by Barton (1975) and included in Table 3, and hence the values given for δ_h should not be compared directly with those for δ_a and δ_b.

It was pointed out that the overall δ values should be similar for solvent and solute in order to achieve maximum solubility. The same applies for the individual parameters δ_d, δ_p, and δ_h. Subdivision of δ in this way allows differences in solubility and solvent selectivity to be anticipated for solvents with similar polarities and similar overall values of δ.

The use of δ_a and δ_b, as distinct from δ_h, can be a useful guide when selecting a solvent for a particular extraction process. Because maximum solubility is promoted by strong hydrogen bonding between solvent and solute molecules, solution should be greatest when the product of δ_a (for solvent) and δ_b (for solute) (or vice versa) is a maximum, rather than when the values are equal for both solvent and solute.

SOLUBILIZATION OF MACROMOLECULES

The mechanisms in the dissolution of neutral polymers and polyelectrolytes are different. At low pH values (~pH 4) H^+-exchanged humic substances have many of the physicochemical properties of neutral polymers having substantial amounts of hydrogen bonding. As the pH is raised the acid groups in the humic substances dissociate and the macromolecules assume the properties of polyelectrolytes. Therefore, it is appropriate to consider some of the features involved in dissolving neutral polymers and polyelectrolytes.

Solubilization of Neutral Polymers

The most widely used theory of polymer solutions is that initiated independently by Flory and by Huggins (see Flory, 1953). This theory is based on the evaluation of the free energy change of mixing ($\Delta_m G_p$) of a liquid and polymer to form a solution, and culminates with an equation of the form

$$\Delta_m G/RT = x_A \ln\phi_A + x_B \ln\phi_B + \chi\phi_A\phi_B(x_A + V_B x_B/V_A) \qquad (6)$$

where x is the mole fraction, ϕ is the volume fraction, subscripts A and B denote solvent and polymer, respectively, and V refers to molar volume. The term χ is called the Flory–Huggins interaction parameter for the particular solvent–polymer pair A–B. This parameter is dimensionless and regarded by Flory as the sum of the *reduced enthalpy of dilution* χ_H and of the *reduced residual partial molar entropy* χ_S. Molyneaux (1975) has outlined succinctly the basis of the derivation of Equation (6) in his treatment of water–polymer systems. The first two terms in Equation (6) are always negative and result from the configurational energy of mixing. For the polymer to dissolve ($\Delta_m G$ negative) χ must be negative, or small if positive. When V_B/V_A is large, that is, for high-molecular-weight materials, the critical value for χ is ≤ 0.5 if complete solution is to be achieved. Because χ_S is usually between 0.2 and 0.6 χ_H must be small.

There are some who question the usefulness of the Flory–Huggins solubility parameter for problems related to the solubilization of polymers, although it is agreed that it is useful for study of the thermodynamics of dilute solutions. Barton (1975) has referred to literature that cites its shortcomings as a practical criterion of solubility. Some of these are:

1. It is concentration dependent.
2. It is difficult to evaluate experimentally.
3. It is influenced by hydrogen bonding.
4. It is inconvenient for multicomponent systems because interactions between each pair of components must be known.

Despite these shortcomings, evaluations of polymer–solvent parameters are very widely used.

The Hildebrand–Scatchard equation (see Hildebrand and Scott, 1951), which is derived on semiempirical grounds, is

$$\Delta_m U_v = (x_A V_A + x_B V_B)(\delta_A - \delta_B)^2 \phi_A \phi_B \tag{7}$$

where $\Delta_m U_v$ is the change in the molar internal energy on mixing at constant volume, δ refers to the solubility parameters, and the rest of the symbols are as defined in Equation (6). Several assumptions were made in deriving this equation:

1. The constant-pressure change of volume on mixing is zero. When such volume changes occur there is a large disparity between $\Delta_m H_v$ and the constant-pressure enthalpy of mixing, $\Delta_m H_p$, which can be measured experimentally.
2. The interaction forces are additive, and the interactions between each pair of molecules are not influenced by the presence of other molecules.
3. The mixing is random and no nearest-neighbor associations are favored.

It is assumed that the value of $-E/V$ [Equation (4)] is additive, which it appears to be even for polymers. The use of the equation relies, of course, on the availability of data for the solubility parameters of solute and solvent.

Similarities in the solvent–solute solubility parameters allow a more negative Gibbs energy of mixing. If it is assumed that the solution is regular ($\Delta_m S$ ideal)

$$\Delta_m A_v = \Delta_m U_v - T\,\Delta_m S_v \tag{8}$$

where A refers to the molar Helmholtz free energy at constant volume, and the other symbols are as defined above. The value for $\Delta_m U_v$ may be obtained from Equation (7), and if $\Delta_m S$ is zero, $\Delta_m A_v$ can be estimated.

Solubilization of Polyelectrolytes

Mention was made of how humic substances expand in water as the result of repulsions between conjugate bases (anions) formed after the release or dissociation of the counterions. The higher the charge density the greater the repulsion between the charges, and the greater the hydrodynamic volume. Electrically charged, ionic species readily solvate or hydrate, and in so doing they bind solvent or water molecules because of strong monopole–dipole interactions. Smaller ions have stronger electric fields at their peripheries, and these bind solvent molecules more strongly. Such strongly bound molecules tend to be localized, which implies a loss of entropy. This loss, however, is compensated to some extent by the breakdown of water structure.

Richards (1980, p. 209) presents a description of the ion atmosphere for a polyelectrolyte in solution. He gives a clear interpretation of the electrical double-layer effects and of the mechanisms by which the presence of excess salts can depress the electrical potential and cause the highly charged polyanions to have many of the physicochemical properties of neutral molecules.

EXTRACTION OF HUMIC SUBSTANCES FROM SOIL

A short account is given in the second major section of the composition and properties of humic substances. That account indicates how the influences of charge-neutralizing hydrogen ions and of divalent and polyvalent cations render these polydisperse polyelectrolytes insoluble in water. Because such ions are undissociated, or only weakly dissociated from the anionic functional groups, they give rise to a type of cross-linking effect from inter- and intrastrand hydrogen bonding, and from bridging in the cases of the divalent and polyvalent cations. Polyvalent cations also form bridges between the macromolecules and the inorganic colloidal components of the soil, and this also complicates the processes of extraction.

Water is the solvent most often used for isolating humic substances. Organic solvents have been used occasionally, but there are not enough data from systematic research to allow a comprehensive evaluation of their effectiveness.

Criteria for Solvents and Methods for the Extraction of Soil Humic Substances

Whitehead and Tinsley (1964) have proposed four criteria for solvents for humic substances. In their view, effective solvents should have:

1. A high polarity and a high dielectric (or permittivity) constant to assist the dispersion of the charged molecules.
2. A small molecular size to penetrate into the humic structures.
3. The ability to disrupt the existing hydrogen bonds and to provide alternative groups to form humic–solvent hydrogen bonds.
4. The ability to immobilize metallic cations.

Stevenson (1982) has also listed four criteria for the ideal extraction method:

1. The method leads to the isolation of unaltered materials.
2. The extracted humic substances are free of inorganic contaminants, such as clay and polyvalent cations.
3. Extraction is complete, thereby ensuring representation of fractions from the entire molecular weight range.
4. The method is universally applicable to all soils.

Structure and Some Solution Properties of Water

Any consideration of water for use as a solvent or in a solvent system should take account of its structure and of some of its unique characteristics. Detailed information about the properties of water is contained in the several volumes on *Water—A Comprehensive Treatise,* edited by F. Franks (1975), and referenced in Franks (1983). The following discussion is intended only to review briefly some of the properties of water which are important for the solubilization of humic substances.

The Bjerrum four-point-charge model provides a useful, though older, concept of the structure of the water molecule. In this model the oxygen atom is at the center of a regular tetrahedron and the fractions of charge are placed at the corners at a distance of 1 Å from the center. The corners with positive charges correspond to the positions of the two hydrogens, and the other two corners are occupied by the lone pairs of electrons. This four-point-charge model has been subjected to adjustments based on quantum mechanical and spectroscopic data which show that the structure is not that of a regular tetrahedron, but that the H—O—H angle is 104°27′ and the O—H bond length is 0.958 Å.

Calculations based on theory, where concepts of the structure of the isolated water molecule are used, suggest that the most stable form of interaction between water molecules involves the linear hydrogen bond. The estimated molar dissociation energies for such hydrogen bonded structures range between 20 and 35 kJ, and the equilibrium O—O distances found are between 2.6 and 3.0 Å (Franks, 1983, p. 15).

Hydrogen bonding confers a considerable degree of structure to liquid water. In the "flickering" cluster model, the individual water molecules tend to form clustered structures by what is known as a *cooperative* interaction. It can be supposed that liquid water is composed of clusters of varying sizes in equilibrium with each other and with free unstructured water molecules. Local energy fluctuations allow the formation and degradation of hydrogen bonds and of clusters in a dynamic process in what is known as the "flicker" effect.

Water is highly structured in ice. The crystal structure of ice-1h, for instance, suggests a tetrahedral molecular packing in which each molecule is hydrogen bonded to four nearest neighbors in a hexagonal lattice arrangement. Such structures have considerable empty space, and this explains why water expands in going from the liquid to the highly ordered ice phase.

When soluble ions or highly polar compounds are introduced into water solvation occurs because of associations, through electrostatic effects, between the solvent and solute species. The ordering of the solvent around the solute leads to a decrease in entropy, but this is compensated by the increased entropy from the breakdown of water cluster structure.

Nonpolar molecule–water interactions are different. When such molecules enter spaces between water clusters they interact more strongly with each other than with the nonclustered water molecules in the void spaces. However, the guest molecules disrupt the hydrogen bonds between the resident, nonclustered water molecules. This raises the energy of the nonbonded water which then tends to associate with the clusters. Thus the clusters grow at the expense of the free water, and the energy and entropy are decreased as the free water becomes localized in the clusters through hydrogen bonding. The result is that the nonpolar solutes tend to associate with themselves, and this effect has given rise to the concept of the hydrophobic bond between nonpolar molecules in the presence of water (Franks, 1975, 1983).

Extraction of Humic Substances in Water and in Aqueous Salt Solutions

Fulvic acids, in the H^+-exchanged form, are, by definition, soluble in water, but humic acids are not. As was pointed out, both acids can be regarded as neutral macromolecules at pH values of 4 and below. (In the presence of strong acids, however, where protonation of appropriate functional groups can take place, the humic substances can be considered polycations.) The easy explanation for the differences between the solubilities of the H^+-fulvic acids and H^+-humic acids is that the fulvic acids contain more polar groups per unit of macromolecule. This should be regarded only as a part of the explanation; the configurations and conformations of similar structures can be very important in considerations of their solubilities. For example, cellulose, a β-(1 $\rightarrow$ 4) linked polyglucose is not soluble in water whereas amylose, an α-(1 $\rightarrow$ 4) linked polyglucose is. One reason for the difference is that the β-conformation allows the sugar monomer units in one polymer strand to have intimate contact with those on another strand to form a linear helix structure stabilized through hydrogen bonding. The α-conformation does not allow such intermolecular intimacy and the polymer is soluble in water. It is interesting to note, though, that the β-D-glucopyranose molecule (chair conformation) can replace almost exactly the chair conformation of water molecules inherent in the ice structure giving rise to glucose–water hydrogen bonds instead of water–water hydrogen bonds (Suggett, 1975). This affinity, however, is not sufficient to overcome the extensive hydrogen bonding between the polymer strands of cellulose.

There is good reason to consider that fulvic acid molecules exhibit hydrogen bonding when dried. However, lack of regularity in the arrangements of the functional groups would suggest that the macromolecules have insufficient order to allow regular and close sequences of inter- and intramolecular hydrogen bonding. Many fulvic acids contain sugar moieties, but it is evident that the molecular weights of these components, and/or the conformations of the sugars and the configurations of their linkages, do not allow sufficient hydrogen bonding between the molecules to prevent them from hydrating.

Table 1 in Chapter 15 shows that water extracted 2.8% of an H^+-exchanged organic histosol as fulvic acids. When the residual soil was sequentially extracted, exhaustively, with aqueous solutions of sulfolane, dimethylsulfoxide (DMSO), pyridine, and ethylenediamine (EDA) or diaminoethane, additional amounts of fulvic acids were isolated. This suggests that the fulvic acids which were not extracted in water initially were adsorbed, or in some way bound to insoluble humic substances or to the other soil components, and were released by the solvents used subsequently in the extraction processes. It is possible, of course, that the additional fulvic substances resulted from the breakdown of humic acid moieties to smaller, water-soluble components.

In their search for nonalkaline solvents for soil humic acids Bremner and Lees (1949) showed that up to about 30% of soil organic matter could be extracted as humic substances by sodium and potassium salts of inorganic and of organic acids. Of these, a 0.1*M* solution of sodium pyrophosphate, neutralized with phosphoric acid, was best, and was followed in order by solutions of the sodium salts of fluoride (NaF), hexametaphosphate $(NaPO_3)_6$, orthophosphate (Na_3PO_4), borate ($Na_2B_4O_7$), NaCl, NaBr, and NaI. The order of decreasing extraction efficiencies for the organic acid salts was oxalate $[(COONa)_2]$, citrate $[(NaO_2CCH_2C(OH)CO_2Na)CH_2CO_2Na]$, tartarate $[NaO_2CCH(OH)CH(OH)CO_2Na]$, malate $[NaO_2CCH(OH)CH_2\text{-}CO_2Na]$, salicylate (*o*-$OHC_6H_4CO_2Na$), benzoate ($C_6H_5CO_2Na$), succinate ($NaO_2CCH_2CH_2CO_2Na$), sodium-4-hydroxybenzenecarboxylate, and ethanoate (CH_3CO_2Na).

The best extractants in the above list of salts form complexes with the polyvalent metals that neutralize charges on the humic substances and link them to the inorganic soil colloids. These polyvalent ions are replaced by sodium ions from the salts. The efficiency of each solvent system will depend on the extent to which the resident cations are exchanged and removed from humic structures. Diffusion of the salts to the interior of solid humic substances is slow. Some channeling can take place, but extensive penetration would probably require the opening up from the outside of the macromolecular structures. It would be necessary for these structures to remain open to allow exchange from the interior to take place.

Mechanisms for expansion of humic molecules after the exchange of divalent and polyvalent cations by sodium ions have been discussed. Available evidence suggests that the more highly oxidized humic substances and those in the lower-molecular-weight ranges are removed by the salt solutions. The more the substances are oxidized the greater will be their charge densities, and the combination of high charge and relatively low hydrodynamic volume increases the possibilities for solution to take place.

The extent of dissolution is influenced by the amounts of low-molecular-weight electrolyte present. Excess electrolyte increases the ionic strength outside the humic structures relative to that inside, and it decreases the thickness of the electrical double layer causing the macromolecules to contract. At high salt concentrations hydration is curtailed and dissolution may not take place.

TABLE 4. Yields and ESR Data for Humic Acids (HA) and Fulvic Acids (FA) Extracted by Different Extractants from a H^+-Exchanged Sapric Histosol[a]

	Yield (%)			ESR [(Spins/g) × 10^{-16}]		
Solvent	HA	FA	Total	HA	FA	HA + FA
2.5*M* EDA (pH 12.6)	49.0	14.0	63.0	15.0	27.5	—
EDA (anhydrous)	2.0	3.0	5.0	6.4	12.8	—
0.5*M* NaOH	58.0	2.0	60.0	4.6	0.4	—
0.1*M* $Na_4P_2O_7$ (pH 7)	13.7	0.8	14.5	4.5	1.9	—
1*M* Na^+-EDTA	12.5	3.8	16.3	0.3	0.3	—
Pyridine[b]	34.0	2.0	36.0	—	—	2.1
DMF[b]	16.0	2.0	18.0	—	—	1.4
Sulfolane[b]	10.0	12.0	22.0	—	—	1.0
DMSO[b]	17.0	6.0	23.0	—	—	4.2

[a] *From Hayes et al. (1975).*
[b] *Extraction with the solvent was followed by exhaustive extraction with water. EDA = diaminoethane or ethylenediamine; DMF = N,N-dimethylformamide; DMSO = dimethylsulfoxide; EDTA = ethylenediaminetetraacetic acid.*

It is not necessary to hydrogen saturate humic substances before extraction with neutral salt solutions containing divalent and polyvalent metals which can form complexes with the organic matter. Alexandrova (1960) showed that serosems (soils with high calcium contents) could be effectively extracted with pyrophosphate without prior exchange with H^+. In fact the acidification process inherent in the H^+ exchange depresses the amounts extracted. The low yields of humic substances obtained by Hayes et al. (1975) for exhaustive extraction with neutral sodium pyrophosphate and EDTA (see Table 4) solutions probably arose because the extractants were not buffered and the acidity of the H^+-exchanged soil significantly lowered the pH of the solvents in contact with the soil. Hence, ionization of the acid groups on the macromolecule was suppressed. It is of interest to note that the carbon contents of the humic acids extracted by these solvents (Table 5) were lower than for the other extractants and the oxygen contents of the pyrophosphate extracts were highest. This indicates that the more highly oxidized materials were extracted in these instances.

Extraction Under Basic Conditions

The data in Table 4 show that the basic solvent systems, 0.5*M* NaOH and 2.5*M* ethylenediamine (EDA), extracted more humic substances from a H^+-exchanged organic soil than did the other solvents. Because of the highly alkaline conditions which prevailed (the pH of the soil suspension in the aqueous EDA system was 12.6) all the acid groups in the macromolecules

TABLE 5. Carbon, Oxygen, Hydrogen, Nitrogen, and Sulfur Contents, on a Dry Ash-Free Basis, of Humic Acids (HA) and Fulvic Acids (FA) Isolated by Solvents Described in Table 4[a]

	Carbon		Oxygen		Hydrogen		Nitrogen		Sulfur	
Solvent	HA	FA	HA	FA	HA	FA	HA	FA	HA	FA
2.5*M* EDA	56.8	51.2	29.2	30.3	5.9	5.7	6.4	11.1	—	—
Pyridine	55.9	47.1	32.9	39.9	5.1	5.3	4.4	6.0	—	—
DMSO	55.0	55.0	35.5	37.1	4.2	4.4	3.3	2.2	2.0	1.3
Sulfolane	54.4	53.2	35.2	37.4	4.8	4.4	3.2	3.3	2.4	1.7
DMF	54.3	52.3	36.8	38.8	4.6	4.1	2.6	3.2	1.7	1.6
0.5*M* NaOH	53.1	45.0	36.3	43.0	6.0	6.0	2.9	4.3	—	—
Na^+-EDTA	52.1	48.4	—	—	4.1	4.2	—	—	—	—
0.1*M* $Na_4P_2O_7$ (pH 7)	50.9	37.3	41.1	50.9	3.3	5.1	3.0	5.0	—	—

[a] *After Hayes et al. (1975).*

were dissociated and the repulsion of charge gave the fully expanded random coil structure. Thus the anionic and polar sites could be readily solvated with water molecules.

The yields of humic substances (humic acids plus fulvic acids) were slightly higher from the exhaustive extractions with aqueous EDA than from those with sodium hydroxide (Table 4) (for the exhaustive extractions soil samples were repeatedly equilibrated with solvent until the supernatant solutions had negligible color). Data in Table 4 suggest that less humic acid and more fulvic acid substances were extracted in the EDA than in the sodium hydroxide solutions. However, it was found that considerable amounts of the substances classified as fulvic acids (components remaining in solution when the aqueous EDA extracts were acidified to pH 1) were precipitated in the sacs during dialysis against distilled water. This would indicate that some humic acids were solubilized by EDA salts formed on acidification, and that these acids were precipitated as the salts were lost during dialysis.

Use of aqueous solutions of sodium hydroxide for the extraction of humic substances was first described by Achard (1786). Such solutions have been the solvents of choice by most workers since that time. Comparison of elemental analysis data (Table 5) for the 2.5*M* EDA- and for sodium hydroxide-extracted humic substances show that the carbon and nitrogen contents of the EDA-soluble substances were significantly higher than for those extracted with sodium hydroxide. The reverse was true for the oxygen contents of these substances.

There are a number of explanations that can be offered for the observed differences in the carbon and hydrogen compositions of these humic substances. One is that the substances extracted by the two solvents are significantly different. Another, and more likely, explanation is that some oxidation occurs during extraction with sodium hydroxide. Bremner (1950) has

TABLE 6. The Influence of pH on the Uptake of Oxygen by Organic Matter Extracted for 7 Hours with Solvents in Media of Different pH Values[a]

Reagent	O_2 Uptake (mm^3/0.2 g)	
	Soil 1	Soil 2
0.5*M* NaOH	896	712
0.5*M* Na_2CO_3, pH 10.5	56	71
0.2*M* Na citrate, pH 7.0	39	58
0.1*M* $Na_4P_2O_7$		
pH 7.0	7	37
pH 8.0	12	—
pH 9.0	31	52

[a] *Stevenson (1982) from data by Bremner (1950).*

shown that solution in aqueous (0.5*M*) sodium hydroxide increased oxygen uptake by soil organic materials (Table 6), and more recently Swift and Posner (1972) have demonstrated that some breakdown of humic acids can take place in alkaline conditions in the presence of oxygen. It is evident from Table 6 that uptake of oxygen (oxidation) is small under mildly alkaline conditions. Data are awaited for the uptake of oxygen by humic substances in sodium hydroxide solutions at pH values below 11, and in solution in EDA.

The enrichment in nitrogen of humic substances extracted in EDA (Table 5) suggests that solvents containing this compound, or any other primary amine, would not be appropriate for use with humic substances. EDA readily reacts with carbonyl groups to form Schiff base structures. Such reactions are rapid when the protonated carbonyl group reacts with the nonprotonated amine. Thus the maximum reaction rate occurs when the product of the concentrations of the protonated carbonyl groups and free amines is maximum. The reaction continues, but at a rate which decreases in an exponential manner, as the pH is raised or lowered from the optimum value for the reaction.

Reference was made to the formation of carbon to nitrogen bonds between amine groups and carbon atoms *alpha* to the keto groups in quinone structures. Such bonds readily resist cleavage in the acid-wash procedure used by Hayes et al. (1975) in their attempts to lower the nitrogen contents of the humic substances extracted by EDA. Dryden (1952) and Rybicka (1959) have observed also that the nitrogen contents of coal materials extracted with EDA were raised. Incorporation of the solvent organic molecules into the humic structures would enhance the carbon and nitrogen contents at the expense of the oxygen lost in the water released in the condensation reactions. The high free radical contents of the humic and fulvic acids isolated when aqueous EDA was used (Table 4), as measured by the electron spin resonance procedure, provides further evidence of denaturation in this solvent system.

Damage through oxidation can be partially avoided when humic solutions in alkaline conditions are handled under dinitrogen gas. Choudri and Stevenson (1957) initiated an approach using this procedure when they extracted their soil humic acids with aqueous sodium hydroxide in an atmosphere of N_2 and in the presence of stannous chloride as an antioxidant. For the isolation of its collection of standard soil humic substances (see Chapter 1), the International Humic Substances Society has recommended equilibrating the soils with 1*M* hydrochloric acid, then neutralizing to pH 7 with 1*M* sodium hydroxide, and adding under dinitrogen gas 0.1*M* sodium hydroxide to give a liquid to soil ratio of 10 : 1.

Extraction with Organic Solvents

Except for extracts with anhydrous EDA, data in Table 4 were obtained for humic substances isolated from an air-dried H^+-exchanged humic histosol soil. For extractions with pyridine, *N,N*-dimethylformamide (DMF), dimethylsulfoxide (DMSO), and sulfolane, soils (60 g) were thoroughly mixed with the appropriate solvent (250 cm^3). After centrifugation the residues were repeatedly extracted with water until the supernatants were only faintly colored. Supernatants for each of the solvent systems were combined and the pH values of the solutions were adjusted to 1.0 using 5*M* hydrochloric acid. Humic and fulvic acids were separated by centrifugation.

For extraction with anhydrous ethylenediamine (EDA) H^+-exchanged soil was dried in vacuo at 75°C over phosphorus pentoxide and then thoroughly mixed with EDA dried by distillation from solid sodium hydroxide. Extraction was repeated twice.

In the absence of water, EDA was a very poor solvent for humic acids (Table 4). This suggests that the solute–solvent interactions were not sufficiently energetic and/or numerous to overcome the strong hydrogen bonding forces between the humic acid macromolecules. This is consistent with the suggestion that the more hydrophobic or less polar components of humic substances orientate toward the exteriors of the structures as shrinking takes place during drying. This effect would give rise to a shielding of the more polar and readily solvated groups in the interiors of the macromolecules.

Inevitably some acid groups in the dry humic acids are contacted by the EDA. These protonate one or both of the amine groups of the solvent, and the conjugate acid structure is then held by ion exchange to the conjugate base structures ($EDAH^+$) of the dissociated acids. However, the size of the EDA molecule and of its protonated derivative inhibit penetration into the tightly hydrogen bonded macromolecular matrix, and so, extensive solvation does not take place. Thus, in the absence of water, humic acids cannot be expected to swell readily and to solvate in EDA or other organic amine solvents. The yields of fulvic acids (Table 4) from extraction with anhydrous EDA were similar to those for extraction with sodium hydroxide. Because fulvic acids are generally smaller and more polar than humic acids, it is probable that sufficient groups which could be solvated were available to the EDA solvent to allow solution to take place.

Pyridine, followed by water, extracted more humic substances than did the DMF, DMSO, or sulfolane systems (Table 4). The enhanced solubilization by pyridine could be attributed partially to a pH effect. When pyridine (eight parts) was diluted with water (one part) to simulate the composition of the solvent in the air-dried soil, the pH of the mixture was 11.6. Because of the low buffering capacity of the solvent system, the pH of the extract was only 4.2 (Swift, 1968). Theory suggests that substantially more humic materials can be brought into solution by pyridine if the pH of the medium is maintained at 9.0 (the pK_b for pyridine is 8.96).

At pH values greater than 9.0 pyridine molecules predominate in the medium and these are then involved in solvating functional groups on the humic structures. There is a logarithmic increase in pyridinium ions as the pH falls, and the organocations formed are held by ion exchange to the conjugate bases of the acidic functional groups on the macromolecules. Although humate–pyridinium salts are less dissociated than those of the humate–monovalent inorganic cation species, they are significantly more dissociated than the H^+- and divalent and polyvalent cation-exchanged materials. This would explain the author's observation of enhanced solubility in water of humic substances from H^+-exchanged humic histosols first extracted with pyridine.

Electron spin resonance (ESR) data suggest that solution in pyridine does not significantly enhance the content of free radicals in humic substances. Pyridine does, however, significantly enhance their nitrogen contents (Table 5), especially in the cases of the fulvic acids, as was found also for extracts with EDA. Sorption, by ion-exchange and possibly by charge-transfer processes, might provide plausible explanations for the enrichments.

The amounts of humic substances extracted into the dipolar aprotic solvents DMF, DMSO, and sulfolane depend entirely on the extents to which these solvents can solvate the macromolecules. ESR data indicate that DMF and DMSO do not generate free radicals in the humic extracts. There is a slight enhancement of sulfur in the humic acids extracted in sulfolane, and this might be attributable to some adsorption of the solvent by the macromolecules. Otherwise the elemental analysis data of the humic substances extracted by these two solvents are similar, and the data for the humic acids isolated in the solvents and in 0.5M sodium hydroxide are comparable. However, fulvic acids from the basic solution have lower carbon and higher oxygen contents (Table 5), and this suggests that some uptake of oxygen occurred under the alkaline conditions. Because some colored materials precipitated during dialysis of the fulvic-type substances from the organic liquid extracts, it is probable that the true fulvic acids are contaminated by humic acids solvated by the residual solvents in the acidified mixtures. These acids precipitate when the solvating molecules are replaced with water during dialysis.

In the experiments of Hayes et al. (1975) DMSO was marginally better than DMF or sulfolane for dissolving humic substances (Table 4). In the ESR there was evidence of a higher free radical concentration in DMSO than in either DMF or sulfolane. Because DMSO would not be expected to generate free radicals, it is reasonable to infer from the ESR data that humic components, which are insoluble in the DMF– and sulfolane–water systems, were dissolved in this solvent. Elemental contents were similar for the humic and fulvic acids of the DMSO extracts, and these data infer that the major difference between the two fractions was one of molecular size. However, some fulvic acid materials were observed to precipitate during dialysis, as was noted for the DMF and sulfolane systems.

In a further set of experiments Hayes et al. (1975) exhaustively extracted an H^+-exchanged sapric histosol with water, and then exhaustively with the series DMF–, sulfolane–, DMSO–, pyridine–, and EDA–water, 1 : 1 (v/v) mixtures in that order. The cumulative amounts of humic substances extracted for any succession of solvents were the same as the amounts extracted by the last solvent in the series without the aid of the others. This would suggest that the materials dissolved in the less efficient solvents were dissolved also in the more efficient members of the series. The analytical data for the different fractions closely resemble those in Table 5 for the substances extracted by the single solvent systems.

The extraction performances of pyridine and of the dipolar aprotic solvents in Table 4 prompted Hayes et al. (1983) to compare the abilities of these and of other solvents to dissolve humic acids. It was presumed that any solvent and treatment which would dissolve such macromolecules could be adapted for the extraction of humic substances from the soil. Samples of humic acids, isolated by the procedure recommended by the International Humic Substances Society from a Florida (Glade) sapric histosol (supplied by Dr. R. L. Malcolm of the U.S. Geological Survey, Denver), were extracted with the solvents (0.2% w/v) listed in Table 7. Swelling was allowed to take place overnight, and after centrifugation each supernatant solution was diluted with the solvent used for extraction until an absorbance reading against the solvent blank could be obtained at 400 nm in each case. The absorbance values quoted (Table 7) represent the product of the reading obtained and the dilution factor used.

Data in Table 7 show that acetonitrile, dioxane, ethanol, and water are very poor solvents for humic acids. Humic acids are, by definition, insoluble in water, although traces of the H^+-exchanged substances are invariably dissolved in it. Similar trace amounts were dissolved in ethanol. Pyridine and formic acid (90%) were also poor solvents for humic substances. The extent of dissolution in pyridine was less than would be predicted from the data quoted for the H^+-exchanged soil in Table 4. However, the Glade humic

TABLE 7. Comparison of Absorbance Values for Solutions Obtained by Mixing H^+-Exchanged Humic Acids (0.2% w/w) with Aqueous and Organic Solvents

Solvent	Absorbance Value
Water	1.0
Dioxane	0.0
Acetonitrile	0.0
Ethanol	1.0
Formic acid (90%)	4.0
Pyridine	5.0
N,N-Dimethylformamide (DMF)	18.0
Formamide	19.0
Dimethyl sulfoxide (DMSO)	21.0
0.5*M* NaOH, pH 9.2	23.0
0.5*M* NaOH	24.0

acids were significantly drier than the soil, and it is concluded that pyridine in the absence of water does not readily solvate humic acids, and the results had similarities to those for anhydrous EDA.

Comparison of results with those obtained by Sinclair and Tinsley (1981) suggests that solvation by the 90% formic acid is disappointing. However, Sinclair and Tinsley used anhydrous formic acid. This acid has many of the desirable characteristics of a good solvent for humic substances, as listed by Whitehead and Tinsley (1964) and outlined in this chapter. It is concluded that the presence of water in the solvent used by Hayes et al. (1983) depressed the ability of formic acid to solvate humic acid molecules.

Formamide, DMSO, and DMF were good solvents for the Glade humic acids and solvation by these approached that by 0.5*M* sodium hydroxide. Swelling was slow in the organic solvents, and this contrasted with the behavior in the aqueous base.

Some of the properties listed for the solvents in Table 1 help to explain the differences in solubilities of the Glade humic acids in the solvents listed in Table 7. For the purpose of the present discussion, solvents giving absorbance values less than 10 and greater than 15 in Table 7 are regarded as poor and as good solvents, respectively, for the H^+-exchanged humic acids. Each of the good solvents has an electrostatic factor (EF) value greater than 140, but so also has acetonitrile, a poor solvent. However, on the basis of the data in Tables 1 and 7 it is appropriate to consider that organic solvents with EF values greater than 140 and with pK_{HB} (a measure of the strength of a solvent as an acceptor in hydrogen bonding) values greater than 2 should be good solvents for humic acids. The low pK_{HB} value for acetonitrile indicates that it is incapable of breaking the hydrogen bonds of such H^+-exchanged acids. DMF and DMSO fulfill these requirements and are seen to be good solvents in Table 7. *N*-methyl-2-pyrolidone has the required pK_{HB} value to be a good solvent but the hydrogen-bonding parameter, δ_h, to be discussed below, is unfavorable, and the high molar volume (Table 1) value suggests that steric constraints could hinder diffusion to solvation sites within the nonexpanded H^+-exchanged humic structures. Law (1984) shows that this compound is a poor solvent for humic substances in a H^+-exchanged sapric histosol, but it performs significantly better in the Ca^{2+}-exchanged soil.

Solubility parameter data (Table 3) provide less clear-cut differences between the good and poor solvents in Table 7. Mention was made of the need for δ, δ_d, δ_p, and δ_h parameters for solvent and solute to be similar for dissolution to take place. There are no measured solubility parameter data available for humic substances, and so accurate predictions cannot be made based on parameter values for solutes and solvents.

The dispersion force (δ_d) parameters are similar for the organic solvents used to provide the data in Table 7, but there are differences in the polar (δ_p) and hydrogen bonding (δ_h) parameters of these solvents. Comparison of data in Tables 3 and 7 suggests that the best solvents for H^+-exchanged humic acids have δ_p, δ_h, and δ_b (proton acceptor) parameters of the order of, or greater than, 6, 5, and 5, respectively. Water satisfies all these criteria, although it is a poor solvent for H^+- and divalent and polyvalent cation-exchanged humic acids. It was pointed out earlier in a previous section that solution is greatest when the products δ_a (solvent) $\times$ δ_b (solute), or vice versa, are maximum. The very large values of δ_h, δ_a, and δ_b indicate the extents of self-association through hydrogen bonding of water molecules. It is clear that the δ_a or δ_b values of the H^+-exchanged humic acids are not sufficient to disrupt these attractive forces.

Formamide (Table 7) was a good solvent for the H^+-exchanged humic acids, and this might not be readily predicted from comparisons of its solubility parameter data with those for ethanol and water. However, although the values for the δ_a and δ_b are large, it is clear that hydrogen bonding must take place between the solvent and the macromolecules. The value of δ_h is similar to that for ethanol, a poor solvent, but the high δ_p parameter, which distinguishes formamide from the other solvents in Table 3, is central to the performance of the solvent as an acceptor in hydrogen bonding with humic acids.

The H^+-exchanged humic acid materials studied were mixtures of polydisperse macromolecules. If it is assumed that molecular size was the only major difference between the components, it is reasonable to suggest that the smaller materials were solubilized preferentially. Hayes et al. (1975) have shown that less of the high-molecular-weight components were extracted in the neutral organic solvent than in the aqueous alkaline solvent systems (see also Hayes and Swift, 1978). On the other hand, differences in the compositions, as well as molecular sizes present additional complications. Therefore, in order to apply solubility parameter theory, which has a more sound theoretical basis, it is desirable to work with humic fractions that are relatively homogeneous with regard to size and composition. This involves a fractionation procedure similar to that employed by Cameron et al. (1972b). Solubility parameter values can then be obtained for the fractions, and such data might suggest solvent systems for the isolation from soil of different humic fractions, or the design of solvent mixtures for more complete extraction of all soil humic components.

Applications of polymer solution theory to the studies of the dissolution of humic acids and of their extraction from soils suffer most because interactions between each pair of components must be known (criterion 4, p. 343). Unfractionated humic and fulvic acids and humic substances in the soil are parts of multicomponent systems, and interactions between the different components are unknown.

The closest approach to successful applications of polymer solution theory to humic substances was made by Chiou et al. (1983) when studying the binding of small organic chemicals by soils. They considered the soil sorbent substances to be amorphous macromolecular humic substances, and they adapted the Flory–Huggins theory to a study of the sorbate species solubilized in the amorphous macromolecules.

Determination of the Flory–Huggins interaction parameter (χ) for solvent–polymer pairs again requires careful fractionation of the humic macromolecules. In the view of this author, much can be learned about the ways in which humic substances are associated through determination of Flory–Huggins and solubility parameters of carefully fractionated humic substances, and through applications of empirical equations such as those of Flory–Huggins [Equation (6)] and Hildebrand–Scatchard [Equation (7)].

Extraction with Solvent Mixtures

Parker (1962) has cited references for the solubilization of neutral polymers such as polyacrylonitrile, nitrocellulose, cellulose acetate, and wood products in DMSO, DMF, and sulfolane. Unlike cellulose, these products are not strongly hydrogen bonded. The review by Suggett (1975) suggests that the addition of urea (10%) to DMSO effectively breaks the hydrogen bonds in cellulose and allows solution to take place. With this in mind Hayes et al. (1983) added urea (10% w/v) to DMSO, and to 0.025*M* sodium tetraborate ($Na_2B_4O_7$; pH 9.2) and 0.1*M* sodium pyrophosphate ($Na_4P_2O_7$; adjusted to pH 7 with H_3PO_4) solutions. Solution of the Glade soil H^+-exchanged humic acids in borate was depressed by the presence of urea, although urea did not influence extraction in the pyrophosphate. Aqueous 10% urea was a very poor solvent for the H^+-exchanged humic acids, but solubilization was almost complete in 5*M* urea. Complete solution of the humic acid was observed when water (5% of the total weight) was added to the urea (10%)–DMSO solution. However, solution was gradually depressed as additional increments of water were added. Although urea–water mixtures greatly improved the efficiency of dissolution and extraction of humic substances from soil their uses cannot be recommended because the raised nitrogen contents of

TABLE 8. Solubilization of H^+-Exchanged Fen Humic Acids (H^+-HA) and Humic Substances from a H^+-Exchanged Sapric Histosol Soil (H^+-SHisl) in 0.1*M* NaOH, DMSO, and Acidified DMSO

Sample and Weight (mg)	Solvent Mixtures: DMSO (cm^3)	Solvent Mixtures: Other Solvents (cm^3)	Absorbance at 465 nm (E_4)	Absorbance at 665 nm (E_6)	E_4/E_6
H_+-HA (1.27)	0.0	0.1*M* NaOH (5.2)	1.212	0.240	5.05
H^+-HA (1.27)	5.2	—	1.045	0.206	5.07
H^+-HA (1.27)	5.18	H_2O (0.02)	1.042	0.196	5.31
H^+-HA (1.27)	5.0	H_2O (0.2)	1.042	0.196	5.31
H^+-HA (1.27)	5.18	HCl (0.02)[b]	0.910	0.110	8.27
H^+-HA (1.27)	5.0	HCl (0.20)[b]	0.890	0.108	8.24
H^+-SHisl (11.7)[a]	0.0	0.5*M* NaOH (5.2)	2.06	0.348	5.92
H^+-SHisl (11.7)[a]	11.7	—	1.71	0.328	5.21
H^+-SHisl (11.7)[a]	11.0	H_2O (0.6); HCl (0.10)[b]	1.90	0.226	8.41
H^+-SHisl (11.7)[a]	11.6	HCl (0.10)[b]	1.98	0.232	8.53

[a] *On dry-weight basis.*
[b] *Concentrated acid.*

the dialyzed extracts suggest that urea interacts with the humic substances (Law, 1984).

Tables 8 and 9 present data for extraction of humic substances from a sapric histosol and from two tropical soils using various combinations of DMSO, acid, and water. For comparison, some data are presented for solution in sodium hydroxide and neutral sodium pyrophosphate solutions. Use is made of E_4/E_6 ratios to indicate differences in the solution conformations and/or compositions of the humic substances in the different solvent systems.

All the samples of the Fenland H^+-exchanged humic acids were completely dissolved in the solvent systems used in Table 8. The data show that the absorbance value at 465 nm for the sodium hydroxide solution was greater than that for the same concentrations of solutes in the DMSO preparations. Although DMSO was stored over calcium hydride to prevent excessive water uptake no attempt was made to fully dry the solvent, and it is assumed that it contained some water. Additional water (up to about 4% of the total volume) did not significantly alter the solution properties of the DMSO. Addition of concentrated hydrochloric acid, even to the extent of only 0.4% of the total volume of the mixture, significantly altered the absorbances and the E_4/E_6 ratios.

The changes observed from the additions of acid can be attributed to alterations of the solution conformations of the humic molecules, as was suggested by Chen et al. (1977), and by Ghosh and Schnitzer (1979). All the humic acid materials were dissolved, whether or not hydrochloric acid was

TABLE 9. Extraction of Humic Substances from a Tropical Rain Forest Soil, Untreated (TRS) and H^+-Exchanged (H^+-TRS), and from a Tropical Savanna Soil, Untreated (TSS) and H^+-Exchanged (H^+-TSS) Using Aqueous Base, Aqueous Neutral Pyrophosphate, DMSO, and Acidified DMSO[a]

Soil[b]	Solvent Mixtures: DMSO (cm^3)	Solvent Mixtures: Other Solvents (cm^3)	Absorbance at 465 nm (E_4)	Absorbance at 665 nm (E_6)	E_4/E_6
TRS	0.0	0.1*M* NaOH (11.7)	0.763	0.117	6.52
H^+-TRS	0.0	0.1*M* NaOH (11.7)	2.145	0.419	5.12
TSS	0.0	0.1*M* NaOH (11.7)	0.433	0.098	4.42
H^+-TSS	0.0	0.1*M* NaOH (11.7)	1.326	0.363	3.65
TRS	0.0	0.5*M* NaOH (11.7)	0.503	0.051	9.86
H^+-TRS	0.0	0.5*M* NaOH (11.7)	1.670	0.272	6.14
TSS	0.0	0.5*M* NaOH (11.7)	0.222	0.042	5.28
H^+TSS	0.0	0.5*M* NaOH (11.7)	2.104	0.648	3.25
TRS	0.0	0.1*M* $Na_4P_2O_7$ (11.7)[c]	0.785	0.107	7.34
H^+-TRS	0.0	0.1*M* $Na_4P_2O_7$ (11.7)	0.902	0.124	7.27
TSS	0.0	0.1*M* $Na_4P_2O_7$ (11.7)	0.720	0.189	3.81
H^+-TSS	0.0	0.1*M* $Na_4P_2O_7$ (11.7)	0.953	0.232	4.11
TRS	11.7	—	0.074	—	—
H^+-TRS	11.7	—	0.419	0.042	9.98
TSS	11.7	—	0.071	0.031	2.29
H^+-TSS	11.7	—	0.640	0.144	4.44
TRS	11.0	H_2O (0.6); HCl (0.1)[d]	1.107	0.131	8.45
H^+-TRS	11.0	H_2O (0.6); HCl (0.1)	1.538	0.181	8.49
TSS	11.0	H_2O (0.6); HCl (0.1)	1.151	0.232	4.96
H^+-TSS	11.0	H_2O (0.6); HCl (0.1)	1.331	0.291	4.57
TRS	10.8	H_2O (0.4); HCl (0.5)[d]	1.518	0.168	9.03
H^+-TRS	10.8	H_2O (0.4); HCl (0.5)	1.444	0.157	9.20
TSS	10.8	H_2O (0.4); HCl (0.5)	1.209	0.240	5.04
H^+-TSS	10.8	H_2O (0.4); HCl (0.5)	1.315	0.263	5.00

[a] *From Fagbenro (1984).*
[b] *0.1 Percent soil suspensions in each case.*
[c] *At pH 7.0.*
[d] *Concentrated acid.*

added to the DMSO. Decreases in the E_4 and E_6 values as a result of additions of acid, and the increases in the E_4/E_6 ratios are therefore consistent with shrinking and the suppressions of the hydrodynamic volumes of the humic acid molecules. An increase in the negative slope of the plot of log absorbance versus wavelength (as observed by Chen et al., 1977) was observed (Law, 1984) when acid was added and this is regarded as further evidence for changes of the solution conformations of the macromolecules in the presence of acid.

Table 8 also compares the extraction of organic materials from a H^+-exchanged Fenland (Norfolk, England) sapric histosol soil. In the H^+-exchanged form this soil contained only about 5% ash. The data show that the E_4/E_6 ratios for extracts in $0.5M$ sodium hydroxide and DMSO are similar, but that the E_4 and E_6 absorbance values are slightly higher for the base extract, and this is in keeping with the trends observed for the H^+-exchanged humic acids. The addition of acid predictably increased the E_4/E_6 ratio, and it can be concluded that it also increased the amount of organic matter extracted because of the increased E_4 values. It is not possible to compare quantitatively the amounts of organic matter extracted by the different solvent systems, but when the depressed absorbance values in DMSO (compared with sodium hydroxide) are taken into account, extraction with acidified DMSO nearly matches in efficiency that in the base.

Data for the tropical soils provide information about differences in the behavior of DMSO (Table 9). The adsorbance values show, as would be predicted, that sodium hydroxide extracted most organic materials from the H^+-exchanged soils, and that $0.1M$ base was generally better than $0.5M$ base. H^+ exchanging had little influence, as might be predicted, and excess salt suppressed solubilization. Again, the addition of acid to DMSO increased solubilization of the colored humic substances. Furthermore, the addition of acid dispenses with the need to H^+-exchange the soils prior to extraction with the solvent.

Anions are very sparingly solvated in DMSO, but cations are much more readily solvated (Martin and Hauthal, 1975, p. 131). Hence exchangeable metal cations that neutralize the charges on the humic macromolecules in the non-H^+-exchanged tropical soils would be solvated by DMSO. The conjugate bases (carboxylates and phenolates) would not be solvated, however, until neutralized with hydrogen ions. A plausible representation of associations between these acids and DMSO might be the following:

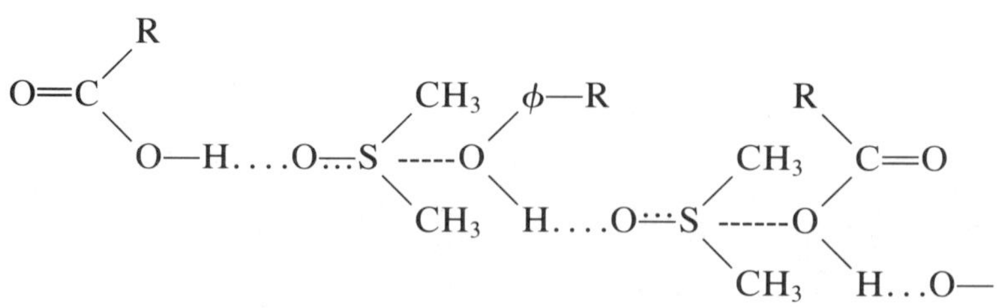

(9)

where R represents the remainder of the macromolecule in each case, and ϕ is an aromatic C_6H_4 (with possibilities for further substitutions in the ring) group.

Association–dissociation equilibria of the type

$$(CH_3)_2SO + HCl \rightleftharpoons (CH_3)_2SO \cdot HCl \rightleftharpoons [(CH_3)_2\overset{+}{S}OH]Cl^- \rightleftharpoons (CH_3)_2\overset{+}{S}OH + Cl^- \quad (10)$$

are known (Martin and Hauthal, 1975, p. 79). This could lead to binding by ion exchange under acid conditions, but this mechanism would not lead to dissolution of the humic molecules. Of more importance is the fact that there are strong interactions between DMSO and water, carboxyl, and phenolic hydroxyl groups (Martin and Hauthal, 1975, p. 77). DMSO–H_2O interactions are stronger than the associations between the water molecules, and the organic molecule associates preferentially with two molecules of water. It can also be expected to associate with the phenolic and carboxyl groups in the humic structures and to break the inter- and intrastrand hydrogen bonds which render the macromolecules insoluble in water. Because of the strong association between DMSO and water the presence of excess water in the medium inhibits interactions between the $>$S=O group and the acid groups in the macromolecule. The nonpolar moieties of this dipolar aprotic solvent are also important in solution processes because these can solvate the hydrophobic structures in the humic substances.

SUMMARY

Water is an excellent solvent for humic substances when the acid groups on the macromolecules are dissociated, and the structures are expanded to allow solvation of the conjugate bases of the acid groups and the polar moieties in the structures. This dissociation is best accomplished with the use of strongly basic aqueous solutions of sodium or potassium hydroxide, but these can degrade the macromolecules and lead to the formation of artifacts. When the charges on the humic molecules are neutralized by divalent and polyvalent cations, dissolution in water is inhibited because the cations are only weakly dissociated and tend to condense the structures by forming short bridges between charges on adjacent strands of the macromolecules. Neutral salt solutions which complex and remove divalent and polyvalent cations allow some degree of solvation of the humic substances in water.

Formamide, *N*-methylformamide, possibly anhydrous formic acid, and the dipolar aprotic solvents *N,N*-dimethylformamide (DMF), and especially dimethyl sulfoxide (DMSO) are the best of the organic solvents for the dissolution of humic substances. Their solvation action can be partially explained by the abilities of these solvents to provide polar and nonpolar moieties for solvation, and to act as acceptors in hydrogen bonding systems. DMSO is the best of the organic solvents tested for humic acids, and its effectiveness is improved by the presence of small amounts of concentrated acid. Swelling of humic substances in DMSO and other organic solvents is slow (equilibration overnight, with vigorous agitation is recommended), and in addition recovery of the classic humic fractions from such solvents is difficult. Because of the low solvation of anions in DMSO the humic substances can be recovered by making the extract alkaline. This, however, would negate the advantage of using a mild solvent. Much systematic research is needed in order to devise means of handling extracts in the organic solvents and to understand fully the mechanisms of solvation of the macromolecules in these solvents.

Note Added to Proof

Humic substances may be recovered from solution in DMSO-HCl mixtures by use of XAD-8 resin columns (H^+-form). The humic materials are sorbed to the resin as the DMSO is washed through in 0.1*M* HCl. Elution with sodium tetraborate (0.01*M*) and sodium chloride (0.1*M*), pH 8.9, releases the humic materials. Any residues remaining may be displaced with 0.05*M* NaOH.

CHAPTER FOURTEEN

Isolation and Concentration Techniques for Aquatic Humic Substances

GEORGE R. AIKEN

ABSTRACT

A critical review is presented of the methods commonly used to isolate and concentrate aquatic humic substances. The important steps in any extraction scheme are filtration, concentration, isolation, and preservation. Various filtration options are outlined and compared; filtration is important for obtaining low-ash humic substances. Advantages and disadvantages of commonly used concentration and isolation techniques are discussed, and sorption on the synthetic macroporous resins XAD-8, a nonionic resin, and Duolite A-7, a weak anion-exchange resin, are recommended as the most efficient methods. Two established extraction schemes utilizing these resins are discussed.

INTRODUCTION

The study of humic substances in water dates back to the Swedish scientist Berzelius, who investigated colored waters of a mineral spring. He later isolated colored organic compounds from swamp water and iron-containing waters by precipitation with added iron, and studied the chemistry of these compounds (Kononova, 1966). Emphasis in the study of humic substances since Berzelius' time has centered on the chemistry and importance of these compounds in soil. In the last 25 years interest in organic compounds in water has grown. Increased awareness concerning the chemical quality of water has stimulated development in the analysis of organic compounds in water, and interest in aquatic humic substances has never been greater than at present.

Organic matter in natural waters can be divided into dissolved and particulate organic carbon. No natural cutoff exists between these two fractions and the distinction is operational. Filtration through a 0.45 μm (micron) filter has been arbitrarily established as the standard procedure for separating dissolved and particulate components (Danielsson, 1982). The distribution of dissolved versus particulate organic carbon varies depending on the type of aquatic sample. Particulate organic carbon in seawater ranges from 0.7% of the total organic carbon in the North Central Pacific to 24% in the Arctic Ocean (Sharp, 1973). Leenheer (1982) reports particulate organic carbon ranging from 7 to 24% of total organic carbon for rivers in the United States.

Research on humic substances in water has been almost exclusively concerned with the dissolved fraction. Suspended organic carbon, very important in so called "black waters" such as tropical rivers, swamps, and bogs where concentrations of humic substances are high, has been little studied, and much research is needed in this area.

Dissolved organic carbon can be divided into six fractions: hydrophobic acids, bases, and neutrals; and hydrophilic acids, bases, and neutrals (Leenheer, 1981). Dissolved organic carbon fractionation data for a variety of freshwater systems are presented in Table 1. Humic substances are the major component of the hydrophobic acid fraction. These compounds range in concentration from 20 μg/L (micrograms per liter) in groundwater to over

TABLE 1. Dissolved Organic Carbon Fractionation Data (mg C/L) for Select Freshwater Systems of the United States

			Hydrophobics			Hydrophilics		
Sample	Date	DOC	Acids	Bases	Neutrals	Acids	Bases	Neutrals
Black Lake, North Carolina	11/81	8.3	3.4(41)[a]	0.1(1)	1.5(19)	0.7(8)	2.6(31)	0.0
Ohio River at Cincinnati, Ohio	6/81	3.7	1.2(32)	0.0	0.9(25)	1.3(36)	0.3(7)	0.0
Missouri River at Sioux City, Iowa	8/81	3.4	0.7(19)	0.8(22)	0.0	1.7(51)	0.3(8)	0.0
Suwannee River at Fargo, Georgia	12/82	38.2	16.0(42)	0.2(1)	1.0(2)	19.2(50)	1.4(3)	0.6(1)

[a] *Numbers in parentheses are the percent of the total DOC of that fraction.*

30 mg/L (milligrams per liter) in surface water (Thurman and Malcolm, 1981). Before their properties can be thoroughly defined, aquatic humic substances must be isolated from other organic compounds and background inorganic species and then be concentrated.

The final product should be free from chemical impurities, which hinder characterization of the isolated material, and should be in a form that can resist biological and chemical degradation. Many options are available for isolating and concentrating these compounds. Each method has advantages and disadvantages, and best results can be obtained by adopting a combination of techniques. It is the purpose of this chapter to critically discuss and evaluate various methods commonly used to isolate and concentrate aquatic humic substances.

FILTRATION

Isolation and concentration of aquatic humic substances begin by separating the sample into dissolved and particulate fractions (Danielsson, 1982). Filtration through a 0.45 μm filter is the accepted method of separation; however, numerous problems accompany this seemingly simple task. A number of filter types are available (Table 2), and many of these have been compared for the isolation of particulate organic carbon (Wangersky and Hincks, 1978; Sheldon, 1972; Quinn and Meyers, 1971) and for the study of dissolved inorganic species (Danielsson, 1982). The effects of filtration on the dissolved organic carbon fraction and on dissolved humic substances have not been addressed.

TABLE 2. Advantages and Disadvantages of Matrix Types Suitable for Filtration of Water for Isolation of Aquatic Humic Substances

Matrix Type	Advantages	Disadvantages
Silver membrane	1. Uniform pore size. 2. Bactericidal properties.	1. Slow flow characteristics. 2. Expensive. 3. Slight sorption of certain organic compounds, such as mercaptans.
Glass fiber	1. Good flow characteristics. 2. Economical.	1. Particles larger than nominal pore size can pass through filter. 2. Slight sorption of certain organic compounds.
Organic membrane Cellulose acetate/ cellulose nitrate	1. Uniform pore size. 2. Economical.	1. Wetting agents contaminate filtrate. 2. Sorption of organics to matrix.

Filtration not only separates dissolved and suspended organic carbon, but also isolates dissolved organic carbon from suspended clay minerals in the sample. These minerals interact with humic substances (Preston and Riley, 1982) and are concentrated along with the compounds of interest, resulting in high ash concentrations in the final product. Clay minerals are similar to humic substances in exchange capacity, metal uptake, water-retention capacity, and heat capacity. Clay mineral impurities in the sample hinder the characterization of humic substances by elemental analysis, infrared spectroscopy, titrimetry, and measurement of metal complexing ability (Malcolm, 1976). Therefore, it is desirable to obtain samples that are low in these minerals. Clay–mineral colloids can also interfere with the concentration of organic compounds by sorption methods. Pores of resin sorbents can be clogged by clay colloids, and hydrous oxides of aluminum, iron, and manganese can be sorbed on or precipitated within the beads (Kunin and Suffet, 1980).

The 0.45 μm cutoff between dissolved and particulate fractions is arbitrary (Danielsson, 1982), and many researchers have commented on the inadequacy of this standard for removal of colloidal species. Sharp (1973) points out that colloids are of the approximate size range of 0.001–1.0 μm, and Kennedy et al. (1974) report that clay minerals of the types found in stream sediments can be much smaller than 0.45 μm. The concern of Kennedy et al. (1974) is that such material can pass the 0.45 μm filter in sufficient quantities to seriously influence the analysis of dissolved inorganic species. These authors recommend a 0.1 μm filter pore size to remove clay colloids more effectively. The presence of these clay minerals is also of concern in the study of aquatic humic substances; however, decrease in pore size from 0.45 to 0.1 μm is accompanied by a decrease in flow rate through the filter, which is a major disadvantage when hundreds of liters of water are to be processed. Use of 0.45 μm filters represents a compromise between flow rate and rejection of clay minerals.

A variety of filters with a 0.45 μm pore size is available. These filters vary in uniformity of pore size, chemical composition, and flow characteristics. Each of these factors affects the performance of the filter; and some filters are better suited for the study of aquatic humic substances than others. The ideal filter would be inert relative to compounds of interest, exhibit good flow characteristics, have a uniform pore size, and be reasonably priced. An evaluation of the suitable filters with respect to pore size, chemical composition, and flow characteristics follows.

Pore Size

Filters can be divided into two types: membrane (screen) filters and depth filters. Membrane filters, such as silver-membrane filters or cellulose-acetate and cellulose-nitrate filters, physically sieve and retain particles on their surfaces. These filters have uniform pore sizes and are rated for absolute

TABLE 3. Variation of Dissolved Organic Carbon Concentration as a Function of Filter Plugging for Silver-Membrane and Glass-Fiber Filters

		Dissolved Organic Carbon Concentration (mg C/L)	
Sample	Filter	Unplugged Filter	Plugged Filter
Suwannee River at Fargo, Georgia	Silver membrane[a]	30	30
	Glass fiber[b]	25[c]	25
Ogeechee River at Grange, Georgia	Silver membrane	5.1	4.9
	Glass fiber	4.8	4.6

[a] *Selas Flotronics, 0.45 μm.*
[b] *Balston Microfiber, 0.3 μm.*
[c] *This sample is not for the same data as silver-membrane filter sample.*

retention; all particles larger than the pore size are retained. Depth filters, such as glass-fiber filters, consist of a matrix of fibers that form a tortuous maze of flow channels. The particulate fraction becomes entrapped by this matrix. These filters do not have a uniform pore size, and it is not possible to rate them for absolute retention. They are rated according to nominal pore size, which is determined by the particle size that is retained by the filter to a predetermined percentage. This percentage is usually given as 98% retention; however, it can be as low as 90%.

Glass-fiber filters with nominal pore sizes will pass some particles that are larger than the stated pore size, thus including additional clay colloids in the filtrate. Particles smaller than the stated pore size can also be trapped within the fiber network. While this is a problem common to all filters, it is more acute with glass-fiber filters. Silver-membrane filters have a more uniform, well-defined pore size, and have been recommended for the study of particulate organic carbon (Wangersky and Hincks, 1978). However, it should be pointed out that problems do arise when using the silver-membrane filters, as the effective pore size of the filters decreases with increased loading (Sheldon, 1972). When large samples or samples high in concentration of particulates are filtered, average retention size is less than stated pore size, and flow rate drops dramatically. The effect on dissolved organics is not significant (Table 3) for waters low in dissolved organic carbon; however, for waters high in colloidal organic matter, such as black water lakes and bogs, some colloidal material smaller than 0.45 μm could be excluded.

Chemical Composition

Organic-membrane filters, such as cellulose-acetate and cellulose-nitrate filters, are often conditioned with organic surfactants to assist in wetting the filter. These surfactants will be leached from the filters, contaminate the filtrate, and will be isolated and concentrated with the humic substances of interest (Hwang et al., 1979). Bleed from these filters can be minimized by leaching the filter with distilled water immediately prior to use; however, even small quantities of impurity will be further concentrated, and use of these filters for the isolation of humic substances is not recommended.

Filters can also sorb certain organic compounds possibly resulting in reduced recovery and fractionation of compounds of interest. Cranston and Buckley (1972) compared organic-membrane, glass-fiber, and silver-membrane filters for the sorption of humic acid. Organic-membrane filters sorbed 15%, glass-fiber filters 5%, and silver-membrane filters 3% of the humic acid in the solution passed through the filters.

Quinn and Meyers (1971) report the sorption of long-chain fatty acids from aqueous solution on glass-fiber filters. Gordon and Sutcliffe (1974) report no apparent sorption of organic carbon on silver-membrane filters. Organic compounds that can interact with silver, such as mercaptans, would be expected to sorb onto these filters. These compounds are not found in significant concentrations in natural surface waters, but, are present in reducing groundwaters and soil-pore waters.

Silver-membrane and glass-fiber filters are recommended for studying particulate organic carbon, because organic contamination of the sample is avoided (Cranston and Buckley, 1972). These filters are also recommended for the isolation of aquatic humic substances, because interactions with the filter by dissolved organic compounds are minimized, and the potential for organic contamination of the sample is slight.

Flow Characteristics

Large volumes of water often need to be processed to obtain sufficient quantities of aquatic humic substances; filtration is the slowest step in this process. In a comparison of filter flow rates, Cranston and Buckley (1972) found filtering times for 47 mm (millimeter) diameter filters increased in this order; glass-fiber filters < silver-membrane filters < organic-membrane filters (cellulose-acetate and cellulose-nitrate). They also report that substantial variation exists between different silver filters from the same manufacturer. This is caused by variation in permeability (number of pores per unit area), not pore size, and did not occur with the other filters studied.

Excellent flow rates can be obtained with glass-fiber filters in the form of filter tubes. These filters are supported by an inert screen, and water is pumped through the filter. The efficiency of Balston's* microfiber filter tubes (with nominal pore size of 0.3 μm at 98% retention) over the 0.45 μm Selas Flotronic silver-membrane filter is shown in Figure 1. Water from the Ogeechee River in Georgia, a river high in clays, was filtered through each filter at 30 psi (pounds per square inch). Even though the filters are of comparable

* The use of brand names in this report is for identification purposes only and does not constitute endorsement by the U.S. Geological Survey.

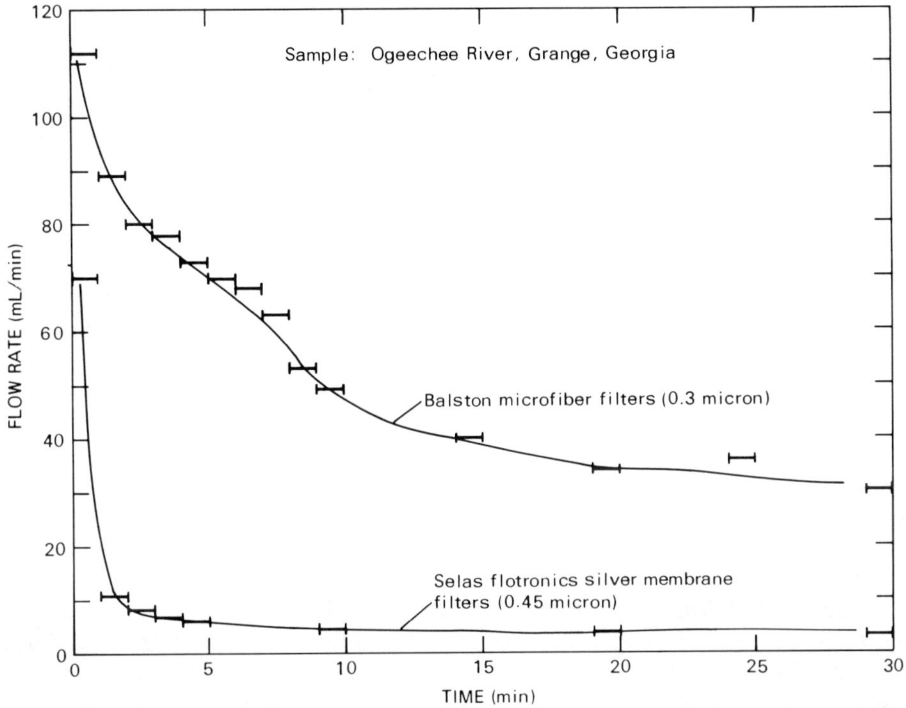

FIGURE 1. Variation of flow rate with filtration time for silver-membrane and glass-fiber filters at a pressure of 30 psi for Ogeechee River water.

size, the glass-fiber tubes have 2.4 times more surface area, based on surface-area calculations from filter dimensions. In reality, the glass-fiber filters have an even greater surface area than the silver-membrane filters, because of the macroreticular nature of depth filters. Flow rate for the silver-membrane filters dropped off abruptly, while flow rate on the glass-fiber filters dropped off gradually. The difference in flow rate resulted in 1500 mL of water being filtered by the glass-fiber filters compared to 200 mL by the silver-membrane filters in 30 minutes of filtration time.

At most sampling sites, many hundreds of gallons of water must be processed to obtain sufficient quantities of aquatic humic substances. The cost of filtration can become prohibitive both in time and in number of filters used. The superior flow rate of the glass-fiber filter tubes over the silver-membrane filters is a definite advantage. Glass-fiber filter tubes are also much less expensive than the silver filters.

SAMPLE PRESERVATION

Samples of naturally occurring organic matter are subject to both biological and chemical degradation. It is important to process the sample as soon as possible after collection to prevent degradation; in particular, it is imperative to filter samples immediately.

Filtration through a 0.45 μm filter effectively removes organisms as small as bacteria. Biological activity can be further suppressed by the addition of a biocide to the filtered sample. Silver, in concentrations as low as 10 ppb (parts per billion), has been found to be an effective bactericide in water (Woodward, 1963). Silver can be added to the solution as $AgNO_3$, or samples can be filtered through silver-membrane filters, resulting in the solubilization and addition of silver to the filtrate. Values as high as 260 ppb silver have been measured in this laboratory after silver filtration of samples that initially contained no silver.

Adjustment of sample pH is another effective measure to minimize biological degradation. Afghan et al. (1974) report that *Pseudomonas* bacteria, responsible for degradation of phenols, is destroyed by both high and low pH values. High pH values increase the chemical oxidation rate of humic substances, and should be avoided (Stevenson, 1982); low pH values can result in precipitation of humic acid if the solution is sufficiently concentrated. Generally, the concentration of humic acid is low and precipitation is not a problem.

Biological degradation can be greatly retarded by chilling the sample with ice or by refrigeration. If practical, samples should be filtered immediately, chilled, and processed as soon as possible. An inorganic biocide, such as Ag(I), should be added if expeditious sample processing is not possible.

Humic substances can also degrade chemically. Such preservation techniques as ultraviolet irradiation or storage at high temperature should be avoided because of chemical alteration of humic material. Chemical oxidation will also cause serious structural alterations of humic substances. Oxidation is enhanced in the presence of NaOH (Stevenson, 1982), and if NaOH is used in the concentration procedure, care should be taken to keep the solution under nitrogen. Contact time with NaOH should be kept to a minimum, and the solution pH adjusted with mineral acid as soon as possible.

CONCENTRATION METHODS

Before humic substances from water can be properly characterized, they need to be further concentrated and isolated from other solutes. The problem of concentrating these substances from water has hindered their study in the past. However, in the last 20 years, methods have been developed to improve the efficiency of the concentration process. Humic substances are now easily isolated and concentrated from waters with very low concentrations of organic carbon such as groundwater (see Chapter 4). Numerous concentration methods are available (Table 4). The more commonly used methods are evaluated below.

Vacuum Distillation

Humic substances have been concentrated by vacuum distillation. This method is carried out at low temperatures, which avoids decomposition and chemical reactions within the sample (Jolley et al., 1975), and is faster than freeze-drying (Katz et al., 1972). All solutes are concentrated by this method, and coprecipitated organic matter must be further extracted from precipitated inorganic salts (Katz et al., 1972). Vacuum distillation of humic substances to dryness is not recommended because it results in a dense product that is not easily removed from the drying vessel and may be difficult to dissolve.

Freeze-Drying (Lyophilization)

Freeze-drying is a gentle method for concentrating humic substances (Malcolm, 1968). High concentration factors are possible, and samples can be taken to dryness. The method is slow and not suitable for processing large volumes of water (Katz et al., 1972). However, small samples can be directly preconcentrated with this method (Glaze et al., 1981). With the exception of volatile organics, all solutes are concentrated by freeze-drying and humic substances must be further isolated from the freeze-dried residue. It is difficult to remove large concentrations of salt after drying the sample (Watts et al., 1981), and prior treatment of a sample to remove inorganic salts is a disadvantage of the method (Milanovich et al., 1975).

Freeze-drying is commonly used in conjunction with other concentration methods as the final step in isolating humic substances from water (Beck et al., 1974; Deinzer et al., 1975; Aldridge et al., 1976; Jolley et al., 1979; Thurman and Malcolm, 1981), and it is here that freeze-drying is most efficient. Samples of aquatic humic substances should be considerably concentrated and desalted prior to freeze-drying. The solid product obtained by freeze-drying can be easily handled and stored without fear of chemical degradation.

Freeze Concentration

Freeze concentration is another method that concentrates all solutes present including volatiles and neutral polar solutes. Further sample treatment is required to separate humic substances from other organic solutes concentrated. Efficiency of concentration is dependent on sample ionic strength, with a high salt content inhibiting efficiency (Baker, 1970), Black and Christman (1963a,b) report that some organic matter can be lost with this method. Freeze concentration is also slow and unsuitable for processing large volumes of water (Shapiro, 1961; Black and Christman 1963b). However, this method is mild (Baker, 1967), inexpensive, and simple (Shapiro, 1961).

TABLE 4. Methods Commonly Used to Isolate and Concentrate Aquatic Humic Substances

Method	Advantages	Disadvantages
Vacuum distillation	1. Low temperatures.	1. All solutes concentrated.
Freeze-drying (Lyophilization)	1. Mild. 2. High concentration factors. 3. Sample taken to dryness.	1. Method is slow. 2. All solutes with the exception of volatiles are concentrated.
Freeze concentration	1. Mild. 2. Inexpensive. 3. Simple.	1. Method is slow. 2. All solutes concentrated.
Coprecipitation	1. Inexpensive. 2. Effective for waters high in DOC.	1. Efficiency dependent on initial DOC. 2. Inefficient on large volumes of water. 3. Isolated organic matter must be separated from inorganic salts.
Ultrafiltration	1. Organic solutes fractionated by molecular size. 2. Large volumes can be processed.	1. Interactions with membrane possible. 2. Fouling of membrane possible.
Reverse osmosis	1. Ambient conditions, mild. 2. Large volumes can be processed.	1. All solutes concentrated. 2. Efficiency dependent on concentration.
Solvent extraction	1. Inorganic salts effectively excluded.	1. Humic substances insoluble in many solvents. 2. Method is slow.
Sorption		
Alumina	1. Organic acids readily sorb to basic adsorbent. 2. Mild eluents.	1. Inefficient desorption. 2. Structural alterations of organic matter possible.
Nylon and polyamide powder	1. Efficient adsorption.	1. Irreversible sorption probable.
Carbon	1. Inexpensive. 2. Simple procedure. 3. Large volumes of water can be readily processed. 4. Organic blanks are low.	1. Irreversible sorption possible. 2. Slow elution rates. 3. Slow sorption rates with high-molecular-weight species. 4. Chemical alteration of organic solutes possible.
Anion exchange (a) Strong-base resins	1. Method is simple. 2. Large volumes can be processed. 3. High capacities for macroporous resins.	1. Irreversible sorption probable. 2. Fouling of resins possible. 3. Resin bleed. 4. All anions concentrated.
(b) Weak-base resins on amphoteric matrix	1. Method is simple. 2. Large volumes can be processed. 3. High capacities for macroporous resins. 4. Efficient desorption. 5. Inorganic salts removed.	1. All organic anions concentrated. Humic substances must be isolated from hydrophilic acids. 2. Extensive cleanup of resin required. 3. Resin bleed. 4. Desorption with NaOH.

TABLE 4. continued

Method	Advantages	Disadvantages
Nonionic macroporous sorbents	1. Method is simple. 2. Resins easily regenerated. 3. Large volumes can be processed. 4. High capacities. 5. Efficient desorption of acrylic ester resins.	1. Irreversible sorption possible on styrene divinylbenzene resins. 2. Desorption with NaOH. Precautions required to prevent oxidation of humic substances. 3. Resin bleed. 4. pH adjustment to pH 2 prior to adsorption.

Coprecipitation

Humic substances have been isolated from water by coprecipitation with $CaCO_3$, $Mg(OH)_2$, $Fe(OH)_3$, $Pb(NO_3)_2$, and $FeCl_3$. This is inexpensive and effective for waters with high dissolved organic carbon (Sridharan and Lee, 1972). However, the efficiency of the method depends on initial concentrations of organic matter (Otsuki and Wetzel, 1973). Recoveries of organic carbon ranging from 16 to 63% have been reported (Williams and Zirino, 1964), indicating that the method is not quantitative for humic substances.

Reagents, such as $FeCl_3$, may contain significant quantities of organic impurities that can be difficult to remove prior to use. In addition, separation of inorganic salts from isolated organic matter is difficult and cumbersome (Jeffrey, 1969). Where large volumes of water are processed, coprecipitation is an impractical method of isolating humic substances.

Ultrafiltration

Ultrafiltration is a valuable fractionation method that has been used successfully to isolate humic substances from water (Milanovich et al., 1975). In this method, dissolved solutes are separated by a membrane according to molecular size. Large volumes of water can be efficiently processed with spiral wound membranes. In theory, aquatic humic substances can be separated from inorganic solutes and low-molecular-weight organic solutes (Michaels, 1968). The disadvantages of ultrafiltration are fouling of membranes and membrane–solute interactions (Buffle et al., 1978). Ultrafiltration has been used primarily to determine the molecular size distribution of aquatic organic compounds, which is further discussed in Chapter 15 by Swift and Chapter 19 by Wershaw and Aiken.

Reverse Osmosis

As a method of concentrating organic matter from water, reverse osmosis has the advantage of utilizing ambient conditions to minimize the possibility of destructive chemical reactions (Deinzer et al., 1975). In addition, large volumes of water can be processed easily. The method concentrates all solutes with the exception of certain organic compounds, such as phenols, which have negative retentions. Kopfler et al. (1975) report up to 85% of DOC present in drinking water can be retained. Low-molecular-weight organic compounds and inorganic salts need to be further separated from humic substances.

Odegaard and Koottatep (1982) report that higher-molecular-weight fractions are excluded well at low concentrations, but movement across the membrane occurs at higher concentrations. These authors report 80–100% removal of humic substances, as determined by color removal. However, reverse osmosis is an expensive, equipment-intensive method.

Solvent Extraction

The insolubility of humic substances in nonpolar organic solvents has limited the use of solvent extraction as a method of isolating humic substances from water. The most effective method for solvent extraction was reported by Eberle and Schweer (1974). Humic acid was efficiently extracted with trioctylamine/chloroform at pH 5 and was recovered by back-extracting with water at pH 10 or above. Butanol has been used to extract freeze-concentrated humic substances; however, not all the material was extracted (Shapiro, 1957). Another method involves acidification of a sample with acetic acid, followed by extraction with isoamyl alcohol. Humic acid precipitates at the interface (Martin and Pierce, 1975). This method is slow; 5 hours were required to extract 100 mL of sample. No data on the behavior of fulvic acid in this solvent extraction were presented.

One advantage of solvent extraction is that inorganic salts can be effectively separated from organic matter (Shapiro, 1957). However, poor extraction efficiencies and slow extraction rates outweigh this advantage.

Sorption Methods

Column chromatographic methods using a wide variety of sorbents have been effective in isolating humic substances. Advantages of these methods include easy handling of large volumes of water, high-concentration factors for isolated solutes, fractionation of dissolved organic solutes according to sorption characteristics, and regeneration of sorbent. The major problem with sorption is the presence of sites that can sorb via different mechanisms, resulting in irreversible sorption. In addition, small pores on the sorbent can exclude large molecules and thereby lower the capacities. These pores can also trap large molecules and hinder elution. Development of synthetic macroporous resins, in both nonionic and anion-exchange forms, has helped solve these problems, and the efficiency of isolating and concentrating humic substances from water has increased. The more widely used sorbents are described below.

Alumina

Alumina is well suited for sorption of acids. The presence of oxide groups on the surface provides alumina with basic binding sites, and weak acids sorb to alumina relative to other sorbents such as silica (Snyder, 1968); strong acids chemisorb to alumina. This sorbent also has acidic binding sites and electron acceptor sites, capable of charge-transfer interactions (Snyder, 1968). These sites reduce desorption efficiency of humic-like molecules from alumina. In addition, many organic compounds react on the alumina surface resulting in structural alterations (Laitinen and Harris, 1975). Moed (1970) reports the isolation of lake organic matter on alumina. Based on absorbance at 270 nm 98% of the soluble, yellow organic matter was sorbed; desorption with 0.008M and 0.3M NaH_2PO_4 buffer was inefficient, with recoveries of 66–80% reported. This sorbent does not require organic solvents or strongly acidic or basic eluents.

Nylon and Polyamide

Nylon, in the form of white nylon stockings, has been found effective for isolation of humic material (Gelbstoff) from seawater. Sieburth and Jensen (1968) report that 70% of the Gelbstoff of seawater could be concentrated on nylon. Elution efficiencies with 0.1N NaOH were high with some irreversible sorption: 8% of Gelbstoff was not eluted; concentration factors of 10,000 were attainable. These authors also report that polyamide powders were less efficient sorbents, with irreversible sorption seriously affecting the isolation of Gelbstoff. Less than half of the sorbed Gelbstoff was eluted from these sorbents. Irreversible sorption of humic substances on polyamide powders is probably due to strong hydrogen bonding between phenolic hydroxyl groups and amide bonds. Additional strong attractions exist for dicarboxylic acids, aromatic carboxylic acids, and quinones (Endres and Hormann, 1963).

Carbon

As a sorbent for isolating humic substances from water, carbon has the advantages of being simple and inexpensive. Large volumes of water can be processed, organic bleed from the carbon is low, and fulvic acid can be quantitatively sorbed (Kerr and Quinn, 1975). Certain disadvantages must be noted. Carbon is a sorbent capable of numerous sorption mechanisms. Irreversible sorption of organic compounds on carbon has been attributed to the presence of surface oxides (Modell et al., 1980). Carbon is also capable of charge-transfer complexation and ion exchange. Kerr and Quinn (1975) report that desorption of seawater fulvic acid from carbon is incomplete; Modell et al. (1980) report slow elution rates and poor recoveries of phenol on granular-activated carbon. These authors point out that elution efficiencies can be greatly improved by using supercritical fluids as the eluent. This is a recent development to improve elution efficiencies that should be investigated further.

The effect of pore size on the kinetics of sorption of large molecules on carbon has been studied. Rapid breakthrough and low-concentration factors of organic compounds have been attributed to slow sorption kinetics (Youssefi and Faust, 1980). McCreary and Snoeyink (1980) report that sorptive capacity decreased with increasing-molecular-weight fractions, and that humic acid was slower to attain equilibrium than the smaller fulvic acid. Slow sorption kinetics particularly hamper column-concentration methods, and the choice of proper flow rate is important.

High ash contents for yellowish-brown organic matter extracted from seawater have been attributed to inorganic impurities in the carbon sorbent used (Jeffrey and Hood, 1958). These authors recommend extraction of carbon with phenol prior to use to remove these inorganic impurities. A more serious problem is the possibility of chemical alteration of organic matter on the surface of carbon (Jeffrey, 1969).

Ion Exchange

Ion exchange has been used routinely to remove organic matter from sugar liquors, pharmaceutical broths, and chemical-process streams, and for water treatment (Tilsley, 1979). The method is simple, resins can be regenerated easily, and large volumes of water can be processed. The major disadvantages are that resins must be extensively cleaned to minimize organic bleed, and that all organic anions are concentrated necessitating further separation of humic substances. Numerous exchange resins are available, and the efficiency of isolating organic matter from water can be maximized by judicious choice of sorbent. Resins are available with a variety of polymer matrices (Fig. 2). Ionizable functional groups, which have mobile ions that can react with or be replaced by other ions, are chemically bonded to the hydrocarbon polymer matrix. Behavior on an ion-exchange resin is determined by the nature of these functional groups (Khym, 1974). In addition, these resins are available in both microporous and macroporous forms. High surface area, macroporous resins are desirable for isolating humic substances from water (Tilsley, 1979).

Two types of anion-exchange resins are commonly used to isolate humic substances from water: strong-base resins that have quaternary ammonium groups, and weak-base resins that have secondary amine groups (Fig. 2). In the macroporous forms, these resins have high capacities for humic substances. Irreversible sorption is a disadvantage with strong-base exchangers. Kim et al. (1976) report that irreversible sorption occurs on strong-base

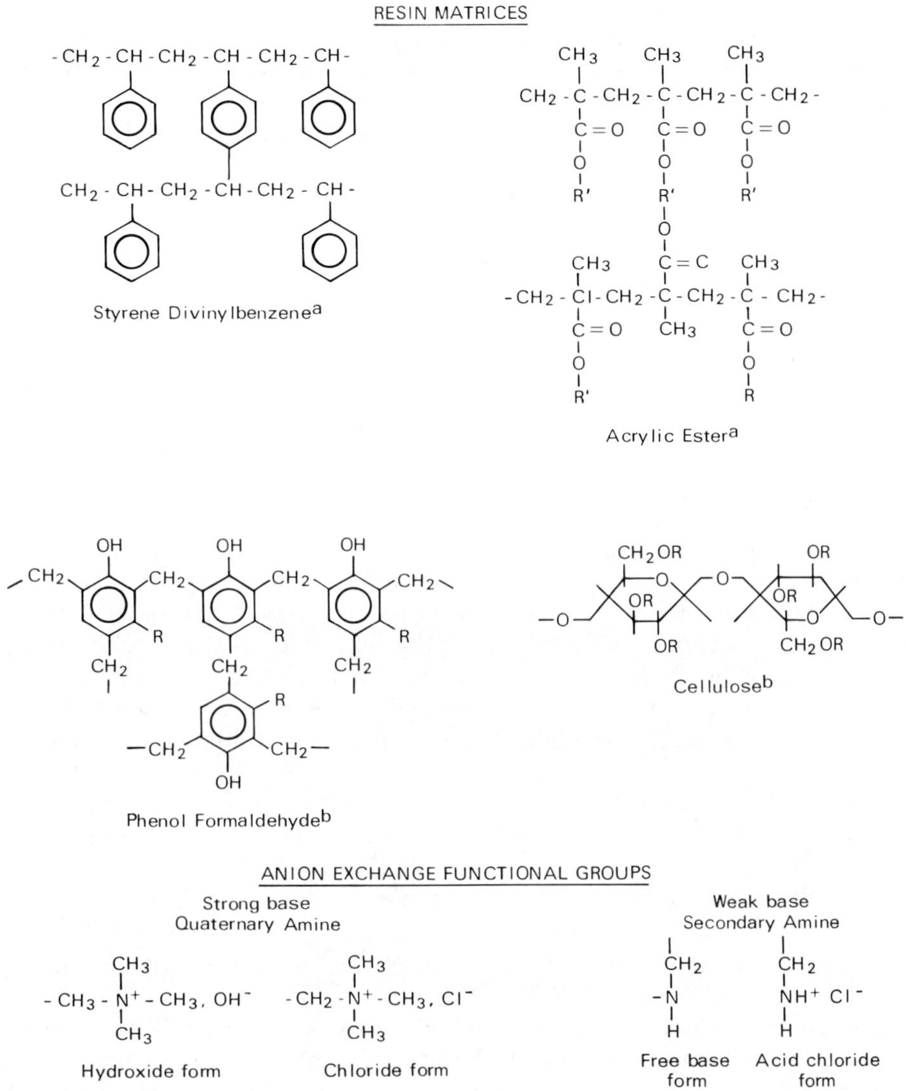

a. Kun and Kunin, 1968
b. Craig, 1953
c. Kim, et al, 1976

FIGURE 2. Structural components of macroporous resin sorbents suitable for concentrating aquatic humic substances.

resins with organic compounds such as phenols and alkylbenzene sulfonates, and that it is caused by the high affinity of these compounds for quarternary ammonium sites. Interaction of organic solutes with the styrene divinylbenzene matrix of strong-base exchangers can cause elution problems (Abrams, 1969). Large molecules, such as humic substances, also diffuse more slowly from the macroporous structures of these resins resulting in eventual fouling (Kunin and Suffet, 1980). In addition, strong-base resins will concentrate all inorganic and organic anions. Use of strong-base resins for isolation of humic substances from water is not recommended.

Phenol-formaldehyde weak-base resins combine both weak-base secondary-amine functional groups with a more hydrophilic matrix than styrene divinylbenzene. The charge- and ion-exchange ability of weak-base exchangers is a function of pH. Secondary-amine functional groups are protonated and positively charged below pH 5.5. Above pH 10, these resins are negatively charged, and anions are repulsed. The resin is neutral between pH 5.5 and pH 10. Kim et al. (1976) report that sorption of organic anions on these resins is pH dependent, with maximum sorption occurring in the pH region in which both resin and solute are uncharged. Organic acids are desalted during the concentration step, because the mechanism of sorption is hydrogen bonding (Abrams, 1969; Sirotkina et al., 1974; Kim et al., 1976) and not ion exchange. Efficient desorption is due to charge exclusion attained by ionizing both the anionic organic solutes and the anionic resin matrix. All organic anions are concentrated on weak-base exchange resins, and humic substances must be isolated from low-molecular-weight hydrophilic acids.

Criteria necessary for an efficient anion sorbent of aquatic humic substances are weak-base functional groups, a macroporous structure, and a hydrophilic matrix that is negatively charged at pH 10 (Abrams and Breslin, 1965). Diamond Shamrock's A-7, a phenol-formaldehyde weak-base resin, has been reported by Leenheer (1981) to be an effective resin for isolation and concentration of aquatic humic substances. This resin has excellent elution characteristics, with 100% recovery of colored organic solutes from water reported when loading was limited to one-half to two-thirds of resin capacity (Abrams and Breslin, 1965). When the resin is loaded to capacity, resin performance is greatly decreased, with premature breakthrough and only 70% recovery of the dissolved organic anions (Abrams and Breslin, 1965; Kunin and Suffet, 1980).

Similar success has been reported for the concentration of aquatic humic substances on diethylaminoethyl cellulose (DEAE cellulose) (Sirotkina et al., 1974; Miles et al., 1983). DEAE cellulose is a weak anion exchanger, with tertiary amine functional groups bonded to a hydrophilic matrix. Miles et al. (1983) report recoveries of 85% and higher for humic substances isolated from four rivers of high DOC. Like Duolite A-7, organic acids are desalted during the concentration step, because sorption of inorganic ions on DEAE cellulose is minimal in the pH range 6.7–7 (Sirotkina et al., 1974).

Humic substances must be further isolated from other organic substances. Major disadvantages are that DEAE cellulose has low exchange capacity relative to other resin exchangers and poor flow characteristics.

Sorption of humic substances from water on cation-exchange resins is extremely limited. MacCarthy and O'Cinneide (1974) report that a cationic fraction of fulvic acid from a bog peat could be isolated with cation-exchange resin. However, aquatic humic material is strongly anionic, and sorption on cation-exchange resins is poor.

Nonionic Macroporous Sorbents

In recent years, it has been found that high recoveries of organic compounds from water are possible with nonionic macroporous sorbents such as the Amberlite XAD resin series. XAD-1 and XAD-2 have been used as sorbents for the isolation of humic substances from seawater (Mantoura and Riley, 1975; Stuermer and Harvey, 1977b); XAD-2 and XAD-8 have also been used to isolate these substances from fresh, surface, and ground water (Weber and Wilson, 1975; Thurman and Malcolm, 1981). These resins are an improvement over such sorbents as carbon, alumina, nylon, and polyamide powder because of high adsorption capacities and ease of elution (Mantoura and Riley, 1975). Bleeding of organic polymer material by nonionic macroporous resins is a disadvantage of this method. Bleed contamination is minimized by extensive Soxhlet extraction of the resin with organic solvents prior to use.

XAD resins are nonionic macroporous copolymers with large surface areas. The "hydrophobic effect" is the principal driving force for sorption on these resins. Sorption of organic acids such as humic substances is determined by the solute's aqueous solubility and solution pH (Thurman et al., 1978) (Fig. 3). At low pH, weak acids are protonated and adsorbed on the resin; at high pH, weak acids are ionized and desorption is favored. Samples are generally acidified with mineral acid, such as HCl, and passed through a column of XAD resin. Adsorbed organic acids are recovered by eluting the column with a basic solution, usually $0.1N$ NaOH. Ammonium hydroxide can also be used as an eluent, however, NH_4^+ can strongly interact with humic substances (Stevenson, 1982) and will be difficult to eliminate from the final product, leading to erroneously high nitrogen contents of the isolated material.

Comparisons of commonly used XAD resins have been published for the isolation of both fulvic acid (Aiken et al., 1979) and humic acid (Cheng, 1977) from water. These resins differ in pore size, surface area, polymer composition, and polarity (Table 5) (Kunin, 1977). As with anion-exchange resins, hydrophobic styrene-divinylbenzene resins (XAD-1, XAD-2, XAD-4) were found more difficult to elute than hydrophilic acrylic-ester resins (Table 6). This is due to hydrophobic interactions, and possible π–π interactions with the aromatic resin matrix of styrene-divinylbenzene resins. In addition, ki-

TABLE 5. Properties of XAD Resins Studied

Resin	Composition[a]	Average Pore Diameter A[a]	Specific Surface Area (m^2/g)[a]	Specific Pore Volume (cm^3/g)[a]	Solvent Uptake, g per g of Dry Resin[b]
XAD-1	Styrene-divinylbenzene	200	100	0.69	—
XAD-2	Styrene-divinylbenzene	90	330	0.69	0.65–0.70
XAD-4	Styrene-divinylbenzene	50	750	0.99	0.99–1.10
XAD-7	Acrylic ester	80	450	1.08	1.89–2.13
XAD-8	Acrylic ester	250	140	0.82	1.31–1.36

[a] *Kunin (1974).*
[b] *Parish (1977).*

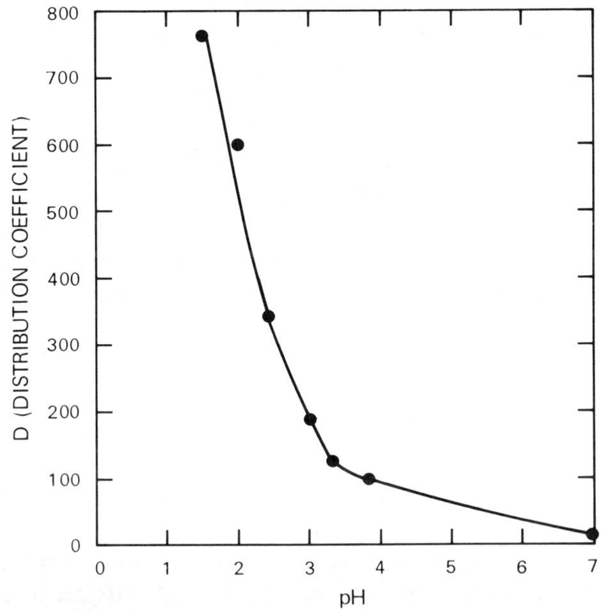

FIGURE 3. pH dependence of the distribution coefficient of fulvic acid on XAD-8.

netics of sorption of fulvic acid on these resins is slow, with diffusion into the resin being the rate-controlling step.

Acrylic-ester resins (XAD-7 and XAD-8) are more hydrophilic, wet more easily, and adsorb more water than styrene-divinylbenzene resins. Kinetics of sorption are much faster, and equilibrium is attained more rapidly. In addition, these resins have higher capacities and are more efficiently eluted than styrene-divinylbenzene resins when fulvic acid is the solute of interest. Because of serious bleed problems of XAD-7 with NaOH (Aiken et al., 1979), XAD-8 is preferred over XAD-7 for the isolation of fulvic acid.

TABLE 6. Distribution Coefficients and Elution Efficiency of Fulvic Acid on XAD Resins

Resin	Distribution Coefficient[a] K_D	Elution Efficiency (%)
XAD-1	475	70
XAD-2	515	75
XAD-4	332	70
XAD-7	1480	98
XAD-8	604	98

[a] *As measured at pH 2 by batch experiment.*

Cheng (1977) found XAD-12, a very hydrophilic XAD resin with weak-base functional groups, to be the best sorbent for humic acid. Because of precipitation of humic acid at low pH, pH 5 was found best for sorption. Fulvic acid, however, adsorbs more strongly at lower pH (Fig. 3), and pH 2 is recommended. Humic acid constitutes only about 5% of dissolved humic substances in water; for this reason, solution pH should be adjusted to pH 2 when isolating aquatic humic substances on XAD resins (Aiken et al., 1979).

The macroporous XAD resins, XAD-8 in particular, are excellent sorbents for humic substances. With the weak anion-exchange resins, such as Duolite A-7, these resins are the sorbents of choice for isolating and concentrating humic substances from water.

EXTRACTION SCHEMES

In isolating aquatic humic substances, it is more efficient to employ a variety of methods in order to yield a high-quality product, free of inorganic salts and low-molecular-weight organic acids, and in a form that will resist degradation. When used alone, none of the methods discussed in the previous section can yield this product. Used in combination, they can be powerful tools.

Any extraction scheme designed should incorporate the following steps to ensure a low-ash product:

1. *Filtration*. Sample should be filtered (≤0.45 μm) to separate dissolved humic substances from particular organic carbon and colloidal clays.
2. *Concentration*. Humic substances should be concentrated by an efficient method, such as sorption on XAD-8 or Duolite A-7.
3. *Isolation*. Humic substances should be isolated from inorganic salts and other organic solutes.
4. *Preservation*. Isolated humic substances should be freeze-dried to yield a stable, easy to handle product with good physical properties.

A scheme devised by Thurman and Malcolm (1981) uses XAD-8 to concentrate and isolate aquatic humic substances. According to this scheme, the sample is first filtered through 0.45 μm silver-membrane filters and acidified. After concentration on XAD-8 resin, the humic acid fraction is precipitated at pH 1. Both humic acid and fulvic acid fractions are hydrogen saturated by passing the sample through a cation-exchange resin in the H-form. These fractions are then freeze-dried to yield low-ash samples of aquatic humic and fulvic acids. This extraction scheme is outlined in Table 7. These authors successfully isolated humic substances from a number of surface and groundwaters. Even samples with DOC values of 0.7 mg C/L could be proc-

TABLE 7. Extraction Scheme Using XAD-8 to Concentrate Aquatic Humic Substances

1. Filter sample through 0.45 μm silver-membrane filter and lower pH to 2.0 with HCl.
2. Pass acidified sample through column of XAD-8; aquatic humic substances adsorb to resin.
3. Elute XAD-8 resin in reverse direction with 0.1*N* NaOH; acidify immediately to avoid oxidation of humic substances.
4. Reconcentrate on smaller XAD-8 column until DOC is greater than 500 mg C/L.
5. Adjust pH to 1.0 with HCl to precipitate humic acid. Separate humic and fulvic acids by centrifugation. Rinse humic acid fraction with distilled water until $AgNO_3$ test shows no Cl^- in washwater. Dissolve humic acid in 0.1*N* NaOH and hydrogen saturate by passing solution through cation-exchange resin in H-form.
6. Reapply fulvic acid fraction at pH 2 to XAD-8 column. Desalt fulvic acid by rinsing column with 1-void volume of distilled water to remove HCl and inorganic salts; elute fulvic acid by back-elution with 0.1*N* NaOH.
7. Hydrogen saturate fulvic acid fraction by immediately passing 0.1*N* NaOH eluate through cation-exchange resin in H-form. Continue cation-exchange process until final concentration of Na^+ is less than 0.1 part per million.
8. Freeze-dry humic acid and fulvic acid fractions.

essed by using multiple adsorption–desorption cycles on XAD-8. Ash contents of the isolated materials were low, typically 1% or less. This scheme is specific for hydrophobic organic acids in water, the majority of which are humic substances. It is straightforward and simple and is highly recommended by this author.

A more general scheme that concentrates and fractionates all the organic constituents from water has been outlined in detail by Leenheer and Noyes (1983). This method also combines filtration, adsorption chromatography, ion exchange, and lyophilization (Fig. 4). Aquatic humic substances are concentrated with other hydrophilic organic acids and inorganic salts. The method presented by the authors for the fractionation of organic acids in the A-7 eluate yields five fractions, which include two humic acid fractions and one fulvic acid fraction. However, this method is complicated, combining rotary evaporation, centrifugation, ion exchange, and adsorption chromatography to separate each fraction. This procedure could be simplified if humic substances were the only solutes of interest.

One particular advantage of this extraction scheme is that large volumes of the sample can be processed on site with no sample manipulation. The sample is pumped through a 0.3 μm Balston microfiber filter tube (glass-fiber filter) directly onto the column array. This scheme fractionates the organic matter present in the water into hydrophobic and hydrophilic acid, base, and neutral fractions. Suspended sediment is also retained, and an extraction procedure for this material is presented. Organic matter from surface water

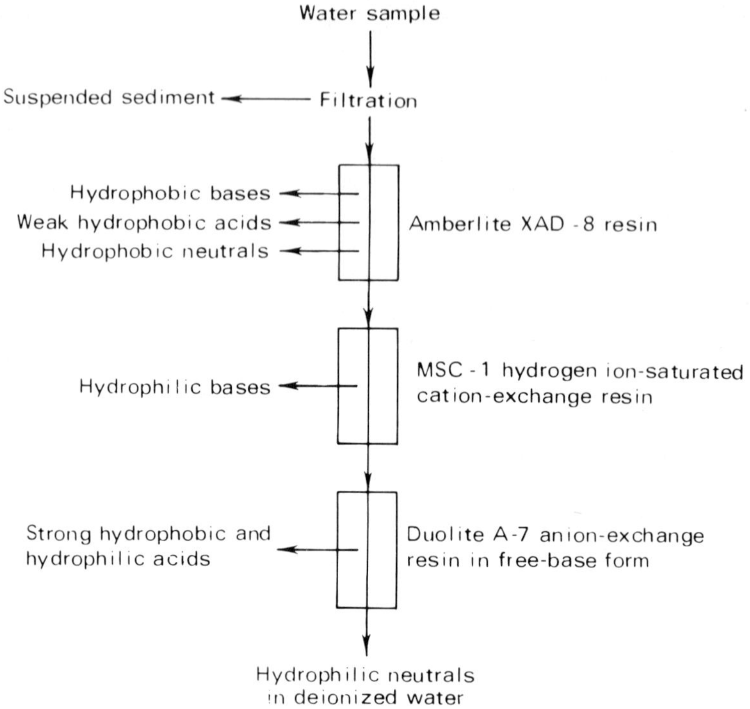

FIGURE 4. Fractionation of organic solutes in water by the method of Leenheer and Noyes (in press).

and groundwater has been extracted using this procedure with good results (Leenheer and Noyes, 1984). It is particularly useful to those interested in the comprehensive study of organic compounds in water.

CONCLUSIONS

During the last 15 years, interest in the study of aquatic humic substances has increased. The problems associated with isolating and concentrating this material from aqueous solution largely have been overcome, and humic substances can be easily extracted from any aquatic sample. Humic substances have been successfully isolated from waters with very low DOC values, such as seawater and groundwater, as well as more concentrated systems.

Advantages and disadvantages of the commonly used methods to isolate and concentrate aquatic humic substances have been presented in this chapter. For most waters, the process of producing low-ash humic material involves filtration, concentration, isolation of humic substances from inorganic and other organic solutes, and lyophilization. Development of synthetic, macroporous resins, both nonionic resins and weak anion-exchange resins, has increased the efficiency of concentrating and isolating aquatic humic substances. XAD-8 and Duolite A-7, in particular, are highly recommended by this author.

ACKNOWLEDGMENTS

Thanks are due to the staff of the U.S. Geological Survey Library, Lakewood, Colorado, for their assistance in locating references, and to other Geological Survey staff members in the preparation of this chapter.

CHAPTER FIFTEEN

Fractionation of Soil Humic Substances

ROGER S. SWIFT

ABSTRACT

Successful fractionation of humic substances extends our knowledge of their molecular properties, assists in their characterization, and aids in the meaningful application of analytical techniques.

A wide range of procedures has been used to achieve fractionation. Classical methods of fractionation involve the adjustment of pH and the addition of salts, organic solvents, or metal ions. More recently, very good fractionations have been carried out on the basis of molecular size differences, using gel permeation chromatography, ultrafiltration, and centrifugation. Similarly, electrophoresis isoelectric focusing and isotachophoresis have been used to produce fractionations on the basis of electrical charge. Adsorption onto a variety of media followed by selective desorption has also proved a useful technique.

In most instances, recent developments in techniques or in the support media available have led to improvements in the fractionations which can be achieved. While there is always room for improvement and refinement, the currently available procedures produce good results if properly understood and industriously applied.

INTRODUCTION

Fractionation Versus Purification

Following the extraction of humic substances from soil media it is necessary to purify the humic substances by separating them from the nonhumic substances. Used in this sense purification is the removal of materials such as carbohydrates, proteins, lipids, low-molecular-weight compounds, and so on which have been co-extracted with the humic substances. This whole process is called isolation and is dealt with in Chapter 13 by Hayes.

Fractionation, on the other hand, is the subdividing of humic substances according to some property related to their molecular composition. Because humic substances are ill defined there will inevitably be some confusion between these processes of purification and fractionation, especially since the same or very similar techniques are used in both cases. Nevertheless, the distinction between the two should be clearly made and understood and adhered to by research workers in this field.

There are some who still approach the fractionation of humic substances with the objective of being rewarded by the isolation of one or more, pure, identifiable compounds. Historically this was a reasonable and laudable objective and many eminent researchers joined in the search. Nowadays, insofar as soil humic substances are concerned, that band is largely made up of the uninitiated. In any case their search, if they choose to continue it, is almost certain to be futile.

Among experts in the field it is now generally accepted that the term humic substances is a generic name referring to a family of macromolecular substances which, although they have a similar origin, structure, and composition, exhibit a wide range of molecular properties. Thus, there is a broad spectrum of related molecules, each one differing almost imperceptibly from the next in terms of one or other of its properties. If this is truly the case, then it is wrong to expect the isolation of a pure compound. It should also be clear that the most that can be expected of a fractionation procedure is to decrease the heterogeneity of the system as much as possible.

The reader should be aware that this chapter is not intended to be a comprehensive review of the literature; rather I have attempted to distill and condense my own views gained from carrying out research into and reading the literature on this subject over a number of years. For coverage of much of the literature, use has been made of the relevant sections of the excellent selection of books, reviews, and monographs on humic substances which have appeared in the last decade or so. Those unfamiliar with the literature will find any number of references by consulting these works (Dubach and Mehta, 1963; Kononova, 1966; Stevenson and Butler, 1969; Schnitzer and Khan, 1972; Flaig et al., 1975; Hayes and Swift, 1978; Schnitzer, 1978; Stevenson, 1982).

Reasons for Fractionating

A significant outcome of the application of fractionation procedures has been the finding that humic substances tend to exhibit a range of values for any given molecular property. Consequently, one of the main reasons for carrying out a fractionation is to determine the range of variation found for properties such as molecular weight, functional group content, elemental composition, and so on. In some instances (e.g., molecular weight) the extent of variation of a property has been found to be so great that an average or mean value for that property conveys little information as to the true situation within the system as a whole.

A second reason is that the measurement of many chemical or physical parameters is made very difficult if the molecules being studied exhibit a wide range for the particular property being measured. For instance, the measurement of molecular weight by colligative properties to give a number average value is a good example. The presence of even a moderate amount of unwanted low-molecular-weight impurity or contaminant can greatly influence the result obtained and give a misleadingly low value. These "impurities" may be low-molecular-weight inorganic compounds, nonhumic organic compounds, or may even be small fragments of humic molecules formed as artifacts during extraction. As a consequence and in order to obtain more meaningful values, it is necessary to remove contaminants and to carry out measurements on well-fractionated samples that exhibit a much narrower variation in the property being measured.

Fractionation procedures have been used as a preliminary step to spectral measurements, elemental analyses, functional group analyses, measurements of charge, charge density, viscosity, and so on. However, adequate fractionation has not as yet been extensively applied prior to carrying out chemical degradation reactions.

A third application for fractionation procedures is their use as characterization or fingerprint techniques to monitor the effect of some other chemical or biological treatment. In this regard gel permeation chromatography has been particularly useful.

Physical and Chemical Properties Used for Fractionation

The use of fractionation techniques has greatly enhanced our knowledge and understanding of humic substances and given greater reliability to and confidence in the data obtained.

As with other biological macromolecules a wide range of techniques has been used for fractionation but generally these exploit physicochemical differences in solubility, reactions with metal ions, molecular size, charge or charge density, and adsorption characteristics. The list of properties exploited for fractionation is unlikely to change greatly but the refinement and sophistication of the techniques will almost certainly continue to improve.

FRACTIONATION ON THE BASIS OF SOLUBILITY AND PRECIPITATION

Use of pH

Given the relatively unsophisticated nature of the techniques available to them, it is not surprising that early workers experienced so much difficulty in coming to terms with humic substances. Indeed, it is perhaps surprising that they made as much progress as they did. Inevitably the fractionation procedures adopted by these early workers were based on solubility properties and utilized precipitation techniques, particularly those based on adjustment of pH or the addition of metal ions. This early work has been extensively reviewed by Kononova (1966) and more recently has been concisely and astutely summarized by Stevenson (1982).

It was during the period 1780 to 1930 that the terms used to name fractions of humic substances were coined. Among those appearing in the literature are (in no particular order) humic acid, fulvic acid, humin, ulmic acid, ulmin, crenic acid, apocrenic acid, hymatomelanic acid, gray humic acid, brown humic acid, and so on (see Stevenson, 1982, Chapter 2, for a discussion on nomenclature). The multiplicity of names reflected the belief of many early workers that they were dealing with and searching for a number of discrete, identifiable, individual compounds. It is in turn a reflection of our greater understanding of humic substances that most of these names are not now used. Many workers now use only the terms fulvic acid, humic acid, and humin and often in practice only the first two of these are required. In certain countries historical loyalty to other fractions persist so that the terms gray and brown humic acid can still be found in the literature.

As is now well known to anyone acquainted with studies of humic substances the definitions relating to the main fractions, particularly when they are obtained by alkaline extraction of a soil, are as follows:

Humin. Insoluble in alkali, insoluble in acid.

Humic Acid. Soluble in alkali, insoluble in acid.

Fulvic Acid. Soluble in alkali, soluble in acid.

A flow diagram outlining the interrelationship of these three fractions is shown in Figure 1 which is taken from Hayes and Swift (1978). It should be clear from the statements made in the first section of this chapter that these are rather arbitrary delineations that provide us with no more than a gross first-stage fractionation.

Although the definitions of humic acid and of other fractions were initially based on extraction of soil with alkaline reagents, the same terminology is used when the extractants are neutral, acidic, or organic. In these cases the term humin means "not extracted," humic acid means "soluble in the ex-

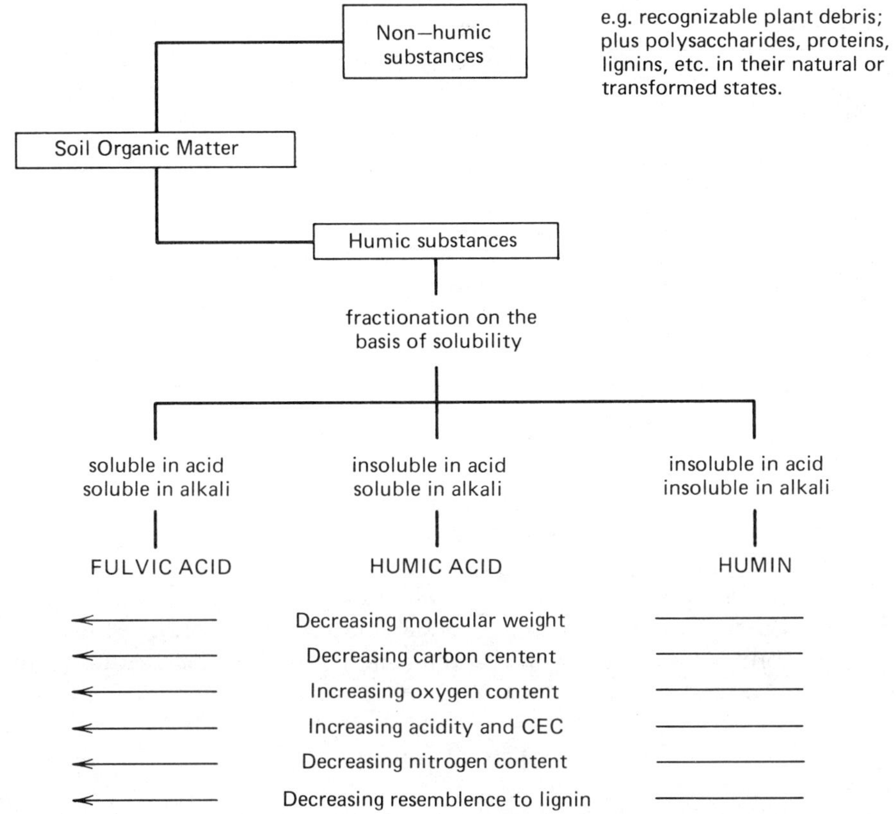

FIGURE 1. Fractionation of soil organic matter and humic substances, showing some property variations (from Hayes and Swift, 1978).

tractant used but precipitated on adjustment to pH = 1," and fulvic acid means "remains soluble at pH = 1." This use of the terminology does not conflict too greatly with the classical distinctions because the materials obtained in this way seem to fit largely with those definitions also. However, it is as well to be aware of the distinction.

Normally the humic acid fraction is precipitated at pH = 1.0 but other workers (Flaig et al., 1975) have chosen pH values of 1.5 or 2.0 in order to decrease the acidity of the precipitation medium. It is also possible to fractionate humic acid by precipitating material at intermediate pH values, for example, pH = 4.8 (Hobson and Page, 1932; Waksman, 1936). The fractionation achieved does not appear to be particularly good probably because of the amount of co-precipitation of one fraction with another.

Solubility can also be used in the reverse way, that is, to gradually extract materials sequentially with increasingly powerful extractants. This has been done by changing pH and the nature of the extractant anion (Posner, 1966) or by using a range of solvents such as that employed by Hayes et al. (1975).

TABLE 1. Data for Successive Extraction of Humic Substances from a Soil[a]

Extractant	Yield (% of Total OM) Humic Acid	Fulvic Acid	Cumulative Total	pH Value of Extractant
Water	0.0	2.8	2.8	—
DMF	15.0	2.2	20.0	6.8
Sulfolane	4.1	1.0	25.1	3.7
DMSO	0.7	0.2	26.0	5.9
Pyridine	14.8	0.6	41.4	11.6
EDA	23.2	6.3	70.9	13.0

[a] *From Hayes et al. (1975).*

Such procedures are covered in more detail in Chapter 13 by Hayes in this book but some of the results obtained are shown in Table 1.

The manipulation of pH as a technique for purification and crude fractionation is likely to remain popular since it allows one to handle relatively large amounts of material rapidly and gain a substantial fractionation. However, because of problems of co-precipitation it is never likely to be used to obtain well-defined fractions with a narrow range of properties.

Salting-Out

Like any other charged macromolecule (polyelectrolyte) the behavior of humic substances in solution is strongly influenced by the presence and concentration of background electrolyte (salt). The observed effects are largely attributable to the way in which the structure of the diffuse double layer of charged ions surrounding the polyelectrolyte changes with background salt concentration. At very low concentrations of background electrolyte the diffuse double layers extend some distance from the surface of the charged macromolecule. When the macromolecules in solution approach one another, intermolecular charge repulsion forces predominate, the molecules repel one another and the polymer remains in dispersion.

If the background electrolyte concentration is increased the extension of the double layer is suppressed much closer to the surface of the molecule. This allows the macromolecules to approach one another more closely so that intermolecular attractive forces predominate and coagulation or precipitation can occur. Suppression of charge interaction to the extent where precipitation occurs is referred to as *salting out*. The general theory of the behavior of charged macromolecules in electrolyte solutions (Tanford, 1961) is important to understanding their behavior in different environments and is vital to the interpretation of many physicochemical measurements. Thus, many of the techniques used for studying and fractionating humic substances

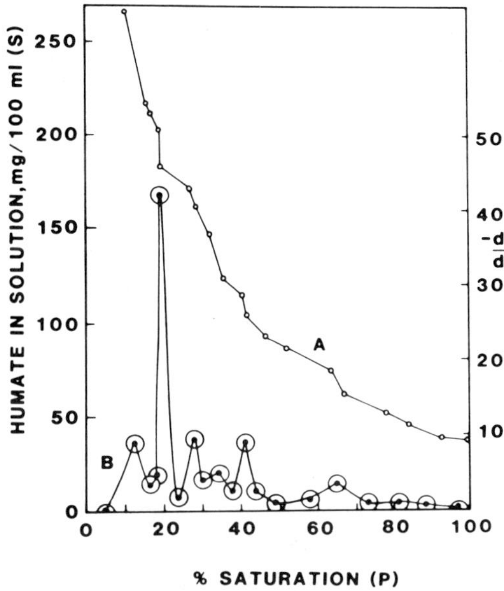

FIGURE 2. Relationship between percentage saturation with ammonium sulfate and the amount of humate remaining in solution: (A) salting-out curve, pH 7.0; (B) differential salting-out curve. [Adapted by Stevenson (1982) from Theng et al. (1968).]

will be better appreciated with an adequate knowledge of polyelectrolyte behavior.

The best known use of salting-out as a method of fractionation is the splitting of humic acid into gray and brown humic acid fractions by the addition of a salt, usually KCl, to a solution of humic acid (Springer, 1938). Theng et al. (1968) obtained a useful fractionation by salting-out using ammonium sulfate at pH = 7. The results obtained are shown in Figure 2. In general, salting-out is unlikely to be used to produce fine fractionations because of the indistinct nature of the boundary conditions and co-precipitation problems.

Use of Metal Ions

It has been known since the earliest studies that humic substances formed insoluble salts with a wide range of metal ions. Many di-, tri-, or tetravalent ions will bring about the precipitation of a greater or lesser amount of humic substances from solution. An early example of this was the use of copper to precipitate apocrenic acid and crenic acid (Berzelius, 1839). In modern times the technique has been used with some success by Dubach et al. (1961) and Sowden and Deuel (1961).

Perhaps because of the ease of precipitating humic acid by the adjustment of pH, the use of metal ions for fractionation has not received great attention. The fractional precipitation of humic acid by utilizing either gradually increasing concentrations of the metal ion causing precipitation or by changing the identity of the precipitating ion may well warrant further investigation. Perhaps, like fractional precipitation using pH adjustment and salting-out procedures, the use of metal ions would give rather gross fractionations. A more exciting prospect is use of chromatographic media onto which are bound metals such as Zn(II), Cu(II), and Ni(II) (e.g., onto Sepharose 6B, produced by Pharmacia). Such materials are almost certain to react with humic substances and could provide a useful means of fractionation if they could be successfully re-eluted under mild conditions.

Use of Organic Solvents

Organic solvents have long been used for extraction and sequential extraction, which is fractionation of a sort (Flaig et al., 1975; Schnitzer, 1978). While the direct use of organic solvents in fractionation has not been widespread, nonetheless, the technique has received some attention. For instance, the separation of hymatomelanic acid from precipitated humic acid is obtained by extraction with ethanol (Oden, 1919). Ethanol has been used to bring about fractional precipitation by addition to alkaline solutions of humic acid (Kyuma, 1964; Kumada and Kawamura, 1968). There is no reason why other water-miscible solvents such as acetone and methanol should not be used in this way. Solvents that are highly immiscible with water (e.g., hexane and benzene) do not appear to remove any substantial fraction of humic substances. These are perhaps best used to remove nonhumic substances (such as fats and waxes) prior to extraction. However, recent work by Allen and MacCarthy (personal communication) has shown that more polar water-immiscible solvents, such as methyl isobutyl ketone and diethyl ether, can be used successfully to purify and fractionate humic substances.

Any fractionation obtained by the use of organic solvents is again likely to be rather crude and the method is not likely to find great favor as a preparative technique. Nonetheless, solubility in nonaqueous solvents can be a very useful adjunct to the chemists armory when physicochemical measurements (such as molecular weight and viscosity) need to be made. In this context, lower-molecular-weight, hydrogen-ion saturated humic substances are most likely to be useful.

FRACTIONATION ON THE BASIS OF MOLECULAR SIZE

Measurement of the molecular weights of humic substances has been the subject of a considerable amount of work and this is dealt with in more detail by Wershaw and Aiken in Chapter 19 of this book. In this section the emphasis will be placed on the application of techniques that use molecular size as a basis for fractionation rather than the determination of molecular weight itself.

Gel Permeation Chromatography

Since its introduction some years ago gel permeation chromatography has become a powerful tool in the study of naturally occurring polymers. While primarily devised and used for studying proteins, the technique has been applied to a wide variety of materials and has been used in the study of humic substances since the early 1960s. Gel permeation chromatography is a rapid, cheap, and very versatile technique. It can be used as a method for separation, purification, and fractionation as well as for determinations of molecular weights and molecular weight distributions of polymer systems. A review of the principles and applications of the technique is provided by Fisher (1969).

Although inorganic materials, such as porous glass beads, have been used, the gels most commonly employed consist of cross-linked polymers (e.g., polysaccharides, polystyrene, and polyamides) in the form of small beads or granules. The gel structure is perfused by a system of pores and the size of these is determined by the degree of cross-linking in the polymer. These pores enable the gel to act as a chromatographic medium giving separations based on differences in molecular size. When a solution, containing a mixture of molecules of varying sizes, is applied to the top of a gel column and eluted with solvent, those molecules which cannot enter the pores in the beads will pass between the beads and will be eluted first from the column. Molecules smaller than the pore sizes of the gel will enter the pores and their passage through the column will be retarded. The extent to which this occurs will depend on the actual size and shape of the molecule, but the net result is that the solute molecules are eluted from the column in order of decreasing molecular size and, for a given polymer, decreasing molecular weight. Gels are available which operate over different molecular weight ranges and have different exclusion limits. By utilizing a range of gels it is possible to determine molecular weight values ranging from several thousands to millions (Table 2).

Gel permeation chromatography has been extensively and successfully applied to studies of humic substances. However, a number of problems are encountered which, if not overcome, can invalidate the results. For instance, the gel material should be inert to the solute molecules so that there are no chemical or physical interactions between gel and solute. When any adsorption of the applied polymer molecules by the gel takes place the observed retention by the column is not solely caused by penetration into the pores, and the resulting separation cannot be entirely attributed to molecular weight differences. Because of their chemical composition, humic substances tend to be readily adsorbed by gel materials. Adsorption behavior

TABLE 2. Range of Sephadex and Sepharose Gels Manufactured by Pharmacia Showing Their Fractionation Range and Exclusion Limit

Gel Type and Grade	Fractionation Range[a] (Molecular Weight) Proteins	Polysaccharides
Sephadex		
G-10	– 700	– 700
G-15	– 1500	– 1500
G-25	1000– 5000	100– 5000
G-50	1500– 30,000	500– 10,000
G-75	3000– 80,000	1000– 50,000
G-100	4000–150,000	1000–100,000
G-150	5000–300,000	1000–150,000
G-200	5000–600,000	1000–200,000
Sepharose		
2B	7×10^4–40×10^6	1×10^5–20×10^6
4B	6×10^4–20×10^6	3×10^4– 5×10^6
6B	1×10^4– 4×10^6	1×10^4– 1×10^6

[a] *In each case the upper figure represents the exclusion limit for the gel.*

shows up typically as a peak or a substantial amount of sample being eluted after the total column volume (Swift and Posner, 1971). Such behavior is illustrated by the elution patterns shown in Figures 3 and 4. Figure 3 shows that the use of sodium chloride solution as eluent leads to a substantial amount of reversible adsorption as indicated by the large amount of material eluted after the total column volume (V_t). The final peaks, eluted after V_t in each pattern shown in Figure 4, indicate that reversible adsorption occurs

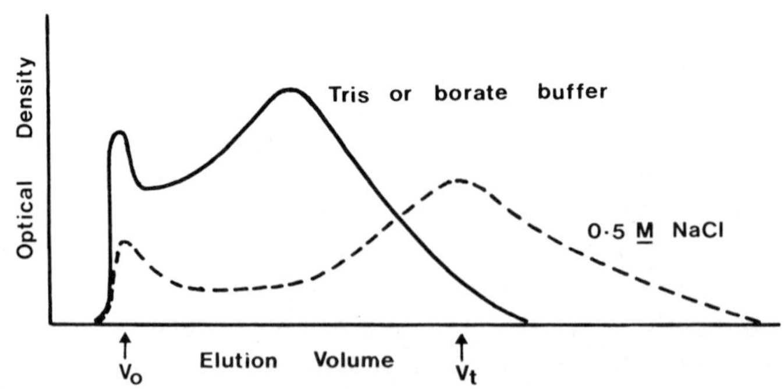

FIGURE 3. Gel permeation chromatography of humic acid on Sephadex G-100 showing the effect of eluent on adsorption.

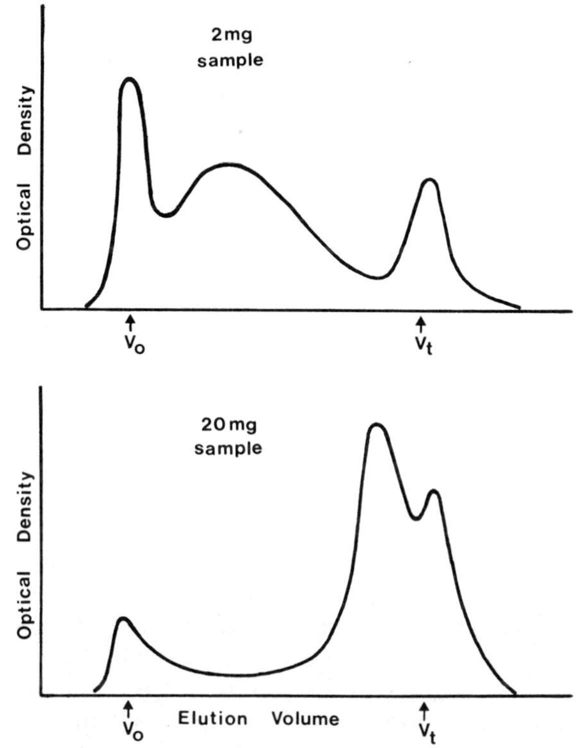

FIGURE 4. Gel permeation chromatography of sodium humate on Sephadex G-100 using water as eluent showing the effect of sample size on the elution pattern.

when water is used as eluent. In both of the systems cited above, irreversible adsorption was also observed (Swift and Posner, 1971).

A further problem arises as the result of charge interactions between residual charged groups on the gel and those on the humic substances leading to attractive or repulsive forces between the gel and charged humic substances. If not suppressed these charges interfere with the separation which again would not take place solely on the basis of molecular size differences. This type of interaction is most likely to occur when water is used as eluent, and typical behavior is shown in Figure 4. It can be seen that at low sample concentration the charge repulsion between the gel and humic material leads to a large amount of the sample being excluded at the void volume (V_0). When the sample size is increased the charge repulsion effects between gel and humic material are somewhat suppressed, and the amount of sample excluded is decreased. The elution patterns shown in Figure 4 are the result of a complex interaction of reversible adsorption, charge interaction, and molecular size fractionation.

In addition, when water is used as eluent and the sample applied contains electrolyte, then a "salt-boundary" effect can occur (Posner, 1963). A typi-

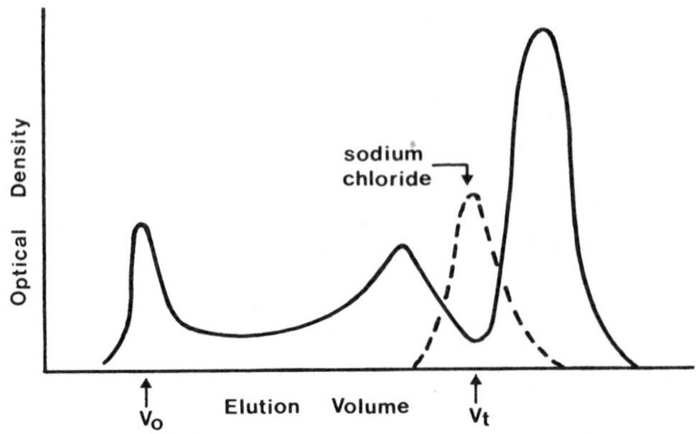

FIGURE 5. Gel permeation chromatography of sodium humate on Sephadex G-100 using water as eluent but with NaCl added to the sample showing the salt-boundary effect.

cal elution pattern is shown in Figure 5, and while fractionation occurs, it is not entirely on the basis of molecular size. In particular, the final peak, occurring after the salt boundary and after the total column volume, consists of material effectively trapped behind the salt layer at the commencement of fractionation. If this peak is collected and reapplied to the column, it will not be eluted at the same position, and therefore it was not initially subjected to fractionation on the basis of molecular size. This technique has been used by a number of workers without a full understanding of the processes taking place.

Swift and Posner (1971) discuss these problems fully and show that they can be largely overcome by careful selection of the gel matrix and by the use of appropriate buffer solutions. Use of a buffer containing a large organic cation such as tris [2-amino-2(hydroxymethyl) propane-1,3-diol], or alternatively borate or some other suitable buffer, is recommended. Even when such procedures are used, there is some indication that a small amount of interaction between gel and solute can still take place, particularly in the cases of the very high-molecular-weight, less-soluble humic acid fractions. Despite these handicaps gel chromatography has proved to be a particularly useful technique for the purification, fractionation, and determination of the molecular size of humic substances. Typical elution patterns obtained by the proper application of gel permeation chromatography (e.g., Dubach et al., 1964; Swift and Posner, 1971) should be consulted. The elution pattern, shown in Figure 3, using borate or tris buffer is an example of the type of curve that should be obtained.

When selecting gels for fractionation work, consideration should be given to the molecular weight ranges for which the gels are suitable and, in particular, to their upper exclusion limits. The manufacturers of gels supply figures for these properties and such values have been quoted extensively when reporting values for the gel chromatography of humic substances. The manufacturers' values are obtained by calibrating the gels with proteins or polysaccharides of known molecular weights (Table 2). It has been shown, however, that the calibrations from humic acid fractions (Cameron et al., 1972a) differ significantly from those based on proteins which tend to have tightly coiled, globular molecular configurations. There was reasonable agreement with some calibrations obtained using polysaccharides which tend to have less compact, randomly coiled molecular configurations.

Many studies using gel permeation chromatography have confirmed the polydisperse nature of humic substances and showed that they cover a wide range of molecular weight values. Cameron et al. (1972b) using gel permeation chromatography separated humic acid fractions ranging in molecular weight from 2000 to 1,500,000 and showed that the most abundant portion of the molecular weight distribution for a sodium hydroxide-extracted humic acid was around 100,000. This wide range of molecular weights within a single sample can present difficulties in choosing a suitable gel. A large-pore gel with a high molecular exclusion limit (e.g., Sepharose 2B) retains most of the sample, but the resolution at lower-molecular-weight values will be very poor. Conversely, a small-pore gel with a lower exclusion limit (e.g., Sephadex G-25) will exclude a major portion of the sample. This problem can be overcome by successively using gels of various exclusion limits and reapplying the excluded or included portion from a given gel to another with a higher or lower exclusion limit (Schnitzer and Skinner, 1968b). The author has found the range of Pharmacia gels Sephadex G-25, G-75, and G-200 and Sepharose 6B to be a particularly useful series for work with humic substances (Table 2).

That many recent studies on humic substances utilize gel permeation chromatography as a central or supporting technique (see, e.g., Kolesnikov, 1978; Chakraborty et al., 1979; Goh and Williams, 1979; Danneberg, 1981; Dawson et al., 1981; Gonzalez et al., 1981; Ruggiero et al., 1981) attests to its usefulness in this type of work. Because of its simplicity, and since it can be used both as a preparative and analytical technique, gel chromatography is certain to remain a useful tool in studies of humic substances.

Ultrafiltration

Another recent advance in the handling of biological macromolecules has been the development of membrane filters. Using a variety of polymer materials and manufacturing processes, filters can be prepared which have a known, controlled pore size ranging from several micrometers to a few nanometers in diameter. Membranes with pore sizes at the larger end of the range are used conventionally to arrest the passage of small particles or microorganisms; this is referred to as microfiltration. Membranes with pore sizes at the lower end of the range can be used to filter molecules in solution on the basis of molecular size and this process is referred to as ultrafiltration. Membranes are manufactured by a variety of companies (e.g., Amicon, Millipore, Sartorius) with nominal molecular weight cut-off values ranging from 50 to 1,000,000 with a large number of cut-off values in-between. The pore size within these membranes is not completely uniform so that the molecular weight cut-off is not as sharp as might be imagined. A convention which appears to have been adopted by manufacturers is that a membrane with quoted molecular weight cut-off will retain 90% or more of spherical, uncharged solute molecules of that molecular weight. As well as this uncertainty in the accuracy of the cut-off point, the actual molecular weight value at which it operates for a given substance will depend upon the charge and molecular configuration of that substance. It has been observed that charge–charge interactions between the solute and the membrane can interfere with the filtration process so that it is no longer based solely on molecular size. Given the highly charged nature of humic substances and unresolved doubts about their molecular configuration, the cut-off values quoted by the manufacturers should be used with caution.

Experimentally, ultrafiltration is a very simple technique. Typically a solution of the sample is placed in a pressure cell with a membrane at the bottom. Pressure is applied to the cell by means of an inert gas and the solution is stirred by means of a magnetic stirrer bar suspended just above the membrane (Figure 6). This prevents concentration polarization and clogging of the membrane which can result if solute molecules are allowed to accumulate at the surface of the membrane. Nevertheless, leakage of high-molecular-weight material tends to occur as the solute concentration increases during ultrafiltration (Buffle et al., 1978).

Ultrafiltration is a most useful technique which, by the use of a suitable series of membranes, allows the rapid fractionation of relatively large quantities of humic substances. The fractionation can be carried out either in order of ascending or descending molecular weight. By choosing a membrane with a low-molecular-weight cut-off value, ultrafiltration can be used for desalting and for concentration. As such it is far superior to dialysis.

While ultrafiltration has been used extensively by water chemists in the isolation of humic substances (Schnitzer, 1978), its use with soil materials has been more limited (Cameron et al., 1972b; Wake and Posner, 1967). From the foregoing discussion it should be clear that, as a technique for preparative fractionation and desalting, ultrafiltration is very attractive and will rival gel permeation chromatography in this type of work. Only time and a considerable amount of effort will tell which is the superior technique.

Ultrafiltration does not lend itself so readily to the determination of molecular weight and molecular weight distribution, and gel permeation chromatography will certainly retain its role as an analytical tool to measure these particular properties of a sample.

Centrifugation

When used properly, ultracentrifugation continues to be our major means of determining molecular weight values for humic substances using sedimentation velocity and other techniques. Indeed, centrifugation studies with humic substances have usually centered upon molecular weight measurements.

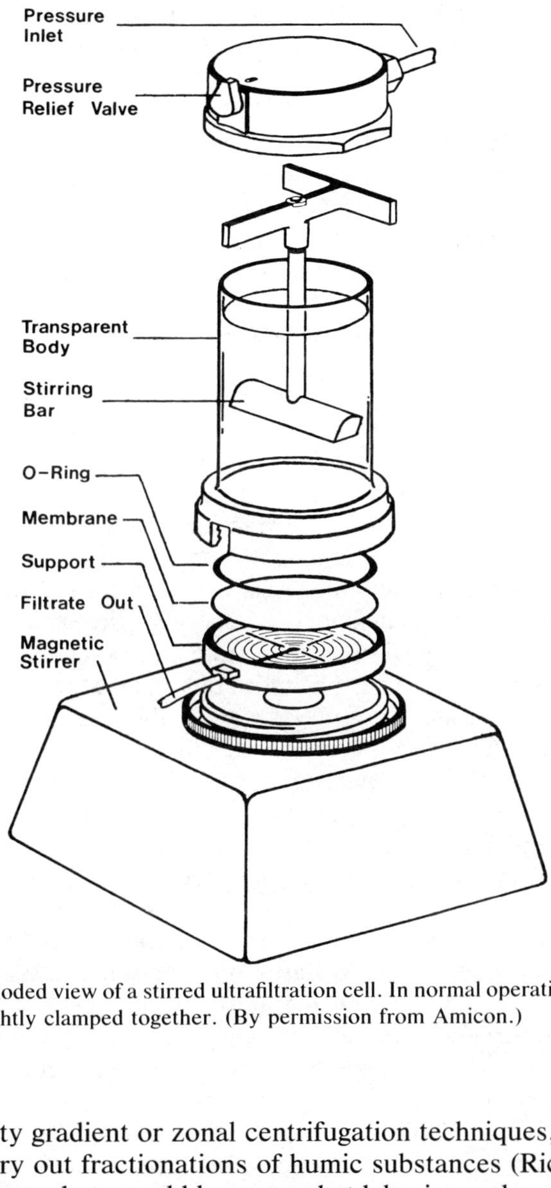

FIGURE 6. Exploded view of a stirred ultrafiltration cell. In normal operation the cell components would be tightly clamped together. (By permission from Amicon.)

By using density gradient or zonal centrifugation techniques, however, it is possible to carry out fractionations of humic substances (Rickwood, 1978). Although the procedure would be somewhat laborious when compared with gel chromatography and ultrafiltration it would be most useful to obtain fractions by a completely different technique to enable us to assess more reliably the authenticity of results obtained by simpler methods.

In any centrifugation study with humic substances the suppression of intermolecular charge repulsion by the addition of electrolyte is essential (Cameron et al., 1972b; Hayes and Swift, 1978). Any ultracentrifugation studies, analytical or preparative, where this has not been properly done should be disregarded.

FRACTIONATION ON THE BASIS OF CHARGE CHARACTERISTICS

The presence of charge resulting from the ionization of functional groups is a fundamental property of humic substances. To some extent, it is this property that is exploited when humic acid is precipitated by acidification. The same property can be more exquisitely exploited by means of ion-exchange and electrophoretic techniques, and these techniques have been extensively used in the fractionation of humic substances.

Electrophoresis and Electrofocusing

Electrophoresis is the term used to describe the movement of charged solute molecules in an electric field. Simple electrophoretic systems in which biological polymers were dissolved in buffer systems, placed on a variety of support media, and subjected to several hundred volts potential difference have yielded excellent results in the fractionation of proteins and polysaccharides.

As a consequence, the technique was taken up with some enthusiasm by those working with humic substances. Generally, the sample is dissolved in an alkaline buffer and placed on a support medium such as cellulose or glass paper in flat beds and glass or gel beads in column systems. The earlier work is well reviewed by Flaig et al. (1975). As with other fractionation procedures discrete fractions are not obtained, but rather there is a gradation of properties with one fraction merging into another. In general, high-molecular-weight gray humic acids migrate very slowly, brown humic acids migrate more quickly, and fulvic acids migrate more rapidly still. This electrophoretic behavior supports the view that the observed sequence is composed of molecules of increasing charge densities and decreasing molecular weights.

Many workers have observed fluorescent areas or fractions during electrophoresis experiments (e.g., Waldron and Mortensen, 1961), and this behavior is usually associated with the more mobile materials. It is not clear whether the fluorescence is due to the presence of closely associated nonhumic components, or is an innate characteristic of particular fractions of humic substances. It is possible that all fractions do in fact fluoresce, but that this fluorescence is masked by the intense light absorption of the gray-brown components in fractions which do not emit measurable fluorescence.

In general, the fractionation obtained by this traditional type of electrophoresis is rather disappointing and usually inferior to that which could be obtained on the basis of molecular weight by the techniques outlined in the previous section of this chapter. As a result, the use of electrophoresis with humic substances has declined somewhat in popularity.

More recently there have been significant developments in electrophoretic techniques and interest in the application of these techniques to humic substances is likely to be rekindled. The techniques in question are polyacrylamide gel electrophoresis (PAGE), isoelectric focusing (IEF), and isotachophoresis (ITP).

In PAGE a polyacrylamide gel is used as the support medium; the sample is simultaneously subjected to fractionation on the basis of charge by electrophoresis and on the basis of molecular weight by gel permeation chromatography. The experimental conditions can be modified by altering the buffer, the exclusion limit of the gel, by changing the composition of the gel (and thereby its exclusion limit) along the path of the sample, by using mixed gels (usually polyacrylamide and agarose), by using swamping amounts of charged or uncharged detergents, and by having a series of buffers partitioned or stacked along the path of the sample. Thus, there are a wide number of variables that can be altered to optimize the fractionation obtained.

In isoelectric focusing a pH gradient is set up within the gel support by incorporation of a range of relatively low-molecular-weight amphoteric substances (usually mixtures of synthetic polyaminopolycarboxylic acids with molecular weights in the range 300–600) called ampholines. Having formed the pH gradient, the ampholines themselves settle at their own isoelectric point and do not migrate any further. In this technique a gel support medium with a very large pore dimension is chosen so that fractionation on the basis of molecular weight does not occur. Assuming that the correct pH range has been chosen, then when a macromolecule is subjected to electrophoresis in the system, it will migrate to the pH of its isoelectric point and then cease to move. It should be noted, however, that there is a possibility of interactions between humic substances and the ampholine molecules. This could influence the nature of the fractionation obtained.

Isotachophoresis is similar to isoelectric focusing but includes the use of additional ampholines or multiphasic buffer systems to act as spacers to improve the separation and resolution. Each of these techniques can be run using columns, tubes, thin layers, or slabs and are comprehensively dealt with by Andrews (1981).

All these techniques have recently been applied to soil humic substances (Cacco et al., 1974; Castagnola et al., 1978; Gonzalez et al., 1981, Kasparov et al., 1981; Curvetto and Orioli, 1982; Orioli and Curvetto, 1982), often in association with gel permeation chromatography. Some useful fractionations have been obtained by these workers (as illustrated in Fig. 7) although one can still get the impression that some workers continue to be disappointed at not isolating discrete compounds. It can be argued whether the concept of an isoelectric point is tenable for humic substances and consequently whether the molecules are subject to isoelectric focusing or simply precipitating at the given pH value.

Recent experience with these techniques by the author indicates that they are a very promising addition to the fractionation armory. At the present time they are more likely to be used analytically for characterization or fingerprinting or to monitor changes arising from other treatments or proce-

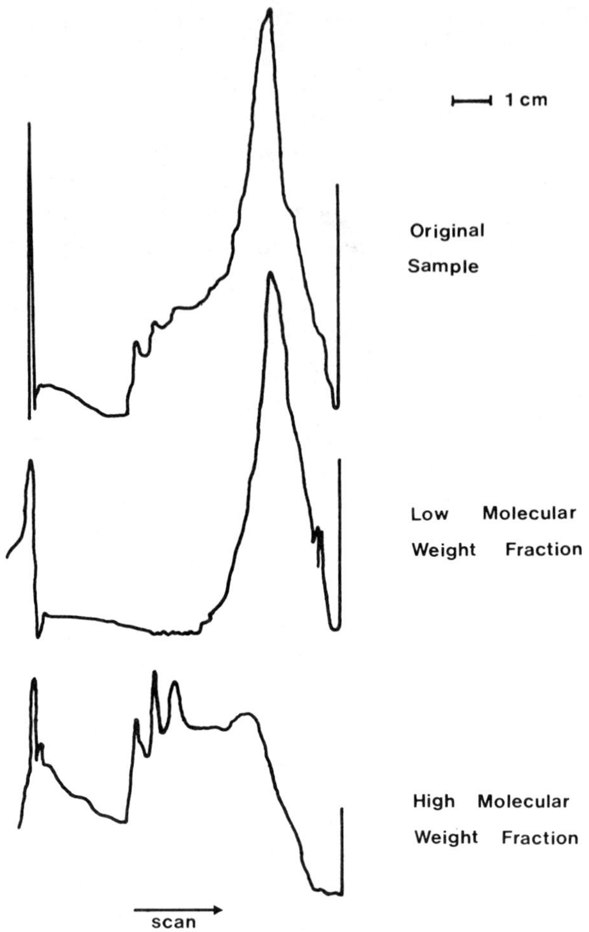

FIGURE 7. Densitometric traces of the isotachophoretic separation of a humic acid sample and two molecular weight fractions obtained from it (from Curvetto and Orioli, 1982).

dures rather than as preparative techniques. However, if problems of heating and convection can be overcome, there is no reason why satisfactory preparative procedures could not be developed. Wider application of PAGE, IEF, and ITP to humic substances offers promising avenues for future research.

Ion-Exchange Media

Anion-exchange resins have been used (Wright and Schnitzer, 1960) in an attempt to fractionate soil humic substances. Some of the humic material is readily retained and a fractionation can be achieved by elution with a salt gradient and/or an alkaline reagent (usually NaCl and NaOH, respectively). In theory the anion-exchange technique should work well, but in practice the

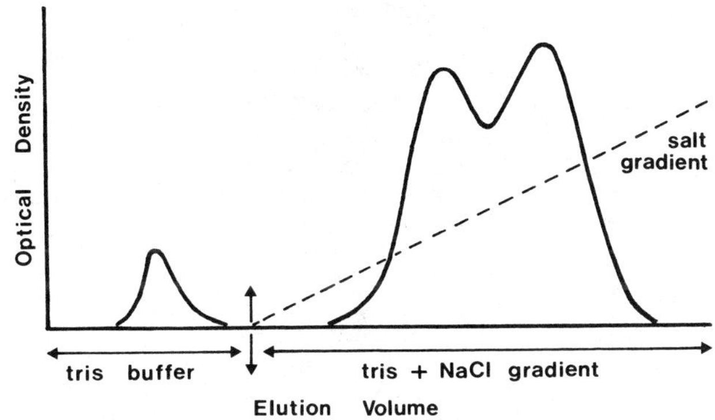

FIGURE 8. Fractionation of a fulvic acid sample on DEAE–cellulose using tris buffer plus a superimposed salt gradient as eluents.

fractionation obtained is rather crude. This is probably due, in part, to the fact that most anion-exchange resins consist of solid (i.e., nonporous) polystyrene beads. This structure greatly restricts their surface area which in turn limits the ability of the resin beads to interact with all charged sites on the humic molecules. Consequently, the humic polyelectrolytes are unable to exhibit fully their charge characteristics, so that charge differences between the molecules will be less well defined, and the fractionation will lose resolution. However, polystyrene-based cation-exchange resins have proved very useful for changing the cation associated with humic substances (Schnitzer, 1978).

Better fractionation results have been obtained when porous ion-exchange media such as anion-exchange cellulose and anion-exchange gels are used (Roulet et al., 1963; Barker et al., 1967). Again, after adsorption of the humic acid onto the gel, a fractionation can be achieved by eluting with buffer solutions and salt gradients, and then, if necessary, with an alkaline reagent. Figure 8 shows a typical ion-exchange fractionation of fulvic acid carried out by the author. The initial peak was removed by eluting with tris buffer alone and the following two peaks with tris buffer plus a salt gradient. A good fractionation can be obtained in this way, but quite often some of the humic material is held so strongly that it is difficult to recover it all.

In the author's view the potential of these materials has not received the attention that they warrant, particularly since they offer a relatively simple but sensitive means of fractionation based primarily on properties of electrical charge rather than molecular weight. In addition, they lend themselves very readily to preparative work.

FRACTIONATION BASED ON ADSORPTION

Fractionation based on adsorption properties has been used particularly with fulvic acid fractions. Due to the nature of the extraction procedures used this

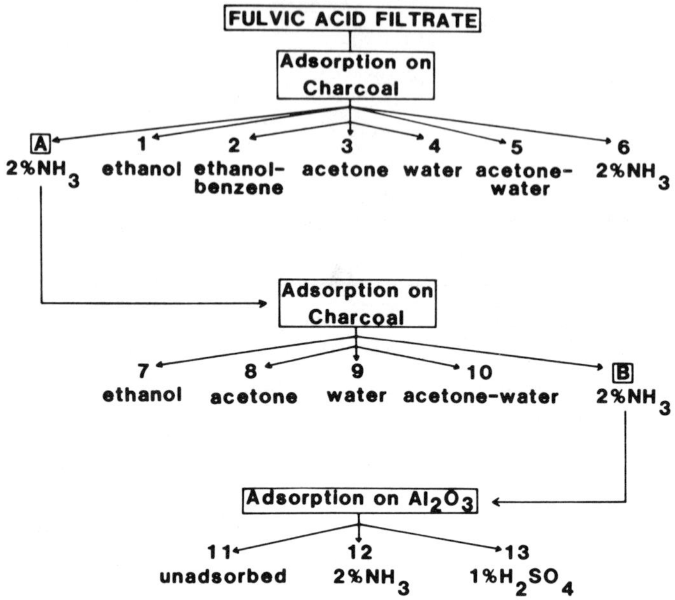

FIGURE 9. Fractionation for fulvic acids. [Adapted by Stevenson (1982) from Dragunov and Murzakov (1970).]

particular fraction tends to contain a considerable amount of nonhumic impurities, and in some cases, it is difficult to distinguish between fractionation and purification. Use has been made of a number of adsorbent media such as charcoal (Forsyth, 1947), alumina (Dragunov and Murzakov, 1970), and gels (Swincer et al., 1969). Desorption has been achieved by a variety of organic solvents and acidic and basic reagents. A rather complex fractionation scheme which illustrates the principles used is shown in Figure 9. Procedures such as those described above are probably best used for separation of humic substances from polysaccharides rather than for the fractionation of humic substances themselves. One reason for this is that adsorption of humic substances is often so strong that they cannot be desorbed without the use of rather strong and potentially damaging reagents.

In this regard the macroporous methylmethacrylate resin XAD-8 is a relatively weak adsorbent and has been shown to be well suited to use in the purification of humic substances from aqueous environments (Aiken et al., 1979; Chapter 14 in this book). It might also be expected that this and other adsorption materials will be suitable for fractionation of soil humic substances. For instance, a wide range of affinity chromatography and metal chelating materials are now manufactured by Pharmacia (e.g., CH— and AH—Sepharose 4B) and the list of products is constantly growing. It is very likely that one or other of these materials could prove a useful medium to produce a fractionation of humic substances based on a property other than molecular weight or electrical charge. Investigation of the properties and effectiveness of these resins should be encouraged as a potential area for research.

SUMMARY

There is now general acceptance that humic substances are a family of related compounds exhibiting a wide range of values with respect to any given property. As a consequence the use of fractionation techniques for the isolation of pure, identifiable compounds has now largely been abandoned. Instead, fractionation of soil humic substances is now used in order to:

1. Decrease the heterogeneity of these materials to allow the meaningful application of various chemical and physical techniques.
2. Allow us to obtain information about the range of molecular properties encountered within the spectrum of humic substances.
3. Characterize or fingerprint samples in order to monitor changes resulting from the application of some treatment or other.

A very wide range of fractionation procedures has been used and a fractionation of some sort is always obtained. The quality of fractionation is variable, and research in this area has largely centered on the improvement and refining of existing techniques, as much as on the search for new ones.

Classical fractionation procedures usually involve precipitation by adjustment of pH, adjustment of salt concentration, addition of organic solvents, or addition of metal ions. The fractionations produced are rather crude, but are generally quick and easy, and the manipulation of pH still remains a popular method today.

The introduction of gel permeation chromatography materials and ultrafiltration membranes has provided powerful techniques which give very good fractionations on the basis of molecular size. Both techniques are now widely used, and although less popular, centrifugation is available as an alternative independent technique.

Fractionations on the basis of charge by classical electrophoretic techniques proved to be rather disappointing. However, the recent introduction of a range of new electrophoretic techniques, such as polyacrylamide gel electrophoresis, isoelectric focusing, and isotachophoresis, has greatly improved the fractionations achieved and given renewed impetus to the use of electrophoresis. Modern ion-exchange media also offer great potential for alternative methods of fractionation based on charge properties.

Adsorption chromatography as a means of fractionation has not been as widely exploited as it deserves, possibly due to initial poor results and cumbersome procedures. Again the recent introduction of new types of adsorption media may provide materials well suited to the fractionation of humic substances.

Although fractionation work is tedious and painstaking, the rewards available in making more meaningful measurements and obtaining greater understanding make it well worthwhile. The armory available for fractionation is more extensive and powerful than it has ever been before. Despite the progress so far, there is still much to be achieved by the enterprising and diligent research worker; a fractionation can be carried as far as one's patience will allow.

CHAPTER SIXTEEN

Fractionation Techniques for Aquatic Humic Substances

JERRY A. LEENHEER

ABSTRACT

A review of current chemical and physical fractionation techniques for aquatic humic substances is presented in this chapter. Factors that hinder the fractionation of aquatic humic substances into individual compounds by conventional approaches include their polyfunctional character, which causes conformational and particle-size changes due to intra- and intermolecular weak-bonding mechanisms, multiple interactions with fractionating media, and high molecular weights that prevent fractionation by gas chromatography. Theoretically, it should be possible to fractionate aquatic humic substances into individual compounds by liquid chromatography. The most promising chromatographic approaches include normal-phase liquid chromatography on weak-base substrates, or reverse-phase liquid chromatography where humic solutes are disaggregated by heating the mobile phase, use of highly polar mobile phases, or by use of polar supercritical-fluid mobile phases. Prior chemical derivatization of polar interacting functional groups to less polar groups should aid in the liquid-chromatographic separations. Methods for forming the methyl and trifluoroethyl esters of carboxyl groups, the acetyl ester and trifluoroethyl ether of hydroxyl groups, and the reduced alcohol of the carbonyl group are presented. Lastly, analyt-

ical and preparative fractionation procedures were formulated with the goal of obtaining pure compounds from aquatic humic substances for structural studies.

INTRODUCTION

Fractionation techniques for aquatic humic substances have not been developed to the same extent as concentration and isolation techniques. Many organic fractionation techniques presuppose a concentrated sample, but aquatic humic substances exist naturally at dilute concentrations in the presence of greater suspended sediment and inorganic solute concentrations. Now that efficient preparative concentration and isolation techniques for aquatic humic substances have been developed, as reported by Aiken in Chapter 14 of this book, renewed emphasis can be given to group fractionations with the ultimate hope that the chromatographic separation of aquatic humic substances into individual compounds can be achieved.

There is no clear distinction between humic and nonhumic substances in water. Humic substances can be associated with nonhumic substances, such as proteins and polysaccharides, through covalent bonding, hydrogen bonding, and electrostatic interactions. The chemical conditions used in the isolation and fractionation procedures will determine the degree of separation of humic from nonhumic substances. Fractionation procedures cannot be clearly distinguished from isolation procedures, because most isolation procedures, such as adsorption chromatography, are also crude fractionation procedures that partly fractionate aquatic humic substances. Therefore, those researchers studying aquatic humic substances need to contend with definitions and procedures dependent on conditions of analytical operations and need to recognize that distinctions between isolates or fraction usually are not well defined. The discussion of this chapter will emphasize fractionation procedures of previously isolated aquatic humic substances, but some discussion of isolation procedures and nonhumic substances will be presented because of the overlap and relationships of these ancillary topics.

Application of conventional fractionation and chromatographic techniques used for hydrocarbons and monofunctional organic compounds to aquatic humic substances has met with little success because of the complex chemical and physical properties of aquatic humic substances. Fractionation techniques developed for polymeric, multifunctional biological substances such as proteins and polysaccharides sometimes can be successfully applied to aquatic humic substances, but the limited set of monomeric units and the regularity in chemical and physical properties found in proteins and polysaccharides are not features of the more heterogeneous aquatic humic substances. Consequently, fractionation techniques developed for biopolymers may not be applicable for aquatic humic substances because of their greater heterogeneity, which results in more diverse and irregular interactions with the fractionating medium.

Most concentration and isolation techniques, except for evaporative and freeze-concentration techniques, are also the first steps in chemical and physical fractionation of aquatic humic substances. This chapter will concentrate primarily on techniques used to subfractionate and chromatographically separate aquatic humic substances previously isolated as crude fractions.

Macro- as well as microfractionation techniques need to be developed, as necessary steps in attaining the goal of organic structure elucidation. Now that several hundred grams of aquatic humic substances can be isolated from water at reasonable time and cost, it is not unrealistic to plan to process this quantity of material to obtain milligram quantities of pure substances for structural studies. Once some structures have been determined, it should be possible to miniaturize the fractionations and use mass spectroscopy for structure identification.

The purposes of this chapter are (1) to present an overview of fractionation methods that have been successfully applied to aquatic humic substances; (2) to examine chemical and physical fractionation mechanisms in the light of what is known about aquatic humic substance properties and structure; (3) to postulate new fractionation approaches that hopefully will result in more homogeneous fractions and, ultimately, pure compounds that comprise aquatic humic substances.

GENERAL CONSIDERATIONS IN DESIGNING FRACTIONATION METHODS

Trial and error approaches seldom have been successful for fractionations of aquatic humic substances. However, fractionations can be designed using the following general considerations about the nature of aquatic humic substances.

Considerations of Aquatic Humic Substance Structure

A conceptual model of aquatic humic substance structure is represented by ORSMAC (organic solute macromolecule) (Fig. 1). By itself, ORSMAC is a solute and is not especially large. Reuter and Perdue (1981) found a number-average molecular weight of approximately 600 for an aquatic fulvic acid isolated from the Satilla River, Georgia, and Thurman et al. (1982) found that aquatic fulvic acids isolated from a variety of surface water and groundwaters generally had molecular weights less than 2000.

Although ORSMAC primarily is an acid because of the predominance of carboxylic acid functional groups, ORSMAC has amphoteric properties from a hydrogen-bonding standpoint. Carboxyl, hydroxyl, and enol are proton-donating groups, whereas keto, ether, and amide are proton-accepting groups. If these groups are stronger conjugate acids and bases, respectively, than water, hydrogen-bonding aggregation will occur between these hydro-

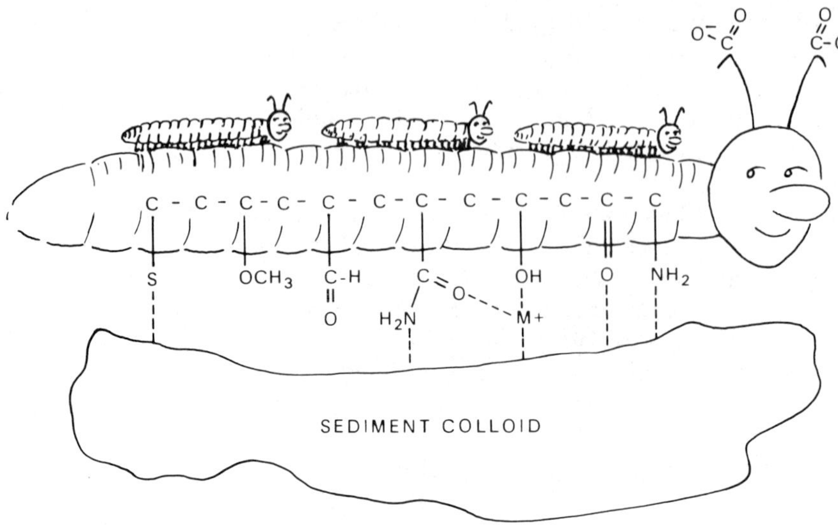

FIGURE 1. ORSMAC, the organic solute macromolecule.

gen-bonding groups and inorganic solutes such as silicic acid and boric acid, or with silica and alumina surfaces on sediment. Metals in solution or on sediment surfaces also will form complexes with various ORSMAC functional groups. Therefore, ORSMAC frequently is aggregated with other ORSMACS, with soluble silica or boric acid, and adsorbed on mineral surfaces.

Lastly, ORSMAC has weakly amphipathic properties, which are characterized by its surface activity and detergent properties. However, the hydrophobic parts of the ORSMAC structure are not large enough and the concentrations are not sufficiently great to cause the formation of micellar structures in aquatic systems.

Fractionation Before or After Concentration

A persuasive argument for fractionating aquatic humic substances before concentration is the minimization of aggregation resulting from intermolecular interactions discussed in the previous section. The resin-adsorption concentration procedures discussed by Aiken in Chapter 14 of this book also accomplish compound group fractionation at ambient concentrations. Most fractionation procedures can be performed at ambient concentrations if detection of the analyte is sufficiently sensitive or if the fractionating medium concentrates the analyte. However, preconcentration needs to be used if a preparative fractionation is desired where the analyte is not concentrated on the fractionating medium.

Hydrogen-bonding effects causing aggregation need to be minimized for successful fractionation of concentrated aquatic humic substances. Techniques used for disaggregation include pH adjustment, ionic-strength adjustment, temperature increase, addition of ion-pairing and hydrogen-bonding reagents to the solvent, solvation in solvents of greater polarity than the functional groups of the solute, and chemical derivatization of polar functional groups to less polar forms.

Preservation of Aggregates and Complexes During Fractionation

Many studies of aquatic humic substances seek to determine the nature and properties of these materials as they exist in the environment. One approach for the study of aquatic humic aggregates and complexes which does not involve fractionation is to measure a constituent *in situ,* such as a trace metal in the presence of the unfractionated aquatic humic substances by specific-ion electrode. Alternatively, one might fractionate the sample and study the individual fractions. Most chemical fractionation techniques and certain physical fractionation techniques cause reversible and irreversible changes in structure, break or reform hydrogen-bonded aggregates and metal complexes, and may even cleave covalent-bond linkages in the structure. Physical fractionation techniques such as ultrafiltration, gel permeation chromatography, and ultracentrifugation are the preferred methods for fractionating aggregates and complexes.

Sediment, Colloidal, and Molecular Size Fractionations

Because of the aggregating and sorptive tendencies of aquatic humic substances, any size fractionation is defined operationally where and when performed, and additional aggregation, as evidenced by precipitation after filtration, is a frequent occurrence. Gjessing (1973) compared ultrafiltration with gel permeation chromatography using aquatic humic substances and found that aggregation and disaggregation that occurred during the fractionations caused significant interchange of material between fractions. Smith (1976) found that increased salinity in the estuary of the Ogeechee River in Georgia caused aggregation of high-molecular-weight humic substances (molecular weight determined by ultrafiltration) to the point of precipitation and sedimentation of the high-molecular-weight fraction.

Size fractionations of aquatic humic substances definitely need to be performed on-site to minimize changes in aggregation during sample preservation, transport, and storage if the size fractionation data are to be related to environmental conditions. If additional aggregation or precipitation or both occur after an on-site size fractionation, the sample should not be refractionated, but needs to be treated as if the on-site fractionation is valid.

Desirable and Undesirable Interactions with Fractionating Medium

It is usually desirable to fractionate aquatic humic substances by only one interaction mechanism operating at a time. Unfortunately, the complexity of aquatic humic substance structures and properties usually causes multiple interactions with the fractionating medium. Examples of undesirable interactions that cause problems with fractionation procedures include adsorptive interactions of aquatic humic substances on Sephadex* gels used for size-exclusion chromatography (Wershaw and Pinckney, 1973b), irreversible sorption of humic substances on the hydrophobic matrix of an anion-exchange resin (Abrams and Breslin, 1965), irreversible adsorption of quinone functionalities of aquatic humic substances on polyamide adsorbents (Endres and Hormann, 1963), and minimal recoveries of humic substances adsorbed from seawater on activated carbon (Kerr and Quinn, 1975).

Examples of successful fractionations of aquatic humic substances where only one fractionation mechanism was operative include utilization of the hydrophobic properties of XAD resins (Mantoura and Riley, 1975; Aiken et al., 1979), hydrogen bonding of weak-acid functionalities of humic constituents to weak-base anion-exchange resins (Kim et al., 1976), and use of ion-exchange celluloses for ion-exchange fractionation of aquatic humic substances without hydrophobic matrix adsorption (Sirotkina et al., 1974). These examples of successful fractionations demonstrate the potential for chromatography of aquatic humic substances when fractionations are designed carefully to avoid undesirable interactions.

Operational Definitions of Fractions

In the absence of definitive chromatographic procedures that separate aquatic humic substances into pure constituents, current fractionation procedures only separate aquatic humic substances into more homogeneous groups of compounds that are defined operationally by the mechanisms operative in the fractionation procedure. An example of operational definitions are the hydrophobic acid, base, and neutral, and hydrophilic acid, base, and neutral compound groups of the dissolved organic carbon (DOC) fractionation procedure (Leenheer and Huffman, 1979). Present compound-group fractionation procedures also separate predominately by major characteristic differences in polyfunctional molecules so that minor characteristic differences will appear in multiple fractions. For example, aliphatic, monocarboxylic acids and aliphatic amines greater than nine-carbon chain length, and aromatic monocarboxylic acids and aromatic amines of three or more rings fractionate into the hydrophobic neutral class of DOC fractionation because the neutral hydrocarbon characteristics of these compounds outweigh their acid–base characteristics. Other examples are base or "cationic" characteristics found in predominately acidic humic substances (MacCarthy and O'Cinneide, 1974). Additional subfractionation procedures that use different forms of chromatography can generate successively purer forms of aquatic humic substances that are homogeneous with respect to

* Use of trade names in this report is for identification purposes only and does not constitute endorsement by the U.S. Geological Survey.

multiple independent properties (hydrophobicity, acidity, and basicity) instead of being homogeneous just with respect to their major characteristic (Leenheer and Noyes, 1983).

Preservation of Fractions

It is usually preferable to complete fractionation procedures in a short time so that preservation of fractions between various fractionating procedures is not necessary. However, samples usually have to be preserved and stored after fractionation and, sometimes, between fractionation steps as well. Short-term storage of aquatic humic substances (days to weeks) is best accomplished by leaving the fraction in the solvated state, refrigerated in a dark bottle under nitrogen. Alkaline solvents need to be avoided during storage because of various hydrolytic and oxidative degradations that occur in alkaline solvents.

Freeze-drying of aquatic humic substances has been recommended as the method of choice for preservation and storage of natural organic substances. However, this author (J. A. Leenheer, unpublished data, 1983) has found that a minor problem of freeze-drying of hydrogen-saturated aquatic humic substances is the formation of ester and lactone linkages as evidenced by proton nuclear magnetic resonance and infrared spectral data. Freeze-drying of neutralized salts of the humic substances is a possible solution to prevent esterification during drying. Freeze-drying definitely is a good method to render a sample relatively inert to biological and photochemical degradations. For further discussion of preservation of aquatic humic substances, see discussion by Aiken (Chapter 14).

CHEMICAL FRACTIONATION METHODS

Precipitation Methods

Acidification of aqueous concentrates and extracts to pH near 1 is the standard procedure to precipitate humic from fulvic acid, and this procedure also has been applied to aquatic humic substances (Thurman and Malcolm, 1981). Aquatic humic substances that interact significantly with metal ions can be precipitated from water by addition of lead(II) nitrate (Klocking and Mucke, 1969). Co-precipitation of aquatic humic materials with aluminum, copper, iron, and magnesium hydroxides has been used to recover aquatic humic substances from various types of water (Jeffrey and Hood, 1958; Williams and Zirino, 1964; Zeichmann, 1976). Humic acids can also be precipitated from an unconcentrated water sample by adding acetic acid and isoamyl alcohol to a sample contained in a separatory funnel, and after shaking, humic acid precipitates at the alcohol–water interface (Martin and Pierce, 1971). Precipitation methods are among the crudest of fractionation methods applied to aquatic humic substances because of intermolecular aggregation that occurs during precipitation; they are generally more useful for isolation and concentration than for fractionation of humic materials.

Solvent Extraction

Partitioning of humic substances into acid, base, and neutral groups by solvent extraction accompanied by pH adjustment is not possible because humic substances are not solvent extractable. The formation of dark films at the liquid–liquid interface or the formation of emulsions demonstrates the amphipathic character of aquatic humic substances whereby the hydrophobic part of the molecule is attracted to the organic solvent surface, but the polar part of the molecule does not cross the interface from the aqueous to organic liquid phase. However, it is possible to form a hydrophobic, extractable ion-pair by addition of tertiary or quaternary long-chain alkyl amines. Eberle and Schweer (1974) developed a procedure whereby trioctylamine dissolved in chloroform efficiently extracted aquatic humic substances and lignosulfonic acids at pH 5. The chloroform extract was back-extracted with water at pH 10 or greater to recover aquatic humic substances. Competition of inorganic anions and nonhumic acids for trioctylamine ion-pair sites causes these substances to be co-extracted with aquatic humic substances. Variation of alkyl chain length of the amine coupled with a liquid–liquid fractionating procedure like counter-current distribution might lead to a useful adaptation of the ion-pair solvent extraction method to chromatographic fractionation of aquatic humic substances.

Adsorption Chromatography

Fractionation of aquatic humic substances by adsorptive interactions has been the most successful method for fractionation as well as concentration and isolation. Humic solutes readily interact with various adsorptive surfaces without the requirement of crossing the interface surface as is necessary with solvent partitioning or absorptive interactions.

Hydrophobic interactions operative in reverse-phase liquid chromatography have been used to concentrate and isolate aquatic humic substances (Mantoura and Riley, 1975; Aiken et al., 1979), but chromatography of aquatic humic substances by reverse-phase, high-performance liquid chromatography has not produced well resolved component chromatograms. Broad, trailing peaks indicative of solute–solute or solute–sorbent, direct-phase interactions are produced by reverse-phase chromatography of aquatic humic substances. A number or combination of different approaches may produce successful fractionations of aquatic humic substances by reverse-phase liquid chromatography. Intermolecular hydrogen-bonding effects responsible for aggregation may be minimized by heating the mobile phase, adjusting the mobile phase pH to an optimum level, adding hydrogen-bonding reagents (alkyl amines) to the mobile phase, using highly polar solvents other than water in the mobile phase, or using supercritical fluids (ammonia, sulfur dioxide, carbon dioxide) as the mobile phase. Because of the large number of theoretical plates being attained by state-of-the-art reverse-phase liquid chromatography, a concerted effort needs to be made to adapt this form of chromatography for fractionation of aquatic humic substances.

Hydrogen-bonding interactions used in normal-phase liquid chromatography have as much potential utility in humic substance fractionations as reverse-phase liquid chromatography. Normal-phase chromatography using organic solvents as mobile phases and silica, alumina, or magnesia as adsorbents has not been successful with aquatic humic substances because of limited solubility of these solutes in organic solvent systems and irreversible interactions of the solutes with the sorbents. However, normal-phase chromatography with aqueous mobile phases has been found to fractionate aquatic humic substances according to the nature of their polar functional group content (Thurman and Malcolm, 1983; Jennings and Ekeland, 1983). Sephadex, a dextran polymer used for gel permeation chromatography, also interacts via hydrogen-bonding mechanisms with weakly basic aromatic amines (Gelotte, 1960) and weakly acidic polyphenols (Woof and Pierce, 1967) to fractionate these materials. Wershaw and Pinckney (1973) attributed the fractionation of aquatic humic substances on Sephadex gels in the absence of ion-pair buffers primarily to adsorptive interactions.

Considerable study has been performed on hydrogen-bonding interactions of aquatic organic solutes with reverse-osmosis membranes (Sourirajan, 1977). Polar parameters designated $\Delta\nu_s$ (acidity) and $\Delta\nu_s$ (basicity) were used to quantify hydrogen bonding by infrared stretching-frequency shifts (in reciprocal centimeters) of proton-donating and proton-accepting functional groups. A listing of the polar parameters for functional groups, solvents, adsorbents, and inorganic solutes important for hydrogen-bonding effects in systems containing aquatic humic substances is shown in Table 1 (Sourirajan and Matsuura, 1977). If an acid or base functional group has a smaller value for $\Delta\nu_s$ (acidity) or $\Delta\nu_s$ (basicity) than water, these weakly hydrogen-bonding groups will be preferentially associated with water because of its relative abundance compared to solvated constituents. However, acid or base functional groups with $\Delta\nu_s$ (acidity) and $\Delta\nu_s$ (basicity) values greater than water can preferentially interact in the presence of water because of their greater hydrogen-bond energies. Therefore, solute–solute or solute–sorbent hydrogen-bonding interactions should occur between phenol (or enol), silicic acid and boric acid, weak acids, and ketone and ether weak bases in aquatic humic substances. Alicyclic ethers and ketones are the weak bases that are slightly stronger than water. Quinones and ketones that enolize are also stronger bases than water. In fulvic acids, the nitrogen content generally is too small to cause significant hydrogen bonding because of its basic properties, but in humic acid, the nitrogen content (with consequent hydrogen-

TABLE 1. Hydrogen Bonding Infrared Frequency Shifts for Various Compounds[a]

Compound	$\Delta\nu_s$ (acidity) (cm^{-1})	$\Delta\nu_s$ (basicity) (cm^{-1})
Water	250	~80
Aliphatic alcohol	120–160	—
Phenols	250–300	—
Esters	—	20–50
Ketones	—	50–90
Ethers	—	60–100
Aromatic amines	—	150–270
Silicic acid and boric acid	—	250–300

[a] *Data obtained from Sourirajan and Matsuura (1977).*

bonding aggregation) is significantly greater. Functional groups sufficiently acidic or basic to ionize in water interact through an ionic mechanism rather than by hydrogen bonding.

Polyamide adsorbents (Endres and Hormann, 1963), weak-base exchange resins (Kim et al., 1976), and ion-exchange celluloses (Sirotkina et al., 1974) all have been used for concentration, isolation, and general compound-group fractionations of aquatic humic substances by hydrogen-bonding interactions; Shapiro (1957) obtained as many as nine fractions of aquatic humic substances using paper chromatography, and Sieburth and Jensen (1968) obtained additional fractionation of aquatic humic substances using two-dimensional paper chromatography.

Moderate- to high-resolution normal-phase liquid chromatography has not been attained yet for aquatic humic substances. Recently, Jennings and Ekeland (Montana State University, unpublished data, 1983) have found that both soil and aquatic fulvic acids were fractionated into several components by normal-phase, high-performance liquid chromatography using a silica packing bonded with organic amines. Thurman and Malcolm (1983) found that weak-base resins retained aquatic humic substances containing phenolic hydroxyl groups at pH 7 in a sodium bicarbonate buffer, whereas humic substances not containing phenolic groups did not interact. As more data become available on the polar functional groups of aquatic humic substances responsible for hydrogen-bonding interactions, it should be possible to design various types of affinity chromatography to separate weakly acid or basic functional groups in aquatic humic substance mixtures.

Fractionation of aquatic humic substances by ion-exchance mechanisms has been limited severely by undesirable matrix interactions of the exchange medium. Hydrophobic matrix-exchange resins also interact with aquatic humic substances by hydrophobic effects (Abrams and Breslin, 1965) and hydrophilic matrix-exchange gels also interact with polyfunctional solutes by hydrogen bonding (Sirotkina et al., 1974). Ligand-exchange chromatography of aquatic humic substances is subject to the same undesirable matrix interactions as ion-exchange chromatography.

Various types of adsorptive chromatography can be combined into analytical-separation schemes. Hydrophilic ion-exchange cellulose adsorbents and Sephadex were used by Sirotkina et al. (1974) to systematically analyze for organic solute distributions in natural waters. The fractionation scheme is shown in Figure 2. Diethylaminoethyl (DEAE) cellulose was used in the free-base form, and carboxymethyl (CM) cellulose was used in the acid form. Sorption of natural organic acids on DEAE cellulose and organic bases on CM cellulose was by hydrogen bonding as well as by ion-exchange, because the adsorbent exchange groups were only slightly ionic at neutral pH where adsorption occurred. Irreversible sorption of hydrophobic solutes was not a problem because of the hydrophilic nature of cellulose sorbents. Natural waters were first concentrated by freeze concentration, and the samples were passed through the adsorbent sequence without pH adjustment. Gel filtration on Sephadex was used to desalt the samples and broadly fractionate into low- and high-molecular-weight components. Recovery studies based on standard additions determined that organic solute losses during fractionation did not exceed 10%, and the fractionation procedure was applied to five different river water samples.

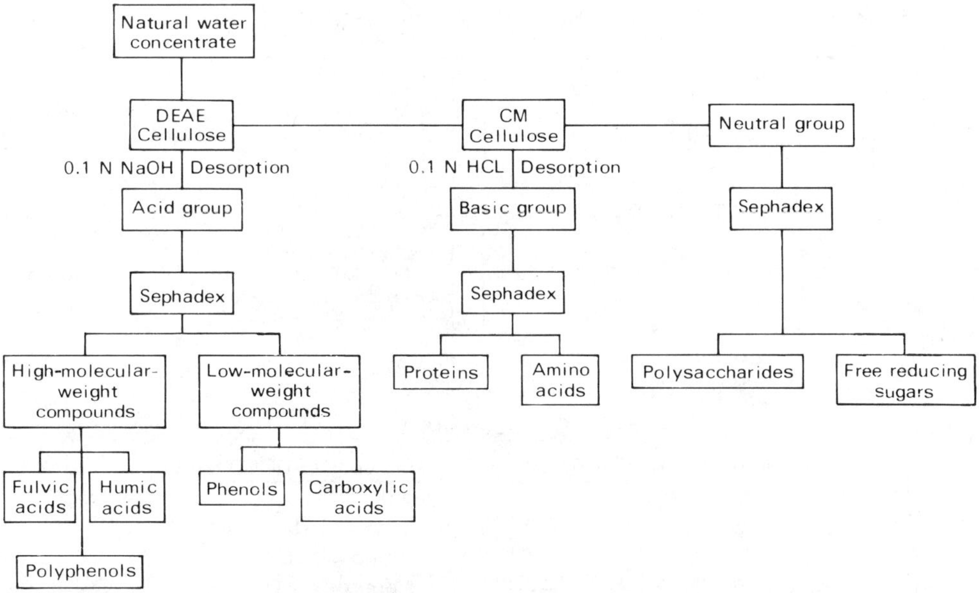

FIGURE 2. Fractionation of organic solutes in water by ion-exchange celluloses and Sephadex. Reprinted with permission from Sirotkina et al., *Zhurnal Analiticheskoi Khimii* **29,** 1626–1632. Copyright © 1974 by Plenum Publishing Corporation.

Leenheer and Huffman (1976) have developed a fractionation procedure for aquatic organic solutes called dissolved organic carbon (DOC) fractionation. The procedure for the analytical DOC fractionation is shown in Figure 3. Hydrophobic solutes are first removed from water by adsorption on Amberlite XAD-8 resin, hydrophilic bases in the effluent are removed by cation-exchange resins, and hydrophilic acids in the effluent are removed by anion-exchange resins. Aquatic humic substances occur primarily in the

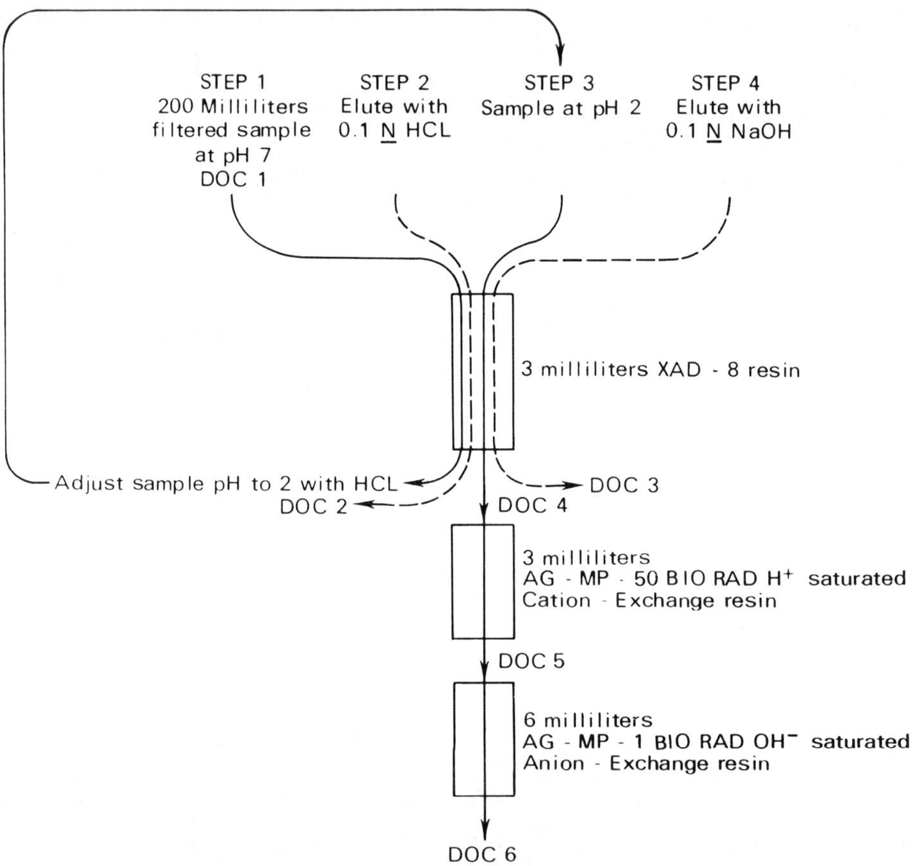

CALCULATIONS

Hydrophobic DOC (mg/L)

Total = DOC 1 − DOC 4

$$\text{Bases} = \text{DOC 2} \times \frac{\text{eluate volume}}{\text{sample volume}}$$

$$\text{Acids} = \text{DOC 3} \times \frac{\text{eluate volume}}{\text{sample volume}}$$

Neutrals = Total − Bases − Acids

Hydrophilic DOC (mg/L)

Total = DOC 4

Bases = DOC 4 − DOC 5

Acids = DOC 5 − DOC 6

Neutrals = DOC 6

FIGURE 3. Analytical scheme for dissolved organic carbon (DOC) fractionation.

hydrophobic acid fraction. The analytical DOC fractionation procedure was not satisfactory as a preparative procedure because of irreversible adsorption of hydrophilic acids on the strong-base anion-exchange resin. Therefore, a preparative DOC fractionation procedure (Fig. 4) was developed (Leenheer, 1981; Leenheer and Noyes, 1983). The recycle step through XAD-8 resin in the analytical procedure was omitted in the preparative procedure to facilitate sample throughput for on-site fractionations, and the Duolite A-7 weak-base anion-exchange resin was substituted because of its efficient desorption of aquatic humic substances when the charge of the adsorbent was reversed at pH > 10. Aquatic humic substances are recovered from the Duolite A-7 resin by infusing 10*N* sodium hydroxide into recycled column water until the pH of the recycled water attains pH 11.5; then the recycle loop is interrupted and sodium salts of aquatic fulvic acids, hydrophilic acids, and inorganic anions are eluted from the column with distilled water. Aquatic fulvic acids are separated from hydrophilic acids and inorganic salts by acidification of the concentrate and readsorption of aquatic fulvic acids on XAD-8. Procedures for purification of the hydrophilic acid fraction include removal of chloride by precipitation of silver chloride on a silver-saturated cation-exchange resin, and removal of sulfate by crystallization of sodium sulfate decahydrate on dilution of the concentrate with ethanol.

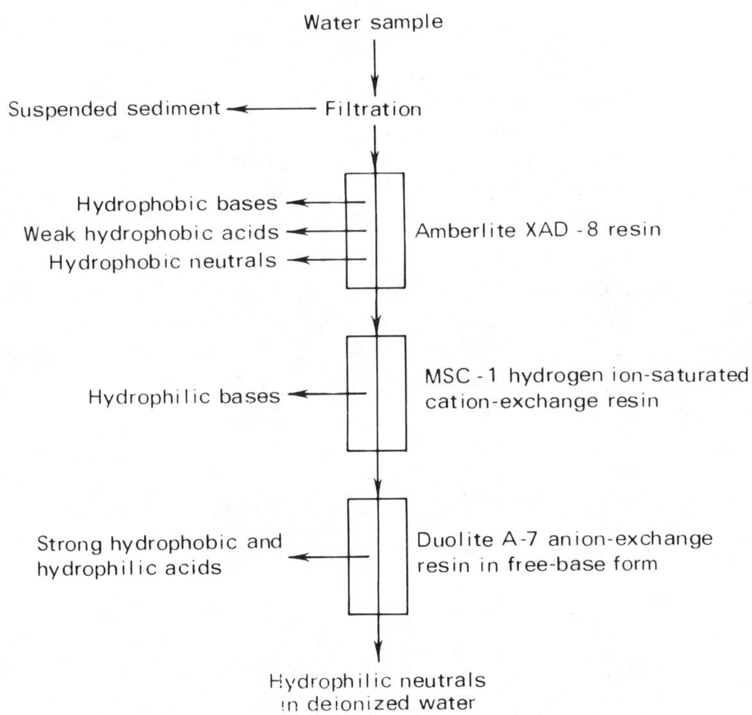

FIGURE 4. Preparative scheme for dissolved organic carbon (DOC) fractionation.

Adsorption chromatography of aquatic humic substances is still in its developmental stage. For progress to be made, aggregation phenomena and undesirable adsorbent interactions will have to be minimized. However, the rapid advances being made in the field of liquid chromatography undoubtedly will have applications to the fractionation of aquatic humic substances.

Derivatives for Chromatography

The author recently synthesized a number of chemical derivatives of aquatic humic substances for nuclear magnetic resonance and mass spectroscopy studies. Attempts to fractionate these derivatives by gas chromatography were not successful, but encouraging preliminary results were obtained for both reverse-phase and normal-phase liquid chromatography. The key to a useful chemical derivative of aquatic humic substances is to eliminate hydrogen bonding between functional groups by modifying or converting all the functional groups to forms with similar properties. For example, a combination of carboxyl group methylation and hydroxyl group acylation converts acidic groups to esters that do not hydrogen bond with ether and ketone functional groups at other positions in the molecule. Homogeneity in functional group properties also enables selection of an adsorbent and mobile phase optimized for those properties.

Methylation

Methylation of carboxyl groups has been found to be very useful for obtaining stable derivatives of aquatic humic substances; however, permethylation of weakly acidic hydroxyl groups has not resulted in sufficient recoveries nor complete derivatization using procedures listed in Blau and King (1978). Carboxyl groups were methylated with diazomethane (Wershaw and Pinckney, 1978), which does not react with ester and lactonic groups in humic substance structures, and by boron trifluoride/methanol, which results in complete conversion of esters and lactones to methyl esters (Blau and King, 1978). Methylation of aquatic humic substances (represented by ORSMAC, Fig. 1) is shown by reactions 1 and 2:

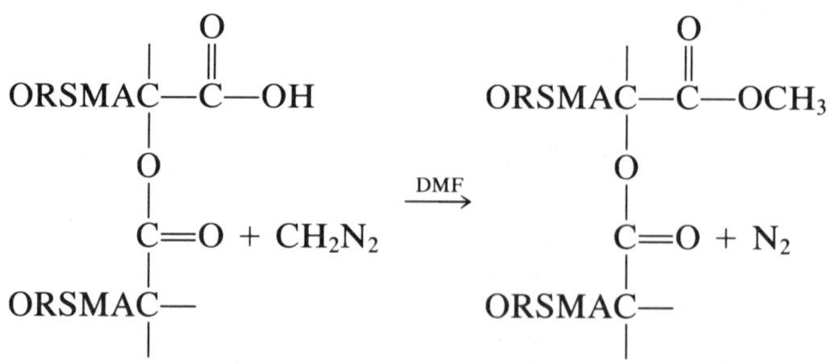

(1)

and

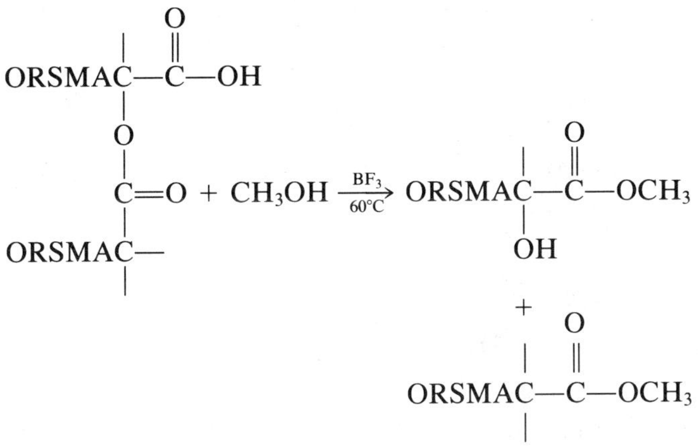

(2)

Reaction (2) is more of a degradative procedure than reaction (1) because boron trifluoride-catalyzed methylation will cleave ester linkages and open up lactone rings. Isolation of the reaction (1) product is accomplished by vacuum drying of the reaction mixture; isolation of the reaction (2) product is more complicated. Reaction (2) is accomplished by heating 10–50 mg of sample in 1 mL of 14% boron trifluoride/methanol at 60°C for 1 hour. Then after cooling, 5 mL of saturated sodium bicarbonate solution is added to the reaction mixture, and any precipitated product is removed by centrifugation. The precipitate is washed twice with 5 mL of distilled water, the supernatant removed by centrifugation, and all supernatant washings are combined. Methylated products dissolved in the reaction mixture and the supernatant washings are isolated by adsorption on a 20 mL column of XAD-8 resin after acidification of the reaction mixture with hydrochloric acid and the supernatant washings to pH 2. A 50 mL rinse of distilled water follows passage of the acidified sample solution through the column. The adsorbed product is desorbed with methanol, and after combination of the precipitated product with the methanol eluate, the product is isolated by a combination of vacuum rotary evaporation and vacuum drying.

Acylation

Acylation of aquatic humic substances with acetic anhydride in a variety of solvents was found to be unsatisfactory because of dehydration and π-bond formation. The mild acylating reagent, *N*-acetylimidazole, did not give complete acylations of hindered hydroxyl groups. Acetyl chloride was tested under a variety of conditions and gave complete derivatization of primary and secondary hydroxyls under the conditions shown in reaction (3):

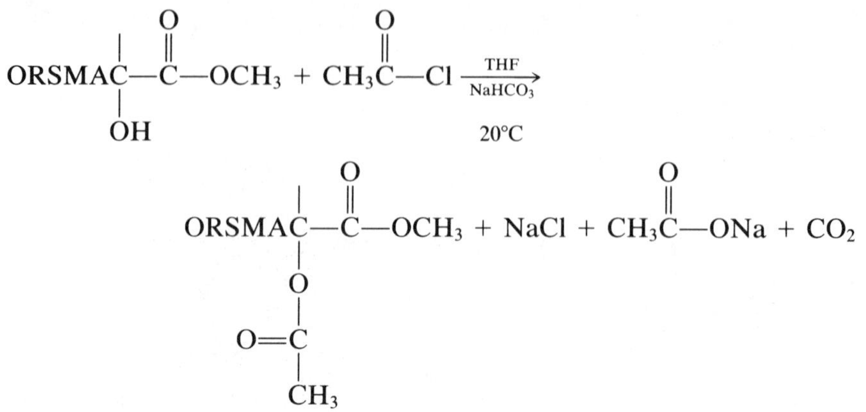 (3)

The methylated sample is dissolved in 50 mL of tetrahydrofuran and 5 g of sodium bicarbonate is added. The suspension is stirred and 5 mL of acetyl chloride is added. Stirring is continued overnight with the reaction mixture at room temperature. The solvent and excess acetyl chloride are removed by vacuum rotary evaporation at 30°C, and the product is partitioned between methylene chloride and water. Reaction by-products partition into water and the product partitions into methylene chloride. Emulsions may form during solvent extraction but they can be broken by addition of water saturated with sodium chloride. The methylene chloride is back-extracted twice with equivalent volumes of water, dried over sodium sulfate, and evaporated from the product.

Primary, secondary, and unmethylated phenolic groups are acylated. Tertiary aliphatic hydroxyl, benzylic hydroxyl, and allylic hydroxyl are less reactive and may be substituted with chlorine if the reaction medium becomes sufficiently acidic during hydrochloric acid evolution.

Trifluoroethylalkylation

The trifluoroethyl (CF_3CH_2—) derivative of organic acids and alcohols has been found to be useful for distinguishing carboxyl from primary hydroxyl from secondary hydroxyl functional groups by ^{19}F nuclear magnetic resonance spectroscopy (Koller and Dorn, 1982). This derivative is formed under conditions shown in reaction (4):

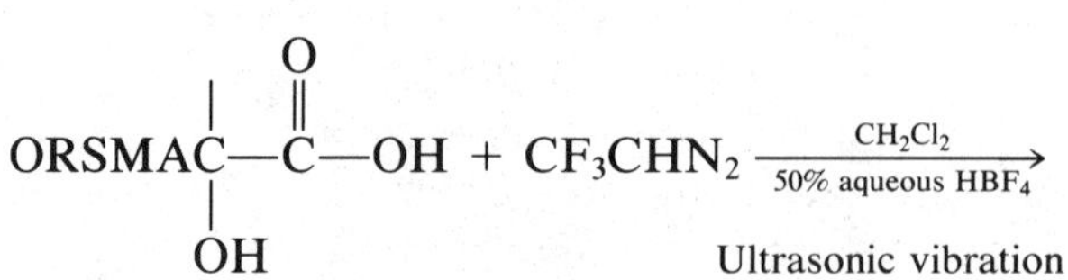

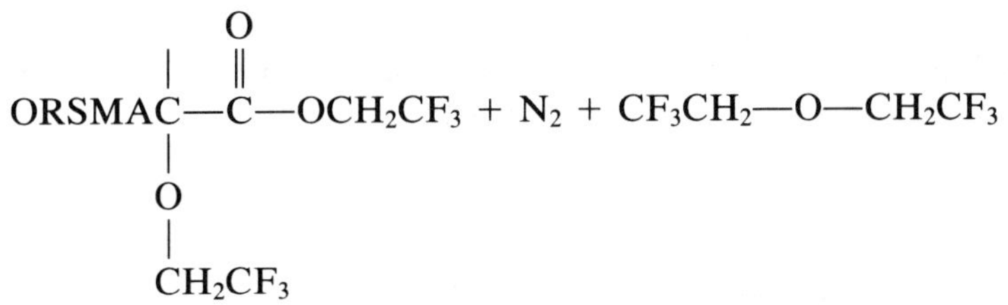

(4)

Reaction conditions are identical to the procedure given by Koller and Dorn (1982) except that ultrasonic vibration was used to disperse the sample suspension and fluoroboric acid catalyst in the methylene chloride while the trifluorodiazoethane gas is being bubbled through the suspension. Reaction is complete when the sample is dissolved in methylene chloride and nitrogen gas evolution ceases. The product is partitioned between methylene chloride and saturated aqueous sodium bicarbonate, which extracts the fluoroboric acid catalyst. After two additional washings with water, the product is dried over sodium sulfate, and the methylene chloride and excess trifluorodiazoethane are removed by evaporation.

The advantage of this reaction is derivatization of both carboxyl and hydroxyl groups in one step. Trifluorodiazoethane is easy to generate by diazotization of trifluoroethylamine, and trifluorodiazoethane does not polymerize in the presence of Lewis acid catalysts for weak-acid peralkylation as does diazomethane. Disadvantages of this derivative are that dehydration and π-bond formation occasionally have been observed in the presence of fluoroboric acid, and the moderate dipole moment of the derivative may result in hydrogen-bonding interactions during chromatography.

Sodium Borohydride Reduction

Conversion of keto and quinone groups to hydroxyl and phenol groups by sodium borohydride reduction is a possible approach toward decreasing hydrogen bonding by conversion of basic functional groups to weak-acid groups that become the predominant characteristic after reduction. Sodium borohydride reduction is performed under conditions shown in reaction (5):

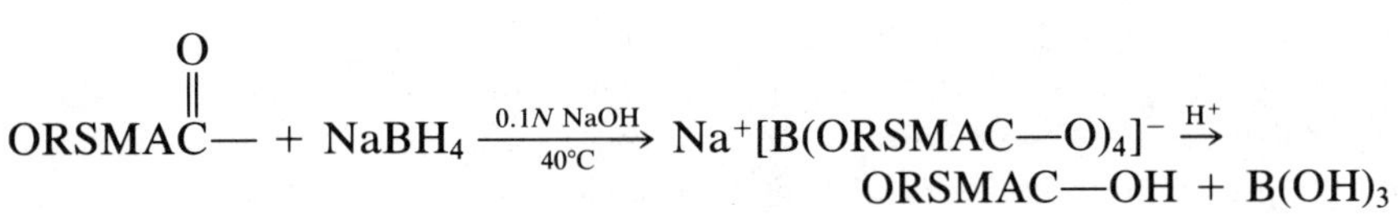

(5)

Equal weights of sample and sodium borohydride dissolved in aqueous $0.1N$ sodium hydroxide are heated at 40°C for 1 hour. The solution then is cooled, acidified to pH 2 with trifluoroacetic acid, and passed through a 20 mL XAD-8 column followed by a 50 mL rinse of $0.1N$ trifluoroacetic acid.

The sample is desorbed from the XAD-8 column with 0.1*N* sodium hydroxide, which immediately is removed by passage of the base eluate through a hydrogen-saturated cation-exchange column connected in series with the XAD-8 column. Most of the trifluoroacetic acid and water are removed from the sample by vacuum rotary evaporation of the sample to a moist residue. Lastly, the sample is diluted to 20 mL in distilled water and freeze-dried. The sodium borohydride reduction product is readily soluble in water, methanol, tetrahydrofuran, dimethylsulfoxide, and dioxane.

PHYSICAL FRACTIONATION METHODS

Electrophoresis

Electrophoretic separations of solutes are determined primarily by the mass-to-charge ratio of the solute. However, certain electrophoretic separations use gel or paper supports for the electrolyte, and adsorptive and molecular-size fractionations may be as significant as electrophoretic fractionation in these systems. Hall (1970) found that polyacrylamide gel electrophoresis of aquatic humic substances gave size fractionations corresponding to Sephadex gel fractionation; however, individual fractions were poorly resolved in polyacrylamide gel electrophoresis. Clapp (1957) and Mortensen and Schwendinger (1963) used electrophoresis with electrolyte supporting curtains to separate colorless polysaccharides from humic substances in soil–water extracts.

Free-flow electrophoresis is accomplished by a laminar flow of unsupported electrolyte between glass plates; the absence of a supporting medium nullifies adsorption and filtration interactions, and the free flow enables relatively large quantities of sample to be processed. Free-flow electrophoresis not only fractionates dissolved charged materials, but also has the capability to fractionate suspended particulate material on the basis of charge as described by Strickler (1967).

Free-flow electrophoresis was applied to dissolved organic materials isolated from three black-water rivers (Leenheer and Malcolm, 1973). The results of free-flow electrophoresis were not superior to curtain-supported or gel electrophoresis with respect to resolution of fractions. An uncharged fraction consisting mainly of soluble polysaccharides and a broad, negatively charged band consisting of aquatic humic solutes were obtained for all three isolates. Significant quantities of polysaccharide were also conjugated with aquatic humic solutes as evidenced by their migration in the electric field. The only advantage of free-flow electrophoresis as opposed to supported-electrolyte electrophoresis is the greater preparative capability of free-flow electrophoresis. Results from the free-flow electrophoresis experiments provide additional evidence that aquatic humic substances are probably aggregated in aqueous solution so that high-resolution fractionations are not possible without altering the nature of the interacting sites.

Ultrafiltration, Gel Permeation Chromatography and Ultracentrifugation

Ultrafiltration, gel permeation chromatography, and ultracentrifugation are also discussed by Swift in Chapter 15 of this book, and most of the physical fractionation procedures applied to soil humic substances can also be applied to aquatic humic substances. Dissolved aquatic humic substances are already dissolved in water in contrast to soil systems, and physical fractionations can be performed with or without preconcentration. If possible, these physical fractionation procedures need to be carried out at ambient concentrations to avoid intermolecular aggregation after concentration.

Gel permeation chromatography was applied to natural waters and accepted as a fractionation technique before ultrafiltration. Gjessing (1965) and Gjessing and Lee (1967) related the fractionation obtained by gel permeation chromatography on Sephadex of natural organic matter in lakes and streams to estimated molecular weights, chemical oxygen demand, color, and organic nitrogen. Since this pioneering work, many other studies too numerous to mention have used gel permeation chromatography for fractionation of aquatic humic substances. Gardner and Landrum (1983) recently used high-resolution, size-exclusion liquid chromatography at ambient concentrations to resolve dissolved organic matter into four distinct components using distilled water as the mobile phase. High-speed centrifugation, applied to the samples prior to size-exclusion chromatography, showed that most of the organic matter was truly dissolved rather than colloidal. Addition of interacting cations caused major changes in the fractionation pattern. Gel filtration of aquatic humic substances was directly compared to ultrafiltration by Gjessing (1973). There was little correlation in molecular-size results between the two methods. Gjessing attributed this to the adsorption and disaggregation of humic materials during gel filtration.

Ultrafiltration is now favored by many investigators rather than size-exclusion chromatography for molecular-size fractionations of aquatic humic substances. A discussion of ultrafiltration is also presented in Chapter 19 by Wershaw and Aiken. Ultrafiltration membranes are also subject to adsorptive interactions, and polar neutral solutes pass through ultrafiltration membranes more readily than charged ionic solutes of comparable size. Adsorption effects are much less for ultrafiltration than gel permeation chromatography because of greater flow rates past the adsorbent surface and smaller adsorbent surface area. As with size-exclusion chromatography, the surface chemistry of various ultrafiltration membranes is an important variable affecting adsorptive interactions.

Molecular-size distributions in a natural river and estuary water were determined by ultrafiltration in a study by Smith (1976). Chian and DeWalle

(1977) applied ultrafiltration to a study of organic acids in a landfill leachate. In both studies, the majority of aquatic humic substances appeared in the 1000–10,000 molecular weight range.

Regrettably, the most promising size-fractionating technique, analytical ultracentrifugation, has not been applied to aquatic humic substances. Matrix interactions are practically absent with ultracentrifugation and detectors are sufficiently sensitive to allow fraction detection without preconcentration. The cost and lack of availability of ultracentrifugation equipment has deterred most investigators, but a few studies using ultracentrifugation need to be performed to confirm or disprove the results and interpretations of size-exclusion chromatography and ultrafiltration of aquatic humic substances.

CONCLUSIONS

With the rapid development of both analytical and preparative fractionation techniques presented in this chapter, in addition to the relatively low molecular weight of dissolved aquatic humic substances, it is reasonable to seek to develop an approach to separate dissolved aquatic humic substances into their component compounds. In order to reach our ultimate goal of identification of all the organic solutes isolated from a natural water, a multistep fractionation scheme will be required (see Fig. 5). The initial fractionation of organic solutes from water can be performed by the preparative dissolved organic carbon fractionation procedure (Leenheer and Noyes, 1983). A further analytical fractionation scheme needs to be developed to obtain separations of discrete compounds from the DOC fractions. Next, medium-resolution chromatography (reverse- and normal-phase column and thin-layer chromatography) needs to be used for compound-group subfractionations of underivatized samples. Lastly, high-resolution liquid chromatography needs to be used on the fractions from medium-resolution chromatography after chemical derivatization breaks up loosely bonded aggregates.

Perhaps the use of supercritical fluids as mobile phases in liquid chromatography will permit sufficiently high resolution to allow circumvention of subfractionation procedures. Interfacing of the final high-resolution chromatography with mass spectrometers or infrared spectrometers will give much structural information, but greater quantities of pure compounds will be needed for nuclear magnetic resonance in more definitive structural studies.

The preparative fractionation scheme will apply the chromatography developed for the analytical scheme to larger samples. From 10 to 20 mg of the pure compound isolated from aquatic humic substances needs to be the objective of the preparative fractionation. The preparative fractionation might have to begin with as much as a kilogram of initial material. After structural determination of a number of aquatic humic substance components, the analytical fractionation, combined with spectral determinations,

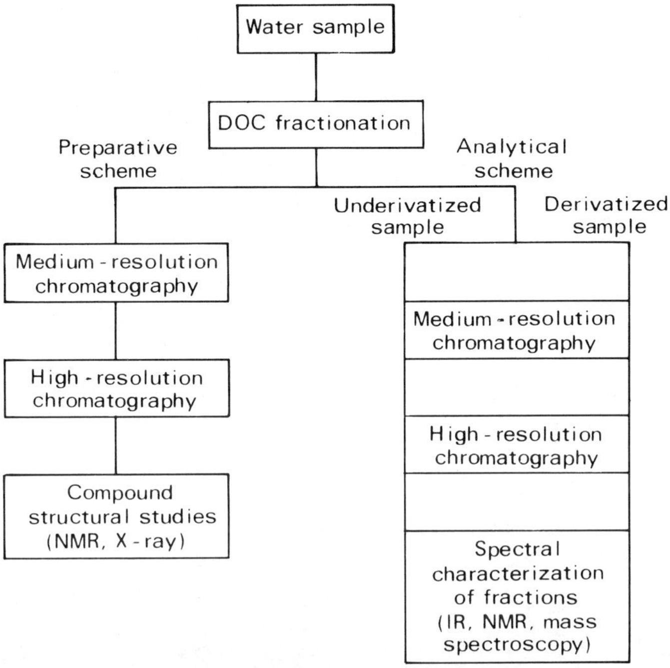

FIGURE 5. Proposed fractionation scheme for dissolved aquatic humic substances.

can be used to monitor reactivity and fate of dissolved aquatic humic substances in the environment.

After chemical structural studies have been successfully applied to dissolved aquatic humic substances, more attention can be given to colloidal and particulate aquatic humic substances. With knowledge of the structure of low-molecular-weight aquatic humic substances, mechanisms of aggregation and polymerization can be postulated and tested by physical fractionation methods, or by chemical structure studies that use the chromatographic methods developed previously. Ultimately, both chemical and physical studies of aquatic and soil humic substances should converge as methods are developed for a large range of molecular weights.

SECTION THREE

CHARACTERIZATION OF HUMIC SUBSTANCES

CHAPTER SEVENTEEN

Analytical Methodology for Elemental Analysis of Humic Substances

E. W. D. HUFFMAN, JR. and HAROLD A. STUBER

ABSTRACT

The applicability of the usual methods of elemental analysis was studied by comparing results obtained by four different laboratories on three different humic and fulvic acid preparations. Identical freeze-dried specimens of each sample were analyzed for carbon, hydrogen, oxygen, nitrogen, sulfur, ash, chlorine, phosphorus, and water content. The results obtained for carbon and hydrogen were within the usually accepted precision ranges of ±0.3% for organic elemental analysis but only when corrected using an accurately determined water content. There were large variations between laboratories in the water content determined as weight loss at 60°C under vacuum, which resulted in unacceptable differences in results between laboratories, especially for hydrogen. Water determined by the Karl Fischer method showed greater precision and it is recommended that water in humic substances be determined by this method. A study of the weight loss under vacuum of the three samples at room temperature, 40, 60, 80 and 110°C and the consequent weight regain on exposure to air was conducted. Rapid regain of moisture indicates these materials should be analyzed on an equilibrated basis rather than dried and that water be determined separately. Results obtained for oxygen, nitrogen, sulfur, ash, chlorine, and phosphorus showed greater variations than commonly accepted, indicating that caution should be used in interpreting the percentage of these components determined in humic substances.

INTRODUCTION

Elemental analysis is probably the most commonly used tool in the characterization of humic substances. Since most humic preparations are mixtures of many components, elemental analysis does not provide an absolute molecular formula but it does provide general compositional information and sets limits as to possible molecular composition. As more highly refined humic substances are produced, the elemental analysis data will become more important in establishing molecular formula information. Elemental analysis is also very useful in establishing the processing efficiency and purity of humic preparations. The elemental composition, along with the element ratios derived from it, such as C/H, O/C, and C/N, are fundamental quantities used in describing and understanding the geochemistry of these substances. Often the elemental data are the single most useful indicator of the unique nature of a given humic or fulvic acid. Elemental composition is often useful in distinguishing different classes of humic substances. For example, there are differences between aquatic and soil fulvic acids and the oxygen to carbon ratios are quite different between groundwater and surface water fulvic acids (Stevenson, 1982; see also Malcolm, Chapter 7 in this book). Elemental analysis provides a cornerstone for our understanding of the diagenesis of humic substances in sediments and is useful in placing soil humic substances in the progressive series between living matter and the peats and coals. In short, it is critically important in interpreting the geochemistry of humic substances. Details relating to the interpretation of results of elemental analysis are presented by Steelink in Chapter 18.

Given the importance of elemental data in interpretive studies, it is necessary to ask questions about the accuracy of the reported results. Very often, elemental data are not obtained by the investigator studying the sample but, rather, by a separate in-house or commercial laboratory. It is imperative that the investigators know the limitations of the results they interpret and how to obtain the best possible results from their laboratories. The purpose of this study was to evaluate the elemental analysis of humic substances as accomplished by the currently common methods. We intended to learn if there are any special problems in attaining accurate data for humic substances, to establish some of the precision limits being attained by laboratories, and, finally, to make recommendations about optimum, or perhaps unacceptable, approaches for elemental analysis of these substances. We know of no published evaluation of the elemental analysis of humic substances.

A basic premise of this discussion is that only limited amounts of sample are available for analysis. We have somewhat arbitrarily assumed that no more than 100 mg of sample are available for a complete elemental analysis. This sample limitation dictates that microchemical methods of analysis be used and that procedures requiring larger amounts of sample will not be considered.

A review of the usual methods of elemental analysis with emphasis on particular problems relative to humic substances is presented. To evaluate the overall reliability of elemental analysis, we conducted a four-laboratory analysis of three carefully prepared, freeze-dried humic substances. In addition to this interlaboratory study, an investigation of the weight loss during vacuum drying of these samples at various temperatures was conducted. We also studied the regain of water on exposure to air. These measurements were made in order to learn more about the water content and hygroscopic nature of the samples, which could then be used as a guide for the proper handling of humic substances in the analytical laboratory.

The sources and handling of samples chosen for this study were known to be carefully and accurately documented. They include one aquatic fulvic acid with a low ash content, one high ash aquatic humic acid, and one low ash soil humic acid. The laboratories chosen include three widely known commercial laboratories in the United States which specialize in organic elemental analysis and one government laboratory with extensive experience in elemental analysis of humic substances.

REVIEW OF METHODS OF ELEMENTAL ANALYSIS

The purpose of this section is to provide a brief review of the methods and techniques commonly used in the elemental analysis of humic substances with special emphasis placed on areas that may cause difficulties in their analysis. There are few specific references to methods of elemental analysis of humic substances. A computerized search of *Chemical Abstracts* since 1966 revealed no references to techniques of elemental analysis when elemental analysis was cross-referenced with humic or fulvic acid materials. In general, the methods of analysis have been developed to be applicable to a wide range of organic materials. However, it should be pointed out that most methods have been validated on the basis of the analysis of stable, nonhygroscopic, nonvolatile, pure compounds and not heterogeneous mixtures.

For more details on these methods, there is a variety of references that may be consulted. The classical methods are given by Pregl (1930) and Niederl and Niederl (1938). Steyermark (1961) and Ingram (1962) have rather thoroughly presented the noninstrumental methods of elemental analysis. The more recent methods, especially the instrumental methods, have been presented in books by Ehrenberger and Gorbach (1973), Belcher (1977), Ma and Rittner (1979), Bance (1980), and Kirsten (1983). Multivolume series such as the *Treatise on Analytical Chemistry* (Kolthoff and Elving 1959–1981) and *Comprehensive Analytical Chemistry* (Wilson and Wilson, 1959) are useful references. A review of papers in this field is presented biyearly (Ma et al., 1982). Most of the methods in the widely used monograph, *Methods of Soil Analysis* (Black, 1965), are not directly applicable to humic substances because of large sample requirements and the fact that much of the emphasis on soil analysis is based on destruction of the inorganic matrix.

Sample Preparation and Handling

Proper preparation and handling of the analytical sample is probably the most important, although most often overlooked, step in analysis. The results of any analysis can be no better than allowed by the sample presented. With heterogeneous materials, it is especially important that great care be used in selecting subsamples for analysis.

It is beyond the scope of this chapter to provide a discussion of techniques involved in isolating and preparing humic substances for analysis. However, one should keep in mind what effects sample preparation techniques will have on subsequent analysis (Walton and Hoffman, 1959).

It is most important to homogenize the material so that representative subsamples can be taken for analysis. Homogeneity is often obtained by exhaustive grinding and mixing of the sample. In principle, since properly prepared freeze-dried materials are prepared by removing the water from immobilized solutes, they should be relatively homogeneous. However, one should ensure that the material does not segregate due to size or density differences and one must recognize that the high surface area of freeze-dried materials may promote absorption of water and/or decomposition reactions.

In general, the greater the purity of the sample material, the better the analysis. Purity is often indicated by ash or, in some cases, heteroatom content. Extractions with mineral acids such as HCl and HF are often used to remove inorganic materials (Stevenson, 1982). It should be recognized that some inorganic materials may be chemically bound to the organic fraction and that demineralization procedures may affect organic constituents.

The moisture content of the sample is an important consideration. If the material is too wet, it rapidly loses water and is difficult to handle. Conversely, if it is too dry, it tends to rapidly gain water from the atmosphere. It is generally desirable to have the sample in approximate equilibrium with the moisture of the laboratory atmosphere. Air sensitive and dry hygroscopic samples can be handled in dry boxes and other inert gas handling systems but these procedures require considerable skill and introduce increased risks of error in the handling. Hydrogen and oxygen determinations are especially difficult using these techniques because the sample must be transferred into the analytical apparatus without gain of water. With other elements, moisture gain after the sample is weighed does not normally affect the result.

The physical nature of the sample is important to the analyst. Freeze-dried materials are particularly difficult to handle. Their low density often makes it difficult to place them in the usual sample containers. The large surface area may cause rapid water gain or loss and the material is often strongly affected by electrostatic charges. Glass or plastic apparatus used to transfer the sample in an inert, dry atmosphere may develop sufficient electrostatic charge to attract sample particles from the sample container to the walls of the transfer apparatus. We have found it desirable to compress freeze-dried material before handling. This compression is most easily accomplished by pressing the material between sheets of aluminum foil. Static problems can be minimized by working with a moist (air equilibrated) sample and exposing the sample to ionizing radiation (usually polonium or a piezoelectric source) to dissipate electrostatic charges. Because of water gain problems, the sample should be allowed to stand on the balance for several minutes to be sure that a stable weight is indicated.

Moisture Determination

The two common approaches to the determination of moisture are to measure the loss in weight when the sample is dried or to measure water directly by the Karl Fischer titration.

The main variables in the loss on drying approach are temperature, pressure, and dessicant. The modified Abderhalden pistol drier (Styermark, 1961) is an efficient drying technique. The sample end of the pistol is placed in a drying oven at specified temperatures. The dessicant, usually P_2O_5, is connected to the drying chamber but is maintained at room temperature. The entire assembly is evacuated, usually to a pressure of 0.1–2 mm of mercury. Vacuum ovens are also commonly used to dry samples. Some systems use a stream of dry gas to sweep moisture from the samples.

The most difficult parameter to optimize is the drying temperature. In general, one wants to use as high a temperature as possible to maximize the water loss without causing decomposition of the sample. In fact, there may be no temperature that will cause the complete removal of water without causing sample decomposition. This problem is further investigated later in the experimental section of this chapter.

An alternative to the loss on drying approach is to directly measure the water lost from the sample while the sample is heated in a dry carrier gas. The water in the carrier gas can be determined by a variety of techniques such as gravimetry, coulometry, or Karl Fischer titration. Although this approach may help to distinguish between water loss and decomposition, the strategy is not without its problems because water is often a decomposition product.

The titrimetric determination of water using the Karl Fischer reagent (mixture of iodine, sulfur dioxide, pyridine, and methanol) is discussed in detail by Mitchell and Smith (1980). Although the reaction is complex, the net reaction is the reaction of water with iodine in a 2 : 1 ratio in the presence of pyridine:

$$I_2 + 2H_2O + (C_5H_5N)_2 \cdot SO_2 + 2C_5H_5N \rightarrow (C_5H_5N)_2 \cdot H_2SO_4 + 2C_5H_5N \cdot HI$$

The reaction has a wide applicability but is subject to certain interferences. In general, any material that will react with iodine in the reagent mixture will cause an interference. Common organic interferences are active carbonyl compounds, ascorbic acid, quinone, mercaptans, and diacyl peroxides (Mitchell, 1961). Several of these interferences can be eliminated or minimized by appropriate modifications to the method such as prereaction to remove interfering materials and extrapolation of the observed endpoint to the true endpoint based on kinetically slow interfering reactions. Care must also be taken to avoid introducing extraneous moisture from the atmosphere and to be sure that the reaction has reached completion.

Determination of the Elements

Organic elemental analysis usually involves a combustion of the weighed sample, separation of the desired product from potentially interfering materials, conversion (usually quantitative) of the element to a form that can be measured, and final measurement. In some cases these steps are combined and several elements can be determined simultaneously. The two general approaches to final determination are absolute and sample calibrated methods. In general, absolute measurements such as gravimetry, coulometry, and volumetry have certain advantages over sample calibrated measurements such as thermal conductivity and nondispersive infrared analysis. The sample calibrated techniques can correct for errors in the analysis that are common to both standard and sample but in many cases these errors may not be consistent between standard and sample. The absolute measurement allows the analyst to show, by analyzing appropriate standards, that the procedure does not contain significant errors and that recoveries are quantitative. Unfortunately, some absolute techniques, particularly gravimetry, have not been readily adaptable to automation and require such considerable skill on the part of the analyst that calibrated techniques may be advantageous for the usual laboratory.

Most organic analysts and instrument manufacturers quote analysis accuracies on the order of ±0.3% absolute for elemental analysis. The statistical significance of this figure is somewhat obscure but the authors generally take it to represent an absolute error of two standard deviations as determined on standard organic reference materials. When applied to humic substances, this level of accuracy is good for carbon, which will typically run 40–70% but certainly represents poor *relative* accuracy (30%) on a 1% nitrogen level. A major reason for this state of affairs is that accuracies are normally determined by analysis of highly purified organic standards. These materials all contain relatively high levels of the analyte of interest and, largely for convenience, accuracy is calculated on an absolute basis. There is little question that absolute accuracies can generally be improved for lower levels of most elements but, because of a lack of standards, it is difficult to prove what level of accuracy is obtained. Usually this problem is overcome to some degree by preparing mixed standards, where the element of interest is diluted by some other matrix. This approach introduces several problems mostly involved with homogeneity of the mixture and the type and purity of the matrix material. Another approach is to analyze a natural sample by several techniques and use an average value. This has obvious limitations in that the average results are really no better than the techniques used. The analyst must always remember that a given analysis may yield results that have high precision but very poor accuracy.

Carbon and Hydrogen

In organic microanalysis carbon and hydrogen are traditionally determined simultaneously. Nitrogen is often determined simultaneously with carbon and hydrogen, but since it is often determined separately it will be discussed separately. Carbon and hydrogen are determined by combusting a weighed sample in an excess of oxygen. The oxidation of benzoic acid illustrates the combustion and its products:

$$C_7H_6O_2 + 7\tfrac{1}{2}O_2 \xrightarrow{1000°C} 7CO_2 + 3H_2O$$

To ensure that this reaction reaches completion, an excess of oxygen should be used. The combustion gases are swept through a combustion tube filling which consists of catalysts for oxidation of the organic material and scrubbers to remove potentially interfering materials such as sulfur and halogens. This oxidation converts all carbon to carbon dioxide and all hydrogen to water. Classically, the CO_2 is absorbed in a basic material such as sodium hydroxide on asbestos (Ascarite) and weighed. Similarly, the H_2O is absorbed by anhydrous magnesium perchlorate and weighed. Carbon dioxide has also been determined after absorption in an appropriate medium by coulometry, electrical conductivity, or aqueous titration. Alternatively, CO_2 has been determined in the gas phase by thermal conductivity (after chromatography or absorption), by infrared spectroscopy, and by a gasometer. Water from the combustion has been determined by thermal conductivity (after chromatography or absorption), Karl Fischer titration, or coulometry, and is often quantitatively converted to other compounds such as acetylene or CO_2 for quantification.

Some humic samples may contain high ash levels and it is therefore important to use sufficiently high combustion temperatures (usually $>1000°C$) to decompose any carbonates that may be present in the sample or form during combustion. If carbonates are present, the carbon value obtained represents total carbon. The carbonate carbon content of the sample can be determined by measuring the quantity of CO_2 evolved from a weighed sample when it is treated with a strong mineral acid (Huffman, 1977). This determined carbonate carbon value can then be subtracted from the total carbon value to given an organic carbon value. Since carbon is the major component of most organic samples, the sample weighing is very important to the accuracy of the analysis. For example, in considering a sample containing 60% carbon and 1% nitrogen, a 1% weighing error would cause a 0.6% error in the carbon value and only 0.01% error in the nitrogen value.

Nitrogen

The commonly used methods for the determination of total nitrogen involve variations of the Dumas or Kjeldahl techniques. The Dumas approach involves oxidation of the nitrogen to elemental or oxide form by combustion of the sample, reduction of nitrogen oxides, and determination of the elemental nitrogen. The Dumas oxidation is usually represented by the following example for acetanilide:

$$2C_8H_9NO + 19\tfrac{1}{2}O_2 \xrightarrow{1000^\circ C} 16CO_2 + 9H_2O + N_2$$

It should be noted that in initial oxidation some nitrogen oxides may be formed but these are reduced to N_2 by reaction with metallic copper at 500–750°C. This determination can be performed simultaneously with the carbon and hydrogen determinations by using an inert carrier gas to pass the combustion products over copper to remove oxygen and reduce nitrogen oxides to nitrogen gas followed by detection of nitrogen. Thermal conductivity detection, either after chromatographic separation of the nitrogen or by comparing the original carrier gas to the sample carrier gas after removal of other constituents, is used. An absolute nitrogen measurement is made using a nitrometer, where the CO_2 carrier gas is absorbed by KOH and the remaining nitrogen gas is determined volumetrically. The classical Dumas procedure used only copper oxide as a combustion oxidant. Modern methods use combustion in an excess of oxygen to eliminate many of the shortcomings of the classical procedure (Merz, 1968; Trutnovsky, 1974). A well maintained Dumas system will give good recovery of virtually all nitrogen species, with the possible exception of slightly low recoveries on materials containing highly oxidized nitrogen species (e.g., dinitrobenzene, sodium nitrate).

The Kjeldahl nitrogen method is based on the fact that, upon digestion with sulfuric acid and catalysts, the organic material is destroyed and most forms of nitrogen are converted to ammonia. When the digestion mixture is made alkaline, the ammonia is released. The ammonia is removed by distillation, absorbed, and titrated. These reactions can be represented as follows:

$$\text{Digestion:}\quad \text{Organic N} \xrightarrow[\text{heat}]{H_2SO_4} 3CO_2 + 4\tfrac{1}{2}H_2O + NH_4HSO_4$$

$$\text{Distillation:}\quad NH_4HSO_4 + 2NaOH \rightarrow NH_3 + Na_2SO_4 + 2H_2O$$

$$\text{Titration:}\quad NH_3 + HCl \rightarrow NH_4Cl$$

Numerous variations to the classical Kjeldahl procedure, such as various digestion catalysts and various methods of determinations of ammonia, have been proposed and are summarized in the reviews of Ma et al. (1982). To determine nitrogen in compounds containing N—N, N=N, NO, and NO_2, modifications such as zinc iron reduction must be made to the usual procedure (Styermark, 1961). Certain forms of nitrogen such as triazoles cannot be determined by the Kjeldahl method.

A recent approach is the use of combustion followed by chemiluminescent detection of nitric oxide produced in the combustion (Drushell, 1977). The senior author's experience is that response of this detector apparently is dependent on the composition of the sample, and that therefore the instrument must be standardized with materials very similar to the unknown (Schuchardt, 1980).

Oxygen

Unlike most other elements, the determination of oxygen involves a reductive pyrolysis in an inert gas. The pyrolysis products are passed over carbon, which converts oxygen in the pyrolysis products to carbon monoxide. The general reaction is as follows:

$$\text{Organic O} \xrightarrow[\text{carbon}]{1120°\text{C}} \text{CO} + \text{CH}_4 + \text{other pyrolysis products}$$

Several investigators have proposed using catalysts in the carbon such as platinum (Oita and Conway, 1954) or nickel (Kirsten, 1983) to lower the required pyrolysis temperature. The senior author has found that noncatalyzed carbon at 1120°C has superior performance to catalyzed carbon. The pyrolysis gases are scrubbed, usually with caustic preparations, to remove interfering materials and the carbon monoxide is then determined. The carbon monoxide from the pyrolysis may be determined directly using a gas chromatographic separation prior to thermal conductivity, infrared absorption spectroscopy, or iodometry. Alternatively, the CO may be oxidized to CO_2 and determined by gravimetry, coulometry, or one of the other methods discussed under carbon determinations.

Oxygen results must be carefully evaluated to interpret their applicability to various sample types. Since water contains 88.89% oxygen, the moisture content of the sample must be accurately known. Any gain or loss of moisture after weighing the sample is by far the most critical with oxygen determinations. The first step of the oxygen procedure is to use the dry carrier gas to backsweep air (oxygen) introduced with the sample out of the system. This backsweep will also remove loosely bound moisture from the sample. In some analyzers it is possible to minimize this moisture loss by cooling the sample to dry ice–acetone temperatures during the backsweep.

The oxygen procedure is generally applicable to organically bound oxygen. Oxygen associated with metals is only determined if the metal is reduced or the oxygen is displaced by another element. Kapron and Brandt

(1961) have used carbon and cuprous chloride to reduce metal oxides. Merz (1970) has prepared an additive mixture containing hexamethylenetetramine, ammonium chloride, and silver chloride to reduce and/or replace oxygen from metals. In general, oxygen associated with refractory metal oxides (e.g., silicon, aluminum, boron) will not be determined unless a high-temperature induction furnace is used for the pyrolysis. However, the senior author has observed that the induction furnace may give low results with organic compounds.

Fluorine causes an absolute interference by nonstoichiometrically replacing oxygen in the pyrolysis tube and the method is not applicable when the sample contains significant levels of fluorine. High levels (>30%) of sulfur and phosphorus may also cause errors in the procedure.

Depending in part on the preparation method, humic substances may be subject to these problems. However, since oxygen is a major constituent of humic substances, it is important to know its true value. As discussed above, careful accounting of the moisture content of the sample is critical. One should also have information on the composition of the ash so that its effect on the oxygen analysis can be considered. If HF has been used to remove ash constituents, the sample should be analyzed for fluorine residues.

Oxygen is commonly determined by difference (e.g., subtracting all other determined values from 100%). Two major drawbacks to this approach are: (1) the calculated value includes the sum of the errors in all other determinations, and (2) the ash may contain elements already determined (e.g., sulfates carbonates, etc.) so that when the ash content is subtracted, these elements are, in effect, subtracted twice. In regard to this last point, it is common for an elemental analysis to total over 100%, because oxygen determined directly is also included in the ash.

Oxygen may also be determined by neutron activation analysis (Anders and Briden, 1964). This approach determines inorganic oxygen as effectively as organic oxygen. Fluorine presents a serious interference with this method. Of course, the considerations of moisture are still important.

Sulfur

A wide variety of techniques are used to decompose the sample and to determine the resulting sulfur species. The reviews by Heinrich et al. (1961) and Alcino et al. (1965) summarize a large selection of the methods used for the determination of sulfur. The general reactions for the decomposition of the sample are as follows:

$$\text{Organic S} + H_2 \xrightarrow[\text{catalyst}]{\text{heat}} H_2S + CH_4$$

$$\text{Organic S} + \text{metal} \xrightarrow[\text{fusion}]{} \text{metal sulfide}$$

$$\text{Organic S} + O_2 \xrightarrow[\text{heat}]{} SO_3 + SO_2 + H_2O + CO_2$$

$$SO_2 \xrightarrow[\text{catalyst, [O]}]{} SO_3$$

The sulfide generated by the first two reactions can be detected with a high degree of sensitivity, making this approach particularly valuable for trace analysis. Many laboratories do not like to use hydrogen or to work with relatively toxic H_2S and therefore the reduction procedures are not widely used.

Many procedures are used for the oxidation of organic compounds for the determination of sulfur. Methods which are described in the above reviews include open tube combustion, oxygen flask combustion, oxygen bomb combustion, and fusion with various salts including sodium peroxide and sodium carbonate. The sulfate resulting from the sample decomposition is usually determined either gravimetrically as barium sulfate or by titration with barium using either thorin or sulfanazo III indicators. Low levels of sulfate can be determined by nephelometry (Ma and Rittner, 1979) or by ion chromatography (Small et al., 1975; Fritz et al., 1982).

Several commercial sulfur analyzers are based on the fact that, in high-temperature combustion in oxygen, sulfur is predominantly converted to sulfur dioxide, which can readily be measured by infrared spectroscopy (LECO Corp., St. Josephs, Michigan), thermal conductivity (Kirsten, 1983), or titration (Mansfield and Gibboney, 1977). The drawback to this approach is that the conversion to SO_2 is not complete and the extent of conversion is dependent on sample matrix. In addition, these analyzers generally require 10–100 mg or more of sample.

Combustion of the sample in an oxygen combustion flask (Alcino et al., 1965) followed by ion chromatography is an attractive method for the analysis of sulfur in humic substances. Not only does the method require relatively little sample and provide high sensitivity, but it also allows simultaneous determination of halogens. One must be sure that sulfur is not rendered insoluble by ash constituents such as calcium.

Halogens (Chlorine, Bromine, Iodine)

Ingram (1962) has discussed many of the methods used for the determination of halogens. In general, the decomposition methods are similar to those used for sulfur, and in many cases sulfur and halogens may be determined on aliquots of a single sample decomposition. The general reactions are as follows: (X represents either Cl, Br, or I)

$$\text{Organic X} + H_2 \xrightarrow[\text{catalyst}]{\text{heat}} HX + CH_4$$

$$\text{Organic X} + \text{metal} \xrightarrow[\text{fusion}]{} \text{metal X}$$

$$\text{Organic X} + O_2 \xrightarrow{\text{heat}} HX + HXO_n + H_2O + CO_2$$

$$HXO_n \xrightarrow[\text{reducing agent}]{} HX$$

$$HXO_n \xrightarrow[\text{oxidation}]{} HXO_4$$

The first four reactions all result in the -1 valence state of the halogen. The halide can be determined by argentometric titration, mercurimetric titration, turbidimetry (as AgX), colorimetery, or ion chromatography. The use of potentiometric titration or ion chromatography allows determination of the individual halides. Bromine and iodine may be determined individually by using selective oxidation techniques as shown by the last general equation above. After the oxidation to the highest oxidation state, the bromine or iodine is titrated iodometrically. For humic substances which normally contain relatively low levels of halogens, the oxygen flask combustion followed by ion chromatography is perhaps the method of choice. This method provides fairly high sensitivity and, as discussed above, allows the simultaneous determination of sulfur.

Phosphorus

Ma and Rittner (1979) conclude that acid digestion is the method of choice for decomposition of organic samples for the determination of phosphorus. The senior author has found that it is important to use perchloric acid at the end of the digestion to convert all phosphorus to orthophosphate. The reaction is as follows:

$$\text{Organic P} \xrightarrow[HNO_3,\ H_2SO_4,\ HClO_4]{\text{heat}} H_3PO_4 + CO_2 + H_2O$$

The orthophosphate can be determined by gravimetry, titrimetry, or colorimetry.

For humic substances which normally are expected to contain small amounts of phosphorus, the molybdenum blue colorimetric method (Ma and Rittner, 1979) is recommended. The blue color provides a high sensitivity, allowing relatively small samples to be used for the determination.

Ash

In the analysis of pure metal organic compounds, ashing may be used to determine the quantity of a specific metal present in the sample. In the case of heterogeneous materials such as humic substances, ashing is generally used as an indication of total inorganic content. The basic concept of ashing—that is, to burn off the organic material and leave the inorganic residue behind—is a simple concept. However, there are a number of factors that may affect the value obtained. The following equations show the reactions that occur under various conditions for a calcium-containing organic material:

$$\text{Organic Ca} + O_2 \rightarrow CaCO_3 + H_2O + CO_2$$

$$\text{Organic Ca} + O_2 \rightarrow CaO + H_2O + CO_2$$

$$\text{Organic Ca} + O_2 \xrightarrow[H_2SO_4]{} CaSO_4 + H_2O + CO_2$$

Nitric acid (1 : 5) is occasionally added to aid in sample decomposition. Sulfuric acid (1 : 5) is commonly added to convert alkalies to sulfates, as shown in the last equation above. If sulfuric acid is not added, elements such as sodium, potassium, calcium, and magnesium may form mixtures of sulfates, chlorides, oxides, carbonates, and so on, depending on the composition of the sample and the ashing conditions. Formation of sulfate greatly enhances the weight of ash (e.g., 1 mg of MgO is converted to 2.99 mg $MgSO_4$) and care must be used in interpreting results of sulfate ashing.

The temperature of ashing commonly varies from low temperatures using an oxygen plasma to 1000°C using a muffle furnace. The usual ashing temperature ranges from 700 to 800°C. Higher temperatures decompose carbonates and ensure complete oxidation of organic matter but may result in losses of more volatile metals.

For microash procedures, the ashing is often performed in oxygen, whereas, macroash procedures usually use air. The use of oxygen may help to ensure complete combustion and enhance the decomposition of carbonates. The important factors influencing these reactions include the nature of the original sample, the temperature of ashing, and the addition of acid to the sample.

For humic substances, the Coombs–Alber (Styermark, 1961) microashing boat and sleeve are recommended. The sleeve helps to prevent spattering losses and may aid in the decomposition of metal organics. The senior author has observed ashes where the metal organic vaporized from the boat but decomposed and left the ash residue in contact with the platinum sleeve. The researcher must determine if a sulfated ash is desired and what temperature is best for the expected ash composition. Ingram (1962) presents an especially useful element by element treatment of ashing. The analyst should recognize that good technique is important in ash determination. Common sources of error include not allowing the ash vessel to come to temperature equilibrium with the balance or not protecting hygroscopic ashes from moisture gain. It should be recognized that phosphorus is not completely volatilized in ashing and will be included in the ash.

Other Elements

It is beyond the scope of this chapter to consider analysis of specific metals and other trace elements. In many humic preparations the ash content represents either contamination from the mineral matrix or contamination from reagents used in processing the preparation. In those cases it may be useful to semiquantitatively determine the elemental content of the ash by subjecting the ash to emission spectroscopy or X-ray fluorescence spectrometry.

Ingram (1962) and Ma and Rittner (1979) have summarized many techniques for specific metal analysis. Since humic substances are generally low in ash content, the most applicable approaches are to determine metals by atomic absorption or inductively coupled plasma spectroscopy after appropriate dissolution of the sample.

PREPARATION OF SAMPLES AND EXPERIMENTAL METHODS

The three humic substances used in this study were obtained from the laboratory of Dr. Ronald Malcolm of the U.S. Geological Survey, Denver, Colorado. Sample No. 1, isolated from Coal Creek, a tributary of the Yampa River in Colorado, is an aquatic fulvic acid which was prepared according to the method described by Thurman and Malcolm (1981). The water sample was filtered using 0.45 μm silver membrane filters, acidified to pH 2, and passed through an XAD-8 resin column (see Aiken, Chapter 14). The column was eluted with 0.1*N* NaOH. The isolated humic and fulvic acids were acidified and reconcentrated using XAD-8. The final concentrate was acidified to pH 1.5 to precipitate the humic acid, which was removed by centrifugation after 24 hours. The soluble fulvic acid was neutralized with NaOH, hydrogen saturated by passage through a cation-exchange resin in the H-form, and freeze-dried.

Sample No. 2, an aquatic humic acid isolated from the Ogeechee River in Georgia, was prepared in the same manner as sample No. 1, except that a Balston glass fiber filter (nominally 0.3 μm) was used instead of silver membrane filters to filter the original sample. This sample had a high ash content because of fine clay which passed through the filters and was ultimately isolated with the humic acid. Analysis of the ash showed it to be largely aluminum, silicon, iron, and oxygen (George Aiken, US Geological Survey, Denver, personal communication).

Sample No. 3 is a soil humic acid which was isolated from an Alaskan soil (Malcolm, 1976). The soil was extracted for 12 hours under N_2 using 0.1*N* NaOH. The humic acid was precipitated at pH 1, isolated by centrifugation, and freeze-dried.

The samples were divided into 100 mg portions and submitted to four laboratories for analysis. The laboratories included three commercial analytical laboratories that specialize in organic microanalysis and one government laboratory that specializes in analysis of humic substances. The following ranges of expected content were supplied in a cover letter to each laboratory.

Carbon	20–70%
Hydrogen	2–8%

Oxygen	20–40%
Nitrogen	0.1–5%
Sulfur	0.1–5%
Ash	0.1–60%
Chlorine	0.005–5%
Phosphorus	0.005–2%

We included the suggestion that the dried materials were extremely hygroscopic and were best analyzed on air equilibrated samples, as supplied. The laboratories were also advised that no additional sample was available for analysis and that all analyses must be completed on the supplied sample.

The drying studies were carried out using a vacuum oven. The air inlet and vacuum outlet were protected with tubes containing approximately 500 g of indicating anhydrous calcium sulfate. A manometer connected to the vacuum line indicated a vacuum of approximately 1 mm of mercury. The samples were weighed into soft glass weighing bottles (pigs) fitted with ground glass stoppers. The pigs were kept stoppered except when the samples were in the oven or being exposed to the air. The pigs were allowed to come to room temperature before being weighed on a six-place Mettler model M5 microbalance.

RESULTS AND DISCUSSION OF INTERLABORATORY STUDY

Interlaboratory Study

The results of the analysis of the three humic substances by four laboratories are shown in Tables 1–3. The results in these tables have been corrected for water content based on an average value of the Karl Fischer water determinations. Table 6 summarizes the methods used by the various laboratories in the interlaboratory study. The agreement between the laboratories for the carbon and hydrogen contents of the samples is excellent for all three samples when the average Karl Fischer water composition is used to correct the raw data. The range of hydrogen values is well within the uncertainty of 0.3% absolute that is the usually accepted standard for organic elemental analysis. Carbon values show a range of up to 0.7% absolute (on sample No. 3), but this still represents fairly good precision on a sample that contains 52% carbon. Carbon and hydrogen results indicate these elements are being determined consistently and reproducibly by the laboratories and that the samples are homogeneous with respect to major elements. The excellent precision for carbon in samples No. 1 and No. 2 is impressive and the results from sample No. 2, the high ash sample, show no special problems are occurring for carbon determination on a high ash sample. Hydrogen was also determined quite precisely on this high ash sample, although the standard deviation and range are somewhat higher than for the low ash samples.

Before discussing the analysis of any of the other elements it is necessary to discuss the critical importance of the determination of water in these samples. The water determination is crucial to the correct determination of carbon, hydrogen, and oxygen, the most abundant elements in humic substances, and there are different approaches to the determination of water from which to choose.

The results in Tables 1, 2, and 3 show that the weight loss on vacuum drying at 60°C was very inconsistent between laboratories on all three samples. The numbers vary by an unacceptable factor of 2 on samples No. 1 and No. 2 and by 30% on sample No. 3. Whatever the cause of this variation, it is clear that large discrepancies would result in the carbon, hydrogen, and oxygen values reported by these laboratories if the raw carbon, hydrogen, and oxygen data were corrected based on these reported moisture values.

Table 4 shows the "as reported" and "corrected" data for carbon, hydrogen, and oxygen from laboratories 1 through 4. Note that the carbon values reported by laboratories 1 and 2 are quite close to the corrected values calculated using an average Karl Fischer water value. This is because both labs calculated these results based on water values close to that average. The results for laboratories 3 and 4 are mostly quite different from the corrected values, and use of these data might therefore suggest there were serious problems with the carbon analysis. We do not feel the carbon analyses were faulty, but rather, that their moisture content determinations by vacuum drying were unreliable.

TABLE 1. Elemental Analysis Data on Sample 1 Corrected for 6.56%[a] Water (All Values are in Weight Percent)

	Values Found						
Determination	Lab 1	Lab 2	Lab 3	Lab 4	$\bar{X}^b$	S^c	RSD[d]
Carbon	54.10	53.69	53.69	54.00	53.87	0.21	0.39
Hydrogen	4.22	4.09	4.23	4.20	4.22	0.013	0.30
Oxygen	37.30	36.55	31.14	—	35.00	3.36	9.6
Nitrogen	1.10	0.88	1.27	0.81	0.99	0.20	20.5
Sulfur	0.43	0.56	0.28	0.98	0.56	0.30	53.7
Chlorine	0.23	0.16	0.097	—	0.16	0.067	41.9
Phosphorus	0.02	0.15	<0.0005	—	0.07	—	—
Ash	1.13	1.44	0.53	0.31	0.85	0.52	61.5
Total	98.44	97.52	91.237	69.30	95.94	—	—
Weight loss at 60°C	7.02	7.21	4.64	3.41	5.57	1.85	33.3
Moist (KF)	6.71	6.41	—	—	—	—	—

[a] *The average Karl Fischer moisture content as determined by Laboratory 1 and Laboratory 2.*
[b] *$\bar{X}$ average*
[c] *S standard deviation*
[d] *RSD relative standard deviation*

TABLE 2. Elemental Analysis Data on Sample 2 Corrected for 5.59%[a] Water (All Values are in Weight Percent)

Determination	Values Found Lab 1	Lab 2	Lab 3	Lab 4	$\bar{X}$	S	RSD
Carbon	20.34	20.30	20.13	20.20	20.24	0.095	0.47
Hydrogen	2.58	2.51	2.74	2.69	2.63	0.10	3.96
Oxygen	18.45	19.31	—	—	18.88	0.61	3.22
Nitrogen	1.31	1.30	0.82	1.28	1.18	0.24	20.2
Sulfur	0.40	0.58	0.26	0.94	0.55	0.29	53.5
Chlorine	0.95	0.76	0.56	—	0.76	0.20	25.66
Phosphorus	0.23	0.19	0.18	—	0.20	0.026	13.23
Ash	57.50	55.22	55.63	46.64	53.74	4.85	9.03
Total	101.76	100.17	83.32	80.75	98.18	—	—
Weight loss at 60°C	4.98	6.01	2.60	3.27	4.21	1.56	37.0
Moist (KF)	5.50	5.69	—	—	—	—	—

[a] *The average Karl Fischer moisture content as determined by Laboratory 1 and Laboratory 2. Symbols have same meaning as in Table 1.*

The hydrogen and oxygen values reported by laboratories 3 and 4 show very large discrepancies from the corrected data. Note the close agreement of the two labs (3 and 4) on hydrogen and the large discrepancies with labs 1 and 2. The reason for this great discrepancy is that neither laboratory 3 or 4 made the necessary correction for the hydrogen and oxygen present in the

TABLE 3. Elemental Analysis Data on Sample 3 Corrected for 7.41%[a] Water (All Values are in Weight Percent)

Determination	Values Found Lab 1	Lab 2	Lab 3	Lab 4	$\bar{X}$	S	RSD
Carbon	52.55	52.47	52.05	51.81	52.22	0.35	0.67
Hydrogen	3.75	3.69	3.83	3.72	3.75	0.060	1.61
Oxygen	34.80	34.36	30.81	—	33.32	2.19	6.57
Nitrogen	3.23	3.14	1.45	2.68	2.63	0.82	31.16
Sulfur	0.32	0.41	0.28	0.94	0.49	0.31	62.56
Chlorine	0.49	0.20	0.17	—	0.29	0.18	60.94
Phosphorus	0.50	0.50	0.39	—	0.46	0.064	13.81
Ash	3.25	1.88	2.14	1.68	2.24	0.70	31.28
Total	98.89	96.65	91.40	69.83	95.40	—	—
Weight loss at 60°C	7.78	7.55	7.33	9.55	8.05	1.02	12.6
Moist (KF)	7.63	7.20	—	—	—	—	—

[a] *The average Karl Fischer moisture content as determined in Laboratory 1 and Laboratory 2. Symbols have same meaning as in Table 1.*

TABLE 4. Comparison of Carbon, Hydrogen, and Oxygen Results as Reported by the Individual Laboratories and as Corrected for Moisture[a] (All Values are in Weight Percent)

Sample	Lab 1		Lab 2		Lab 3		Lab 4	
	Corr.	As Reported	Corr.	As Reported	Corr.	As Reported	Corr.	As Reported
Carbon								
No. 1	54.10	54.19	53.69	53.60	53.69	53.61	54.00	52.24
No. 2	20.34	20.32	20.30	20.32	20.13	19.51	20.20	19.72
No. 3	52.55	52.68	52.47	52.35	52.05	52.01	51.81	53.04
Hydrogen								
No. 1	4.22	4.16	4.09	4.10	4.23	4.86	4.20	4.77
No. 2	2.58	2.56	2.51	2.50	2.74	3.26	2.69	3.24
No. 3	3.75	3.67	3.69	3.71	3.83	4.66	3.72	4.66
Oxygen								
No. 1	37.30	37.22	36.55	36.62	31.14	38.76	—	—
No. 2	18.45	18.52	19.31	19.24	—	—	—	—
No. 3	34.80	34.68	34.36	34.45	30.81	37.36	—	—

[a] *Corrected for the average Karl Fischer moisture determined by Laboratory 1 and Laboratory 2.*

moisture in the sample when calculating the hydrogen and oxygen content of the sample. This is a serious error, which might not have been detected if this were not a multilaboratory study.

The greatest absolute deviation was in oxygen results. Much of this is probably due to the problems with moisture as discussed above and the inherent difficulties with the oxygen method. The elemental totals for samples No. 1 and No. 3 determined by lab 3 are approximately 9% short of adding to 100%. Since the carbon and hydrogen values are all in good agreement, and it is difficult to believe that any other element or elements not determined could account for 9% of the total, the oxygen values reported by lab 3 appear to be grossly in error. Laboratory 3 felt that an oxygen determination was not appropriate on the high ash sample (No. 2). Laboratory 4 does not perform direct oxygen determinations. There was fair agreement between lab 1 and lab 2 on all oxygen results. The low elemental totals on samples No. 1 and No. 3 are most likely caused by low oxygen values, due to loss of moisture during the backsweep phase of the analysis. The high total obtained by lab 1 on sample No. 2 may be due to measuring a portion of the oxygen twice (directly and in ash).

With the exception of lab 3, which was markedly low on two samples and high on the other, the nitrogen values were in good general agreement.

Laboratory 4 obtained consistently higher sulfur values. The results of the other labs are within 0.16% absolute of their mean sulfur value.

Based on the data obtained, it is difficult to evaluate the results for chlorine and phosphorus. Laboratory 3 obtained consistently lower results on both elements and lab 1 tended to have the highest chlorine values.

The ash values showed considerable variation between laboratories. Table 5 summarizes the ashing procedures used by the various laboratories. Although lab 4 obtained consistently low ash values, in general, higher ash values are obtained where lower temperatures, spattering protection, and platinum containers are used.

It is valuable to use the total of the elemental analyses plus ash to evaluate the overall reliability of analysis. Since laboratory 4 did not run all elements, especially oxygen, their total has little meaning. The low totals obtained by lab 3 suggest a major error in the analysis. The totals obtained by labs 1 and 2 suggest minor errors, as discussed under oxygen above, or that some ele-

TABLE 5. Summary of Methods Used by the Various Laboratories in the Interlaboratory Study

Determination	Lab 1	Lab 2	Lab 3	Lab 4
Carbon and hydrogen	Modified Coulometrics, Inc.	Perkin Elmer model 240 elemental analyzer	Modified Pregel	Coleman model 33
Nitrogen	Modified Dumas (Merz)	Perkin Elmer model 240 elemental analyzer	Dumas	Coleman nitrogen analyzer (Dumas)
Oxygen	Modified Coulometrics, Inc. model 5060 oxygen analyzer	Carlo–Erba oxygen analyzer	Perkin Elmer model 240	—
Sulfur	Oxygen flask combustion followed by ion chromatography	Oxygen flask combustion followed by ion chromatography	Potassium fusion and methylene blue determination of evolved H_2S	Oxygen flask combustion followed by barium titration
Chlorine	Oxygen flask combustion followed by ion chromatography	Oxygen flask combustion followed by ion chromatography	Colorimetric	—
Phosphorus	Acid digest yellow phosphomolybdate colorimetric	Acid digest yellow phosphomolybdate colorimetric	Gravimetric	—
Ash	Platinum boat inside platinum sleeve in tube with oxygen flow heated to 750°C to constant weight	Ceramic crucible at 750°C in ashing furnace to constant weight	In crucible over burner at 890°C	Muffle furnace for 4 hours at 800°C. Did not use microbalance

ment has not been determined. Low elemental totals can be caused by low ash values, possibly caused to some extent by volatilization of organometallic compounds.

Effect of Temperature on Weight Loss on Drying

The results of a study of weight loss on drying of three humic substances are shown in Table 6 and in Figure 1. Duplicate samples of each of the humic substances were vacuum dried to constant weight at 20, 40, 60, 80 and 110°C over P_2O_5. The weight loss at each temperature was determined. The dried samples were then exposed to air and the weight regain was recorded after 1, 5, and 10 minutes and after 6 hours.

All three samples show a gradual, consistently increasing weight loss with increasing temperatures up to 80°C. Presumably, this represents the loss of increasingly more tightly bound water as more energy is available for desorption. Two of the samples (No. 1 and No. 3) show a sharply increased loss at 110°C. The samples also were observed to darken in color, suggesting that some structural decomposition is occurring at this temperature. Sample No. 2, however, does not show a large additional loss at 110°C, but rather a small, consistent increase compared to lower temperatures. These results imply that there is no "correct" temperature for a moisture determination

TABLE 6. Moisture Regain as a Function of Drying Temperature and Time

Sample	Drying Temperature °C	Loss On Drying (%)	% Regained at Time After Air Exposure			
			1 min	5 min	10 min	6 hours
1	20	5.01	0.04	0.57	1.16	3.88
2	20	3.22	0.33	1.06	1.58	2.84
3	20	5.52	0.19	1.02	1.78	5.08
1	40	6.24	0.12	0.60	1.14	4.41
2	40	4.23	0.13	0.73	1.36	3.36
3	40	7.10	0.16	0.86	1.67	5.96
1	60	7.02	0.18	0.69	1.24	4.64
2	60	4.98	0.08	0.75	1.37	3.82
3	60	7.78	0.24	1.07	1.94	6.14
1	80	8.33	0.13	0.53	1.04	4.00
2	80	5.48	0.21	0.82	1.28	3.38
3	80	8.75	0.10	0.69	1.38	5.79
1	110	12.92	0.17	0.66	1.09	3.06
2	110	6.50	0.10	0.75	1.35	2.61
3	110	12.48	0.24	0.92	1.61	4.58

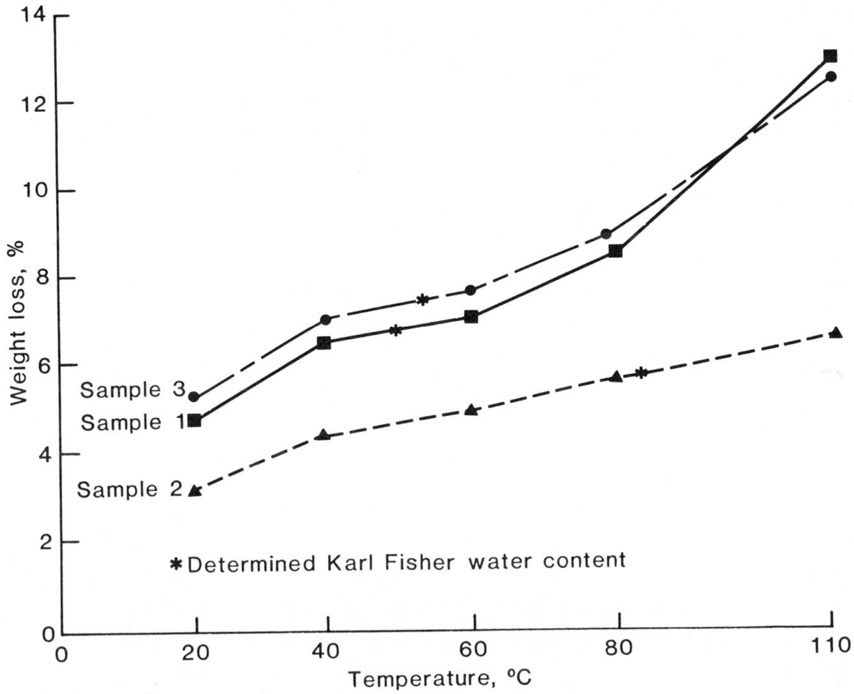

FIGURE 1. Weight loss on drying for three humic substances.

based on weight loss, but suggest a greater risk of decomposition above 80°C. Normally, this determination is made at 60°C and, for these samples, this appears to be a safe temperature. As shown in Figure 1, the loss on drying at this temperature roughly corresponds to the moisture determined by the Karl Fischer titration.

A further indication that decomposition occurs at higher temperatures can be found in the data on the weight regained on exposure to air. Samples heated to 110 and 80°C actually regained less of their lost weight than did samples heated to only 60°C. Note that samples No. 1 and No. 3, heated to 110°C, regained much less of their initial weight after 6 hours than when dried at 60°C. For all samples, weight regain increased up to 60°C, then decreased. This suggests, although does not prove, that decomposition may be increasing above 60°C.

These humic substances are extremely hygroscopic when vacuum dried and Table 6 shows they rapidly absorb moisture from the air. This can create problems for accurate elemental analyses if dried samples are used. For example, after drying at 60°C, these samples regained between 0.6 and 1.2% of their initial weight after only 5 minutes exposure to air. To obtain an accurate weight and to transfer such a substance for analysis would certainly require a drybox or glovebag. For this reason, we suggest that these samples be weighed and analyzed when equilibrated with the moisture in the air and that the water content be determined separately and a correction be made on the raw results for this water content.

The Karl Fischer water determination is a widely used technique in analytical chemistry (Kolthoff and Elving, 1961; Mitchell and Smith, 1980) and has been successfully used on many complex materials such as carbohydrates and whole grains (Mitchell and Smith, 1980). The method works well on other complex organic substances and it may work well on humic substances.

However, there are some reasons to be cautious about applying the Karl Fischer method to humic and fulvic acids. Some of the reasons for caution are: (1) the reaction is quite complex, (2) the reagents involved are subject to interferences due to both oxidation and reduction, and (3) some organic functional groups can cause interferences. An example of a potential negative interference is the oxidation of HI by quinones to yield I_2. The presence of quinones has been reported in humic substances (Stevenson, 1982). Active carbonyl compounds, such as aldehydes, can combine with the solvent, methanol, forming acetals and water, thereby causing a positive interference. This reaction occurs at a slow rate and its contribution can be isolated (because of its relative slowness) and subtracted. We feel that the Karl Fischer method can be useful for humic substances but that those interpreting the results need to be aware of the possible interferences.

This study of methods to determine the moisture content of humic substances has yielded no absolute values for moisture content. The loss on drying shows no end point for water loss and the potential interferences of the Karl Fischer titration prevent definition of absolute values. Certainly, additional research needs to be done in this area. For example, techniques such as NMR and IR might be utilized to define and correlate the moisture status of these substances. In many cases it is sufficient to have a reference point to bring these substances to for comparison purposes. Based on Figure 1, loss on drying at 60°C should be the most consistent reference point, however, the interlaboratory study results showed very poor agreement. Based on this limited study we conclude that the Karl Fischer method gives more consistent results on these samples.

CONCLUSIONS

The data obtained in this study suggest that care should be used in evaluating results of elemental analyses. The results of the interlaboratory study range from rather good agreement for carbon and hydrogen to poor agreement for other elements and water. Since no reference humic substances were available, the results of this study can only be evaluated on a relative basis. The establishment of standard reference humic substances by the International Humic Substances Society (see Chapter 1) should be a great aid in helping to validate the applicability of methods and laboratories for the analysis of humic substances. One should use a reputable analytical laboratory and critically evaluate the results obtained. A complete elemental analysis has the additional advantage of allowing calculation of a mass total which should be near 100%.

The moisture status of the samples must be carefully determined. Although the Karl Fischer water determination is subject to certain limitations, it appeared to give the most consistent results on humic substances. With the possible exception of oxygen determination, it appeared desirable to perform analysis on an air equilibrated sample and correct the results to a dry basis using the determined moisture value. Two of the labs in the interlaboratory study neglected to correct for the hydrogen and oxygen in water when correcting the analytical results.

ACKNOWLEDGMENTS

We wish to thank Dr. Ronald Malcolm and George Aiken of the U.S. Geological Survey for supplying the samples used in this study. We also wish to thank Dale Raines of Huffman Laboratories who carried out the experimental work on the determination of moisture.

CHAPTER EIGHTEEN

Implications of Elemental Characteristics of Humic Substances

CORNELIUS STEELINK

ABSTRACT

Elemental composition is an important chemical property which can be used to establish the nature and source of humic substances. When the percent composition data are displayed as the atom ratios H/C, O/C, and N/C, some general characteristics become visible. Soil, coal, marine, and aquatic humates may be distinguished, one from the other. Structural trends may be identified in specific environments, such as lake sediments and soil profiles. Nonhumate contaminants can be detected. Atom ratios may also aid the investigator in proposing hypothetical structures for humic and fulvic acids and serve as a guide in the synthesis of artificial humates.

Although the elemental composition is a useful guide, its validity is dependent on the purity of the sample and the operational definition of a humic substance. This requires a consensus among investigators on the methodology for extracting humates from their environments and also a consensus on what fraction of this extracted material will be labeled "humic" or "fulvic." Fortunately, significant progress has been made toward this consensus during the past 5 years. In the near future, it will be possible to obtain reproducible analytical data on a rigorously defined humic fraction.

INTRODUCTION

Elemental analysis is the cornerstone of all chemical inquiry. Almost before one asks "what is it?", one asks "what is its percent composition?". From such data, a chemist writes the first empirical formula, and later a molecular formula. If the substance is very similar to a known compound, the chemist may well infer that it is a simple modification of the molecular formula. Or, if the unknown substance is extracted from a plant or microorganism, an educated guess about its structure may be made, based on the formulas of other compounds extracted from the same source.

In any event, after the elemental content is established, the elucidation of the structure is carried forward by the full panoply of analytical methods which are available in a modern laboratory. Elemental analysis is also an indicator of purity. If repeated, nondegradative manipulations or fractionations change the composition, then one knows that one is working with a mixture.

Such is the accepted dogma (or methodology) of establishing chemical structures. What does elemental composition contribute to the elucidation of the structures of *humic* substances? Here, one faces a range of analytical values instead of precise percentages for each element. For soil humic and fulvic acids, the most commonly accepted percentages are shown in Table 1 (Schnitzer and Khan, 1978).

Reference to a composition table would enable a chemist to supply a formula of $C_{10}H_{12}O_5N$ for humic acid and $C_{12}H_{12}O_9N$ for fulvic acid, disregarding sulfur. Other simple empirical formulas could be devised, depending on which carbon, hydrogen, oxygen, and nitrogen values within the ranges were chosen.

The absurdity of such empirical formulas is illustrated by just two examples: the same empirical formula for fulvic acid could also be written for whole wood (a complex mixture) or an oligomer of maleic anhydride, a formula already proposed for humic acid (Anderson and Russell, 1976). The "average" elemental percentage composition would accommodate both substances.

TABLE 1. Average Values for Elemental Composition of Soil Humic Substances

	Humic Acids	Fulvic Acids
Carbon	53.8–58.7%	40.7–50.6%
Hydrogen	3.2–6.2	3.8–7.0
Oxygen	32.8–38.3	39.7–49.8
Nitrogen	0.8–4.3	0.9–3.3
Sulfur	0.1–1.5	0.1–3.6

Clearly then, the elemental composition is not unique to humic substances nor does it define humic substances. However, it does set boundaries for probable chemical structures, which are confirmed by other properties, such as:

Molecular weight	Adsorbing properties
Functional groups	Solubility
Absorption spectra	Exchange capacity
Magnetic resonance spectra	Paramagnetism
Complexation ability	Fluorescence
Wetting properties	Titration curves

Thus, the compound HEXAPUS, a cyclotriveratrylene derivative (Menger et al., 1981) has many properties remarkably similar to fulvic acid and humic acid. In particular, it has the ability to chelate metals and entrap nonpolar substances in aqueous solution. However, it has 70% carbon, which rules it out as a natural humic acid.

A systematic approach to the chemical structure of humic or fulvic acid could be patterned on the historical approach to the structure of lignin. Like humic acid, the composition of lignin varies with its source. By common agreement, lignin chemists have adopted Norway spruce as a reference source for native or milled wood conifer lignin. Not only were proposed structural studies based on this lignin material itself, but also on its precursors. Many important structural features were gleaned from a study of precursors and their polymerization products.

In an analogous manner, the structure of humic or fulvic acid could be investigated. One could choose a well-defined forest podzol whose origins are established. The structure of the humic substance could be inferred from its "distance" from lignin, a factor determined by its age, oxidizing environment, methoxyl content, and elemental analysis (Flaig, 1966a; Christman and Oglesby, 1971). A full discussion of this approach is treated in a later section.

The most serious problem in the study of humic substances is the lack of reproducibility of analytical results. One would expect soil humates to vary with soil type, aquatic humates to vary with water sources, and coal humates to vary with coal rank. But even within one well-defined source, the elemental composition will vary between samples, depending on extraction and fractionation procedures. There are cases in which the same authors have used the same source and the same extraction procedure and have obtained significantly different elemental analyses. Before any meaningful structural conclusions can be deduced from elemental analysis, a rational definition of humic substances will have to be established (MacCarthy, 1976; Malcolm and MacCarthy, 1979).

DEFINITION OF HUMIC SUBSTANCES

The classic definitions of soil humic and fulvic acids are based on solubility (Schnitzer and Khan, 1972). Thus, humic acid is the alkali-soluble material in soil, which is precipitated at pH 1. The material which remains soluble in the extract at pH 1 is fulvic acid. A more recent definition for aquatic humic substances is given by Thurman and Malcolm (1981). Here the material which adsorbs on an XAD column from an acid aqueous solution is defined as aquatic humus. That part of the adsorbed material which is soluble in acid and base is fulvic acid; the portion insoluble in acid is humic acid. Another definition of an aquatic humic substance is based on adsorption by DEAE–cellulose columns (Miles et al., 1983).

Obviously, a large number of compounds would seem to fit these two definitions. (Salicylic acid and galacturonic acid, for example, would fit the solubility parameters for soil humic and fulvic acids, respectively.) However, the added restrictions of elemental analysis (see Table 1)—dark color (e.g., absorption spectra) titration data, and molecular weight—would narrow the possible chemical structures.

Molecular weight is an important criterion for defining humic substances. Soil humic substances appear to have higher molecular weights than their aquatic counterparts. Molecular weight measurements on humic substances are highly dependent on the method used, as well as pH, concentration, and ionic strength (Ghosh and Schnitzer, 1980b). Humic acids have reported number average molecular weight ranges of 3000–1,000,000; fulvic acids range from 500 to 5000 in molecular weight (Stevenson, 1982, pp. 295–307). Fractionation of humates on Sephadex indicates a mixture of macromolecular materials whose elemental composition varies with molecular weight (Povoledo et al., 1975; Chen and Schnitzer, 1978).

SOURCES OF HUMIC SUBSTANCES

Humic and fulvic acids are presumed to arise by two classical natural processes. Terrestrial humates are found in the following pathway: plants → soil humates → peat → coal. Aquatic humates start with soil leachates or marine phytoplankton and go through a sequence: sediments → kerogen → petroleum. There are conditions which mix the two processes as well. As a result, there are a host of names and symbols applied to these compounds, such as peat humic acid, coal fulvic acid, soil humic acid, and so on. Depending on their oxidation state, they may be heavily bound to metal ions. Within each class of humic acid, there are subclassifications, such as Podzol B_h humic acid, lignite fulvic acid. Other types are classified by geological age, depth in a sediment, and type of aquatic environment. The following discussion will attempt to relate elemental composition to these broad classes of humates.

Soil Humic and Fulvic Acids

Soils have been classified into 10 orders (Stevenson, 1982). Within each order are soil horizons, such as A_1, B_2, and so on, which indicate biological weathering patterns. In some categories, carbonic acid is the principal weathering agent, and in other categories organic chelates from decomposing vegetation are the agents. Thus, a multiplicity of soil environments exists, each with different substrates to manufacture humic substances (Schnitzer and Khan, 1978).

A few surveys of elemental analyses of soil humates and fulvates have been made for different soil types. Schnitzer and Khan (1978) have recorded values for a large number of humates extracted from arctic, temperate, subtropical, and tropical soils. The values center around 54–56% carbon, 4–5% hydrogen, and 34–36% oxygen. The neutral soils show narrow ranges of carbon, hydrogen, and oxygen values, the acid soils show broader ranges. About the only significant deviation from these values occurs with the acid soils, which show broad ranges and higher oxygen contents than the other soils. Hatcher (1980) has observed the same ranges for another group of soil humates.

Sedimentary Humic Substances

The composition of humic substances in lake sediments appears to be a function of depth. Sampling of cores from recent lake sediments in Japan (Koyama, 1966) shows that total carbon increased while total oxygen decreased with depth. Ishiwatari (1975a) isolated humic and fulvic acids from lake sediments and found a similar trend in the elemental composition. He summarized the changes in humic acid composition which occur with increased depth as follows:

1. Increase in carbon content.
2. Increase in C/H ratio.
3. Increase in C/N ratio.
4. Increase in color intensity.
5. Decrease in proteinaceous material.
6. Decrease in carbohydrate-like material.

One should expect such changes in an anaerobic environment. The humic acids all had compositions within the accepted range (see Table 1), but the trend with depth was unmistakable.

Coal Humic Acids

Humic acids in coal may have at least two origins. Humic acid remaining in coal may be the residue of the geological precursors of coal, or, the humic

TABLE 2. Elemental Analyses of Some Coal Humic Acids

Source	C (%)	H (%)	O (%)	N(%)	Reference
Brown coal	59.7	3.90	35.3	0.70	Camier et al. (1981)
Lignite soft coal	61.2	4.1	34.8	—	Flaig (1966a)
Subbituminous coal	69.2	4.5	26.3	—	Flaig (1966a)

acid content may be the result of a reoxidation of coal. The commonly accepted genesis of brown coal humic acids is shown in the following sequence (Camier et al., 1981):

$$\text{Plant material} \xrightarrow[\text{cellulose}]{} \text{lignin monomer} \xrightarrow{\text{polymerization}}$$
$$\text{humic acid} \xrightarrow[H_2O]{} \text{humic acid gel} \xrightarrow{\text{cross-linking}} \text{brown coal}$$

Some brown coals may contain up to 30% humic acid. Typical compositions are shown in Table 2. Bituminous coal contains no humic acids.

Lignin-like substances appear in low rank coals (Hayatsu, 1981) as determined by CuO-NaOH degradation products. The elemental analyses of humates from these coals support these findings.

Aquatic Humic and Fulvic Acids

Aquatic humic substances may be found in groundwater, river water, lakes, marshes, bogs, swamps, and seawater. The source of the humates may be autochthonous or allochthonous: that is, the humates may be formed from phytoplankton in the water or they may be leached into the aquatic environment from terrestrial plants, leaf litter, soil, or subsurface deposits. Relatively undisturbed marine environments have humic and fulvic acids formed almost entirely from native phytoplankton; inland surface waters contain major contributions from allochthonous sources. Mixing of the two types of materials may occur, as in the estuaries of rivers.

Marine humic substances and groundwater humates form two distinct groups, both of which differ significantly from each other and from inland surface waters. Marine humic acids are distinguished by high H/C ratios (Ertel and Hedges, 1983) and O/C ratios of around 0.50. This indicates high aliphatic character. This has been confirmed by ^{13}C NMR analysis (Gillam and Wilson, 1983; Hatcher et al., 1983a). The data suggest that marine diatoms or algae may be the major source of these humates and that uronic acids may be the major structural units in marine fulvates.

On the other hand, groundwater humates are characterized by low O/C (0.3–0.4) atomic ratios and H/C ratios of around 1.0. This implies a kerogen or Trona acid origin or contamination by these materials.

The surface waters (marshes, bogs, rivers, streams) contain humates similar to their soil counterparts, possessing only sightly higher O/C ratios. Aquatic fulvates, on the other hand, have lower O/C ratios than their soil counterparts; no doubt this is a reflection of the lower carbohydrate content of the aquatic fulvates (Thurman and Malcolm, 1983). Sedimentary fulvic acids appear to be modified uronic acids.

All comparisons of the elemental composition of soil humates with aquatic humates must be viewed with caution. In recent years, the extraction methodology has changed rapidly. A soil sample extracted with base in the classical manner may have up to 20% carbohydrate and up to 10% ash, while a sample extracted with XAD resins may have less than 2% carbohydrate and 2% ash. To compare two samples extracted by two different methods would be akin to comparing apples and oranges. Fortunately, a body of literature is accumulating which reports the properties of humates extracted by a single, well-defined method.

Synthetic Humic Acids from Chemical and Microbiological Reactions

Synthetic humic acids have been prepared by a number of investigators over the past 50 years. Most of these preparations are based on a preference for a specific hypothesis for the origin of humic substances: (1) partial degradation of lignin; (2) chemical polymerization of microbial phenols or lignin monomers and amino acids; (3) cell autolysis products; (4) microbial synthesis (see Stevenson, 1982, Chapter 8 for review).

Thus Flaig (1966a) has examined hypothesis 1 by analyzing the humic products of rotting straw. Other workers have oxidized mixtures of model phenols and quinones to support hypothesis 2. Martin and Haider (1980) have treated mixtures of phenol and amino acids with fungal and bacterial enzymes to simulate hypothesis 4. Even artificial landfills have produced humates (Fuller and Artiola-Fortuny, 1982).

From all these methods have come dark-brown materials, many of which fit the general range of elemental composition of soil humates. Table 3 lists some selected data.

Schnitzer and Mathur (1978) conclude that the best substrates for the chemical preparation of humates are catechol, pyrogallol, or hydroquinone (with or without added nitrogen); that is, the elemental composition of the humic products fits that of an "ideal" soil humic acid. This is not entirely fortuitous. All three substrates are possible degradation products of lignins or tannins. Under strong oxidizing conditions, they could undergo cleavage or condensation reactions to produce phenolic carboxylic acids. All have O/C atomic ratios which would, upon oxidative polymerization or oxidative cleavage, yield the correct O/C ratios for the average soil humic acid. The amount of nitrogen incorporated by these polymers would depend on the soil environment. By the use of such substrates and with a knowledge of the environment in which the humic acid is naturally formed, a chemist should be able to synthesize a reasonable model humic acid. Ellwardt et al. (1981)

TABLE 3. Elemental Composition of Synthetic Humic Acids

Source	C (%)	H (%)	O (%)	N (%)	References
p-benzoquinone	59.0	1.64	39.4	—	Flaig (1960b)
o-benzoquinone	59.1	3.7	37.2	—	Murphy and Morre (1960)
Catechol/NH_3/ oxidation	56.2	4.36	35.7	3.5	Schnitzer and Mathur (1978)
Aspergillus metabolites	51.0	5.6	40.6	2.8	Christman and Oglesby (1971)
Penicillium metabolites	45.2	6.1	45.9	2.8	Christman and Oglesby (1971)
		Reference Humic Acids			
Source	C	H	O	N	References
Peat humic acid	58.3	4.97	32.2	2.62	Christman and Oglesby (1971)
"Ideal" soil humic acid	56.2	4.70	35.5	3.2	Schnitzer and Mathur (1978)

used artificial lignin derived from coniferyl alcohol as substrates for enzymatic oxidation to humic-like substances. The recent synthesis of the compound HEXAPUS (Menger et al., 1981) suggests another approach. This substance lowers the surface tension of water, binds to nonpolar compounds, and chelates metals. Although its carbon content is too high for a natural humic acid, one could make analogues that would fit within the range of natural humates. The starting material could be a benzyl alcohol derived from a lignin model compound. A carboxylic acid side chain could be attached, whose structure would fit the C/H/O ratio of any specified humate.

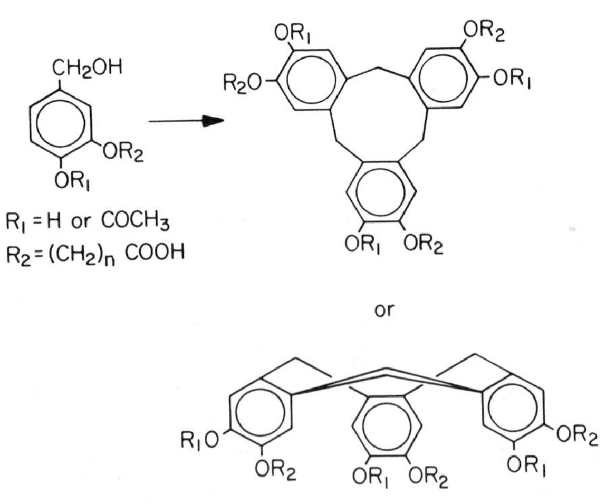

FIGURE 1. Synthesis of HEXAPUS.

Acid catalyzed condensation would then form the trimer or higher polymer. Compounds of this type have been synthesized (Collet et al., 1981). In this synthetic approach, one would obtain a discrete compound whose structure is exactly known and whose physicochemical properties could be tailored to molecular constitution (see Fig. 1).

ATOMIC RATIOS OF CARBON, HYDROGEN, NITROGEN, AND OXYGEN CONTENT AS A GUIDE TO HUMIC SUBSTANCES IDENTIFICATION

The simplest way to display information about the elemental composition of a substance is to use atomic ratios. The ratios of H/C, O/C, and N/C can prove very useful in the following ways:

1. To identify types of humic substances.
2. To monitor the structural changes of humates in soils and sediments.
3. To devise structural formulas for humates.
4. To eliminate the role of ash content.

Van Krevelen (1961) has developed a system of chemical structures for coals, based on atomic ratios. Many other investigators have applied this system to humic and fulvic acids (Visser, 1983a).

In Table 4 are listed atomic ratios for a large number of humic and fulvic acids. Many of these are averages, taken from a wide variety of sources. Methods of extraction differ from one to another. Even with all these variables, one can still make some generalizations:

1. The O/C ratio is the clearest indicator of humic types. Soil humic acid ratios cluster around 0.50 while soil fulvic acids center around 0.70. Aquatic humic acids are close to the soil ratio of 0.50; however, aquatic fulvic acids possess consistently lower ratios of around 0.60. The difference may reflect different extraction procedures or it may represent lower carbohydrate contents in the water environment (Thurman and Malcolm, 1983). Another indication of the effect of carbohydrate content on the O/C ratio comes from the analysis of sediments (Ishiwatari 1975a). With increasing depth, the carbohydrate content of fulvates increases while the carbohydrate content of humates decreases. This is revealed by the changes in the O/C ratios. These ratios may indicate serious contamination from other sources. Thus, groundwater humates show abnormally low O/C ratios; this may be due to contamination by kerogens or Trona acids.

Comparisons are always more significant if they are obtained from a group of humates from related sources, such as a soil profile or a sediment profile.

TABLE 4. Atomic Ratios of Elements in Humic and Fulvic Acids and Related Compounds

Source	H/C	O/C	N/C	References
		Soil Fulvic Acids		
Average of many samples	1.4	0.74	0.04	Schnitzer and Khan (1978)
Average of many samples	0.83	0.70	0.06	Ishiwatari (1975a)
Average of many samples	0.93	0.64	0.03	Malcolm et al. (1981)
		Soil Humic Acids		
Average of many samples	1.0	0.48	0.04	Schnitzer and Khan (1978)
Average of many samples	1.1	0.50	0.02	Ishiwatari (1975a)
Neutral soils, average	1.1	0.47	0.06	Hatcher (1980)
Aldrich humic acid	0.8	0.46	0.01	Steelink et al. (1983)
Amazon HA/FA = 84/16	0.97	0.57	0.04	Leenheer (1980)
		Sedimentary Humic Acids		
Lake, 0–8 cm deep	1.42	0.57	0.10	Ishiwatari (1975a)
Lake, 20–24 cm deep	1.34	0.51	0.09	Ishiwatari (1975a)
Lake, 20–30 cm deep	0.80	0.46	0.06	Povoledo et al. (1975)
		Sedimentary Fulvic Acids		
Lake	1.53	0.80	0.08	Ishiwatari (1975a)
		Aquatic Fulvic Acids		
Suwannee River	0.81	0.54	0.01	Thurman and Malcolm (1981)
Suwannee River	1.0	0.63	0.01	Thurman and Malcolm (1983)
River (high OCH_3)	0.99	0.57	0.02	Plechanov et al. (1983)
River (low OCH_3)	1.05	0.59	0.03	Plechanov et al. (1983)
Surface	0.90	0.56	0.06	Malcolm et al. (1981)
Groundwater (Biscayne)	0.90	0.47	0.03	Thurman and Malcolm (1981)
Groundwater (Fox Hill)	1.3	0.34	0.001	Thurman and Malcolm (1981)
Groundwater (average of 5 samples)	1.2	0.39	0.01	Thurman and Malcolm (unpublished)
		Aquatic Humic Acids		
River (high OCH_3)	1.02	0.50	0.02	Plechanov et al. (1983)
Suwannee River	0.82	0.51	0.06	Thurman and Malcolm (1981)
Many rivers	1.11	0.59	0.03	Thurman and Malcolm (unpublished)
Spring (XAD-8)	0.81	0.59	0.01	Leenheer (1980)
Spring HA/FA = 35/65	0.92	0.66	0.01	Leenheer (1980)
Groundwater (Biscayne)	0.69	0.39	0.09	Thurman and Malcolm (1981)
Groundwater (Fox Hill)	0.98	0.28	0.04	Thurman and Malcolm (1981)
		Sedimentary Marine Humic Acid		
Average of 4 samples	1.27	0.51	0.08	Ertel and Hedges (1983)

TABLE 4. Continued

Source	H/C	O/C	N/C	References
	Miscellaneous			
Peat humic acid	1.0	0.41	0.04	Christman and Oglesby (1971)
Peat humic acid	1.2	0.43	0.02	Ishiwatari (1975a)
Kerogen	1.56	0.082	0.03	Fester and Robinson (1966)
Trona acid	1.56	0.10	0.03	Fester and Robinson (1966)
Humic acid from subbituminous coal	0.78	0.28	—	Flaig (1966a)
Humic acid from lignite soft coal	0.80	0.43	0.01	Flaig (1966a)
Polygalacturonic acid	1.30	1.0		
Mugeneic acid	1.64	0.65	0.16	Ohfune et al. (1981)
Polymaleic acid	1.0	0.75		Anderson and Russell (1976)
Periodate lignin	1.2	0.37		Hatcher et al. (1982)
Brauns native lignin	1.16	0.35		Lai and Sarkanen (1971)

2. H/C ratios are clustered around 1.0 for most soil and aquatic humates and fulvates. Lake and marine sedimentary humic substances have somewhat higher H/C ratios than their soil or water counterparts (Ishiwatari, 1975a). A plot of E_4/E_6 ratios* versus H/C ratios shows a direct correlation for terrestrial humic acids (Ertel and Hedges, 1983). The magnitude of the E_4/E_6 ratio is inversely proportional to the degree of condensation or the molecular weight (Chen et al., 1977). Ratios above 1.3 indicate that the material may be a nonhumic substance.

EFFECT OF FRACTIONATION ON ELEMENTAL ANALYSIS

What is a "pure" humic acid? Or a "real" fulvic acid? Such questions have plagued soil scientists ever since they noted that these brown substances were not simple, discrete compounds. Did humic acid consist of a "core" attached to peripheral carbohydrates, proteins, fatty acids, and phenolics? Were fulvic acids mixtures of monomeric compounds held together by hydrogen bonds? Could humates co-precipitate with other compounds or complex with nonpolar substances? The answers to all these questions may be yes.

Through the years, many fractionation schemes have been employed to try to answer these questions (Kononova, 1961). In many of these studies, elemental analysis of the fractions was used to identify the humic substance

* $E_4/E_6 \text{ ratio} = \dfrac{\text{Absorbance of alkaline humate solution at 465 nm}}{\text{Absorbance of alkaline humate solution at 665 nm}}.$

in the fraction. Frequently, other chemical and physical properties (such as spectral and functional group) were used. Unfortunately, no definite answers have emerged from these studies. The following account can only describe some of the fractionation techniques and propose some generalizations.

Organic Solvent Extraction

Schnitzer and Barton (1963) separated a soil fulvic acid into a methanol-insoluble and -soluble fraction. The methanol-soluble fraction had less carbon than the original substance. Steelink et al. (1983) treated Aldrich and Fluka humic acids by exhaustively extracting them with methanol. The methanol-soluble fraction revealed methyl ester functions, as detected by

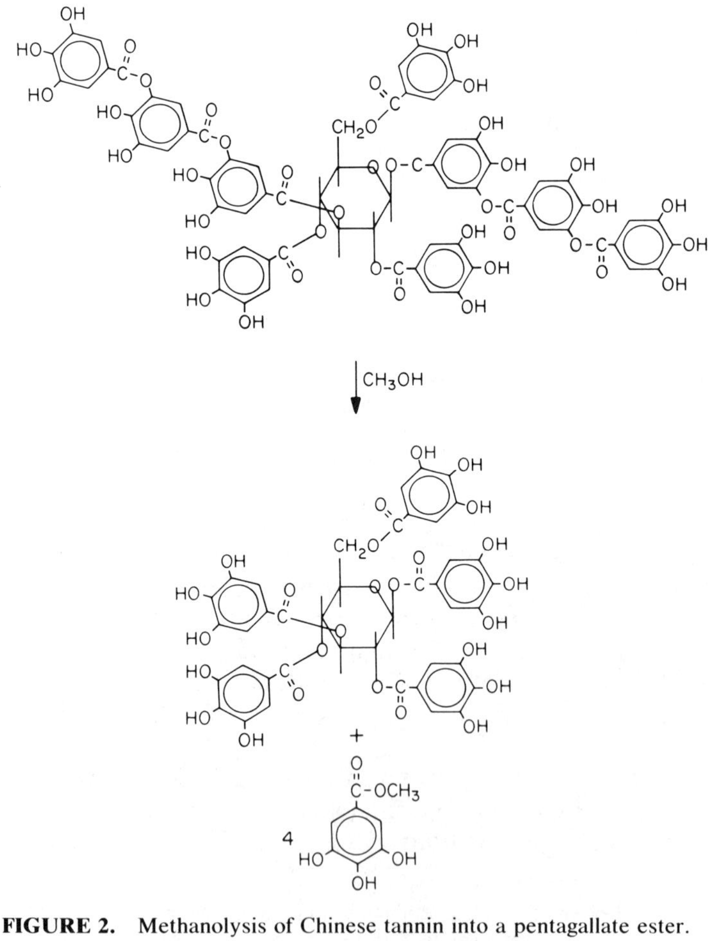

FIGURE 2. Methanolysis of Chinese tannin into a pentagallate ester.

^{13}C NMR. This reaction was reminiscent of studies with hydrolyzable tannins, in which methanol caused chemical degradation. The phenomenon evokes a provocative question: Does humic acid consist of a basic core attached to smaller ester functions? Solvent extraction might provide an answer. The analysis of hydrolyzable tannins provides a good example. Treatment of Chinese tannin with methanol yields gallic acid derivatives plus a large tannin molecule (see Fig. 2). Further treatment of the large fragment with methanol continued to yield gallic acid derivatives plus a smaller tannin molecule (Armitage et al., 1961). Here, the "neutral" solvent is actually hydrolyzing ester bonds (methanolysis) to form methyl esters. What, then, is the "true" tannin: the nonagallate or the pentagallate?

A true tannin is operationally defined as a substance that will precipitate proteins from aqueous solutions. The pentagallate will not fit this description but a nonagallate will.

Another case in point is the solvent extraction of lignins (Nimz, 1966). Are the propylbenzene oligomers found in the extract degradation products of the original lignin polymer, or are they precursors co-occurring with the polymer? In any case, their identities have helped establish the structure of native lignin.

Even with condensed tannins, one finds that very mildly acidic conditions can cause anthocyanin groups to split off the polymer. Size exclusion chromatography of condensed tannins (Thompson et al., 1972) always reveals mixtures of oligomers. What is the "true" tannin molecule?

Still, it can be said that the various extraction schemes for tannins and lignins have provided valuable insights into the structures of these natural polymers. From the many monomeric compounds which co-occur with the polymers, scientists have inferred, correctly, that the polymer was constructed from many of the monomers.

Chromatographic Separations

Resins

In the past 10 years, a wide variety of resins have been used to fractionate humates. Interest in these resins originates from a search for methods to concentrate materials from natural waters. It is necessary to mention the first "resin" used to fractionate humates: charcoal (Forsyth, 1947). Carbohydrates and proteins were preferentially separated by this method. Anderson and Russell (1976) purified fulvic acid by a charcoal separation. Elemental analysis showed this fulvic acid to be identical to a direct citric acid extract of the soil.

A variety of new methods have been developed using XAD resins. These are too numerous to summarize, but the work of Thurman and Malcolm (1983) and co-workers represents the state of the art. One of the most significant effects of XAD resin treatment is the removal of carbohydrates (Thurman and Malcolm, 1983). By a combination of XAD resins, Enzacryl gel chromatography, and cation/anion-exchange resins, they have managed to exclude inorganic ions (ash) and nonacidic organic contaminants. The result of all these procedures is a "pure", low ash humic substance from a natural water. An example of the elemental analysis of a fulvic acid from such a procedure is shown below:

Source	C	H	O	N	Ash	Reference
Suwannee River	51.3	4.32	42.9	0.56	0.005	Thurman and Malcolm (1983)
Suwannee River	54.6	3.71	39.29	0.47	0.95	Thurman and Malcolm (1981)

Some problems still remain, as these data indicate. Here, a fulvic acid from the same source isolated by the same procedure by the same investigators shows quite different elemental analyses. Are these two substances the same compound?

Aquatic humic substances have also been isolated from aquatic sources by DEAE–cellulose chromatography (Miles et al., 1983).

Gel Permeation

Separation of molecular weight fractions of humates by gel permeation chromatography has been used for some time (Schnitzer and Khan, 1978; De Haan, 1983). In one report, Povoledo et al. (1975) showed that carbon, hydrogen, and nitrogen contents all increased with increasing molecular weight of the fractions. Stevenson (1982) confirms these observations.

HPLC Fractionation

Little has been published in this area in relation to humic substances. Humates tend to have retention times close to, or before, the solvent front in most reverse-phase columns. In one report (Rodgers et al., 1981), a fulvic material was separated into seven fractions on a silanized BioSil column. The fractions were analyzed by infrared spectroscopy. One of the fractions was patently not a fulvic acid, although it co-precipitated with fulvic acid.

Electrophoresis

Leenheer and Malcolm (1973) separated a soil fulvic acid into four fractions by electrophoresis. They found a dramatic change in elemental composition, which correlated with carbohydrate content. Thus, the high polysaccharide fraction had high H/C and O/C ratios, while the low carbohydrate fractions had low H/C and O/C ratios. All fractions, however, fitted into the accepted ranges for soil fulvic acids, even though the atomic ratios varied by factors of more than 50%. Earlier electrophoresis experiments had also shown dramatic differences among fractions (Kononova, 1961).

EFFECT OF HYDROLYSIS ON ELEMENTAL ANALYSIS

Acid hydrolysis has been used extensively to remove ash, protein, and carbohydrate from humic substances (see reviews by Schnitzer and Khan, 1972, 1978; Stevenson, 1982; Kononova, 1961). In theory, this method should remove peripheral groups and co-precipitated materials and leave the "core" of humic or fulvic acid. What actually happens may be quite disconcerting. Structural changes may take place, as well as the removal of contaminants. Depolymerization may occur, but acid-catalyzed condensation may also take place. These changes can be seen by the loss of hydrogen content and by a decrease in the E_4/E_6 ratios. All of these events are best summarized in Table 5 (Stevenson, 1982).

As can be seen from Table 5, the carbon content increases while the hydrogen and nitrogen contents decrease, a trend which would be predicted from loss of protein and carbohydrate. Decarboxylation and condensation would also account for these changes. Humic acid appears to suffer only minor changes. This is especially apparent in the small loss of carboxyl and change in E_4/E_6 ratios. One would have to agree with Stevenson (1982) that soil humic acids may be recoverable with acid hydrolysis, but soil fulvic acids are changed too much to be considered related to the original compound. The acid hydrolysis method becomes less feasible for aquatic humates, since these compounds appear to be more closely related to soil fulvates than to soil humates.

Can any meaningful information be gleaned from acid hydrolysis? Cer-

TABLE 5. Effects of 6*N* HCl Hydrolysis on Analytical Characteristics of Humic and Fulvic Acids

	Humic Acid		Fulvic Acid	
Characteristics	Untreated	After Hydrolysis	Untreated	After Hydrolysis
Carbon (%)	57.2	58.1	49.5	53.9
Hydrogen (%)	4.4	3.9	4.5	4.1
Nitrogen (%)	2.4	0.7	0.8	0.1
Total acidity (meq/g)	8.1	7.0	12.4	8.9
Carboxyl (COOH) (meq/g)	4.8	4.3	9.1	4.9
Phenolic OH (meq/g)	3.3	2.7	3.3	4.0
E_4/E_6	4.3	4.0	7.1	4.3
Ash (%)	7.9	0.3	2.0	0

tainly, the identities of the amino acids or the monosaccharides or the monomeric acids can be determined. But the manner in which these monomers are linked to one another and to the "core" humic acid is not revealed. Perhaps milder methods of hydrolysis will have to be developed. In the field of lignin chemistry, 0.2*N* HCl/Dioxane/water hydrolysis has been used (Nimz, 1966) to strip away portions of the lignin molecule. Tannin chemists have used mild acid hydrolysis coupled with a nucleophilic trapping agent (Thompson et al., 1972) to selectively remove monomers from the tannin polymer.

ELEMENTAL COMPOSITION AND ITS RELATION TO ELECTRON SPIN RESONANCE PROPERTIES

Ever since the early work of Steelink and Tollin (1967), there have been attempts to relate ESR spectra to chemical structures for humates. Steelink (1966) provided evidence that at least two types of stable radicals existed in humic substances; both of them were resistant to acid hydrolysis and fractionation and were therefore not artifacts or contaminants. The results of later investigators are reviewed by Stevenson (1982) and Schnitzer and Khan (1978).

The general conclusions of all investigators is that a phenol–quinone complex is responsible for the free radical species. Schnitzer and Levesque (1979) proposed that the ESR signal height, a measure of total radical content, is dependent on the concentration of phenolic groups. It appears from Table 6 that free radical content increases with humification. In turn, this correlates with greater aromaticity, as measured by lower E_4/E_6 values (Ertel and Hedges, 1983). Since most phenolic free radicals absorb in and around 600 nm, this relationship is not unexpected.

The following conclusion appears justifiable: a low H/C atomic ratio correlates with a high spin density, a low E_4/E_6 ratio, a high degree of humification, and a high phenolic content. However, one must remember that there

TABLE 6. Free Radical Content of Humic Acids and Salts[a]

Humic Acid Source	Acid Form (spins/g) $\times 10^{17}$	Sodium Salt (spins/g) $\times 10^{17}$
English podzol	2.0	100
Swiss muck soil	6.0	240
Wisconsin podzol, humic	7.0	180
Wisconsin podzol, fulvic	4.0	7.0
NALCO Leonardite	4.0	66
New Mexico lignite humic acid	28.0	300
Artificial catechol-amine humic acid	23.0	540

[a] *Steelink (1966).*

are factors which can affect free radical content. Among the most potent factors which contribute to spin content is the pH of a humate solution. A change of two or three pH units may cause a 20-fold change in radical content. Salts and metal complexes also affect the concentration of radicals.

STRUCTURAL FORMULAS BASED ON ELEMENTAL ANALYSIS

One of the favorite pastimes of soil organic chemists is the construction of hypothetical formulas for humic acid. The Introduction to this chapter stated that elemental analysis was the first step in such a process. It is fitting that we conclude with a structural formula.

The nature of the structure depends on the assumptions one uses as to the probable precursors for the humate. Lignin is a favorite precursor; others include tannins, quinones, microbiological metabolites, sugar–amine condensation products, and other products (Stevenson, 1982).

Flaig (1966a) has reported elemental analysis for a humic acid formed from decaying straw (Table 7). Here, the prime source was ligniferous plant tissue. A humic acid of typical elemental composition was obtained from a ligniferous source. More recent evidence supports this type of biodegradation, based on direct ^{13}C NMR monitoring (Ellwardt et al., 1981). A number of soil humic acids fit these data.

Can one propose a reasonable chemical structure from elemental data? The answer is "yes" if titration data are included. A typical soil humic acid might have the following characteristics:

$$H/C = 1.0 \qquad \text{COOH content} = 4.4 \text{ meq/g}$$
$$O/C = 0.5 \qquad \text{Phenol content} = 2.6 \text{ meq/g}$$

An empirical formula (excluding nitrogen) should have the formula $CHO_{0.5}$. If one assumes the phenylpropane backbone of lignin as the carbon skeleton, then the simplest molecular formulas should be $C_{18}H_{18}O_9$. At this point, the functional group content must be taken into account. A carboxyl content of 4.4 meq/g translates into 1.5 COOH groups per mole of $C_{18}H_{18}O_9$ or 3 COOH groups per $C_{36}H_{36}O_{18}$. Therefore, a tetramer of phenylpropane

TABLE 7. Elemental Analysis of Products of Degradation of Straw

Source	C	H	O	N	OCH_3
Bjorkman lignin from straw	60.68	5.79	33.11	0.4	16.74
Humic acid from straw after 270 days	57.53	4.58	36.94	0.95	—

units of molecular weight 756 is the smallest unit that will accommodate 3 COOH groups and 2 phenol groups. That leaves 10 oxygen atoms to be distributed in ether, hydroxyl, or carbonyl functions. The nature of these groups can be inferred from ^{13}C NMR studies (Steelink et al., 1981; Thurman and Malcolm, 1983; Ellwardt et al., 1981).

The tetramer can be functionalized in a number of ways to correspond to the properties of various humates. Oligomers such as a tetramer are probably products of microbiological degradation of lignin. If these tetramers contain muconic acids, they can be polymerized by oxidative enzymes or free radicals in the soil to form larger molecules in a manner analogous to butadiene Further oxidation to produce more COOH groups would lead to typical fulvic acid structures, with loss of more aromaticity. The positions of the major oxygen functionalities would have to account for the complexation and encapsulation properties, but there are many reasonable possibilities.

The question of carbohydrate content raises an interesting issue. If one assumes that the "true" humic acid is not bound to carbohydrate, then the tetramer model is acceptable. If humic acid is considered a carbohydrate-bound molecule, then one must take into account the actual percent carbohydrate in the structure. This would mean assigning lower O/C and H/C ratios for the noncarbohydrate portion: in other words, making it more condensed or aromatic.

Assuming 3 COOH, 2 phenol, 1 quinoid, 2 ketone, and 6 hydroxyl groups in the tetramer, one could propose a basic core of four units (see Fig. 3).

These four units might be connected by aryl–aryl, β–C–C– or benzyl–aryl linkages, the most resistant linkages to microbiological attack (see Fig. 4). The proximity of quinone and catechol groups would favor charge transfer or radical formation and give rise to the absorption in the visible spectrum at 665 nm. Chelation sites are plentiful. The ratios of aliphatic to aromatic carbon atoms are consistent with the results of NMR studies (Hatcher et al., 1983a). It should be noted that treatment of a compound like the tetramer with strong acid or strong base would cause further condensation. This would lead to oxidative degradation products composed of highly substituted benzene carboxylic acids, which are artifacts of the degradation analysis.

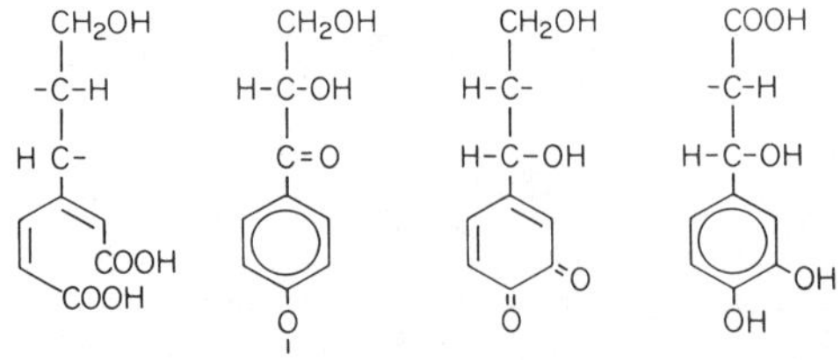

FIGURE 3. Proposed basic units in humic acids.

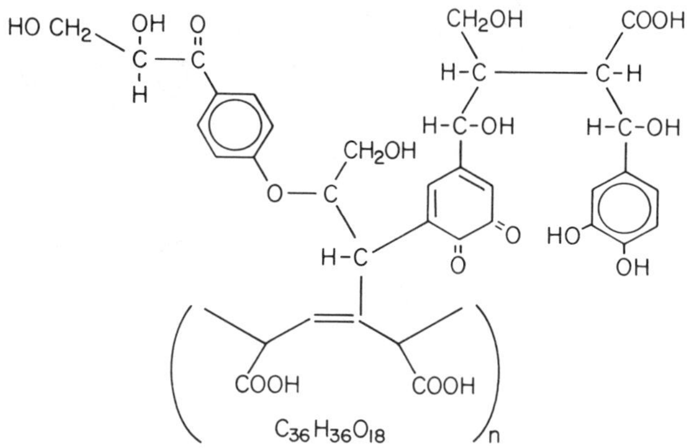

FIGURE 4. Proposed tetramer structure for humic acid.

CONCLUSION

Throughout this chapter, an attempt has been made to characterize humic substances by elemental composition. Minimal references have been made to other properties of humates. This has been a difficult assignment, since so much information about humates is derived from other analytical measurements.

Nevertheless, it is surprising how often one refers to elemental analysis to confirm the presence of a humic or fulvic acid. It is useful in determining whether a brown, macromolecular acidic material is from a coal, soil, marine sediment, or kerogen. It is most useful in characterizing structural trends in a specific environment, such as in sediments or soil profiles. Atomic ratios, especially O/C ratios, are the simplest way to display elemental composition of humates. They also help one devise hypothetical structures for humates. As a guide in the synthesis of artificial humic substances, they are invaluable. In addition, atomic ratios help the investigator identify nonhumate contaminants.

In the final analysis, the validity of elemental composition depends on the purity of the sample. The term "purity" is an operational concept. It requires a consensus among investigators on the methodology for extracting humates from their environments and also a consensus on what fraction of the extracted material will be labeled "humic." Fortunately, real progress has been made toward these two goals in the past 5 years, and there is evidence that a system of classifying humates will be established. When that happens, the implications of elemental characterizations of humic acids will be more meaningful.

NOTE ADDED TO PROOF

The term humate(s), in this chapter, is used synonymously with humic substances or humic acid(s); the term fulvate(s) is used synonomously with fulvic acids.

CHAPTER NINETEEN

Molecular Size and Weight Measurements of Humic Substances

ROBERT L. WERSHAW and GEORGE R. AIKEN

ABSTRACT

A brief overview of the most commonly used methods for determining molecular weight and size of humic substances is presented. Methods discussed for measuring molecular size are: gel permeation chromatography, ultrafiltration, scattering of electromagnetic radiation, and electron microscopy. Methods discussed for measuring molecular weight are: ultracentrifugation, viscometry, and colligative-property measurements. Emphasis is placed on the limitations and problems of each method. The chemical properties of humic substances that interfere with molecular weight determination by these methods and result in questionable data are discussed.

INTRODUCTION

The determination of molecular weight has long been a valuable tool used to help identify and understand the basic chemistry of unknown organic compounds. Knowledge of molecular weights in the study of humic substances is important for several reasons: (1) to establish proximate molecular formulas in conjunction with data provided by other methods of characterization; (2)

to help establish stoichiometric relationships between humic substances and other chemical species (e.g., trace metals) through the conversion of weight to molar concentrations; and (3) to aid in the comparison of humic substances extracted from various environments.

The task, however, of determining molecular weights for humic substances has not been simple. Humic substances comprise one of the most widely distributed classes of natural products on Earth. Defining the chemistry and understanding the nature of humic substances have long been hampered because most humic substances are not discrete chemical entities, but are a complex mixture of organic substances (Kononova, 1966) with a wide range of molecular sizes (polydisperse).

Problems associated with determining molecular weights of mixtures have long been recognized. Lansing and Kraemer (1935) point out that the usual methods of determining molecular weight yield "average molecular weights," and depending on the methods used, these averages are not directly comparable. Failure to consider this fact is one source of confusion in comparing molecular weight data for humic substances. Three different types of molecular weight averages are commonly used:

1. Number-average molecular weight ($\overline{Mn}$) is generated by physicochemical methods that determine the total number of molecules present, regardless of size. It is expressed mathematically as

$$\overline{Mn} = \frac{\sum_i n_i M_i}{\sum_i n_i} \tag{1}$$

where n_i is the number of molecules of molecular weight M_i. Colligative-property measurements (e.g., freezing-point lowering, vapor-pressure osmometry, and membrane equilibria), which depend on the number of molecules in solution, give number-average molecular weights. In the case of a polydisperse system, number-average molecular weight emphasizes the lower molecular weights (Moore, 1972).

2. Weight-average molecular weight ($\overline{Mw}$) is generated by methods such as light scattering and sedimentation which depend on the masses of material in different fractions (Moore, 1972). Weight-average molecular weight is expressed as

$$\overline{Mw} = \frac{\sum_i n_i M_i^2}{\sum_i n_i M_i} \tag{2}$$

Weight-average molecular weight tends to favor the heavier-molecular-weight species of a mixture, resulting in higher-molecular-weight values than $\overline{Mn}$.

3. z-Average molecular weight ($\overline{Mz}$) also may be calculated from data generated by ultracentrifugation, and is expressed as

$$\overline{Mz} = \frac{\sum_i n_i M_i^3}{\sum_i n_i M_i^2} \tag{3}$$

A complete theoretical treatment of z-average molecular weight has been developed by Lansing and Kraemer (1935), who introduced the concept. For a monodisperse system, $\overline{Mn} = \overline{Mw} = \overline{Mz}$; for a polydisperse system, this is not the case. The ratio $\overline{Mw}/\overline{Mn}$ can be used as an indication of polydispersity (Stevenson, 1982).

In addition to methods that yield average molecular weights, there are a number of methods (gel filtration, ultrafiltration, small-angle X-ray scattering) that measure molecular size. In these methods, model compounds of known molecular weight and composition are used to estimate the molecular weight of humic substances. Problems can arise if the model compounds are not sufficiently similar to the humic substances of interest. Choice of appropriate model compounds is hampered by lack of detailed information about the chemical structures of humic substances.

Data presented in other reviews of molecular weight of humic substances (Stevenson, 1982; Schnitzer and Khan, 1972; Thurman et al., 1982) point out the degree of confusion surrounding this important method of characterization. Molecular weight values range from 500 daltons for some aquatic humic substances to more than 10^6 daltons for soil humic acids, and there is little agreement between the various methods used to determine these weights. It is the purpose of this chapter to briefly examine the common methods used to determine molecular weight and size of humic substances. Special emphasis has been placed on discussing the limitations of each method.

METHODS FOR MEASURING MOLECULAR SIZE

Gel Permeation Chromatography

Gel permeation or gel filtration chromatography has been extensively applied to the determination of the molecular sizes of humic and fulvic acids.

Most of these studies have used Sephadex* gels for this purpose although a few studies have used other gels. Molecular weight data can only be estimated by calibrating the gel with appropriate standards.

Sephadex is a cross-linked dextran polysaccharide, which is prepared by reacting dextran with epichlorohydrin. It is available in several different grades depending on the degree of cross-linking. These different grades of Sephadex allow one to carry out molecular size fractionation experiments over a wide range of molecular sizes.

Porath and Flodin (1959) were the first to make a systematic study of gel permeation chromatography. They studied the fractionation of saccharides and serum proteins using a Sephadex cross-linked dextran gel column. The dextran used in their experiments was in bead form (100–200 mesh in the dry form); these beads swelled when soaked in water. They determined that small molecules were retarded in their passage through the column, whereas large molecules passed through the column unretarded. This behavior was attributed to migration of the small molecules into the swollen beads, whereas the molecules larger than a certain size were excluded from the beads. The amount of retardation of molecules of intermediate size was inversely related to the size of the molecules. Porath and Flodin (1959) pointed out that the molecular exclusion size could be altered by changing the degree of cross-linking of the dextran.

In order to measure accurate molecular sizes by gel permeation chromatography, the only interaction between the analyte molecule and gel must be a size-dependent interaction. All other physical and chemical interactions, such as adsorption and electrostatic interactions, must be absent or very weak (Yau et al., 1979). However, as Swift and Posner (1971) have reported, humic acids do interact with the most commonly used gels such as Sephadex, both electrostatically and by adsorption. The adsorption of humic acid molecules is most pronounced at low pH when they are unionized, whereas, the electrostatic interactions are greatest in solutions of low ionic strength. In order to suppress these effects and to obtain accurate molecular size distributions, Swift and Posner (1971) recommended that the measurements be made in a basic buffer of relatively high ionic strength, such as 1*M* tris buffer at pH 9. Under these conditions they found that they could obtain what appeared to be a ". . . continuous fractionation on the basis of molecular weight." The criteria used to determine that this type of fractionation was taking place on their columns were: (1) The elution volume of the analyte was ". . . largely independent of sample concentration . . .", and (2) the entire sample was eluted from the column within the total column volume. In addition, the column must be calibrated with standards that have similar chemical and physical properties to the analytes in order to obtain a measure of the molecular sizes of the compounds being fractionated. Thus,

* Use of trade names is for descriptive use only and does not constitute an endorsement by the U.S. Geological Survey.

globular-protein standards are used for globular-protein analytes and polysaccharides are used for polysaccharide analytes. In the case of humic acids, Posner and Creeth (1972) attempted to measure the molecular weights of several humic acids by equilibrium ultracentrifugation. However, all the samples examined were polydisperse and, therefore, not suitable as standards for the calibration of Sephadex.

Another complicating factor that causes humic acids to act in nonideal fashion in a gel permeation column is that humic acids form molecular aggregates in solution. Wershaw and Pinckney (1973b) have determined that humic acids and their fractions isolated from Sephadex columns aggregate in solution, and that the degree of aggregation is a function of both pH and concentration. Ritchie and Posner (1982) found similar behavior in molecular weight fractions of humic acid isolated during gradient ultracentrifugation. More recent work (Wershaw and Pinckney, unpublished data) also indicates that apparently monodisperse fulvic acids undergo conformational changes as a function of solution ionic strength. The term monodisperse is used here to indicate a system in which the results of molecular size measurements are similar to those which would be obtained with a well-defined system of uniform particle size. In the case of humic substances, this probably only indicates a relatively narrow distribution of particle sizes.

Reuter and Perdue (1981) collected the fractions that were excluded from various grades of Sephadex and measured the number-average molecular weight of each of the fractions by vapor-pressure osmometry. They found that the molecular weights of the excluded molecules decreased as expected from Sephadex G-50 to Sephadex G-10 but that the number-average molecular weights, which are biased toward a low molecular weight (Moore, 1972), measured by osmometry, were much lower than the exclusion limits for globular proteins on the various grades of Sephadex. For example, the number-average molecular weight of the humic acid fraction excluded from G-50 was 1231 daltons, whereas the published globular-protein exclusion limit for this grade is 30,000 daltons. They do not explain this departure, but it is not unreasonable to ascribe it to the fact that humic acids undergo molecular aggregation (Wershaw and Pinckney, 1973b). In a system where a dynamic equilibrium exists between aggregated and disaggregated particles, one would not expect two totally different measurement methods based on different physical principles to yield similar results. A discussion of aggregating systems is beyond the scope of this chapter; for further discussion, the reader is referred to Cann and Fink (1983).

Saito and Hayano (1979) published the results of a preliminary study of the fractionation of humic and fulvic acids with a Toyo Soda high-performance gel. They found that humic and fulvic acids were completely excluded from the gel pores by electrostatic effects when distilled water was used as an eluent. When the column was eluted with $0.1M$ NaCl, this exclusion was not observed, and they found no evidence of adsorption of the humic substances on the gel. From these results, the authors suggested that this gel might be useful for molecular size fractionation of humic substances. However, much more work needs to be done to establish that true molecular size fractionation is indeed taking place.

Ultrafiltration

Ultrafiltration is a method of separating macromolecules according to molecular size, by filtration under an applied hydrostatic pressure through a membrane. This method is similar to reverse osmosis and differs only in the size of particles allowed to pass the membrane. Reverse osmosis separates particles with molecular dimensions similar to the solvent; ultrafiltration separates particles in the range of 10 times the size of the solvent molecules to approximately 0.5 μm. A number of different membrane types with a wide variety of nominal molecular weight cutoffs (500 to 10^6 daltons) are available. Even though the membranes are classified by the manufacturer according to molecular weight cutoff, it is emphasized that solute molecules are separated according to molecular size in ultrafiltration. Molecular weight data for humic substances can be estimated only by comparison of fractionation results with those obtained using suitable standards.

In theory, ultrafiltration is a rather simple process (Cross and Strathmann, 1973; Michaels, 1968). Under hydrostatic pressure, solute molecules, within the molecular weight cutoff of the membrane, pass, along with solvent, through the micropores of the membrane. Larger solutes are retained and concentrated. The separation is made in a cell (see Fig. 6, Chapter 15) with constant stirring to prevent concentration polarization and clogging of the membrane.

In practice a number of problems should be noted:

1. Because the size of the micropores is not uniform, the molecular weight cutoffs given by the manufacturer are not as sharp as would be expected. Nominal molecular weight cutoffs usually represent the particle size that will be 90% retained. In addition, ultrafiltration separates solutes according to molecular size, which will be dependent on molecular charge and configuration.

2. The separation process in ultrafiltration is dependent both on pressure and concentration gradient (Cross and Strathmann, 1973). As the volume of solvent in the cell decreases, the concentration of larger-molecular-size solutes in the cell increases, resulting in a breakthrough of larger-molecular-size solutes. Buffle et al. (1978) recommend that the filtration volume never exceeds 90% of the initial total volume.

3. Buffle et al. (1978) also note that the reactivities of humic and fulvic acids with other dissolved species or colloidal particles and with each other to form aggregates seem to be the most important factors in obtaining irreversible results with ultrafiltration. The possibility of interactions of humic substances in solution increases with the increased concentration of large-molecular-size material in the cell.

4. Interactions between humic substances and the membrane are possible. Care needs to be taken in choosing appropriate membranes to make a molecular size fractionation. To illustrate the importance of this point, two commonly used membranes with a molecular weight cutoff of 10,000 daltons, Amicon's UM-10 and PM-10, were compared using an aquatic fulvic acid from the Suwannee River, Georgia (Table 1). The molecular weight of the fulvic acid had previously been estimated as 1000–1500 daltons by small-angle X-ray scattering. Filtration volumes did not exceed 90% of initial volume. The membranes differ only in chemical composition. The UM-10 membrane retained 30–50% of the fulvic acid and was found to be discolored, indicating sorption of fulvic acid on the membrane. The PM-10 membrane retained none of the fulvic acid and behaved as expected. Use of standards such as Vitamin B_{12} (1357 daltons) or cytochrome-*c* (12,600 daltons) may not be sufficient to standardize the membranes for use with humic substances, because of the lack of chemical similarity of these materials with humic substances. Serious errors concerning the molecular size distribution of organic matter can result if this problem is not addressed.

Ultrafiltration is a versatile technique which has been used to concentrate organic compounds from water (Chapter 14), to fractionate humic substances by molecular size (Chapter 15) and to estimate molecular weights of organic matter (Gjessing, 1970; Ogura, 1974; Buffle et al., 1978). It has been used mostly to study aquatic humic substances although a few researchers have applied the method to soil humic material. Care needs to be exercised in using the method and in studying the data of others.

Scattering of Electromagnetic Radiation

When electromagnetic radiation impinges on a molecule, the electrons and protons of the molecule begin to vibrate. These vibrating, charged particles of the molecule are new sources of radiation, radiating in all directions. In this discussion, we are concerned only with scattered radiation of the same frequency as the impinging radiation. It can be shown that the angular distribution of the scattered intensity is a function of the size and shape of the scattering particles. Both light scattering and small-angle X-ray scattering

TABLE 1. Retention of Suwannee River Fulvic Acid (Molecular Weight 1500 daltons) by Ultrafiltration Membranes of 10,000 dalton Cutoff

Membrane	Milligrams of Carbon per Liter			Percent Retained
	Initial Solution	Concentrate	Filtrate	
UM-10	100	700	30	30
	20	100	10	50
PM-10	20	20	20	0

have been used for determination of molecular size and molecular weight (see Kerker, 1968, for a detailed discussion of light scattering and Guinier and Fournet, 1955, for a discussion of small-angle X-ray scattering). A brief discussion of the theory of small-angle X-ray scattering is given below in order to allow one to understand the experimental results.

Guinier and Fournet (1955) determined that for an ensemble of randomly oriented, identical scattering particles in which there is no long-range order the scattered intensity, $I(h)$, may be represented to a close approximation by the equation

$$I(h) = I_e(h)Nn^2e^{-h^2R^2/3} \tag{4}$$

where $I_e(h)$ is the scattered intensity that would result if a single electron were substituted for one of the scattering particles; N is the total number of particles in the ensemble; n is the number of electrons per particle; R is the radius of gyration of one of the particles; and $h = (2 \sin 2\theta)/\lambda$, where λ is the wavelength of the X-radiation. The radius of gyration of the particle, which is defined as the root-mean-square distance of the electrons in the particle from the center of charge, is a useful general characteristic of comparative molecular or particle size. Rewriting Equation (4) yields

$$\ln I(h) = -\frac{h^2R^2}{3} + \text{constant} \tag{5}$$

From Equation (5), it is seen that the radius of gyration may be calculated from the slope of a plot of ln $I(h)$ versus h^2; this is the so-called Guinier plot or curve. In a system in which all the scattering particles are of equal size, the Guinier plot will be a straight line.

In a polydisperse system, the Guinier plot, which is actually a summation of many Guinier plots, is no longer a straight line, but is concave upward. If only a few widely different-sized particles are in the system, it may be possible to discern discrete straight-line segments of the curve, and these segments may be used to calculate radii of gyration of the different particle sizes. In the general case, a Guinier plot of a polydisperse system of particles of uniform electron density will yield only the range of particle sizes present in the system, not the distribution of sizes (Kratky, 1963).

Wershaw et al. (1967) were the first to use small-angle X-ray scattering for determination of particle sizes in humic acid solutions. They determined that the unfractionated sodium humate, isolated from a North Carolina sandy soil, was polydisperse in solution. Further studies (Wershaw and Pinckney, 1973a) on humic acid fractions isolated by adsorption chromatography on Sephadex indicated that these fractions formed molecular aggregates in solution, and that the degree of aggregation was a function of both pH and concentration. Three different types of aggregation have been detected: (1) increased aggregation below pH 3.5 with little disaggregation above pH 3.5;

(2) disaggregation up to pH 7 and then reaggregation above pH 7; and (3) a continual decrease in aggregation with increasing pH. Wershaw and Pinckney (1973a) pointed out that the changes in degree of aggregation of humic acid salts in solution probably reflect the interaction of several different bonding mechanisms, including hydrogen bonding and coulombic interactions.

Lindquist (1970) determined that a soil sodium humate which was polydisperse in solution before acid hydrolysis became monodisperse after hydrolysis. The measured radius of gyration of the hydrolyzed sample was between 18 and 19 Å. It is well known that acid hydrolysis of humic substances releases both amino acids and simple sugars. These compounds probably are present as proteins and polysaccharides in most humic acid preparations. It is possible some of the large particles that Lindquist (1970) detected in his unhydrolyzed sample were these proteins and polysaccharides. But it seems more likely that the proteins and polysaccharides were associated by weak bonding mechanisms to the humic acid particles, perhaps even causing some of the aggregation. In this regard, Wershaw and Pinckney (1980) determined that one of the humic acid fractions isolated by adsorption chromatography on Sephadex is a clay–humic complex in which the humic acid is bound to the clay mineral particles by proteins. Deamination of the complexes releases these humic acids from the clays.

In a more recent study, Thurman et al. (1982) used small-angle X-ray scattering to study aquatic fulvic and humic acids. They found that many aquatic fulvic acids are monodisperse and that their radii of gyration are generally less than 10 Å. The humic acids, on the other hand, are polydisperse with some much larger particles present.

As far as we have been able to ascertain, only one light scattering study of a humic acid has been published. Orlov and Gorshkova (1967), as quoted in Orlov et al. (1971), found that the weight-average molecular weight of a chernozem humic acid was 66,200 daltons, and that of a sod-podzolic humic acid was 65,800 daltons. As Orlov et al. (1971) pointed out, it is difficult to compare these results with those obtained by other methods, because of the polydispersity of most humic acids, and the fact that they undergo aggregation reactions. Another problem not noted by Orlov and co-workers is that humic substances strongly absorb radiation and fluoresce so that it is difficult to measure accurately the amount of scattered radiation.

Electron Microscopy

Two different types of electron microscopy have been applied to humic substances: transmission electron microscopy (TEM) and scanning electron microscopy (SEM). The transmission electron microscope functions in much the same way as the light microscope except the wavelength of the radiation impinging on the samples in electron microscopy is about 10^{-5} that of the light microscope. Under the most favorable circumstances, one obtains a resolution of 0.1 or 0.2 nm in a transmission electron microscope, whereas the best one can hope for with a light microscope is 150 or 200 nm (Goodhew, 1975).

One of the major problems in observing organic materials by TEM is that of obtaining sufficient contrast to allow one to distinguish the important structural features of the specimen. Contrast in TEM is mainly caused by differences in electron scattering; for example, if one region scatters very little and an adjoining region scatters more, then the boundary between the regions can be distinguished. In the case of organic samples, it generally is necessary to enhance contrast by a staining technique. A heavy metal salt is used either as a positive or negative stain. In positive staining, stain penetrates the section and deposits preferentially at certain features. In negative staining, heavy metal particles surround the feature of interest. Another technique that has been used widely to allow viewing of the features of a sample is preparation of a carbon replica.

The scanning electron microscope is based on a somewhat different principle than the transmission electron microscope. In the scanning electron microscope, the viewed image is formed by the secondary electrons emitted from the sample surface when an electron beam is scanned across this surface. These secondary electrons are detected by a suitable detector and counted. An image is then formed on a cathode ray tube in which the brightness of the raster spot is proportional to the number of electrons emitted at each point on the sample surface. In order to prevent charging of the surface as the electron beam is scanned across it, the surface is coated with a conductor such as gold.

The first study of humic substances by electron microscopy was done by Flaig and Beutelspacher (1951) (see Flaig, 1967). They determined that the humic acid particles aggregate into a "ramified" structure at pH 3.5 but disaggregate at pH 8 to a more or less monodisperse system. They reported particle diameters for a representative chernozem humic acid to be about 50–100 Å. These samples were prepared by freeze-drying small droplets of humic acid solution on a supporting film and then increasing the contrast by shadow coating with platinum (Flaig et al., 1975).

Schnitzer and Kodama (1975) used TEM and Chen and Schnitzer (1976) used SEM to study the effect of pH on the structure of humic substances. In both studies, the authors determined that a gradual change occurred from a fibrous structure at low pH to a more sheet-like structure at higher pH. They also determined that the particles decreased in size with increasing pH.

However, in a later study, Stevenson and Schnitzer (1982), using a technique similar to that described by Flaig et al. (1975), determined that changes in pH had no effect on the structures of humic substances observed by TEM. The authors deposited drops of dilute aqueous solutions of humic and fulvic acids at various pH values on freshly cleaned mica sheets. The drops were frozen rapidly and freeze-dried, and the dried humic substance particles were replicated with a platinum and carbon film. In this way they hoped to obtain an accurate characterization of the morphology of the humic and fulvic acid particles. Five major types of structures common to humic and fulvic acids were determined: (1) small discrete spheroids 5–9 nm in diameter; (2) flattened aggregates of spheroids 100–200 nm in diameter; (3) linear chains of these aggregates; (4) flattened filaments 15–150 nm in width; and (5) perforated sheets. The initial concentration of the humic substances and the time of freeze-drying determined the type of structure predominant in the deposit; varying the pH had no effect.

Wershaw and Vader (unpublished data) observed by TEM particles with a diameter of approximately 20 Å in a humic acid fraction that had been negatively stained with sodium phosphotungstate. A similar size for these particles was measured by small-angle X-ray scattering. The particles measured in the other studies referenced before were substantially larger than these, and probably represent aggregates of smaller particles.

Orlov et al. (1975) reported that the features of humic acids observed in electron photomicrographs were not representative of the shape or ordering of humic acid particles in solution. They reached this conclusion from two different experiments: (1) They found by X-ray diffraction that humic acid particles are more ordered in an aqueous gel than in solid state, and (2) they found that the particles they observed in their electron photomicrographs were spherical, whereas their viscometry studies indicated that the particles were ellipsoidal, with an axial ratio of 1 : 6.

It also is very unlikely that the size and shape of humic substances observed in electron photomicrographs accurately represent the size and shape of these materials in soils because they have been dissolved and deposited under much different conditions than their formation in soils.

METHODS FOR MEASURING MOLECULAR WEIGHT

Ultracentrifugation

Two different types of ultracentrifugation generally are used to determine molecular weights: equilibrium ultracentrifugation and sedimentation velocity ultracentrifugation. Equilibrium ultracentrifugation, as the name implies, is an equilibrium measurement and can be described by using the formalism of thermodynamics. The sedimentation-velocity experiment, on the other hand, is a transport measurement that is much less well understood than the equilibrium experiment. As Eisenberg (1976) pointed out, the transport of molecules in a gravitational field is a complex phenomenon that is difficult to represent mathematically.

The normal treatment, although admittedly imperfect, is that given by Schachman (1959). The molecular weight of the sedimentary particles in a sedimentation-velocity experiment may be calculated from the well-known Svedberg equation:

$$M = \frac{RTS}{D_0(1 - \bar{v}\rho)} \tag{6}$$

where D_0 is the diffusion coefficient at infinite dilution, ρ is the density of the solution, $\bar{v}$ is the partial specific volume of the solution $\partial v/\partial m|_{(T,p)}$, and S is the Svedberg constant.

As Williams (1972) pointed out, this equation is valid only in a two-component system, in which S and D have been measured in the same solution, at the same temperature, and have been extrapolated to infinite dilution. If the particles that are sedimenting are charged, then a background electrolyte needs to be added to prevent the establishment of an electric potential gradient in the cell by the charged particles.

In the equilibrium ultracentrifugation experiment the partial molar free energy (or chemical potential) for a charged solute i, μ_i, in a gravitational field, $\varphi = -\frac{1}{2}\omega^2 r^2$, can be represented by the equation

$$\mu_i = \mu_i' + M_i\varphi_i + \nu\mathcal{F}\psi, \tag{7}$$

where M_i is the molecular weight of species i, ν is the charge number of species i, $\mathcal{F}$ is the Faraday constant, ψ is the electrostatic potential, and μ_i' is the partial molar free energy of i in the absence of gravitational and electrostatic fields (see Eisenberg, 1976).

In a two-component system of an electrostatically neutral solute and a solvent in a centrifuge cell at a distance r from the center of rotation of the cell, Eisenberg (1976) showed that the solution of Equation (7) yields the well-known equilibrium ultracentrifugation equation:

$$\frac{d \ln C_i}{dr^2} = \frac{\omega^2}{2RT}(1 - \bar{v}\rho)M_i \tag{8}$$

where C_i is the concentration of the solute, $\bar{v}$ is the apparent partial specific volume of a solute molecule, R is the gas constant, T is the temperature, and ρ is the solution density. If the solute is a polyelectrolyte, then a supporting electrolyte needs to be added to the solute to maintain electroneutrality.

Most ultracentrifuge studies of humic substances have used the sedimentation-velocity method. Flaig et al. (1975) have reviewed this work. In general, the results of this work indicated that all the solutions studied were polydisperse; as we have discussed previously, in a polydisperse system, the ordinary sedimentation-velocity equation cannot be applied because each different particle size has a different diffusion coefficient and different sedimentation constant.

Cameron et al. (1972b) determined that whole humic acid samples usually are too polydisperse to yield reliable molecular weight measurements by sedimentation-velocity ultracentrifugation. They carefully fractionated a soil humic acid by ultrafiltration and gel permeation chromatography in an attempt to obtain monodisperse fractions. The fractions they obtained were more homogeneous than the unfractionated solution, but still were not monodisperse. The molecular weights for their fractions ranged from 2600 to 1,360,000 daltons. This range is so broad that even if it were possible to measure accurately some sort of average molecule, it would have little meaning.

Posner and Creeth (1972) attempted to use equilibrium ultracentrifugation to measure the number-average, weight-average, and z-average molecular weights of soil humic acids in solution. They were able to get some indication of the degree of polydispersity of their sample by determining the ratios of these various molecular weight averages. For example, the $\overline{Mn}/\overline{Mz}$ ratio for a soil humic acid was 20, indicating a significant degree of polydispersity. Fractions isolated from this humic acid by ultrafiltration were less polydisperse as indicated by smaller ratios.

Ritchie and Posner (1982) recently published the results of a detailed sedimentation-velocity study of humic acid fractions that were isolated by density-gradient ultracentrifugation. Their work clearly demonstrated the necessity of performing molecular weight studies on samples with narrow molecular weight ranges, because of variation of diffusion coefficients and sedimentation coefficients with molecular weight. They pointed out that the use of unfractionated humic acids in molecular weight measurements could be the reason for variation in results between various workers. In addition, the molecular weight and size of their fractions decreased with increasing pH.

Viscometry

Ghosh and Schnitzer (1980b) used viscometry to determine the molecular weight and the shape of unfractionated humic and fulvic acids in solution. In addition to measuring molecular weights of humic and fulvic acids in solution, they concluded from their results that humic and fulvic acids change configuration in solution as a function of concentration, pH, and ionic strength.

Ghosh and Schnitzer (1980b) calculated molecular weights from viscosity measurements using the Staudinger equation:

$$[\eta] = KM^{\alpha} \tag{9}$$

where $[\eta]$, the intrinsic viscosity, is related to molecular weight by two adjustable parameters. These parameters are determined for a homologous polymer series with standards of known molecular weight (Eisenberg, 1976).

Ghosh and Schnitzer (1980b) have attempted to estimate these parameters; however, their estimates are questionable because well-defined standards were not used.

Colligative-Property Measurements

By definition, a colligative property is a thermodynamic property that depends on the number of particles in solution, and not on the nature of these particles. The colligative properties of a solution are: vapor-pressure lowering, freezing-point lowering (cryoscopy), boiling-point elevation, and osmotic pressure. Each of these properties can be shown to be related to the lowering of vapor pressure of the solvent by the addition of solute. In the limit of infinite dilution, each of these properties is proportional to the number of molecules of solute present. The theory and derivation of the mathematical relationships for each of the colligative properties can be found in many physical chemistry texts such as Maron and Prutton (1969). Reports by Bonnar et al. (1958) and Glover (1975) are excellent works describing the determination of molecular weight by measurement of these properties.

The molecular weight obtained by measurement of a colligative property is a number-average molecular weight. Two serious problems need to be addressed in determining number-average molecular weights of humic substances:

1. A molecular weight value less than the actual molecular weight is obtained for a dissociated organic acid (Hansen and Schnitzer, 1969) because the dissociated protons are "counted" as solute molecules, increasing the value for the number of solute molecules present. Dissociation can be suppressed by making measurements in organic solvents; however, humic substances are insoluble or only partly soluble in most organic solvents. In addition, dimerization can occur in some solvents (Shapiro, 1964). It is possible to mathematically correct for dissociation of the acid. The method of Hansen and Schnitzer (1969) simply applies a correction technique which assumes that the dissociation of humic substances can be described by

$$H_nA = nH^+ + A^{n-}$$

This method only requires the experimentally determined molecular weight, $\overline{Mn}$, and solution pH. Wilson and Weber (1977) have proposed a method of correcting for dissociation which assumes that fulvic acid is comprised of two types of carboxylic acid groups. This correction method uses $\overline{Mn}$, solution pH, and the experimentally determined equilibrium constants for the two carboxylic acid types. Reuter and Perdue (1981) critically discuss the methods of Hansen and Schnitzer (1969) and Wilson and Weber (1977), and propose a rigorous method of their own that requires knowledge of the instrument readout and pH. This method makes no assumptions about the chemistry of dissociation of humic substances. Gillam and Riley (1981) also present a correction method that does not require dissociation constant data. Their method requires knowledge of solution pH and the apparatus constant (K_{app}), which is the appropriate colligative-property constant corrected for errors that result from imperfect solvent purity and manipulation of the instrument (Bonnar et al., 1958).

2. Number-average data of a polydisperse system are of little value. It is important to use other methods, such as small-angle X-ray scattering, to determine the extent of polydispersity of the system before making conclusions from colligative-property measurements.

The most widely used colligative-property measurements for the molecular weight determination of humic substances are cryoscopy and vapor-pressure osmometry. In the cryoscopic experiment, the freezing temperature of a solution of the sample is compared with that of the pure solvent. $\overline{Mn}$ values for polymer solutions as much as 50,000 daltons have been determined by the cyroscopic method (Glover, 1975). For humic substances, reported molecular weights range from as low as 300 daltons for fulvic acids to 74,000 daltons for peat humic acids (Stevenson, 1982). Sulfolane, which has a greater cryoscopic constant than water, has been used as a solvent for the cryoscopic determination of molecular weights of humic substances (Wood et al., 1961; Schnitzer and Desjardins, 1962). Although this solvent appears to solve the problems of dissociation, it requires extensive purification (Gillam and Riley, 1981).

In vapor-pressure osmometry, the vapor-pressure difference between solution and solvent is determined by measuring the temperature increase associated with the condensation of vapor into the solution (Glover, 1975). This method is rapid, and measurements can be made with a variety of organic solvents, as well as water. The upper molecular weight limit is 40,000 daltons or less. For humic substances reported values range from 300 to more than 2000 daltons (Schnitzer and Khan, 1972).

SUMMARY AND CONCLUSIONS

The determination of molecular sizes and weights for humic substances is a task complicated by the nature of these materials. A brief overview of the methods most commonly used to determine molecular sizes and weights of humic substances has been presented in this chapter. Many of the methods discussed are powerful techniques that can yield information about the molecular size, shape, and weight of humic substances. No single method alone, however, is sufficient to provide a complete understanding of these molecular characteristics. Meaningful and accurate conclusions can only be made by using the data provided by different methods of analysis.

The fine details of theory, methodology, and application are beyond the scope of this chapter. An effort has been made, however, to address the limitations of each of the methods presented. Occasionally, in our enthusiasm to obtain new insights into the chemistry of humic substances, we pay too little heed to the limitations of the methodology, thereby diminishing the value of our conclusions. In determining molecular size and molecular weight data, recognizing these limitations is critical for the advancement of our understanding of the chemistry of humic substances.

CHAPTER TWENTY

Acidic Functional Groups of Humic Substances

E. MICHAEL PERDUE

ABSTRACT

A detailed discussion of the factors that affect the acidities of organic acids is presented, with emphasis on statistical, electrostatic, and delocalization effects. From such considerations and the very high degree of oxygen substitution in humic substances, it can be concluded that humic substances must contain a complex mixture of nonidentical acidic functional groups. Given this view of humic substances, quantitative methods of functional group analysis and mathematical models of proton binding have been evaluated.

Although all potentiometric methods of functional group analysis are inherently operational, the total acidity can be determined by the barium hydroxide method, if appropriate precautions are taken to remove all colored material from the reaction mixtures prior to the final titration. Any method that purportedly distinguishes between different classes of functional groups should be used with full awareness of the operational nature of the results that will be obtained. If potentiometric data are augmented with calorimetric and/or spectroscopic data, more accurate estimates of functional group concentrations may possibly be obtained.

Until recently, most mathematical models that were used to describe proton binding by humic substances were far too simple, and, consequently, served only as empirical curve-fitting equations. More recently, several authors have used mathematical models that treat humic substances as having a continuous distribution of acidic functional groups. These more realistic models are highly recommended and are expected to find greater acceptance in the near future.

INTRODUCTION

To properly describe the acid–base properties of humic substances, it is essential that (1) the identification and quantification of acidic functional groups be accomplished in a rigorous, reproducible manner and (2) the range of pK_a values that exists in humic substances be described by a suitable model that is as rigorous as possible. Despite years of research, neither of these objectives has been satisfactorily met. Consequently, it is not generally possible to compare results obtained by different scientists if different methods and/or humic samples have been used.

This chapter examines the fundamental structural properties of organic molecules that directly or indirectly affect pK_a values of organic acids, making it possible to estimate theoretical upper limits for concentrations of common acidic functional groups, to assess methods of quantitative analysis of acidic functional groups in humic substances, and to examine several models that have been proposed for the description of proton binding by humic substances. Some of the concepts that appear obvious in the context of this chapter have been included for the simple reason that the literature on acidity of humic substances reveals that these points have frequently been either misunderstood or neglected.

ACIDIC FUNCTIONAL GROUPS IN ORGANIC COMPOUNDS

Both the abundances and the pK_a values of acidic functional groups in humic substances are constrained and controlled by the compositional and structural properties of these substances. It is therefore important to be familiar with the principal structural properties that determine the acidities of organic acids. In this section of the chapter, compositional constraints on the abundances of acidic functional groups and a general discussion of statistical, electrostatic, and delocalization effects on the acidities of organic acids are presented. Specific applications to humic substances are given in the following sections of the chapter.

Analytical Constraints on Acidic Functional Group Concentrations

The commonly encountered acidic functional groups in organic compounds include carboxylic acids, phenols, ammonium ions, alcohols, and thiols. To a lesser extent, sulfonic acids and "active methylene" compounds (those containing the —CO—CH_2—CO— structural moiety) are also encountered.

Because humic substances generally contain low amounts of nitrogen and sulfur (see next section), the remaining discussion will focus on the properties of oxygen acids (carboxylic acids, phenols, alcohols) and "active methylene" compounds.

The abundance of oxy-acid functional groups in an organic substance or mixture of such substances is potentially limited by the oxygen content of the substance, but those theoretical limits depend very much on whether the oxy-acids are predominantly carboxylic acids or other functional groups such as phenols, alcohols, and enols. For example, on the basis of its oxygen content alone, an organic compound or mixture of such compounds containing 48% oxygen could contain 15 mmol/g of carboxyl groups or 30 mmol/g of the other types of oxy-acids.

Other structural constraints can impose much lower limits on the potential concentrations of acidic functional groups. Both carboxyl and phenolic groups are associated with unsaturation (e.g., π bonds and/or rings), and thus the overall degree of unsaturation in an organic compound or mixture of such compounds may actually limit the abundances of oxygen acids. Perdue (1984) shows that the amount of unsaturation in a mixture of organic compounds (ϕ_{total}, in mmol/g) can be rigorously calculated as

$$\phi_{total} = C_{total} + N_{total}/2 - H_{total}/2 + 1000/\overline{Mn} \qquad (1)$$

where C_{total}, H_{total}, and N_{total} are elemental abundances in mmol/g and $\overline{Mn}$ is the number-average molecular weight of the mixture. Fortunately, for substances with relatively high molecular weights such as humic substances, the calculation of ϕ_{total} is relatively insensitive to $\overline{Mn}$, making it possible to estimate that parameter if it has not been independently determined (as is often the case). The combined unsaturation of aromatic rings, carboxyl groups, carbonyl groups, alkenes, esters, amides, cycloalkyl groups, and so on cannot exceed ϕ_{total}, as given in Equation (1). Inasmuch as phenolic hydroxyl groups are more or less directly limited by the aromaticity of the mixture, it is evident that the concentrations of both carboxyl and phenolic groups are potentially limited by structural constraints on unsaturation.

Statistical Effects on Acidity

Whenever an organic acid contains two or more chemically identical (i.e., stereochemically equivalent) functional groups, statistical factors that originate in the entropy of formation of the acid and/or its conjugate base contribute to the variation of thermodynamic dissociation constants with the degree of dissociation of the acid. Such statistical effects are implicitly included in equations that are often used to describe acid–base equilibria in synthetic and natural polymers. Because those equations have frequently been applied to proton binding by humic substances, a brief discussion of statistical effects on acid–base equilibria is presented here. The reader is referred to Hine (1975) and Tanford (1961) for a more thorough discussion.

Following the approach of Tanford (1961), it is useful to examine the ionization of a hypothetical diprotic acid (HA-BH). Thermodynamically, the acid–base properties of this acid are completely described by two acid dissociation constants (K_1 and K_2). However, at the molecular level, it is clearly more appropriate to define four dissociation constants as depicted in Equation (2):

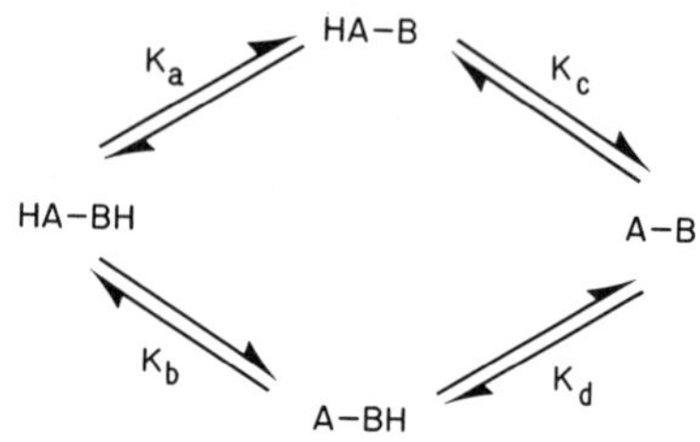

(2)

Whether or not A and B represent identical ligands, the products K_aK_c and K_bK_d must be equal. The relationship between K_a and K_b or between K_c and K_d depends on whether or not A and B are identical. With complete generality, it can be shown that the thermodynamic acid dissociation constants (K_1 and K_2) are related to the microscopic constants by the following expressions:

$$K_1 = K_a + K_b \qquad K_2 = K_cK_d/(K_c + K_d) \tag{3}$$

In the case where A and B are identical, K_a equals K_b and K_c equals K_d. It follows that K_1 equals $2K_a$ and K_2 equals $0.5K_c$. Therefore, the thermodynamic dissociation constants differ from the "intrinsic" dissociation constants, with the ratio K_1/K_2 equal to $4(K_a/K_c)$. Whether or not there are electrostatic interactions that affect K_1 and K_2 (discussed in the following sections), the statistical factor of 4 will always be present. A comparison of K_1/K_2 for a series of symmetrical dicarboxylic acids HOOC—$(CH_2)_n$—COOH is given in Figure 1. As n increases, electrostatic effects should approach zero and K_a/K_c should approach unity, leaving only a statistical factor of 4, as is evident in this figure.

If all acidic functional groups in a molecule are absolutely identical and are physically so far apart that there are no electrostatic effects on pK_a values, the titration curve of a polyprotic acid can be represented by the titration curve of a suitable monoprotic acid whose K_a value is the statistically corrected "intrinsic" K_a for the polyprotic acid. For instance, in the previous example, if there were no electrostatic effects to be considered and A and B were identical, then K_a would equal K_c. The thermodynamic con-

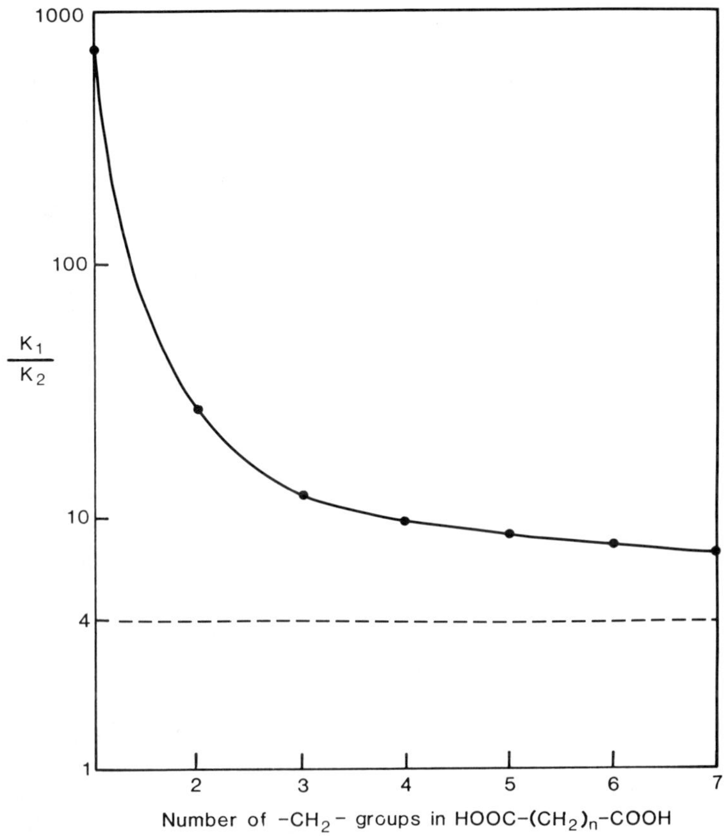

FIGURE 1. Electrostatic effects on pK_a values of symmetrical dicarboxylic acids.

stants would be $K_1 = 2K_a$ and $K_2 = 0.5K_a$ and the titration curve could be exactly described in terms of either 1 mole of a diprotic acid (K_1 and K_2) or 2 moles of an intrinsic monoprotic acid ($K_a = 0.5K_1 = 2K_2$).

Tanford (1961) also applies the concept of an "intrinsic" constant to polymeric organic acids in which all acidic functional groups are assumed to be identical, with apparent pK_a differences being attributed to statistical effects and electrostatic effects that increase with increasing degree of ionization of the polymer. Empirical terms can be introduced to "correct" for electrostatic effects arising from formally charged sites in a molecule and statistical effects are handled as previously described. The "intrinsic" K_a value corresponds to the apparent K_a value when the polymer is uncharged. Such models are most relevant in regular polymers whose repeating monomer units exist in identical electrostatic environments when the polymer is uncharged.

Electrostatic Effects on Acidity

The acidity of an ionizable functional group depends greatly on its electrostatic environment (i.e., the spatial distribution of proximate dipolar groups and formally charged groups in the molecule). The quantitative treatment of the electrostatic interactions of acidic functional groups with both charged and dipolar substituents is thoroughly discussed by Hine (1975), King (1965), and Stock (1972), so only an overview is given here. The examples and discussion do not include delocalized systems, which are treated separately in the next section of the chapter.

Several theories have been proposed to account for substituent-induced changes in the acidities of organic acids. Substituent effects are most clearly illustrated in a symmetrical proton transfer reaction such as

$$HA_1 + A_2 \rightleftharpoons A_1 + HA_2 \tag{4}$$

where HA_1 and HA_2 are two structurally similar organic acids differing only by a charged or dipolar substituent, and A_1 and A_2 are the respective conjugate bases.

Electrostatic Field Models

In an analysis of the influence of charged substituents on the pK_a values of carboxylic acids, Bjerrum (1923) suggested that, for symmetrical proton transfer reactions such as Equation (4), pK_a differences can be accounted for in terms of purely electrostatic field effects (and "statistical effects," whenever they are warranted). If the proton and charged substituent are treated as point charges separated by a distance r in a solvent continuum whose dielectric constant D is that of the bulk solvent, the work required to transfer the proton from r to infinity in the electrostatic field of the charged substituent can be calculated from classical electrostatic theory. The pK_a difference between the unsubstituted and substituted acids (ΔpK) is then given by

$$\Delta pK = \frac{Nze^2}{2.303RTDr} \tag{5}$$

where R is the gas constant, T is the absolute temperature, N is Avogadro's number, ze is the charge of the substituent, and e is the proton charge.

Bjerrum's equation can be extended to include dipolar substituents if the dipole moment (μ) of the substituent is treated as a point dipole located at a distance r from the acidic proton (see Fig. 2). The pK_a difference between the unsubstituted and substituted acids is then given by

$$\Delta pK = \frac{Ne\mu \cos \theta}{2.303RTDr^2} \tag{6}$$

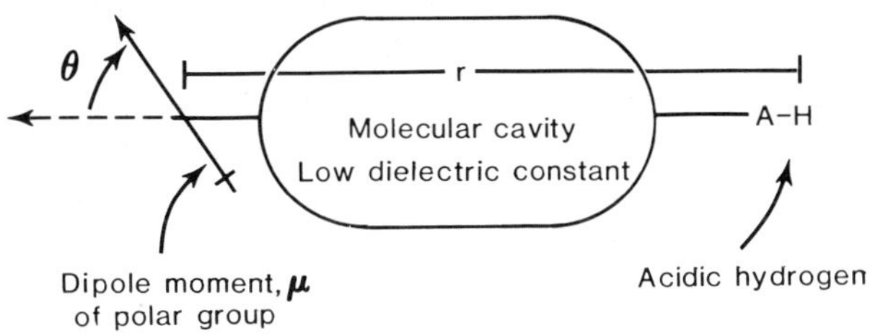

FIGURE 2. Electrostatic model of dipolar substituent effects.

where θ is the angle between the dipole moment vector and a line connecting the point dipole with the acidic proton, and all other symbols are the same as those previously defined. It is very important to note that the effect of a dipolar group is acid-strengthening when $\theta < 90°$, acid-weakening when $\theta > 90°$, and approaches zero as θ approaches 90°. In a flexible molecule that can readily assume a conformation that minimizes electrostatic repulsions, the polar groups will assume an orientation that places the positive end of the dipole nearer to the negatively charged anion that forms upon ionization, thus lowering the Gibbs free energy for ionization of the acidic functional group. Consequently, dipolar groups are almost always acid-strengthening. Equations (5) and (6) usually yield qualitatively correct but quantitatively low estimates of the actual effects of charged and dipolar substituents.

In a major contribution to the understanding of transmission of substituent effects in organic molecules, Westheimer and co-workers (Kirkwood and Westheimer, 1938; Westheimer and Kirkwood, 1938; Westheimer and Shookhoff, 1939; Westheimer et al., 1942) pointed out that the electrostatic field of a charged or dipolar substituent is at least partially transmitted through the molecule itself. Because the dielectric constant of an organic molecule is much smaller than that of a polar solvent such as water, substituent effects are attenuated with increasing distance to a much smaller extent than predicted by Equations (5) and (6). Kirkwood and Westheimer described methods for computation of the "effective" dielectric constant for a molecule–solvent system. The use of the effective dielectric constant (D_{eff}) in place of the solvent's dielectric constant (D) in Equations (5) and (6) greatly improved the quantitative capabilities of electrostatic models of substituent effects.

The relative extent to which dipolar and charged substituent effects decrease with increasing distance from an acidic functional group can be obtained by dividing Equation (5) into Equation (6):

$$\delta \Delta \mathrm{p}K = \frac{\mu \cos \theta}{zer} \tag{7}$$

where $\delta\Delta pK$ is the relative effect of dipolar and charged groups on pK_a values and the other terms are as previously defined. Using the group dipole moment of a carboxyl group ($\mu = 1.65$), the optimum angle of orientation ($\theta = 0$), and $z = 1$, the magnitude of $\delta\Delta pK$ can be calculated at selected values of r. Using $r = 2$–4 Å, $\delta\Delta pK$ is predicted to fall between 0.09 and 0.17. In all likelihood, each uncharged, polar group that is in proximity to a carboxyl group will exert an acid-strengthening effect that is more than 10% as strong as the acid-weakening effect that would be exerted by a negatively charged group at the same distance from the acidic group.

Inductive Model

Another model of transmission of polar substituent effects (often described in introductory organic chemistry texts) is the so-called inductive model, in which it is proposed that a charged or dipolar group modifies the pK_a of an acidic functional group by successive polarization of all the intervening σ

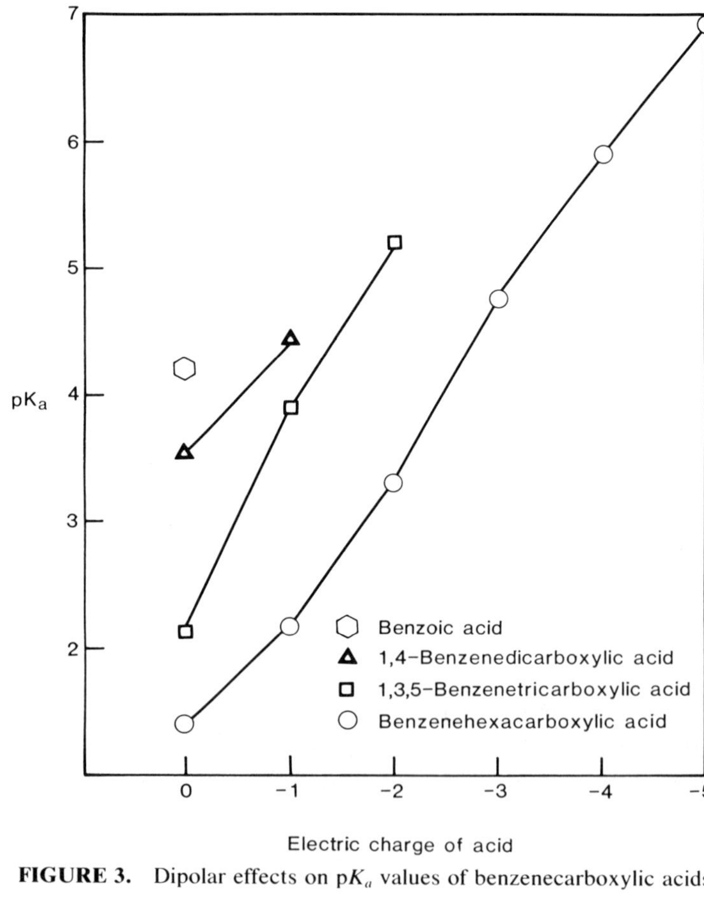

FIGURE 3. Dipolar effects on pK_a values of benzenecarboxylic acids.

and π bonds between the functional group and the substituent. Numerous researchers have synthesized and studied acid systems that were designed to compare the predictions of the inductive and electrostatic field models. Stock (1972) has summarized the results of many such investigations and, in general, much better agreement between theoretical and experimental results was obtained using the electrostatic field model of Kirkwood and Westheimer.

Electrostatic Effects in Simple Organic Acids

As a simple illustration of the importance of dipolar effects on acid strength, the pK_a values of selected benzenecarboxylic acids are presented in Figure 3. Using benzoic acid as a reference, it is clear that the first ionization constants of all the other acids are greater, even though all the acids are uncharged. A small part of the increased acidity is due to statistical effects, but most of the effect is attributable to dipolar stabilization of the monoanion in the polyprotic acids. The dipolar effect is so pronounced in these molecules that the K_a values of several of the anions are greater than that of benzoic acid.

The effect of charge-dipole separation (r in Eqs. (5) and (6)] on the magnitude of dipolar substituent effects is illustrated in Figure 4. Several homologous series of organic acids containing some of the types of polar substituents that might be present in humic substances are included. To eliminate the statistical factor, $K_1/2$ values are used for the dicarboxylic acid series.

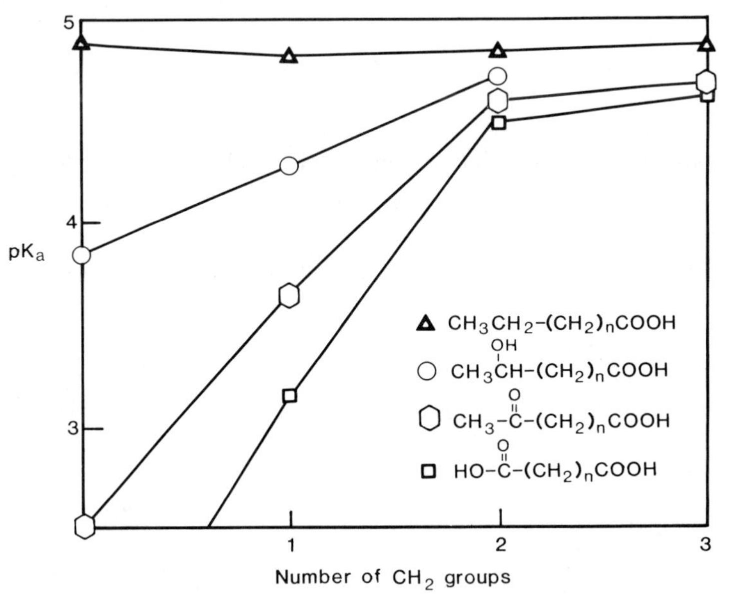

FIGURE 4. Dipolar effects on pK_a values of aliphatic carboxylic acids.

The simple alkanoic acids are included as a point of reference. It is evident that all the acid series yield similar results. If polar groups are within a distance of 1 to 3 C—C bonds from an acidic functional group, pK_a values are lowered by 0.1–2.0 pK units.

Delocalization Effects on Acidity

The delocalization of electron density in the conjugate base of an acid enhances its acidity, as indicated by the relative acidities of alcohols and carboxylic acids. While O—H bonds are heterolytically broken in both cases, only carboxylic acids yield delocalized anions. This subject is presented in adequate detail in most organic chemistry texts (Hine, 1975; March, 1977; Lowry and Richardson, 1981), so only a few examples that illustrate the significance of this phenomenon are presented here.

The pK_a values of selected benzoic acids and phenols are given in Table 1. Most of the compounds listed and many similar compounds have been identified as oxidation products of lignin (Hedges et al., 1982) and of humic substances (Reuter et al., 1983; Liao et al., 1982). The data cannot be completely explained in terms of statistical and electrostatic effects. While the dipolar —CHO and —$COCH_3$ groups are acid-strengthening in both classes of acids, their effect is much more pronounced in the phenols. The negatively charged —CO_2^- group is acid-weakening, as expected, in the benzoic acid, but is quite unexpectedly acid-strengthening in the phenol. Both the stronger dipolar effects and the "reverse" effect of —CO_2^- in the substituted phenols are easily understood in terms of delocalization of negative charge in the phenoxide anions. It is not possible to write "resonance" structures for benzoate anions that delocalize the charge onto either the benzene ring or onto para-substituents, so the principal effect of a charged or polar group on the acidity of a substituted benzoic acid is electrostatic. In contrast, the negative charge of a phenoxide ion is delocalized onto the benzene ring and onto the para-substituents that are included in Table 1. Some of the pertinent

TABLE 1. pK_a Values of Selected Phenols and Benzoic Acids

	pK_a	
X Group	X— 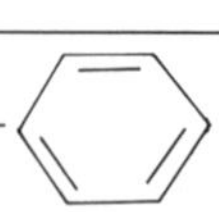 —OH	X— 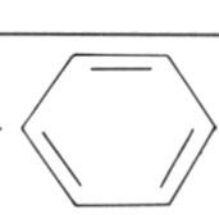 —COOH
—H	9.98	4.20
—CHO	7.62	3.76
—$COCH_3$	8.05	3.70
—COO^-	9.46	4.46

resonance structures are given below (Y = H, CH_3, or O^-):

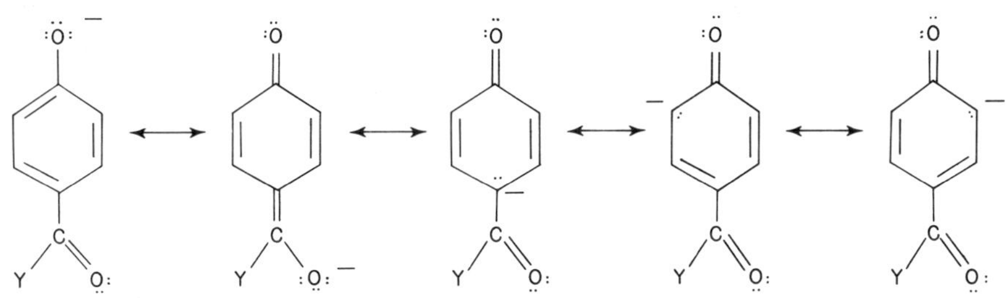

The ability of the —CHO and —$COCH_3$ groups to delocalize charge in phenoxide ions complements the acid-strengthening electrostatic effects of those substituents, resulting in a larger enhancement of acidity than in similarly substituted benzoic acids. The delocalization of phenoxide charge by the —CO_2^- substituent completely overrides the acid-weakening electrostatic effect of that substituent, making that anionic phenol even more acidic than phenol itself.

Acid Dissociation Constants in Simple Organic Acids

To illustrate the overall result of statistical, electrostatic, and delocalization effects on the acidities of organic acids, pK_a values for *all* the organic acids containing only carbon, hydrogen, and oxygen that are tabulated in two critical compilations (567 pK_a values; Christensen et al., 1976; Martell and Smith, 1977) are summarized as frequency histograms in Figure 5. The use of pK_a as the independent variable is warranted because pK_a values are directly proportional to standard Gibbs free energies of ionization and are directly related to the charge or polarity of a substituent [see Eqs. (5) and (6)].

The range of acid types encompasses everything from monoprotic to hexaprotic acids, with acidic functional groups consisting mainly of carboxylic acids (409), phenols (95), and a number of alcohols and β-dicarbonyl carbon acids (63). Therefore, the observed distribution of pK_a values is a result of structural, statistical, and electrostatic effects. It is apparent from Figure 5 that the frequency of occurrence of the carboxyl groups is approximately a Gaussian distribution with a mean pK_a value of about 4.5. The phenolic sites are similarly distributed around a mean pK_a of approximately 10. The remaining acidic functional groups (mainly β-dicarbonyl compounds, enols, and alcohols) occur throughout the pK_a range of 0–13. The small cluster of peaks in the pK_a 12–13 range is composed of the alcoholic hydroxyl groups of monosaccharides.

There is clearly some overlap of pK_a ranges for the classes of acidic functional groups. While the distribution of pK_a values in humic substances

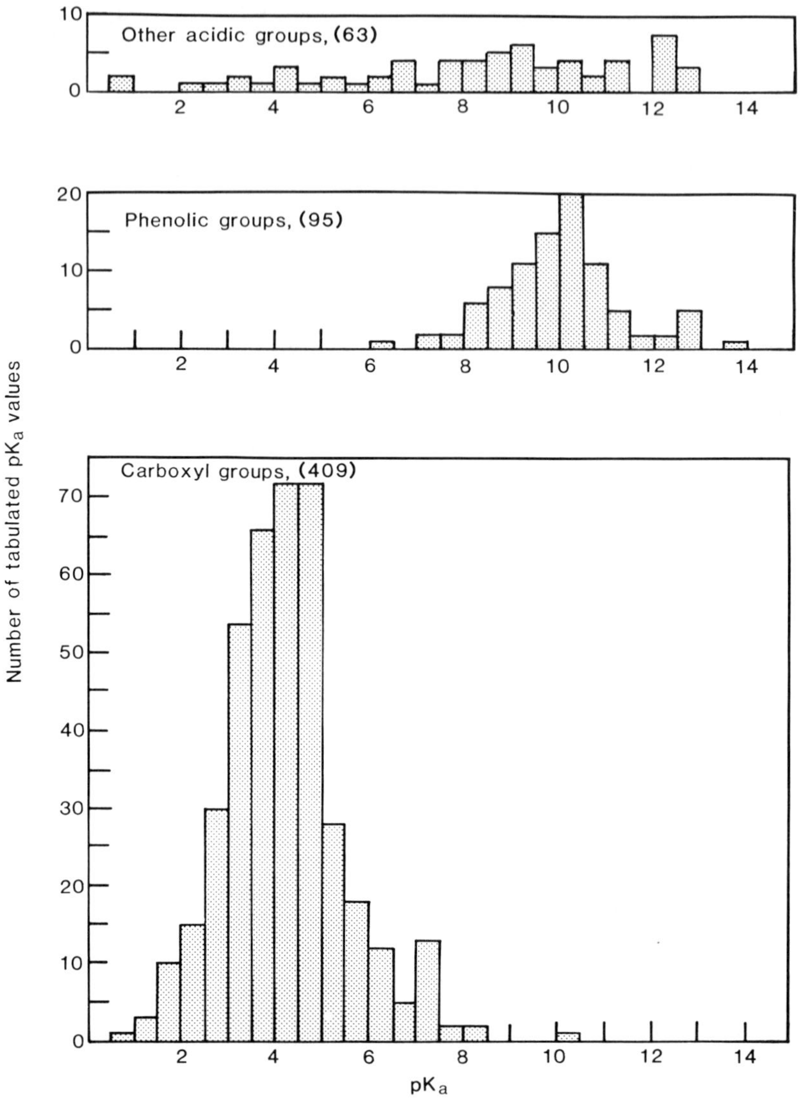

FIGURE 5. Distribution of pK_a values in simple organic acids.

will undoubtedly differ from the statistical sample of simple organic acids given in Figure 5, it would be unlikely that any sort of pK_a-dependent method of functional group analysis could unambiguously distinguish carboxylic acids from the other classes of acids that could be present. The operational nature of such functional group methodologies is discussed at length by Perdue et al. (1980) and is reviewed later in this chapter.

ACIDIC FUNCTIONAL GROUPS IN HUMIC SUBSTANCES

Analytical Constraints on Acidic Functional Group Concentrations

Before examining the methodologies of functional group analysis and the modeling of acid–base equilibria in humic substances, it is worthwhile to consider how functional group concentrations, the choice of proton binding models, and so on may be constrained by the chemical composition of humic substances. First of all, humic substances are indisputably a highly complex mixture that has thus far been essentially unresolvable into significant amounts of pure components. Nevertheless, several properties of humic substances are, in fact, well defined and experimentally measurable. These include elemental composition and, in some cases, number-average molecular weights (Reuter and Perdue, 1981). These parameters can be used to calculate the amount of unsaturation in a humic substance sample [see Eq. (1)], which, in turn, can often be used to place upper limits on the concentrations of carboxyl and phenolic functional groups.

Most aquatic humic and fulvic acid samples contain less than 1% sulfur. For such samples, the total concentration of sulfur acids cannot exceed 0.3 mmol/g. The nitrogen content of these samples is also quite low (<2%). This fact, coupled with the fact that more than half of the nitrogen is usually recoverable as amino acids (presumably from polypeptide residues, in which the amide nitrogen is not basic), limits the concentration of nitrogen acids to about 0.7 mmol/g. The theoretical limits for sulfur acids and nitrogen acids would be higher in humic acids, which may contain more sulfur and nitrogen than aquatic humic or fulvic acids. However, even in such cases, sulfur acids and nitrogen acids are relatively unimportant in comparison with oxygen acids.

The fulvic acid sample of Hatcher et al. (1981c), which has the elemental composition 50.9% carbon (42.4 mmol/g), 3.6% hydrogen (36.0 mmol/g), 1.0% nitrogen (0.7 mmol/g), 0.6% sulfur (0.2 mmol/g), and 43.9% oxygen (27.4 mmol/g), will be used for illustrative purposes in subsequent discussion of analytically imposed constraints on the structural features of humic substances. Hatcher et al. (1981c) report an aromatic carbon content of 28% for this sample. The number-average molecular weight of this material is probably near 1000 g/mol (Reuter and Perdue, 1981). The reported "wet chemical" values for total acidity, carboxyl, phenolic, and carbonyl content in this humic sample are 12.4, 9.1, 3.3, and 3.1 mmol/g, respectively. (As discussion in a subsequent section of this chapter reveals, total acidity is probably the most accurately known "functional group" parameter.)

The aromatic carbon content of this sample (28% of 42.4 mmol/g or 11.9 mmol/g) is equivalent to 2.0 mmol/g of benzene rings. If the phenolic hydroxyl content of this sample were to exceed 2.0 mmol/g (as indicated by the wet chemical data), the *average* aromatic ring in the sample would have to contain more than one phenolic hydroxyl group (a highly unlikely occurrence, given the preponderence of nonhydroxylated and monohydroxylated aromatic compounds in proteins and lignins and the virtual absence of polyhydroxylated aromatic oxidative degradation products of humic substances, lignin, etc.). Hatcher et al. (1981c, 1983a) report that very little phenolic content is indicated in the ^{13}C NMR spectra of most humic substances that they have examined.

Substitution of the elemental composition of this sample into Equation (1) yields a total unsaturation of 25.8 mmol/g. The aromatic carbon accounts for 7.9 mmol/g of the unsaturation (4 mmol of unsaturation per 6 mmol of aromatic carbon) in the sample, leaving 17.8 mmol/g of unsaturation in other groups. Such a large amount of nonaromatic unsaturation makes it virtually impossible to estimate the carboxyl content of this material, because, even if the total acidity of the sample was entirely attributable to carboxyl groups, those groups could not account for all the nonaromatic unsaturation. At least 17.8 − 12.4 or 5.4 mmol/g of additional unsaturation is present in the sample, possibly as C=O groups, esters, amides, cycloalkyl groups, pyranose and furanose rings, and so on. This information, while not germane to the discussion of acidic functional groups, is definitely relevant in discussions of the origin and fate of humic substances in soil and aquatic environments.

Acid Dissociation Constants of Functional Groups in Humic Substances

Not only do the elemental composition, $\overline{Mn}$, and aromaticity of a humic substance constrain the concentrations of acidic functional groups but they also give a crude picture of the spatial distribution of dipolar groups and formal charges in the vicinity of an acidic functional group. Such information is especially useful in evaluating the structural soundness of the multitude of models that are used to describe the acid–base properties of humic substances.

The fulvic acid sample of Hatcher et al. (1981c) contains about nine carboxyl groups per average molecule (assuming an $\overline{Mn}$ of 1000 g/mol). In the unionized state, these carboxyl groups themselves are quite polar and are generally acid-strengthening, as are carbonyl groups, hydroxyl groups, methoxyl groups, and so on. As pH increases, carboxyl groups are converted into negatively charged carboxylate ions, which are almost always acid-weakening. The other polar groups, however, continue to be acid-strengthening at all pH values.

The magnitude of the effect of polar groups on pK_a values in humic substances depends on the proximity of such groups to acidic functional groups. If ester groups can be neglected, the average number of noncarboxyl carbons per polar functional group is calculated from the elemental composition and carboxyl content as $(C_{total} - [COOH])/(O_{total} - [COOH])$, where all parameters are in mmol/g. Statistically, the fulvic acid of Hatcher et al. (1981c) contains either a carboxyl group or an oxygen-containing substituent (carbonyl, alcohol, or ether) for every 1.8 noncarboxyl carbons. In other words, the sample is *highly substituted with polar groups*. Whether or not those polar groups are uniformly distributed within and between molecules, it is clearly the case that many of the carboxyl groups are sterically quite close to other polar groups (either carboxyl or other oxygen-containing groups).

At an average site density of one polar group per 1.8 noncarboxyl carbons, the average distance between polar groups in this humic substance sample is about 2–4 Å. It is therefore possible to anticipate that humic substances must contain carboxyl groups that are far more acidic than acetic or benzoic acids. As those acidic groups are converted to carboxylate ions at higher pH values, the accumulation of negative charge will eventually overcome the shorter-range dipolar effects due to unionized carboxyl groups and other polar groups, and pK_a values of carboxyl groups and weaker classes of acidic functional groups should increase above those of simple reference acids.

In light of the above discussion, the spectrum of acidic functional groups in humic substances is expected to be at least as "continuous" and complex as the distribution of simple organic acids whose pK_a values are displayed in the form of a frequency histogram in Figure 5, especially if the acidic properties of humic substances are affected by configurational effects, Donnan potential effects, and so on. Humic substances almost certainly contain a highly complex mixture of *nonidentical* functional groups with pK_a values that span the entire possible range that is determined by the leveling effect of H_2O on the strengths of acids (no acid stronger than H_3O^+ can exist as a major chemical species in aqueous solutions and no acid weaker than H_2O will ionize appreciably in aqueous solutions). This view of humic substances is reflected in the following sections of the chapter, where methods of functional group analysis and mathematical models of acid–base equilibria in humic substances are addressed. The ultimate method or model has to be consistent with the anticipated complexity of the mixture of acidic functional groups in humic substances.

DETERMINATION OF ACIDIC FUNCTIONAL GROUP CONCENTRATIONS

Numerous descriptions of methods of analysis of the acidic functional groups of humic substances have been published (e.g., Stevenson and Butler, 1969; Schnitzer and Khan, 1972; Stevenson, 1982). The published methods include direct titrations, discontinuous titrations, indirect titrations, indirect titrations coupled with either a distillation or ultrafiltration step, thermometric titration methods, nonaqueous titrations, irreversible reactions of acidic hydrogens with various organic and inorganic reagents, and so on. The two most commonly described methods are the barium hydroxide total acidity determination and the calcium acetate exchange method for determination of carboxyl groups, both of which are classified as indirect potentiometric titration methods. Rather than examine the experimental protocol of each of these published methods, the theoretical limitations, inherent assumptions, and ambiguities of some of the potentiometric and thermometric techniques are examined. The reader is referred to other sources such as Stevenson (1982) or Schnitzer and Khan (1972) for more experimental details.

Total Acidity Determinations

The phrase "total acidity" conveys a sense of certainty and definition, suggesting to the reader that *all* acidic hydrogens in humic substances are included in this category. Upon reflection, however, one must conclude that the total acidity of a sample includes only those acidic hydrogens that react with a specified reagent under a specified set of experimental conditions. It is apparent that a good total acidity method must involve equilibration of the sample with a reagent of very high pH, so that even the weakest acids will react. This requirement presents two serious experimental problems that must be overcome to achieve good results. First, it is difficult to accurately determine how much base has reacted with the humic substances in the presence of the extreme excess of base that is needed to reach an appropriately high pH. Second, equilibration should be achieved as quickly as possible to avoid base-catalyzed side reactions that might alter the apparent total acidity of the sample (e.g., hydrolysis of esters and peptides, which releases additional carboxylic acids). Some common potentiometric methods are discussed below.

Direct Titrations

The amount of base consumed in a direct titration approaches the total acidity if sufficiently high pH values are achieved. Several factors, however, limit the usefulness of this approach. First, many authors report a pronounced tendency for pH values to decrease with time in alkaline solutions (Dempsey and O'Melia, 1983; see Flaig et al., 1975 for a review of earlier work) although the reasons for this behavior are a matter of speculation. One possibility is the aforementioned hydrolysis reactions, including the hydrolysis of metal–organic complexes (Martin and Reeve, 1958). In addition, in the presence of oxygen, it is believed that this phenomenon can be attributed to both chemically and bacterially mediated oxidation reactions that produce acids (Swift and Posner, 1972). Borggaard (1974) has found that alkaline solutions of humic substances are stable only under nitrogen in sterile, sealed ampoules. Further research is needed to elucidate the cause of pH drift in direct titrations of humic substances. Another problem encountered in direct titrations is the difficulty of making accurate pH measurements at high pH, due in part to electrode interferences arising from the very high concentrations of alkali metal ions. Consequently, it is difficult to determine the amount of base that has reacted at a particular high pH. Until these problems are solved, total acidities cannot be accurately determined by direct titrations.

Barium Hydroxide Method

This method was originally designed for use with coals and was adapted for humic substances research by Schnitzer and co-workers (e.g., Schnitzer and Khan, 1972). In the barium hydroxide method, a 50–100 mg sample of humic substances in 20 ml of 0.1*M* barium hydroxide and a comparable reagent blank are allowed to equilibrate for 24 hours in a N_2 atmosphere. The sample "suspension" is then filtered through an unspecified type of filter, the residue is carefully rinsed, and the filtrate plus washings are titrated to pH 8.4 with standard HCl (see Fig. 6). The difference in sample and blank titrations is used to calculate "total" acidity. Because the pH is greater than 13 in this reagent solution, most of the acidic hydrogens in a sample of humic substances would be ionized and it should be possible to reasonably estimate total acidity.

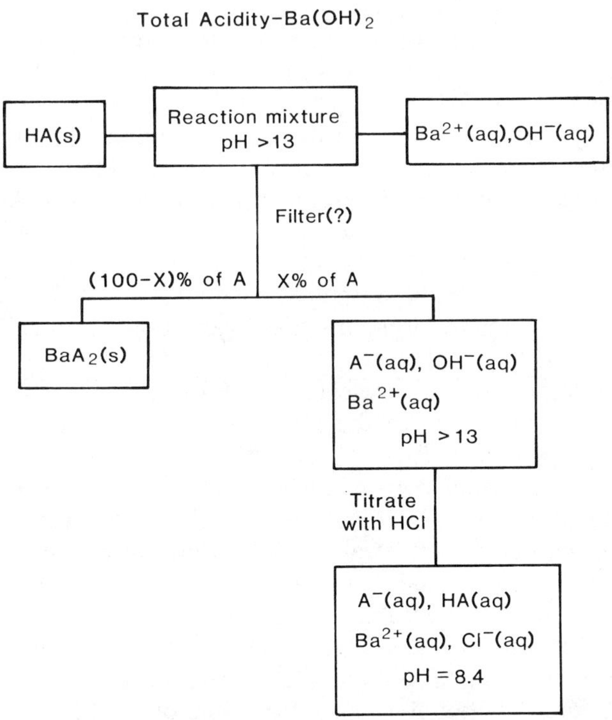

FIGURE 6. The barium hydroxide total acidity method.

As other scientists have adopted this procedure for studying a variety of humus-like materials, the *critical importance of the filtration step* has been recognized (Davis, 1982). As indicated in Figure 6, if any of the *weakly acidic* ionized groups (A^-) in the reaction mixture remain in solution rather than precipitating as barium salts, those groups would be re-protonated during titration to pH 8.4. The total acidity of the sample would be underestimated by a corresponding amount. Davis (1982) reports this problem in a study of dissolved organic matter from lake water, for which measured total acidities were even less than estimates obtained from direct titrations. It is probably reasonable to conclude that total acidities obtained on obviously colored filtrates should be viewed with skepticism.

Nonaqueous Titrations

Nonaqueous titrations have been used to estimate total acidity (e.g., Wright and Schnitzer, 1959). This approach utilizes an aprotic solvent such as pyridine or dimethylformamide, in which a very strong base such as ethoxide ion can be used to react with the acidic functional groups of the humic substance sample. The comparison of "nonaqueous total acidity" and "aqueous total acidity" should be tempered with caution and any efforts to subdivide nonaqueous total acidity into functional group classes by analogy with pK_a values of acidic groups in aqueous solution are theoretically unfounded. Some reasons for concern are presented in the next paragraph.

The ionization of a carboxylic acid, phenol, enol, or alcohol in a solvent (S) [$AH + S = A^- + SH^+$] always results in the net production of ions (ionogenic reaction), while the ionization of an ammonium ion [$AH^+ + S = A + SH^+$] is an isoelectric reaction. Because of the greater stability of ions in water than in organic solvents, ionogenic reactions are particularly sensitive to the nature of the solvent. While the pK_a values of ammonium ions are similar in aqueous and nonaqueous solvents, carboxylic acids, phenols, and so on are much weaker acids in organic solvents.

As discussed previously, ion stabilities are also affected by electrostatic and delocalization effects. The effect of a charged or dipolar substituent on the pK_a of an acidic group is expected to be even greater in most organic solvents than in water, due to the lower dielectric constants of common organic solvents [see D in Eqs. (5) and (6)]. Delocalization effects can be quite spectacular in organic solvents. For example, picric acid (whose conjugate base is a highly delocalized phenoxide ion) is a stronger acid than HBr in dimethylformamide (Sears et al., 1956). Both the absolute and relative pK_a values of acidic functional groups are thus quite different in nonaqueous solvents than in aqueous solutions.

Because this chapter is intended to examine only potentiometric and thermometric methods of functional group analysis, the irreversible methods of analysis of acidic functional groups [reaction of acidic H with CH_2N_2, $LiAlH_4$, or $(BH_3)_2$] are not discussed. The most likely potentiometric procedure to yield unambiguous results may well be the barium hydroxide method, modified to include a filtration step (possibly even ultrafiltration) to remove all colored compounds from the "suspension" prior to final titration.

Carboxyl Group Determinations

A major portion of the total acidity in humic substances is believed to be attributable to carboxyl groups, and several potentiometric methods for determination of carboxyl content have been proposed. All these methods are ultimately operational, effectively defining carboxyl content as the concentration of functional groups that are neutralized at some explicitly or implicitly specified pH. Some of the more common methods will be briefly discussed in the following paragraphs.

Direct Titrations

The aforementioned problems with direct titration methods are much less significant in carboxyl group determinations because extremely high pH values are not used in this analysis. Many investigators have simply used normal potentiometric titrations to estimate carboxyl content, even though no distinct equivalence points are apparent in the resulting titration curves. Commonly, an arbitrarily selected pH is used to effectively define the carboxyl content of a humic substance sample. Because the equivalence point for titration of any weak acid must lie above pH 7, that pH has occasionally been used to define carboxyl content in direct titrations. Other authors (e.g., Gamble, 1972) have attempted to modify the Gran functions for simple titrations to analyze titration curves of humic substances. The resulting values for carboxyl content are clearly operational and should be used accordingly. The reader is referred to Flaig et al. (1975) for a more thorough review of direct titration methods.

Calcium Acetate Exchange Method

The most commonly used method of determination of carboxyl content is the calcium acetate exchange reaction, in which 50–100 mg of humic substances in 50 mL of 0.1*M* $Ca(CH_3COO)_2$ and a comparable reagent blank are equilibrated for 24 hours in a N_2 atmosphere. After filtering the suspension through an unspecified type of filter and carefully washing the residue, the filtrate and washings are titrated to pH 9.8 with standard NaOH (see Fig. 7). Carboxyl content is calculated from the difference in sample and blank titrations. The intent of this technique is to convert the complex mixture of acids in humic substances into an equivalent amount of CH_3COOH by the reaction

$$2RCOOH + Ca(CH_3COO)_2 \rightleftharpoons Ca(RCOO)_2 + 2CH_3COOH \qquad (8)$$

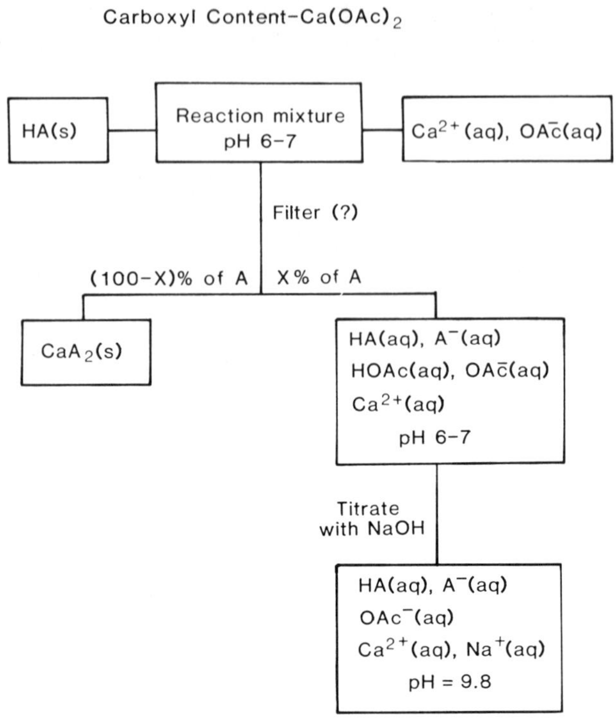

FIGURE 7. The calcium acetate exchange method for carboxyl groups.

Several problems are inherent in this method, so the results should be considered as quite operational (Dubach et al., 1964; van Dijk, 1966; Stevenson and Goh, 1972; Holtzclaw and Sposito, 1979; Perdue, 1979; Perdue et al., 1980). First, unlike the barium hydroxide reagent used for total acidity, $0.1M$ calcium acetate is poorly buffered. The equilibrium pH, which determines the extent to which the acidic functional groups of humic substances will react, is dependent on the amount of humic substances added to the 50 mL of calcium acetate. Perdue et al. (1980) demonstrate that the binding of Ca^{2+} to the humic substance sample displaces additional protons that do not react if sodium acetate or pyridine is used as the exchange base. It cannot be assumed that those "excess" protons are derived exclusively from carboxyl groups.

Second, an even more serious problem with this method is the *critical nature of the filtration step* (Holtzclaw and Sposito, 1979; Perdue, 1979; Perdue et al., 1980). Any moderately acidic functional groups (pK_a 7–10) that are soluble in the equilibrium mixture (pH 6–7) will be titrated along with acetic acid when the filtrate is titrated to pH 9.8 (see Fig. 7). All the functional groups detected in that titration are simply assumed to be carboxyl groups. Thus, while the use of high concentrations of humic substances can lead to underestimation of carboxyl content, the complexation of Ca^{2+} and the incomplete filtration of humic substances in the reaction mixture tend to result in potential overestimation of carboxyl content. It is obviously not possible to know how to interpret the results of the calcium acetate exchange reaction.

Two alternative modifications have been proposed to eliminate the filtration error in the calcium acetate method. Holtzclaw and Sposito (1979) use steam distillation to separate the acetic acid from the nonvolatile humic substances in the reaction mixture. This approach would gradually shift the equilibrium in Equation (8) to the right, theoretically yielding a higher carboxyl content than the method of Perdue et al. (1980), who separate acetic acid from essentially all the humic substances in the reaction mixture by ultrafiltration using Amicon UM-2 membranes prior to titration with NaOH. This modification results in a carboxyl content that corresponds to the concentration of functional groups that were dissociated at the equilibrium pH of the reaction mixture (typically pH 6–7). Weakly acidic carboxyl groups would not be included in the carboxyl content. Both methods yield lower values for carboxyl content than the original method of Schnitzer and Gupta (1965).

The relative results obtained by direct titration and by the calcium acetate reaction (unmodified and the method of Perdue et al., 1980) are given in Figure 8. In order to compare the results of these different methods, the titration data have been corrected for dilution, blank titrations, and so on, so

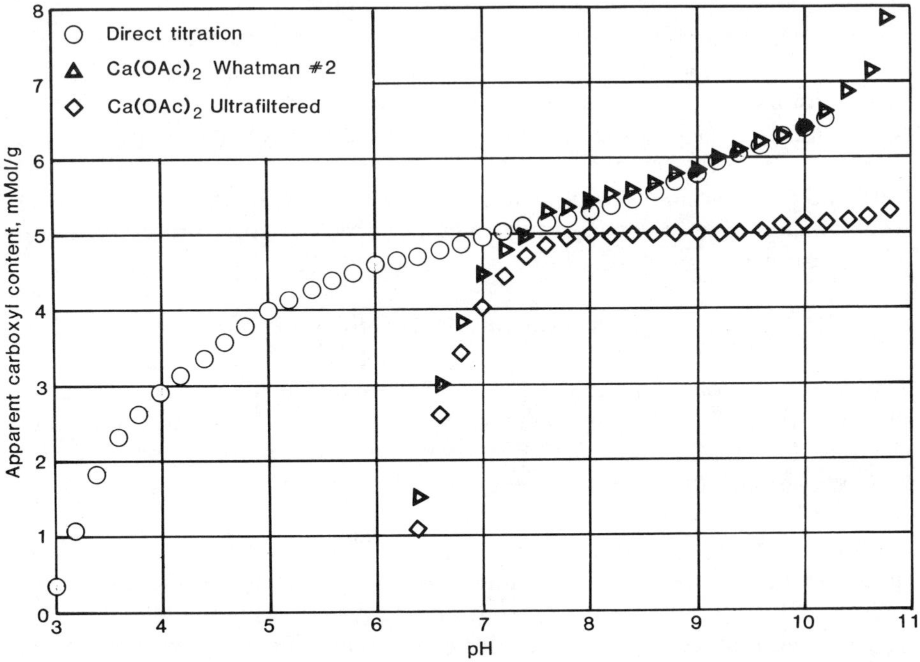

FIGURE 8. Direct and indirect estimates of carboxyl content.

that the actual amount of base consumed is given as a function of pH. Such a plot should level off after all carboxyl groups are titrated, if only carboxyl groups are detected by the method. The direct titration in $0.033\,M$ $Ca(NO_3)_2$ indicates that the slope of carboxyl content versus pH never reaches zero, although the slope is smallest in the pH 6–8 range. This titration curve clearly indicates the ambiguity of carboxyl content that is simply defined as the amount of base neutralized at an arbitrarily selected pH. The results of the calcium acetate reaction also illustrate the filtration problem. The reaction mixture that was filtered through a Whatman No. 42 paper contains virtually all the original aquatic humus sample and the titration curve of that reaction mixture merges with the direct titration curve above the endpoint for titration of acetic acid. In contrast, the ultrafiltered reaction mixture yields a relatively constant estimate of carboxyl content that is lower than the other estimates at all pH values. Even though it is possible to eliminate the filtration error, the fact that carboxyl contents are still either explicitly or implicitly defined in terms of the pH of the reaction mixture should not be overlooked.

Phenolic Hydroxyl Group Determinations

There is no potentiometric method that can be used to determine the phenolic hydroxyl content of humic substances. Consequently, phenolic groups are often simply assumed to be the difference between total acidity and carboxyl content. This author prefers the use of "weakly acidic groups" to represent these functional groups, as recommended by Stevenson (1982). While some of the weakly acidic groups are probably phenols, weaker carboxyl groups, alcoholic groups in carbohydrate entities, and enols may be of some importance (see Fig. 5). In an earlier section of the chapter, it was pointed out that the phenolic hydroxyl content of humic substances is probably limited by the amount of aromatic carbon in the sample. Before all the weakly acidic groups are assumed to be phenols, the structural implications of such an assumption should be examined.

Thermometric Titrations

The neutralization of an acidic functional group is usually accompanied by the evolution of heat. For the reaction of OH^- with an acidic functional group with $pK_a < 9$, the reaction is essentially quantitative (reacted base = added base) and the heat evolved is directly proportional to the quantity of added base. The slope of heat versus moles of reacted base is the enthalpy of neutralization of the acidic functional group. If the acidic functional group is rather weak ($pK_a > 9$), the reaction with added OH^- is incomplete and the amount of heat evolved is not proportional to the quantity of added base. For simple, well-defined organic acids of this type, nonlinear regression methods can be used to simultaneously determine both the enthalpy of neutralization and the pK_a of the acid (Christensen et al., 1968; Christensen et al., 1972; Eatough et al., 1972a,b). Because the enthalpies of ionization of carboxylic acids and phenols are usually quite different, even when their pK_a values happen to be in the same range, thermometric titrations can be used to more accurately identify the types of functional groups that undergo reaction at a specified pH (see Perdue, 1978, 1979).

The technique of titration calorimetry has been successfully used to determine the nature and abundances of a variety of acidic functional groups in proteins (Jespersen and Jordan, 1970). Several authors (Ragland, 1962; Khalaf et al., 1975) have attempted to use calorimetry to detect the acidic functional groups of humic substances. Much more rigorous attempts have been made by Choppin and Kullberg (1978) and by Perdue (1978, 1979). Perdue (1978, 1979) has found that the enthalpy of neutralization of the acidic functional groups of humic substances is constant over most of a titration and equals the expected value for neutralization of carboxylic acids. At higher levels of added base, the heat evolved is no longer proportional to the amount of added base because weakly acidic groups are being titrated. The amount of base consumed in the linear portion of the titration curve equals the absolute lower limit for carboxyl content (Perdue et al., 1980). The more weakly acidic groups that are subsequently neutralized can be carboxyl, phenolic, enolic, and/or alcoholic hydroxyl groups and cannot be quantified by this method. Perdue (1978, 1979) simplistically modeled humic substances as a mixture of a moderately acidic and a weakly acidic compound; however, the fitting parameters of the thermometric titration curve are of no chemical significance in the complex mixture of organic acids that is undoubtedly present in humic substances. Now that it is known that most of the carboxyl groups in humic substances have essentially the same enthalpy of ionization, the technique could be modified to examine the pK_a distribution of the carboxyl groups; however, the prohibitively large sample size required by Perdue's calorimetry system has prevented further experiments from being carried out.

EQUILIBRIUM MODELS OF PROTON BINDING BY HUMIC SUBSTANCES

Many mathematical models have been used to describe proton binding by humic substances; however, in every case, the models were initially developed for other purposes, often for the description of proton binding by proteins, acidic polymers, ion exchange resins, and so on. The assumptions and approximations that were inherent in the original models have often been overlooked or forgotten when those models are applied to humic substances. In this section, several common models are examined to evaluate their applicability to humic substances, with due consideration for the complexity of this mixture of nonidentical organic acids.

At the outset, the reader should recognize that, as a mixture becomes more complex, less detail can be obtained from titration curves, which tend to become rather featureless. Given a fairly smooth, featureless titration curve, almost any mathematical model with several adjustable fitting parameters can be used to empirically fit the data, making it impossible to use goodness-of-fit to determine whether the mathematical model is also a sound chemical model. That judgment must be reached primarily by chemical intuition and secondarily by goodness-of-fit. If the objective of the modeling effort is simply curve-fitting experimental data, there is little reason for choosing one model over another. However, the temptation to attribute chemical significance to demonstrably empirical fitting parameters should be carefully avoided. The following review assumes that all models are approximately equivalent insofar as curve-fitting of data is concerned. The focus is therefore on the chemical soundness of the models that are commonly used to describe proton binding by humic substances.

Mathematical Properties of Multiligand Equilibria

In recent years, largely through the efforts of Gamble and co-workers (Gamble, 1970, 1972; Burch et al., 1978) and MacCarthy and co-workers (MacCarthy, 1977; MacCarthy and Smith, 1979), the more important mathematical properties of multiligand mixtures have been presented. Perdue and Lytle (1983a,b) have extended these concepts and reviewed the mathematical details of determination of stability "constants" from laboratory data. Only a brief overview of the mathematics of proton binding is presented in this chapter.

The first modeling question that must be addressed is: Can the acidic functional groups of humic substances be treated as a mixture of monoprotic acids, even though more complex acids are undoubtedly present? The conditions under which this approach is valid have been outlined by Simms (1926a,b). This work and a simple example presented by Perdue et al. (1984) strongly suggest that, on the basis of pH titration data alone, it is not generally possible to distinguish a polyprotic acid from a mixture of monoprotic acids. Therefore, the neglect of the polyprotic character of some of the acids in humic substances should not be of any consequence.

The binding of a proton (H) by a single ligand (L_i) can be described by a dissociation constant (K_i).

$$HL_i = H + L_i \tag{9}$$

$$K_i = \frac{[H][L_i]}{[HL_i]} \tag{10}$$

In a complex mixture of monoprotic acids, the overall degree of ionization of acidic functional groups (α) can be calculated from the electroneutrality equation at any point in the titration of the acid mixture with strong base.

$$C_B + [H^+] - [OH^-] = \alpha C_L \tag{11}$$

where the total stoichiometric concentrations of acidic groups and added base are C_L and C_B, respectively. Strictly speaking, C_L must correspond to the total acidity of a humic sample, because the concentrations of individual classes of functional groups cannot be unambiguously determined. In actual practice, however, operationally defined concentrations of functional group classes are often used in Equation (11), with only a portion of the titration curve being analyzed. Occasionally, in order to treat an entire titration curve, Equation (11) is modified to express the total organic anion concentration in terms of the concentrations of two "classes" of functional groups:

$$C_B + [H^+] - [OH^-] = \alpha_I C_I + \alpha_{II} C_{II} \tag{12}$$

While this equation is easily written, it contains two totally unknown pH-dependent variables (α_I and α_{II}), in addition to the ambiguous separation of C_L into C_I and C_{II}, and cannot be rigorously solved. If it is assumed that the pK_a ranges of the two "classes" of functional groups do not overlap significantly, the values of α_I and α_{II} can be estimated. As earlier discussion has indicated, however, it is improbable that such a convenient distribution of acidic functional groups would exist in humic substances.

It is useful to express α [as defined in Eq. (11)] in terms of the summation of individual functional groups,

$$\alpha = \frac{\Sigma[L_i]}{C_L} = \sum \left(\frac{K_i}{K_i + [H^+]}\right)\left(\frac{C_i}{C_L}\right) \tag{13}$$

where C_i/C_L is the mole fraction of the ith functional group. Likewise, an apparent average dissociation "constant" can be calculated at any point in a titration.

$$\bar{K} = \frac{[H^+]\Sigma[L_i]}{\Sigma[HL_i]} = [H^+]\left[\frac{\alpha}{1 - \alpha}\right] = \frac{[H^+](C_B + [H^+] - [OH^-])}{(C_L - C_B - [H^+] + [OH^-])} \tag{14}$$

The attractiveness of this approach is undoubtedly enhanced by the fact that $\bar{K}$ can be directly calculated from experimental data. However, if Equations (10) and (14) are combined and all terms in the summations are divided by $[HL_r]$, the concentration of an arbitrarily selected reference ligand, $\bar{K}$ can be expressed as

$$\bar{K} = \frac{\Sigma K_i([HL_i]/[HL_r])}{\Sigma([HL_i]/[HL_r])} \tag{15}$$

The average dissociation "constant" ($\bar{K}$) is a weighted average of many K_i values and cannot be a constant unless all the weighting factors ($[HL_i]/$

$[HL_r]$) are constants. Because the ligands probably have a range of K_i values, HL_i and HL_r will not remain in a constant ratio as base is added to a protonated ligand mixture. For example, if the strongest proton binding ligand (the weakest acidic functional group) in the mixture is chosen as the reference ligand, then all the $[HL_i]/[HL_r]$ values will be greatest at very low levels of added base and will steadily decrease as base is added. It follows that $\bar{K}$ *values must be functions rather than constants,* and that $\bar{K}$ values will decrease as base is added during a titration of a protonated ligand mixture. An analogous set of equations and conclusions regarding metal binding by humic substances was published by MacCarthy (1977) and MacCarthy and Smith (1979).

Simple Binding Site Models

The simplest and least satisfactory models are those that treat humic substances as a mixture of a few simple acids. Such models can only be used as simple curve-fitting equations. A frequent conceptual error in metal complexation models that occasionally appears in proton binding models is the assumption that the pK_a values and binding site concentrations obtained from discrete curve-fitting models are the "average" values for several "classes" of ligands. For instance, if a titration curve can be fit by assuming that humic substances consist of three monoprotic acids, the authors might conclude that the humic sample contains three "classes" of binding sites whose average pK_a values and concentrations correspond to the values that were obtained for the hypothesized three monoprotic acids. The previous discussion in this section has clearly shown that "average constants" do not exist in multiligand mixtures of nonidentical ligands, so the "pK_a" and "ligand concentration" values that are obtained by fitting experimental data to discrete ligand models must be regarded simply as curve-fitting parameters with no chemical significance.

One example of this approach has been published by Sposito and co-workers (Sposito and Holtzclaw, 1977; Sposito et al., 1977), who have published proton binding data and a discrete ligand model for fulvic acids derived from sewage sludge. The experimental data, which appear to have been carefully obtained, contain some peculiar anomalies that are difficult to explain. For instance, when low fulvic acid concentrations are titrated with strong base, the low pH region of the titration curve indicates that some functional groups are reprotonated as base is added. Sposito and co-workers have attributed this phenomenon to counterion condensation. The same experimental observation was also reported by Perdue et al. (1980).

Sposito and Holtzclaw (1977) also reported that the acidic functional groups of humic substances appeared to become weaker at lower fulvic acid concentrations, which led them to suggest a complex proton binding model that not only treats fulvic acid as a mixture of four "mean fulvic acid units" but also specifically includes a hydrogen-bond-mediated aggregation of "mean fulvic acid units" (Sposito et al., 1977). It can be shown, however, that the Δn_{OH} function that was used to analyze their fulvic acid data will produce the same apparent phenomenon when it is used to analyze the titration data of a simple monoprotic acid such as acetic acid. It is not possible to use Δn_{OH} to support the aggregation hypothesis. In a more recent paper, Dempsey and O'Melia (1983) show that there is no difference in normalized proton binding curves for fulvic acid solutions varying in concentration from about 110 to 1330 mg/L. Likewise, Burch et al. (1978) found the normalized titration curves of two fulvic acids to be independent of fulvic acid concentration.

If the aggregation component of the model of Sposito et al. (1977) is overlooked, the model reduces to a simple discrete four ligand model. There are two circumstances under which the model could be chemically appropriate: (1) if the fulvic acid contained only four functional groups or (2) if the fulvic acid contained four classes of functional groups, each class consisting of a group of absolutely identical ligands that do not interact electrostatically with one another. In the latter case, the four pK_a values would be statistically corrected intrinsic constants (see earlier discussion of statistical effects). Such a model is inappropriate for the complex mixture of nonidentical ligands that is expected to exist in humic substances (i.e., while it has enough adjustable parameters to adequately fit titration data, those parameters cannot be attributed with chemical significance).

Intrinsic Binding Site Models

The so-called intrinsic binding site models are more appropriate than simple discrete models because they provide a means for modeling a more-or-less continuous distribution of binding sites, if the differences in pK_a values can be attributed solely to the electrostatic effects of charged substituents. Many of the models that have been used to describe proton binding by humic substances assume that all functional groups of a particular structural type are inherently identical, with the observed range of pK_a values being entirely attributable to the formation of negatively charged conjugate base ions as the degree of ionization of the humic substances increases. Accordingly, those models utilize some type of extrapolation procedure to obtain an "intrinsic" pK_a value for the average "uncharged" molecule. It is implicit in such models that all acidic functional groups are in identical electrostatic environments in the absence of formal charges. The electrostatic properties of dipolar groups, which are abundant in humic substances, are totally ignored. Chemical intuition and the elemental composition of humic substances both suggest that this type of model is not likely to be appropriate. A rigorous presentation of the intrinsic binding site models is given by Tanford (1961) and the most careful study of the applicability of those equations to humic substances is that of Dempsey and O'Melia (1983).

A number of authors have used intrinsic pK_a models to describe proton binding by humic substances, using one or more of a group of related equations that attempt to account for the increase in apparent pK_a with increasing degree of ionization (α) of a particular class of acidic functional groups of humic substances. Only the carboxyl portion of titration curves is usually fitted to these equations. One commonly used equation is

$$pH = pK_{int} + n[\log(\alpha/(1 - \alpha))] \quad (16)$$

where pK_{int} is the intrinsic pK_a value that would apply to all acidic functional groups in the uncharged molecule and n is a variable that reflects the extent to which pK values are modified by electrostatic effects. In a molecule in which the acidic groups are so far apart that there is essentially no interaction between groups, n approaches a value of unity.

Equation (16) is a modified Henderson–Hasselbalch equation and is usually attributed to Katchalsky and Spitnik (1947). Some of the authors who have applied this equation to humic substances include Pommer and Breger (1960), Huizenga and Kester (1979), Dempsey and O'Melia (1983), and Varney et al. (1983). While experimental data for humic substances are fitted reasonably well by this equation, the fundamental assumption that all carboxyl groups are inherently identical in the absence of formally charged groups must not be overlooked.

A very similar equation was introduced by Katchalsky et al. (1954), in which an attempt is made to use the Hermans–Overbeek equation (Hermans and Overbeek, 1948) to compute the actual electrostatic free energy change that can be attributed to the accumulation of negative charge in a flexible linear polyelectrolyte. This equation is of the general form:

$$pH = pK_{int} + \log(\alpha/(1 - \alpha)) - 0.868wn\alpha \quad (17)$$

where n is the average number of carboxyl groups per humic substance molecule and w is a composite term that depends on the ionic strength of the solution, the dielectric constant of the solvent, and the apparent size of the humate anion. This equation has been applied to humic substances by Posner (1964), Wilson and Kinney (1977), Dempsey and O'Melia (1983), Plechanov et al. (1983), and Varney et al. (1983).

Posner (1964) pointed out that graphically estimated values of w did not change with ionic strength in the expected fashion, leading him to suggest that humic substances contained many different acidic functional groups on many different molecules and was not similar to a truly polymeric acid. As was the case with Equation (16), Equation (17) is based on the assumption that all carboxyl groups are identical in the absence of electrical charge, so the values of pK_{int} and w that are derived from titration data should not be accepted as rigorously defined chemical parameters.

More recently, Marinsky and co-workers (Marinsky et al., 1980; Marinsky et al., 1982a,b) have pointed out that Equations (16) and (17) are not applicable to cross-linked resins, gels, and so on, and have presented a fundamentally sound method of correcting for Donnan potential terms in such systems. The model assumes the existence of a "gel phase" in which all acidic functional groups have the same intrinsic pK_a value. As was the case in the simpler electrostatic models, however, it seems highly unlikely that all functional groups in humic substances would have the same pK_a value. Nevertheless, it is very important to note that the concept of a gel phase is quite plausible for higher-molecular-weight fractions of humic substances. None of the other mathematical models discussed in this chapter consider the potential complications in modeling acid–base equilibria in a two-phase system.

Continuous Distribution Models

The intrinsic binding site models represent a very simplified approach toward modeling a continuous distribution of proton binding ligands. As previously indicated, those models are severely limited by their assumption that only statistical effects and electrostatic effects of charged groups have any effect on the acidities of functional groups.

More general modeling approaches have recently evolved that incorporate all types of structural effects without attempting to separate those effects into the components (statistical, electrostatic, and delocalization) that were discussed previously. These continuum models include a rather rigorous model (Gamble, 1970, 1972; Burch et al., 1978) that is considered as a benchmark in the understanding of the acid–base properties of humic substances. The model uses either graphical or numerical methods to estimate the "instantaneous" dissociation constant of the group that is reacting at a particular degree of ionization of the humic substance. Shuman et al. (1983) have described an affinity spectrum model which approximates the actual distribution of ligands by applying mathematical methods that were originally developed to model dynamic relaxation in viscoelastic materials. Finally, Perdue and Lytle (1983b) have presented a model that assumes that the distribution of functional groups of humic substances may resemble a mixture of the distributions that are given in Figure 5 (a nearly Gaussian distribution of carboxyl groups and a similar distribution of phenolic and weaker acidic groups). Each of these models will be briefly examined in this section.

Gamble's Method

In a thorough paper that discusses many of the important properties of multiligand mixtures in the context of acid–base equilibria, Gamble (1970) has provided a direct method for estimating the distribution of pK_a values in humic substances. Gamble assumes that a continuous distribution of nonidentical binding sites exists in humic substances and that the stoichiometric concentration of each class of acidic functional groups can be independently determined. As earlier discussion has indicated, the unambiguous separation of total acidity into its structural components (e.g., carboxyl and phenolic groups) is not readily accomplished. Only carboxyl groups are generally treated (Gamble 1970, 1972; Burch et al., 1978), but the method itself could be extended to the weaker acids as well.

By assuming that the dissociation constant of a particular functional group is functionally related to the overall degree of ionization of the mixture (α), it is possible to estimate the "instantaneous" dissociation constant of the group that is ionizing at a particular pH by either graphical or numerical differentiation of appropriate equations. In choosing α as the independent variable, the model conceptually parallels the "intrinsic" binding site models that were discussed in the previous section. In fact, although the method is ultimately unaffected, Gamble attributes the variation in dissociation constants to the increased negative charge that accompanies the increased degree of ionization of the mixture. To apply this method, the variable $\alpha[H^+]$ is plotted versus α. The instantaneous dissociation constant at a particular value of α is given by

$$K = -\frac{d(\alpha[H^+])}{d\alpha} \tag{18}$$

The derivative is either graphically estimated at the desired value of α or the experimental points are fitted to a high-order polynomial equation whose first derivative is then evaluated at the desired value of α. No attempt is made to use any sort of "chemical" equation to fit the data, so, even though the method is the most rigorous one that has been published to date, the results cannot readily be incorporated into multicomponent chemical equilibrium models. This fact has lessened the impact of the method in the scientific community.

The above synopsis of Gamble's method has glossed over some of the features of this model by using symbols that were previously defined in this chapter to represent somewhat different variables in Gamble's paper. Specifically, Gamble uses mass action quotients, which are conditional stoichiometric dissociation constants at constant ionic strength, instead of equilibrium constants, so that ambiguities regarding activity coefficients can be avoided. Accordingly, he uses H^+ molality instead of pH or H^+ activity. These differences do not significantly affect the conceptual aspects of Gamble's method as described here.

Affinity Spectrum Method

Hunston and co-workers (Klotz and Hunston, 1971; Hunston, 1975; Thakur et al., 1980) have presented an approximation technique for describing the binding of small molecules to proteins. The method exploits the formal similarity between the appropriate chemical equations and the equations that are used to describe the dynamic response of viscoelastic materials as presented by Ferry (1970). The reader is referred to these references for a more thorough development of the method. Shuman et al. (1983) have adapted this method to describe metal complexation by humic substances. It is easily modified further to describe the distribution of proton binding ligands in humic substances.

The model assumes the existence of a continuum of binding sites in which the stoichiometric concentration of binding sites with a particular pK_a value is functionally related to the pK_a value. In other words, the probability of occurrence of a binding site depends on the Gibbs free energy of dissociation of that site. The nature of the hypothesized distribution function is, of course, unknown; however, the affinity spectrum technique is specifically designed to numerically estimate that function from observable titration data.

The summation in Equation (13) is replaced by an integral over the range of pK_a values:

$$\alpha = \frac{1}{n_0} \int \frac{N(K)K}{[\mathrm{H}^+] + K}\, d(\mathrm{p}K) \tag{19}$$

where $N(K)$ is the hypothesized distribution function that can be used to calculate the probability of occurrence of a ligand with a particular pK_a value, and n_0 is the integral of $N(K)d(\mathrm{p}K)$ over all pK_a values. To use this method, the experimental data are numerically smoothed so that α values can be obtained at any specified value of $[\mathrm{H}^+]$. Then, $N(K)$ values are calculated from Equation (20) using α values that are obtained either graphically or numerically at $[\mathrm{H}^+]$ values of K/a, Ka, K/a^2, and Ka^2, where a is an empirical constant (a = 1.585) which affects the resolving power of the technique. The function $N(K)$ is approximately given by

$$N(K) = \frac{n_0}{2\log(a)} \left(f_1 - \left[\frac{a}{(a-1)^2} \right] (f_2 - 2f_1) \right) \tag{20}$$

where

$$f_1 = \alpha(K/a) - \alpha(Ka) \quad \text{and} \quad f_2 = \alpha(K/a^2) - \alpha(Ka^2) \tag{21}$$

The variables f_1 and f_2 are computed from four α values, as indicated in Equation (21). This technique appears to locate the K values of some simple ligand mixtures, but it does not properly recognize the discrete nature of those ligand mixtures. From the limited number of attempts this author has made to test this approach, the method always predicts the existence of rather broad distributions of binding sites, even when only a few discrete sites are present. Nevertheless, when a complex mixture of proton binding sites is present, the method may be useful for estimating the actual nature of the binding site distribution.

Gaussian Distribution Method

The difficulty of implementing Gamble's method and the inability of the affinity spectrum method to distinguish between simple discrete ligands and continuous distributions of ligands must inevitably limit the usefulness of those modeling approaches. Another alternative approach would be to assume a particular distribution function that could be substituted for $N(K)$ in Equation (19). Many probability distribution functions are known and it is even possible to write computer software that selects the most appropriate function from a family of distribution functions (e.g., the family of Pearson distributions). The simplest distribution function is the normal or Gaussian distribution, in which the probability of occurrence of a given ligand is assumed to be described by the symmetrical Gaussian distribution function. The complete distribution of ligands can then be described by the mean (μ) and variance (σ^2) of the distribution. The frequency histograms given in Figure 5 for a large number of carboxyl and phenolic functional groups are obviously approximately symmetrical about mean pK_a values. It would seem reasonable that a simple mixture of normal distributions of carboxyl and phenolic hydroxyl groups would provide a good approximation of the distribution of acidic functional groups in humic substances. Such a model has been used to describe proton binding by humic substances (Posner, 1964; Perdue and Lytle, 1983b; Perdue et al., 1984), so only an overview is presented here.

In a simple Gaussian distribution of proton binding ligands,

$$\frac{C_i}{C_L} = \frac{1}{\sigma\sqrt{2\pi}} \exp\left[-\frac{1}{2}\left[\frac{\mu - \mathrm{p}K_i}{\sigma}\right]^2\right] d\mathrm{p}K \tag{22}$$

where C_i/C_L is the mole fraction of ligands in the interval $d\mathrm{p}K$ whose *dissociation* constant for proton binding is expressed as a negative logarithm (pK_i), and σ is the standard deviation for the distribution of pK_i values about the mean pK value (μ) for the mixture of ligands. Combining Equation (22) with (13) and converting from a discrete summation to a continuous integral, the overall degree of ionization of acidic functional groups in a mixed normal distribution of carboxyl and phenolic groups is given by

$$\alpha = \alpha_{\mathrm{I}}(\mathrm{C_I}/\mathrm{C_L}) + \alpha_{\mathrm{II}}(\mathrm{C_{II}}/\mathrm{C_L}) \tag{23}$$

where

$$\alpha_{\mathrm{I}} = (\sigma\sqrt{2\pi})^{-1} \int_a^b \frac{K}{K + [\mathrm{H}^+]} \exp\left[-\frac{1}{2}\left[\frac{\mu - \mathrm{p}K}{\sigma}\right]^2\right] d\mathrm{p}K \tag{24}$$

and a similar term is written for α_{II}.

Although the limits of integration have not been indicated above, they are fixed, in fact, by the leveling effect of H_2O on the strengths of acids and bases. No acid stronger than H_3O^+ can exist as a major species in aqueous solution. Likewise, no acid weaker than H_2O can ionize appreciably in aqueous solution. The lower and upper limits are $a = -1.74$ and $b = 15.74$, respectively. Of course, the Gaussian distribution has infinite range and, if used without regard to limits, does provide an approximation to the underlying ligand distribution. However, improved accuracy can easily be obtained by incorporating the known limits of integration into the distribution model.

The six curve-fitting parameters in Equations (23) and (24) (C_{I}, μ_{I}, σ_{I}, C_{II}, μ_{II}, and σ_{II}) can be determined through use of a nonlinear regression method that minimizes a weighted residual sum of squares (RSS),

$$\text{RSS} = \sum \frac{(\alpha_{\text{calc}} - \alpha_{\text{exp}})^2}{\alpha_{\text{exp}}} \tag{25}$$

where RSS is summed over all the data points (α_{exp}, $[\text{H}^+]$) in the titration. The iterative minimization procedure involves three steps: (1) given initial estimates of C_{I}, μ_{I}, σ_{I}, C_{II}, μ_{II}, and σ_{II}, α_{calc} values are calculated from Equations (23) and (24) at each value of $[\text{H}^+]$; (2) RSS is evaluated from Equation (25); (3) using a nonlinear minimization algorithm, such as steepest descent, Marquardt's, or Fletcher–Powell, improved guesses are generated for the six fitting parameters. These three steps are simply repeated until the values of RSS stabilize, converging to a minimum.

The Gaussian distribution model readily distinguishes a two-ligand mixture from a continuous distribution of ligands and can also distinguish between unimodal and bimodal distributions. Thus, the problems associated with the affinity spectrum technique are not encountered. Furthermore, the model is appropriately formulated to be easily incorporated into existing multicomponent chemical equilibrium computational programs such as MINEQL (Westall et al., 1976). The main weakness to this approach is the *a priori* assumption of the type of distribution of proton binding sites that exists in humic substances. The fact that all the combined contributions of statistical, charge–charge, charge–dipole, and delocalization effects result in approximately normal distributions of carboxyl and phenolic groups in simple, well-defined organic acids suggests that this model should serve as a good first approximation of the most probable distribution of acidic functional groups in humic substances.

SUMMARY AND CONCLUSIONS

The literature on proton binding by humic substances indicates that statistical effects, delocalization effects, and, probably most importantly, the effects of dipolar groups on the acidity of a functional group have generally been ignored. An attempt has been made in this chapter to provide the reader with a rather detailed discussion of the nature of substituent effects on the dissociation constants of organic acids. Statistical, electrostatic, and delocalization effects have been treated separately.

The very high degree of substitution of oxygen-containing groups in humic substances (an average of about one COOH or other group for every two noncarboxyl carbons) and the tremendous complexity of the mixture of organic compounds in humic substances guarantee that the acidity of humic substances can only be attributed to a complex mixture of nonidentical functional groups. Carboxyl groups are clearly abundant; however, structural considerations based on the degree of unsaturation of humic substances and on ^{13}C NMR spectra indicate that phenolic content may be less than previously believed. The alcoholic hydroxyl groups of sugars are sufficiently acidic to react during total acidity determinations, but their contribution to total acidity has not been quantified.

Methods of functional group analysis that are based on pK_a values of the functional groups (all potentiometric methods) can only yield operationally defined estimates of the concentration of a particular class of acidic functional groups. For this reason, an increased reliance on spectroscopic methods of functional group analysis is recommended. Only total acidity, as determined by the barium hydroxide method, appears to be a potentially accurate potentiometric method of analysis, if some type of ultrafiltration technique is used whenever normal filtration fails to remove all color from reaction mixtures.

A number of common methods that have been used to model the acidity of humic substances have been reviewed. The failure of discrete models and intrinsic binding site models to even begin to acknowledge the complexity of the distribution of acidic functional groups in humic substances reduces those models to empirical curve-fitting equations. The "stability constants" and "binding site concentrations" obtained by fitting data to such models are simply empirical curve-fitting parameters with no chemical significance. Several models that treat humic substances as a continuous distribution of acidic functional groups have been proposed. None of these models is entirely satisfactory, but the conceptual approach that is embodied in these models is definitely more consistent with the complexity of humic substances. Further improvements in such models are anticipated.

In the view of this author, much of the confusion and mysticism about the properties of humic substances is directly attributable to the use of chemically naive mathematical models to describe the observed behavior of humic substances. It is incumbent upon those of us who work with humic substances to acknowledge that such a complex mixture of acidic functional groups can only be described by relatively complex mathematical models. Many simple mathematical models, even polynomial equations with absolutely no chemical basis, can accurately fit experimental data in a typical titration, simply because of the large number of adjustable fitting parameters in the models. We are denied the luxury of relying solely on goodness-of-fit as evidence for the chemical soundness of a particular model. Instead, chemical intuition and, secondarily, goodness-of-fit must be used. Only in this manner will good, sound chemical models eventually be developed.

CHAPTER TWENTY-ONE

Spectroscopic Methods (Other Than NMR) for Determining Functionality in Humic Substances

PATRICK MacCARTHY and JAMES A. RICE

ABSTRACT

The applicability of spectroscopic methods (other than NMR) for determining functionality in humic substances is reviewed. Spectroscopic methods, like all other investigational techniques, are severely limited when applied to humic substances. This is because humic substances are comprised of complicated, ill-defined mixtures of polyelectrolytic molecules, and their spectra represent the summation of the responses of many different species. In some cases only a small fraction of the total number of molecules contributes to the measured spectrum, further complicating the interpretation of spectra. The applicability and limitations of infrared spectroscopy, Raman spectroscopy, UV–visible spectroscopy, spectrofluorimetry, and electron spin resonance spectroscopy to the study of humic substances are considered in this chapter. Infrared spectroscopy, while still very limited when applied to humic substances, is by far the most useful of the methods listed above for determining functionality in these materials. Very little information on the functionality of humic substances has been obtained by any of the other spectroscopic methods.

INTRODUCTION

It is evident from Chapter 1 that humic substances have been recognized and studied for a long time. Considering this fact, what do we definitively know about the fundamental chemical nature of humic substances? That, of course, is a formidable question, and in this chapter we address only one segment of that query: namely, what information have researchers been able to definitively acquire concerning the functionality of humic substances by spectroscopic methods?

This chapter deals exclusively with spectroscopic methods involving electromagnetic radiation. Consequently, the application of methods such as mass spectrometry are not considered here. The emphasis of the chapter is on determining the functionality of humic substances. Many electromagnetic spectroscopic methods, such as gamma-ray spectroscopy and atomic absorption spectroscopy, are not used for functional group analysis of *any* type of material and, consequently, are also omitted from consideration in this chapter. The application of nuclear magnetic resonance (NMR) spectroscopy for determining functionality in humic substances is discussed by Wershaw in Chapter 22 of this book and thus NMR is not discussed any further here. In preparing to write this chapter the authors came to the realization that infrared spectroscopy overwhelmingly dominated the manuscript; this revelation, of itself, was instructive in that it forced us to *realize* what we apparently already unwittingly had known: that is, other than infrared spectroscopy (and NMR), spectroscopic methods provide relatively little information on the functionality of humic substances.

In conformance with the theme of this book, stated in Chapter 1, we hope to critically evaluate the applicability and limitations of spectroscopic methods as applied to determining the functionality of humic substances, and in so doing to provide the reader with a renewed perspective on, and comprehension of, this subject. The term functionality will be used in a rather broad sense in that it incorporates more than simply the functional groups alone. For example, the identification of structural features such as those of an aliphatic or aromatic nature, the presence of unsaturation, or quinone/hydroquinone moieties, are considered within the scope of the present chapter. The determination of detailed structural information on humic substances is not within the domain of this chapter. In fact, obtaining detailed structural information about a humic substance is not within the realm of present-day technology as far as the following spectroscopic methods are concerned.

Finally, this is more than simply a chapter on the application of spectroscopic methods to humic substances—it embodies an essay on the fundamental nature of humic substance investigations and addresses the unique scientific approach, indeed the philosophical attitude, which must be adopted in studying these materials.

LIMITATIONS INHERENT IN THE STUDY OF HUMIC SUBSTANCES

Before delving into the applications of spectroscopy to humic substances it is necessary to pause and examine what is generally known about the chemical nature of these materials. There are some simple questions which are worth asking. One such question is: Why is it that the fundamental chemical nature of humic substances is still largely a mystery, despite the extensive investigations that have been carried out on these materials for many years? In order to put this question into perspective, one can compare the major strides made in the study of other very complex systems such as proteins, polysaccharides and nucleic acids. These latter biochemical substances once posed major problems to scientists, but in the interim all these materials have yielded to the chemist's scalpel and their fundamental chemical structures have been elucidated. In fact, in many cases, not only is the primary structure of very complex molecules known but secondary and tertiary structures have also been established. Yet the fundamental chemical structure(s) of humic substances remains unknown! Humic substances are frequently referred to in the literature as polymers; however, no one has yet proven (or, for that matter, disproven) the presence of a sequence of monomeric units in humic substances. In the authors' opinion, such a proof, or disproof, would constitute a major breakthrough in humic substances research. Until that information is ascertained, researchers should be conservative in the use of that terminology in describing humic substances. There is ample justification for referring to humic acids as polyelectrolytes or as mixtures of macromolecules; there is no justification at this time for calling these substances polymers.

Of course there is much that we do know about humic substances; for example, there is considerable information available concerning their role in the environment. Much is known concerning their pedogenic and agronomic importance (Waksman, 1938; Kononova, 1975; Stevenson, 1982), but not all that information is understood. At the chemical level we know the elemental contents, and to a certain degree the functional group contents of various humic samples, but our knowledge of the "backbone" structure of humic substances is considerably more vague. We know *relatively* little about the juxtaposition of the various functional groups in humic substances; what we do know, or what we think we know in this context, is largely inferential. The state of disarray in our understanding of humic substances is epitomized by the diversity of structures that have been proposed over the years for humic acids (e.g., Gillet, 1956; Swain, 1963; Felbeck, 1965a; Kononova, 1966; Manskaya and Drozdova, 1968; Degens and Mopper, 1975; Stevenson, 1982). Admittedly, the humic acids which were the subject of those investigations were derived from different sources and frequently by different extractive techniques, but knowledge of that fact does not ameliorate the situation. All the proposed structures do have certain features in common: all contain essentially the same functional groups and all possess both aromatic and aliphatic character. However, the models differ dramatically in the fundamental structural backbone and in the relative positioning of the various functional groups. While all these structures are consistent with *known* properties of humic acids, for example, acidity or the presence of aromatic and aliphatic character, to the authors' knowledge none of these structures has proven effective in predicting previously *unknown* properties of humic acids—the ultimate test of any scientific hypothesis.

What is it then that has impeded progress in elucidating the fundamental chemical structure of humic substances, compared to the major advances in our understanding of proteins or polysaccharides? A study of the history of chemistry reveals that there is one basic reason for this dichotomy. Virtually all major breakthroughs in structural determinations have resulted from experiments on pure, or essentially pure, compounds. When confronted with a mixture, the chemist or biochemist attempts to separate it into pure components. Following such a separation, the individual components are then subjected to rigorous chemical and physical investigation: hence the pervasive role of separation techniques such as distillation, crystallization, and extraction throughout the history of chemistry and the prominence of the wide variety of chromatographic and other separation techniques in modern-day chemistry. It was only when proteins and other biopolymers could be isolated in pure form that significant advances were made in elucidating their structures. In many cases, purification leads to a crystalline product thereby making its solid state structure amenable to determination by X-ray or neutron diffraction. In the study of humic substances we are confronted with, and must remain constantly cognizant of, the fact that, to date, no satisfactory separation of a humic substance into its pure components has been accomplished. While virtually every separation technique available has been applied to humic substances, no fraction of what could be called a pure humic substance has yet been isolated. Consequently, when working with humic substances, researchers must contend with the inescapable fact that they are working with *mixtures*. With this in mind, two general avenues of attack are identifiable. First, renewed attempts at separating the humic substances into more clearly defined fractions or ideally into pure components can be pursued. Success in separating humic substances into pure components would be the ultimate breakthrough, perhaps, in humic substances research. However, there is no proof that such a separation is possible, and skepticism as to the feasibility of such a separation is evidenced in the literature (Dubach and Mehta, 1963; Dubach et al., 1964; Felbeck, 1965b). Even proof that such a separation is infeasible would be a most significant finding of itself because it would help to focus attention on other approaches to the "humic substances problem."

Second, one can study humic substances by the application of various chemical and instrumental methods, while bearing in mind the limitations imposed on the interpretability of the data resulting from the heterogeneous nature of the sample. This is a point all too often overlooked in the study of humic substances—chemical and physical methods of investigation are severely limited when applied to mixtures rather than to pure substances.

In many cases authors never acknowledge that the subject of their investigation is a mixture even when such information is essential to the interpretation of the data. Somewhat more puzzling are the examples, which persist in the literature, where the multicomponent nature of humic substances is acknowledged but later ignored in the choice of experiments and interpretation of data! *The study of humic substances is the study of complicated, ill-defined mixtures,* and many researchers appear to overlook this basic fact.

A simple example will serve to illustrate the difficulty of dealing with a mixture compared to a pure compound. The classical method for determining structure is chemical degradation followed by separation and identification of the decomposition products. With this information, and an understanding of the chemistry involved in the degradation reactions, the researcher attempts to rationalize what compound could have given rise to the particular combination of products identified. Frequently, a number of different degradation procedures must be employed in order to supply sufficient information to solve the problem. Consider now a mixture consisting of three compounds comprising, say, a tripeptide, a disaccharide, and the ester of a phenolic acid. If a researcher was presented with this sample with the understanding that it was composed of a single substance and asked to determine its structure, formidable problems would be encountered. In the application of degradation techniques the decomposition products from all three components would be intermingled, and establishing the original structure would actually be impossible since a single pure substance did not exist in the first place. The interpretability of other chemical and physical data acquired from mixtures is likewise limited. For example, in applying any spectroscopic method to humic substances one measures the summation of the signals of the numerous components in the mixture which respond to that particular frequency range, with all signals superimposed upon each other. What is measured is the net or average response of these particular components in the assemblage. Deciphering this garbled message into chemically meaningful information is, at best, a formidable task.

A critical examination of chemical and physical methods of investigation reveals that they are generally confined to simple systems, and, once a mixture is involved, one's ability to interpret the data is extremely limited. These complications are aggravated in the case of humic substances which are the epitome of molecular complexity as evidenced by the results of separation attempts to date. The limitations imposed on the interpretability of elemental analysis data and hydrolytic data are discussed by Steelink in Chapter 18 of this book.

The application of various spectroscopic methods to the study of humic substances will now be discussed in light of the above-mentioned limitations inherent in the study of these multicomponent mixtures.

INFRARED SPECTROSCOPY

Brief Introduction to IR Spectroscopy as Applied to Functional Group Analysis

The absorption of infrared (IR) radiation by matter corresponds to vibrational and rotational transitions within the material. In the case of solids and liquids one can generally observe only the vibrational bands, and these are the only bands of relevance in the study of humic substances. There are two general types of vibrations—stretching and bending—as illustrated for the water molecule (Alpert et al., 1964):

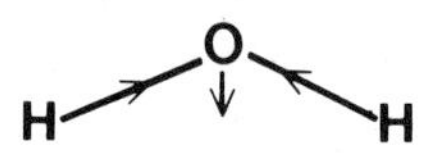

SYMMETRIC STRETCHING VIBRATION

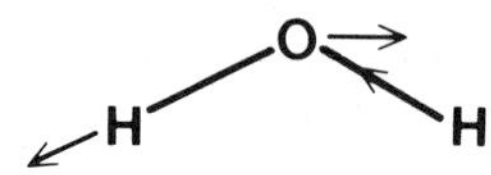

ASYMMETRIC STRETCHING VIBRATION

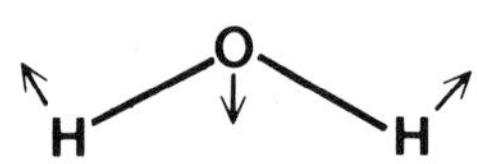

BENDING VIBRATION

In the case of very simple molecules, normal coordinate analysis (i.e., a mathematical correlation of band frequency with structure) can provide a complete structural determination for the compound at hand. However, for other than the simplest molecules such analysis is beyond present-day interpretational skills. But even in the absence of such rigorous interpretational abilities, infrared spectroscopy is still a very powerful tool in chemistry because it can provide information concerning the presence of specific functional groups or other structural entities within a molecule. This is due to the fortuitous fact, that, within limits, the absorption bands corresponding to a particular vibration of a given bond occur at a given frequency. If this absorption is within the spectral region from 4000 to ~1250 cm^{-1}, it is relatively unaffected by the remainder of the molecule. This is the so-called characteristic group frequency region which makes infrared spectroscopy so useful in general, and to the study of humic substances in particular. For example, with methyl (—CH_3) and methylene (—CH_2—) groups the C—H stretching bands typically occur at ~2860 cm^{-1} (symmetric stretch) and ~2920 cm^{-1} (asymmetric stretch), regardless of the nature of the molecule as a whole (Alpert et al., 1964). The O—H stretching bands typically occur in the range between 3500 and 2800 cm^{-1}. The force constant of the O—H bond is more prone to electronic and other influences than is the C—H bond force constant, thereby accounting for the broader range over which O—H absorption bands are found. In contrast to the characteristic group frequency region, absorption bands occurring at frequencies less than ~1250 cm^{-1} (the so-called fingerprint region) "are profoundly affected by the molecular structure as a whole" (Olsen, 1975).

The theory of infrared spectroscopy will be outlined very briefly here insofar as it is required for understanding the spectra of humic substances. The classical equation for the frequency, ν, or wavenumber, $\bar{\nu}$ (= ν/c), of a covalent bond is given by

$$\nu = \frac{1}{2\pi} \sqrt{\frac{k}{M_r}} \tag{1}$$

or

$$\bar{\nu} = \frac{1}{2\pi c} \sqrt{\frac{k}{M_r}} \tag{2}$$

One should note that *both* ν and $\bar{\nu}$ are often referred to as frequency, even though strictly speaking the latter should be called wavenumber and has the dimensions of reciprocal distance rather than reciprocal time. In Equations (1) and (2), c is the velocity of light *in vacuo,* k is the force constant, and M_r is the reduced mass for the two entities of masses m_1 and m_2 involved in a vibration. Reduced mass, M_r, is given by

$$M_r = \frac{m_1 m_2}{m_1 + m_2} \tag{3}$$

Thus, the greater the force constant and the smaller the reduced mass, the greater the frequency or wavenumber of a particular band. Absorption of radiation occurs (provided certain selection rules are obeyed) when the frequency of the incident radiation coincides with the vibrational frequencies as given by Equations (1) and (2), leading to absorption bands, that is, decreased transmittance, at those frequencies. Transmittance, T, is defined by

$$T = \frac{I}{I_0} \tag{4}$$

where I is the intensity of the transmitted radiation and I_0 is the intensity of the incident radiation. The IR spectrum consists of a plot of T versus frequency.

By far the single largest application of infrared spectroscopy is for qualitative analysis as a result of the highly structured nature of infrared spectra. However, infrared spectroscopy can also be used for quantitative analysis, where the absorbance, A,

$$A = -\log T \tag{5}$$

is related to concentration (C) through Beer's law,

$$A = a_\nu bC \tag{6}$$

where a_ν is the absorptivity of the sample at the frequency of measurement ν, and b is the optical path length.

When tackling a complex chemical problem in many areas of chemistry, it is advisable to adopt a variety of methodologies, both chemical and instrumental, in combination, rather than relying on a single approach in attempting to solve the problem. In this context, the utilization of infrared spectroscopy in conjunction with chemical derivatization methods has proven to be a fruitful marriage in the solution of many chemical problems in the past. This is particularly true in the case of humic substances—the utility of infrared spectroscopy has been expanded considerably when used in conjunction with chemical derivatization as will be discussed later.

As seen from Equations (1) and (2), any change in the force constant or the reduced mass of a given system alters the vibrational frequency. Both changes are of relevance to studies of humic substances and each will be illustrated by specific examples in the following two sections.

Effect of Hydrogen Bonding on IR Spectra

When hydrogen is bonded to the more electronegative atoms such as oxygen or nitrogen the bond is polarized, leaving the hydrogen with a partial positive charge; this hydrogen atom can then interact electrostatically with the lone pair of electrons on oxygen or nitrogen atoms in other molecules. This is referred to as hydrogen bonding, and, as illustrated in the following figure, it can occur between functional groups of the same molecule (intramolecular) or between functional groups in different molecules (intermolecular).

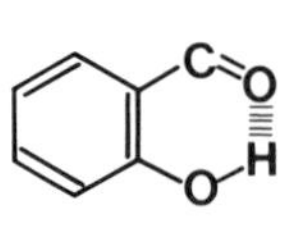

INTRAMOLECULAR HYDROGEN BONDING

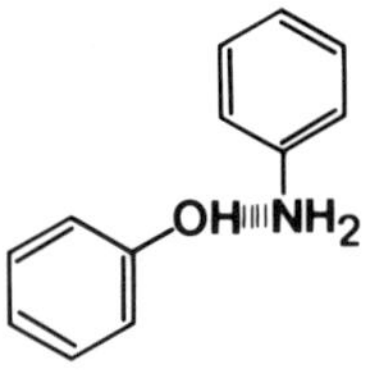

INTERMOLECULAR HYDROGEN BONDING

Hydrogen bonding is indicated by the hatched markings. Hydrogen bonding facilitates the increased separation of the hydrogen atom from the atom to which it is covalently bound, thereby effectively diminishing the force constant of the covalent bond. This results in a diminished frequency of absorption [Eqs. (1) and (2)]. Hydrogen bonding also results in broader bands due to the statistical distribution in the extent of hydrogen bonding in an assemblage of molecules (Bellamy, 1958).

Effect of Isotope Substitution on IR Spectra

In general, isotope substitution involves somewhat laborious chemical procedures. However, in the case of hydrogen attached to the more electronegative atoms such as oxygen, nitrogen, and sulfur, the hydrogen atom is capable of engaging in rapid exchange with other similar hydrogens into which it comes in contact, for example,

$$\mathrm{ROH} + \mathrm{R'OH^*} \rightleftarrows \mathrm{ROH^*} + \mathrm{R'OH} \tag{7}$$

where the asterisk is used to differentiate between the two hydrogen atoms. Such hydrogen atoms are referred to as exchangeable or active hydrogens. If a compound containing active hydrogens is contacted with a solvent containing active deuterium atoms, facile deuterium–hydrogen exchange can occur; for example,

$$\mathrm{ROH} + \mathrm{D_2O} \rightleftarrows \mathrm{ROD} + \mathrm{HOD} \tag{8}$$

By using an excess of deuterating solvent this reaction can be driven essentially to completion. Substitution of deuterium for hydrogen results in a virtual doubling of the reduced mass [Eq. (3)], and a consequent decrease in the absorption frequency by a factor of approximately $\sqrt{2}$, or 1.4 [Eqs. (1) and (2)]. Simple calculations show that such pronounced changes upon isotope substitution do not occur with other types of isotopic substitution (except, of course, for tritium–hydrogen exchange). This topic is reviewed in a recent paper (MacCarthy, 1983).

Sample Preparation in IR Spectroscopy

There are two generally applicable methods of preparing dried humic samples for IR spectroscopy—the alkali halide pressed-pellet method and the mull technique. In the pressed-pellet method approximately 1 mg of the dried sample is thoroughly mixed with about 100 mg of dried alkali halide salt, usually KBr, and compressed into a pellet which is then placed in the sample path of an infrared spectrometer and its spectrum recorded. As KBr is infrared-transparent over the conventional range of 4000 to 400 cm^{-1} only the spectrum of the sample within the KBr matrix is observed. This procedure, with its advantages and limitations, is discussed further in the following references (Stimson, 1962; Fridman, 1967; Parker, 1971; Price, 1972; Ataman and Mark, 1976).

The mull technique involves thoroughly mixing the sample with a low vapor pressure, medium-molecular-weight hydrocarbon, generally known as Nujol. The mull or dispersion is then held between two infrared windows and its spectrum recorded. Since the Nujol itself absorbs radiation (absorption bands at ~2900, ~1460, and ~1375 cm^{-1}), the observed spectrum consists of the superposition of the spectrum of the compound of interest on that of the supporting oil. Other mulling media have also been employed in order to obtain a wider window using the mull technique (Williams and Fleming, 1966). This technique, with its advantages and limitations is discussed in more detail elsewhere (Potts, 1963; Alpert et al., 1964; Parker, 1971; Price, 1972).

A third method of sample preparation, not quite as common as the two described above, is the cast film method. In this technique a solution of the sample is evaporated to dryness and the IR spectrum of the deposited solid film is recorded. This has the advantage that no extraneous matrix is required; however, a satisfactory film may not form in all cases. The cast film method is discussed in more detail in the following references (Alpert et al., 1964; Parker, 1971; Price, 1972).

Methods for measuring the infrared spectra of aqueous solutions or suspensions of humic substances are discussed later in this chapter.

Application of IR Spectroscopy to Humic Substances

With a few exceptions (MacCarthy et al., 1975; MacCarthy and Mark, 1975) all infrared spctra of humic substances have been measured on dried solid samples, and the pressed-pellet method has been used almost exclusively. The mull technique has been used to a very limited extent in the study of humic substances (Ceh and Hadzi, 1956; Orlov et al., 1962; Wagner and Stevenson, 1965), and a few workers have also used the cast film method (MacCarthy and Mark, 1975; Wershaw and Pinckney, 1980). Typical infra-

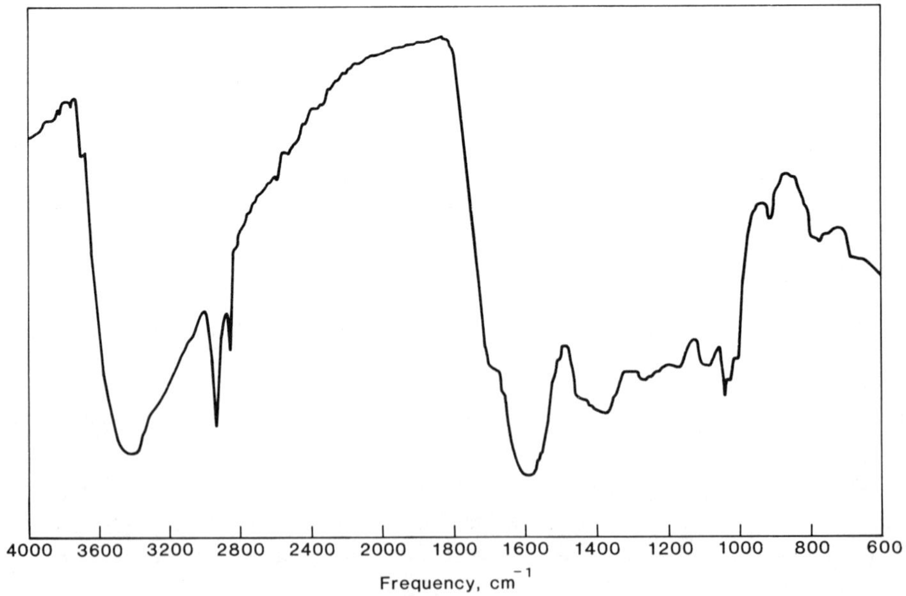

FIGURE 1. Infrared spectrum of a commercial (Pflatz and Bauer) humic acid.

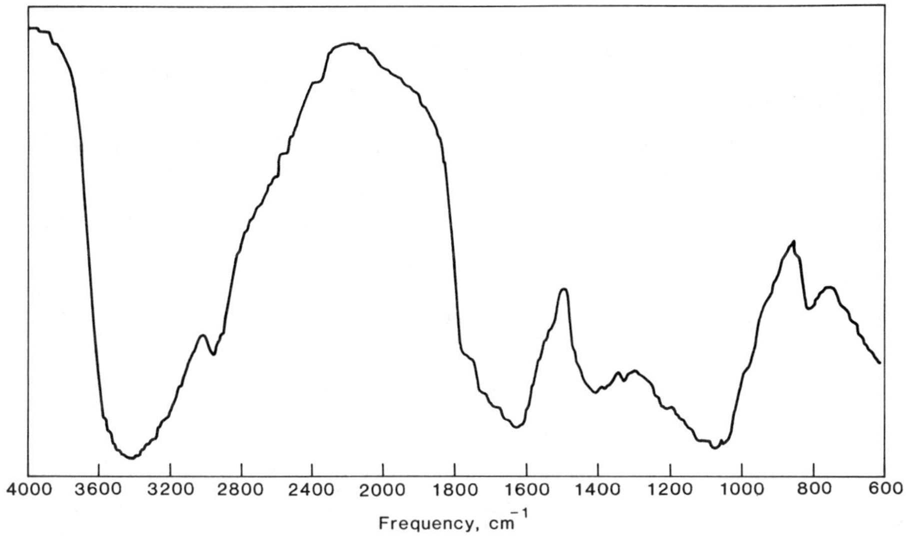

FIGURE 2. Infrared spectrum of a peat fulvic acid.

red spectra of humic substances are shown in Figures 1–3. The most striking feature of the infrared spectra of humic substances, for a person familiar with studying the infrared spectra of pure compounds, is the overall simplicity of the spectra. For comparison purposes the infrared spectrum of a simple molecule, benzoic acid, and of a polymer, polystyrene, are shown in Figures 4 and 5, respectively; these latter spectra are characterized by many narrow, well-defined absorption bands. In contrast, the spectrum of humic acid (Fig. 1) consists of relatively few bands that are very broad. This simplicity is more apparent than real because the broadness of the bands results from the fact that one is dealing with a complex mixture in the case of humic substances. A particular type of functional group in a humic substance can exist in a wide variety of chemical environments each characterized by slightly different force constants for its bonds. Since humic substances are

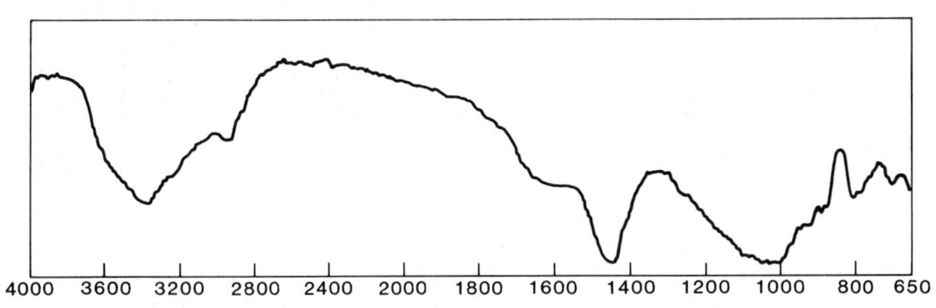

FIGURE 3. Infrared spectrum of humin (in salt form) isolated from a stream sediment.

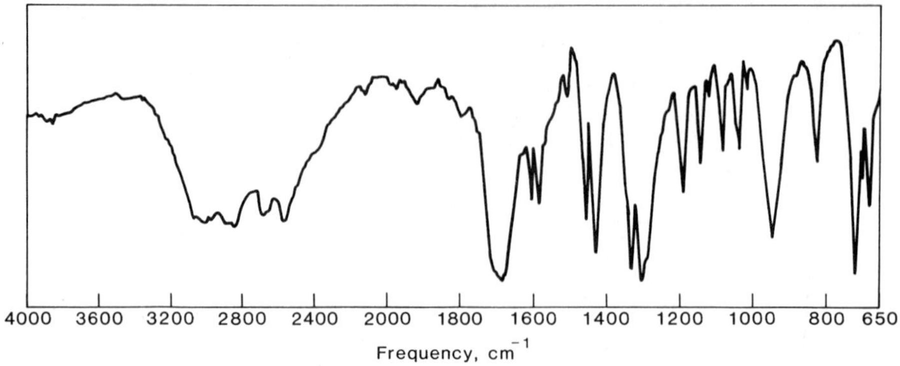

FIGURE 4. Infrared spectrum of benzoic acid.

comprised of complex mixtures, as evidenced by the fact that no satisfactory separation of humic substances has yet been achieved, each type of functional group probably exists in a wide diversity of chemical environments. As a result, there is severe overlapping of absorption bands from the individual constituents in the complex mixture thus accounting for the broad bands and the *apparent* simplicity of the spectra. Other factors can also contribute to the broadening of bands in the infrared spectra of humic substances as will be discussed later.

While there are differences in the infrared spectra of humic substances derived from different sources (Stevenson, 1982), the overall similarity in the spectra of humic substances of diverse origin is more noteworthy than the differences. This similarity must be interpreted with caution, however! It is not uncommon in the humic literature to find the statement that since

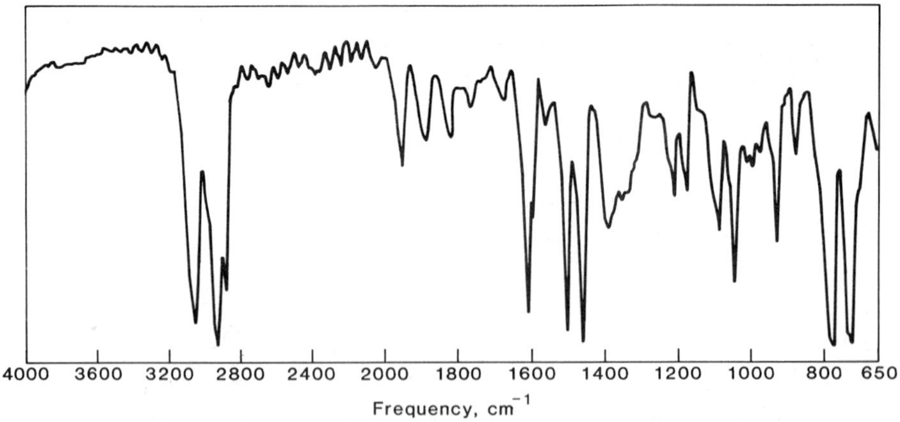

FIGURE 5. Infrared spectrum of a polystyrene film.

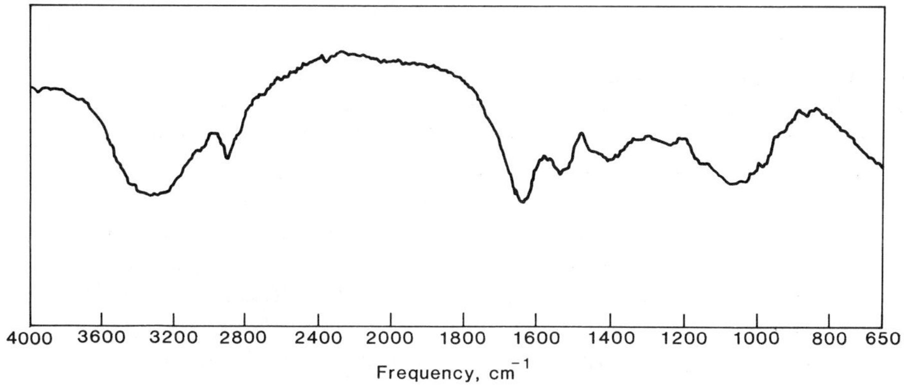

FIGURE 6. Infrared spectrum of urease.

humic substances from diverse sources display similar infrared spectra the materials must have similar structures (Manskaya and Drozdova, 1968; Schnitzer, 1971). A more extensive review of the infrared spectra of a wide variety of materials, however, shows that the situation is not as clearcut as implied above; for example, the infrared spectrum of urease, a discrete enzyme, is shown in Figure 6. This spectrum possesses broad bands that are relatively few in number. This is a spectrum of a discrete biopolymer, and while it is not identical to that of a humic substance, the similarities are striking! Such a conclusion of structural similarity is not justified—the similarity in the infrared spectra indicates only that the net functional group content in each of the various samples may be similar.

An advantage of IR spectroscopy in general, and as applied to humic substances in particular, is the small quantity of sample required. For example, the alkali halide pressed-pellet technique, as normally carried out, requires about 1–2 mg of sample, which is considerably less than that required for the other most useful spectroscopic method, NMR, in the study of humic substances (see Chapter 22). However, microsampling methods have been developed which can record IR spectra on less than 0.01 μg of sample (Alpert et al., 1964; Parker, 1971; Price, 1972; Griffiths and Block, 1973). These special techniques may have advantage in the study of humic samples isolated only in minute quantities, or for investigating small samples obtained in the fractionation of humic substances.

Interpretation of IR Spectra of Humic Substances

In this section the major absorption bands of humic substances are discussed. Different workers report the bands at slightly different frequencies, and the cited values should be regarded as approximate.

The 3400 cm^{-1} Region

This is the region where the absorption due to OH stretching occurs (Wagner and Stevenson, 1965) and its broadness is generally attributed to hydrogen bonding (Theng et al., 1966; Juo and Barber, 1969). This assignment is substantiated by observing a decrease in the absorption intensity upon methylation and acetylation (Wagner and Stevenson, 1965). Wagner and Stevenson (1965) have shown that when methylated fulvic acid was saponified, the absorption in the 2500–3100 cm^{-1} region increased to its original value.

In light of the prior discussion on the effects of hydrogen bonding, it is interesting to examine the 3400 cm^{-1} band in more detail. The OH absorption band in the spectra of humic substances is not extraordinarily broad when compared to that of some pure compounds such as benzoic acid (compare Fig. 1 with Fig. 4). In fact, considering the ill-defined nature of humic substances and the fact that the OH groups presumably occur in a wide variety of chemical environments, it is surprising that the OH absorption band of a humic substance is not considerably broader than what is usually observed.

It is generally claimed that humic substances engage in pronounced hydrogen bonding (Cannon and Sutherland, 1945; Ceh and Hadzi, 1956; Schnitzer, 1965b; Stevenson and Goh, 1971, 1972; Flaig et al., 1975; Schnitzer, 1978; Ruggiero et al., 1978; Stevenson, 1982). As discussed previously, a consequence of hydrogen bonding is the diminished force constant for the OH bond and concomitant lowering of the absorption wavenumber [Eq. (2)]. The wavenumber of the OH absorption in a humic substance (the band centered at ~3400 cm^{-1}) is substantially greater than that in many simple carboxylic acids (compare Figs. 1, 4, and 7). In light of the view which many people hold, that humic substances have a major carboxylic acid character, it is surprising that the frequency of the OH absorption band in humic substances is not decreased further than what is observed. Instead, the OH absorption band in humic substances occurs in a region closer to that characteristic of phenolic and alcoholic OH groups. These questions have not been raised or discussed in the literature to the authors' knowledge.

The 2920 and 2860 cm^{-1} Bands

The absorption bands at 2920 and 2860 cm^{-1} are evident in the spectra of most humic substances, usually superimposed on the shoulder of the broad O—H stretching band. They are generally more pronounced in humic acids than in fulvic acids. These bands are attributed to the asymmetric and symmetric stretching vibrations, respectively, of aliphatic C—H bonds in methyl and/or methylene units (Theng et al., 1966). This assignment is consistent with the observed increase in absorbance of these bands upon methylation of the humic substance (Wagner and Stevenson, 1965; Wershaw et al., 1981).

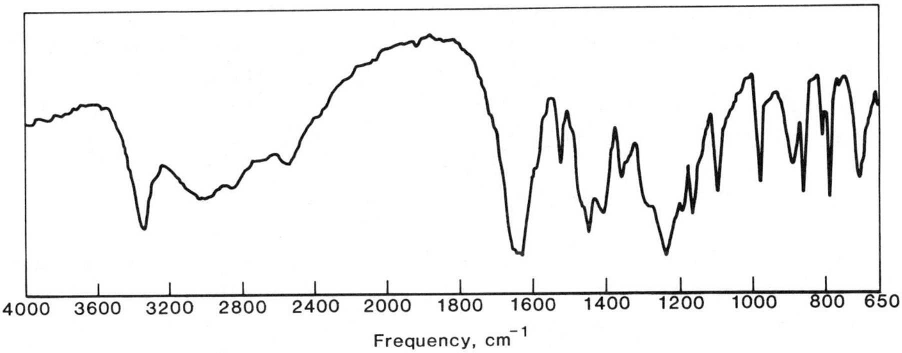

FIGURE 7. Infrared spectrum of 2,4-dihydroxybenzoic acid.

The 1720 cm^{-1} Band

The 1720 cm^{-1} band in humic substances is generally attributed to the C=O stretching vibration, due mainly (though not completely) to carboxyl groups. This is one of the more easily assigned bands in humic substances, in that titration to ~pH 7.0 causes it to largely disappear with concomitant appearance of a band at ~1600 cm^{-1} and intensification of the absorption in the 1400 cm^{-1} region. Similar changes occur upon neutralization of simple carboxylic acids (Bellamy, 1968). This type of behavior is universally reported for humic substances (Schnitzer and Skinner, 1963; Wagner and Stevenson, 1965; Theng et al., 1966). Theng and Posner (1967) have shown that the absorptivity of a band at ~1720 cm^{-1} in humic substances could be correlated with the exchange capacity (i.e., COOH content).

The 1600–1650 cm^{-1} Band

The C=C bond in benzene is infrared inactive; however, in benzene derivatives, which decrease the symmetry of the molecule, this band is infrared active. The band at ~1650 cm^{-1} in humic substances has been assigned to aromatic C=C "double bonds" conjugated with C=O and/or COO^- (Schnitzer and Skinner, 1968a). Theng and Posner (1967) attributed this band to $\alpha\beta$-, or $\alpha\beta$-$\alpha'\beta'$-unsaturated ketones which are known to absorb in this region. The bending vibration of water is centered at 1640 cm^{-1} and may contribute to the IR absorption in this region if the sample is not thoroughly dry.

The 1510 cm^{-1} Band

Juo and Barber (1969) attribute the band at 1510 cm^{-1} to stretching vibrations of aromatic C=C bonds. A small, well-defined peak at 1510 cm^{-1} was reported in a sample of peat fulvic acid evaporated to dryness at pH 2.0; this peak was absent when the sample was dried at pH 7.0. This band was assigned to bending vibrations of $—NH_3^+$ groups (MacCarthy and O'Cinneide, 1974). Confirmatory evidence for this assignment was obtained from deamination and ion-exchange studies.

The 1380–1440 cm^{-1} Band (Shoulder)

The 1450 cm^{-1} band has been attributed to the bending vibration of aliphatic C—H groups (Juo and Barber, 1969) as occurs in simple compounds (Meloan, 1963).

The 1400 cm^{-1} Band

The absorption at 1400 cm^{-1} may be due to the O—H bending vibrations of alcohols or carboxylic acids, as occurs with simple model compounds.

The 1220 cm^{-1} Band

This band has been assigned to the C—O stretching vibration and OH bending deformations, due mainly to carboxyl groups since it disappears on producing the salt form (Juo and Barber, 1969) and as occurs with simple carboxylic acids (Meloan, 1963).

IR Spectra of Chemically Altered Humic Substances

The assignment of infrared absorption bands of humic substances is facilitated by chemical modification of the material. In particular, the assignment of the 3400 cm^{-1} band to OH stretching vibrations has been substantiated by methylation of OH groups (Wagner and Stevenson, 1965).

However, it is worth noting that many of the infrared spectra of methylated samples of humic materials still display rather pronounced absorption bands in the 3400 cm^{-1} region (Wagner and Stevenson, 1965; Flaig et al., 1975; Stevenson, 1982) even though in some cases it was alleged that the methylation was exhaustive (Ceh and Hadzi, 1956). This suggests that derivatization of those samples was not complete. As discussed in Chapter 16 of this book, Leenheer has developed a technique for derivatizing humic substances resulting in samples that have virtually no absorbance in the 3400 cm^{-1} region (Fig. 8). Wershaw et al. (1981) have described the methylation of fulvic acid in dimethylformamide using diazomethane followed by reaction with sodium hydride/methyl iodide. They state that infrared analysis "has shown that complete methylation was obtained" after additional permethylation with methyl iodide and sodium hydride in dimethylformamide. The use of these techniques which result in exhaustive derivatization, as evidenced from the IR spectra, warrants more complete investigation in the future.

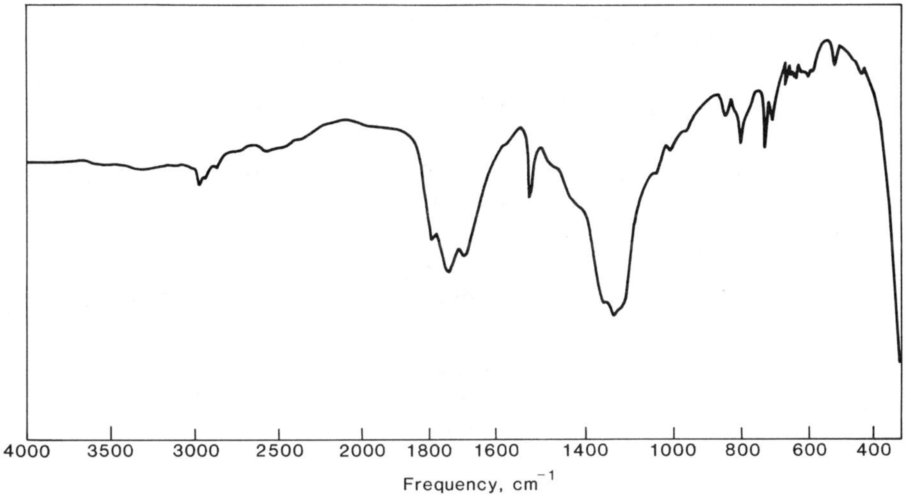

FIGURE 8. Infrared spectrum of methylated and trifluoroacetylated humic acid (after Leenheer, Chapter 16, this book).

Methylation and acetylation provide confirmatory evidence for the assignment of various bands to aliphatic components in the humic substances (Wagner and Stevenson, 1965; Flaig et al., 1975; Stevenson, 1982). The presence of amine groups in a peat fulvic acid has been confirmed by observing the changes in the infrared spectra of the sample following reaction with nitrous acid (MacCarthy and O'Cinneide, 1974).

IR SPECTRA OF HUMIC SUBSTANCES IN THE AQUEOUS STATE

In the natural environment humic substances occur almost invariably in the wet state. For this reason it is of interest to study humic substances in the aqueous environment. However, there are a number of problems associated with measuring the IR spectra of aqueous solutions or suspensions, the most serious being the intense absorption bands of the water which is present in large excess. The two most intense bands of water are broad and are centered at 3400 and 1640 cm^{-1}; these bands obscure major areas of interest in the spectra of humic substances. Two approaches have been used to record the IR spectra of humic substances in the aqueous environment: (1) measurement of spectra in D_2O media using conventional (i.e., dispersive) spectrometers, or (2) measurement of spectra in H_2O media using a Fourier transform IR spectrometer. When studying the IR spectra of aqueous samples, it is necessary to work with concentrated solutions or slurries due to the very short path-length cells required, typically 10–20 μm. These path lengths are about a thousandfold shorter than the common path lengths used in UV–visible spectroscopy and are necessitated by the strong absorption due to the water itself.

IR Spectra of Humic Substances in Deuterium Oxide

The infrared spectra of H_2O, D_2O, and an approximately 1 : 1 mixture of H_2O and D_2O are shown in Figure 9 (MacCarthy and Mark, 1975). The broad OD absorption bands in D_2O, centered at ~2500 and ~1200 cm^{-1}, correspond to the 3400 and 1600 cm^{-1} bands of H_2O. The additional band at ~1450 cm^{-1} in the spectrum of the D_2O/H_2O mixture is due to the bending vibration of the

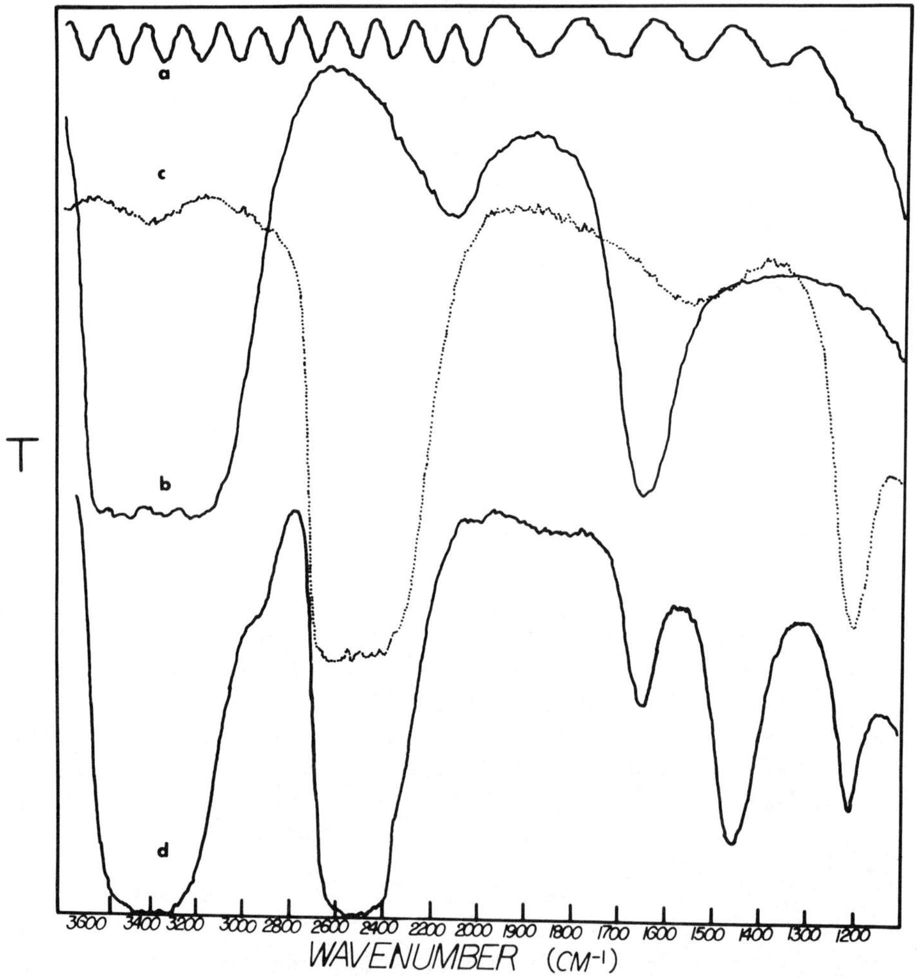

FIGURE 9. (a) Transmittance of CaF_2 cell (25 μm spacer); infrared spectra of (b) H_2O, (c) D_2O, and (d) an approximately 1 : 1 mixture of H_2O and D_2O. Spectra (b), (c), and (d) measured in 15 μm cells (MacCarthy and Mark, 1975).

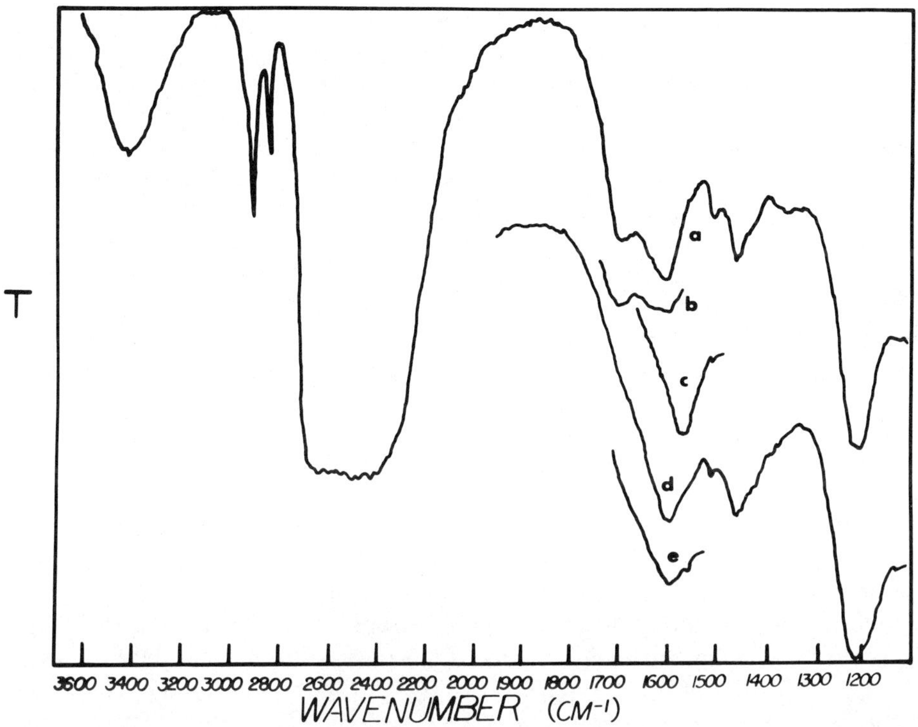

FIGURE 10. Infrared spectra in 15 μm CaF_2 cells of (a) humic acid in D_2O, (b) humic acid in D_2O/DCl, (c) humic acid in $D_2O/NaOD$, (d) Cu^{2+}–humic acid complex in D_2O, and (e) Fe^{3+}–humic acid complex in D_2O (MacCarthy and Mark, 1975).

HOD molecule. As seen from curve c in Figure 9, D_2O is essentially transparent to infrared radiation over regions of interest in the study of humic substances. When humic substances are dissolved or suspended in D_2O the active protons undergo exchange with deuterium atoms as described by Equation (8); thus, one actually records the spectrum of the *deuterated* humic substance in D_2O. The OH bands are replaced by OD bands and the changes in wavenumber can be calculated from Equation (2). Figure 10 shows the infrared spectra of slurries of deuterated humic acid in D_2O and in D_2O/DCl, of deuterated sodium humate in $D_2O/NaOD$, and the spectra of Cu(II) and Fe(III) complexes of humic acid in D_2O. The changes that occur in the 1550–1700 cm^{-1} region, and in the 1375 cm^{-1} region on going from the acid to the salt form, or to the Cu(II) or Fe(III) complexes, are readily observed in D_2O medium. The absorption band at ~1700 cm^{-1}, due to the carbonyl of the unionized carboxyl group, disappears upon neutralization and a corresponding band due to the carboxylate ion appears at ~1374 cm^{-1}.

It is interesting to compare the relative intensities of the bands at ~1700 and ~1600 cm^{-1} in spectra of (1) deuterated humic acid, (2) deuterated humic acid to which DCl has been added, and (3) dried humic acid (hydrogen form) pressed into a pellet. In the sample of deuterated humic acid (in D_2O) the band at ~1600 cm^{-1} is considerably more intense than that at ~1700 cm^{-1}. However, after addition of DCl both bands have approximately the same intensity, similar to that observed in the dried sample. Evidently, the humic acid is partially dissociated at ambient pD values causing a decrease in intensity at ~1700 cm^{-1}, but this dissociation is suppressed by a decrease in pD or by drying. These results demonstrate that this approach allows the *equilibrium* state of the organic matter in the aqueous system to be investigated, unlike the KBr pellet method which alters the equilibrium status of the system under investigation.

In the spectra of the deuterated humic acid and of its salt and complexes (Figure 10; a,c,d,e) the aliphatic C—H stretching bands at 2860 and 2920 cm^{-1} are clearly evident without the severe interference from the broad OH band which generally occurs with the hydrogen form of these substances.

Preliminary experiments on the kinetics of deuterium–hydrogen exchange and of hydrogen–deuterium exchange with humic substances indicated that the reactions were complete within a matter of minutes (MacCarthy and Mark, 1975). This is in contrast to the slow exchange found for some native proteins and nucleic acids but similar to that found for denatured proteins and many other materials (Parker, 1971, Ch. 11). More refined experiments along those lines, performed over shorter time periods, may provide information on the nature of the hydrogen bonding in humic substances.

Fourier Transform IR Spectra of Humic Substances in Water

Fourier transform IR spectroscopy (FTIR) has a number of advantages compared to dispersive IR spectroscopy (Griffiths, 1975). In particular, FTIR is characterized by (1) enhanced resolution, (2) improved signal-to-noise (S/N) ratio, and (3) higher energy throughput. The superior resolution is not of particular value when dealing with humic substances since the spectra are inherently broad as a result of the complicated, multicomponent nature of humic materials, and broad bands will be observed regardless of the resolution of the spectrometer! This is a typical example of the limitations imposed on an instrumental method of investigation by the nature of humic substances per se.

The high throughput (i.e., the amount of energy passing through the spectrometer) is of particular advantage in energy-limited situations, such as when working with highly absorbing samples. Figure 11A shows a segment of the IR spectrum of water and Figure 11B shows the same segment from a spectrum of sodium humate in water (MacCarthy et al., 1975). Curve C of Figure 11 shows the result of ratioing curves A and B (i.e., obtaining the "difference spectrum"). The aliphatic C—H stretching bands (2860 and 2920 cm^{-1}) which occur as slight humps on the shoulders of the broad O—H stretching bands of water in spectrum B become well-defined bands in spec-

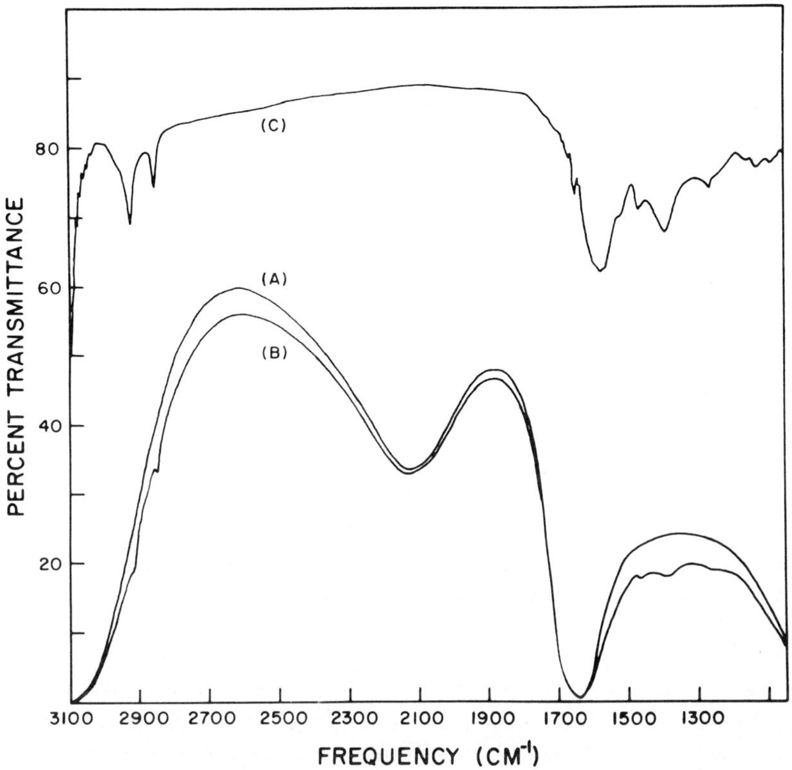

FIGURE 11. Fourier transform infrared spectra of (A) H_2O, (B) solution of sodium humate in H_2O, and (C) spectrum of sodium humate obtained by ratioing (B) with (A) (path length = 17.6 μm; CaF_2 cells) (MacCarthy et al, 1975).

trum C. In addition, whereas spectrum B and spectrum A are virtually superimposed at ~1640 cm^{-1} (the maximum of the HOH bending mode) the ratioed spectrum C reveals spectral information concealed beneath the water band. The bands in the 1050–1550 cm^{-1} region of the humic acid spectrum, which are barely discernible when superimposed on the water background in curve B, are quite distinct in the ratioed spectrum C.

Figure 12 shows the spectra of slurries of humic acid and Cu^{2+}–humate in H_2O. The main difference between the spectrum of sodium humate (Fig. 11C; pH 8.0) and that of humic acid (Fig. 12A) is the occurrence of the band centered at ~1700 cm^{-1} in the latter. Upon neutralization, this band decreases in intensity considerably and "new" bands appear at ~1570 and ~1395 cm^{-1} consistent with the ionization of carboxylic acid groups. The fact that the ~1700 cm^{-1} band has practically disappeared at pH 8.0 indicates that this band is due almost exclusively to carboxyl groups with little contribution from aldehyde and ketone groups. This is consistent with what has been reported by other workers (Wagner and Stevenson, 1965; Theng et al., 1967) using the KBr pellet method on *dried* samples.

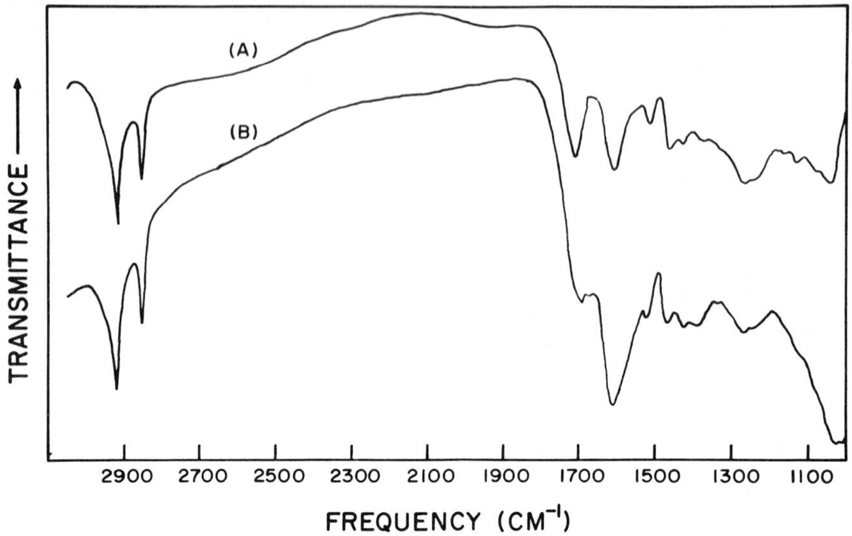

FIGURE 12. Fourier transform infrared spectra of (A) humic acid in H_2O and (B) Cu^{2+}–humate in H_2O (CaF_2 cells) (MacCarthy et al, 1975).

In the spectrum of the Cu^{2+}–humic acid complex (Fig. 12B) the intensity of the ~1710 cm^{-1} band has decreased considerably compared to that of the humic acid. This provides direct evidence for the participation of carboxyl groups in complexation by humic substances in the aqueous environment.

Overall, the spectra of the humic substances recorded in the aqueous state are comparable to those measured by the conventional KBr pellet technique on dried samples. However, FTIR allows the samples to be observed in their native wet state and avoids the shifts in chemical equilibrium which must necessarily accompany the drying process. More detailed information on the ionization of the carboxylic acid functional groups could be obtained by measuring the spectra of the aqueous samples over a wide range of pH values. FTIR may also provide a tool for investigating the kinetics of metal–humic interactions and other reactions of humic substances in the aqueous state.

RAMAN SPECTROSCOPY

Raman spectroscopy is a technique in which monochromatic radiation in the visible range (400–800 nm) is scattered by a sample (Woodward, 1967; Olsen, 1975; Long, 1977). Most of the scattered radiation is due to Rayleigh scattering (scattered radiation of same frequency as incident radiation); however, a small fraction of the scattered radiation is shifted to lower frequencies (Stokes lines) and an even smaller fraction to higher frequencies (anti-Stokes lines). The difference in energy between the Raman-scattered radiation and the incident radiation corresponds to the energy differences of vibrational transitions in the scattering molecule. In other words, the *differences* in energy between the various bands of scattered radiation *in the visible range* correspond to the actual energy of the absorbed radiation in the infrared range. Thus, like infrared spectroscopy, Raman spectroscopy provides information on the vibrational transitions within molecules. A major advantage of Raman spectroscopy is that, since visible radiation is employed, water is a suitable solvent for this technique, and it would thus appear to be the method of choice for studying the vibrational spectra of humic substances in the aqueous state provided the solutions are reasonably concentrated. For dilute samples, such as naturally occurring concentrations of aquatic humic substances, the background signal from water may be a significant interference. Attempts to measure the Raman spectra of humic substances have been unsuccessful to date: the relatively intense fluorescence exhibited by the humic substances obscures any Raman signal that may be present (P. MacCarthy and D. F. Schriver, unpublished results 1971; M. C. Goldberg, U.S. Geological Survey, Denver, personal communication 1982). Techniques for discriminating between Raman and fluorescence signals are not presently applicable to humic substances.

UV–VISIBLE SPECTROSCOPY

Brief Introduction to UV–Visible Spectroscopy

The ultraviolet and visible regions of the electromagnetic spectrum extend from ~10 to 400 nm and from 400 to ~800 nm, respectively. The ultraviolet and visible (UV–visible) regions are generally considered together since both correspond to electronic transitions within the absorbing species (UV–visible spectra are also referred to as electronic spectra). The UV region is classified into the far UV (10–200 nm) and the near UV (200–400 nm). Oxygen absorbs strongly in the far UV, and since evacuation of the spectrometer is the simplest means of alleviating this interference the far UV region is often referred to as the vacuum ultraviolet.

The UV–visible spectra of solutions generally show broad bands that are few in number. The fine structure frequently associated with vapor-phase or solid-state spectra is "lost" in solution spectra due to broadening as a result of random interactions with neighboring molecules.

The absorption of UV–visible radiation can frequently be attributed to a specific segment or functional group within the molecule; such absorbing entities may contain σ, π, or n electrons. Examples of common chromophores are functional groups containing unbonded electrons (e.g., carbonyl), sulfur, nitrogen, or oxygen atoms, and conjugated carbon–carbon multiple bonds.

For a more detailed discussion of the theory and applications of UV–visible spectroscopy the reader is referred to the texts by Jaffé and Orchin (1962), Williams and Fleming (1966), and Olsen (1975).

Application of UV–Visible Spectroscopy to Humic Substances

In many instances UV–visible spectroscopy is a valuable tool in the identification of chromophoric functional groups in discrete organic molecules. But even a simple, two-component mixture may make the interpretation of a UV–visible spectrum difficult from a functional group point of view. The following example from Olsen (1975) illustrates this point well:

> . . . . in confirming the presence of a carbonyl group, a band in the 270 to 290 nm region (with a equal to about 50) might indeed be due to the carbonyl group, but, on the other hand, it might also be due to an impurity having a short conjugated system. Thus, the presence of only 0.25 percent of an impurity with high-intensity absorption ($a_\nu \approx 20{,}000$) will simulate the carbonyl absorption band . . . .

Considering the difficulty in obtaining functional group information from the UV–visible spectra of even relatively simple chemical systems, one can appreciate the enormity of the problem when attempting to extricate functional group information from the UV–visible spectra of humic substances. The UV–visible spectra of humic substances are generally featureless (Kumada, 1955; Zeichmann, 1964; Butler and Ladd, 1969; Schnitzer, 1971, 1978; Schnitzer and Khan, 1972; Bailly, 1973; Hayes and Swift, 1978; Ghosh and Schnitzer, 1979; Stevenson, 1982) with absorptivities increasing toward shorter wavelengths. Figure 13 shows typical UV–visible spectra of several humic acids (Kumada, 1955). It has been stated (Schnitzer, 1971) that while the UV–visible

> . . . absorption spectra . . . do not provide much detailed information on their chemical structure, the similarity of their spectra . . . suggests that we are dealing with compounds with similar basic structures.

However, considering the complicated, multicomponent nature of humic substances, with their variety of chromophores and probable variations in absorptivities, such statements must be viewed with extreme skepticism. A more reasonable inference from examining the UV–visible spectra of humic substances is that they result from the overlap of the absorbances of various chromophores (Hayes and Swift, 1978; Stevenson, 1982). As Hayes and Swift (1978) have pointed out, it is not possible to

> . . . observe or measure one particular chromophore, or to derive definitive information with regard to chemical composition.

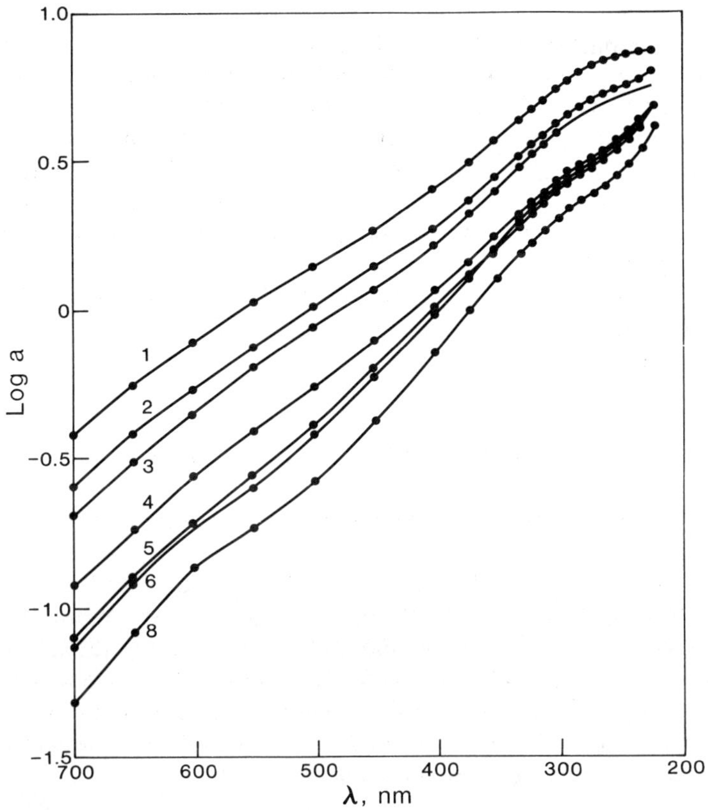

FIGURE 13. UV–visible spectra of several soil humic acids (Kumada, 1955).

Insofar as its applicability to the determination of functionality of a humic substance is concerned, one must conclude that UV–visible spectroscopy is severely limited by the inherent nature of humic substances.

The UV–visible absorptivities of humic substances do vary as a function of pH (Salfield, 1965; MacCarthy and O'Cinneide, 1974) and one can speculate that these changes are due to the ionization of carboxylic and phenolic functional groups. The UV–visible spectra of humic substances do not exhibit an isosbestic point when plotted at various pH values, which is consistent with their multicomponent nature. As with other spectroscopic methods it is not known what proportion of the molecules contribute to the absorbance at a particular wavelength.

Despite the limitations referred to above, UV–visible spectroscopy does have many useful applications for purposes other than determining functionality in humic substances research. The reader is referred to Flaig et al. (1975), Schnitzer (1978), and Stevenson (1982) for reviews of the applications of UV–visible spectroscopy to the study of humic substances. Examples of these applications are in estimating the degree of humification using E_4/E_6 ratios (i.e., the ratio of the absorbance at 465 nm to that at 665 nm) (Chen et al., 1977) and for determining the concentration of dissolved humic substances based on Beer's law plots for the particular substance under study. UV–visible spectroscopy is also used for studying the interaction of metal ions with fulvic acid (Schnitzer and Khan, 1972; Stevenson, 1982); however, caution must be observed to ensure that scattering of the radiation is not a problem (MacCarthy and Mark, 1976).

In conclusion, UV–visible spectroscopy has little value for studying functionality in humic substances and cannot be used for the direct determination of functional groups in these materials.

FLUORESCENCE SPECTROSCOPY

Brief Introduction to Fluorescence Spectroscopy

The absorption of UV–visible radiation raises a molecule from the ground electronic and vibrational states to excited electronic and vibrational states. Most molecules return to the ground state by dissipating the additional energy in the form of heat. In some molecules, however, only a fraction of this energy is dissipated as heat and the residual energy is emitted as electromagnetic radiation of longer wavelength than the incident radiation. This is fluorescence and may be depicted as follows:

$$X + h\nu \rightarrow X^* \rightsquigarrow X^{**} \rightarrow X + h\nu' \tag{9}$$

where X represents the ground state of a species capable of fluorescing, h is Planck's constant, X^* represents an electronically and vibrationally excited state of X, X^{**} represents an electronically excited state of X following radiationless loss of energy from X^*, and ν and ν' are the frequencies of the incident and fluorescent radiation, respectively.

Relatively few aliphatic and alicyclic molecules exhibit fluorescence in the UV–visible region (Wehry and Rogers, 1966; Olsen, 1975). In order for a molecule to fluoresce, it must possess chromophores which absorb the incident radiation. Among the structural entities that give rise to fluorescence are those containing conjugated double bonds or aromatic rings. Groups capable of donating electrons, such as —OH or —NH_2, enhance fluorescence. Electron-withdrawing groups, such as —COOH, tend to diminish fluorescence in aromatic compounds (Wehry and Rogers, 1966).

The intensity of fluorescence, I_f, is given by

$$\begin{aligned} I_f &= kI_0 a_\nu bC \\ &= kI_0 A \end{aligned} \tag{10}$$

for low-absorbance solutions where I_0 represents the intensity of the incident radiation and k is a proportionality constant; a_ν, b, C, and A have the same meanings as before [see Eqs. (5) and (6)].

There are two types of fluorescence spectra. An excitation spectrum is obtained by scanning the incident radiation over a range of wavelengths while monitoring the fluorescence at a fixed wavelength (usually at λ_{max} for fluorescence). Theoretically, the excitation spectrum of a molecule should have the same shape as the absorption spectrum as evident from Equation (10). An emission spectrum results from irradiating the sample with light of a fixed wavelength (usually λ_{max} for absorption), and scanning the emission (fluorescence) spectrum. For a more detailed discussion of fluorescence the reader is referred to Hercules (1966) or Olsen (1975).

Application of Fluorescence Spectroscopy to Humic Substances

Humic substances are known to fluoresce (Seal et al., 1964; Schnitzer, 1971; Flaig et al., 1975; Hayes and Swift, 1978; Stevenson, 1982). In order to fluoresce, some molecules in the humic substances must first absorb the incident UV–visible radiation. As such, fluorescence suffers from all the limitations of UV–visible spectroscopy as regards determining functionality in humic substances. It is known that, in general, only a small fraction of molecules that absorb radiation actually undergo fluorescence. Since humic substances are comprised of a multicomponent mixture, it is likely that the fluorescence spectra represent an even smaller fraction of the total molecules than those responsible for the UV–visible absorption spectra of a humic material. It would be of interest to compare the shapes of the *excitation* spectra with those of the UV–visible absorption spectra to acquire further information in this regard; as is evident from the discussion in connection with Equation (10) above, both spectra would have the same shape only if all the absorbing molecules fluoresced. To the authors' knowledge such comparisons have not been made to date.

Schnitzer (1971) and Schnitzer and Khan (1972) review the literature on the use of spectrofluorimetry in the study of humic substances. The inapplicability of fluorescence to the direct determination of functionality in humic substances is apparent from those reviews. Figure 14 (from Plechanov et al., 1983) shows the fluorescence spectra of several aquatic humic and fulvic acids. The relatively well-defined structure typical of the spectra of simple molecules such as anthracene (Fig. 15) is absent from the humic acid spectra. Ewald et al. (1983) report corrected fluorescence spectra for humic substances, but these spectra provide no further insight into the functionality of these materials.

As with UV–visible absorption spectroscopy, the fluorescence spectra of humic substances represent the summation of the signals from many different fluorescing molecules, and it is not possible, at present, to provide any detailed interpretation of these spectra. Functionality has been *inferred* from

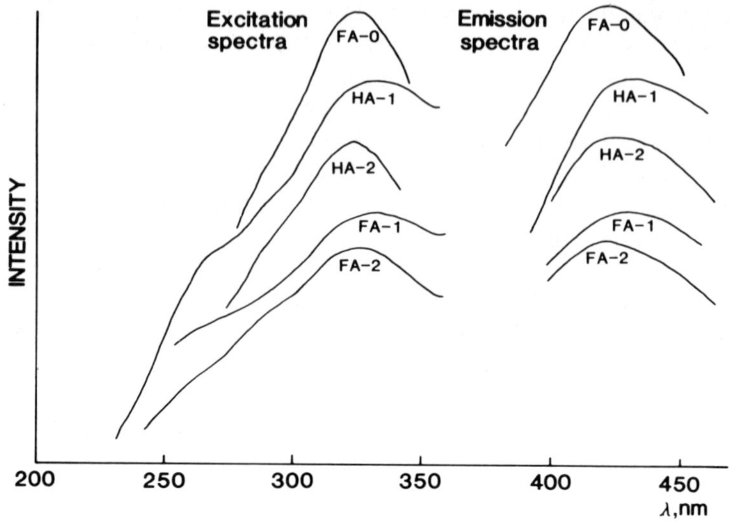

FIGURE 14. Excitation and emission fluorescence spectra of humic and fulvic acids from various soil and aquatic sources (Plechanov et al., 1983).

fluorescence studies. Seal et al. (1964) suggested a "sequence of polymers" to explain their observations. As discussed earlier in this chapter, any interpretation of humic substances data in terms of polymers is not justified on the basis of currently available information. Visser (1983b) attempted to correlate the phenolic content of a humic material to the wavelength of the main excitation peak. However, functional groups could not be directly identified from the fluorescence spectra in any of the above studies.

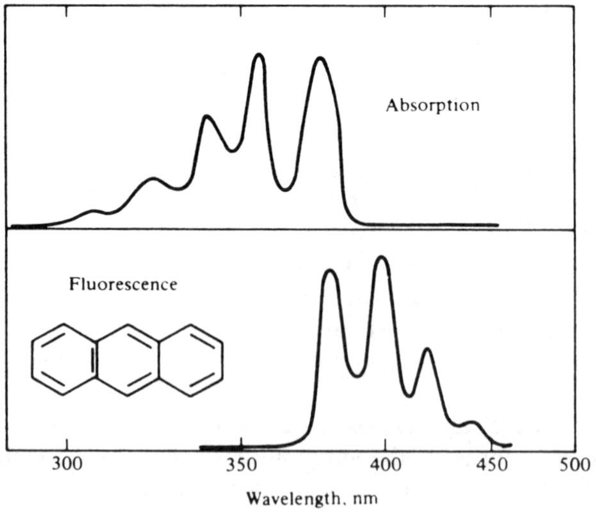

FIGURE 15. Absorption spectrum and emission fluorescence spectrum of anthracene (Whiffen, 1966).

One must conclude that fluorescence spectroscopy cannot be used for the direct determination of functionality in humic substances.

ELECTRON SPIN RESONANCE (ESR) SPECTROSCOPY

Brief Introduction to ESR Spectroscopy

When molecules containing unpaired electrons are placed in a magnetic field, the energy level of each electron is split into two discrete states through interaction of the magnetic moment of the electrons with the applied field. These molecules can be excited from the lower to the higher energy level by absorption of electromagnetic radiation in the microwave region (about 6–9 Hz). This is referred to as electron spin resonance (ESR) spectroscopy or as electron paramagnetic resonance (EPR) spectroscopy. Thus, ESR spectroscopy is used specifically for the study of species with unpaired electrons, that is, free radicals.

The splitting of the energy levels of unpaired electrons by means of a magnetic field is referred to as the Zeeman effect. The energy of an unpaired electron in a magnetic field is given by

$$E = -g\beta M_z H_0 \tag{11}$$

where g is the Landé or spectroscopic splitting factor which has a value of 2.0023 for a "free" or isolated electron, β is the Bohr magneton, M_z is the component of the spin angular momentum in the direction of the z axis of the applied magnetic field and may have discrete values of $+\frac{1}{2}$ or $-\frac{1}{2}$, and H_0 is the magnetic field strength. For a given value of H_0, the energy difference (ΔE) between the two discrete spin states of the electron is given by

$$\Delta E = g\beta H_0 \tag{12}$$

Absorption (or stimulated emission) of radiation occurs when

$$\Delta E = h\nu \tag{13}$$

where, as previously, h and ν represent Planck's constant and frequency, respectively.

When a free radical is placed in a magnetic field, equilibrium will be reached with a slightly greater number of electrons in the lower energy level than in the upper energy level. When radiation of a frequency satisfying Equation (13) is incident on the sample, electrons in the lower energy state are excited to the upper energy state with concomitant absorption of energy, and electrons in the upper energy level are stimulated to emit radiation by a transition to the lower energy level. Since there are more molecules in the lower energy state there is a net absorption of radiation. For a more detailed discussion of ESR spectroscopy the reader is referred to Steelink and Tollin (1967) or Olsen (1975).

Application of ESR Spectroscopy to Humic Substances

The first application of ESR spectroscopy to a humic substance was by Rex (1960) who studied the alkali extract of peat, thus discovering the free radical nature of humic substances. In the following decade, a number of papers describing the application of ESR spectroscopy to humic substances was published, among them being papers by Steelink and Tollin (1962, 1967), Steelink et al. (1963), Atherton et al. (1967), Schnitzer and Skinner (1969), and Riffaldi and Schnitzer (1972).

The free radicals in humic substances appear to be remarkably stable with respect to time and chemical attack (Steelink and Tollin, 1962; Steelink et al., 1963; Steelink, 1964). This stability has led some workers to suggest that humic substances comprise a free radical, or mixture of free radicals, of the semiquinone type (Steelink and Tollin, 1962; Steelink et al., 1963; Atherton et al., 1967). The increase in spin concentration upon converting humic acids to the sodium salts led Steelink (1964) to propose that the radical species in humic acid is a semiquinone co-existent with a quinhydrone species. Steelink (1964) also suggests that a free radical is an integral part of the humic macromolecule. The stability of this radical has been attributed to the delocalization of the unpaired electron over an aromatic system (Theng and Posner, 1967), or a shielding effect on the free radical due to the macromolecular network (Steelink, 1964).

A typical ESR spectrum for humic acid is presented in Figure 16 (Steelink and Tollin, 1967). This is similar to a spectrum reported for a soil fulvic acid by Schnitzer and Skinner (1969). The ESR spectrum of humic acid or fulvic acid consists of a single line identified by its position and width. In general, the ESR spectra reported for humic substances are devoid of hyperfine

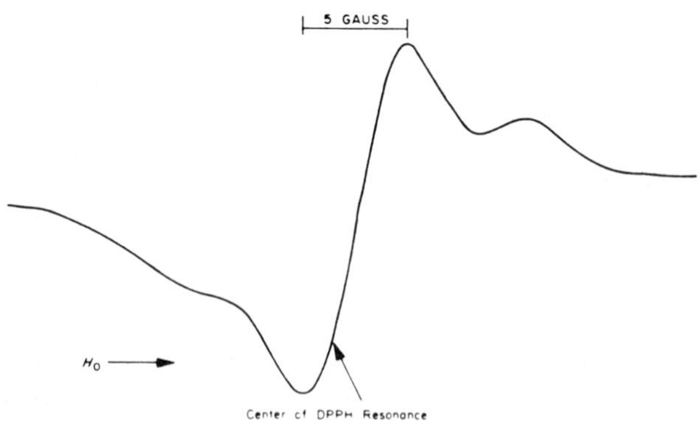

FIGURE 16. Electron spin resonance spectrum of a soil humic acid (Steelink and Tollin, 1967).

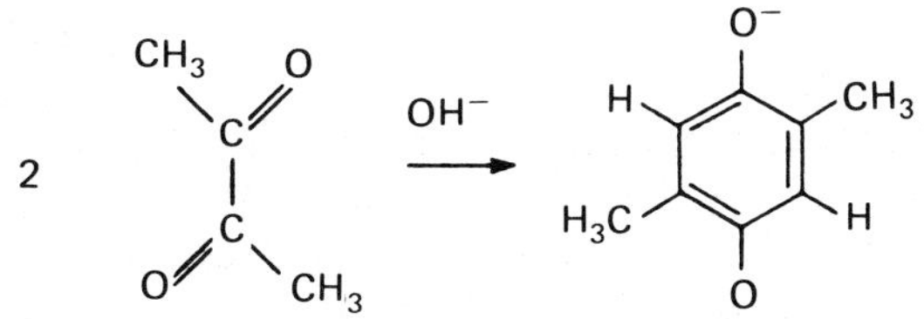

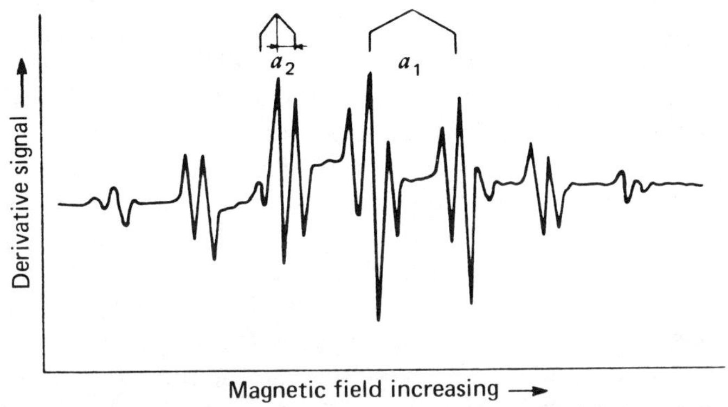

FIGURE 17. Electron spin resonance spectrum of 2,5-dimethylquinone (Williard, Merritt and Dean, 1974).

splitting, although Atherton et al. (1967) have reported hyperfine splitting in spectra of sodium humate (0.1*N* NaOH). Those spectra showed four lines which were attributed to the interaction of the unpaired electron with two nonequivalent protons. Senesi et al. (1977a) reported hyperfine splitting in the spectrum of fulvic acid following oxidation with H_2O_2.

The ESR spectrum of a simple free radical, 2,5-dimethylquinone, is shown in Figure 17. This spectrum has a total of 21 lines (seven triplets), and since each line is characterized by two parameters (position and width) there are a total of 42 bits of data from which to deduce information about this one relatively small molecule. In contrast, the ESR spectra of humic substances are characterized, in general, by only a single line, and one is then left with only two pieces of data from which to deduce information, functional or structural, concerning the nature of the complicated, multicomponent mixtures in humic substances. This, again, puts into perspective, the limitations imposed on various investigational methods when applied to humic substances.

Another type of information obtainable from ESR spectra is the spin concentration in the sample. The spin concentrations of humic acid range from 1.4×10^{17} to 1.2×10^{19} spins/g (Steelink and Tollin, 1967; Schnitzer, 1978). The spin concentrations of fulvic acids have been reported to range from 3×10^{17} to 1.3×10^{19} spins/g (Steelink, 1964; Schnitzer and Skinner, 1969). Spin concentrations for humin are not widely available in the literature, but in one instance range from 5.6×10^{17} to 1.7×10^{18} spins/g (Riffaldi and Schnitzer, 1972).

The reported spin concentrations correspond to one free radical per 1100 molecules (number average molecular weight of 951) for an untreated fulvic acid in the work of Schnitzer and Skinner (1969) and one free radical per 250 molecules of humic acid (1.4×10^{18} spins/g, with a molecular weight of 20,000) in the work of Steelink (1964). Consequently, the ESR spectrum is providing data on only a small fraction of the total molecules in the humic mixture, as pointed out by Riffaldi and Schnitzer (1972). Hayes et al. (1975) found that the free radical contents of humic and fulvic acids vary with the solvent used in the extraction, suggesting that the free radical in a humic substance may be an artifact of the extractive procedure. With the severity of these limitations in mind, any generalization concerning the functionality or structural nature of humic substances based on ESR data must be conservative.

In conclusion, the ESR spectra of humic substances contain relatively little data from which to deduce any detailed information concerning the nature of these materials. The ESR spectrum results from only a small fraction of the total number of molecules that comprise the humic or fulvic acid, further complicating any attempts to interpret the spectra on a microscopic basis.

CONCLUSIONS

On the basis of the information presented in this chapter one must conclude that the various spectroscopic methods are severely limited insofar as their applicability for determining functionality in humic substances is concerned. These limitations are not unique to spectroscopic methods, but apply, in general, to all techniques for studying humic substances. The difficulty in trying to interpret data on humic substances in molecular terms resides in the inherent nature of these materials. Humic substances are comprised of complicated mixtures of polyelectrolytes which have, to date, defied all attempts at fractionation into discrete components. While the multicomponent nature of humic substances is well established, and generally acknowledged in the literature, it is surprising how frequently the heterogeneity of humic materials is ignored in interpreting the data obtained by various experimental techniques.

Of the various methods evaluated, infrared spectroscopy is by far the most useful for determining functionality of humic substances. However, even this technique is severely limited compared to its usefulness when dealing with discrete compounds. It has been pointed out that the OH stretching band in the IR spectra of humic substance is *not* particularly broad, in fact, when compared to the OH stretching bands in the spectra of discrete compounds. Furthermore, in view of the emphasis given to the hydrogen bonding in humic substances it is surprising that the OH stretching band has not been shifted to lower wavenumbers. This point warrants further investigation in the future. There has been relatively little work carried out on the IR spectra of humic substances in the aqueous state which would allow the molecules to be observed in the equilibrium state, and this area also warrants further study. Some of the more recently developed techniques for exhaustively derivatizing humic substances have the potential for considerably enhancing the quality of the IR spectra obtained and for facilitating the interpretation of the adsorption bands in humic substances. If the fluorescence problem could be resolved, Raman spectroscopy would be particularly useful in studying the vibrational spectra of humic substances. However, this is a formidable problem and there is no indication that it will be solved in the very near future.

The ill-defined nature of the UV–visible and fluorescence spectra of humic substances mitigates against their use for providing detailed information concerning the functionality of humic substances. No direct information relating to the functionality of humic substances can be obtained from these methods and only inferential information, based, for example, on the variation of the spectra as a function of pH, can be acquired.

It is equally difficult to arrive at concrete conclusions concerning the functionality of humic substances based on their ESR spectra. When compared to the ESR spectra of discrete free radical molecules, the ESR spectra are seen to be exceedingly crude. In addition, the ESR spectrum results from only a very small fraction of the molecules present in the system, further complicating any interpretation. The ESR spectra of humic substances have been interpreted in terms of the presence of semiquinone moieties. While such interpretations are reasonable, and consistent with what is known about these substances, no definitive proof of the nature of the free radical entities in humic substances has yet been provided.

In conclusion, it should be restated that spectroscopic methods have many useful applications in humic substances research; however, these methods are extremely limited when applied to determining the functionality of humic substances. Upon surveying this subject, one is struck by the realization that no major development in the application of the various spectroscopic methods discussed here has been made in recent years. Perhaps the most important contribution of spectroscopy to humic substances research over the past decade has been the application of the various NMR techniques, as reviewed by Wershaw in Chapter 22.

CHAPTER TWENTY-TWO

Application of Nuclear Magnetic Resonance Spectroscopy for Determining Functionality in Humic Substances

ROBERT L. WERSHAW

ABSTRACT

A wide variety of chemical and spectroscopic techniques has been used to determine functionality in humic substances. Although nuclear magnetic resonance (NMR) spectroscopy has been used for a much shorter period of time than most other techniques for determining functional group concentrations, this technique has provided far more definitive information than all other methods combined. However, substantially more work must be done to obtain the quantitative data that are necessary for both structural elucidation and geochemical studies. In order to increase the accuracy of functional group concentration measurements, the effect of variations in nuclear Overhauser enhancement (NOE) and relaxation times must be evaluated. Preliminary results suggest that spectra of fractions isolated from humic substances should be better resolved and more readily interpreted than spectra of unfractionated samples.

INTRODUCTION

General Theory

Nuclear magnetic resonance (NMR) spectroscopy, like other spectroscopic techniques, is dependent on the interaction of electromagnetic radiation with nuclear, atomic, or molecular species. In general, the atomic spectrum of an element consists of a series of lines which correspond to energy-state transitions of the various orbital electrons of the element. If a spectrometer of sufficient resolution is used, it is possible to observe splitting of the principal lines of an atomic spectrum. This splitting is the so-called fine structure of the spectrum, and it led early spectroscopists to use four quantum numbers to completely describe the state of an electron in an atom. Careful examination of each line of the fine structure revealed that some of these lines could be resolved into two or more lines. Pauli was the first to attribute this splitting to a magnetic interaction between the nucleus of the atom and its moving electrons.

The discovery of this interaction led to the postulation that an atomic nucleus possesses a spin-angular momentum represented by the spin angular momentum vector $\mathbf{I}\hbar$, where $\mathbf{I}$ is the nuclear spin and $\hbar$ is Planck's constant, h, divided by 2π. It has been found experimentally that $\mathbf{I}$ is an odd integer multiple of $\frac{1}{2}$ for nuclei of odd atomic mass numbers (isotope number), zero for nuclei of even atomic mass numbers and even nuclear charges (atomic number), and an integer for nuclei of even atomic mass numbers and odd nuclear charges. The nuclei that we are concerned with here, ^{1}H, ^{13}C, and ^{19}F, have an $\mathbf{I}$ of $\frac{1}{2}$.

In an analogous fashion to classical mechanics, where a spinning charged body possesses a magnetic moment, a spinning nucleus possesses a magnetic moment, $\boldsymbol{\mu}_n$ given by the equation

$$\boldsymbol{\mu}_n = \gamma_n \hbar \mathbf{I}$$

where γ_n is a proportionality constant called the magnetogyric ratio, which is a constant for a given nucleus. If a group of spinning nuclei of a given species is placed in a magnetic field, $\mathbf{H}$, the magnetic moments of each of the nuclei will interact with the field in such a way that the total energy, E, of the system will be $-\gamma_n H I_z$, where I_z is the component of $\mathbf{I}$ in the direction of the field.

Quantum mechanics requires that the nuclear spin status represented by the nuclear spin quantum number, m_I, be quantized in such a way that m_I assumes one of the values in the set $I, (I - 1) \cdot \cdot \cdot -I$. Thus, for a spin of $\frac{1}{2}$, m_I can be either $+\frac{1}{2}$ or $-\frac{1}{2}$.

When an assembly of nuclei is placed in a magnetic field, those nuclei with a spin of $\frac{1}{2}$ will align themselves so that their spin magnetic moment vectors will be parallel to the field vector and those with spin of $-\frac{1}{2}$ will be antiparallel. Those nuclei with a spin of $-\frac{1}{2}$ will have slightly higher energy than those with a spin of $+\frac{1}{2}$. At thermal equilibrium, according to the Boltzmann distribution law, the number of atoms of spin $\frac{1}{2}$, $N_{1/2}$, divided by the number of atoms of spin $-\frac{1}{2}$, $N_{-1/2}$, is given by the equation

$$\frac{N_{1/2}}{N_{-1/2}} = e^{-\Delta E/kT}$$

where ΔE, the energy difference between the two states, is equal to $\gamma\hbar H$. From this relationship we see that the energy difference is a function of the magnetic field; the higher the field, the greater the energy difference.

Transitions between the two different energy states may be brought about by superimposing on the stationary magnetic field an oscillating electromagnetic field, the magnetic vector of which is perpendicular to the steady field **H**. The frequency of the oscillating field must satisfy the resonance condition:

$$\nu = \frac{\gamma_n H I_z}{\pi}$$

In a nuclear magnetic resonance experiment one places a sample in a uniform magnetic field and applies an oscillating magnetic field perpendicular to it. Either the steady field or the frequency of the oscillating field is varied until the resonance condition is met. At resonance, the nuclei will absorb energy from the oscillating field and undergo transition to the higher energy state. If absorption of energy is to continue some of the nuclei in the higher energy state must give up some of their energy as they undergo transition from the higher to the lower energy state. This loss of energy is called relaxation. If relaxation does not take place then eventually the populations in these two spin states will become equal and no more energy will be absorbed from the oscillating field (a condition called saturation). This requirement results from the fact that there is an equal probability the oscillating field will cause nuclear spin transitions from the higher to lower energy as from the lower state to the higher state. Clearly, in order to obtain information from the nuclear magnetic resonance experiment the systems cannot be in a state of saturation. Therefore, if thermal equilibrium is to be maintained as some nuclei absorb energy and are promoted to the higher quantum state an equal number of nuclei must lose energy and decay to the lower quantum state.

In general, there are two different types of relaxation encountered in NMR spectroscopy which are characterized by two different relaxation times, T_1 and T_2. The spin–lattice relaxation time, T_1, is characteristic of the relaxation of the component of the nuclear magnetization that is parallel to

H, while T_2, the transverse relaxation time, is characteristic of the relaxation of the component of the nuclear magnetization that is transverse to **H**. The spin–lattice relaxation is brought about by loss of energy from the excited nuclear spins to the surrounding molecules (molecular lattice). The transverse or spin–spin relaxation, on the other hand, is caused by interchange of energy between different nuclear spins.

Both relaxation times are functions of the thermal motion of the molecules, of chemical-exchange relaxations, and other chemical and physical interactions. Therefore, T_1 and T_2 measurements may be used to give a relative measure of these interactions in different systems. For example, they may be used to evaluate the change in the activity coefficient of counterions, such as sodium ions, brought about by the presence of humic or fulvic acid polyions in solution. A further discussion of this is beyond the scope of this chapter, but it is mentioned here to emphasize the versatility of NMR in the study of macromolecular interactions.

Chemical Shift

Up to now we have been discussing the nuclear magnetic resonance phenomenon in more or less isolated nuclei. Of more interest to the chemist is the magnetic resonance of nuclei in chemical compounds. When a nucleus that possesses a spin, such as a hydrogen (^1H) or carbon (^{13}C) atom, exists in a chemical compound, the spinning nucleus is partially shielded by the surrounding atom from the external magnetic field. Therefore, the effective magnetic field impinging upon the nuclear spin is altered and this in turn requires that either the stationary magnetic field or the frequency of the oscillating field be changed in order to obtain resonance. If we suppose that the stationary field is fixed as is the case in all Fourier transform spectrometers, then the frequency of the oscillating field is the parameter that will be altered in order to achieve resonance.

The shielding of a proton, for example, in an organic compound is a function of the chemical environment in the vicinity of the proton. Thus, the resonance frequencies of the various protons in the chemical compound give information on the chemical structure of the compound. And, in fact, ^1H and ^{13}C NMR spectroscopy are some of the most powerful tools for the elucidation of the chemical structure of organic compounds.

The NMR frequency of a given nucleus is generally measured relative to a suitable standard. This type of measurement gives rise to the so-called chemical shift, δ, defined by the equation

$$\delta = \frac{(\nu_{\text{sample}} - \nu_{\text{reference}})}{\nu_{\text{oscillator}}} \times 10^6$$

This difference measurement is used in order to obtain the best precision possible.

NMR Spectroscopy on Liquid Samples

The nuclear magnetic resonance spectra that are obtained for organic compounds in solution generally consist of very sharp, well-defined lines. The extreme sharpness of these lines allows one to detect very small differences in the magnetic environments of nuclei.

The two most important types of nuclear magnetic interactions that take place in different magnetic environments are (1) the nuclear Zeeman interaction, which gives rise to the chemical shift, and (2) the nuclear spin–spin coupling. The magnetic field at a nucleus consists of two components: (1) the externally applied stationary field, **H**, altered by the nuclear shielding, and (2) the magnetic field induced by the other spinning nuclei in the molecule (spin–spin coupling). As we have pointed out previously, the first term gives rise to the chemical shift. The second term leads to the spin–spin splitting of the chemically shifted lines.

There are several different spin–spin coupling mechanisms. The strongest one is the dipole–dipole coupling, caused by the direct interaction of the magnetic moments of spinning nuclei that are close together. This coupling can cause substantial line broadening, but in liquids the motion of molecules averages out the dipolar effect so that the net dipolar splitting is zero. The most important mechanism for spin–spin coupling in liquid systems is caused by internuclear coupling via the bonding electrons. This coupling is much weaker than dipole–dipole coupling and leads to very fine line splitting; however, this splitting is detectable because of the high resolution inherent in NMR measurements in liquids.

NMR Spectroscopy on Solid Samples

Recently there has been a great deal of interest in obtaining ^{13}C NMR spectra of solids. This interest has developed because there are a large number of organic substances, such as coal, kerogen, and soil humus, which are not readily soluble in organic solvents. In the past it has not been possible to obtain high-resolution ^{13}C NMR spectra of solids, but some relatively recent advances have greatly improved the situation.

NMR spectra of solids in general consist of much broader lines than those of liquid samples. This situation is due to anisotropic interactions in solids. The major anisotropic interaction, as we have pointed out above, is dipole–dipole coupling. The line broadening of ^{13}C solid-state spectra due to dipole–dipole coupling can be eliminated to a large extent by high-power proton decoupling. Proton decoupling is accomplished by irradiating the protons at their resonant frequencies so that they become activated and are no longer coupled to the ^{13}C nuclei. However, this does not eliminate broadening due to chemical-shift anisotrophy. This may be eliminated by the so-called magic-angle spinning technique (Andrew, 1971). Enhanced sensitivity is also accomplished by cross polarization.

By combining all of the above techniques into the so-called cross-polarization magic-angle spinning experiment (CP/MAS), relatively high-resolution solid-state NMR spectra can be obtained. It should be pointed out, however, that the best solid-state spectra are still of substantially lower resolution than routine liquid-state spectra, and therefore additional information not present in the solid-state spectra can be obtained from liquid spectra.

Fourier Transform NMR

The success of an NMR experiment is dependent on obtaining an adequate signal at the resonant frequency of nuclei of interest. The signal intensity, if we ignore line-broadening effects, is a function of two factors: (1) the nuclear moment (μ) of the nuclear species of interest, and (2) the abundance of the species in the sample. The proton, which has the highest nuclear magnetic moment of any nucleus, is normally used as the standard to which the nuclear magnetic moments of all other nuclei are compared. In general, this comparison is calculated as a relative sensitivity which is the ratio of the magnetic moment of the nucleus of interest to that of the proton. In the case of ^{13}C this relative sensitivity is 0.016, and for ^{19}F it is 0.83.

The low relative sensitivity of ^{13}C is coupled with a low natural abundance of ^{13}C of 1.1%, so that the overall effect is to greatly reduce the signal intensity that is obtained in natural abundance ^{13}C NMR measurements compared with those obtained in proton NMR experiments.

Adequate proton NMR spectra can be obtained by continuous wave measurements; however, routine ^{13}C spectra could not be obtained until the development of Fourier transform spectrometers which allow one to accumulate a large number of spectra in a relatively short period of time and to average these spectra to increase the signal-to-noise ratio of the measurements.

The first NMR spectrometers developed were continuous-wave (CW) instruments and they are still in use for proton and fluorine NMR spectroscopy. In these instruments the irradiation frequency is fixed and the magnetic field strength of the magnet is slowly and continuously changed. When the correct magnetic field for the fixed frequency is reached for a proton in a particular chemical environment, then an absorption peak appears in the recorder of the instrument. The area of this absorption peak is a function of the number of protons in the sample that are in this particular chemical environment.

The inherent low sensitivity of ^{13}C NMR requires that the results of at least a few hundred, and normally several thousand, spectra be averaged in order to be able to detect the absorption peaks above the noise in a normal sample. Since a single continuous-wave spectrum takes between 100 and 500 seconds, simple signal averaging in order to obtain satisfactory signal-to-noise ratios is generally prohibitive.

Greatly increased sensitivity is obtained in the Fourier transform experiment because all absorption frequencies in a fixed magnetic field are excited at the same time by a radiation pulse of short duration. This pulse is generated by rapidly turning on and off a signal of a discrete frequency. The rapid turning on and off of the pulse causes it to be no longer a single frequency but to be composed effectively of a range of frequencies with a band width of approximately $1/t$, where t is the duration of the pulse.

The NMR signal obtained from the resonating nuclei after the sample has been irradiated by the pulse is the so-called free-induction decay curve. This curve consists of peaks and valleys. The spectrometer samples the free-induction decay curve at set time intervals and records the data, which are in a time domain. NMR spectra, however, are normally given in terms of frequency and therefore the spectrum must be transformed by use of the Fourier transform pairs:

$$f(t) = \int_{-\infty}^{\infty} g(\omega)e^{i\omega t}d\omega$$

$$g(\omega) = \frac{1}{2\pi}\int_{-\infty}^{\infty} f(t)e^{-i\omega t}dt$$

where $f(t)$ is the function measured in the time domain and $g(\omega)$ is the corresponding function in the frequency domain. Very efficient algorithms have been written so that this transform can be carried out rapidly by a digital computer.

Pulse NMR techniques allow one to not only measure the NMR spectrum of the sample but also to evaluate the relaxation times of the nuclei present. The relaxation information is not readily obtainable from the continuous-wave experiment, and, therefore, in addition to being much faster, the pulse experiment also yields more information than the continuous-wave experiment.

NMR SPECTROSCOPY OF UNDERIVATIZED HUMIC SUBSTANCES

Proton NMR

A representative proton NMR spectrum of a soil fulvic acid is given in Figure 1. The various spectral regions of interest are indicated in the figure. These spectral regions were recognized by the early workers in the field and the interpretations they have made of these data, although differing in some details, have not changed in later studies.

Oka et al. (1969), made the first proton NMR measurements on an underivatized humic acid. They studied three different peat humic acids extracted with a solution consisting of 1% sodium hydroxide, 3% sodium ace-

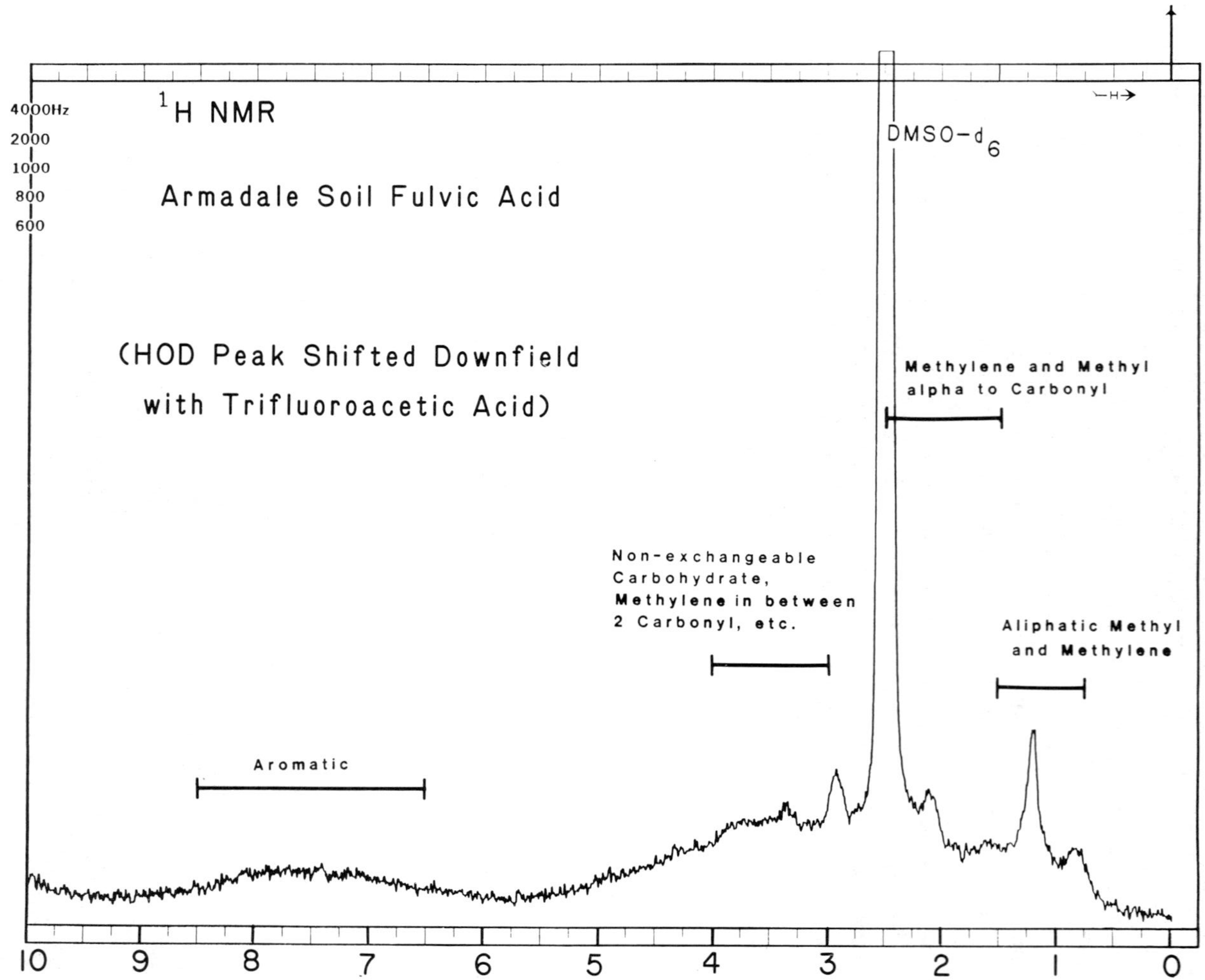

FIGURE 1. Representative proton NMR spectrum of a fulvic acid.

tate, and 1.8% sodium pyrophosphate. The spectra of the three humic acids were very similar, each spectrum consisting of the series of broad bands given in Table 1.

In general the assignments in Table 1 are in agreement with later studies except for the region between 4.0 and 5.5 ppm where Oka et al. (1969) and Lakatos and Meisel (1978) assigned it to lactone protons. However, other workers have assigned this to exchangeable protons [see discussion of work by Ruggiero and co-workers (1979c), below]. We have found in our own studies that this peak disappears upon methylation of hydroxyl and carboxyl groups.

Lüdemann et al. (1973) and Lentz et al. (1977) used proton NMR to study a number of soil humic acid fractions dissolved in D_2O. The spectra had broad bands in the following regions: 1–2.5 ppm, 3.8 ppm, and 6.5–8.5 ppm. An intense DOH peak was present at about 5.2 ppm. The aliphatic region from 1 to 2.5 ppm was composed of several broad bands; a single methoxyl band was present at 3.8 ppm and a single very broad band was present in the aromatic region. They calculated the percentage of aromatic protons to the total protons by integration under the peaks. The highest percentage of aromatic protons was 35% and the lowest was 19%.

Stuermer and Payne (1976) compared the ^{1}H and ^{13}C NMR spectra of a Sargasso seawater fulvic acid to terrestrial fulvic acids. The proton spectrum of the Sargasso Sea fulvic acid that they isolated by adsorption on XAD-2 resin had broad bands in the following regions (their assignment of the protons is shown in parentheses): 1.0–1.7 ppm (aliphatic protons), 1.7–2.5 ppm (protons on carbon atoms adjacent to functional groups such as carbonyl groups), and 7.3–7.9 ppm (aromatic protons). The positions of these bands were measured relative to tetramethylsilane (TMS) as 0. The relative areas of these bands were 15 : 10 : 1. Stuermer and Payne (1976) pointed out that

TABLE 1. Major Proton Resonances of Humic Materials[a]

δ (relative to TMS as 0)	Assignment
13.0 ppm	Carboxylic acid protons
10.0 ppm	Hydroxyl protons
6.0–7.5 ppm	Aromatic protons
4.0–5.5 ppm	Lactone protons
3.7 ppm	Methoxyl protons
2.6 ppm	Aliphatic protons attached C atom α to a benzene ring[b]
1.3 ppm	Aliphatic protons β to a benzene ring[b]
0.9 ppm	Aliphatic protons γ to a benzene ring[b]

[a] *Spectra measured in deuterodimethylsulfoxide.*

[b] $—C_{\alpha}H_2—C_{\beta}H_2—C_{\gamma}H_2—$

the relative concentration of aromatic protons in this sample was much lower than that detected by others in terrestrial (soil) fulvic acids. Their ^{13}C spectra gave a similar result. They pointed out that the low concentration of aromatic protons "may reflect the lack of abundant aromatic precursors in the marine environment."

The spectrum that Wilson et al. (1978) obtained from a Wakanui silt loam soil humic material showed a much stronger aromatic band than that obtained from the Sargasso Sea fulvic acid. These authors estimate from the ^{1}H NMR spectral data that 65% of carbon in this humic material was in aromatic or carboxylic acid groups. In addition to a broad band in the aromatic region between 6 and 8.5 ppm the Wakanui spectrum had well-defined peaks at 0.8, 1.2, 1.9, 2.1, 2.5, 3.3, 3.6, and 3.9 ppm. The band at 0.8 ppm was one of the stronger and more distinct bands in the spectrum; Wilson et al. (1978) attributed it to CH_2 groups γ or further from an aromatic ring. This band and the sharp bands near it indicate that alkyl chains are present in this material. They point out that polymethylene is present in their sample and that its presence is probably due to its low biodegradability.

Several previous studies also indicated that alkyl chains are present both in aquatic and soil fulvic materials (Wilson and Goh, 1977a,b; Grant, 1977). The Wilson and Goh papers which report on ^{13}C NMR spectra will be dealt with in the section of this chapter dealing with ^{13}C. Grant developed an exhaustive extraction procedure which allowed him to extract "essentially all the organic matter from the soil." This extraction procedure consists of alternate extraction with acetone and either formic acid or hydrochloric acid. Some of the fractions that Grant (1977) isolated in this way gave NMR spectra that were very similar to what one would expect from polymethylene chains. Grant states that up to 30% of the organic matter from the soils he extracted is composed of polymethylene chains, and that his fraction 1 which is extracted with acetone may contain even higher amounts. Sciacovelli et al. (1977) also detected polymethylene groups in different solvent extracts of soils. In both studies, however, it is difficult to relate the results to studies of humic and fulvic acids because the extraction procedures used are much different than those that are normally used for humic and fulvic acids.

Ruggiero and co-workers (Ruggiero et al., 1978, 1979a,b, 1980, 1981) have conducted a series of ^{1}H and ^{13}C NMR studies on underivatized, derivatized, and degraded humic and fulvic acids. Ruggiero's group was the first to make a concerted effort to eliminate the strong exchangeable proton absorption band whose chemical shift is a function of relative concentrations of carboxylic acid and alcohol hydroxyl groups in the sample. The exchangeable proton band generally obscures the region between approximately 4 and 5 ppm but it can also distort the aromatic region.

Ruggiero's group used two different techniques for elimination of the obscuring of the important regions of the spectra by the exchangeable proton band: (1) exhaustive drying of the sample solution in deuterodimethylsulfoxide (DMSO) with molecular sieves and (2) shifting of the exchangeable proton peak by adding acid to the sample.

Wilson et al. (1983) used another technique for elimination of the band due to exchangeable protons. They saturated these protons by irradiating them at their resonant frequency. This allowed them to obtain greatly improved definition in other regions of the spectra. Unfortunately, however, this irradiation also eliminates other bands in the same region as the exchangeable protons.

Harvey et al. (1983) have proposed that marine fulvic and humic acids are cross-linked triglycerides. They relied heavily on proton NMR data in reaching this conclusion. However, they made no attempt to eliminate the exchangeable proton bands in their spectra, which means that any bands in the region obscured by the exchangeable proton band have not been taken into account in their proposed structures. Their spectra either entirely lacked evidence of aromatic protons or had a relatively weak band in the aromatic region. From this they concluded that aromatic structures were relatively unimportant in marine fulvic and humic acids. However their data do not rule out the presence of highly substituted phenolic structures in marine fulvic and humic acids. These structures could well have very few aromatic protons.

The use of the molecular sieve technique allowed Ruggiero's group to eliminate the effect of exchangeable protons from the aromatic region and thereby get an accurate measure of aromatic protons in the sample. In all samples they found appreciable concentrations of aromatic protons. They have concluded from this work that "aromatic structures are significant constituents of humic substances. . . ." They further state that proton NMR, because of its greater sensitivity, is better than ^{13}C NMR for estimating the aromaticity of humic substances. In this regard they have entered into a controversy with Wilson and Goh (see Ruggiero et al., 1981, and Wilson and Goh, 1981) over the interpretation of the results of a ^{13}C NMR study which Wilson and Goh (1977a) published. Ruggiero et al. (1981), apparently felt that Wilson and Goh did not give proper weight to the aromatic structures in the samples they examined by ^{13}C NMR. It appears that the difference in opinion arises from the fact that Ruggiero and his group believe that their results on humic material from Italian soils and those of some published papers indicate that most humic substances have significant concentrations of aromatic structures, while Wilson and Goh feel that there is a wide range of aromaticities and that generalizations such as those made by Ruggiero and co-workers are unwarranted. As we shall see, the evidence seems to bear out the position of Wilson and Goh.

Ruggiero et al. (1978, 1981) have also examined the ^{1}H NMR spectra of different molecular weight fractions of humic and fulvic acids isolated by adsorption and gel permeation chromatography. They found that there are differences in the aliphatic and aromatic regions of the spectrum between the different fractions. Their data are fragmentary and substantially more work must be done before generalizations can be made. However they have shown that differences can be detected by NMR and that it should be a sensitive tool for studying the geochemical processes that give rise to various humic and fulvic fractions.

The variability of the ^{1}H and ^{13}C NMR spectra of humic acids from different environments is graphically shown by the work of Dereppe et al. (1980). They extracted humic acid from three marine sediments, a podsol soil, and a peat moss. One of the marine sediments that came from the Norway sea was derived from plants of low lignin content. Another of the marine sediments was mostly of planktonic origin and the third marine sediment was both marine and terrestrial in origin. The three marine sediment humic acids gave very similar spectra. Each had broad bands in the aliphatic region and in the region between 2.8 and 4.1 ppm which the authors assigned to protons alpha to aromatic rings or carboxyl groups. In contrast to the podsol and peat moss humic acids, aromatic bands were absent from these three samples. The peat humic acid had a much larger aromatic band than the podsol humic acid.

The peat humic acid sample that Hatcher et al. (1980b) examined by proton NMR also had a relatively high aromaticity (23%), determined by integration of peak areas; however, aromaticities of the podsol soil humic acids ranged as high as 35%. A higher aromaticity of 33% was obtained for the peat sample when it was determined by ^{13}C NMR. This difference between the aromaticities measured by proton and ^{13}C NMR is what is to be expected if some of the aromatic protons have been replaced by other elements or functional groups. The high probability that at least some of the aromatic rings in humic materials are highly substituted means that more accurate aromaticities are probably obtained from ^{13}C NMR than from proton NMR. Hatcher et al. (1980b) also measured the ^{1}H NMR spectra of several sediment humic acid samples from the New York Bight and one mangrove lake humic acid. These were more aliphatic and less aromatic than the soil and peat humic acids.

Hatcher et al. (1980b) pointed out that there is a high percentage of methyl protons in all their samples as evidenced by a strong band at 0.9 ppm. This indicates that the aliphatic groups are highly branched. They interpreted the peaks at 1.3 ppm as methylene protons and those at 1.6 ppm as methine protons. Some of their samples showed fine structure in the aromatic region with peaks at 6.5, 6.9, and 7.2 ppm. They pointed out that the bands at 6.9 and 6.5 ppm may indicate single and multiple substitutions of electron-donating groups on aromatic rings.

The New York Bight humic acids which Hatcher et al. (1980c) examined have a markedly different origin than any of the humic acids discussed up to now because they were isolated from sediment derived from sewage sludge. In spite of this, their NMR spectra were generally similar to other humic acids in that aliphatic bands were at least as strong as other bands in the spectra. This was not the case, however, for the spectrum of a fulvic acid which Sposito et al. (1978) isolated from an anaerobically digested sewage sludge. The most prominent peak in the ^1H NMR spectrum of this sample is a broad band centered at 3.8 ppm. The area under this band is at least three times that of any of the other bands in the spectrum. The authors have assigned this band to protons in polysaccharide decomposition products and it is not unreasonable considering its source.

One must be cautious in interpreting data on carbohydrates, hydroxyl acids, and uronic acids in humic substances. As Thurman and Malcolm (1983) have pointed out, the amount of these materials in a fulvic acid is a function of the way that it was isolated. They have shown that nonassociated carbohydrates, uronic acids, and hydroxyl acids may be separated from fulvic acid by adsorption chromatography on XAD resins. For example, a prairie soil fulvic acid contained 20% carbohydrate before XAD adsorption chromatography and only 5% after chromatography. In the fulvic acid isolated by Sposito et al. (1978) adsorption chromatography was not used in the purification process and therefore some of the carbohydrate that they report may not be an integral part of the fulvic acid structure.

Saito and Hayano (1981) have also interpreted the presence of a band between 3.3 and 4.6 ppm to indicate that there are polysaccharide ether structures in some of their samples. They found that this band was stronger in fulvic acids from marine sediments than the corresponding humic acids. The marine sediment fulvic acids were higher in oxygen than marine sediment humic acids. Aldrich humic, which is presumably terrestrial in origin, has a still lower oxygen content but does not have a band in this region. These data led Saito and Hayano to conclude that their marine sediment fulvic acids have a "polysaccharide character."

Hatcher et al. (1981) pointed out that the aliphatic region of terrestrial humic acids is very similar to that of marine humic acids and that the only difference is the presence of aromatic bands in the terrestrial humic acid spectra. In previous work, Hatcher (1980) and Hatcher et al. (1980b) concluded from the H/C ratio of 1.5 and presence of a strong terminal methyl band at 0.9 ppm that marine humic acids have highly branched and cross-linked paraffinic carbon atoms. These structures appear to arise from algal and microbial lipids. The similarity in the aliphatic region in terrestrial humic acids suggests that soil microbial lipids may be the source of the aliphatic structures in terrestrial humic acids.

Carbon-13 NMR

Liquid State Theory

Carbon-13 NMR spectroscopy of humic materials presents some particular problems that are not encountered in proton spectroscopy (see Wilson, 1981). The most important of these problems are (1) the low sensitivity of carbon in relation to protons, (2) the low abundance of ^{13}C, (3) the highly variable relaxation times of carbon, and (4) variable nuclear Overhauser enhancement (NOE). We have already dealt with the first two problems in the Introduction. It is appropriate here to discuss the two other problems which are closely related to each other.

The discussion of variable T_1 values will be limited to the dipolar relaxation component of T_1, T_1^D, because dipolar relaxation is generally the major relaxation mechanism for carbon atoms attached to protons. The dipolar relaxation time T_1^D for a given carbon atom may be calculated to first approximation by the equation

$$\frac{1}{T_1^D} = h\gamma_C^2\gamma_H^2 \sum_i r_i^{-6}\,\tau_c,$$

where γ_C and γ_H are the magnetogyric ratios of ^{13}C and ^1H, respectively, r_i is the length of the C—H bond of the ith proton attached to the carbon atom, and τ_c is the correlation time of the C—H bond. This correlation time is a measure of how rapidly the bond is undergoing reorientation in the magnetic field. The relaxation time, T_1^D, is therefore an inverse function of the number of protons attached to a carbon atom and of the correlation time. This relationship shows that there will be a wide range of T_1^D values in organic compounds. As Wilson (1981) has pointed out, T_1 values for ^{13}C nuclei in organic compounds can vary from less than 1 second to several minutes. This wide variation is particularly troublesome in Fourier transform (FT) NMR studies where the atoms are excited by a short pulse of electromagnetic radiation followed by a period in which the nuclei are allowed to relax to the ground state. Ideally, all nuclei should relax to the ground state before the next pulse. If this condition is not met for all the carbon atoms in the molecule of interest then the absorption lines for those atoms that have not completely relaxed will be diminished in size compared to the more rapidly relaxing atoms.

The ^{13}C NMR spectra of organic molecules are normally strongly split by proton coupling. This splitting may be eliminated by irradiating the sample with a strong radio-frequency (rf) field which is tuned to the proton resonant frequency. In order to decouple all protons in the sample it is necessary to irradiate all of them at their respective resonant frequencies. This may be accomplished by tuning the proton rf field to the center of the proton region and modulating the field with an audio "noise" signal with a band pass of about 100 Hz. This irradiation is equivalent to simultaneously irradiating all protons in the sample at their resonant frequencies. When this is done it is found that not only do the carbon multiplets collapse, but that the increase in intensities of the resulting peaks is generally much greater than one would expect from contributions of these split peaks. This additional increase in intensity is called the nuclear Overhauser enhancement (NOE). The maximum signal increase due to the NOE of 2.987 is only obtained when the carbon atoms relax exclusively by dipolar relaxation. If any other mechanism contributes to the relaxation there is a concomitant decrease in the NOE. In the extreme case of a carbon atom that is not attached to a proton there is no NOE. These differences in relaxation times and NOE values may be overcome in some cases by the use of an appropriate paramagnetic relaxation reagent.

Solid-State Theory

Up to this point in the discussion we have been concerned mainly with spectra measured in the liquid state; however, ^{13}C NMR spectra are now determined routinely for solid samples as well as for liquid samples. However, the solid-state NMR experiment poses some particular instrumental problems which must be solved in order to obtain satisfactory results. We shall discuss these only very briefly here, before entering into a detailed comparison of the results of ^{13}C NMR measurements of humic materials in both the solution and solid states. For a more complete discussion of solid-state NMR theory the reader is referred to Schaefer and Stejskal (1979), Palmer and Maciel (1982), and the references contained therein.

The linewidth that one would obtain if one tried to measure ^{13}C NMR spectra of solid samples in the same way as liquid samples would be approximately 20,000 Hz. This linewidth is approximately equivalent to the total ^{13}C chemical shift range for most organic compounds. Therefore it is apparent that it is not possible to obtain useful chemical shift information in this way.

This extreme line broadening is due to two effects: (1) static dipolar interactions of the ^{13}C atoms with adjacent protons and (2) chemical shift anisotropy. The value of the dipolar interaction tensor of two interacting nuclei is a function of the orientation of the vector connecting the two nuclei with the applied static magnetic field. This dipolar coupling interaction causes a splitting of the ^{13}C NMR bands. In solid samples where the nuclei are fixed in space and the ^{13}C–^{1}H vectors have all possible orientations, the accompanying splitting of a given ^{13}C chemical shift causes extreme line broadening. In liquid samples this does not occur because the rapid movement of the molecules results in an averaging of the dipolar interactions to zero (see Carrington and McLachlan, 1967).

The dipolar coupling in the solid state can be eliminated by irradiating the protons with a strong signal at their resonant frequency. This double resonance experiment is similar to the double resonance technique used to remove similar ^{13}C–^{1}H coupling in liquid samples except that much higher powers are required.

The chemical shift anisotropy broadening arises from the fact that the chemical shift of a given ^{13}C atom in a molecule will vary to some extent as a function of the orientation of the molecule in the magnetic field. In liquid samples this variation is averaged out to a single isotropic value; however, in amorphous solids and powders the true linewidth of a given chemical shift will be markedly increased by the anisotropy. This effect can be eliminated by spinning the sample at the so-called magic angle (54.74°) as first suggested by Lowe (1959).

A very brief description of the theory of the cross-polarization (CP) experiment will be given here; for a complete discussion the reader is referred to Pines et al. (1973). The cross-polarization experiment is dependent on the Hartmann–Hahn condition: $\gamma_I \mathbf{H}_I = \gamma_S \mathbf{H}_S$. In this equation γ_I is the magnetogyric ratio of the more abundant spins (^{1}H), γ_S is the magnetogyric ratio of the less abundant spins (^{13}C), $\mathbf{H}_I$ is the rf magnetic field at the resonant frequency of the I nuclei and $\mathbf{H}_S$ is the field at the resonant frequency of the S nuclei. When the condition is obtained the energy level difference for a nuclear spin transition is the same for the S and I nuclei and energy can be freely transferred between the two systems. This transfer of energy is called cross-polarization and it takes place for a period of time called the cross-polarization contact time which is dependent on experimental conditions. After transfer of energy, the number of spins in the less abundant species approaches that of the more abundant species and therefore the sensitivity of the less abundant species is enhanced.

The cross-polarization time, T_{CH}, for a given carbon atom is a function of the distance between the carbon atom and adjacent protons. In order that quantitative results be obtained from the cross-polarization experiment, these T_{CH} values for all carbon atoms in the sample must be less than the experimental contact time, which in turn must be less than all T_1 values, that is, the amount of time in the experiment when the Hartmann–Hahn condition exists must be less than the T_1 values of all ^{13}C and ^{1}H atoms (see Palmer and Maciel, 1982, for a more complete discussion). If these conditions do not apply, then the integrated band areas will not be representative of the concentration of the various carbon atoms present in the sample.

Our discussion of the ^{13}C NMR spectra of humic materials may be conveniently divided in three parts: (1) the liquid-state NMR spectra of underivatized samples, (2) the solid-state spectra of underivatized materials, and (3) the liquid-state spectra of derivatized samples.

Liquid-State Results

Wilson (1981) has reviewed much of the literature of the ^{13}C NMR spectroscopy of soil organic materials. He has shown that the major ^{13}C resonances of humic substances are as listed in Table 2. Although these data were obtained mainly from soil humic substances, the same resonances are encountered in aquatic humic substances, the only difference being that relative intensities of the bands are generally different.

The basic problem in the interpretation of ^{13}C NMR spectra of humic substances is that for quantitation, as we have pointed out above, the integrated area under a given band in a ^{13}C NMR spectrum is not only a function of the number of carbon atoms resonating at that frequency, but is also a

TABLE 2. Major ^{13}C Resonances of Humic Substances

Chemical Shift (ppm)	Assignment
190–200	Carbons in aldehydes, ketones, and C=S groups
160–190	Carboxyl carbons in carboxyl, ester and amide groups
110–160	Aromatic carbons and olefinic carbons
90–110	Acetal carbons
50–70	CO carbons—alcohols esters, ethers, carbohydrates, amines
0–50	Alkyl carbons

function of the relaxation time of the carbon atoms and the NOE of the atoms. Newman et al. (1980) were the first to address this problem. They performed a series of progressive saturation experiments on a soil humic acid in solution to evaluate the effect of relaxation time on the peak areas (Freeman and Hill, 1971), and gated decoupling experiments to eliminate the NOE. In this gated coupling experiment the proton decoupler is turned on only during the time that the carbon spectrum is being acquired, but is shut off during the rest of the time. In this way the proton splitting of the carbon bands is eliminated but the NOE is not obtained.

In the normal progressive saturation experiment a series of 90° pulses is applied to the sample. Each pulse is separated from the pulse before it and after it by a time delay τ. The steady-state magnetization (NMR signal) of a given carbon atom is measured. This pulse sequence is repeated for a series of different τ values, where all τ values are less than 5 times the T_1 value of the particular carbon atom. Under these conditions the carbon atom will not have time to completely relax before the next 90° pulse occurs. After a short period the magnetization will reach a steady state, M_z, which will be a function of τ. It can be shown that a plot of ln $(M_z - M_z^0)$ versus τ yields a straight line, the slope of which is $-1/T_1$. In this equation M_z^0 is the equilibrium magnetization, that is, the magnetization of the fully relaxed carbon atoms.

Newman et al. (1980) have modified this experiment and have plotted the integrated intensities of all the different types of carbon atoms in the sample against τ in an attempt to find the optimum pulse spacing. They found, not surprisingly, that their plot of total magnetization of all the carbon atoms in the sample versus τ is not exponential as one would expect for a single type of carbon atom. Under these circumstances they tried to choose the τ value that would yield the best-resolved spectrum. They also found that the aromatic signal accounted for 28% of the total integrated area with a τ of 36 msec and 25% for a τ of 1.5 sec. They pointed out that this difference is within their experimental error. They concluded that these data suggest that the relaxation curves for the various carbon atoms in the sample are similar and that long pulse delays are not necessary to obtain at least an estimate of the relative abundances of the various carbon atoms in the sample.

The gated dicoupling experiment performed by Newman et al. (1980) on a New Zealand soil humic acid showed that the amount of nuclear Overhauser enhancement for all resonances except that attributed to carboxyl carbon is the same within experimental error. The authors state the results of the progressive saturation and gated decoupling experiments "suggest" that solution ^{13}C NMR spectroscopy may be used to compare the aromaticities of humic acids, that is, the fraction of the total carbon of the sample that is aromatic. The authors reached this conclusion on the basis of results from only one sample, and even in that sample a substantial error is introduced into their calculations by the fact that the NOE for the carboxyl carbons is different from that of the other carbons.

Solid-State Results

Hatcher et al. (1983a) have recently written a comprehensive review of the application of solid-state ^{13}C NMR to the analysis of sedimentary humic substances. In this review, Hatcher et al. (1983a) state that the cross-polarization magic angle spinning techniques (CP/MAS) provide a quantitative measure of the aromatic, paraffinic, carboxylic acid, and ether groups in humic and fulvic acids. In general, Hatcher and other workers in the field have calculated the relative concentrations of the various groups by integrating the areas of the corresponding peaks in an NMR spectrum. Examination of the spectra in the paper by Hatcher et al. (1983a) indicates generally that the peaks are poorly resolved and that there is substantial overlap of some peaks. Under these circumstances one must estimate the peak shapes in regions of overlap. For this reason Hatcher and others estimate that the relative errors in the peak areas are between 5 and 10%. Although this is probably a reasonable estimate for many of the spectra, the peaks in some of the spectra overlap so much that the errors are probably substantially more.

Hatcher et al. (1983a) discuss in some detail other errors that can arise from (1) incomplete relaxation of the spins of the carbon atoms during the experiment, (2) isolation of carbon atoms from protons so that incomplete transfer of polarization takes place between the protons to the ^{13}C atoms (see Alemany et al., 1983), and (3) unequal distribution of free radicals in the sample.

A reduction in intensities can result from incomplete cross-polarization. Hatcher et al. (1983a) have results from a number of experiments on coals to show that these effects are probably not important in humic substances; however, the transfer value of the coal data is questionable because of the marked difference in chemical structure between coal and humic materials. They state that in humic substances there is less chance of intensity distortion due to the isolation of carbon atoms from protons since humic substances are less aromatic than coals. However, one could also conclude that since humic substances have a larger variety of structural elements than coals there is a greater chance of differences in relaxation times and cross-polarization efficiencies between different ^{13}C atoms in humic substances than in coals.

Hatcher et al. (1983a) obtained potentially more meaningful results from experiments in which they varied the contact times and repetition rates for a number of humic and fulvic acid samples. Unfortunately, they do not present detailed results of these experiments and therefore it is difficult to evaluate them, although they state that a contact time of 1 msec and a repetition rate of 1.5 sec gave quantitative results. They also compared liquid-state and solid-state spectra and state that the results are comparable within their limits of error. However, this does not prove that they are obtaining quantitative results because some of the same effects could be distorting both the liquid- and solid-state spectra in the same way. For example, if the relaxation time for a given functional group in a sample is markedly different from those of the other functional groups in the sample, then the peak areas of this functional group in both the liquid- and solid-state spectra will not be representative of its concentration.

Another problem in comparing CP/MAS spectra to liquid-state spectra is that there may be a difference between the chemical shift of a given group in the liquid state and that in the solid state. This is particularly pronounced in the case of β-ketones where Imashiro et al. (1982) found that the chemical shift of the carbonyl and enol carbons in a CP/MAS spectrum may be as much as 20 ppm downfield from the corresponding shift measured in DMSO solution. They attributed this to strong intermolecular hydrogen bonding in the solid.

In the most comprehensive study to date, Hatcher et al. (1983a) have attempted to measure the functional group concentrations in humic acids, fulvic acids, and humins from a number of different freshwater and marine sediments, soils, and plants. They found that the aromaticities of humic acids and humins were between about 20 and 70% with the aromaticities of the corresponding fulvic acids generally being lower. They suggest that high aromaticity indicates vascular plant origin. Aliphatic structures are major components of most humic substances and predominate in humic substances from submerged sediments such as peats, poorly drained soils, algal sapropels, and marine sediments. Hatcher and his co-workers assume that these aliphatic structures indicate contributions from algae and other microorganisms to the humic material. Gillam and Wilson (1983) have come to a similar conclusion for dissolved marine humic substances. They found that the extractable material from a marine diatom culture (*Phaeodactylum tricornutum*) gives a ^{13}C NMR spectrum similar to that of dissolved marine humic material. They point out that this suggests that marine diatoms are important contributors to marine humic substances.

Both Hatcher et al. (1983a) and Preston and Ripmeester (1982) found that fulvic acids are mainly polysaccharide in structure. Hatcher et al. (1983a) points out that a good indicator of the presence of polysaccharide is the NMR peak at 105 ppm which they assigned to anomeric carbons in polysaccharides. As we have pointed out previously, care must be exercised in interpreting evidence of polysaccharides in fulvic acids unless precautions have been taken to separate associated from unassociated polysaccharides. Hatcher et al. (1983a) used ultrafiltration through an Amicon PM/O* membrane (10,000 daltons exclusion) for some of their samples to separate the fulvic acids from low-molecular-weight polysaccharides. Preston and Ripmeester (1982) did not attempt to separate the fulvic acids from the unassociated polysaccharides. The separation technique used by Hatcher et al. (1983a) is a molecular size fractionation and as such will not necessarily separate polysaccharides from what is traditionally called fulvic acid. Preston and Ripmeester (1982) found that solid-state spectra of acid hydrolysis residues of organic soils are much more aromatic than the original soils or humic substances isolated from the soils. This suggests that hydrolysis of soil removes amino acids, proteins, carbohydrates, and low-molecular-weight phenols. Preston and Ripmeester are generally more cautious than Hatcher and co-workers in using peak areas in NMR spectra as a measure of functional group concentration. They point out that aromatic peaks in both liquid- and solid-state spectra may be reduced in intensity by line broadening caused by coordination of paramagnetic ions to aromatic or phenolic structures.

NMR SPECTROSCOPY OF DERIVATIZED HUMIC SUBSTANCES

Wershaw et al. (1981) and Mikita et al. (1981) used a different approach in measuring relative amounts of carboxyl, alcoholic, phenolic, and carbohydrate functional groups in fulvic and humic acids by ^{13}C NMR. They have permethylated humic and fulvic acids with ^{13}C-enriched reagents and then measured the intensities of the ^{13}C NMR peaks of the samples in the region between 50 and 62 ppm. They found that the nuclear Overhauser enhancement (NOE) for all OCH_3 groups in this region was uniform and therefore integration of the peak areas should give an accurate representation of relative abundance of functional groups if there is no distortion due to relaxation effects. One would expect that to a first approximation all OCH_3 groups would have similar T_1 values; however, this should be verified experimentally for each sample. The chemical shift assignments for the various func-

* Any use of trade names is for identification purposes only and does not constitute endorsement by the U.S. Geological Survey.

TABLE 3. ^{13}C NMR Chemical Shift Assignments for Methylated Compounds

Chemical Shift of Methoxyl Band (ppm)	Assignments
51.2	Aliphatic carboxyl
52.5	Aromatic carboxyl
55–56	Phenol
57.5	Aliphatic OH
58–60	Aliphatic or carbohydrate OH
61–62	Aliphatic or carbohydrate OH
45.6	Amino nitrogen

tionalities determined by model compound studies are given in Table 3. At the present time data on only a few samples have been published using derivatization procedures and only relative concentrations of functional groups have been measured.

Leenheer et al. (1983) have recently shown that ketone functional groups can be determined by one of the following techniques: (1) preparation of the methoxime followed by proton or ^{13}C NMR spectroscopy of the derivative; (2) reduction of the ketone carbonyl group with sodium borohydride followed by methylation of the resultant alcohol and NMR analysis of the methyl ethers.

Leenheer et al. (1983) have also prepared the trifluoroethyl ethers of humic materials by reaction of a sample suspended in methylene chloride with trifluorodiazoethane. Proton, ^{13}C, and ^{19}F NMR of this derivative provide information on the concentration of aliphatic hydroxyl groups.

CONCLUSIONS AND RECOMMENDATIONS FOR FUTURE WORK

Nuclear magnetic resonance spectroscopy is a powerful new tool for the determination of functional groups in humic substances. However, substantially more work must be done in order to obtain the type of quantitative data necessary for both structural and geochemical studies.

At the present time the measurement of functional group concentrations in humic substances by both ^{13}C and ^{1}H NMR spectroscopy in both liquid and solid samples is at best semiquantitative. In order to increase the accuracy of the concentration measurements on functional groups the effect of variations in NOE in liquid samples must be evaluated by gated decoupling experiments on every sample to be measured. In addition, the relaxation times of all functional groups must be measured for each sample and the delay times must then be chosen to allow for adequate relaxation of the group with the longest T_1. In order to reduce the effects of paramagnetic ions the samples must be carefully purified using ion-exchange techniques. Contact times and repetition rates should be optimized for each solid sample run by the CP/MAS technique (see Hatcher et al., 1983a).

Another area that must be explored is that of the NMR spectra of fractionated humic and fulvic acids. Preliminary studies in our laboratory indicate that the humic acid fractions obtained by adsorption chromatography on Sephadex gels have different NMR spectra. Fractionation of humic substances may therefore allow one to measure NMR spectra of more homogeneous samples. These spectra should be better resolved and be more readily interpretable.

A rigorous comparison study should be undertaken of all NMR techniques presently used. All spectra should be measured using the optimum parameters as determined from relaxation and gated decoupling experiments, and experiments to determine optimum parameters for CP/MAS should be conducted. This study would provide a basis for determining how to best measure the concentrations of functional groups in humic substances.

Recent studies by Leenheer, Wershaw, and Thorn in the U.S. Geological Survey laboratory in Denver indicate that it should be possible to develop derivatization techniques to measure keto groups, lactones, and other functionalities in fulvic and humic acids. Therefore, it appears that further work should be done on derivatization as a means of measuring concentration of groups that presently cannot be measured.

Preliminary results from experiments performed at the University of Arizona (Thorn and Steelink, oral communication) suggest that ^{29}Si NMR spectroscopy on silylated humic substances should provide additional information on oxygen-containing functional groups. The chemical shift range of silyl esters and ethers is approximately 10 ppm and therefore adequate separations of the oxygen-containing functional groups can probably be attained.

In a recent study Preston et al. (1982) have demonstrated that ^{15}N NMR spectra can be obtained from "synthetic humic acids." This work suggests that ^{15}N NMR spectroscopy might be useful for elucidating the mechanisms of binding of ammonia, amino acids, and nitrates to humic substances, and that derivatization of humic substances with ^{15}N-enriched reagents might be fruitful.

Concluding Remarks

This book has been divided into three sections: Geochemistry, Isolation and Fractionation, and Characterization. These are not isolated topics, but are intimately interconnected. For example, the characterization data depend on how the samples were isolated and fractionated; conversely, a logical choice of fractionation methodology is influenced by our knowledge of the characteristics of humic substances. Similarly, we cannot hope to understand the geochemical roles of humic substances without the analytical data acquired by fractionation and characterization techniques.

The chapters in Section I of this book, dealing with the geochemistry of humic substances in diverse environments, illustrate the fact that we are just beginning to understand the function of humic substances in natural systems. Although at the present time our ability to quantitatively describe any of the reactions of humic substances is limited, it is clear that many of the chemical and biochemical reactions that take place in soils, sediments, terrestrial surface waters, estuaries, and oceans result in or are strongly influenced by the presence of humic substances. What we have learned up to this time about the geochemistry of humic substances provides us with a tantalizing prelude of what is to come as our understanding of these ubiquitous materials increases.

There has been an enormous explosion in knowledge in the fields of protein and nucleic acid biochemistry in the last three or four decades. The knowledge gained in these fields has in turn provided a much deeper understanding of the life processes in all living organisms. Indeed, modern biology has evolved into molecular biology in which practically all organismic processes are described at the molecular level. If evolution of the study of the geochemistry of humic substances follows a similar path, and we believe it must, then future progress in humic substance geochemistry will require a more detailed understanding of the molecular constituents of humic substances and of their physical and chemical properties.

In Section II of this book techniques for isolating humic substances from natural soils and waters, and for fractionating these extracts into less heterogeneous mixtures, are reviewed. Despite the fact that attempts at fractionating and purifying humic substances have constantly pervaded research in humic substances, we are confronted with one sobering realization—*no one has yet succeeded in isolating a pure humic substance,* and consequently we are still constrained to work with mixtures if we are to pursue research on humic substances. This "mixture problem" must have a major influence on our approach to the study of chemistry and geochemistry of humic substances and on our interpretation of the data. If and when pure humic substances are obtained, then, and only then, can the tools of conventional chemistry be directly transferred to the study of humic substances.

Section III of this book is devoted to some of the more recent methods developed to characterize humic substances. These methods have yielded information that is important in understanding the geochemical roles of humic substances. For example, the recent finding that aquatic fulvic acids generally have a molecular weight between 700 and 1500 daltons has changed our concept of fulvic acid from a high-molecular-weight polyelectrolyte to an oligoelectrolyte having 8–12 acidic functional groups per molecule. However, studies using these methods have also raised questions more numerous than the answers they have provided. Using the example of the acidic properties of aquatic fulvic acid, which has important geochemical significance, we see that there are many approaches being taken to modeling the acidic behavior of aquatic fulvic acid. One of the leading approaches is to assume that fulvic acid is a complex mixture of a wide variety of different types of acidic organic compounds. However, spectroscopic characterization, especially ^{13}C NMR spectroscopy of derivatized samples, and elemental composition would suggest that the heterogeneity of aquatic fulvic acids may not be as great as assumed in this modeling approach. Many other inconsistencies in underlying assumptions and approaches will be noted by the careful reader of this book. These inconsistencies provide a challenge for all of us to develop new ideas and methodologies, which should then be extensively tested and applied.

In conclusion, what do we, the Editors, hope to have accomplished in this book? Our goal was to compile critical and comprehensive reviews on three areas of humic substances research and to delineate "what we know, what we don't know, and what we think we know" in these subject areas. In the past, a clear distinction has not always been made between what is fact and what is conjecture in the discussion of humic substances. It is our hope that this book serves to focus more critical attention on these distinctions.

Bibliography

Abelson, P. H. (1967). Conversion of biochemicals to kerogen and *n*-paraffins. In *Researches of Geochemistry,* vol. 2 (P. H. Abelson, ed.). Wiley, New York, pp. 63–85.

Abelson, P. H. (1978). Organic matter in the earth's crust. *Ann. Rev. Earth Planet. Sci.* **6,** 325–351.

Abelson, P. H. and Hare, P. E. (1970). Uptake of amino acids by kerogen. *Carnegie Inst. Washington Yearb.* **68,** 297–303.

Abelson, P. H. and Hare, P. E. (1971). Reactions of amino acids with natural and artificial humus and kerogens. *Carnegie Inst. Washington Yearb.* **69,** 327–334.

Abrams, I. M. (1969). Removal of organics from water by synthetic resinous absorbents. *Chem. Eng. Prog.* **65,** 106–112.

Abrams, I. M. and Breslin, R. P. (1965). Recent studies on the removal of organics from water. *Proc. Int. Water Conf.,* Pittsburgh, 7 pp.

Achard, F. K. (1786). Chemische Utersuchung des Torfs. *Crell's Chem. Ann.* **2,** 391–403.

Acton, C. J., Paul, E. A., and Rennie, D. A. (1963). Measurements of the polysaccharide content of soils. *Can. J. Soil Sci.* **43,** 141–150.

Adams, R. S., Jr. (1973). Factors affecting soil adsorption and bioactivity of pesticides. *Residue Rev.* **47,** 1–54.

Afghan, B. K., Belliveau, P. E., LaRose, R. H., and Ryan, J. F. (1974). An improved method for determination of trace quantities of phenols in natural waters. *Anal. Chim. Acta* **71,** 355–366.

Aiken, G. R., Thurman, E. M., Malcom, R. L., and Walton, H. F. (1979). Comparison of XAD macroporous resins for the concentration of fulvic acid from aqueous solution. *Anal. Chem.* **51,** 1799–1803.

Alcino, J. F., Cohen, A. I., and Everhard, M. E. (1965). Organic analysis: Sulfur. In *Treatise on Analytical Chemistry,* Pt. II, Vol. 12 (I. M. Kolthoff and P. J. Elving, eds.). Wiley-Interscience, New York, pp. 57–115.

Aldridge, M. H., Pressley, T. A., Chapin, C., and Welebir, A. (1976). Characterization of nonvolatile organic material during physicochemical treatment of the District of Columbia raw wastewater. *Water Resources Research Center Report* Number 9, Washington Technical Institute, 28 pp.

Alemany, L. B., Grant, D. M., Pugmire, R. J., Alger, T. D., and Zilm, K. W. (1983). Cross-polarization and magic-angle sample spinning NMR spectra of model organic compounds. 2. Molecules of low or remote protonation. *J. Am. Chem. Soc.* **105,** 2142–2147.

Alexandrova, L. N. (1960). The use of sodium pyrophosphate for isolating free humic substances and their organic-mineral compounds from the soil. *Soviet Soil Sci.* **2,** 190–197.

Allen, H. L. (1976). Dissolved organic matter in lake water: Characteristics of molecular weight size fractions and ecological implications. *Oikos* **27,** 64–74.

Alpert, N. L., Keiser, W. E., and Szymanski, H. A. (1964). *IR Theory and Practice of Infrared Spectroscopy*. Plenum, New York, 380 pp.

American Public Health Association (1971). *Standard Methods* 13th ed. American Public Health Association, Washington, D.C., 769 pp.

Anders, O. U. and Briden, D. W. (1964). A rapid nondestructive method of precision oxygen analysis by neutron activation. *Anal. Chem.* **36,** 287–292.

Anderson, A. H., Berrow, M. L., Farmer, V. C., Hepburn, A., Russell, J. D., and Walker, A. D. (1982). A reassessment of podzol formation processes. *J. Soil Sci.* **33,** 125–136.

Anderson, A. H. and Russell, J. D. (1976). Possible relationship between soil fulvic acids and polymaleic acid. *Nature* **260,** 597.

Anderson, G. (1957). Nucleic acid derivatives in soil. *Nature* **180,** 287–288.

Anderson, G. (1958). Identification of derivatives of deoxyribonucleic acid in humic acid. *Soil Sci.* **86,** 169–174.

Anderson, G. (1961). Estimation of purines and pyrimidines in soil humic acid. *Soil Sci.* **91,** 156–161.

Anderson, M. A. and Morel, F. M. M. (1982). The influence of aqueous iron chemistry on the uptake of iron by the coastal diatom *Thalassiosira weissflogii*. *Limnol. Oceanogr.* **27,** 789–813.

Andreux, F., Golebiowska, D., and Metche, M. (1979). Polymerisation oxydative deso-diphenols en presence ou non d'amino-acides. Cas des systemes catechol-glycocolle et catechol-diglycylglycine. In *C. R. Assoc. Gen. Groupe Polyphenols, Logrono,* pp. 178–188.

Andrew, E. R. (1972). The narrowing of NMR spectra of solids by high speed specimen rotation and the resolution of chemical shift and spin multiplet structures of solids. In *Progress in NMR Spectroscopy,* Vol. 8, Pt. 1 (J. W. Emsley, J. Feehez, and L. H. Satcliffe, eds.). Pergamon, New York, pp. 1–39.

Andrews, A. T. (1981). *Electrophoresis: Theory, Techniques and Biochemical and Clinical Applications*. Clarendon, Oxford.

Antweiler, R. C. and Drever, J. I. (1983). The weathering of a late tertiary volcanic ash. Importance of organic solutes. *Geochim. Cosmochim. Acta,* **47,** 623–629.

Aramentano, T. V. (1979). The role of organic soils in the world carbon cycle; problem definition and research needs. The Institute of Ecology, 37 pp.

Ardakani, S. M. and Stevenson, F. J. (1972). A modified ion-exchange technique for the determination of stability constants of metal–soil organic matter complexes. *Soil Sci. Soc. Am. Proc.* **36,** 884–890.

Armitage, R. G. S., Gramshaw, J. W., Haslam, E., Haworth, R. D., Jones, K., Rogers, H. J., and Searle, T. (1961). Gallotannins, Part III. The Constitution of Chinese, Turkish, Sumach, and Tara Tannins. *J. Chem. Soc.,* 1842.

Armstrong, D. E. and Konrad, J. G. (1974). Nonbiological degradation of pesticides. In *Pesticides in Soil and Water* (W. D. Guenzi, ed.). American Society of Agronomy, pp. 123–131.

Aschan, O. (1908). Die Bedeutung der Wasserloslichen Humusstoffe (Humussole) fur die Bildung der See-und Sumpferze. *J. Prakt. Chem.* (N.F. 77) **185,** 172–188.

Aschan, O. (1932). Water humus and its role in the formation of marine iron ore. *Ark. Kemi Mineral. Geol.* **10A,** No. 15, 1–143.

Ataman, O. Y. and Mark, H. B., Jr. (1976). Ion exchange between ammonium zeolite and the supporting matrix in KBr pellets. *Anal. Lett.* **9,** 1135–1141.

Atherton, N. M., Cranwell, P. A., Floyd, A. J., and Haworth, R. D. (1967). Humic acid. I. ESR spectra of humic acids. *Tetrahedron* **23,** 1653–1667.

Bada, J. L., Hoopes, E., and Ho, H. (1982). Combined amino acids in Pacific Ocean waters. *Earth Planet. Sci. Lett.* **58,** 276–284.

Bailly, J. P. (1973). Spectroscopie UV–visible des acides humiques non fractionnes et de certaines de leur fractions obtenues sur gels de dextranes. *Plant Soil* **38,** 281–297.

Baker, R. A. (1967). Trace organic contaminant concentration by freezing. I. Low inorganic aqueous solutions. *Water Res.* **1,** 61–77.

Baker, R. A. (1970). Trace organic contaminant concentration by freezing. IV. Ionic effects. *Water Res.* **4,** 559–573.

Baker, W. E. (1973). The role of humic acids from Tasmanian podzolic soils in mineral degradation and metal solubilization. *Geochim. Cosmochim. Acta* **37,** 269–281.

Ballard, T. M. (1971). Role of humic carrier substances in DDT movement through forest soil. *Soil Sci. Soc. Am. Proc.* **35,** 145–147.

Bance, S. (1980). *Handbook of Practical Organic Microanalysis.* Halsted, New York.

Barber, M., Bordoli, R. S., Sedgwick, R. D., and Tyler, A. N. (1981). Fast atom bombardment of solids as an ion source in mass spectrometry. *Nature* **293,** 270–275.

Barber, R. T. (1968). Dissolved organic carbon resists microbial oxidation. *Nature* **220,** 274–275.

Barker, S. A., Hayes, M. H. B., Simmonds, R. G., and Stacey, M. (1967). Studies on soil polysaccharides. I. *Carbohydr. Res.* **5,** 13–24.

Barnes, M. A. and Barnes, W. C. (1978). Organic compounds in lake sediments. In *Lakes (Chemistry, Geology, Physics)* (A. Lerman, ed.). Springer-Verlag, New York, pp. 127–152.

Barth, E. F. and Acheson, N. H. (1962). High-molecular-weight material in tap water. *J. Am. Water Works Assoc.* **54,** 959–964.

Bartholomew, W. V. and Clark, F. E. (eds.) (1965). *Soil Nitrogen.* American Society of Agronomy. Madison, WI, 615 p.

Barton, A. F. M. (1975). Solubility parameters. *Chem. Rev.* **75,** 731–753.

Bartuska, V. J., Maciel, G. E., Schaefer, J., and Stejskal, E. O. (1977). Prospects for carbon-13 nuclear magnetic resonance analysis of solid fossil fuel materials. *Fuel* **56,** 354–358.

Batley, G. E. and Florence, T. M. (1974). An evaluation and comparison of some techniques of anodic stripping voltammetry. *J. Electroanal. Chem.* **55,** 23–43.

Baxter, R. M. and Carey, J. H. (1982). Reactions of singlet oxygen in humic waters. *Freshwater Biol.* **12,** 285–292.

Beck, K. C., Reuter, J. H., and Perdue, E. M. (1974). Organic and inorganic geochemistry of some coastal plain rivers of the southeastern United States. *Geochim. Cosmochim. Acta* **38,** 341–364.

Belanger, A. and Hamilton, H. A. (1979). Determination of Disolfoton and Permethrin residues in an organic soil and their translocation into lettuce, onion, and carrot. *J. Environ. Sci. Health* **B14,** 213–226.

Belanger, A. and Mathur, S. P. (1983). Persistence of fonofos in two organic soils each containing four levels of copper. *J. Environ. Sci. Health, Part B,* **18,** 713–723.

Belanger, A., Mathur, S. P., and Maretl, P. (1982). The effect of fonofos and carbofuran on microbiological population and persistence of fonofos in four organic soils infested with onion maggot. *J. Environ. Sci. Health* **17B,** 171–182.

Belcher, R. (ed.) (1977). *Instrumental Organic Elemental Analysis.* Academic, New York, 299 pp.

Bellamy, L. J. (1958). *The Infra-Red Spectra of Complex Molecules,* 2nd ed. Methuen, London, 423 pp.

Bellamy, L. J. (1960). *The Infra-Red Spectra of Complex Molecules,* 2nd ed. Wiley, New York, 425 pp.

Bellamy, L. J. (1968). *Advances in Infrared Group Frequencies.* Methuen, London, 304 pp.

Bellar, T. A., Lichtenberg, J. J., and Kroner, R. C. (1974). The occurrence of organohalides in chlorinated drinking water. *J. Am. Water Works Assoc.* **66,** 703–706.

Benoit, G. J., Turekian, K. K., and Benninger, L. K. (1979). Radiocarbon dating of a core from Long Island Sound. *Estuarine Coastal Mar. Sci.* **9,** 171–180.

Bergmann, W. (1978). Zur Strukturaufklaerung von Huminsaeuren aus Abwasser. Ph.D. Thesis, 136 pp.

Berner, R. A. (1982). Burial of organic carbon and pyritic sulfur in the modern ocean: Its geochemical and environmental significance. *Am. J. Sci.* **282,** 451–473.

Berzelius, J. J. (1839). *Lehrbuch der Chemie,* 3rd ed. (translation by Wöhler). Dresden and Leipzig.

Birge, E. A. and Juday, C. (1922). The inland lakes of Wisconsin. The plankton. I. It's quantity and chemical composition. *Bull. Wis. Geol. Nat. Hist. Surv.,* **64** (Sci. ser. 13), 222 pp. 402, 405, 406, 420, 421, 422, 432.

Birge, E. A. and Juday, C. (1926). Organic content of lake water. *Bull. U.S. Bur. Fish.* **42,** 185–205.

Birge, E. A. and Juday, C. (1934). Particulate and dissolved organic matter in inland lakes. *Ecol. Monogr.* **4,** 440–474.

Bjerrum, N. (1923). Dissoziationskonstanten von mehrbasichen Sauren und ihre Anwendung zur Brechnung molekularer Dimensionen. *Z. Phys. Chem.* **106,** 219–242.

Black, A. P. and Christman, R. F. (1963a). Characteristics of colored surface waters. *J. Am. Water Works Assoc.* **55,** 753–770.

Black, A. P. and Christman, R. F. (1963b). Chemical characteristics of fulvic acids. *J. Am. Water Works Assoc.* **55,** 897–912.

Black, C. A. (ed.) (1969). *Methods of Soil Analysis.* American Society of Agronomy, Madison, WI, 1572 pp.

Blau, K. and King, G. H. (eds.) (1978). *Handbook of Derivatives for Chromatography.* Heyden & Son, London.

Blom, L., Edelhausen, L., and van Krevelen, D. W. (1957). Chemical structure and properties of coal. XVIII. Oxygen groups in coal and related products. *Fuel* **36,** 135–153.

Bloom, P. R., McBride, M. B., and Weaver, R. M. (1979). Aluminum organic matter in acid soils: Buffering and solution aluminum activity. *Soil Sci. Soc. Am. J.* **43,** 488–493.

Blutstein, H. and Smith, J. D. (1978). Distribution of species of Cu, Pb, Zn, and Cd in a water profile of the Yarra river estuary. *Water Res.* **12,** 119–125.

Boehm, P. D. and Quinn, J. G. (1973). Solubilization of hydrocarbons by the dissolved organic matter in seawater. *Geochim. Cosmochim. Acta* **37,** 2459–2477.

Bollag, J. M. (1983). Cross-coupling of humus constituents and xenobotic substances. In *Aquatic and Terrestrial Humic Materials* (R. F. Christman and E. J. Gjessing, eds.). Ann Arbor Science, Ann Arbor, MI, pp. 126–141.

Bollag, J. M., Liu, S. Y., and Minard, R. D. (1980). Cross-coupling of phenolic humus constituents and 2,4-dichlorophenol. *Soil Sci. Soc. Am. J.* **44,** 52–56.

Bolt, G. H. (ed.) (1982). *Soil Chemistry. B. Physico-Chemical Models.* Elsevier, New York.

Bonnar, R. V., Dimbat, M., and Stross, F. H. (1958). *Number Average Molecular Weights.* Wiley-Interscience, New York, 310 pp.

Borggaard, O. K. (1974). Experimental conditions concerning potentiometric titration of humic acid. *J. Soil Sci.* **25,** 189–195.

Boudou, J. P. (1981). Diagenese organique de sediments organique de se deltaiques (Delta de la Mahakam, Indonesie). These, Universite d'Orleans, France.

Bourbonniere, R. A. (1979). Geochemistry of humic matter in Holocene Great Lakes sediments. Ph.D. thesis, University of Michigan, 373 pp.

Bourbonniere, R. A. and Meyers, P. A. (1978). Characterization of sedimentary humic matter by elemental and spectroscopic methods. *Can. J. Spectrosc.* **23,** 35–41.

Bourbonniere, R. A. and Meyers, P. A. (1979). Geochemistry of humic matter in a freshwater sediment. Presented at 9th International Meeting on Organic Geochemistry.

Boyd, S. A., Sommers, L. E., and Nelson, D. W. (1979). Infrared spectra of sewage sludge fractions: Evidence for an amide metal binding site. *Soil Sci. Soc. Am. J.* **43,** 893–899.

Boyle, E. A., Edmond, J. M., and Sholkovitz, E. R. (1977). The mechanism of iron removal in estuaries. *Geochim. Cosmochim. Acta* **41,** 1313–1324.

Bracewell, J. M. and Robertson, G. W. (1976). A pyrolysis-gas chromatography method for discrimination of soil humus types. *J. Soil Sci.* **27,** 196–205.

Bramryd, T. (1980). The role of peatlands for the global carbon dioxide balance. *Proc. 6th Intl. Peat Congr.,* pp. 9–11.

Breger, I. A. (1950). The role of lignin in coal genesis. Ph.D. dissertation, Massachusetts Institute of Technology, 160 pp.

Breger, I. A. (1960). Diagenesis of metabolites and a discussion of the origin of petroleum hydrocarbons. *Geochim. Cosmochim. Acta* **19,** 297–308.

Breger, I. A. (1963). Origin and classification of naturally occurring carbonaceous substances. In *Organic Geochemistry* (I. A. Breger, ed.). Pergamon, Elmsford, N.Y., pp. 50–86.

Breger, I. A. (1976). Geochemical interrelationships among fossil fuels, marsh, and landfill gas. In *The Future Supply of Nature-made Petroleum and Gas Technical Reports* (R. F. Meyer, ed.). Pergamon, Elmsford, N.Y., pp. 913–918.

Breger, I. A. and Brown, A. (1962). Kerogen in the Chattanooga Shale. *Science* **137,** 221–224.

Breger, I. A., Hatcher, P. G., Romankiw, L. A., Miknis, F. P., and Maciel, G. E. (1983). Upper Devonian black shales of the eastern United States: Organic geochemical studies—past and present. In *Geochemistry and Chemistry of Oil Shales* (No. 230) (F. P. Mikins and J. F. McKay, eds.). American Chemical Society Symposium Series, American Chemical Society, Washington, D.C., pp. 181–198.

Breger, I. A., Krasnow, M., and Chandler, J. C. (1978). Peat from the Everglades of Florida: A study of the origin of coal and of natural gas. *Tenth Intl. Congr. Sedimentology,* July 1978, Abstracts 1, pp. 85–86.

Bremner, J. M. (1950). Some observations on the oxidation of soil organic matter in the presence of alkali. *J. Soil Sci.* **1,** 198–204.

Bremner, J. M. (1967). Nitrogenous compounds. In *Soil Biochemistry,* Vol. 1 (A. D. McLaren and G. H. Peterson, eds.). Marcel Dekker, New York, pp. 19–66.

Bremner, J. M. (1968). The nitrogenous constituents of soil organic matter and their role in soil fertility. *Pontif. Accad. Sci. Scripta Varia* **32,** 143–193.

Bremner, J. M. and Lees, H. (1949). Studies of soil organic matter: II. The extraction of organic matter from soil by neutral reagents. *J. Agric. Sci.* **39,** 274–279.

Bresnahan, W. T., Grant, C. L., and Weber, J. H. (1978). Stability constants for the complexation of copper(II) ions with water and soil fulvic acids measured by an ion selective electrode. *Anal. Chem.* **50,** 1675–1679.

Brezonik, P. L., Brauner, P. A., and Stumm, W. (1976). Trace metal analysis by anodic stripping voltammetry: Effect of sorption by natural and model organic compounds. *Water Res.* **10,** 605–612.

Brown, F. S., Baedecker, M. J., Nissenbaum, A., and Kaplan, I. R. (1972). Early diagenesis in a reducing fjord, Saanich Inlet, British Columbia. III. Changes in organic constituents of sediment. *Geochim. Cosmochim. Acta* **36,** 1185–1203.

Brown, M. (1975). High-molecular-weight material in Baltic seawater. *Mar. Chem.* **3,** 253–258.

Brown, M. (1977). Transmission spectroscopy examinations of natural waters. C. Ultraviolet spectral characteristics of the transition from terrestrial humus to marine yellow substance. *Estuarine Coastal Mar. Sci.* **5,** 309–317.

Buffle, J. (1977). Les substances humiques et leurs interactions avec les ions mineraux. In *Conference Proceedings de la Commission d'Hydrologie Appliquee de l'A.G.H.T.M.* L'Universite d'Orsay, pp. 3–10.

Buffle, J., Deladoey, P., and Haerdi, W. (1978). The use of ultrafiltration for the separation and fractionation of organic ligands in freshwaters. *Anal. Chim. Acta.* **101,** 339–357.

Buffle, J., Greter, F. L., and Haerdi, W. (1977). Measurement of complexation properties of humic and fulvic acids in natural waters with lead and copper ion-selective electrodes. *Anal. Chem.* **49,** 216–222.

Burch, R. D., Langford, C. H., and Gamble, D. S. (1978). Methods for the comparison of fulvic acid samples: The effects of origin and concentration on acidic properties. *Can. J. Chem.* **56,** 1196–1201.

Burns, R. G. (ed.) (1978). *Soil Enzymes.* Academic, New York, 380 pp.

Burns, R. G. (1982). Microbial adhesion to soil surfaces: Consequences for growth and enzyme activity. Chapter 14, *Symposium of the Society of Chemical Industries and Society for General Microbiology,* University of Reading, pp. 248–262.

Butler, J. H. A. and Ladd, J. N. (1969). Effect of extractant and molecular size on the optical and chemical properties of soil humic acids. *Aust. J. Soil Res.* **7,** 229–239.

Cacco, G., Maggioni, A., and Ferrari, G. (1974). Electrofocusing: A new method for the characterization of soil humic matter. *Soil Biol. Biochem.* **6,** 145–148.

Cameron, R. S., Swift, R. S., Thornton, B. K., and Posner, A. M. (1972a). Calibration of gel permeation chromatography materials for use with humic acid. *J. Soil Sci.* **23,** 342–349.

Cameron, R. S., Thornton, B. K., Swift, R. S., and Posner, A. M. (1972b). Molecular weight and shape of humic acid from sedimentation and diffusion measurements on fractionated extracts. *J. Soil Sci.* **23,** 394–408.

Camier, R. J., Siemon, S. R., Battaerd, H. A. J., and Stanmore, B. R. (1981). The nature and possible significance of particulate structure in alkali-treated brown coal. *Coal Structure (Adv. Chem. Ser.)* **192,** 311–336.

Campbell, C. A., Paul, E. A., Rennie, D. A., and McCallum, K. J. (1967a). Factors affecting the accuracy of the carbon-dating method in soil humus studies. *Soil Sci.* **104,** 81–85.

Campbell, C. A., Paul, E. A., Rennie, D. A., and McCallum, K. J. (1967b). Dynamics of soil humus utilizing C dating techniques. Soil Sci. **104,** 217–224.

Cane, R. F. (1976). The origin and formation of oil shale. In *Oil Shale* (T. F. Yen and G. V. Chilingarian, eds.). Elsevier, New York, pp. 25–60.

Cann, J. R. and Fink, N. H. (1983). Effect of microheterogeneity on the sedimentation behavior of self-associating proteins. *Biophys. Chem.* **17,** 29–34.

Cannon, C. G. and Sutherland, G. B. B. M. (1945). The infra-red absorption spectra of coals and coal extracts. *Trans. Faraday Soc.* **41,** 279–288.

Carlberg, G. E. and Martinsen, K. (1982). Adsorption/complexation of organic micropollutants and comparison of two analytical methods for analysing organic pollutants in humus waters. *Sci. Tot. Environ.* **25,** 245–254.

Carlson, D. J. and Mayer, L. M. (1980). Enrichment of dissolved phenolic material in the surface microlayer of coastal waters. *Nature* **286,** 482–483.

Carlson, D. J. and Mayer, L. M. (1983). Relative influences of riverine and macroalgal phenolic materials on UV absorbance in temperate coastal waters. *Can. J. Fish. Aquat. Sci.* **40,** 1258–1263.

Carlson, D. J., Mayer, L. M., Brann, M. L., and Mague, T. H. (unpublished). Binding of monomeric organic compounds to macromolecular dissolved organic matter in seawater.

Carrington, A. and McLachlan, A. D. (1967). *Introduction to Magnetic Resonance with Applications to Chemistry and Chemical Physics*. Harper and Row, New York, 266 pp.

Carter, C. W. and Suffet, I. H. (1982). Binding of DDT to dissolved humic materials. *Environ. Sci. Technol.* **16,** 735–740.

Casagrande, D. J., Gronli, K., and Sutton, N. (1980). The distribution of sulfur and organic matter in various fractions of peat: Origins of sulfur in coal. *Geochim. Cosmochim. Acta* **44,** 24–32.

Castagnola, M., Heras, R. G. de las, and Marini-Bettolo, G. B. (1978). The effect of urea on electrophoretic pattern of soil humic acids., *J. Chromatogr.* **147,** 438–442.

Ceh, M. and Hadzi, D. (1956). Infrared spectra of humic acids and their derivatives. *Fuel* **35,** 77–83.

Chakraborty, G., Banerjee, S. K., and Gupta, S. K. (1979). Molecular weights and total acidities of humic acids extracted from soil aggregates of different sizes. *J. Indian Soc. Soil Sci.* **27,** 129–132.

Chen, Y. and Schnitzer, M. (1976). Scanning electron microscopy of a humic acid and of a fulvic acid and its metal and clay complex. *Soil Sci. Soc. Am. J.* **40,** 682–686.

Chen, Y. and Schnitzer, M. (1978). Surface tension of aqueous solutions of soil humic substances. *Soil Sci.* **125,** 7–15.

Chen, Y., Senesi, N., and Schnitzer, M. (1977). Information provided on humic substances by E_4/E_6 ratios. *Soil Sci. Soc. Am. J.* **41,** 352–358.

Cheng, K. L. (1977). Separation of humic acid with XAD resins. *Microchim. Acta* **1977-II,** 389–396.

Cheshire, M. V., Berrow, M. L., Goodman, B. A., and Mundie, C. M. (1977). Metal distribution and nature of some Cu, Mn, and V complexes in humic and fulvic acid fractions of soil organic matter. *Geochim. Cosmochim. Acta* **41,** 1131–1138.

Cheshire, M. V., Cranwell, P. A., Falshaw, C. P., Floyd, A. J., and Haworth, R. D. (1967). Humic acid. II. Structure of humic acids. *Tetrahedron* **23,** 1669–1682.

Chian, E. S. K. and DeWalle, F. B. (1977). Characterization of soluble organic matter in leachate. *Environ. Sci. Technol.* **11,** 158–163.

Chiou, C. T., Peters, L. J., and Freed, V. H. (1979). A physical concept of soil–water equilibria for nonionic organic compounds. *Science* **206,** 831–832.

Chiou, C. T., Porter, P. E., and Schmedding, D. W. (1983). Partition equilibria of nonionic organic compounds between soil organic matter and water. *Environ. Sci. Technol.* **17,** 227–231.

Choppin, G. R. and Kullberg, L. (1978). Protonation thermodynamics of humic acid. *J. Inorg. Nucl. Chem.* **40,** 651–654.

Choudri, M. B. and Stevenson, F. J. (1957). Chemical and physicochemical properties of soil colloids. III. Extraction of organic matter from soils. *Soil Sci. Soc. Am. Proc.* **21,** 508–513.

Chrambach, A. and Rodbard, D. (1971). Polyacrylamide gel electrophoresis. *Science* **172,** 440–451.

Christensen, D. and Blackburn, T. H. (1982). Turnover of ^{14}C-labelled acetate in marine sediments. *Mar. Biol.* **71,** 113–119.

Christensen, J. J., Hansen, L. D., and Izatt, R. M. (1976). *Handbook of Proton Ionization Heats and Related Thermodynamic Quantities*. Wiley, New York, 269 pp.

Christensen, J. J., Ruckman, J., Eatough, D. J., and Izatt, R. M. (1972). Determination of equilibrium constants by titration calorimetry. I. Introduction to titration calorimetry. *Thermochim. Acta* **3,** 203–218.

Christensen, J. J., Wrathall, D. P., Oscarson, J. O., and Izatt, R. M. (1968). Theoretical evaluation of entropy titration method for calorimetric determination of equilibrium constants in aqueous solution. *Anal. Chem.* **40,** 1713–1717.

Christman, R. F. (1970). Chemical structures of color producing organic substances in water. in *Proceedings on Symposium on Organic Matter in Natural Waters* (D. W. Hood, ed.). University of Alaska, Institute of Marine Sciences, Sept. 1968, pp. 181–198.

Christman, R. F. and Gjessing, E. T. (eds.) (1983). *Aquatic and Terrestrial Humic Materials*. Ann Arbor Science, Ann Arbor, MI, 538 pp.

Christman, R. F., Johnson, J. D., Pfaender, F. K., Norwood, D. L., Webb, M. R., Haas, J. R., and Babenrieth, M. J. (1980). Chemical identification of aquatic humic chlorination products. In *Water Chlorination: Environmental Impact and Health*, Vol. 3 (R. L. Jolley, W. A. Brungs, and R. B. Cumming, eds.). Ann Arbor Science, Ann Arbor, MI, pp. 75–84.

Christman, R. F. and Minear, R. A. (1971). Organics in lakes. In *Organic Compounds in Aquatic Environments* (S. J. Faust and J. V. Hunter, eds.). Marcel Dekker, New York, pp. 119–143.

Christman, R. F., Norwood, D. L. Millington, D. S., Johnson, J. D., and Stevens, A. A. (1983). Identity and yields of major halogenated products of aquatic fulvic acid chlorination. *Environ. Sci. Technol.* **17,** 625–628.

Christman, R. F. and Oglesby, R. T. (1971). Microbiological degradation and formation of humus. In *Lignin: Occurrence, Formation, Structure and Reactions* (K. Sarkanen and C. H. Ludwic, eds.). Wiley-Interscience, New York, pp. 780–795.

Clapp, C. E., Jr. (1957). High molecular weight water-soluble muck; isolation and determination of constituent sugars of a borate complex-forming polysaccharide employing electrophoretic techniques. Ph.D. dissertation, Cornell University *Dissert. Abstr.* **17,** 963–964.

Clapp, C. E., Jr. (1969). High molecular weight water-soluble muck: Isolation and determination. University Microfilms, University Microfilms Limited, 156 pp.

Clarke-Whistler, K. and Rowsell, J. A. (1983). Peatlands as fish and wildlife habitat. *Proc. Symposium '82, Can. Natl. Comm. Int. Peat Soc.,* pp. 170–191.

Cloern, J. E. (1982). Does the benthos control phytoplankton biomass in South San Francisco Bay? *Mar. Ecol.—Prog. Ser.* **9,** 191–202.

Clymo, R. S. (1983). Peat. In *Ecosystems of the World. 4A. Mires, Swamp, Bog, Fen and Moor. General Studies* (A. J. P. Gore, ed.). Elsevier, New York, pp. 159–224.

Collet, A., Gabard, J., Jacques, J., Cesario, M., Guilhem, J., and Pascard, C. (1981). Synthesis and absolute configuration of chiral (C_3) cyclotriveratrylene derivatives. *J. Chem. Soc. Perkin I,* 1630–1634.

Collienne, R. H. (1983). Photoreduction of iron in the epilimnion of acidic lakes. *Limnol. Oceanogr.* **28,** 83–100.

Cortez, J. and Schnitzer, M. (1979a). Purines and pyrimidines in soils and humic substances. *Soil Sci. Soc. Am. J.* **43,** 958–961.

Cortez, J. and Schnitzer, M. (1979b). Nucleic acid bases in soils and their association with organic and inorganic soil components. *Can. J. Soil Sci.* **59,** 227–286.

Craig, P. N. (1953). Synthesis of ion exchange resins. *Ann. N.Y. Acad. Sci.* **57,** 67–78.

Cranston, R. E. and Buckley, D. E. (1972). The application and performance of microfilters in analysis of suspended particulate matter. Unpublished manuscript, *Bedford Institute of Oceanography Report Series* B1-R-72-7, 14 pp.

Cranwell, P. A. (1982). Lipids of aquatic sediments and sedimenting particulates. *Prog. Lipid Res.* **21,** 271–308.

Cronin, J. R. and Morris, R. J. (1982). The occurrence of high molecular weight humic material in recent organic-rich sediment from the Manbian Shelf. *Estuarine Coastal Shelf Sci.* **15,** 17–27.

Crosby, D. G. (1970). The nonbiological degradation of pesticides in soils. In *Proceedings of the International Symposium on Pesticides in Soil.* Michigan State University, East Lansing, pp. 86–94.

Cross, R. A. and Strathmann, H. (1973). Barrier separation processes. In *An Introduction to Separation Science* (B. L. Karger, L. R. Snyder, and C. Horvath, eds.). Wiley-Interscience, New York, pp. 469–496.

Crosser, M. L. and Allen, H. E. (1977). Determination of complexation capacity of soluble ligands by ion exchange equilibrium. *Soil Sci.* **123,** 176–181.

Curvetto, N. R. and Orioli, G. (1982). Electrophoretic subfractions of low and high molecular weight humic acid fractions. *Plant Soil* **66,** 205–215.

Dack, M. R. J. (1976). The influence of solvent on chemical reactivity. In *Techniques of Chemistry*. Vol. VIII (A. Weissberger, ed.), *Solutions and Solubilities,* Part II (M. R. J. Dack, ed.). Wiley, New York, pp. 95–157.

Danielsson, L. G. (1982). On the use of filters for distinguishing between dissolved and particulate fractions in natural waters. *Water Res.* **16,** 179–182.

Danneberg, O. H. (1981). Possibilities for separation of humic and non-humic substances. *Bodenkultur* **32,** 93–104.

Datsko, V. G. (1940). Photometric investigation of the color reaction of sulfuric acid with organic substances in natural water. *J. Appl. Chem. (SSR)* **13,** 384–389.

Davis, A. and Spackman, W. (1964). The role of the cellulosic and lignitic components of wood in artificial coalification. *Fuel* **43,** 215–224.

Davis, J. A. (1982). Adsorption of natural dissolved organic matter at the oxide/water interface. *Geochim. Cosmochim. Acta* **46,** 2381–2393.

Davis, J. A. and Gloor, R. (1981). Adsorption of dissolved organics in lake water by aluminum oxide. Effect of molecular weight. *Environ. Sci. Technol.* **15,** 1223–1229.

Dawson, H. J., Hrutfiord, B. F., Zasoski, R. J., and Ugolini, F. C. (1981). The molecular weight and origin of yellow organic acids. *Soil Sci.* **132,** 191–199.

Day, J. (1968). The classification of organic soils in Canada. *Proc. 3rd Int. Peat. Congr.,* pp. 80–84.

Deb, B. C. (1949). The movement and precipitation of iron oxides in podzol soils. *J. Soil Sci.* **1,** 112–122.

Debyser, Y. and Gadel, F. (1977). Etude geochimique des composes humiques et des kerogenes. In *Orgon I, mer de Norvege, Geochimie Organique des sediments marins profonds.* Editions du CNRS, Paris, pp. 247–268.

Debyser, Y. and Gadel, F. (1979). Geochimie des kerogenes dans les sediments. In *Orgon III, Geochimie organique des sediments marins profonds.* Editions du CNRS, Paris, pp. 375–404.

Debyser, Y. and Gadel F. (1981). Geochimie des kerogenes dans les sediments. In *Orgon IV, Geochimie organique des sediments marins profonds.* Editions du CNRS, Paris, pp. 447–482.

Debyser, Y. and Gadel, F. (1983). Geochimie des kerogenes. In *Geochimie organique des sediments marins—D'Orgon a Mesidor.* Editions du CNRS, Paris, pp. 421–449.

Debyser, Y., Gadel, F., Leblond, C., and Martinez, M. J. (1978). Étude des composes humiques, des kerogenes et de la fraction hydrolysable dans les sediments. In *Orgon II, Geochimie organique des sediments marins profonds.* Editions du CNRS, Paris, pp. 339–354.

Degens, E. T. and Mopper, K. (1975). Early diagenesis of organic matter in marine sediments. *Soil Sci.* **19,** 65–72.

Degens, E. T., Reuter, J. H., and Shaw, K. N. F. (1964). Biochemical compounds in offshore California sediments and sea water. *Geochim. Cosmochim. Acta* **28,** 45–66.

De Haan, H. (1972a). Molecule-size distribution of soluble humic compounds from different natural water. *Freshwater Biol.* **2,** 235–241.

De Haan, H. (1972b). The biological transformation of soluble humic substances in Tjeukemeer, the Netherlands: A preliminary report. In *Humic Substances, Their Structure and Func-*

tion in the Biosphere (H. L. Golterman and D. Povoledo, eds.). Pudoc, Wageningen, The Netherlands, pp. 63–69.

De Haan, H. (1974). Effect of a fulvic acid fraction on the growth of a *Pseudomonas* from Tjeukemeer (The Netherlands). *Freshwater Biol.* **4,** 301–310.

De Haan, H. (1977). Effect of benzoate on microbial decomposition of fulvic acids in Tjeukemeer (The Netherlands). *Limnol. Oceanogr.* **22,** 38–44.

De Haan, H. (1983). Use of ultraviolet spectroscopy, gel filtration, pyrolysis/mass spectrometry and numbers of benzoate-metabolizing bacteria in the study of humification and degradation of aquatic organic matter. In *Aquatic and Terrestrial Humic Material* (R. F. Christman and E. T. Gjessing, eds.). Ann Arbor Science, Ann Arbor, MI, pp. 165–182.

De Haan, H. and De Boer, T. (1978). A study of the possible interactions between fulvic acids, amino acids and carbohydrates from Tjeukemeer, based on gel filration at pH 7.0. *Water Res.* **12,** 1035–1040.

De Haan, H. and De Boer, T. (1979). Seasonal variations of fulvic acids, amino acids, and sugars in Tjeukemeer, The Netherlands. *Arch. Hydrobiol.* **85,** 30–40.

De Haan, H., De Boer, T., and Halma, G. (1979). Curie point pyrolysis/mass spectrometry of fulvic acids from Tjeukemeer, The Netherlands. *Freshwater Biol.* **9,** 315–317.

De Haan, H., De Boer, T. and Haogveld, H. L. (1981a). Metal binding capacity in relation to hydrology and algal periodicity in Tjeukemeer, The Netherlands. *Arch. Hydrobiol.* **92,** 11–23.

De Haan, H., Halma, G. De Boer, T., and Haverkamp, J. (1981b). Seasonal variations in the compositions of fulvic acids in Tjeukemeer, The Netherlands, as studied by Curie-point pyrolysis–mass spectrometry. *Hydrobiologia* **78,** 87–95.

Deinzer, M., Melton, R., and Mitchell, D. (1975). Trace organic contaminants in drinking water; their concentration by reverse osmosis. *Water Res.* **9,** 799–805.

Delhez, R. (1960). Ultraviolet absorption spectrum of water stored in polyethylene bottles. *Chemist-Analyst* **49,** 20–21.

Demmering, W. (1936). The use of absorption spectral analyses for the investigation and routine checking of contaminated surface water. *Vom Wasser* **11,** 220–237.

Dempsey, B. A. and O'Melia, C. R. (1983). Proton and calcium complexation of four fulvic acid fractions. In *Aquatic and Terrestrial Humic Materials* (R. F. Christman and E. T. Gjessing, eds.). Ann Arbor Science, Ann Arbor, MI, pp. 239–273.

Dereppe, J. M., Boudou, J. P., Moreaux, C., and Durand, B. (1983). Structural evolution of a sedimentologically homogeneous coal series as a function of carbon content by solid-state ^{13}C NMR. *Fuel* **62,** 575–580.

Dereppe, J. M., Moreau, C., and Debyser, Y. (1980). Investigation of marine and terrestrial humic substances by ^{1}H and ^{13}C nuclear magnetic resonance and infrared spectroscopy. *Org. Geochem.* **2,** 117–124.

Dickinson, C. H. (1983). Microorganisms in peatlands. In *Ecosystems of the World. 4A. Mires: Swamps, Bog, Fen, and Moor. General Studies* (A. J. P. Gore, ed.). Elsevier, New York, pp. 225–246.

Djuricic, M., Murphy, R. C., Vitorovic, D., and Biemann, K. (1971). Organic acids obtained by alkaline permanganate oxidation of kerogen from the Green River (Colorado) shale. *Geochim. Cosmochim. Acta* **35,** 1201–1207.

Döbereiner, J. W. (1822). Zur Pneumatischen Chemie. III. *Phytocheime* **1822,** 64–74.

Dorsch, J. E. and Bidleman, T. F. (1982). Natural organics as fluorescent tracers of river–sea mixing. *Estuarine Coastal Shelf Sci.* **15,** 701–707.

Dragunov, S. S. and Murzakov. B. G. (1970). Heterogeneity of fulvic acids of a chernozem. *Pochvovedenie* **3,** 115–121.

Drake, E. H. and Motto, H. L. (1982). An analysis of the effect of clay and organic matter content on the cation-exchange capacity of New Jersey soils. *Soil Sci.* **133,** 281–288.

Drushell, H. V. (1977). Determination of nitrogen in petroleum fractions by combustion with chemiluminescent detection of nitric oxide. *Anal. Chem.* **49,** 932–936.

Dryden, J. G. C. (1952). Solvent power for coals at room temperature. Chemical and physical factors. *Chem. Ind. (London),* 502–508.

Dubach, P. and Mehta, N. C. (1963). The chemistry of soil humic substances. *Soils and Fertilizers* **26,** 293–300.

Dubach, P., Mehta, N. C., and Deuel, H. (1961). Extraktion von Huminstoffen aus dem B-Horizont eines Podsols mit ADTE. *Z. Pflanzenernahr Dung. Bodenk.* **95,** 119–123.

Dubach, P., Mehta, N. C., Jakab, T., Martin, F., and Roulet, N. (1964). Chemical investigations on soil humic substances. *Geochim. Cosmochim. Acta* **28,** 1567–1578.

Durand, B. (ed.) (1980). *Kerogen.* Editions Technip, Paris, 519 pp.

Durand, B. and Monin, J. C. (1980). Elemental analysis of kerogens. In *Kerogen* (B. Durand, ed.). Editions Technip, Paris, pp. 113–142.

Durand, B. and Nicaise, G. (1980). Procedures for kerogen isolation. In *Kerogen* (B. Durand, ed.). Editions Technip, Paris, pp. 35–54.

Durand, B. and Paratte, M. (1983). Oil potential of coals: A geochemical approach. In *Petroleum Geochemistry and Exploration of Europe* (J. Brooks, ed.). Blackwell Scientific, Oxford, pp. 255–265.

Eaker, D. and Porath, J. (1967). Sorption effects of gel filtration. I. A survey of amino acid behavior on Sephadex G-10. *Sep. Sci.* **2,** 507–550.

Eatough, D. J., Christensen, J. J., and Izatt, R. M. (1972a). Determination of equilibrium constants by titration calorimetry. II. Data reduction and calculation techniques. *Thermochim. Acta* **3,** 219–232.

Eatough, D. J., Izatt, R. M., and Christensen, J. J. (1972b). Determination of equilibrium constants by titration calorimetry. III. Application of method to several chemical systems. *Thermochim. Acta* **3,** 233–246.

Eberle, S. H. and Beuerstein, W. (1979). On the pK spectrum of humic acid from natural waters. *Naturwissenschaften* **66,** 572–573.

Eberle, S. H. and Schweer, K. H. (1974). Bestimmung von Huminsaure und Ligninsulfonsaure im Wasser durch Flussig-Flussingextraktion. *Vom Wasser* **41,** 27–44.

Eckert, J. M. and Sholkovitz, E. R. (1976). The flocculation of iron, aluminum, and humates from river water by electrolytes. *Geochim. Cosmochim. Acta* **40,** 847–848.

Eggelsmann, R. (1976). Peat consumption under influence of climate, soil condition, and utilization. *5th Intl. Peat Congr. of IPS* **3,** 233–247.

Ehrenberger, F. and Gorbach, S. (1973). *Methoden der Organischen Elementar-und Spurenanalyse.* Verlag Chemie, Weinheim.

Eisenberg, Henryk (1976). *Biological Macromolecules and Polyelectrolytes in Solution.* Clarendon, Oxford, 272 pp.

Ellwardt, P-C., Haider, E. K., and Ernst, L. (1981). Studies on the microbiological degradation of lignin by NMR spectroscopy with specifically labelled DHP lignin from coniferyl alcohol. *Holzforschung* **35,** 103–109.

Elwood, J. W., Newbold, J. D., O'Neill, R. V., and Van Winkle, W. (1983). Resource spiraling: An operational paradigm for analyzing lotic ecosystems. In *Dynamics of Lotic Ecosystems* (T. D. Fontaine III and S. M. Bartell, eds.). Ann Arbor Science, Ann Arbor, MI, pp. 3–27.

Emerson, W. W. (1959). The structure of soil crumbs. *J. Soil Sci.* **10,** 235–244.

Endres, H. and Hormann, H. (1963). Preparative and analytical separation of organic compounds by means of chromatography on polyamide. *Agnew. Chem. Inst. Ed. Eng.* **2,** 254–260.

Ennis, M. T. and Brogan, J. C. (1961). The availability of Cu from copper humic acid complexes. *Irish. J. Agric. Res.* **1,** 35–42.

Ernst, W. (1966). Nachweis, Identifizierung und quantitative Bestimung von Uronsauren im Polysaccharidfraktionen mariner Sedimente der Deutschen Bucht. *Veroff. Inst. Meeresforsch. Bremerh.* **10,** 81–92.

Ertel, J. R. and Hedges, J. I. (1983). Bulk chemical and spectroscopic properties of marine and terrestrial humic acids, melanoidins, and catechol-based synthetic polymers. In *Aquatic and Terrestrial Humic Materials* (R. F. Christman and E. T. Gjessing, eds.). Ann Arbor Science, Ann Arbor, MI, pp. 143–162.

Espitalie, J., Laporte, J. L., Madec, M., Marquis, F., Leplat, P., Paulet, J., and Boutefeu, A. (1977). Methode rapide de caracterisation des roches meres, de leur potentiel petrolier et de leur degre devolution. *Rev. Inst. Fr. Petr.* **XXXII** (1), 23–42.

Ewald, M., Belin, C., Berger, P., and Weber, J. H. (1983). Corrected fluorescence spectra of fulvic acids isolated from soil and water. *Environ. Sci. Technol.* **17,** 501–504.

Exarchos, C. C. (1976). Studies of the fate of cell wall polymers of higher plants in peat: A contribution to the geochemistry of coal. Ph.D. dissertation, Pennsylvania State University, 157 pp.

Fagbenro, J. A. (1984). Extraction of humic substances from tropical soil. Personal Communication, Chemistry Dept., University of Birmingham.

Farmer, V. C., Skjemstad, J. O., and Thompson, C. H. (1983). Genesis of humus B horizons in hydromorphic humus podzols. *Nature* **304,** 342–344.

Farnham, R. S. (1978). Wetlands as energy sources. In *Proceedings of the National Symposium on Wetlands.* American Water Research Association, pp. 661–672.

Farnham, R. S. and Finney, H. R. (1965). Classification and properties of organic soils. *Adv. Agron.* **17,** 115–162.

Feder, G. L. and Lee, R. W. (1981). Water-quality reconnaissance of Cretaceous aquifers in the southeastern coastal plain. *U.S. Geol. Surv. Open File Report No. 81-696,* 10 pp.

Felbeck, G. T., Jr. (1965a). Studies on the high pressure hydrogenolysis of the organic matter from a muck soil. *Soil Sci. Am. Proc.* **29,** 48–55.

Felbeck, G. T., Jr. (1965b). Structural chemistry of soil humic substances. *Adv. Agron.* **17,** 327–368.

Felbeck, G. T. (1971). Chemical and biological characterization of humic matter. In *Soil Biochemistry,* Vol. 2 (A. D. McLaren and J. Skujins, eds.). Marcel Dekker, New York, pp. 36–59.

Fenchel, T. and Blackburn, T. H. (1979). *Bacteria and Mineral Cycling.* Academic, New York, 225 pp.

Ferry, J. D. (1970). *Viscoelastic Properties of Polymers,* Second Edition. Wiley, New York, Chap. 4.

Fester, J. I. and Robinson, W. E. (1966). Oxygen functional groups in Green River oil shale kerogen and trona acids. *Coal. Sci. (Adv. Chem. Ser.)* **55,** 22–24.

Figura, P. and McDuffle, B. (1980). Determination of labilities of soluble trace metal species in aqueous environmental samples by anodic stripping voltammetry and Chelex column and batch methods. *Anal. Chem.* **52,** 1433–1439.

Filip, Z., Haider, K., Beutelspacher, H., and Martin, J. P. (1974). Comparison of IR-spectra from melanins of microscopic soil fungi, humic acids, and model phenol polymers. *Geoderma* **11,** 37–52.

Fischer, F., and Schrader, H. (1921). The origin and chemical structure of coal. *Brennstoff-Chem.* **2,** 37–45.

Fisher, L. (1969). An introduction to gel chromatography. In *Laboratory Techniques in Biochemistry and Molecular Biology,* Vol. 1, Pt. II (T. S. Work and E. Work, eds.). North Holland, Amsterdam, pp. 157–390.

Fisher, N. S., Bjerregaard, P., Huynh-Ngoc, and Harvey, G. R. (1983). Interactions of marine plankton with transuranic elements. II. Influence of dissolved organic compounds on americium and plutonium accumulation in a diatom. *Mar. Chem.* **13,** 45–56.

Flaig, W. (1960a). Chemie der Humusstoffe. *Soumen Kem.* **A33,** 229–231.

Flaig, W. (1960b). Comparative chemical investigations on natural humic compounds and their model substances. *Sci. Proc. Roy. Dublin Soc.* **4,** 49–62.

Flaig, W. (1966a). Chemistry of humic substances in relation to coalification. *Coal Sci.* (Adv. Chem. Ser.) **55,** 64–65.

Flaig, W. (1966b). The chemistry of humic substances. In *The Use of Isotopes in Soil Organic Matter Studies.* Report of FAO/IAEA Technical Meeting, Pergamon, Elmsford, NY, pp. 103–127.

Flaig, W. (1967). Chemical composition and physical properties of humic substances. In *Studies About Humus, Transactions of the International Symposium "Humus Et Planta IV"* (B. Noval and V. Rypacek, eds.). Prague, pp. 81–112.

Flaig, W. (1972). Biochemical factors in coal formation. In *Advances in Organic Geochemistry* H. R. von Gaertner and H. Wehner, eds.). Pergamon, Elmsford, N.Y., pp. 197–232.

Flaig, W. and Beutelspacher, H. (1951). Zur kenntnis der huminsauern. *Z. Pflanzenernahr. Dung. Bodenk.* **52,** 1–21.

Flaig, W. and Beutelspacher, H. (1954). Physikalische Chemie der Huminsauren, Landbouwk. *Tijdschr.* **66,** 306–336.

Flaig, W., Beutelspacher, H., and Rietz, E. (1975). Chemical composition and physical properties of humic substances. In *Soil Components: Volume 1. Organic Components* (J. E. Gieseking, ed.) Springer-Verlag, New York, pp. 1–211.

Flaig, W. and Breyhan, T. (1956). Uber das Vorkommen von Indolverbindungen in Schwarzerde-Huminsauren. *Z. Pflanzenernahr. Dung. Bodenk.* **75,** 132–135.

Flory, P. J. (1953). *Principles of Polymer Chemistry.* Cornell University Press, Ithaca, NY.

Forsyth, W. G. C. (1947). Studies on the more soluble complexes of soil organic matter. I. A method of fractionation. *Biochem. J.* **41,** 176–181.

Forsman, P. (1963). Geochemistry of kerogen. In *Organic Geochemistry* (I. A. Berger, ed.). Pergamon, Elmsford, NY, pp. 148–182.

Foster, P. and Morris, A. W. (1971). The seasonal variation of dissolved ionic and organically associated copper in the Menai Straits. *Deep-Sea Res.* **18,** 231–236.

Fotiyev, A. V. (1970). The nature of aqueous humus. *Dokl. Akad. Nauk USSR* **199,** 193–195.

Fox, L. E. (1981). The geochemistry of humic acid and iron during estuarine mixing. Ph.D. Thesis, University of Delaware, 219 pp.

Fox, L. E. (1983a). The removal of dissolved humic acid during estuarine mixing. *Estuarine Coastal Shelf Sci.* **16,** 431–440.

Fox, L. E. (1983b). Geochemistry of humic acid during estuarine mixing. In *Aquatic and Terrestrial Humic Materials* (R. Christman and E. T. Gjessing, eds.). Ann Arbor Science, Ann Arbor, MI, pp. 407–427.

Francko, D. A. and Heath, R. T. (1979). Functionally distinct classes of complex phosphorus compounds in lake water. *Limnol. Oceanogr.* **24,** 463–473.

Francko, D. A. and Heath, R. T. (1982). UV-sensitive complex phosphorus: Association with dissolved humic material and iron in a bog lake. *Limnol. Oceanogr.* **27,** 564–569.

Franks, F. (1975). The hydrophobic interaction. In *Water, A Comprehensive Treatise. Aqueous Solutions of Amphiphiles and Macromolecules,* Vol. 4 (F. Franks, ed.). Plenum, New York, pp. 1–94.

Franks, F. (1983). *Water.* The Royal Society of Chemistry, London.

Fraser, D. C. (1961). Organic sequestration of copper. *Econ. Geol.* **56,** 1063–1078.

Freedman, B. and Hutchinson, T. C. (1980). Effect of smelter pollutants on forest leaf litter decomposition near a nickel–copper smelter at Sudbury, Ontario. *Can. J. Bot.* **558,** 1722–1736.

Freeman, Ray, and Hill, J. D. W. (1971). Fourier transform study of NMR spin–spin relaxation. *J. Chem. Phys.* **55,** 1985–1986.

Frercks, W. and Puffe, D. D. (1959). Comparative studies on "soil respiration" versus CO_2 evolution from various peat soils. *Z. Pflanzenernahrung Dung. Bodenk.* **87,** 108–118.

Fridman, S. A. (1967). Pelleting techniques in infrared analysis—A review and evaluation. In *Progress in Infrared Spectroscopy,* Vol. 3. Plenum Press, New York (H. A. Szymanski, ed.). pp. 1–23.

Frimmel, F. H., Immerz, I., Niedermann, H., and Sattler, D. (1980). Komplexierung von Metallionen durch Bewaesserhumistoffe. II. Charakterisierung der isolierten Humistoffe. *Z. Wasser Abwasser Forsch.* **13,** 119–124.

Frimmel, F. H. and Sattler, D. (1982). Komplexchemische Charakterisierung isolierter Gewaesserhumistoffe nach gelchromatographischer Fraktionierung. *Vom Wasser* **59,** 335–350.

Fritz, J. S., Gjerde, P. T., and Pohlandt, C. (1982). *Ion Chromatography.* Verdoy, New York.

Fuchs, W. (1930). Huminsauren. *Kolloid Z.* **52,** 248–252; **53,** 124–126.

Fuchs, W. (1931). *Die Chemie der Kohle.* Springer, Berlin.

Fuchsman, C. H. (1980). *Peat. Industrial Chemistry and Technology.* Academic, New York, 279 pp.

Fukushima, K. (1982). Vacuum pyrolysis of recent sedimentary humic acids and kerogens. *Geochem. J.* **16,** 43–49.

Fuller, W. H. and Artiola-Fortuny, J. (1982). Humic substances in landfill leachates: I. Humic acid extraction and identification. *J. Environ. Quality* **11,** 663–669.

Fuoss, R. M. and Strauss, U. M. (1948). Electrostatic interaction of polyelectrolytes and simple electrolytes. *J. Polym. Sci.* **3,** 602–603.

Gagosian, R. B. and Lee, C. (1981). Process controlling the distribution of biogenic organic compounds in seawater. In *Marine Organic Chemistry* (E. K. Duursma and R. Dawson, eds.). Elsevier, New York, pp. 91–123.

Gagosian, R. B. and Stuermer, D. H. (1977). The cycling of biogenic compounds and their diagenetically transformed products in seawater. *Mar. Chem.* **5,** 605–632.

Galimov, E. M. (1980). $^{13}C/^{12}C$ in kerogen. In *Kerogen* (B. Durand, ed.). Editions Technip, Paris, pp. 271–300.

Gamble, D. S. (1970). Titration curves of fulvic acid: The analytical chemistry of a weak acid polyelectrolyte. *Can. J. Chem.* **48,** 2662–2669.

Gamble, D. S. (1972). Potentiometric titration of fulvic acid: Equivalence point calculations and acidic functional groups. *Can. J. Chem.* **50,** 2680–2690.

Gamble, D. S. (1973). Na^+ and K^+ binding by fulvic acid. *Can. J. Chem.* **51,** 3217–3222.

Gamble, D. S. and Schnitzer, M. (1974). The chemistry of fulvic acid and its reaction with metal ions. In *Trace Metals and Metal–Organic Interactions* (P. Singer, ed.). Ann Arbor Science, Ann Arbor, MI, pp. 225–302.

Gamble, D. S., Schnitzer M., and Hoffman, I. (1970). Cu^{2+}–fulvic acid chelation equilibrium in 0.1*M* KCl at 25°C. *Can J. Chem.* **48,** 3197–3204.

Gamble, D. S., Schnitzer, M., and Skinner, D. S. (1977). Mn(II)–fulvic acid complexing equilibrium measurements by electron spin resonance spectrometry. *Can. J. Soil Sci.* **57,** 47–53.

Gardner, W. S. and Landrum, P. F. (1983). Characterization of ambient levels of ultraviolet-absorbing dissolved humic materials in natural waters by aqueous liquid chromatography. In *Aquatic and Terrestrial Humic Materials* (R. F. Christman and E. T. Gjessing, eds.). Ann Arbor Science, Ann Arbor, MI, pp. 203–217.

Garroway, A. N., Moniz, W. B., and Resing, H. A. (1976). ^{13}C NMR in cured epoxies: Benefits of magic angle rotation. In *Organic Coatings and Plastics*. Preprints, 172nd ACS meeting, Vol. 36, p. 133.

Geering, H. R., Hodgson, J. F., and Sdano, C. (1969). Micronutrient cation complexes in soil solution: IV. *Soil Sci. Soc. Am. Proc.* **33,** 81–85.

Gelotte, G. (1960). Studies on the gel filtration sorption properties of the bed material "Sephadex." *J. Chromatogr.* **3,** 330–342.

Ghassemi, M. and Christman, R. F. (1968). Properties of the yellow organic acids of natural waters. *Limnol. Oceanogr.* **13,** 583–597.

Ghosh, K. and Schnitzer, M. (1979). UV and visible absorption spectroscopic investigations in relation to macromolecular characteristics of humic substances. *J. Soil Sci.* **30,** 735–745.

Ghosh, K. and Schnitzer, M. (1980a). Fluorescence excitation spectra of humic substances. *Can. J. Soil Sci.* **60,** 373–379.

Ghosh, K. and Schnitzer, M. (1980b). Macromolecular structures of humic substances. *Soil Sci.* **129,** 266–276.

Gierasch, L. M., Frey, M. H., Hexem, J. G., and Opella, S. J. (1982). Solid state NMR of linear and cyclic peptides. In *NMR Spectroscopy: New Methods and Applications, American Chemical Society Symposium Series,* Vol. 191 (G. C. Levy, ed.) American Chemical Society, Washington, D.C., pp. 233–248.

Gilbert, E. (1980). Einwirkung von Ozon auf Huminšauren in wäßrigen Lösungen. *Vom Wasser* **55,** 1–13.

Gillam, A. H. and Riley, J. P. (1981). Correction of osmometric number-average molecular weights of humic substances for dissociation. *Chem. Geol.* **33,** 355–366.

Gillam, A. H. and Wilson, M. A. (1983). Application of ^{13}C NMR spectroscopy to the structural elucidation of dissolved marine humic substances and their phytoplankton precursors. In *Aquatic and Terrestrial Humic Materials* (R. F. Christman and E. T. Gjessing, eds.) Ann Arbor Science, Ann Arbor, MI, pp. 25–36.

Gillet, A. (1956). Von der Cellulose zum Anthrazit Die Entwicklung der Humine aus holziger Materie. II. Desoxydierende Hydrierung durch Disproportionierung. *Brennstoff Chemie* **37,** 395–402.

Given, P. H. (1972). Biological aspects of the geochemistry of coals. In *Advances in Organic Geochemistry 1981* (H. R. von Gartner and H. Wehner, eds.). Pergamon, Elmsford, NY, pp. 69–92.

Given, P. H., and Dickinson, C. H. (1975). Biochemistry and Microbiology of peat. In *Soil Biochemistry,* Vol. 3 (E. A. Paul and A. D. McLaren, eds.). Marcel Dekker, New York.

Gjessing, E. T. (1965). Use of "Sephadex" gel for the estimation of molecular weight of humic substances in natural water. *Nature* (*London*) **208,** 1091–1092.

Gjessing, E. T. (1970). Ultrafiltration of aquatic humus. *Environ. Sci. Technol.* **4,** 437–438.

Gjessing, E. T. (1973). Gel and ultramembrane filtration of aquatic humus: A comparison of the two methods. *Schweiz, A. Hydrol.* **35,** 286–294.

Gjessing, E. T. (1976). *Physical and Chemical Characteristics of Aquatic Humus.* Ann Arbor Science, Ann Arbor, MI, 120 pp.

Gjessing, E. T. (1981). The effect of aquatic humus on the biological availability of cadmium. *Arch. Hydrobiol.* **91,** 144–149.

Gjessing, E. T. and Berglind, L. (1981). Absorption of PAH to aquatic humus. *Arch. Hydrobiol.* **92,** 24–30.

Gjessing, E. T. and Gjerdahl, T. G. (1970). Influence of ultraviolet radiation on aquatic humus. *Vatten* **26,** 144–145.

Gjessing, E. T. and Gjerdahl, T. G. (1975). Electromobility of aquatic humus: Fractionation by use of the isoelectric focusing technique. *Proceedings of the International Meeting on Humic Substances, Nieuwerslius, 1972*. Poduc, Wageningen, pp. 43–51.

Gjessing, E. T. and Lee, G. F. (1967). Fractionation of organic matter in natural waters on Sephadex columns. *Environ. Sci. Technol.* **1,** 631–638.

Glaze, W. H., Jones, P. C., and Saleh, F. W. (1981). Size exclusion, reverse phase, and weak anion exchange chromatography of natural organics in water. In *Advances in the Identification and Analysis of Organic Pollutants in Water,* Vol. 1 (L. H. Keith, ed.). Ann Arbor Science, Ann Arbor, MI, 120 pp.

Gloor, R., Leidner, H., Wuhrmann, K., and Fleischmann, T. H. (1981). Exclusion chromatography with carbon detection. A tool for further characterization of dissolved organic carbon. *Water Res.* **15,** 457–462.

Glover, C. A. (1975). Absolute colligative property measurements. In *Polymer Molecular Weights,* Pt. 1 (P. E. Slade, Jr., ed.). Marcel Dekker, New York, pp. 79–159.

Gocke, K., Dawson, R., and Liebeweit, G. (1981). Availability of dissolved free glucose to heterotrophic microorganisms. *Mar. Biol.* **62,** 209–216.

Goh, K. M. and Williams, M. R. (1979). Changes in molecular weight distribution of soil organic matter during soil development. *J. Soil Sci.* **30,** 747–755.

Gonzalez, N. M. de, Castagnola, M., and Rossetti, D. (1981). Humic characterization of Columbian soil by disc electrophoresis and infrared spectroscopy following gel filtration. *J. Chromatogr.* **209,** 421–431.

Goodhew, P. J. (1975). *Electron Microscopy and Analysis.* Wykeham Publications, London, 191 pp.

Goodman, B. A. (1980). Exchange of comments on the simulation of electron paramagnetic resonance spectra of copper–fulvic acid complex. *Anal. Chem.* **52,** 1770–1771.

Goodman, B. A. and Cheshire, M. V. (1972). A Mossbauer spectroscopic study of the effect of pH on the reaction between iron and humic acid in aqueous media. *J. Soil Sci.* **30,** 85–91.

Goodman, B. A. and Cheshire, M. V. (1973). Electron paramagnetic resonance evidence that copper is complexed in humic acid by the nitrogen of porphyrin groups. *Nature* (*London*) **244,** 158–159.

Goodman, B. A. and Cheshire, M. V. (1975). The bonding of vanadium in complexes with humic acid: An electron paramagnetic resonance study. *Geochim. Cosmochim. Acta* **39,** 1711–1713.

Goodman, B. A. and Cheshire, M. V. (1976). The occurrence of copper–porphyrin complexes in soil humic acids. *J. Soil Sci.* **27,** 337–347.

Goodman, B. A. and Cheshire, M. V. (1982). Reduction of molybdate by soil organic matter: EPR evidence for formation of both Mo(V) and Mo(III). *Nature* **299,** 618–620.

Gordon, D. D. and Sutcliffe, W. H. (1974). Filtration of seawater using silver filters for particulate nitrogen and carbon analysis. *Limnol. Oceanogr.* **19,** 989–993.

Gore, A. J. P. (ed.) (1983). *Ecosystems of the World. Mires: Swamp, Bog, Fen, and Moor. 4A. General Studies. 4B. Regional Studies.* Elsevier, New York, pp. 440–479.

Goryunova, S. V. (1952). Characteristics of dissolved organic substances in water of Glubokae Lake. *Tr. Inst. Mikrobiol. Akad. Nauk SSSR* **3,** 166–179.

Goryunova, S. V. (1954). Characterization of dissolved organic substances in water of Beloe Lake. *Tr. Inst. Mikrobiol. Akad. Nauk SSSR* **3,** 185–193.

Grant, D. (1977). Chemical structure of humic substances. *Nature* **270,** 709–710.

Greenland, D. J. (1965a). Interaction between clays and organic compounds in soils. Part I. Mechanisms of interaction between clay and defined organic compounds. *Soils and Fertilizers,* **38 (no. 5),** 415–532.

Greenland, D. J. (1965b). Interactions between clays and organic compounds in soils. Part II. Absorption of Soil Organic Compounds and its effect on soil properties. *Soils and Fertilizers* **28 (no. 6),** 521–532.

Greenland, D. J. (1971). Interactions between humic and fulvic acids and clay. *Soil Sci.* **111,** 34–41.

Greenland, D. J. and Mott, C. J. B. (1978). Surfaces of soil particles. In *The Chemistry of Soil Constituents* (D. J. Greenland and M. H. B. Hayes, eds.). Wiley, New York, pp. 321–353.

Greter, F. L., Buffle, J., and Haerdi, W. (1979). Voltammetric study of humic and fulvic substances. Part I. *J. Electroanal. Chem.* **101,** 211–229.

Griffin, R. A. and Chian, E. S. K. (1980). Attenuation of water-soluble polychlorinated biphenyl by earth materials. U.S. Environmental Protection Agency EPA-600/2-80-028, Cincinnati.

Griffith, S. M., Silver, J., and Schnitzer, M. (1980). Hydrazine derivatures at Fe^{3+} sites in humic materials. *Geoderma* **23,** 299–302.

Griffiths, P. R. (1975). *Chemical Infrared Fourier Transform Spectroscopy.* Wiley, New York, 340 pp.

Griffiths, P. R. and Block, F. (1973). Considerations for infrared ultramicrosampling. *Appl. Spectrosc.* **27,** 431–435.

Guinier, A. and Fournet, G. (1955). *Small Angle X-Ray Scattering.* Wiley, New York, 268 pp.

Guy, R. D. and Chakrabarti, C. L. (1976). Analytical techniques for speciation of heavy metal ions in the aquatic environment. *Chem. Can.* **28,** 26–29.

Hackmann, R. H. and Todd, A. R. (1953). Some observations on the reaction of catechol derivatives with amines and amino acids in the presence of oxidizing agents. *Biochem. J.* **55,** 631–637.

Hair, M. E. and Bassett, C. R. (1973). Dissolved and particulate humic acids in an East Coast estuary. *Estuarine Coastal Mar. Sci.* **1,** 107–111.

Hall, K. J. (1970). Natural organic matter in the aquatic environment. Ph.D. dissertation, University of Wisconsin.

Hall, J. K. and Lee, G. F. (1974). Molecular size and spectral characterization of organic matter in a meromictic lake. *Water Res.* **8,** 239–251.

Hallsworth, E. C. and Wilkinson, G. K. (1958). The contribution of clay and organic matter to the cation-exchange capacity of the soil. *J. Agric. Sci.* **51,** 1–3.

Hama, T. and Handa, N. (1980). Molecular weight distribution and characterization of dissolved organic matter from lake waters. *Arch. Hydrobiol.* **90,** 106–120.

Handa, N. (1972). Organogeochemical studies of a 200-meter core sample from Lake Biwa. I. The determination of carbohydrate and organic carbon. *Proc. Japan Acad.* **48,** 510–515.

Handa, N. (1973). Organogeochemical studies of a 200–meter core sample from Lake Biwa. II. The determination of protein and lipid. *Proc. Japan Acad.* **49,** 265–270.

Hansen, C. M. (1967). *Three-Dimensional Solubility Parameter and Solvent Diffusion Coefficient.* Danish Technical Press, Copenhagen.

Hansen, E. H. and Schnitzer, M. (1966). The alkaline permanganate oxidation of Danish illuvial organic matter. *Soil Sci. Soc. Am. Proc.* **30,** 745–748.

Hansen, E. H. and Schnitzer, M. (1969). Molecular weight measurements of polycarboxylic acids in water by vapor pressure osmometry. *Anal. Chim. Acta* **46,** 247–254.

Harris, R. F., Chesters, G., and Allen, O. N. (1966). Dynamics of soil aggregation. *Adv. Agron.* **18,** 107–169.

Harrison, P. G. and Mann, K. H. (1975). Detritus formation from eelgrass (*Zostera marina L.*). The relative effects of fragmentation, leaching, and decay. *Limnol. Oceanogr.* **20,** 924–934.

Harvey, G. R. and Boran, D. A. (1982a). The synthesis of marine humic substances. *Trans. Am. Geophys. U.* **63,** 62.

Harvey, G. R. and Boran, D. A. (1982b). Physical, chemical, and PMR spectroscopic properties of dissolved humic substances from the Gulf of Mexico. *Trans. Am. Geophys. U.* **63,** 62.

Harvey, G. R., Boran, D. A., Chesal, L. A., and Tokar, J. M. (1983). The structure of marine fulvic and humic acids. *Mar. Chem.* **12,** 119–132.

Hassett, J. P. and Anderson, M. (1979). Association of hydrophobic organic compounds with dissolved organic matter in aquatic systems. *Environ. Sci. Technol.* **13,** 1526–1629.

Hatcher, P. G. (1978). The organic geochemistry of Mangrove Lake, Bermuda. *NOAA Prof. Paper* **10,** 90–92.

Hatcher, P. G. (1980). The origin, composition, chemical structures and diagenesis of humic substances, coals and kerogens as studied by nuclear magnetic resonance. Ph.D. dissertation, University of Maryland, 283 pp.

Hatcher, P. G., Breger, I. A., Dennis, L. W., and Maciel, G. E. (1982a). Chemical structures in coal: NMR studies and a geochemical approach. In *American Chemical Society Division of Fuel Chemistry Preprints,* Vol. 27, Nos. 3–4. Kansas City, Missouri, pp. 172–183.

Hatcher, P. G., Breger, I. A., Dennis, L. W., and Maciel, G. E. (1983a). Solid-state ^{13}C NMR of sedimentary humic substances: New revelations on their chemical composition. In *Aquatic and Terrestrial Humic Materials* (R. F. Christman and E. T. Gjessing, eds.). Ann Arbor Science, Ann Arbor, MI, pp. 37–82.

Hatcher, P. G., Breger, I. A., and Earl. W. L. (1981a). Nuclear magnetic resonance studies of ancient buried wood. I. Observations on the origin of coal to the brown coal stage. *Org. Geochem.* **3,** 49–55.

Hatcher, P. G., Breger, I. A., and Mattingly, M. A. (1980a). Structural characteristics of fulvic acids from continental shelf sediments. *Nature* **285,** 560–562.

Hatcher, P. G., Breger, I. A., Szeverenyi, N. M., and Macial, G. E. (1982b). Nuclear magnetic resonance studies of ancient buried wood. II. Observations on the origin of coal from lignite to bituminous coal. *Org. Geochem.* **4,** 9–18.

Hatcher, P. G., Lyons, P. C., Thompson, C. L., and Brown, F. W. (1982c). Organic matter in a coal ball: Peat or coal? *Science* **217,** 831–833.

Hatcher, P. G., Maciel, G. E., and Dennis, L. W. (1981b). Aliphatic structure of humic acids: A clue to their origin. *Org. Geochem.* **2,** 43–48.

Hatcher, P. G., Rowan, R., and Mattingly, M. A. (1980b). ^{1}H and ^{13}C NMR or marine humic acids. *Org. Geochem.* **2,** 77–85.

Hatcher, P. G., Schnitzer, M., Dennis, L. W., and Maciel, G. E. (1981c). Aromaticity of humic substances in soils. *Soil Sci. Soc. Am. J.* **45,** 1089–1094.

Hatcher, P. G., Spiker, E. C., Szeverenyi, N. M., and Maciel, G. E. (1983b). Selective preservation and origin of petroleum-forming aquatic kerogen. *Nature* **305,** 498–501.

Hatcher, P. G., VanderHart, D. L., and Earl, W. L. (1980c). Use of solid-state ^{13}C NMR in structural studies of humic acids and humin from Holocene sediments. *Org. Geochem.* **2,** 87–92.

Havlicek, S. C., Reuter, J. H., and Ghosal, M. (1979). Identification of major and minor classes of natural organic substances found in drinking water. *U.S. Environmental Protection Agency Report* No. 68-10-4480, U.S. Government Printing Office, Washington, D.C.

Haworth, R. D. (1971). Chemical nature of humic acid. *Soil Sci.* **111,** 71–79.

Hayatsu, R., Winans, R. E., McBeth, R. L., Scott, R. G., Moore, L. P., and Studier, M. H. (1981). Structural characterization of coal: Lingin-like polymers in coals. In *Coal Structure* (*Avd. Chem. Ser.*) **192,** 149–150.

Hayes, M. H. B. (1970). Adsorption of triazine herbicides on soil organic matter including a short review on organic matter chemistry. *Residue Rev.* **32,** 131–174.

Hayes, M. H. B., Law, I., and Tuck, J. J. (1983). Dissolution of humic substances in organic and aqueous solvents. Unpublished data, Chemistry Department, University of Birmingham.

Hayes, M. H. B. and Swift, R. S. (1978). The chemistry of soil organic colloids. In *The Chemistry of Soil Constituents* (D. J. Greenland and M. H. B. Hayes, eds.). Wiley-Interscience, New York, pp. 179–230.

Hayes, M. H. B., Swift, R. S., Wardle, R. E., and Brown, J. K. (1975). Humic materials from an organic soil: A comparison of extractants and of properties of extracts. *Geoderma* **13,** 231–245.

Hayes, M. H. B. and Swift, R. S. (1981). Organic colloids and organo-mineral associations. *Int. Soc. Soil Sci. Bull.* **60,** 67–74.

Hedges, J. (1976). The formation and clay mineral reactions of melanoidin. *Carnegie Inst. Washington Yearb.* **75,** 792–800.

Hedges, J. I., Ertel, J. R., and Leopold, E. B. (1982). Lingin geochemistry of a Late Quaternary sediment core from Lake Washington. *Geochim. Cosmochim. Acta* **46,** 1869–1877.

Heinrich, B. J., Grimes, M. D., and Puckett, J. E. (1961). Sulfur. In *Treatise On Analytical Chemistry,* Vol. 7, Pt. 2 (I. M. Kolthoff and P. J. Elving, eds.). Wiley-Interscience, New York, pp. 1–135.

Helling, S. C., Chesters, G., and Corey, R. B. (1964). Contribution of organic matter and clay to soil cation-exchange capacity as affected by the pH of the saturating solution. *Soil Sci. Soc. Am. Proc.* **28,** 517–520.

Herbin, J. P. and Deroo, G. (1979). Étude sedimentologique de la matiere organizue dans les argiles noires cretacees de l'Atlantique sud. *Docum. Lab. Geol. Fac. Sci. Lyon,* No. 75, 71–87.

Hercules, D. M. (1966). *Fluorescence and Phosphorescence Analysis.* Wiley, New York, 258 pp.

Hermann, R. (1845). Bermerkungen zu Mulder's Untersuchungen uber Modersubstanzen and Ackererde. *J. Prakt. Chem* **34,** 156–163.

Hermans, J. J. and Overbeek, J. Th. G. (1948). The dimensions of charged long-chain molecules in solutions containing electrolytes. *Rec. Trav. Chim.* **67,** 761–776.

Hickel, B. (1978). Phytoplankton population dynamics in Plusse (East-Holstein, Germany). *Verh. Gesell Oekol.* **7,** 119–129.

Hildebrand, J. H. Prausnitz, J. N., and Scott, R. L. (1970). *Regular and Related Solutions.* Van Nostrand–Reinhold, New York.

Hildebrand, J. H. and Scott, R. L. (1951). *Solubility of Non-Electrolytes,* 3rd ed. Reinhold, New York.

Hildebrand, J. H. and Scott, R. L. (1962). *Regular Solutions.* Prentice-Hall, Englewood Cliffs, NJ.

Hine, J. (1975). *Structural Effects on Equilibria in Organic Chemistry.* Wiley, New York, 347 pp.

Hobson, R. P. and Page, H. J. (1932). Studies on the carbon and nitrogen cycles of the soil. *J. Agric. Sci.* **22,** 297–299; 497–515; 516–526.

Hodgson, J. F. (1963). Chemistry of micronutrient elements in soils. *Adv. Agron.* **15,** 119–159.

Hodgson, J. F., Lindsay, W. L., and Trierweiler, J. F. (1966). Nutrient cation complexing in soil solution. II. *Soil Sci. Soc. Am. Proc.* **30,** 723–726.

Hoering, T. C. (1973). A comparison of melanoidin and humic acid. *Carnegie Inst. Washington Yearb.* **72,** 682–690.

Holty, J. G. and Heilman, P. E. (1971). Molecular sieve fractionation of organic matter in a podsol from southeastern Alaska. *Soil Sci.* **112,** 351–356.

Holtzclaw, K. M. and Sposito, G. (1979). Analytical properties of the soluble, metal-complexing fractions in sludge–soil mixtures: IV. Determination of carboxyl groups in fulvic acid. *Soil Sci. Soc. Am. J.* **43,** 318–323.

Hong, J. K. and Yamane, I. (1981). Distribution of inositol phosphate in the molecular size fractions of humic and fulvic acid fractions. *Soil Sci. Plant Nutr.* **27,** 295–303.

Hongve, D., Skogheim, O. K., Hindar, A., and Abrahamsen, H. (1980). Effects of heavy metals in combination with NTA, humic acid, and suspended sediment on natural phytoplankton photosynthesis. *Bull. Environ. Contam. Toxicol.* **25,** 594–600.

Honjo, F. (1980). Material fluxes and modes of sedimentation in the mesopelagic and bathipelagic zones. *J. Mar. Res.* **38,** 53–97.

Höpner, T. and Orliczek, C. (1978). Humic matter as a component of sediments in estuaries. In *Biogeochemistry of Estuarine Sediments*. UNESCO, Paris, pp. 70–74.

Hoppe-Seyler, F. (1889). Uber Huminsubstanzen, ihre Emstenung and ihre Eigenschaften. 2. *Physiol. Chem.* **13,** 66–121.

Horner, L. and Sturm, K. (1957). Zur kenntnis der o-chinone. 10. Modellreaktionen zur Melaninbildung. *Liebigs Ann. Chem.* **608,** 128–139.

Hsu, T. S. and Bartha, R. (1974). Biodegradation of chloroaniline–humus complexes in soil and in culture solutions. *Soil Sci.* **118,** 213–220.

Huang, W. H. and Keller, W. D. (1971). Dissolution of clay minerals in dilute organic acids at room temperature. *Am. Mineral.* **56,** 1082–1095.

Huc, A. Y. and Durand, B. (1973). Etude des acides humiques et de l'humine de sediments recents consideres comme precurseurs des kerogenes. In *Advances in Organic Geochemistry* (B. Tissot and F. Bienner, eds.). Editions Technip, Paris, pp. 53–72.

Huc, A. Y., and Durand, B. M. (1977). Occurrence and significance of humic acids in ancient sediments. *Fuel* **56,** 73–80.

Huc, A. Y., Durand, B., and Jacquin, F. (1974). Caracterisation des acides humiques de sediments marins recents et comparaison avec leurs homologues terrestres. *Bull. ENSAIA Nancy* **16,** 59–75.

Huc, A. Y. (1980). Origin and formation of organic matter in recent sediments and its relation to kerogen. In *Kerogen* (B. Durand, ed.). Editions Techniq, Paris, Chapter 14, pp. 445–474.

Huffman, E. W. D., Jr. (1977). Performance of a new automatic carbon dioxide coulometer. *Microchem. J.* **22,** 567–573.

Huizenga, D. L. and Kester, D. R. (1979). Protonation equilibria of marine dissolved organic matter. *Limnol. Oceanogr.* **24,** 145–150.

Hunston, D. L. (1975). Two techniques for evaluating small molecule–macromolecule binding in complex system. *Anal. Biochem.* **63,** 99–109.

Hunter, K. A. (1983). On the estuarine mixing of dissolved substances in relation to colloid stability and surface properties. *Geochim. Cosmochim. Acta* **47,** 467–473.

Huntsman, S. A. and Barber, R. T. (1977). Primary production off northwest Africa: The relationship of wind to nutrient conditions. *Deep-Sea Res.* **24,** 25–33.

Hwang, C. P., Lackie, T. H., and Munch, R. R. (1979). Correction for total organic carbon, nitrate, and chemical oxygen demand when using the MF-Millipore filter. *Environ. Sci. Technol.* **13,** 871–872.

Ihnatowice, A. (1952). Investigations on the oxygen groups in coal. *Glowny Instytut Gornictwa Kommun.* No. 125, Katowice.

Imashiro, F., Maeda, S., Takegoshi, K., Terao, T., and Saika, A. (1982). Intermolecular hydrogen-bonding effects on ^{13}C NMR shielding for enol forms of diketones in the solid state. *Chem. Phys. Lett.* **92,** 642–644.

Incze, L. S., Mayer, L. M., Sherr, E. B., and Macko, S. A. (1982). Carbon inputs to bivalve mollusks: A comparison of two estuaries. *Can. J. Fish. Aquat. Sci.* **39,** 1348–1352.

Ingram, G. (1962). *Methods of Organic Elemental Microanalysis*. Reinhold, New York.

Ishiwatari, R. (1966). Structural characteristics of humic substances in recent lake sediments. In *Advances in Organic Geochemistry* (G. D. Hobson and G. C. Speers, eds.). Pergamon Press, Oxford, pp. 283–311.

Ishiwatari, R. (1967a). Infrared absorption band at 1540 cm^{-1} of humic acid from a recent lake sediment. *Geochem. J.* **1,** 61–70.

Ishiwatari, R. (1967b). Elementary composition of humic acids from lake sediments. *Chikyukagaku (Geochemistry)* **1,** 7–14.

Ishiwatari, R. (1969a). An estimation of the aromaticity of a lake sediment humic acid by air oxidation and evaluation of it. *Soil Sci.* **107,** 53–57.

Ishiwatari, R. (1969b). Fractionation and characterization of humic acid from a lake sediment. *Geochem. J.* **2,** 175–184.

Ishiwatari, R. (1970). Structural characteristics of recent lake sediments. In *Advances in Organic Geochemistry 1966.* (G. D. Hobson and G. C. Speers, eds.). Pergamon, Elmsford, NY, pp. 285–311.

Ishiwatari, R. (1971). Molecular weight distribution of humic acids from lake and marine sediments. *Geochem. J.* **5,** 121–132.

Ishiwatari, R. (1973). Chemical characterization of fractionated humic acids from lake and marine sediments. *Chem. Geol.* **12,** 113–126.

Ishiwatari, R. (1974). Electron spin resonance of sedimentary humic acids in relation to their aromatic character. *Geochem. J.* **8,** 97–102.

Ishiwatari, R. (1975a). Chemical nature of sedimentary humic acid. In *Humic Substances, Their Structure and Function in the Biosphere* (D. Povoledo and H. L. Golterman, eds.). Centre for Agricultural Publications and Documentation, Wageningen, pp. 87–107.

Ishiwatari, R. (1975b). Transformation of sedimentary humic acid: facts and speculation. In *Humic Substances, Their Structure and Function in the Biosphere* (D. Povoledo and H. L. Golterman, eds.). Centre for Agricultural Publications and Documentation, Wageningen, pp. 109–121.

Ishiwatari, R. (1977). Organic geochemistry of a 200-meter core sample from Lake Biwa. III. Isolation and characterization of kerogen and humic acid. *Proc. Japan Acad.* **53,** Ser. B, 47–50.

Ishiwatari, R. (1979). Insoluble organic matter in sediments and its geochemical role. In *Recent Progress of Natural Sciences in Japan* Vol. 4. Science Council of Japan, pp. 103–110.

Ishiwatari, R. and Hanya, T. (1965). Infrared spectroscopic characteristics of humic substances extracted from some recent sediments. *Nippon Kagaku Zasshi (J. Chem. Soc. Japan, Pure Chem. Sec.)* **86,** 1270–1274.

Ishiwatari, R., Ishiwatari, M., Rohrback, B. G., and Kaplan, I. R. (1977). Thermal alteration experiments on organic matter from recent marine sediments in relation to petroleum genesis. *Geochim. Cosmochim. Acta* **41,** 815–828.

Ishiwatari, R. and Kawamura, K. (1981). Abundance of humic compounds in the 200-meter sediment core sample from Lake Biwa. In *Paleolimnology of Lake Biwa and the Japanese Pleistocene,* Vol. 9 (S. Horie, ed.). Kyoto University, pp. 103–111.

Ishiwatari, R., Kosaka, M., and Hanya, T. (1966). Compositional and optical characteristics of humic substances extracted from recent lake sediments. *Nippon Kagaku Zasshi (J. Chem. Soc. Japan, Pure Chem. Sec.)* **87,** 557–566.

Ishiwatari, R. and Machihara, T. (1982). Structural changes of young kerogen on laboratory heating as revealed by alkaline potassium permanganate oxidation. *Geochim. Cosmochim. Acta* **46,** 825–831.

Ishiwatari, R. and Machihara, T. (1983). Early stage incorporation of biolipids into kerogen in a lacustrine sediment: Evidence from alkaline potassium permanganate oxidation of lipids and humic matter. *Org. Geochem.* **4,** 179–184.

Ishiwatari, R., Ogura, K., and Horie, S. (1980a). Organic geochemistry of a lacustrine sediment (Lake Haruna, Japan). *Chem. Geol.* **29,** 261–280.

Ishiwatari, R., Yamamoto, S., and Fukushima, K. (1980b). Possible interactions of insoluble organic geopolymers (kerogen) with other organic compounds in the aquatic environment. I. Interaction with fatty acid methyl esters. In *Advances in Organic Geochemistry 1979 Physics and Chemistry of the Earth,* Vol. 12 (A. G. Douglas and J. R. Maxwell, eds.). Pergamon, Elmsford, NY, pp. 495–503.

Ivarson, K. C. and Schnitzer, M. (1979). The biodegradability of the "unknown" soil nitrogen. *Can. J. Soil Sci.* **59,** 59–67.

Jackson, K. S., Jonasson, I. R., and Skippen, G. B. (1978). The nature of metal–sediment–water interactions in freshwater bodies, with emphasis on the role of organic matter. *Earth Sci. Rev.* **14,** 97–146.

Jackson, T. A. (1975). Humic matter in natural waters and sediments. *Soil Sci.* **119,** 56–64.

Jackson, T. A. and Hecky, R. E. (1980). Depression of primary productivity by humic matter in lake and reservoir waters of the boreal forest zone. *Can. J. Fish. Aquat. Sci.* **37,** 2300–2317.

Jaffé, H. H. and Orchin, M. (1962). *Theory and Applications of Ultraviolet Spectroscopy.* Wiley, New York, 624 pp.

Jamison, V. C. (1945). The penetration of irrigation and rain water into sandy soils in central Florida. *Soil Sci. Soc. Am. Proc.* **10,** 25–29.

Jeffrey, L. M. (1969). Development of a method for isolation of gram quantities of dissolved organic matter from seawater and some chemical and isotopic characteristics of the isolated materials. Ph.D. Thesis, Texas A&M University, College Station, 152 pp.

Jeffrey, L. M. and Hood, D. W. (1958). Organic matter in seawater: An evaluation of various methods for isolation. *J. Mar. Res.* **17,** 203–224.

Jennings, P. W. and Ekeland, R. (1983). Fractionation of fulvic acid using a bonded NH_2 column. Unpublished research, Montana State University.

Jerlov, N. (1955). Factors influencing the transparency of the Baltic waters. *Rept. Oceanogr. Inst. Gothenburg* **25,** 18 pp.

Jespersen, N. D. and Jordan, J. (1970). Thermometric enthalpy titration of proteins. *Anal. Lett.* **3,** 323–334.

Jocteur-Monrozier, L. (1981). Les composes humiques a l'interface ocean–sediment. *Oceanis* **7** (3), 309–325.

Jocteur-Monrozier, L., Benijoly, M., Pillion, P., Andreux, F., Sochier, B., and Pelet, R. (1983). Distribution of organic matter in grain-size fractions of some recent sediments. In *Advances in Organic Geochemistry 1981* (M. Bjoroy et al., eds.). Wiley, New York, pp. 323–327.

Jocteur-Monrozier, L. and Jeanson, P. (1979). L'azote organique et ammoniacal. Combinaisons et stabilite chemique. In *Orgon III, Geochimie organique des sediments marins profonds.* Editions du CNRS, Paris, pp. 405–424.

Jocteur-Monrozier, L. and Jeanson, P., (1981). Caracteres generaux et evolution de l'azote organique dans les sediments superficiels. In *Orgon IV, Geochimie organique des sediments marins profonds.* Editions du CNRS, Paris, pp. 503–528.

Jolley, R. L., Katz, S., Mrochek, J. E., Pitt, W. W., and Rainey, W. T. (1975). Analyzing organics in dilute aqueous solutions. *Chemtech* **5,** 312–318.

Jolley, R. L., Lee, N. E., Pitt, W. W., Denton, M. S., Thompson, J. E., Hartmann, S. J., and Mashni, C. I. (1979). Effects of chlorine, ozone, and ultraviolet light on nonvolatile organics in wastewater effluents. In *Progress in Wastewater Disinfection Technology* (A. D. Venosa, ed.). EPA 600/9-79-018, pp. 233–245.

Jones, G. B. and Jordan, M. B. (1979). The distribution of organic material and trace metals in sediments from the River Liffey Estuary, Dublin. *Estuarine Coastal Mar. Sci.* **8,** 37–47.

Juo, A. S. R. and Barber, S. A. (1969). Reaction of strontium with humic acid. *Soil Sci.* **108,** 89–94.

Kaila, A. (1956). Determination of the degree of humification in peat samples. *Maataloust Aikak* **28,** 18–35.

Kalle, K. (1938). Zum Problem der Meereswasserfarbe. *Ann. Hydrogr. Mar. Meteorol.* **66,** 1–13.

Kalle, K. (1966). The problem of the Gelbstoff in the sea. *Oceanography and Marine Biology, Annual Review* (H. Barnes, ed.) **4,** 91–104.

Kapron, M. and Brandt, M. (1961). Determination of oxygen in organometallic and inorganic compounds by a modified underzaucher method. *Anal. Chem.* **33,** 1762–1764.

Karavaev, N. M. and Budyak, N. F. (1960). Investigation of so-called humic acids from fresh-water sapropels. *Dokl. Akad. Nauk SSSR* **132,** 192–194.

Karavaev, N. M., Vener, P. A., and Koroleva, K. I. (1964). On the composition and chemical nature of sapropelic acids. *Dokl. Akad. Nauk SSSR* **156,** 877–879.

Karger, B. L., Snyder, L. R., and Eon, C. (1976). An expanded solubility parameter treatment for classification and use of chromatographic solvents and adsorbents. Parameters for dispersion, dipole and hydrogen bonding interactions. *J. Chromatogr.* **125,** 71–88.

Karger, B. L., Synder, L. R., and Horvath, C. (1973). *An Introduction to Separation Science.* Wiley-Interscience, New York, 586 pp.

Karickhoff, S. W., Brown, D. S., and Scott, T. A. (1979). Sorption of hydrophobic pollutants on natural sediments. *Water Res.* **13,** 241–248.

Kasparov, S. V., Tikhomirov, F. A., and Fless, A. D. (1981). Use of disc electrophoresis to fractionate humic acids. *Moscow Univ. Soil Sci. Bull.* **36,** 21–28.

Katchalsky, A., Schavit, H., and Eisenberg, H. (1954). Dissociation of weak polymeric acids and bases. *J. Polym. Sci.* **13,** 69–84.

Katchalsky, A. and Spitnik, P. (1947). Potentiometric titration of polymethacrylic acid. *J. Polym. Sci.* **2,** 432–446.

Katz, S., Pitt, W. W., Scott, C. D., and Rosen, A. A. (1972). The determination of stable organic compounds in waste effluents at microgram per liter levels by automatic high-resolution ion exchange chromatography. *Water Res.* **6,** 1029–1037.

Kaufman, D. D., Plimmer, J. R., Kearney, P. C., Blake, J., and Guardia, F. S. (1968). Chemical versus microbial degradation of amitrole in soil. *Weed Sci.* **16,** 266–272.

Kazakov, G. (1953). Soviet peat resources: A descriptive survey. *Research Program on the U.S.S.R.,* New York.

Keller, R. A., Karger, B. L., and Snyder, L. R. (1971). In *Gas Chromatography* (R. Stock, ed.). Institute of Petroleum, London, 125 pp.

Kaufman, D. D., Plimmer, J. R., Kearney, P. C., Blake, J., and Guardia, F. S. (1968). Chemical versus microbial degradation of amitrole in soil. *Weed Sci.* **16,** 266–272.

Kemp, A. L. W. and Johnston, L. M. (1979). Diagenesis of organic matter in the sediments of Lakes Ontario, Erie, and Huron. *J. Great Lakes Res.* **5**(1), 1–10.

Kemp, A. L. W. and Mudrochova, A. (1973). The distribution and nature of amino acids and other nitrogen-containing compounds in Lake Ontario surface sediments. *Geochim. Comsochim. Acta* **37,** 2191–2206.

Kemp, A. L. W. and Mudrochova, A. (1975). Nitrogen in sedimented organic matter from Lake Ontario. In *Humic Substances, Their Structure and Function in the Biosphere.* (D. Povoledo and H. L. Golterman, eds.). Centre for Agricultural Publications and Documentation, Wageningen, pp. 137–157.

Kemp, A. L. W. and Wong, H. K. T. (1974). Molecular-weight distribution of humic substances from Lakes Ontario and Erie sediments. *Chem. Geol.* **14,** 15–22.

Kennedy, V. C., Zellweger, G. W., and Jones, B. F. (1974). Filter pore-size effects on the analysis of Al, Fe, Mn, and Ti in water. *Water Resour. Res.* **10,** 785–790.

Kerker, Milton (1968). Light scattering. *Industrial and Eng. Chem.* **60,** 31–46.

Kerr, R. A. and Quinn, J. G. (1975). Chemical studies on the dissolved organic matter in seawater. Isolation and fractionation. *Deep-Sea Res.* **22,** 107–116.

Khalaf, K. Y., MacCarthy, P., and Gilbert, T. W. (1975). Application of thermometric titrations to the study of soil organic matter. II. Humic acids. *Geoderma* **14,** 331–340.

Khan, A., Hassett, J. J., Banwart, W. L., Means, J. C., and Wood, S. G. (1979). Sorption of acetophenone by sediments and soils. *Soil Sci.* **128,** 297–302.

Khan, D. V. (1945). A method of isolating the insoluble fraction (humin) from podzolic soils. *Dokl. Acad. Nauk Lenina,* 7–8.

Khan, S. U. (1973). Interaction of humic substances with bipyridylium herbicides. *Can. J. Soil Sci.* **53,** 199–204.

Khan, S. U. (1982). Bound residues in soils and plants. *Residue Rev.* **84,** 1–25.

Khan, S. U., Belanger, A., Hogue, E. J., Hamilton, H. A., and Mathur, S. P. (1976a). Residues of paraquat and linuron in an organic soil and their uptake by onions, lettuce, and carrots. *Can. J. Soil Sci.* **56,** 407–412.

Khan, S. U., Hamilton, H. A., and Hogue, E. J. (1976b). Fonofos residues in an organic soil and vegetable crops following treatment of the soil with the insecticide. *Pestic. Sci.* **7,** 553–558.

Khan, S. U. and Schnitzer, M. (1972). Permanganate oxidation of humic acids, fulvic acids, and humins extracted from A_h horizons of a black chernozem, a black solod, and black solonetz soil. *Can. J. Soil Sci.* **52,** 43–51.

Khan, S. U. and Sowden, F. J. (1971). Distribution of nitrogen in the Black Solonetzic and Black Chernozemic soils of Alberta. *Can. J. Soil Sci.* **51,** 185–193.

Khan, S. U. and Sowden, F. J. (1972). Distribution of nitrogen in fulvic acid fractions extracted from the Black Solonetzic and Black Chernozemic soils of Alberta. *Can. J. Soil Sci.* **52,** 116–118.

Khym, J. X. (1974). *Analytical Ion-Exchange Procedures in Chemistry and Biology.* Prentice-Hall, Englewood Cliffs, NJ, 257 pp.

Kickuth, R. and Scheffer, F. (1976). Konstitutive and nutritive Eigenschaften des sogenannten Pseudo-Amidstickstoffs aus Huminsauren. *Agrochimica* **XX,** 376–386.

Kim, B. R., Snoeyink, V. L., and Saunders, F. M. (1976). Adsorption of organic compounds by synthetic resins. *J. Water Pollut. Control Fed.* **48,** 120–133.

King, E. J. (1965). *Acid–Base Equilibria,* MacMillan, New York.

Kirkwood, J. G. and Westheimer, F. J. (1938). The electrostatic influence of substituents on the dissociation constants of organic acids. I. *J. Chem. Phys.* **6,** 506–512.

Kirsten, W. J. (1983). *Organic Elemental Analysis.* Academic, New York.

Kivinen, E. (1980a). New statistics on the utilization of peatlands in different countries. *Proc. 6th Intl. Peat Congr., Duluth, Minn.,* pp. 48–51.

Kivinen, E. (1980b). Proposal for a general classification of virgin peat. *Proc. 6th Intl. Peat Congr., Duluth, Minn.,* pp. 1–47.

Kivinen, E. and Pakarinen, P. (1980). Peatland areas and the proportion of virgin peatlands in different countries. *Proc. 6th Intl. Peat Congr., Duluth, Minn.,* pp. 52–54.

Klocking, R. and Mucke, D. (1969). Isolierung wasserloslicher huminsauren (fulvosauren) aus ihren blie (11)—chelatuerbindungen. *Z. Chem.* **9,** 453–454.

Klotz, I. M. and Hunston, D. L. (1971). Properties of graphical representations of multiple classes of binding sites. *Biochemistry* **10,** 3065–3069.

Kodama, H. and Schnitzer, M. (1968). Thermal analysis of a fulvic acid–montmorillonite complex. *Proc. Intl. Clay, Conf. Tokyo, 1969,* Israel Universities Press, Jerusalem, pp. 765–774.

Koenings, J. P. and Hooper, F. F. (1976). The influence of colloidal organic matter on iron and iron–phosphorus cycling in an acid bog lake. *Limnol. Oceanogr.* **21,** 684–696.

Kolesnikov, M. P. (1978). Molecular weight distribution of humic acids according to gel chromatography on Sephadexes. *Sov. Soil Sci.* **10,** 174–181.

Koller, K. L. and Dorn, H. C. (1982). Acid-catalyzed reactions of 2,2,2-trifluorodiazoethane for analysis of functional groups by ^{19}F nuclear magnetic resonance spectroscopy. *Anal. Chem.* **54,** 529–533.

Kolthoff, I. M. and Elving, P. J. (1959–1981). *Treatise on Analytical Chemistry.* Wiley-Interscience, New York.

Kononova, M. M. (1961). *Soil Organic Matter, Its Nature, Its Role in Soil Formation and Soil Fertility* (translated from the Russian by T. Z. Nowakowski and A. C. D. Newman). Pergamon, Elmsford, NY.

Kononova, M. M. (1966). *Soil Organic Matter.* Pergamon, Elmsford, NY, 544 pp.

Kononova, M. M. (1975). Humus of virgin and cultivated soils. In *Soil Components—Volume 1, Organic Components* (J. E. Gieseking, ed.). Springer, Berlin, pp. 475–526.

Kononova, M. M. and Belchikova, N. P. (1961). Rapid methods of determining the humus composition of mineral soils. *Sov. Soil Sci.* **10,** 75–87.

Kopfler, F. C., Melton, R. G., Mullaney, J. L., and Tardiff, R. G. (1975). Human exposure to water pollutants. In *Fate of Pollutants in the Air and Water Environments,* Pt. 2 (I. H. Suffet, ed.). Wiley-Interscience, New York, pp. 419–433.

Koyama, T. (1966). Ratios of organic carbon, hydrogen, and nitrogen in recent sediments, *Coal Sci. (Adv. Chem. Ser.)* **55,** 43–44.

Kramer, C. J. M. and Duinker, J. C. (1984). Complexation capacity and conditional stability constants for copper of sea- and estuarine waters, sediment extracts and colloids. In *Complexation of Trace Metals in Natural Waters* (C. J. M. Kramer and J. C. Duinker, eds.). Martinus Nijhoff/Dr. W. Junk, Dordrecht, The Netherlands, pp. 217–228.

Kratky, O. (1963). X-ray small angle scattering with substances of biological interest in diluted solutions. In *Progress in Biophysics,* Vol. 13. Pergamon, Elmsford, NY, pp. 105–173.

Kreulen, D. J. W. (1962). Chemie der Kohle. Sechs Abhandlungen uber Braunkonlen/Lignite: *Freib. Forsch. H.* **A244,** 80 pp.

Kreulen, D. J. W. and Kreulen-van Selus, F. J. (1956). Humic acids and their role in the formation of coal. *Brennstoff-Chem.* **37,** 14–19.

Krom, M. D. and Sholkovitz, E. R. (1977). Nature and reactions of dissolved organic matter in the interstitial waters of marine sediments. *Geochim. Cosmochim. Acta* **41,** 1565–1573.

Kumada, K. (1955). Absorption spectra of humic acids. *Soil Sci. Plant Nutr.* **1,** 29–30.

Kumada, K. (1977). *Chemistry of Soil Organic Matter* (in Japanese). Tokyo Diagaku Shuppankai, 220 pp.

Kumada, K. and Aizawa, K. (1958). Infrared spectra of humic acids. *Soil Plant Food* (Tokyo) **3,** 151–159.

Kumada, K. and Kawamura, Y. (1968). On the fractionation of humic acids by a fractional precipitation technique. *Soil Sci. Plant Nutr.* **14,** 198–200.

Kunin, R. (1977). Polymeric adsorbents for treatment of waste effluents. *Polym. Eng. Sci.* **17,** 58–62.

Kunin, R. and Suffet, I. H. (1980). Removal of humic material from drinking water by anion-exchange resins. In *Activated Carbon Adsorption of Organics from the Aqueous Phase,* Vol. 2 (M. J. McGuire and I. H. Suffet, eds.). Ann Arbor Science, Ann Arbor, MI, pp. 425–442.

Kyuma, K. (1964). A fractional precipitation technique applied to soil humic substances. *Soil Sci. Plant Nutr.* **10,** 33–35.

Laane, R. W. P. M. (1980). Conservative behavior of dissolved organic carbon in the Ems-Dollart Estuary and the Western Wadden Sea. *Neth. J. Sea Res.* **14,** 192–199.

Ladd, J. N. and Butler, J. H. A. (1975). Humus–enzyme systems and synthetic organic polymer-enzyme analogs. In *Soil Biochemistry,* Vol. 4. (E. A. Paul and A. D. McLaren, eds.). Marcel Dekker, New York, pp. 143–194.

Lai, Y. Z. and Sarkanen, K. V. (1971). Isolation and structural studies. In *Lignins: Occurrence, Formation, Structure, and Reactions.* Wiley-Interscience, New York, 178 pp.

Laitinen, H. A. and Harris, W. E. (1975). *Chemical Analysis,* 2nd ed. McGraw-Hill, New York, 611 pp.

Lakatos, B., Koreca, L., and Meisel, J. (1977a). Comparative study of Mossbauer parameters of iron humates and polyuronates. *Geoderma* **19,** 149–157.

Lakatos, B. and Meisel, J. (1978). Biopolymer–metal complex systems. V. PMR investigation of humic substances. *Acta Agron. Acad. Sci. Hung.* **27,** 313–320.

Lakatos, B., Tibai, T., and Meisel, J. (1977b). ESR spectra of humic acids and their metal complexes. *Geoderma* **19,** 319–338.

Lansing, W. D. and Kraemer, E. O. (1935). Molecular weight analysis of mixtures by sedimentation equilibrium on the Svedberg ultracentrifuge. *J. Am. Chem. Soc.* **57,** 1369–1377.

Larson, R. A. and Hufnal, J. M., Jr. (1980). Oxidative polymerization of dissolved phenols by soluble and insoluble inorganic species. *Limnol. Oceanogr.* **25,** 505–512.

Larson, R. A. and Rockwell, A. L. (1980). Fluorescence spectra of water-soluble humic materials and some potential precursors. *Arch. Hydrobiol.* **39,** 416–425.

Law, I. A. (1984). Extraction of humic substances in DMSO. Personal Communication, Chemistry Department, University of Birmingham.

Leenheer, J. A. (1980). Origin and nature of humic substances in the water of the Amazon River basin. *Acta Amazonica* **10,** 513–526.

Leenheer, J. A. (1981). Comprehensive approach to preparative isolation and fractionation of dissolved organic carbon from natural waters and wastewaters. *Environ. Sci. Technol.* **15,** 578–587.

Leenheer, J. A. (1982). United States Geological Survey Data Information Service. In *Transport of Carbon and Minerals in Major World Rivers,* Pt. I (E. Degens, ed.). Report of a workshop, Hamburg, Germany, March 8–12, 1982, SCOPE, pp. 355–356.

Leenheer, J. A. (1983). Unpublished data. U.S. Geological Survey.

Leenheer, J. A. and Huffman, E. W. D., Jr. (1976). Classification of organic solutes in water by using macroreticular resins. *U.S. Geol. Surv. J. Res.* **4,** 737–751.

Leenheer, J. A. and Huffman, E. W. D., Jr. (1979). Analytical method for dissolved-organic-carbon fractionation. *U.S. Geol. Surv. Water-Resour. Invest.,* 79-4.

Leenheer, J. A. and Malcolm, R. L. (1973). Fractionation and characterization of natural organic matter from certain rivers and soils by free-flow electrophoresis. *U.S. Geol. Surv. Water-Supply Paper,* 1817-E.

Leenheer, J. A., Malcolm, R. L., McKinley, P. W., and Eccles, L. A. (1974). Occurrence of dissolved organic carbon in selected groundwater samples in the United States. *U.S. Geol. Surv. J. Res.* **2,** 361–369.

Leenheer, J. A. and Noyes, T. I. in press. Effects of organic wastes from processing of Green River formation oil shale on water quality. *U.S. Geol. Surv. Prof. Paper.*

Leenheer, J. A. and Noyes, T. I. (1984). A filtration and column adsorbent system for onsite concentration and fractionation of organic substances from large volumes of water. *U.S. Geol. Surv. Water-Supply Paper* No. 2230.

Leenheer, J. A., Wershaw, R. L., and Noyes, T. I. (1983). Chemical and spectroscopic characterization of natural organic solutes in surface and groundwaters associated with Green River formation oil shale: Symposium on "Early Diagenesis and Formation of Humic Substances," Geochemistry Division of the 185th National Meeting of the American Chemical Society, March 1983.

Lees, J. (1950). A note on the copper-retaining power of a humic acid from peat soil. *Biochem. J.* **46,** 450–451.

Lentz, H., Lüdemann, H. D., and Ziechmann, W. (1977). Proton resonance spectra of humic acids from the solum of a podzol. *Geoderma* **18,** 325–328.

Leversee, G. J. (1981). Effects of humic acids on bioaccumulation of six polycyclic aromatic hydrocarbons. Presented at the Symposium on Aquatic and Terrestrial Humic Materials, University of North Carolina, Chapel Hill, November 5, 1981.

Levesque, M. P. (1981). Characterization and differentiation of peat materials in the context of peat soils classification. Proceedings of the Organic Soils Mapping and Interpretations Workshop, Fredericton, *N. B. Res. Br. Agric. Can. L.R.R.I. Monogr.* **82-44,** 74–104.

Levesque, M. and Dinel, H. (1977). Fiber content, particle size distribution, and some related properties of four peat materials in eastern Canada. *Can. J. Soil Sci.* **57,** 187–195.

Levesque, M., Dinel, H., and Marcoux, R. (1980a). Evaluation des criteres de differenciation pour la classification de 92 materiaux tourbeaux du Quebec et de l'Ontario. *Can J. Soil Sci.* **60,** 479–486.

Levesque, M. P. and Mathur, S. P. (1979). A comparison of various means of measuring the degree of decomposition of virgin peat materials in the context of their relative biodegradability. *Can. J. Soil Sci.* **59,** 397–400.

Levesque, M. P. and Mathur, S. P. (1982a). The effects of using copper for mitigating Histosol subsidence on: I: The yield and nutrition of oats and lettuce grown on Histosols, mineral sublayers, and their mixtures. *Soil Sci.* **135,** 88–100.

Levesque, M. P. and Mathur, S. P. (1983b). Effect of liming on yield and nutrient concentration of Reed Canarygrass grown in two peat soils. *Can. J. Soil Sci.* **63,** 469–478.

Levesque, M., and Mathur, S. P. (1984). The effect of using copper for mitigating Histosol subsidence on: *III:* The yield and nutrition of onions, minicarrots, and carrots grown on Histosols, mineral sublayers, and their mixtures. *Soil Sci.* **138,** No 2., pp. 127–137.

Levesque, M., Morita, H., Schnitzer, M., and Mathur, S. P. (1980b). The physical, chemical, and morphological features of some Quebec and Ontario peats. *Res. Br. Agric. Can. L.R.R.I. Monogr.* **62,** 70 pp.

Li, G. C. and Felbeck, G. T., Jr. (1972). Atrazine hydrolysis as catalyzed by humic acids. *Soil Sci.* **114,** 201–209.

Liao, W., Christman, R. F., Johnson, J. D., and Millington D. S. (1982). Structural characterization of aquatic humic material. *Environ. Sci. Technol.* **16,** 403–410.

Lindquist, I. (1967). Adsorption effects in gel filtration of humic acid. *Acta Chem. Scand.* **21,** 2564–2566.

Lindquist, I. (1970). A small angle X-ray scattering study of sodium humate solutions. *Acta Chem. Scand.* **24,** 3068–3069.

Livingston, D. A. (1963). Chemical composition of rivers and lakes. *U.S. Geol. Surv. Prof. Paper* 440G, 64 pp.

Loder, T. S. and Liss, P. S. (1982). The role of organic matter in determining the surface charge of suspended particles in estuarine and oceanic waters. *Thal. Jugos.* **18,** 433–447.

Long, D. A. (1977). *Raman Spectroscopy.* McGraw-Hill, New York, 276 pp.

Loughman, F. C. (1969). *Chemical Weathering of Silicate Minerals.* Elsevier, New York, 154 pp.

Lowe, I. J. (1959). Free induction decays of rotating solids. *Phys. Rev. Lett.* **2**(7), 285–287.

Lowe, L. E. (1978). Carbohydrates in soil. In *Soil Organic Matter* (M. Schnitzer and S. U. Khan, eds.). Elsevier, New York, pp. 65–94.

Lowry, T. H. and Richardson, K. S. (1981). *Mechanism and Theory in Organic Chemistry*, 2nd ed. Harper and Row, New York, 991 pp.

Lüdemann, H. D., Lentz, H., and Liechmann, W. (1973). Protonenresonanzpektroskopie von Ligninen und Huminsauren bie 100 Megahertz. *Erdoel Kohle, Erdgas, Petrochem. Brennst. Chem.* **26,** 506–509.

Lytle, C. R. and Perdue, E. M. (1981). Free, proteinaceous, and humic-bound amino acids in river water containing high concentrations of aquatic humus. *Environ. Sci. Technol.* **15,** 224–228.

Ma, T. S. and Rittner, R. C. (1979). *Modern Organic Elemental Analysis*. Marcel Dekker, New York.

Ma, T. S., Wang, C. Y., and Gutterson, M. (1982). Organic elemental analysis. *Anal. Chem.* **54,** 87R–96R.

MacCarthy, P. (1976). A proposal to establish a reference collection of humic materials for interlaboratory comparisons. *Geoderma* **16,** 179–181.

MacCarthy, P. (1977). An interpretation of stability constants for soil organic matter–metal ion complexes under Schubert conditions. *J. Environ. Sci. Health* **A12,** 43–59.

MacCarthy, P. (1983). Novel deuteration techniques for solid samples in infrared spectrometry, including a diversionary discourse on the nature of scientific investigation and serendipitous aspects thereof. *Colo. Sch. Mines Q.* **78,** 9–22.

MacCarthy, P. and Malcolm, R. L. (1978). 11th Congress, International Society of Soil Science, Edmonton, Proposed organizational scheme and experimental methods for establishing a reference collection of humic substances. *Abstracts* **1,** 57–58.

MacCarthy, P. and Malcolm, R. L. (1979). The need to establish a reference collection of humic substances. In *Trace Organic Analysis: A New Frontier in Analytical Chemistry* (S. N. Chesler and H. S. Hertz, eds.). U.S. National Bureau of Standards Special Publication No. 519, Gaitersburg, pp. 793–796.

MacCarthy, P. and Mark, H. B., Jr. (1975). Infrared studies on humic acid in deuterium oxide. I. Evaluation and potentialities of the technique. *Soil Sci. Soc. Am. Proc.* **39,** 663–668.

MacCarthy, P. and Mark, H. B., Jr. (1976). An evaluation of Job's method of continuous variations as applied to soil organic matter–metal ion interactions. *Soil Sci. Soc. Am. J.* **40,** 267–276.

MacCarthy, P. and Mark, H. B., Jr. (1977). A further examination of the Schubert ion-exchange method as applied to soil organic matter–metal ion interactions. In *Biological Implications of Metals in the Environment,* Proceedings of 15th Annual Life Science Symposium, Richmond, WA, pp. 197–212.

MacCarthy, P., Mark, H. B., Jr., and Griffiths, P. R. (1975). Direct measurement of the infrared spectra of humic substances in water by Fourier transform infrared spectroscopy. *J. Agric. Food Chem.* **23,** 600–602.

MacCarthy, P. and O'Cinneide, S. (1974). Fulvic acid: I. Partial fractionation. *J. Soil Sci.* **25,** 420–428.

MacCarthy, P., Peterson, M. J., Malcolm, R. L., and Thurman, E. M. (1979). Separation of humic substances by pH gradient desorption from a hydrophobic resin. *Anal. Chem.* **51,** 2041–2043.

MacCarthy, P. and Smith, G. C. (1979). Stability surface concept: A quantitative model for complexation in multiligand mixtures. In *Chemical Modeling in Aqueous Systems* (E. A. Jenne, ed.). ACS Symp. Ser. No. 93, American Chemical Society, Washington, D.C., pp. 94–114.

MacFarlane, R. B. (1978). Molecular weight distribution of humic and fulvic acids of sediments from a north Florida estuary. *Geochim. Cosmochim. Acta.* **42,** 1579–1582.

Machihara, T. and Ishiwatari, R. (1980). Characterization of young kerogen in a lacustrine sediment by alkaline potassium permanganate oxidation. *Geochem. J.* **14,** 279–288.

Machihara, T. and Ishiwatari, R. (1981). Characteristics of insoluble organic matter in lake sediments as revealed by alkaline potassium permanganate oxidation. *Ver. Internat. Verein. Limnol.* **21,** 244–247.

Machihara, T. and Ishiwatari, R. (in press). Evaluation of alkaline permanganate oxidation method for the characterization of young kerogen. *Org. Geochem.*

Maciel, G. E., Bartuska, V. J., and Miknis, F. P. (1979). Characterization of organic material in coal by proton-decoupled ^{13}C nuclear magnetic resonance with magic angle spinning. *Fuel* **58,** 391–394.

Maillard, L. C. (1912). Formation d'humus et de combustibles mineraux sans intervention de l'oxygene atmospherique, des microorganisms, de haute temperatures ou de forte pression. *C. R. Acad. Sci. Paris* **154,** 66.

Maillard, L. C. (1913). Formation des matieres humiques par action des acides amines sur les sucres. *C. R. Acad. Sci. Paris* **156,** 1159.

Maillard, L. C. (1916). Synthese de matieres humiques par action des acides amines sur les sucres reducteures. *Ann. Chem. Phys.* **5,** 258–317.

Malcolm, R. L. (1968). Freeze-drying of organic matter, clays, and other earth materials. *U.S. Geol. Surv. Prof. Paper,* 211–216.

Malcolm, R. L. (1976). Method and importance of obtaining humic and fulvic acids of high purity. *J. Res. U.S. Geol. Surv.* **4,** 37–40.

Malcolm, R. L. and Durum, W. H. (1976). Organic carbon and nitrogen concentration and annual organic load of six selected rivers of the United States. *U.S. Geol. Surv. Water-Sup. Paper,* 1817-F.

Malcolm, R. L. and MacCarthy, P. (1979). *A Proposal for Implementing a Reference Collection of Humic and Fulvic Acids.* In *Trace Organic Analysis: A New Frontier in Analytical Chemistry.* (S. N. Chesler and H. S. Hertz, eds.). U.S. National Bureau of Standards Special Publications, No 519, 789–791.

Malcolm, R. L., Thurman, E. M., and Aiken, G. R. (1977). The concentration and fractionation of trace organic solutes from natural and polluted waters using XAD-8, a methylmethacrylate resin. In *Trace Substances in Environmental Health—XI.* (D. D. Hemphill, ed.). University of Missouri, Columbia, pp. 307–314.

Malcolm, R. L., Wershaw, R. L., Thurman, E. M., Aiken, G. R., Pinckney, D. J., and Kaakinen, J. (1981). Reconnaissance sampling and characterization of aquatic humic substances at the Yuma Desalting Test Facility, Arizona. *U.S. Geol. Surv. Water Resour. Invest.,* #81-42.

Mansfield, E. B. and Gibboney, D. A. (1977). Method and apparatus for sulfur analysis. U.S. Patent 4,026,665.

Manskaya, S. M. and Drozdova, T. V. (1968). *Geochemistry of Organic Substances.* Pergamon, Elmsford, NY, 345 pp.

Mantoura, R. F. C. (1981). Organo-metallic interactions in natural waters. In *Marine Organic Chemistry* (Duursma and Dawson, eds.). Elsevier, New York, pp. 179–224.

Mantoura, R. F. C., Dickson, A., and Riley, J. P. (1978). The complexation of metals with humic materials in natural waters. *Estuarine Coastal Mar. Sci.* **6,** 387–408.

Mantoura, R. F. C. and Riley, J. P. (1975). The analytical concentration of humic substances from natural waters. *Anal. Chim. Acta* **76,** 97–106.

Mantoura, R. F. C. and Woodward E. M. S. (1983). Conservative behavior of riverine dissolved organic carbon in the Severn Estuary, chemical and geochemical implications. *Geochim. Cosmochim. Acta* **47,** 1293–1309.

March, J. (1977). *Advanced Organic Chemistry: Reactions, Mechanisms, and Structure,* 2nd ed. McGraw-Hill, New York, 1328 pp.

Marinsky, J. A., Gupta, S., and Schindler, P. (1982a). The interaction of Cu(II) ion with humic acid. *J. Colloid Interface Sci.* **89,** 401–411.

Marinsky, J. A., Gupta, S., and Schindler, P. (1982b). A unified physicochemical description of the equilibria encountered in humic acid gels. *J. Colloid Interface Sci.* **89,** 412–426.

Marinsky, J. A., Wolf, A., and Bunzl, K. (1980). The binding of trace amounts of lead(II), copper(II), cadmium(II), zinc(II), and calcium(II) to soil organic matter. *Talanta* **27,** 461–468.

Maron, S. H. and Prutton, C. F. (1969). *Principles of Physical Chemistry,* 4th ed. Macmillan, New York, 886 pp.

Martel, Y. A. and Paul, E. A. (1974). The use of radiocarbon dating of organic matter in the study of soil genesis. *Soil Sci. Soc. Am. Proc.* **38,** 501–506.

Martell, A. E. and Calvin, M. (1952). *Chemistry of Metal Chelate Compounds.* Prentice-Hall, Englewood Cliffs, NJ.

Martell, A. E. and Smith R. M. (1977). *Critical Stability Constants. Volume 3. Other Organic Ligands.* Plenum, New York, 495 pp.

Martin, A. E. and Reeve, R. (1958). Chemical studies of podzolic illuvial horizons. III. Titration curves of organic matter suspensions. *J. Soil Sci.* **9,** 89–100.

Martin, D. and Hauthal, H. G. (1975). *Dimethyl Sulphoxide* (translated by E. S. Halberstadt). Van Nostrand–Reinhold, New York.

Martin, D. F. and Pierce, R. A. (1975). A convenient method of analysis of humic acids in freshwater. *Environ. Lett.* **1,** 49–52.

Martin, F., Saiz-Jimenez, C., and Cert, A. (1977). Pyrolysis–Gas Chromatography–Mass Spectrometry of soil humic fractions. I. The low boiling point compounds. *Soil Sci. Soc. Am. J.* **41,** 1114–1118.

Martin, J. P. and Haider, K. (1980). A comparison of the use of phenolase and peroxidase for the synthesis of model humic acid-type polymers. *Soil Sci. Soc. Am. J.* **44,** 983–988.

Martin, J. P., Haider, K., and Bondietti, E. (1975). Properties of model humic acids synthesized by phenoloxidase and autoxidation of phenols and other compounds formed by soil fungi. *Proc. Int. Meet. Humic Substances, Nieuwersluis, 1972,* pp. 171–186, Pudoc, Wageningen.

Mason, C. F. and Standen, V. (1983). Aspects of secondary production. In *Ecosystrems of the World. 4A. Mires: Swamps, Bog, Fen, and Moor. General Studies* (A. J. P. Gore, ed.). Elsevier, New York, pp. 367–382.

Mason, J. (1955). Reactions between quinones and proteins. *Nature* **175,** 771–772.

Mathur, S. P. (1982a). The role of soil enzymes in the degradation of organic matter in the tropics, subtropics, and temperate zone. Symposium Papers of the 12th Congress. *Intl. Soc. Soil Sci.* **1,** 125–136.

Mathur, S. P. (1982b). The inhibitory role of copper in the enzymic degradation of organic soils. *Proceedings of the International Peat Symposium.* Bemidji State University, pp. 191–219.

Mathur, S. P. (1982c). Organic soil subsidence: A scan of conventional wisdom and current research. Proc. Organic Soils Mapping and Interpretations Workshop. *Agric. Can. Res. Br. LLRI Pub.* No. 82-44, 139–156.

Mathur, S. P. (1983). A lack of bactericidal effect of subsidence-mitigating copper in organic soils. *Can. J. Soil Sci.* **63,** 645–649.

Mathur, S. P., Belanger, A., Valk, M., Preston, C. M., Knibbe, E. N., and Sanderson, R. B. (1983). Study of onions grown in microplots on three organic soils each containing four levels of copper. *Can. J. Soil Sci.* **63,** 221–228.

Mathur, S. P., Hamilton, H. A., and Levesque, M. P. (1979a). The mitigating effect of residual fertilizer copper on the decomposition of an organic soil in situ. *Soil Sci. Soc. Am. J.* **43,** 200–203.

Mathur, S. P., Hamilton, H. A., and Preston, C. M. (1979b). The influence of variation in copper content of an organic soil on the mineral nutrition of oats grown in situ. *Commun. Soil. Sci. Plant Anal.* **10,** 1399–1409.

Mathur, S. P. and Levesque, M. P. (1980). Relationship between acid phosphatase activities and decomposition rates of twenty-two virgin peat materials. *Commun. Soil Sci. Plant Anal.* **11,** 155–162.

Mathur, S. P. and Levesque, M. P. (1983a). The effects of using copper for mitigating Histosol subsidence on: II. The distribution of Cu, Mn, Zn, and Fe in an organic soil, mineral sublayers and their mixtures in the context of setting a threshold of phytotoxic soil-Cu. *Soil Sci.* **135,** 166–176.

Mathur, S. P. and Levesque, M. P. (1983b). Effect of liming on the yield, nutrition, and copper status of potatoes, carrots, and onions grown sequentially in peat soils. *Can. J. Soil Sci.* **63,** 229–244.

Mathur, S. P., Levesque, M. P., and Desjardins, J. G. (1979c). The relative immobility of fertilizer and native copper in an organic soil under field conditions. *Water, Air, Soil Pollut.* **11,** 207–215.

Mathur, S. P., McDougall, J. I., and McGrath, M. (1980). Levels of activities of some carbohydrases, protease, lipase, and phosphatase in organic soils of differing copper content. *Soil Sci.* **129,** 376–385.

Mathur, S. P. and Morley, H. V. (1975). A biodegradation approach for investigating pesticide incorporation into soil humus. *Soil Sci.* **120,** 238–240.

Mathur, S. P. and Morley, H. V. (1978). Incorporation of methoxychlor-^{14}C in model humic acids prepared from hydroquinone. *Bull. Environ. Contam. Toxicol.* **20,** 268–274.

Mathur, S. P. and Preston, C. M. (1981). The effect of residual fertilizer copper on ammonification, nutrification, and proteolytic population in some organic soils. *Can. J. Soil Sci.* **61,** 445–450.

Mathur, S. P. and Rayment, A. F. (1977). The influence of trace element fertilization on the decomposition rate and phosphatase activity of a Mesic Fibrisol. *Can. J. Soil Sci.* **57,** 397–408.

Mathur, S. P. and Saha, J. G. (1975). Microbial degradation of lindane-^{14}C in a flooded sandy loam soil. *Soil Sci.* **150,** 301–307.

Mathur, S. P. and Saha, J. G. (1977). Degradation of lindane-^{14}C in a mineral soil and in an organic soil. *Bull. Environ. Contam. Toxicol.* **17,** 424–430.

Mathur, S. P. and Sanderson, R. B. (1978). Relationship between copper contents, rates of soil respiration, and phosphatase activities on some Histosols in an area of southwestern Quebec in the summer and the fall. *Can. J. Soil Sci.* **58,** 125–134.

Mathur, S. P. and Sanderson, R. B. (1980a). The partial inactivation of degradative soil enzymes by residual fertilizer copper in Histosols. *Soil Sci. Soc. Am. J.* **44,** 750–755.

Mathur, S. P. and Sanderson, R. B. (1980b). Relationships between copper contents, rates of biodegradation, and degradative enzyme activities of some cultivated organic soils in Ontario, Canada. *Proc. 6th Intl. Peat Congr.,* 628–632.

Mathur, S. P., Sanderson, R. B., Belanger, A., Valk, M., Knibbe, E. N., and Preston, C. M. (1984). The effect of copper applications on the movement of copper and other elements in organic soils. *Water, Air Soil Pollut.,* **22,** 277–288.

Mathur, S. P. and Schnitzer, M. (1978). A chemical and spectroscopic characterization of some synthetic analogues of humic acids. *Soil Sci. Soc. Am. J.* **42,** 591–595.

Matsuda, K. and Schnitzer, M. (1973). Reactions between fulvic acid, a soil humic material, and dialkyl phthalates. *Bull. Environ. Contam. Toxicol.* **6,** 200–204.

Mayer, L. M. (1982a). Retention of riverine iron in estuaries. *Geochim. Cosmochim. Acta* **46,** 1003–1009.

Mayer, L. M. (1982b). Aggregation of colloidal iron during estuarine mixing. Kinetics, mechanism, and seasonality. *Geochim. Cosmochim. Acta* **46,** 2527–2535.

Mayer, L. M., Macko, S. A., Mook, W. H., and Murray, S. (1981). The distribution of bromine in coastal sediments and its use as a source indicator for organic matter. *Org. Geochem.* **3,** 37–42.

Mayer, L. M., Schick, L. L., and Chang, C. A. (1983). Kinetics of trivalent chromium complexation and adsorption at the river–estuary interface. Implications for reactive pollutant transport. In *Proceedings of the International Conference on Heavy Metals in the Environment,* CEP Consultants, pp. 1112–1115.

Mazumdar, B. K., Chakrabartty, S. K., Saha, M., Anad, K. S., and Lahiri, A. (1959). Further studies on the mechanism of oxidation of coal. *Fuel* **38,** 469–482.

McBride, M. B. (1978). Transition metal bonding in humic acid: An ESR study. *Soil Sci.* **126,** 200–209.

McBride, M. B. (1980a). A comparative electron spin resonance study of VO^{2+} complexation in synthetic molecules and soil organics. *Soil Sci. Soc. Am. J.* **44,** 495–499.

McBride, M. B. (1980b). Influence of pH and metal ion content on vanadyl ion fulvic acid interactions. *Can. J. Soil Sci.* **60,** 145–149.

McCreary, J. J. and Snoeyink, V. L. (1980). Characterization and activated carbon adsorption of several humic substances. *Water Res.* **14,** 151–160.

McKnight, D. M. (1981). Chemical and biological processes controlling the response of a freshwater ecosystem to copper stress: A field study for the $CuSO_4$ treatment of Mill Pond Reservoir, Burlington, Massachusetts. *Limnol. Oceanogr.* **26,** 518–531.

McKnight, D. M., Feder, G. L., Thurman, E. M., Wershaw, R. L., and Westall, J. C. (1983). Complexation of copper by aquatic humic substances from different environments. In *Biological Availability of Trace Metals* (R. E. Wildung, and E. A. Jenne, eds.). Elsevier, New York, pp. 65–76.

McKnight, D. M., Thurman, E. M., Wershaw, R. L., and Hemond, H. (1985b). Biogeochemistry of aquatic humic substances in Thoreau's Bog, Concord, *Ecol.* in press.

McKnight, D. M. and Morel, F. M. M. (1979). Release of weak and strong copper-complexing agents by algae. *Limnol. Oceanogr.* **24,** 823–837.

McKnight, D. M. and Morel, F. M. M. (1980). Copper complexation by siderophores from filamentous blue-green algae. *Limnol. Oceanogr.* **25,** 62–71.

McKnight, D. M., Periera, W. E., Zeazan, C. L., and Wissmar, R. C. (1982). Characterization of dissolved organic materials in surface waters within the blast zone of Mt. St. Helens, Washington. *Org. Geochem.* **4,** 85–92.

McKnight, D. M., Klein, J., and Wissmar, R. C. (1985a). Changes in the organic material in lakes in the blast zone of Mt. St. Helens, Washington, *U.S. Geol. Surv. Circ.* 850.

McLaren, R. G. and Crawford, D. V. (1973). Studies on soil copper. I. *J. Soil Sci.* **24,** 172–181.

McLean, E. O., Reicosky, D. C., and Lakshmanan (1969). Aluminum in soils: VII. *Soil Sci. Soc. Am. Proc.* **29,** 374–378.

Means, J. C. and Wijayaratne, R. (1982). Role of natural colloids in the transport of hydrophobic pollutants. *Science* **215,** 968–970.

Meloan, C. E. (1963). *Elementary Infrared Spectroscopy.* MacMillan, New York, 193 pp.

Melzer, A. and Steinberg, C. (1983). Nutrient cycling in freshwater ecosystems. *Encyclopedia of Plant Physiology,* 12 D. Springer, New York.

Menger, F. M., Takeshita, M., and Chow, J. F. (1981). Hexapus, a new complexing agent for organic molecules. *J. Am. Chem. Soc.* **103,** 5938–5939.

Merz, W. (1968). Automatische Schnellmethode zur Stickstoffbestimmung. *Z. Anal. Chem.* **237,** 272–279.

Merz, W. (1970). Beitrag zur mikrobestimmung von sauerstoff in organischen substanzen in gegenwart von metallen sowie in metallorganischen verbindungen und automatische endbestimmung der kohlensaure. *Anal. Chim. Acta* **50,** 305–313.

Meuzelaar, H. L. C., Haider, K., Nagar, B. R., and Martin, J. P. (1977). Comparative studies of pyrolysis–mass spectra of melanins, model phenolic polymers, and humic acids. *Geoderma* **17,** 239–252.

Meybeck, M. (1982). Carbon, nitrogen, and phosphorus transport by world rivers. *Am. J. Sci.* **282,** 401–450.

Meyer, J. L. and Tate, C. M. (1983). The effects of watershed disturbance on dissolved organic carbon dynamics of a stream. *Ecology* **64,** 33–44.

Meyers, P. A. and Quinn, J. G. (1973). Factors affecting the association of fatty acids with mineral particles in sea water. *Geochim. Cosmochim. Acta* **37,** 1745–1759.

Michaels, A. S. (1968). Ultrafiltration. In *Progress in Separation and Purification,* Vol. 1 (E. S. Perry, ed.). Wiley-Interscience, New York, pp. 297–333.

Midwood, R. B. and Felbeck, G. T., Jr. (1968). Analysis of yellow organic matter from fresh water. *J. Am. Water Works Assoc.* **60,** 357–366.

Mikita, M. A., Steelink, C., and Wershaw, R. L. (1981). Carbon-13 enriched nuclear magnetic resonance method for the determination of hydroxyl functionality in humic substances. *Anal. Chem.* **53,** 1715–1717.

Miknis, F. P., Bartuska, V. J., and Maciel, G. E. (1979a). Cross-polarization ^{13}C NMR with magic angle spinning. Some applications to fossil fuels and polymers. *Am. Labs.* **1979,** 8 pp.

Miknis, F. P., Maciel, G. E., and Bartuska, V. J. (1979b). Cross-polarization magic angle spinning ^{13}C NMR spectra of oil shales. *Org. Geochem.* **1,** 169–176.

Milanovich, F. P., Ireland, R. R., and Wilson, D. W. (1975). Dissolved yellow organics: Quantitative and qualitative aspects of extraction by four common techniques. *Anal. Lett.* **8,** 337–344.

Miles, C. J. and Brezonik, P. L. (1981). Oxygen consumption in humic-colored waters by a photochemical ferrous–ferric catalytic cycle. *Environ. Sci. Technol.* **15,** 1089–1095.

Miles, C. J., Tuschall, J. R., and Brezonik, P. L. (1983). Isolation of aquatic humus with diethylaminoethylcellulose. *Anal. Chem.* **55,** 410–411.

Miller, R. H. and Wilkinson, J. F. (1977). Nature of the organic coating on sand grains of nonwettable golf greens. *Soil Sci. Soc. Am. J.* **41,** 1203–1204.

Mills, G. L., Hanson, A. K., Jr., Quinn, J. G., Lammela, W. R., and Chasteen, N. D. (1982). Chemical studies of copper–organic complexes isolated from estuarine waters using C_{18} reverse-phase liquid chromatography. *Mar. Chem.* **11,** 355–378.

Mingelgrin, U. and Gerstl, Z. (1983). Reevaluation of partitioning as a mechanism of nonionic chemicals adsorption in soils. *J. Environ. Qual.* **12,** 1–11.

Mitchell, J., Jr. (1961). Water. *Treatise on Analytical Chemistry,* Pt. II, Sec. A. Wiley-Interscience, New York, pp. 69–197.

Mitchell, J. and Smith, M. D. (1980). Aquametry. In *A Treatise on Methods for the Determination of Water,* Pt. III, Vol. 5 (P. J. Elving and J. D. Winefordner, eds.). Wiley, New York.

Modell, M., de Filippi, R., and Krukonis, V. (1980). Regeneration of activated carbon with supercritical carbon dioxide. In *Activated Carbon Adsorption of Organics from the Aqueous Phase,* Vol. 1 (I. H. Suffet and M. J. McGuire, eds.). Ann Arbor Science, Ann Arbor, MI, pp. 447–461.

Moed, J. R. (1970). Aluminum oxide as adsorbent for natural water-soluble yellow material. *Limnol. Oceanogr.* **15,** 140–142.

Molyneaux, P. (1975). Synthetic polymers. In *Water, A Comprehensive Treatise Vol. 4. Aqueous Solutions of Amphiphiles and Macromolecules* (F. Franks, ed.). Plenum, New York, pp. 569–754.

Montgomery, J. R. and Santiago, R. J. (1978). Zinc and copper in "particulate" forms and "soluble" complexes with inorganic and organic ligands in the Guanagibo River and Coastal zone, Puerto Rico. *Estuarine Coastal Mar. Sci.* **6,** 111–116.

Moore, E. S. and Maynard, J. E. (1929). Solution, transportation, and precipitation of iron and silicia. *Econ. Geol.* **24,** 506–527.

Moore, L. R. (1968). Cannel coals, bogheads, and oil shales. In *Coal and Coalbearing Strata* (D. C. Murchison and T. S. Westoll, eds.). Oliver and Boyd, Edinburgh, pp. 19–29.

Moore, P. D. and Bellamy, D. J. (1974). *Peatlands.* Springer-Verlag, New York, 221 pp.

Moore, R. M., Burton, J. D., Williams, P. J. LeB., and Young, M. L. (1979). The behavior of dissolved organic material, iron, and manganese in estuarine mixing. *Geochim. Cosmochim. Acta* **43,** 919–926.

Moore, W. J. (1972). *Physical Chemistry,* 4th ed. Prentice-Hall, Englewood Cliffs, NJ, 977 pp.

Morita, H. (1975). Polyphenols in the lime water extractives of peat. *Soil Sci.* **120,** 112–116.

Morris, A. W. (1974). Seasonal variation of dissolved metals in inshore waters of the Menai Straits. *Mar. Pollut. Bull.* **5,** 54–59.

Morris, R. F. and Penny, B. G. (1971). Persistence of linuron residues in soils at Kelligrews, Newfoundland. *Can. J. Plant Sci.* **51,** 242–245.

Morris, R. J. (1975). The amino acid composition of a deep-water marine sediment from the upwelling region Northwest of Africa. *Geochim. Cosmochim. Acta* **39,** 381–388.

Mortensen, J. L. and Schwendinger, R. B. (1963). Electrophoretic and spectroscopic characterization of high molecular weight components of soil organic matter. *Geochim. Cosmochim. Acta* **27,** 201–208.

Mott, C. J. B. (1981). Anion and ligand exchange. In *The Chemistry of Soil Processes* (D. J. Greenland and M. H. B. Hayes, eds.). Wiley, New York, pp. 179–219.

Muenster, U. (1982). Physikalisch-chemische Untersuchungen zur Anreicherung, Fraktionierung und Strukturaufklaerung monomerer und polymerer organischer Substrate im DOC vom Plussee. PhD Thesis, University of Kiel, Germany, 281 pp. Excerpts in this thesis will be published in *Arch. Hydrobiol.*

Mulder, G. J. (1840). Untersuchungen uber Humusartige Materien. *Ann. Chem. Pharm.* **36,** 243–295.

Mulder, G. J. (1862). *Die Chemie der Ackerkrume, Ubers.* Muller, Leipzig.

Murphy, D. and Moore, A. W. (1960). A possible structural basis of natural humic acid. *Sci. Proc. Roy. Dublin Soc.* **4,** 191–195.

Murphy, T. P., Lean, D. R. S., and Nalewjko, C. (1976). Blue-green algae: Their excretion of iron-selective chelators enables them to dominate other algae. *Science* **192,** 900–902.

Musani, Lj., Valenta, P., Nurnberg, H. W., Konrad, Z., and Branica, M. (1980). On the chelation of toxic trace metals by humic acid of marine origin. *Estuarine Coastal Mar. Sci.* **11,** 639–649.

Nelson, N. A., Kelly, R. C., and Johnson, R. A. (1982). Prostaglandins and the arachidonic acid cascade. *Chem. Eng. News,* 30–44, August 16.

Newell, R. C. (1965). The role of detritus in the nutrition of two marine deposit feeders, the prosobranch *Hydrobia ulvae* and the bivalve *Macoma Balthica. Proc. Zool. Soc.* **144,** 25–45.

Newman, R. H., Tate, K. R., Barron, P. F., and Wilson, M. A. (1980). Towards a direct, nondestructive method of characterising soil humic substances using ^{13}C NMR. *J. Soil Sci.* **31,** 623–631.

Niederl, J. B. and Niederl, V. (1938). *Micromethods of Quantitative Organic Elementary Analysis*. Wiley, New York.

Nimz, H. (1966). Oligomere Abbauphenole des Fichtenlignins. *Chem. Ber.* **99,** 2638–2640.

Nissenbaum, A. (1974). The organic geochemistry of marine and terrestrial humic substances: Implications of carbon and hydrogen isotope studies. In *Advances in Organic Geochemistry* (B. Tissot and F. Bienner, eds.). Editions Technip, Paris, pp. 39–52.

Nissenbaum, A. (1979). Phosphorus in marine and non-marine humic substances. *Geochim. Cosmochim. Acta* **43,** 1973–1978.

Nissenbaum, A., Baedecker, M. J., and Kaplan, I. R. (1971). Studies on dissolved organic matter from interstitial water of a reducing marine fjord. In *Advances in Organic Geochemistry* (H. R. V. Gaertner and H. Wehner, eds.). Pergamon, Elmsford, NY, pp. 427–440.

Nissenbaum, A. and Kaplan, I. R. (1972). Chemical and isotopic evidence for the *in situ* origin of marine humic substances. *Limnol. Oceanogr.* **17,** 570–582.

Nriagu, J. O. and Coker, R. D. (1980). Trace metals in humic and fulvic acids from Lake Ontario sediments. *Environ. Sci. Technol.* **14,** 443–446.

O'Brien, B. J. and Stout, J. D. (1978). Movement and turnover of soil organic matter as indicated by carbon isotope measurements. *Soil Bio. Biochem.* **10,** 309–317.

Odegaard, H. and Koottatep, S. (1982). Removal of humic substances from natural waters by reverse osmosis. *Water Res.* **16,** 613–620.

Oden, S. (1914). Zur Kolloidchemie der Humusstoffe. *Kolloid Z.* **14,** 123–130.

Oden, S. (1919). Die Huminsauren. *Kolloidchem. Beih.* **11,** 75–260.

Ogner, G. and Gjessing, E. T. (1975). Permanganate oxidation of methylated and nonmethylated aquatic humus in Norway. *Geoderma* **14,** 139–145.

Ogner, G. and Schnitzer, M. (1970a). The occurrence of alkanes in fulvic acid, a soil humic fraction. *Geochim. Cosmochim. Acta* **34,** 921–928.

Ogner, G. and Schnitzer, M. (1970b). Humic substances: Fulvic acid–diaklyl phthalate complexes and their role in pollution. *Science* **170,** 317–318.

Ogura, N. (1974). Molecular weight fractionation of dissolved organic matter in coastal seawater by ultrafiltration. *Mar. Biol.* **24,** 305–312.

Ohfune, Y., Tomita, M., and Nomoto, K. (1981). Total synthesis of 2′-deoxymugineic acid, the metal chelator excreted from wheat root. *J. Am. Chem. Soc.* **103,** 2409–2410.

Ohle, W. (1934). Uber organische Stoffe in Binnenseen. *Verh. Intl. Verein. Limnol.* **6,** 249–262.

Ohle, W. (1935). Organische Kolloide in ihrer Wirkung auf den Stoffhaushalt der Gewaesser. *Naturwissenschaften* **35,** 480–484.

Ohle, W. (1937). Kollidgele als Naehrstoffregeneratoren der Gewaesser. *Naturwissenschaften* **37,** 471–474.

Oita, I. J. and Conway, H. S. (1954). Direct determination of oxygen in organic substances by modified Schultze method. *Anal. Chem.* **26,** 600–602.

Oka, H., Sasaki, M., Itho, M., and Suzuku, A. (1969). Study on the chemical structure of peat humic acids by high resolution nuclear magnetic resonance spectroscopy. *Nenryo Kyokai Shi (J. Japanese Fuel Soc.)* **48,** 295–302.

Oliver, R. and Thurman, E. M. (1982). The influence of aquatic humic substance properties on trihalomethane potential. *Water Chlorination: Environmental Impact and Health Effects*, Vol. 4, Chap. 16 (Jolley, Brungs, Cotruvo, Cumming, Mattice, and Jacobs, eds.). Ann Arbor Science, Ann Arbor, MI, pp. 231–241.

Olsen, E. D. (1975). *Modern Optical Methods of Analysis*. McGraw-Hill, New York, 629 pp.

Ong, H. L. and Bisque, R. E. (1968). Coagulation of humic colloids by metal ions. *Soil Sci.* **106,** 220–224.

Ong, H. L., Swanson, V. E., and Bisque R. E. (1970). Natural organic acids as agents of chemical weathering. *U.S. Geol. Surv. Prof. Paper* 700C, 130–137.

Orioli, G. A. and Curvetto, N. R. (1982). Evaluation of extractants for soil humic substances. *Plant and Soil* **66,** 353–361.

Orlov, D. S., Ammosova, Ya. M., Glebova, G. I., Gorshkova, Ye. I., Il'in, N. P., and Kolesnikov, M. P. (1971). Molecular weights, sizes, and configuration of humic-acid particles. Translated from *Pochvovedeniye* **11,** 43–56.

Orlov, D. S. (1974). *Humus Acids of Soils* (in Russian). Moscow University Press, Moscow, 332 pp.

Orlov, D. S., Ammosova, Ya. M., and Glebova, G. I. (1975). Molecular parameters of humic acids. *Geoderma* **13,** 211–229.

Orlov, D. S., Rozanova, O. N., and Matyukhina, S. (1962). Infrared absorption spectra of humic acids. *Sov. Soil Sci.* **1,** 15–21.

Ortner, P. B., Kreader, C., and Harvey, G. R. (1983). Interactive effects of metals and humus on marine phytoplankton carbon uptake. *Nature* **301,** 57–59.

O'Shea, T. A. and Mancy, K. H. (1976). Characterization of trace metal–organic interactions by anodic-stripping voltammetry. *Anal. Chem.* **48,** 1603–1607.

Otsuki, A. and Hanya, T. (1967). Some precursors of humic acid in recent lake sediment suggested by infrared spectra. *Geochim. Cosmochim. Acta* **31,** 1505–1515.

Otsuki, A. and Wetzel, R. G. (1973). Interaction of yellow organic acids with calcium carbonate in freshwater. *Limnol. Oceanogr.* **18,** 490–493.

Packham, R. F. (1964). Studies of organic color in natural water. *Proc. Soc. Water Treatment Exam.* **13,** 316–334.

Page, H. J. (1930). Studies on the carbon and nitrogen cycles in the soil. *J. Agric. Sci.* **20,** 455–459.

Palacas, J. G., Swanson, V. E., and Love, A. H. (1968). Organic geochemistry of recent sediments in the Chochawhatchee Bay Area, Florida—A preliminary report. *U.S. Geol. Surv. Prof. Paper* 600C, C97-C106.

Palmer, A. R. and Maciel, G. E. (1982). Relaxation behavior in the carbon-13 nuclear magnetic resonance spectrometric analysis of kerogen with cross-polarization and magic-angle spinning. *Anal. Chem.* **54,** 2194–2198.

Parish, J. R. (1977). Macroporous resins as supports for a chelating liquid-ion exchanger in extraction chromatography. *Anal. Chem.* **49,** 1189–1192.

Parker, A. J. (1962). The effects of solvation on the properties of anions in dipolar aprotic solvents. *Q. Rev. Chem. Soc.* **16,** 163–187.

Parker, F. S. (1971). *Applications of Infrared Spectroscopy in Biochemistry, Biology, and Medicine*. Plenum, New York, 601 pp.

Parsons, J. W. and Tinsley, J. (1975). Nitrogenous substances. In *Soil Components,* Vol. 1 (J. E. Gieseking, ed.). Springer-Verlag, New York, pp. 263–304.

Passer, M. (1957). Analytical procedure for the organic constituents of peat. Report No. 14, Chemical Products from Peat Project, University of Minnesota, 38 pp.

Paul, E. A., Campbell, C. A., Rennie, D. A., and McCallum, K. J. (1964). Investigations of the dynamics of soil humus using carbon dating techniques. *Trans. 8th Intl. Congr. Soil Sci., Bucharest* **III,** 201–208.

Paul, E. A. and Veen, J. A. van (1978). The use of tracers to determine the dynamic nature of organic matter. *Trans. 11th Intl. Congr. Soil Sci. Edmonton* **3,** 61–102.

Pelet, R. (1978). Geochimie organique des sediments marins profouds de l'appareil detritique de l'Amazone et de la fosse de Cariaco: Vue d'ensemble. In *Orgon II, Geochimie organique des sediments marins profonds.* Editions du CNRS, Paris, pp. 371–390.

Pelet, R. (1979). Geochimie organique des sediments marins profonds au large de la Mauritanie et du Senegal: Vue d'ensemble In *Orgon III, Geochimie organique des sediments marins profonds*. Editions du CNRS, Paris, pp. 425–441.

Pelet, R. (1980). Evolution geochimique de la matiere organique. In *Kerogen* (B. Durand, ed.). Editions Technip, Paris, pp. 475–499.

Pelet, R. (1983). Preservation and alteration of present-day sedimentary organic matter. In *Advances in Organic Geochemistry, 1981* (M. Bjoroy et al., ed.). Wiley, New York, pp. 241–250.

Pelet, R. and Debyser, Y. (1977). Organic geochemistry of Black Sea cores. *Geochim. Cosmochim. Acta* **41,** 1575–1586.

Pennanen, V. (1975). Humus fractions and their distribution in some lakes in Finland. *Proc. Intl. Meet. Humic Substances, Mieuwersluis.*, pp. 207–215.

Perdue, E. M. (1978). Solution thermochemistry of humic substances. I. Acid–base equilibria of humic acid. *Geochim. Cosmochim. Acta* **42,** 1351–1358.

Perdue, E. M. (1979). Solution thermochemistry of humic substances. II. Acid–base equilibria of river water humic substances. In *Chemical Modeling in Aqueous Systems,* Vol. 93 (E. A. Jenne, ed.). American Chemical Society, Washington, D.C., pp. 94–114.

Perdue, E. M. (1983). Association of organic pollutants with humic substances: partitioning equilibria and hydrolysis constants. In *Aquatic and Terrestrial Humic Materials* (R. F. Christman and E. T. Gjessing, eds.). Ann Arbor Science, Ann Arbor, MI, pp. 441–460.

Perdue, E. M. (1984). Analytical constraints on the structural features of humic substances. *Geochim. Cosmochim. Acta* **48,** 1435–1442.

Perdue, E. M., Beck, K. C., and Reuter, J. H. (1976). Organic complexes of iron and aluminum in natural waters. *Nature* **260,** 418–420.

Perdue, E. M. and Lytle, C. R. (1983a). A critical examination of metal–ligand complexation models: Application to defined multiligand mixtures. In *Aquatic and Terrestrial Humic Materials* (R. F. Christman and E. T. Gjessing, eds.). Ann Arbor Science, Ann Arbor, MI, pp. 295–313.

Perdue, E. M. and Lytle, C. R. (1983b). A distribution model for binding of protons and metal ions by humic substances. *Environ. Sci. Technol.* **17,** 654–660.

Perdue, E. M., Reuter, J. H., and Ghosal, M. (1980). The operational nature of acidic functional group analyses and its impact on mathematical descriptions of acid–base equilibria in humic substances. *Geochim. Cosmochim. Acta* **44,** 1841–1851.

Perdue, E. M., Reuter, J. H., and Parrish, R. S. (1984). A statistical model of proton binding by humus. *Geochim. Cosmochim. Acta* **48,** 1257–1263.

Petruzzelli, G. and Guidi, G. (1976). Influence of soil organic matter on copper availability to plants. *Z. Pflanzenernaehr. Dung. Bodenkd.* **6,** 679–684.

Petruzzelli, G., Guidi, G., and Sequi, P. (1975). The absorption of copper by roots in relation to chelating properties of soil organic matter. In *Studies about Humus*. Trans. Int. Symp. Humus et Planta VI., pp. 18–22.

Philp, R. P. and Calvin, M. (1976). Possible origin for insoluble organic (kerogen) debris in sediments from insoluble cell-wall materials of algae and bacteria. *Nature* **262,** 134–136.

Philp, R. P., Calvin, M., Brown, S., and Yang, E. (1978). Organic geochemical studies on kerogen precursors in recently deposited algal mats and oozes. *Chem. Geol.* **22,** 207–231.

Piccolo, A. and Stevenson, F. J. (1981). Infrared spectra of Cu^{2+}, Pb^{2+}, and Ca^{2+} complexes of soil humic substances. *Geoderma* **27,** 195–208.

Pimentel, G. C. and McClellan, A. L. (1960). *The Hydrogen Bond.* Freeman, San Francisco.

Pines, A., Gibby, M. G., and Waugh, J. S. (1973). Proton-enhanced NMR of dilute spins in solids. *J. Chem. Phys.* **59,** 569–590.

Piotrowicz, S. R., Harvey, G. R., Boran, D. A., Weisel, C. P., and Springer-Young, M. (1984). Cadmium, copper, and zinc interactions with marine humus as a function of ligand structure. *Mar. Chem.*, **14,** 333–346.

Piotrowicz, S. R., Harvey, G. R., Springer-Young, M., Courant, R. A., and Boran, D. A. (1983b). Studies of cadmium, copper, and zinc interactions with marine fulvic and humic materials in seawater using anodic stripping voltammetry. In *Trace Metals in Seawater* (C. S. Wong, J. D. Burton, K. Bruland, and E. D. Goldberg, eds.). Plenum, New York, pp. 699–717.

Piotrowicz, S. R., Springer-Young, M., Puig, J. A., and Spencer, M. J. (1982). Anodic stripping voltammetry for evaluation of organic–metal interactions in seawater. *Anal. Chem.* **54,** 1367–1371.

Piper, T. J. and Posner, A. M. (1972). Humic acid nitrogen. *Plant and Soil* **36,** 595–598.

Plechanov, N., Josefsson, B., Dyrssen, D., and Lundquist, K. (1983). Investigations on humic substances in natural waters. In *Aquatic and Terrestrial Humic Materials* (R. F. Christman and E. T. Gjessing, eds.). Ann Arbor Science, Ann Arbor, MI, pp. 387–405.

Pocklington, R. (1977). Chemical processes and interactions involving marine organic matter. *Mar. Chem.* **5,** 479–496.

Pocklington, R. and Leonard, J. D. (1979). Terrigenous organic matter in sediments of the St. Lawrence estuary and the Saguenay Fjord. *J. Fish. Res. Bd. Can.* **36,** 1250–1255.

Podolski, J. E. (1979). Cadmium bioaccumulation assays. Their relationship to various ionic equilibria in Lake Superior water. *Environ. Sci. Technol.* **13,** 701–706.

Pommer, A. M. and Breger, I. A. (1960). Potentiometric titration and equivalent weight of humic acid. *Geochim. Cosmochim. Acta* **20,** 30–44.

Ponomareva, V. V. and Ettinger, A. I. (1954). Characterization of the organic substances dissolved in neva water. *Zh. Prikl. Khim* **27,** 774–781.

Porath, J. and Flodin, P. (1959). Gel filtration: A method for desalting and group separation. *Nature* **183,** 1657–1659.

Porrier, M. A., Bordelon, B. R., and Laseter, J. L. (1972). Adsorption and concentration of dissolved carbon-14 DDT by coloring colloids in surface water. *Environ. Sci. Technol.* **6,** 1033–1035.

Porter, L. K. (1975). Nitrogen transfer in ecosystems. In *Soil Biochemistry,* Vol. 4 (E. A. Paul and A. D. McLaren, eds.). Marcel Dekker, New York, pp. 1-30.

Porter, N. A., Lehman, L. S., Weber, B. A., and Smith, K. J. (1981). Unified mechanism for polyunsaturated fatty acid autoxidation. Competition of peroxy radical hydrogen abstraction, b-scission and cyclization. *J. Am. Chem. Soc.* **103,** 6447–6454.

Posner, A. M. (1963). Importance of electrolyte in the determination of molecular weights by "Sephadex" gel filtration with especial reference to humic acid. *Nature* **198,** 1161–1163.

Posner, A. M. (1964). Titration curves of humic acid. In *Proceedings of the 8th International Congress of Soil Science Part II*. Bucharest, Romania.

Posner, A. M. (1966). The humic acid extracted by various reagents from a soil. *J. Soil Sci.* **17,** 65–78.

Posner, A. M. and Creeth, J. M. (1972). A study of humic acid by equilibrium ultracentrifugation. *J. Soil Sci.* **23,** 333–341.

Post, L. von and Granlund, E. (1926). Peat resources in the south of Sweden. *Swed. Geol. Unders.,* Ser. C., No. 335.

Postma, H., Manuels, M. W., and Rommets, J. W. (1976). Breakdown and production of fluorescent substances in Dutch waters. *Neth. J. Sea Res.* **10,** 499–516.

Potts, W. J. (1963). *Chemical Infrared Spectroscopy*. Wiley, New York, 322 pp.

Poutanen, E. L. and Morris, R. J. (1983). The occurrence of high molecular weight humic compounds in the organic-rich sediments of the Peru continental shelf. *Oceanol. Acta* **6**(1), 21–28.

Povoledo, D. and Gerletti, M. (1963). Studies on the sedimentary, acid-soluble organic matter from Lake Maggiore (North Italy). I. Heterogeneity and chemical properties of a fraction precipitated by barium ions. *Mem. Ist. Ital. Idrobiol.* **17,** 115–150.

Povoledo, D. and Gerletti, M. (1968). Recent studies on the sedimentary organic matter from Lake Maggiore (Northern Italy). *Mitt. Intl. Verein. Limnol.* **14,** 145–154.

Povoledo, D., Murray, D., and Pitze, M. (1975). Pigments and lipids in the humic acids of some Canadian lake sediments. In *Humic Substances, Their Structure and Function in the Biosphere* (D. Povoledo and H. L. Golterman, eds.). Centre for Agricultural Publications and Documentation, Wageningen, pp. 233–258.

Povoledo, D. and Pitze, M. (1979). Preparation of low-ash humic acids, and the accompanying losses of organic material, from Canadian lake sediments and a gleysol. *Soil Sci.* **128,** 1–8.

Prakash, A. (1971). Terrigenous organic matter and coastal phytoplankton fertility. In *Fertility of the Sea* (J. D. Costlow, ed.). Gordon and Breach, New York, pp. 351–369.

Prakash, A., Rashid, M. A., and Rao, D. V. S. (1973). Influence of humic substances on the growth of marine phytoplankton: Diatoms. *Limnol. Oceanogr.* **18,** 516–524.

Pregl, F. (1930). *Die quantitative organische Mikroanalyse*. Springer, Berlin.

Premuzic, E. T., Benkovitz, C. M., Gaffeny, J. S., and Walsh, J. J. (1982). The nature and distribution of organic matter in the surface sediments of world oceans and seas. *Org. Geochem.* **4,** 63–77.

Preston, C. M., Mathur, S. P., and Rauthan, B. S. (1981). The distribution of copper, amino compounds, and humus fractions in organic soils of differing copper contents. *Soil Sci.* **131,** 344–352.

Preston, C. M., Rauthan, B. S., Rodger, Charles, and Ripmeester, J. A. (1982). A hydrogen-1, carbon-13, and nitrogen-15 nuclear magnetic resonance study of *p*-benzoquinone polymers incorporating amino nitrogen compounds ("synthetic humic acids"). *Soil Sci.* **134,** 277–293.

Preston, C. M. and Ripmeester, J. A. (1982). Application of solution and solid-state ^{13}C NMR to four organic soils, their humic acids, fulvic acids, humins, and hydrolysis residues. *Can. J. Spectrosc.* **27,** No. 4, 99–105.

Preston, C. M., Dudley, R. L., Fyfe, C. A., and Mathur, S. P. (1984). Effects of variations in contact times and copper contents in a ^{13}C CPMAS NMR study of samples of four organic soils. *Geoderma* **33,** 245–253.

Preston, M. R. (1979). An investigation into the behavior of humic compounds in estuaries. Ph.D. Thesis, University of Liverpool, 142 pp.

Preston, M. R. and Riley, J. P. (1982). The interactions of humic compounds with electrolytes and three clay minerals under simulated estuarine conditions. *Estuarine Coastal Shelf Sci.* **14,** 567–576.

Price, W. J. (1972). Sample handling techniques. In *Laboratory Methods in Infrared Spectroscopy* (R. G. J. Miller and B. C. Stace, eds.). Heyden and Sons, London, pp. 97–128.

Provasoli, L. (1963). Organic regulation of phytoplankton fertility. In *The Sea,* Vol. 2. Wiley-Interscience, New York, pp. 165–219.

Quinn, J. G. and Meyers, P. A. (1971). Retention of dissolved organic acids in seawater by various filters. *Limnol. Oceanogr.* **16,** 129–131.

Ragan, M. A., Smidsrod, O., and Larsen, B. (1979). Chelation of divalent metal ions by brown algal polyphenols. *Mar. Chem.* **7,** 265–271.

Ragland, J. L. (1962). The use of thermometric titrations in soil chemistry studies. *Soil Sci. Soc. Am. Proc.* **26,** 133–137.

Rakovsky, V. (1959). Processes of formation of fuels and the ways of solving the general problem of genesis. In *Symposium on the Nature of Coal*. Central Fuel Research Institute, Jealgora, India, pp. 19–26.

Rashid, M. A. (1971). Role of humic acids of marine origin and their different molecular weights in complexing di- and trivalent metals. *Soil Sci.* **111,** 298–306.

Rashid, M. A. and King, L. H. (1969). Molecular weight distribution measurements on humic and fulvic acid fractions from marine clays on the Scotian Shelf. *Geochim. Cosmochim. Acta* **33,** 147–151.

Rashid, M. A. and King, L. H. (1970). Major oxygen-containing functional groups present in humic and fulvic acid fractions isolated from contrasting marine environments. *Geochim. Cosmochim. Acta* **34,** 193–201.

Rashid, M. A. and Prakash, A. (1972). Chemical characteristics of humic compounds isolated from some decomposed marine algae. *J. Fish. Res. Bd. Can.* **29,** 55–60.

Rebhun, M. and Manka, J. (1971). Classification of organics in secondary effluents. *Environ. Sci. Technol.* **5,** 606–608.

Reddy, M. M. (1978). Kinetic inhibition of calcium carbonate formation by waste-water constituents. In *Chemistry of Wastewater Technology* (A. J. Rubin, ed.). Ann Arbor Science, Ann Arbor, MI, pp. 31–58.

Resing, H. A., Garroway, A. N., and Hazlett, R. N. (1978). Determination of aromatic hydrocarbon fraction in oil shale by ^{13}C NMR with magic angle spinning. *Fuel* **57,** 450–454.

Reuter, J. H. (1977). Inferred configurational changes of aquatic humic substance. In *Abs. Geol. Soc. Am. Ann. Mtg.,* 1140 pp.

Reuter, J. H., Ghosal, M., Chian, E. S. K., and Giabbai, M. (1983). Oxidative degradation studies on aquatic humic substances. In *Aquatic and Terrestrial Humic Materials* (R. F. Christman and E. T. Gjessing, eds.). Ann Arbor Science, Ann Arbor, MI, pp. 107–125.

Reuter, J. H. and Perdue, E. M. (1977). Importance of heavy metal–organic matter interactions in natural waters. *Geochim. Cosmochim. Acta* **41,** 325–334.

Reuter, J. H. and Perdue, E. M. (1981). Calculation of molecular weights of humic substances from colligative data: Application to aquatic humus and its molecular size fractions. *Geochim. Cosmochim. Acta* **45,** 2017–2022.

Rex, R. W. (1960). Electron paramagnetic resonance studies on stable free radicals in lignins and humic acids. *Nature* **188,** 1185–1186.

Reynolds, R. C., Jr. (1978). Polyphenol inhibition of calcite precipitation in Lake Powell. *Limnol. Oceanogr.* **23,** 585–597.

Reynolds, T. M. (1963). Chemistry of nonenzymic browning. I. The reaction between aldoses and amines. *Adv. Food Res.* **12,** 1–52.

Reynolds, T. M. (1965). Chemistry of nonenzymic browning. II. *Adv. Food Res.* Div. of Food Preservation, Commonwealth Scientific and Industrial Research Organization, No. Ryde, New South Wales, Australia. **14,** 167–283.

Rice, D. L. (1982). The detritus nitrogen problem: New observations and perspectives from organic geochemistry. *Mar. Ecol. Prog. Ser.* **9,** 153–162.

Richards, E. G. (1980). *An Introduction to the Physical Properties of Large Molecules in Solution*. Cambridge University Press, Cambridge.

Rickwood, D. (1978). *Centrifugation: A Practical Approach*. Information Retrieval.

Rifai, N. and Bertru, G. (1980). La biodegradation des acides fulviques. *Hydrobiologia* **75,** 181–184.

Riffaldi, R. and Schnitzer, M. (1972a). Electron spin resonance spectrometry of humic substances. *Soil Sci. Soc. Am. Proc.* **36,** 301–305.

Riffaldi, R. and Schnitzer, M. (1972b). Effects of 6*N* HCl hydrolysis on the analytical characteristics and chemical structure of humic acids. *Soil Sci.* **115,** 349–356.

Riley, J. P. and Taylor, D. (1969). The analytical concentration of traces of dissolved organic materials from seawater with Amberlite XAD-1 resin. *Anal. Chim. Acta* **46,** 307–309.

Rinderknecht, H. and Jurd, L. (1958). A novel non-enzymatic browning reaction. *Nature* **181,** 1268–1269.

Ritchie, G. S. P. and Posner, A. M. (1982). The effect of pH and metal binding on the transport properties of humic acids. *J. Soil Sci.* **33,** 233–247.

Ritchie, P. F. and Purves, C. B. (1947). Periodate lignins: Their preparation and properties. *Pulp Paper Mag. Can.* **48,** 74–82.

Robin, P. L., Rouxhet, P. G., and Durand, B. (1977). Caracterisation des kerogenes et de leur evolution par spectroscopie infrarouge: fonctions carbonees. In *Advances in Organic Chemistry 1975* (R. Campos and J. Goni, eds.). Enadisma, pp. 693–716.

Robison, W. E. (1976). Origin and characteristics of Green River oil shale. In *Oil Shale* (T. F. Yen and G. V. Chilingarian, eds.). Elsevier, New York, pp. 61–79.

Rodgers, D. H., Ogle, L. D., and Bombaugh, K. J. (1981). Study of the HPLC separation and characterization of non-volatile organics from drinking water. Radian Corporation Report 200-33 to the Environmental Protection Agency, MERL.

Rook, J. J. (1974). Formation of haloforms during chlorination of natural waters. *Water Treatment Exam.* **23,** 234–243.

Rook, J. J. (1979). Complexation of cadmium(II) with water- and soil-derived fulvic acids: Effect of pH and fulvic acid. *Can. J. Chem.* **57,** 1263–1268.

Ross, D. J. and Speir, T. W. (1979). Biochemical activities of organic soils from subantarctic tussock grasslands on Campbell Island: 2: Enzyme activities. *N. Z. J. Sci.* **22,** 173–182.

Rossknecht, H. (1980). Phosphatelimination durch autochthone Calcitfaellung im Bodensee-Obersee. *Arch. Hydrobiol.* **88,** 328–344.

Rosswall, T. (1982). Microbiological regulation of the biogeochemical nitrogen cycle. *Plant and Soil* **67,** 15–34.

Roulet, N., Mehta, N. C., Dubach, P., and Deuel, H. (1963). Abtrennung von kohlenhydraten und stickstoff-verbindungen aus huminstoffen durch gelfiltration und ionenaustausch-chromatographie. *Z. Pflanzenernahr. Dung. Bodenkd.* **103,** 1–9.

Ruggiero, P., Interesse, F. S., Cassidie, L., and Sciacovelli, O. (1979a). ^{1}H NMR spectra of humic and fulvic acids and their peracetic oxidation products. *Geochim. Cosmochim. Acta* **44,** 603–609.

Ruggiero, P., Interesse, F. S., and Sciacovelli, O. (1979b). ^{1}H and ^{13}C NMR studies on the importance of aromatic structures in fulvic and humic acids. *Geochim. Cosmochim. Acta* **43,** 1771–1775.

Ruggerio, P., Interesse, F. S., and Sciacovelli, O. (1979c). ^{1}H NMR evidence of exchangeable aromatic protons in fulvic and humic acids. *Soil Biol. Biochem.* **12,** 297–299.

Ruggiero, P., Interesse, F. S., Cassidei, L., and Sciacovelli, O. (1981). ^{1}H NMR and IR spectroscopic investigations on soil organic fractions obtained by gel chromatography. *Soil Biol. Biochem.* **13,** 361–366.

Ruggiero, P., Sciacovelli, O., Testini, C., and Interesse, F. S. (1978). Spectroscopic studies on soil organic fractions. II. IR and ^{1}H NMR spectra of methylated and unmethylated fulvic acids. *Geochim. Cosmochim. Acta* **42,** 411–416.

Ruhe, R. V. (1969). Principles for dating pedogenic events in the Quarternary. *Soil Sci.* **107,** 398–402.

Rybicka, S. M. (1959). The solvent extraction of low rank vitrain. *Fuel* **38,** 45–54.

Ryhaenen, R. (1968). Die Bedeutung der Humussubstanzen im Stoffhaushalt der Gewaesser Finnlands. *Mitt. Int. Verein. Limnol.* **14,** 168–178.

Saar, R. A. and Weber, J. H. (1980). Comparison of spectrofluorometry and ion-selective electrode potentiometry for determination of complexes between fulvic acid and heavy-metal ions. *Anal. Chem.* **52,** 2095–2100.

Saar, R. A. and Weber, J. H. (1982). Fulvic acid: Modifier of metal-ion chemistry. *Environ. Sci. Technol.* **16,** 510A–517A.

Sackett, W. M. and Thompson, R. R. (1963). Isotopic organic carbon composition of recent continental derived clastic sediments of the eastern Gulf Coast, Gulf of Mexico. *Bull. Am. Assoc. Petrol. Geol.* **47,** 525–531.

Saint-Paul, C., Monin, J. C., and Durand, B. (1980). Methode de caracterisation rapide des hydrocarbures de C_1 a C_{35} contenus dans les roches sedimentaires et dans les huiles. *Rev. Inst. Fr. Petrole* **XXXV-6,** 1065–1078.

Saito, Y. and Hayano, S. (1979). Application of high-performance aqueous gel permeation chromatography to humic substances from marine sediment. *J. Chromatogr.* **177,** 390–392.

Saito, Y. and Hayano, S. (1981). Characterization of humic and fulvic acids isolated from marine sediments of Sagami and Suruga Bays with ^{13}C and proton nuclear magnetic resonance. *J. Oceanogr. Soc. Japan* **36,** No. 6, 286–292.

Saiz-Jimenez, C., Haider, K., and Muezelaar, H. L. C. (1979). Comparisons of soil organic matter and its fractions by pyrolysis mass spectrometry. *Geoderma* **22,** 25–38.

Saleh, F. Y., Chang, D. Y., and Frye, J. S. (1983). Cross-polarization carbon-13 nuclear magnetic resonance and fast atom bombardment mass spectrometry of fractionated fulvic acid. *Anal. Chem.* **55,** 862–866.

Salfield, J. C. (1965). Die Charakterisierung von huminstoffen durch differensen-spektrogramme. *Mitt. Mtdsch. Bodenkundl. Ges.* **4,** 289–290.

Saussure, T. de (1804). Recherches chemiques sur la vegetation. *Paris Ann.* **12,** 162.

Saville, T. (1917). The nature of color in water. *J. N.E. Water Works Assoc.* **31,** 79–123.

Saxby, J. D. (1976). The significance of organic matter in ore genesis. In *Handbook of Strata-Bound and Strateform Ore Deposits. Vol. 2 Geochemical Studies* (K. H. Wolf, ed.). Elsevier, New York, pp. 111–131.

Schachman, H. K. (1959). *Ultracentrifugation in Biochemistry*. Academic, New York, 272 pp.

Schaefer, J. and Stejskal, E. O. (1979). High resolution of solid polymers. In *Topics in Carbon-13 NMR Spectroscopy* (G. C. Levy, ed.). Wiley, New York, pp. 283.

Scheffer, F. (1954). Neure Erkentnisse in der Humusforschung. *Trans. 5th Intl. Congr. Soil Sci.* **1,** 208–228.

Scheffer, F. and Ulrich, B. (1960). *Humus and Humusdungung*. Ferdinand Enke, Stuttgart, 266 pp.

Schindler, D. W., Lean, D. R. S., and Fee, E. J. (1975). Nutrient cycling in freshwater ecosystems. In *Productivity of World Ecosystems*. National Academy of Science, Washington, D.C., pp. 96–105.

Schindler, J. E., Williams, D. J., and Zimmerman, A. P. (1976). Investigation of extracellular electron transport by humic acids. In *Environmental Biogeochemisty* 1: Carbon, Nitrogen, Phosphorus, Sulfur and Selenium Cycles, (J. O. Nriagu, ed.). Ann Arbor Science Publishers, Ann Arbor, MI, pp. 109–115.

Schloesing, Th. (1876). Sur les echanges d'ammoniaque entre l'atmosphere et la terre vegetale. *C. R. Acad. Sci.* **82,** 1105–1108.

Schnitzer, M. (1965a). Contribution of organic matter to the cation exchange capacity of soil. *Nature* **207,** 667–668.

Schnitzer, M. (1965b). The application of infrared spectroscopy to investigations on soil humic components. *Can. Spec.* **10,** 121–127.

Schnitzer, M. (1967). Humic–fulvic acid relationships in organic soils and humification of the organic matter in these soils. *Can. J. Soil Sci.* **47,** 245–250.

Schnitzer M. (1969). Reaction between fulvic acid, a soil humic compound and inorganic soil constituents. *Soil Sci. Soc. Am. Proc.* **33,** 75–81.

Schnitzer, M. (1971). Characterization of humic constituents by spectroscopy. In *Soil Biochemistry,* Vol. 2 (A. D. McLaren and J. Skujins, eds.). Marcel Dekker, New York, pp. 60–95.

Schnitzer, M. (1976). The chemistry of humic substances. In *Environmental Biochemistry,* Vol. 1 (J. O. Nriagu, ed.). Ann Arbor Science, Ann Arbor, MI, pp. 89–107.

Schnitzer, M. (1977). Recent findings of the characterization of humic substances extracted from soils from widely differing climatic zones. In *Proceedings of the Symposium on Soil Organic Matter Studies, Braunsweig,* pp. 117–131.

Schnitzer, M. (1978). Humic substances: Chemistry and reactions. In *Soil Organic Matter* (M. Schnitzer and S. U. Khan, eds.). Elsevier, New York, pp. 1–64.

Schnitzer, M. (1982). Quo vadis organic matter research? *Trans. 12th Intl. Congr. Soil Sci. New Delhi* **4,** 67–78.

Schnitzer, M. and Barton, D. H. R. (1963). A new experimental approach to the humic acid problem. *Nature* **198,** 217–219.

Schnitzer, M. and Desjardins, J. G. (1962). Molecular and equivalent weights of organic matter of a podzol. *Soil Sci. Soc. Am. Proc.* **26,** 362–365.

Schnitzer, M. and Desjardins, J. G. (1965). Carboxyl and phenolic hydroxyl groups in some organic soils and their relation to the degree of humification. *Can. J. Soil Sci.* **45,** 257–264.

Schnitzer, M. and Desjardins, J. G. (1966). Oxygen-containing functional groups in organic soils and their relation to the degree of humification as determined by solubility in sodium pyrophosphate solution. *Can. J. Soil Sci.* **46,** 236–243.

Schnitzer, M. and Gupta, V. C. (1964). Some chemical characteristics of the organic matter extracted from the O and B_2 horizons of a gray wooded soil. *Soil Sci. Soc. Am. Proc.* **28,** 374–377.

Schnitzer, M. and Gupta, V. C. (1965). Determination of acidity in soil organic matter. *Soil Sci. Soc. Am. Proc.* **29,** 274–277.

Schnitzer, M. and Hansen, E. H. (1970). Organo-metallic interactions in soils: 8. An evaluation of methods for the determination of stability constants of metal–fulvic acid complexes. *Soil Sci.* **109,** 333–340.

Schnitzer, M. and Hindle, D. A. (1980). Effect of peracetic acid oxidation on N-containing components of humic materials. *Can. J. Soil Sci.* **60,** 541–548.

Schnitzer, M. and Khan, S. U. (1972). *Humic Substances in the Environment.* Marcel Dekker, New York, 327 pp.

Schnitzer, M. and Khan, S. U. (eds.) (1978). *Soil Organic Matter.* Elsevier, New York, 319 pp.

Schnitzer, M. and Kodama, H. (1975). An electron microscopic examination of fulvic acid. *Geoderma* **13,** 279–287.

Schnitzer, M. and Kodama, K. (1976). The dissolution of micas by fulvic acid. *Geoderma* **15,** 381–391.

Schnitzer, M. and Levesque, M. (1979). Electron spin resonance as a guide to the degree of humification of peats. *Soil Sci.* **127,** 140–145.

Schnitzer, M., Marshall, P. M., and Hindle, D. A. (1983). The isolation of soil humic and fulvic acid components rich in "unknown" N. *Can. J. Soil Sci.,* **63,** 425–433.

Schnitzer, M. and Mathur, S. P. (1978). A chemical and spectroscopic characterization of some synthetic analogues of humic acids. *Soil Sci. Soc. Am. J.* **42,** 591–595.

Schnitzer, M. and Skinner, S. I. M. (1963). Organo–metallic interactions in soils: 1. Reactions between a number of metal ions and the organic matter of a podzol B_h horizon. *Soil Sci.* **96,** 86–93.

Schnitzer, M. and Skinner, S. I. M. (1968a). Alkali versus acid extraction of soil organic matter. *Soil Sci.* **105,** 392–396.

Schnitzer, M. and Skinner, S. I. M. (1968b). Gel filtration of fulvic acid, a soil humic compound.

In *Isotopes and Radiation in Soil Organic Matter Studies,* International Atomic Energy Agency, Vienna, pp. 41–55.

Schnitzer, M. and Skinner, S. I. M. (1969). Free radicals in soil humic compounds. *Soil Sci.* **108,** 383–390.

Schothorst, C. J. (1977). Subsidence of low moor peat soils in Western Netherlands. *Geoderma* **17,** 265–271.

Schreiner, O. and Shorey, E. C. (1910). Chemical nature of soil organic matter. *U.S. Dept. Agric. Bur. Soils Bull.* 74 (see also Bulletins 53, 70, 77, 80, 83, 87, 89, 90, and 108).

Schreiner, O. and Shorey, E. C. (1914). *U.S. Dept. Agric. Bur. Soils Bull.* 108.

Schuchardt, K. (1980). Stockstoffelementanalysis in mineralolen *Erdol Kohle Erdgas Petrochemie* **33,** 96.

Sciacovelli, O., Senesi, N., Solinas, V., and Testini, C. (1977). Spectroscopic studies on soil organic fractions. I. IR and NMR spectra. *Soil Biol. Biochem.* **9,** 287–293.

Seal, B. K., Roy, K. B., and Mukherjee, S. K. (1964). Fluorescence emission spectra and structure of humic and fulvic acids. *J. Ind. Chem. Soc.* **41,** 212–214.

Sears, P. G., Wolford, R. K., and Dawson, L. R. (1956). Conductances of some acids, bromides, and picrates in dimethylformamide at 25°C. *J. Electrochem. Soc.* **103,** 633–636.

Senesi, N. (1981). Free radicals in electron donor–acceptor reactions between a soil humic acid and photosynthesis inhibitor herbicides. *Z. Pflanzenernahr Dung. Bodenkd.* **144,** 580–586.

Senesi, N., Chen, Y., and Schnitzer, M. (1977a). Hyperfine splitting in electron spin resonance spectra of fulvic acid. *Soil Biol. Biochem.* **9,** 371–372.

Senesi, N., Chen, Y., and Schnitzer, M. (1977b). The role of humic acids in extracellular electron transport and chemical determination of pE in natural waters. *Soil Biol. Biochem.* **9,** 397–403.

Senesi, M., Griffith, S. M., and Schnitzer, M. (1977c). Binding of Fe^{3+} by humic materials. *Geochim. Cosmochim. Acta* **41,** 969–976.

Senesi, N. and Schnitzer, M. (1977). Effects of pH, reaction time, chemical reduction, and irradiation on ESR spectra of fulvic acid. *Soil Sci.* **123,** 224–234.

Shapiro, J. (1957). Chemical and biological studies on the yellow organic acids of lake water. *Limnol. Oceanogr.* **2,** 161–179.

Shapiro, J. (1961). Freezing-out, a safe technique for concentration of dilute solutions. *Science* **133,** 2063–2064.

Shapiro, J. (1964). Effect of yellow organic acids on iron and other metals in water. *J. Am. Water Works Assoc.* **56,** 1062–1082.

Shapiro, J. (1967). Concentration of volatile substances in aqueous solution and production of water-free organic by freezing out. *Anal Chem.* **39,** 280.

Sharp, J. H. (1973). Size classes of organic carbon in seawater. *Limnol. Oceanogr.* **18,** 441–447.

Sharp, J. H., Culberson, C. H., and Church, T. M. (1982). The chemistry of the Delaware estuary. General considerations. *Limnol. Oceanogr.* **27,** 1015–1028.

Sheldon, R. W. (1972). Size separation of marine seston by membrane and glass-fiber filters. *Limnol. Oceanogr.* **17,** 494–498.

Shioya, M. and Ishiwatari, R. (1983). Laboratory thermal conversion of sedimentary lipids to kerogen-like matter. Organic Geochem. **5,** 7–12.

Shmuk, A. A. (1930). The chemical nature of soil organic matter. *Byull. Pochvoveda,* 5–7.

Sholkovitz, E. R. (1976). Flocculation of dissolved organic and inorganic matter during the mixing of river water and seawater. *Geochim. Cosmochim. Acta* **40,** 831–845.

Sholkovitz, E. R., Boyle, E. A., and Price, N. B. (1978). The removal of dissolved humic acids and iron during estuarine mixing. *Earth Planet. Sci. Lett.* **40,** 130–136.

Sholkovitz, E. R. and Copland, D. (1981). The coagulation, solubility, and adsorption properties of Fe, Mn, Cu, Ni, Cd, Co, and humic acids in a river water. *Geochim. Cosmochim. Acta* **45,** 181–189.

Shultz, D. J. and Calder, J. A. (1976). Organic carbon $^{13}C/^{12}C$ variations in estuarine sediments. *Geochim. Cosmochim. Acta* **40,** 381–385.

Shuman, M. S., Collins, G. J., Fitzgerald, P. J., and Olson, D. L. (1983). Distribution of stability constants and dissociation rate constants among binding sites on estuarine copper–organic complexes: Rotated disk electrode studies and an affinity spectrum analysis of ion-selective electrode and photometric data. In *Aquatic and Terrestrial Humic Materials* (F. R. Christman and E. T. Gjessing, eds.). Ann Arbor Science, Ann Arbor, MI, pp. 349–370.

Sieburth, J. McN. and Jensen, A. (1968). Studies on algal substances in the sea. I. Gelbstoff (humic material) in terrestrial and marine waters. *J. Exp. Mar. Biol. Ecol.* **2,** 174–189.

Sieburth, J. McN. and Jensen, A. (1969). Studies on algal substances in the sea. II. The formation of gelbstoff (humic material) by exudates of *Phaeophyta. J. Exp. Mar. Biol. Ecol.* **3,** 275–289.

Siegel, A. (1971). Metal interactions in the marine environment. In *Organic Compounds in Aquatic Environments* (S. J. Faust and J. V. Hunter, eds.). Marcel Dekker, New York, pp. 265–295.

Sigleo, A. C., Helz, G. R., and Zoller, W. H. (1980). Organic-rich colloidal material in estuaries and its alteration by chlorination. *Environ. Sci. Technol.* **14,** 673–679.

Sigleo, A. C., Hoering, T. C., and Helz, G. R. (1982). Composition of estuarine colloidal material: Organic components. *Geochim. Cosmochim. Acta* **46,** 1619–1626.

Simms, H. S. (1926a). Dissociation of polyvalent substances. I. Relation of constants to titration data. *J. Am. Chem. Soc.* **48,** 1239–1250.

Simms, H. S. (1926b). Dissociation of polyvalent substances. II. Relation of constants to chemical structure. *J. Am. Chem. Soc.* **48,** 1251–1261.

Simonson, R. W. (1959). Outline of a generalized theory of soil genesis. *Soil Sci. Soc. Am. Proc.* **23,** 152–156.

Sinclair, A. H. and Tinsley, J. (1981). Studies of soil organic matter using formic acid solvents: Factors affecting composition and degree of formylation of organic isolates. *J. Soil Sci.* **32,** 103–117.

Sirotkina, I. S., Varshall, G. M., Lur'e, Y. Y., and Stepanova, N. P. (1974). Use of cellulose sorbents and Sephadexes in the systematic analysis of organic matter in natural water. *Zh. Anal. Khim.* **29,** 1626–1632.

Skogerboe, R. K. and Wilson, S. A. (1981). Reduction of ionic species by fulvic acid. *Anal. Chem.* **53,** 228–232.

Skopintsev, B. (1972). On the age of stable organic matter–aquatic humus in oceanic waters. In *The Changing Chemistry of the Oceans,* Proceedings of the 12th Nobel Symposium, 1971 (D. Dyrssen and D. Jagner, eds.). Wiley, New York.

Skopintsev, B. (1982). New data on the composition of organic matter from ocean water and the energy bound in it. *Oceanology* **21,** 582–588.

Skopintsev, B. A. and Krylova, L. P. (1955). Optical properties of the organic substance of the aqueous humus present in the upper layers of inland waters. *Gidrokhim. Maerialy* **23,** 31–35.

Small, H., Stevens, T. S., and Bauman, W. C. (1975). Novel ion-exchange chromatographic method using conductimetric detection. *Anal. Chem.* **47,** 1801–1809.

Smart, P. L., Finlayson, B. L., Rylands, W. D., and Ball, C. M. (1976). The relation of fluorescence to dissolved organic carbon in surface waters. *Water Res.* **10,** 805–811.

Smith, D. (1982). Nitrogen fixation. In *Experimental Microbial Ecology* (R. G. Burns and J. H. Slater, eds.), Blackwell Scientific, London, pp. 212–220.

Smith, R. G., Jr. (1976). Evaluation of combined applications of ultrafiltration and complexation capacity techniques to natural waters. *Anal. Chem.* **48,** 74–76.

Snyder, L. R. (1968). *Principles of Adsorption Chromatography*. Marcel Dekker, New York, 413 pp.

Snyder, L. R. (1978). Solvent selection for separation processes. In *Techniques of Chemistry,* Vol. XII. *Separation and Purification,* 3rd ed. (E. S. Perry and A. Weissberger, eds.). Wiley, New York, pp. 25–75.

Soil Taxonomy, *USDA Agric. Handbook* **436,** 211–226.

Sourirajan, S. (ed.) (1977). *Reverse Osmosis and Synthetic Membranes; Theory—Technology—Engineering,* National Research Council.

Sourirajan, S. and Matsuura, J. (1977). Physicochemical criteria for reverse osmosis separations. In *Reverse Osmosis and Synthetic Membranes; Theory—Technology—Engineering,* National Research Council, Chap. 2, pp. 5–43.

Sowden, F. J. (1959). Investigations on the amounts of hexosamines found in various soils and methods for their determinations. *Soil Sci.* **88,** 138–143.

Sowden, F. J. (1969). Effect of hydrolysis time and iron and aluminum on the determination of amino compounds in soil. *Soil Sci.* **107,** 364–371.

Sowden, F. J., Chen, Y., and Schnitzer, M. (1977). The nitrogen distribution in soils formed under widely differing climatic conditions. *Geochim. Cosmochim. Acta* **42,** 1524–1526.

Sowden, F. J. and Deuel, H. (1961). Fractionation of fulvic acids from the B horizon of a podzol. *Soil Sci.* **91,** 44–47.

Sowden, F. J., Griffith, S. M., and Schnitzer, M. (1976). The distribution of nitrogen in some highly organic tropical volcanic soils. *Soil. Biol. Biochem.* **8,** 55–60.

Sowden, F. J. and Schnitzer, M. (1967). Nitrogen in illuvial organic matter. *Can. J. Soil Sci.* **47,** 111–116.

Spackman, W., Cohen, A. D., Given, P. H., and Casagrande, D. J. (1974). The comparative study of the Okefenokee Swamp and the Everglades–mangrove swamp–marsh complex of southern Florida. *Field Guidebook for the Geology Society of America Meeting, Miami,* 265 pp.

Speigt, M. C. D. and Blackith, R. E. (1983). The animals. In *Ecosystems of the World, 4A. Mires; Swamps, Bog, Fen, and Moor. General Studies* (A. J. P. Gore, ed.). Elsevier, New York, pp. 349–366.

Spiker, E. C. and Hatcher, P. G. (1984). Carbon isotope fractionation of sapropelic organic matter during early diagenesis: *Org. Geochem.* **5,** 283–290.

Sposito, G. and Holtzclaw, K. M. (1977). Titration studies on the polynuclear, polyacidic nature of fulvic acid extracted from sewage sludge–soil mixtures. *Soil Sci. Soc. Am. J.* **41,** 330–336.

Sposito, G., Holtzclaw, K. M., and Keech, D. A. (1977). Proton binding in fulvic acid extracted from sewage sludge–soil mixtures. *Soil Sci. Soc. Am. J.* **41,** 1119–1125.

Sposito, G., Holtzclaw, K. M., and LeVesque-Madore, C. S. (1979). Cupric ion complexation by fulvic acid extracted from sewage sludge-soil mixtures. *Soil Sci. Soc. Am. J.* **43,** 1148–1155.

Sposito, G., Schaumberg, G. D., Perkins, T. G., and Holtzclaw, K. M. (1978). Investigation of fulvic acid extracted from sewage sludge using carbon-13 and proton NMR spectroscopy. *Environ. Sci. Technol.* **12,** 931–934.

Sprengel, C. (1826). Uber Pflanzenhumus, Humussaure and Humussaure Salze. *Kastner's Arch. Ges. Naturlehre* **8,** 145–220.

Sprengel, C. (1837). *Die Bodenkunde oder die Lehre vom Boden,* Muller, Leipzig.

Springer, U. (1938). Der Heutige Stand der Humusuntersuchungsmethoden. *Bodenk. Pflanzanernahr.* **6,** 312.

Sridharan, N. and Lee, G. F. (1972). Coprecipitation of organic compounds from lake water by iron salts. *Environ. Sci. Technol.* **6,** 1021–1023.

Srna, R. F., Garrett, K. S., Miller, S. M., and Thum, A. B. (1980). Copper complexation capacity of marine water samples from southern California. *Environ. Sci. Technol.* **14,** 1482–1486.

Stabel, H. H. (1977). Gebundene Kohlenhydrate als stabile Komponenten im Schoehsee und in *Scenedesmus-Kulturen. Arch. Hydrobiol. Suppl.* **53,** 159–254.

Stabel, H. H., Moaledj, K., and Overbeck, J. (1979). On the degradation of dissolved organic molecules from Plussee by oligocarbophilic bacteria. *Arch. Hydrobiol. Beih. Ergebn. Limnol.* **12,** 95–104.

Stabel, H. H. and Steinberg, C. (1976a). Geloeste organische Molekuele im Walchensee. *Gewaesser Abwaesser* **60/61,** 100–112.

Stabel, H. H. and Steinberg, C. (1976b). Cleavage of macromolecular soluble organic matter. *Naturwissenschaften* **63,** 533.

Stach, E., (1975). *Coal Petrology.* Gebruder Borntraeger, Stuttgart, 428 pp.

Stadnikov, G. (1930). *Die Entstehung von Kohle and Erdol.* Enke, Stuttgart, 254 pp.

Steelink, C. A. (1964). Free radical studies of lignin, lignin degradation products, and soil humic acid. *Geochim. Cosmochim. Acta* **28,** 1615–1622.

Steelink, C. A. (1966). Electron paramagnetic resonance studies of humic acid and related model compounds. *Coal Sci. (Adv. Chem. Ser.)* **55,** 80–90.

Steelink, C. (1977). Humates and other natural organic substances in the aquatic environment. *J. Chem. Educat.* **54,** 599–603.

Steelink, C., Mikita, M. A., and Wershaw, R. L. (1981). Carbon-13 enriched NMR method for the determination of hydroxyl functionality in humic substances. *Anal. Chem.* **53,** 1715–1717.

Steelink, C. A., Reid, T., and Tollin, G. (1963). Structures of humic acid. IV. Electron paramagnetic resonance studies. *Biochim. Biophys. Acta* **66,** 444–447.

Steelink, C., Thorn, K., and Mikita, M. A. (1983). Magnetic resonance studies of humates and related model compounds. In *Aquatic and Terrestrial Humic Materials* (R. L. Christman and E. T. Gjessing, eds.). Ann Arbor Science, Ann Arbor, MI, pp. 87–104.

Steelink, C. A. and Tollin, G. (1962). Stable free radicals in soil. *Biochim. Biophys. Acta* **59,** 25–34.

Steelink, C. and Tollin, G. (1967). Free radicals in soil. In *Soil Biochemistry* (A. D. McLaren and G. H. Peterson, eds.). Marcel Dekker, New York, Chap. 6.

Stefan, C. and Jocteur-Monrozier, L. (1983). To be published.

Steinberg, C. (1976). Ueber die Kopraezipitation ninhydrinpositiver Stoffe. *Vom Wasser* **47,** 275–280.

Steinberg, C. (1977a). Schwer abbaubare, stickstoffhaltige geloeste organische substanzen im Schoehsee und in Algenkulturen. *Arch. Hydrobiol./Suppl.* **53,** 48–158.

Steinberg, C. (1977b). Vergleich der geloesten organischen Stoffe verschiedener holsteinischer Seen. *Arch. Hydrobiol.* **80,** 297–307.

Steinberg, C. and Herrmann, A. (1981). Utilization of dissolved metal organic compounds by freshwater microorganisms. *Verh. Intl. Verein. Limnol.* **21,** 231–235.

Steinberg, C. and Baltes, G. F. (1984). Influence of metal compounds of fulvic acid/molybdenum blue reactive phosphate associations. *Arch. Hydrobiol.* **100,** 61–71.

Stephens, J. C. and Spier, W. H. (1970). Subsidence of organic soils in the U.S.A. *Intl. Assoc. Sci. Hydrol. Symp.* No. 89, 523–534.

Stephenson, R. L. and Lyon, G. L. (1982). Carbon-13 depletion in an estuarine bivalve: Detection of marine and terrestrial food sources. *Oecologia* **55,** 110–113.

Stevens, R. J. and Stewart, B. M. (1982). Concentration, fractionation and characterization of soluble organic phosphorus in river water entering Lough Neagh. *Water Res.* **16,** 1507–1519.

Stevenson, F. J. (1972). Organic matter reactions involving herbicides in soil. *J. Environ. Qual.* **1,** 333–343.

Stevenson, F. J. (1974). Nonbiological transformations of amino acids in soil and sediments. In *Advanced in Organic Geochemistry* (B. Tissot and F. Bienner, eds.). Editions Technip, Paris, pp. 143–151.

Stevenson, F. J. (1976a). Stability constants of Cu^{2+}, Pb^{2+}, and Cd^{2+} complexes with humic acids. *Soil Sci. Soc. Am. J.* **40,** 665–672.

Stevenson, F. J. (1976b). Organic matter reactions involving pesticides in soil. *Am. Chem. Soc. Symp. Ser.* **29,** 180–207.

Stevenson, F. J. (1977). Nature of divalent transition metal complexes of humic acid as revealed by a modified potentiometric titration method. *Soil Sci.* **123,** 10–17.

Stevenson, F. J. (1982). *Humus Chemistry; Genesis, Composition, Reactions,* Wiley-Interscience, New York, 443 pp.

Stevenson, F. J. (1983). Trace metal–organic matter interactions in geologic environments. In *Significance of Trace Elements in Solving Petrogenetic Problems* (S. A. Theophrastus, ed.). National Technical University of Athens, Athens.

Stevenson, F. J. and Ardakani, M. S. (1972). Organic matter reactions involving micronutrients in soils. In *Micronutrients in Agriculture.* (J. J. Mortvedt, P. M. Giordano, and W. L. Lindsay, eds.). American Society of Agronomy, pp. 79–114.

Stevenson, F. J. and Butler, J. H. A. (1969). Chemistry of humic acids and related pigments. In *Organic Geochemistry* (G. Eglinton and M. T. J. Murphy, eds.). Springer-Verlag, New York, pp. 534–557.

Stevenson, F. J. and Fitch, A. (1981). Reactions with organic matter. In *Copper in Soils and Plants* (J. F. Loneragan, A. D. Robson, and R. D. Graham, eds.). Academic, New York, pp. 69–95.

Stevenson, F. J. and Goh, K. M. (1971). Infrared spectra of humic acids and related substances. *Geochim. Cosmochim. Acta* **35,** 471–483.

Stevenson, F. J. and Goh, K. M. (1972). Infrared spectra of humic and fulvic acids and their methylated derivatives: Evidence for nonspecificity of analytical methods for oxygen-containing functional groups. *Soil Sci.* **113,** 334–345.

Stevenson, F. J., Krastonav, S. A., and Ardakani, M. S. (1973). Formation constants of Cu(II) complexes with humic and fulvic acids. *Geoderma* **9,** 129–141.

Stevenson, I. L. and Schnitzer, M. (1982). Transmission electron microscopy of extracted fulvic and humic acids. *Soil Sci.* **133,** 179–185.

Stewart, A. J. (1984). Interactions between dissolved humic materials and organic toxicants. In *Synthetic Fossil Fuel Technologies. Results of Health and Environmental Studies* (K. E. Cowser, ed.). Butterworth, London, pp. 505–521.

Stewart, A. J. and Wetzel, R. G. (1980). Fluorescence: Absorbance ratios—a molecular-weight tracer of dissolved organic matter. *Limnol. Oceanogr.* **25,** 559–564.

Stewart, A. J. and Wetzel, R. G. (1981a). Asymmetrical relationships between absorbance, fluorescence, and dissolved organic carbon. *Limnol. Oceanogr.* **26,** 590–597.

Stewart, A. J. and Wetzel, R. G. (1981b). Dissolved humic materials: Photodegradation, sediment effects, and reactivity with phosphate and calcium precipitation. *Arch. Hydrobiol.* **92,** 265–286.

Stewart, A. J. and Wetzel, R. G. (1982). Influence of dissolved humic material on carbon assimilation and alkaline phosphatase activity in natural algalbacterial assemblages. *Freshwater Biol.* **12,** 369–380.

Stimson, M. M. (1962). Sampling for spectrophotometry with special reference to the use of the potassium bromide disc technique. In *Progress in Infrared Spectroscopy,* Vol. 1 (H. A. Szymanski, ed.). Plenum Press, New York. pp. 143–150.

Stobbe, P. C. and Wright, J. R. (1959). Modern concepts of the genesis of Podzols. *Soil Sci. Soc. Am. Proc.* **23,** 161–164.

Stock, L. M. (1972). The origin of the inductive effect. *J. Chem. Educ.* **49,** 400–404.

Strickler, A. (1967). Continuous particle electrophoresis: A new analytical and preparative capability. *Sep. Sci.* **2,** 335–355.

Strome, D. J. and Miller, M. C. (1978). Photolytic changes in dissolved humic substances. *Verh. Intl. Verein. Limnol.* **20,** 1248–1254.

Stone, A. T. (1983). The reduction and dissolution of Mn(III) and Mn(IV) oxides by organics. Ph.D. dissertation, California Institute of Technology, Pasadena, pp. 258–260.

Stuber, H. A., and Leenheer, J. A. (1978). Assessment of a resin based fractionation procedure for monitoring organic solutes from oil shale retorting wastes. In *Establishment of Water Quality Monitoring Programs,* (L. G. Everett and Schmidt, K. D., eds.). American Water Resources Assoc., Minneapolis, MN, pp. 266–272.

Stuermer, D. H. (1975). The characterization of humic substances in seawater. Ph.D. Thesis, Woods Hole Oceanographic Institution, Woods Hole, MA, 187 pp.

Stuermer, D. H. and Harvey, G. R. (1974). Humic substances from seawater. *Nature* **250,** 480–481.

Stuermer, D. H. and Harvey, G. R. (1977a). Structural studies on marine humus: A new reduction sequence for carbon skeleton determination. *Mar. Chem.* **5,** 55–70.

Stuermer, D. H. and Harvey, G. R. (1977b). The isolation of humic substances and alcohol-soluble organic matter from seawater. *Deep-Sea Res.* **24,** 303–309.

Stuermer, D. H. and Payne, J. R. (1976). Investigation of seawater and terrestrial humic substances with carbon-13 and proton nuclear magnetic resonance. *Geochim. Cosmochim. Acta* **40,** 1109–1114.

Stuermer, D. H., Peters, K. E., and Kaplan, I. R. (1978). Source indicators of humic substances and proto-kerogen. Stable isotope ratios, elemental compositions and electron spin resonance spectra. *Geochim. Cosmochim. Acta* **42,** 989–997.

Stumm, W. and Morgan, J. J. (1970). *Aquatic Chemistry.* Wiley-Interscience, New York.

Stumm, W. and Morgan, J. J. (1981). *Aquatic Chemistry,* 2nd ed. Wiley-Interscience, New York.

Styermark, A. (1961). *Quantitative Organic Microanalysis.* Academic, New York.

Suberkropp, K., Godshalk, G. L., and Klug, M. J. (1976). Changes in the chemical composition of leaves during processing in a woodland stream. *Ecology* **57,** 720–727.

Suess, E. (1980). Particulate organic carbon flux in the oceans. Surface productivity and oxygen utilization. *Nature* **288,** 260–263.

Sueur, S., Van Den Berg, C. M. J., and Riley, J. P. (1982). Measurement of the metal complexing ability of exudates of marine macroalgae. *Limnol. Oceanogr.* **27,** 536–543.

Sugai, S. F. and Healy, M. L. (1978). Voltammetric studies of the organic association of copper and lead in two Canadian inlets. *Mar. Chem.* **6,** 291–308.

Suggett, A. (1975). Polysaccharides. In *Water, A Comprehensive Treatise:* Vol. 4, *Aqueous Solutions of Amphiphiles and Macromolecules* (F. Franks, ed.), Plenum, New York, pp. 519–567.

Sunda, W. G. and Guillard, R. R. L. (1976). The relationship between cupric ion activity and the toxicity of copper to phytoplankton. *J. Mar. Res.* **34,** 511–529.

Sunda, W. and Lewis, J. A. M. (1978). Effect of complexation by natural organic ligands on the toxicity of copper to a unicellular alga, *Monochrysis Lutheri. Limnol. Oceanogr.* **23,** 870–876.

Sunda, W. G., Huntsman, S. A., and Harvey, G. R. (1982). Photoreduction of manganese oxides by organic substances in natural waters, effects of manganese geochemistry and plant nutrition. *EOS, Trans. Am. Geophys. U.* **63,** 1010.

Sunda, W. G., Huntsman, S. A., and Harvey, G. R. (1983). Photoreduction of manganese oxides in seawater and its geochemical and biological implications. *Nature* **301,** 234-236.

Suyama, K. (1981). Amino-carbonyl reaction in fatty aldehyde–primary amine system. *Nippon Nogeikagaku Zasshi* **55,** 1131–1138.

Swain, F. M. (1963). Geochemistry of humus. In *Organic Geochemistry* (I. A. Breger, ed.), International Series of Monographs on Earth Sciences, Vol. 16. Pergamon Press Symposium Publication Division I (distributed in Western Hemisphere by MacMillan), pp. 87–147.

Sweet, M. S. and Perdue, E. M. (1982). Concentration and speciation of dissolved sugars in river water. *Environ. Sci. Technol.* **16,** 692–698.

Swift, R. S. (1968). *Physico-Chemical Studies on Soil Organic Matter.* Ph.D. Thesis, University of Birmingham.

Swift, R. S. and Posner, A. M. (1971). Gel chromatography of humic acid. *J. Soil Sci.* **22,** 237–249.

Swift, R. S. and Posner, A. M. (1972). Autoxidation of humic acid under alkaline conditions. *J. Soil Sci.* **23,** 381–393.

Swincer, G. D., Oades, J. M., and Greenland, D. J. (1969). Extraction, characterization and significance of soil polysaccharides. *Adv. Agron.* **21,** 195–235.

Swindale, L. D. and Jackson, M. L. (1956). Genetic processes in some residual podzolized soils of New Zealand. *Trans. 6th Intl. Congr. Soil Sci. E,* 233–239.

Szilagyi, M. (1971). Reduction of Fe^{3+} ion by humic acid preparations. *Soil Sci.* **111,** 233–235.

Tadros, M. S. and Varney, E. A. (1982). The interrelationship between soil arthropods and soil fungi in woodland and field soils. In *New Trends in Soil Biology* (Proc. 8th Intl. Colloq. Soil Zoology). Dieu-Brichart, Louvaine-la-neuve, Belgium, pp. 247–252.

Taft, R. W., Gurka, D., Joris, L., Schleyer, Von R., and Rakshys, J. W. (1969). Studies of hydrogen-bonded complex formation with *p*-fluorophenol. V. Linear free energy relationships with OH reference acids. *J. Am. Chem. Soc.* **91,** 4801–4808.

Tan, K. H. (1980). The release of silicon, alumina, and potassium during decomposition of soil minerals by humic acid. *Soil Sci.* **129,** 5–11.

Tanford, C. (1961). *Physical Chemistry of Macromolecules.* Wiley, New York.

Tate, R. L. III (1978). Cultural and environmental factors affecting the longevity of *Escherichia coli* in Histosols. *Appl. Environ. Microbiol.* **35,** 925–929.

Taylor, L. C. E. (1981). Fast atoms allow MS study of large molecules. *Ind. Res. Devel.* **23,** 124–128.

Teichmuller, M. (1975). Origin of the petrographic constituents of coal. In *Stach's Testbook of Coal Petrology,* Gebruder Borntraeger, Berlin, pp. 1976–2038.

Telang, S. A., Baker, B. L., Costerton, J. W., Lodd, T., Mutch, R., Wallis, P. M., and Hodgson, G. W. (1981). *Biochemistry of Mountain Stream Waters: The Marmot System.* Inland Waters Directorate, Scientific Series #101, p. 101.

Tessler, A., Campbell, P. G. C., and Bissom, M. (1979). Sequential extraction procedure for the speciation of particulate trace metals. *Anal. Chem.* **51,** 844–851.

Thakur, A. K., Musons, P. J., Hunston, D. L., and Rodbard, D. (1980). Characterization of ligand-binding systems by continuous affinity distributions of arbitrary shape. *Anal. Biochem.* **103,** 240–254.

Theis, E. R. (1945). The collagen–quinone reaction. I. Fixation and thermolability as a function of pH values. *J. Biol. Chem.* **157,** 23–33.

Theng, B. K. G. (1976). Interactions between montmorillonite and fulvic acid. *Geoderma* **15,** 243–251.

Theng, B. K. G. (ed.) (1979). *Formation and Properties of Clay–Polymer Complexes.* Elsevier, Amsterdam.

Theng, B. K. G. (1979). Humic substances. In *Formation and Properties of Clay–Polymer Complexes* (B. K. G. Theng, ed.), Elsevier, Amsterdam, pp. 283–327.

Theng, B. K. G. and Posner, A. M. (1967). Nature of the carbonyl groups in soil humic acid. *Soil Sci.* **104,** 191–201.

Theng, B. K. G., and Scharpenseel, H. W. (1975). The adsorption of ^{14}C-labelled humic acid by montmorillonite. In *Proceedings of the ^{14}C International Clay Conference,* Mexico City, 1975 (S. W. Bailey, ed.), Applied Science Publishers, Barking, U.K., pp. 643–653.

Theng, B. K. G., Wake, J. R. H., and Posner, A. M. (1966). The infrared spectrum of humic acid. *Soil Sci.* **102,** 70–72.

Theng, B. K. G., Wake, J. R. H., and Posner, A. M. (1967). The humic acids extracted by various reagents from a soil. II. Infrared, visible, and ultraviolet absorption spectra. *J. Soil Sci.* **18,** 349–363.

Theng, B. K. G., Wake, J. R. H., and Posner, A. M. (1968). Fractional precipitation of soil humic acid by ammonium sulfate. *Plant and Soil* **29,** 305–316.

Thompson, R. S., Jacques, D., Haslam, E., and Tanner, R. J. N. (1972). Plant proanthocyanidins. I. Introduction, isolation, structure and distribution in nature of plant procyanidins. *J. Chem. Soc. Perkin Trans. 1,* 1387–1390.

Thurman, E. M. (1979). Isolation, characterization, and geochemical significance of humic substances from groundwater: Ph.D. Thesis, University of Colorado, Boulder.

Thurman, E. M. (1985). *Organic Geochemistry of Natural Waters,* Martinus Nijhoff/Dr. W. Junk Publishers, Dordrecht, The Netherlands (in press).

Thurman, E. M. and Malcolm, R. L. (1979). Concentration and fractionation of hydrophobic organic acid constituents from natural waters by liquid chromatography. *U.S. Geol. Surv. Water-Supply Paper 1817-G,* **16,** U.S. Government Printing Office, Washington D.C.

Thurman, E. M. and Malcolm, R. L. (1981). Preparative isolation of aquatic humic substances. *Environ. Sci. Technol.* **15,** 463–466.

Thurman, E. M. and Malcolm, R. L. (1983). Structural study of humic substances: New approaches and methods. In *Aquatic and Terrestrial Humic Materials* (R. F. Christman and E. T. Gjessing, eds.). Ann Arbor Science, Ann Arbor, MI, pp. 1–23.

Thurman, E. M., Malcolm, R. L., and Aiken, G. R. (1978). Prediction of capacity factors for aqueous organic solutes on a porous acrylic resin. *Anal. Chem.* **50,** 775–779.

Thurman, E. M., Wershaw, R. L., Malcolm, R. L., and Pinckney, D. J. (1982). Molecular size of aquatic humic substances. *Org. Geochem.* **4,** 27–35.

Tibbetts, T. E. (1981). Peat—A major Canadian resource. In *Proceedings of the Symposium on Peat,* Thunder Bay, Ontario, October 26–28, 1981. pp. 7–23.

Tilsley, G. M. (1979). Interaction of organic matter with anion resins. *Chem. Ind.* **5,** 142–149.

Tipping, E. (1981). The adsorption of aquatic humic substances by iron oxides. *Geochim. Cosmochim. Acta* **45,** 191–199.

Tipping, E. and Cooke, D. (1982). The effects of adsorbed humic substances on the surface charge of Goethite (α-FeOOH) in freshwaters. *Geochim. Cosmochim. Acta* **46,** 75–80.

Tipping, E. and Higgins, D. C. (1982). The effect of adsorbed humic substances on the colloid stability of hematite particles. *Colloid Surf.* **5,** 85–92.

Tipping, E. and Woof, C. (1983). Seasonal variations in the concentrations of humic substances in a soft-water lake. *Limnol. Oceanogr.* **28,** 168–172.

Tipping, E. and Woof, C. (1983b). Elevated concentrations of humic substances in a seasonally anoxic hypolimnion: Evidence for co-accumulation with iron. *Arch. Hydrobiol.* **98,** 137–145.

Tissot, B., Durand, B., Espitalie, J., and Combaz, A. (1974). Influence of nature and diagenesis hydrocarbons in early Toarcian shales. *Bull. Am. Assoc. Petrol. Geol.* **55,** 2177–2193.

Tissot, B., Durand, B., Espitalie, J., and Combaz, A. (1974). Influence of nature and diagenesis of organic matter in formation of petroleum. *Am. Assoc. Petrol. Geol. Bull.* **58,** 499–506.

Tissot, B. P. and Welte, D. H. (1978). *Petroleum Formation and Occurrence*. Springer-Verlag, New York.

Tokar, J. M., Harvey, G. R., and Chesal, L. A. (1981). A gas lift system for large volume water sampling. *Deep-Sea Res.* **28,** 1395–1399.

Tokar, J. M., Harvey, G. R., and Chesal, L. A. (1983). A portable system for extraction of organics from thousand liter volumes of seawater. U.S. National Oceanic and Atmospheric Administration Technical Memorandum ERL AOML-53.

Truitt, R. E. and Weber, J. H. (1981). Copper(II) and cadmium(II) binding abilities of some New Hampshire freshwaters determined by dialysis titration. *Environ. Sci. Technol.* **15,** 1204–1208.

Trutnovsky, H. (1974). A novel combustion system for the microdetermination of nitrogen. *Microchem. J.* **19,** 347–352.

Tryland, Q. and Gjessing, E. T. (1975). The ultrafilterability of iron in humic waters. *Vatten* **1,** 2–6.

Turekian, K. K. (1977). The fate of metals in the oceans. *Geochim. Cosmochim. Acta* **41,** 1139–1144.

Tuschall, J. R., Jr. and Brezonik, P. L. (1983). Complexation of heavy metals by aquatic humus: A comparative study of five analytical methods. In *Aquatic and Terrestrial Humic Materials* (R. F. Christman and E. T. Gjessing, eds.). Ann Arbor Science, Ann Arbor, MI, pp. 275–294.

Tyler, G. (1976). Heavy metal pollution, phosphatase activity and mineralization of organic phosphorus in forest soils. *Soil Biol. Biochem.* **8,** 327–332.

U.S. EPA (U.S. Environmental Protection Agency) (1974). Lower Mississippi River Facility, New Orleans Area Water Supply Study—Draft Analytical Report.

Uzaki, M. and Ishiwatari, R. (1983). Determination of cellulose and non-cellulose carbohydrates in recent sediments by gas chromatography. *J. Chromatogr.* **260,** 487–492.

Vallentyne, J. R. (1962). Solubility and the decomposition of organic matter in nature. *Arch. Hydrobiol.* **58,** 423–434.

van Dijk, H. (1966). Some physico-chemical aspects of the investigation of humus. In *The Use of Isotopes in Soil Organic Matter Studies,* Pergamon, Oxford, pp. 129–141.

van Dijk, H. (1971). Cation binding by humic acids. *Geoderma* **5,** 53–67.

van Krevelen, D. W. (1961). *Coal Typology, Chemistry, Physics and Constitution.* Elsevier, Amsterdam.

van Krevelen, D. W. (1963). Geochemistry of coal. In *Organic Geochemistry.* (I. A. Breger, ed.) Pergamon, Oxford, pp. 183–247.

Vannote, R. L., Minshall, G. W., Cummins, K. W., Sedell, J. R., and Cushing, C. E. (1980). The river continuum concept. *Can. J. Fish. Aquat. Sci.* **37,** 130–137.

Van Vleet, E. S. and Quinn, J. G. (1977). Input and fate of petroleum hydrocarbons entering the Providence River and Upper Narragansett Bay from wastewater effluents. *Environ. Sci. Technol.* **11,** 1086–1092.

Varney, M. S., Mantoura, R. F. C., Whitfield, M., Turner D. R., and Riley, J. P. (1983). Potentiometric and conformational studies of the acid–base properties of fulvic acid from natural waters. In *Trace Metals in Sea Water* Proceedings of the NATO Conference, 1981, pp. 751–772.

Varossieau, W. W. and Breger, I. A. (1952). Chemical studies on ancient buried wood and the origin of humus. In *Comptes Rendus, 3rd Congress on Stratigraphy and Geology on the Carboniferous, Herrlen,* pp. 637–646.

Vila, F. J. G., Lentz, H., and Ludemann, H. D. (1976). FT ^{13}C nuclear magnetic resonance spectra of natural humic substances. *Biochem. Biophys. Res. Commun.* **72,** 1063–1069.

Vinkler, P., Lakatos, B., and Meisel, J. (1976). Infrared spectroscopic investigations of humic substances and their metal complexes. *Geoderma* **15,** 231–242.

Visser, S. A. (1982). Acid functional group content of aquatic humic matter: Its dependence upon origin, molecular weight and degree of humification of the material. *J. Environ. Sci. Health A* **17,** 767–788.

Visser, S. A. (1983a). Application of Von Krevelen's graphical–statistical method for the study of aquatic humic material. *Environ. Sci. Technol.* **17,** 412–417.

Visser, S. A. (1983b). Fluorescence phenomena of humic matter of aquatic origin and microbial cultures. In *Aquatic and Terrestrial Humic Materials* (R. F. Christman and E. T. Gjessing, eds.), Ann Arbor Science, Ann Arbor, MI, 183–202.

Vitorovic, D., Vucelic, D., Gasic, M. J., Juranic, N., and Macura, S. (1978). Analysis of the organic matter of oil shales by nuclear magnetic resonance. *Org. Geochem.* **1,** 89–91.

Volk, B. G. and Schnitzer, M. (1973). Chemical and spectroscopic methods for assessing subsidence in Florida Histosols. *Soil Sci. Soc. Am. Proc.* **37,** 886–888.

Vucelic, D., Juranic, N., and Vitorovic, D. (1979). Potential of proton-enhanced ^{13}C NMR for the classification of kerogens. *Fuel* **58,** 759–764.

Wagner, G. H. and Stevenson, F. J. (1965). Structural arrangement of functional groups in soil humic acid as revealed by infrared analyses. *Soil Sci. Soc. Am. Proc.* **29,** 43–48.

Wake, J. R. H. and Posner, A. M. (1967). Membranes for measuring low molecular weights by osmotic pressure. *Nature* **213,** 692–693.

Waksman, S. A. (1936). *Humus.* Williams and Wilkins, Baltimore, MD.

Waksman, S. A. (1938). *Humus—Origin, Chemical Composition, and Importance in Nature,* 2d ed., Williams and Wilkins, Baltimore, MD, 526 pp.

Waksman, S. A. and Iyer, N. R. N. (1932). Contribution to our knowledge of the chemical nature and origin of humus. *Soil Sci.* **34,** 43–69.

Waldron, A. C. and Mortensen, J. L. (1961). Soil nitrogen complexes. II. Electrophoretic separation of organic components. *Soil Sci. Soc. Am. Proc.* **25,** 29–32.

Wallace, G. T., Jr. (1982). The association of copper, mercury, and lead with surface-active organic matter in coastal seawater. *Mar. Chem.* **11,** 379–394.

Wallis, P. M. (1979). *Sources, transportation, and utilization of dissolved organic matter in groundwater and streams.* Inland Water Directorate scientific series #100, Environment Canada, Ottawa, 49 pp.

Walmsley, M. E. (1973). Physical and chemical properties of peat. In *Muskeg and the Northern Environment in Canada* (N. W. Radforth and C. O. Brawner, eds.). University of Toronto Press, Toronto, pp. 82–129.

Walton, W. W. and Hoffman, J. I. (1959). Principles and methods of sampling. In *Treatise on Analytical Chemistry,* Pt. I, Vol. 1 (I. M. Kolthoff and P. J. Elving, eds.). Wiley-Interscience, New York, pp. 67–97.

Wang, T. S., Li, S. W., and Ferng, Y. L. (1978). Catalytic polymerization of phenolic compounds by clay minerals. *Soil Sci.* **126,** 15–21.

Wangersky, P. J. and Hincks, A. V. (1978). The shipboard intercalibration of filters used in the measurement of particulate organic carbon. Research Council of Canada Publication Number NRCC-16767, 32 pp.

Watts, C. D., Clathorne, B., Crane, R. I., and Fielding, M. (1981). Development of techniques for the isolation and identification of nonvolatile organics in drinking water. In *Advances in the Identification and Analysis of Organic Pollutants in Water,* Vol. 1 (L. H. Keith, ed.). Ann Arbor Science, Ann Arbor, MI, pp. 383–397.

Weber, J. B., Weed, S. B., and Ward, T. M. (1969). Adsorption of *s*-triazines by soil organic matter, *Weed Sci.* **17,** 417–421.

Weber, J. H. (1983). Metal ion speciation studies in the presence of humic materials. In *Aquatic and Terrestrial Humic Materials* (R. F. Christman and E. T. Gjessing, eds.). Ann Arbor Science, Ann Arbor, MI, pp. 295–314.

Weber, J. H. and Wilson, S. A. (1975). The isolation and characterization of fulvic acid and humic acid from river water. *Water Res.* **9,** 1079–1084.

Wehry, R. K. and Rogers, L. B. (1966). Fluorescence and phosphorescence in organic molecules. In *Fluorescence and Phosphorescence Analysis* (D. M. Hercules, ed.). Wiley, New York, pp. 81–150.

Welte, D. (1973). Recent advances in organic geochemistry of humic substances and kerogen. In *Advances in Organic Geochemistry* (B. Tissot and F. Bienner, eds.). Editions Technip, Paris, pp. 3–13.

Wershaw, R. L., Bohner, G. E., Jr. (1969). Pyrolysis of humic and fulvic acids. *Geochim. Cosmochim. Acta* **33,** 757–762.

Wershaw, R. L., Burcar, P. J., and Goldberg, M. C. (1969). Interaction of pesticides with natural organic material. *Environ. Sci. Technol.* **3,** 271–273.

Wershaw, R. L., Burcar, P. J., Sutula, C. L., and Wiginton, B. J. (1967). Sodium humate solution studies with small-angle X-ray scattering. *Science* **157,** 1429–1431.

Wershaw, R. L. and Goldberg, M. C. (1972). Interaction of organic pesticides with natural organic polyelectrolytes. In *Fate of Organic Pesticides in the Aquatic Environment,* Advances in Chemistry Series III (S. D. Faust, ed.). American Chemical Society, Washington, D.C., pp. 149–158.

Wershaw, R. L., Mikita, M. A., and Steelink, C. (1981). Direct ^{13}C NMR evidence for carbohydrate moieties in fulvic acids. *Environ. Sci. Technol.* **15,** 1461–1463.

Wershaw, R. L. and Pinckney, D. J. (1971). Association and dissociation of a humic acid fraction as a function of pH. *U.S. Geol. Surv. Prof. Paper* No. 750-D.

Wershaw, R. L. and Pinckney, D. J. (1973a). Determination of the association and dissociation of humic acid fractions by small angle X-ray scattering. *J. Res. U.S. Geol. Surv.* **1,** 701–707.

Wershaw, R. L. and Pinckney, D. J. (1973b). The fractionation of humic acids from natural water systems. *U.S. Geol. Surv. J. Res.* **1,** 361–366.

Wershaw, R. L. and Pinckney, D. J. (1978). Methylation of humic acid fractions. *Science* **199,** 906–907.

Wershaw, R. L. and Pinckney, D. J. (1980). Isolation and characterization of clay–humic complexes. In *Contaminants and Sediments* Vol. 2 (Robert A. Baker, ed.). Ann Arbor Science, Ann Arbor, MI, pp. 207–219.

Weser, G., Suess, E., Blazer, W., Liebeziet, G., Muller, P. J., Ungerer, C. A., and Zenk, W. (1982). Fluxes of biogenic components from sediment trap deployment in circumpolar waters of the Drake Passage. *Nature* **299,** 145–147.

Westall, J. C., Zachary, J. L., and Morel F. M. M. (1976). MINEQL. A computer program for the calculation of chemical equilibrium composition of aqueous systems. Techn. Note No. 18, Massachusetts Institute of Technology.

Westheimer, F. H., Jones, W. A., and Lad, R. A. (1942). The electrostatic influence of substituents on reaction rates. *J. Chem. Phys.* **10,** 478–485.

Westheimer, F. H. and Kirkwood, J. G. (1938). The electrostatic influence of substituents on the dissociation constants of organic acids. II. *J. Chem. Phys.* **6,** 513–517.

Westheimer, F. H. and Shookhoff, M. W. (1939). The electrostatic influence of substituents on the dissociation constants of organic acids. III. *J. Am. Chem. Soc.* **61,** 555–560.

Wetzel, R. G. (1975). Organic carbon cycle and detritus. In *Limnology*. Saunders, Philadelphia, pp. 538–621.

Wetzel, R. G. (1983). *Limnology*. Saunders College Publishing, Philadelphia, 767 pp.

Wetzel, R. G. and Otsuki, A. (1974). Allochthonous organic carbon of a marl lake. *Arch. Hydrobiol.* **73,** 31–56.

Wetzel, R. G., Rich, P. H., Miller, M. C., and Allen, H. L. (1972). Metabolism of dissolved and particulate detrital carbon in a temperate hot-water lake. *Mem. Ist. Ital. Idrobiol.* **29** (Suppl.), 185–243.

Whiffen, H. (1966). *Spectroscopy*. Wiley, New York.

Whitehead, D. C. and Tinsley, J. (1964). Extraction of soil organic matter with dimethylformamide. *Soil Sci.* **97,** 34–42.

Wilke, R. J. and Dayal, R. (1982). The behavior of iron, manganese, and silicon in the Peconic River estuary, New York. *Estuarine Coastal Shelf Sci.* **15,** 577–586.

Willard, H. H., Merritt, L. L., Jr., and Dean, J. A. (1974). *Instrumental Methods of Analysis*, 5th ed. Van Nostrand, New York, 860 pp.

Willey, J. D. (1984). The effect of seawater mangesium on natural fluorescence during estuarine mixing, and implications for tracer applications. *Mar. Chem.* **15,** 19–45.

Willey, J. D. and Atkinson, L. P. (1982). Natural fluorescence as a tracer for distinguishing between Piedmont and coastal plain river water in the nearshore waters of Georgia and North Carolina. *Estuarine Coastal Shelf Sci.* **14,** 49–59.

Willey, L. M., Kharaka, Y. K., Press, T. S., Rapp, J. B., and Barnes, I. (1975). Short-chain aliphatic acid anions in oil field waters and their contribution to measured alkalinity. *Geochim. Cosmochim. Acta* **39,** 1707–1711.

Williams, D. H. and Fleming, I. (1966). *Spectroscopic Methods in Organic Chemistry*. McGraw-Hill, London.

Williams, J. W. (1972). *Ultracentrifugation of Macromolecules*. Academic, New York.

Williams, P. M. (1968). Stable carbon isotopes in the dissolved organic matter of the sea. *Nature* **219,** 152–153.

Williams, P. M. (1983). Bomb-^{14}C in abyssopelagic fish and crustacea. *Am. Soc. Limnol. Ocean.* 46th Annual Meeting Abstracts, 61 pp.

Williams, P. M. and Zirino, A. (1964). Scavenging of "dissolved" organic matter from seawater with hydrated metal oxides. *Nature* **204,** 462–464.

Williams, R. B. (1958). Characterization of hydrocarbons in petroleum by nuclear magnetic resonance spectrometry. *Am. Soc. Test. Mater. Spec. Publ.*, 1958 (**224**), 168–194.

Williams, V. R. (1914). Soil science. *Pochvovedenie* **1.**

Wilson, A. L. (1959). Determination of fulvic acids in water, *J. Appl. Chem.* **9,** 501–510.

Wilson, C. L. and Wilson, D. W. (1959). *Comprehensive Analytical Chemistry*. Elsevier, New York.

Wilson, D. E. and Kinney, P. (1977). Effects of polymeric charge variations on the proton–metal ion equilibria of humic materials. *Limnol. Oceanogr.* **22,** 281–289.

Wilson, M. A. (1981). Applications of nuclear magnetic resonance spectroscopy to the study of the structure of soil organic matter. *J. Soil Sci.* **32,** 167–186.

Wilson, M. A., Barron, P. F., and Gillam, A. H. (1981a). The structure of freshwater humic substances as revealed by ^{13}C NMR spectroscopy. *Geochim. Cosmochim. Acta* **45,** 1743–1750.

Wilson, M. A., Collin, P. J., and Tate, K. R. (1983). ^{1}H nuclear magnetic resonance study of a soil humic acid. *J. Soil Sci.* **34,** 297–304.

Wilson, M. A. and Goh, K. M. (1977a). Proton-decoupled pulse Fourier-transform ^{13}C nuclear magnetic resonance of soil organic matter. *J. Soil Sci.* **28,** 645–652.

Wilson, M. A. and Goh, K. M. (1977b). ^{13}C nuclear magnetic resonance spectroscopy of humic substances. *Plant and Soil* **46,** 287–289.

Wilson, M. A. and Goh, K. M. (1981). Comment on the paper: "^{1}H and ^{13}C NMR studies on the importance of aromatic structures in fulvic and humic acids" (Ruggiero et al., 1979) *Geochim. Cosmochim. Acta* **45,** 489–490.

Wilson, M. A., Jones, A. J., and Williamson, B. (1978). Nuclear magnetic resonance spectroscopy of humic materials. *Nature* **276,** 487–489.

Wilson, M. A., Pugmire, R. J., Zilm, K. W., Goh, K. M., Heng, S., and Grant, D. M. (1981b). Cross-polarization ^{13}C NMR spectroscopy with magic angle spinning characterizes organic matter in whole soils. *Nature* **294,** 648–650.

Wilson, S. A., Huth, T. C., Arndt, R. E., and Skogerboe, R. K. (1980). Voltammetric methods for determination of metal binding by fulvic acid. *Anal. Chem.* **52,** 1515–1518.

Wilson, S. A. and Weber, J. H. (1977). A comparative study of number-average dissociation-corrected molecular weights of fulvic acids isolated from water and soil. *Chem. Geol.* **19,** 285–293.

Witthauer, J. and Kloecking, R. (1981). Untersuchungen sum Vorkommen und zur Bindung von Kohlenhydraten in Morrwasserhuminsaeuren. *Acta Hydrochim. Hydrobiol.* **9,** 281–295.

Wood, J. C., Moschapedis, S. E., and Elafson, R. M. (1961). Studies in humic acid chemistry. I. Molecular weights of humic acids in sulfolane. *Fuel* **40,** 193–201.

Woodward, L. A. (1967). General introduction to Raman spectroscopy. In *Raman Spectroscopy* (H. A. Szymanski, ed.), Plenum, New York, pp. 1–43.

Woodward, R. B. (1956). Synthesis. In *Perspectives in Organic Chemistry* (A. Todd, ed.). Wiley-Interscience, New York, pp. 155–184.

Woodward, R. L. (1963). Review of the bactericidal effectiveness of silver. *J. Am. Water Works Assoc.* **55,** 881–886.

Woodwell, G. M., Whittaker, R. H., Reiners, W. A., Likens, G. E., Delwiche, C. C., and Botkin, D. B. (1978). The biota and the world carbon budget. *Science* **199,** 141–146.

Woodwell, G. M. and Houghton, R. H. (1977). Biotic influences on the world carbon budget. In *Global Chemical Cycles and Their Alterations by Man* (W. Stumm, ed.). Wiley, New York, pp. 61–72.

Woof, J. B. and Pierce, J. S. (1967). Separation of complex mixtures of polyhydroxyl phenols on columns of Sephadex. *J. Chromatogr.* **28,** 94–103.

Wright, J. R. and Schnitzer, M. (1959). Oxygen containing functional groups in the organic matter of a Podzol soil. *Nature* **184,** 1462–1463.

Wright, J. R. and Schnitzer, M. (1960). Oxygen-containing functional groups in the organic matter of the A_0 and B_h horizon of a podzol. *Proc. 7th Intl. Congr. Soil Sci.,* Madison, Wisconsin, pp. 2, 120–127.

Wright, J. R. and Schnitzer, M. (1961). An estimate of the aromaticity of the organic matter of a Podzol soil. *Nature* **190,** 703–704.

Wynn-Williams, D. D. (1982). Microflora–microfauna interactions in Antarctic moss peat decomposition processes. In *Muskeg and the Northern Environment in Canada* (N. W. Radforth and C. O. Brawner, eds.). University of Toronto Press, Toronto, pp. 82–129.

Yamamoto, S. (1983). Chemical studies on the formation process of organic geopolymers in Recent sediments. Doctoral thesis, Tokyo Metropolitan University.

Yamamoto, S. and Ishiwatari, R. (1981). Chemical characterization of organic geopolymers (kerogens) in river and lake sediments. *Chikyukagaku (Geochemistry)* **15,** 60–69.

Yamamoto, S., Ishiwatari, R., and Hon-nami, H. (1981). Possible interaction of insoluble organic geopolymers (kerogen) with other organic compounds in the aquatic environment. II. Interaction with alkylbenzenesulfonates. *Japan J. Water Pollut. Res.* **4,** 65–72.

Yamazaki, H., Hohda, S., and Nishikawa, Y. (1980). Chemical forms of chromium in natural water—interaction of chromium(III) and humic substances in natural water. *J. Oceanogr. Soc. Japan* **35,** 233–240.

Yau, W. W., Kirkland, J. J., and Bly, D. D. (1979). *Modern Size-Exclusion Liquid Chromatography: Practice of Gel Permeation and Gel Filtration Chromatography.* Wiley, New York.

Yen, R. F., Erdman, J. G., and Saraceno, A. J. (1962). Investigation of the nature of free radicals in petroleum asphaltenes and related substances by electron spin resonance. *Anal. Chem.* **34,** 694–700.

Young, D. K., and Yen, T. F. (1977). The nature of straight-chain aliphatic structures in Green River kerogen. *Geochim. Cosmochim. Acta* **41,** 1411–1417.

Youssefi, M. and Faust, S. D. (1980). Adsorption and formation of light halogenated hydrocarbons and humic acid in water by granular activated carbon. In *Activated Carbon Adsorption of Organics from the Aqueous Phase,* Vol. 1 (I. H. Suffet and M. J. McGuire, eds.). Ann Arbor Science, Ann Arbor, MI, pp. 133–143.

Zeichmann, W. (1964). Spectroscopic investigations of lignin, humic substances, and peat. *Geochim. Cosmochim. Acta* **28,** 1555–1566.

Zeichmann, W. (1976). Huminstoffe in Sudamerikanishen fluss systemen. *Amazoniana* **6,** 135–144.

Zelazny, L. W. and Calisle, V. W. (1974). Physical, chemical, elemental, and oxygen-containing functional group analysis of selected Florida Histosols. In *Histosols—Their Characteristics, Classification, and Use* (A. R. Aandahl, S. W. Buol, O. E. Hill, and H. H. Bailey, eds.). Soil Science Society of America, Special Publications Series No. 6, American Society of Agronomy, Madison, Wisconsin, pp. 63–78.

Zepp, R. G., Baughman, G. L., and Scholtzhauer, P. F. (1981a). Comparison of photochemical behavior of various humic substances in water: I. Sunlight induced reactions of aquatic pollutants photosensitized by humic substances. *Chemosphere* **10,** 109–117.

Zepp, R. G., Baughman, G. L., and Schlotzhauer, P. F. (1981b). Comparison of photochemical behavior of various humic substance in water. II. Photosensitized oxygenations. *Chemosphere* **10,** 119–126.

Zepp, R. G., Wolfe, N. L., Baughman, G. L., and Hollis, R. C. (1977). Singlet oxygen in natural waters. *Nature* **267,** 421–423.

Zilm, K. W., Pugmire, R. J., Grant, D. M., Wood, R. E., and Wiser, W. H. (1979). A comparison of the carbon-13 NMR spectra of solid coal and their liquids obtained by catalytic hydrogenation. *Fuel* **58,** 11–16.

Zimmerman, A. P. (1981). Electron intensity, the role of humic acids in extracellular electron transport and chemical determination of pE in natural waters. *Hydrobiologia* **78,** 259–265.

Zimmerman, J. T. F. and Rommets, J. W. (1974). Natural fluorescence as a tracer in the Dutch Wadden Sea and the adjacent North Sea. *Neth. J. Sea Res.* **8,** 117–125.

Zsolnay, A. (1979). Coastal colloidal carbon. A study of its seasonal variation and the possibility of river input. *Estuarine Coastal Mar. Sci.* **9,** 559–567.

Zunino, H. and Martin, J. P. (1977). Metal-binding organic macromolecules in soils: I. *Soil Sci.* **123,** 65–76.

Zunino, H., Pierano, P., Aguilera, M., and Escobar, I. (1972). Determination of maximum complexing ability of water-soluble complexants. *Soil Sci.* **114,** 414–416.

APPENDIX A

Glossary of Terms

Concepts from many different scientific disciplines are discussed in the different chapters of this book. A glossary of terms is provided to ensure that the terminology associated with each discipline is not an impediment to the reader. Each of the words defined in the Glossary is italicized when it appears in the definition of another word also defined in the Glossary. The definitions are written to present to the reader the essential aspects of a given term in the context in which it was used. These definitions, therefore, are not intended to be comprehensive.

Acetylation. Insertion of an acetyl radical $CH_3\overset{O}{\overset{\|}{C}}\cdot$ into an organic molecule containing OH or NH_2 groups. Acetylation is commonly used to determine the number of hydroxyl groups in fats and oils.

Acylation. Reaction of an acyl group (RCO—) with the active hydrogen sites on functional groups such as —OH or $—NH_2$.

Alicyclic. Of or pertaining to a group of closed ring, nonaromatic organic compounds that may contain saturated carbon–carbon bonds, carbon–carbon double bonds, and/or, carbon–carbon triple bonds, but do not have alternating unsaturated/saturated bonds that allow for electron resonance around the ring.

Aliphatic. Of or pertaining to a broad category of carbon compounds distinguished only by a straight, or branched, open chain arrangement of the constituent carbon atoms. The carbon–carbon bonds may be saturated or unsaturated.

Alkaloid. A class of nitrogen-containing organic compounds, usually derived from the nitrogen-containing ring compounds, such as pyridine. Examples of alkaloids include morphine, nicotine, codeine, cocaine, and caffeine.

Alkane. A saturated (i.e., no double bonds) hydrocarbon compound containing only carbon and hydrogen. Also called paraffins.

Alkanoic acid. A carboxylic acid derived from an *alkane*.

Alkene. An open chain hydrocarbon compound containing at least one carbon–carbon double bond.

Allochthonous. Produced outside the system, for example, produced in a catchment basin and transported to a lake (opposite of *autochthonous*).

Allylic. Containing a propene group, that is, 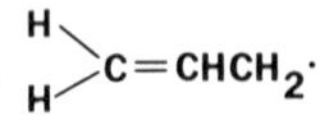

Alum. An aluminum sulfate of some type, commonly used as a flocculent in water treatment.

Amino acid. One of a group of related biochemicals that form the basic subunits of proteins and have the general chemical formula:

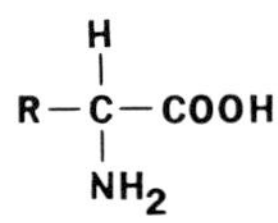

Amino sugar. One of a group of sugars with a hydroxyl group replaced by an amine, for example, glucosamine:

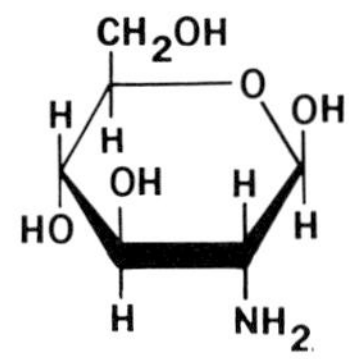

Amphoteric. Capable of acting either as a base or an acid, for example, amino acids which contain both a basic group (NH_2) and an acidic group (COOH).

Anisotropic. Possessing properties with different values when measured along axes in different directions.

Anodic stripping voltammetry. The initial reduction and collection of metals at the surface of an anode, and the subsequent stripping of the metals by oxidation, resulting in a voltammogram that identifies the metals and their respective concentrations.

Anomeric. Referring to a cyclic stereoisomer of the carbohydrate series with isomerism involving only the arrangement of atoms or groups at the aldehyde or ketone position.

Anthocyanin. A class of soluble pigments (blue to red) found in plant tissues; a specific example is leucoanthocyanin.

Aryl. Of or pertaining to organic compounds that contain a benzene ring.

Aromatic. Of or pertaining to organic compounds that resemble benzene in chemical behavior.

Autochthonous. Produced within the system, for example, within a lake (opposite of *allochthonous*).

Autolysis. Self-hydrolysis by tissue-degrading enzymes; self-digestion.

Autoxidation. An oxidation that occurs spontaneously in the presence of air and is self-catalyzed in the sense that the first oxidation reaction promotes the second, and so on. Autoxidation can be initiated by heat, light, metallic catalysts, or free radical generators.

Axenic. Composed of one species and free of all others; commonly used in reference to pure cultures of bacteria.

Benthic. Deriving from or occurring at the bottom of a body of water (n. benthos).

Bentonite. A soft, porous, moisture-absorbing rock composed essentially of clayey minerals, in particular, montmorillonite, in the form of extremely small crystals.

Benzylic. Pertaining to the radical formed by the removal of one hydrogen atom from the side chain of toluene:

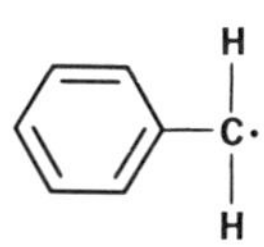

Biodeposition. Incorporation of material into sediment, either as part of the detritus of an organism, or as an indirect result of biological activity.

Biogenic. Of biological origin or in some way resulting from biological activity.

Biogeochemical. A biochemical that is no longer part of a living organism and has undergone, or is undergoing, alteration in waters, soils, or sediments.

Biolith. A biogenic rock, that is, one that is composed of organic remains or is in some way derived from biological activity.

Biome. A complex biological community covering a large geographic area and usually characterized by a dominant, climax-type of vegetation; for example, a grassland is a biome.

Biosphere. That part of the world in which life can exist without artificial means of support.

Biota. A term that encompasses the spectrum of living things within a given area.

Biotite. A member of the mica mineral group; lustrous, with a dark green, brown to black color; approximate composition $K(Mg,Fe)_3(AlSi_3O_{10})(OH)_2$.

Bioturbation. Perturbation of the upper layer of sediment by macroinvertebrates and other organisms.

Bog. An area of wet, soggy ground; an acid *peatland* (Europe).

"Bomb" ^{14}C. Radioactive carbon derived from atomic bomb explosions that was initially introduced into the atmosphere in the 1940s and dramatically increased the background level of ^{14}C present in the environment.

Carboniferous. The combined Mississippian and Pennsylvanian periods of the upper Paleozoic era of the geologic time scale; commenced approximately 345 million years ago.

Catagenesis. The middle stage of evolution of sedimentary organic matter, involving thermal transformation of organic compounds which occurs as a result of the increasing geothermal gradient in sediments with depth. As a general rule, the residual organic material following catagenesis has a lower oxygen and hydrogen content, and is more aromatic (see also *diagenesis, metagenesis*).

Cellulose. A degradation-resistant polysaccharide composed of glucose subunits which is the major structural component of the cell walls of plants, accounting for more than 50% of the organic carbon in the *biosphere*.

Cheluviation. Downward movement in soil of metal ions as soluble organic complexes.

Chernozem. A soil consisting of a black or nearly black, organic-rich A *horizon,* high in exchangeable calcium, underlain by a lighter-colored transitional horizon above a layer of calcium carbonate accumulation; occurs in a cool subhumid climate under tall and midgrass prairie vegetation.

Chitin. A homopolymer of the *amino sugar,* acetyl glucosamine; acts as a structural component of the exoskeleton of insects and crustaceans.

Chromophore. A chemical moiety that absorbs light and contributes color to a compound when exposed to light. Examples of some chromophores are N═N— (in azo dyes) and —C═O (in humic substances).

Clastic. Referring to a rock or a sediment that is a consolidation of broken fragments derived from rocks in other locations; the fragments have been transported mechanically to their place of deposition and eventual consolidation.

Coalification. A process by which organic matter is converted into coal of increasingly higher rank. Brown coal, bituminous coal, and anthracite coal are coal ranks in order of increasing carbon content, from 69 to 95% carbon.

Colligative property. A thermodynamic property that depends on the number of particles in solution, and not on the nature of these particles.

Copropel. The brown or grey pulpy material formed from the dead remains of microscopic plants and animals in the top muds of *eutrophic* lakes and *marshes*.

Coulometric. Pertaining to the quantity of electric charge passed through a chemical system. In coulometric methods of analysis the total quantity of electricity (coulombs) required to exhaustively electrolyze a substance is measured and related to the quantity of that substance present through Faraday's Laws.

Curie point pyrolysis. Rapid pyrolysis of a sample by means of a current induced in a wire of ferromagnetic material. The wire stabilizes at its Curie point, the temperature at which it ceases to be ferromagnetic.

Dalton. One atomic mass unit, equivalent to $\frac{1}{12}$th the mass of the carbon atom (mass number 12).

Dextran. Polymers of glucose which have high molecular weights and chain-like structure; produced from sucrose by certain bacteria.

Diagenesis. The less advanced stage of evolution of sedimentary organic matter, involving a low-temperature transformation of organic compounds typically in low-temperature waters, soils, and relatively shallow sediments. Low-temperature chemical reactions and biological activity (e.g., consumption and production of organic compounds) contribute to diagenesis. Diagenesis is divided into early diagenesis, when organic matter loses mainly nitrogen, and diagenesis, when organic matter loses mainly oxygen (see also *catagenesis* and *metagenesis*).

Diatom. A microscopic single-celled plant (algae) growing in marine or freshwater; develops a siliceous box-like skeleton that may accumulate significantly in sediments.

Dimerization. The combination of two molecules to form a larger molecule.

Dipole. An electric dipole consists of equal and opposite charges ($\pm q$) separated by a small distance (d); the dipole moment is a vector quantity of magnitude qd. A dipole can interact electrostatically with other dipoles or with charged particles, and aligns itself in an electric field.

Dolomite. A sedimentary carbonate mineral, of the formula $CaMg(CO_3)_2$, generally thought to be formed from limestone by the replacement of Ca with Mg.

Dystrophic. Subject to a high loading of *allochthonous* organic matter, but with a low level of *autochthonous* input. Dystrophic lakes are heavily stained ("brown water") and have a high content of *humic substances*.

Ecophysiological. Of or relating to the normal functioning of an organism considered in an ecological context.

Electrical double layer. In surface chemistry, a diffuse layer of primary adsorbed ions together with their counterions of opposite charge, surrounding a particle in suspension. The primary or first layer of ions adsorbs as a result of attraction to ions of opposite charge *within* the particle. The second, or counter layer of ions adsorb as a result of attraction to ions of opposite charge in the primary layer.

Electrophoresis. The differential migration of charged particles through a given medium (of varied composition) due to an imposed potential difference between immersed electrodes; can be used to separate charged polymeric compounds such as proteins.

Eluate. In liquid chromatography, a solution of the formerly sorbed substance in the mobile phase liquid as it emerges from the column.

Eluviation. The transportation of dissolved or particulate soil material from one area or specific *horizon* of the soil to another by the movement of water.

Endmember. In a general context, one extreme of a series.

Enolic. Pertaining to the enol group (also called alkenol), 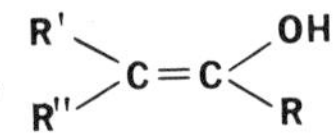

where a carbon atom is double bonded to another carbon and singly bonded to both a hydroxyl group and to another atom.

Eocene. An epoch of the lower Tertiary period of the geologic time scale; commenced approximately 54 million years ago.

Epilimnion. The uppermost layer of a lake during the summer period of lake stratification; this layer is warmer than underlying waters and is generally well mixed. Stratification occurs only in lakes of sufficient depth, as determined by the local temperature and wind conditions (see also *hypolimnion*).

Epiphytic. Growing on *macrophyte* surfaces.

Esterification. The formation of an ester, 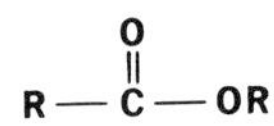 ; a common type of esterification is the reaction of an alcohol with a carboxylic acid.

Eutrophic. Refers to waters high in nutrient loading with high primary production of organic material by algae and/or macrophytes. Excessive deposition of algal materials often results in anoxic conditions in the bottomwaters of eutrophic lakes. Eutrophic waters often exhibit simple *trophic* structure, and a low level of species diversity. Anthropogenic inputs, particularly phosphate enrichment, often induce eutrophication (see also *mesotrophic, oligotrophic*).

Evapotranspiration. The process of losing water from the combination of evaporation from the soil surface and transpiration from plants.

Fen. A *peatland* covered by water, receiving inputs from groundwater and/or surface flow, and generally having a higher mineral content than a *bog*.

Fibric. Least decomposed of organic soil material. Has high fiber content (U.S. system).

Fluvial. Originating in, or derived from, a stream or river.

Fulvic acid. That fraction of *humic substances* that is soluble under all pH conditions.

Gel permeation chromatography. A method of separating molecules by size, usually carried out on columns that are tightly packed with gel and completely filled with solvent. Differential permeation into the gel pores by sample molecules determines the elution order of the sample molecules. Also known as gel filtration chromatography or exclusion chromatography.

Gelbstoff. A complex mixture of natural compounds dissolved in seawater, characterized by light absorbance that increases with decreasing wavelength, giving a yellow color to the water and causing a blue fluorescence when irradiated with long-wave ultraviolet radiation.

Gley. Soil developed under reducing conditions as a result of poor drainage, characterized by a gray and mottled coloration.

Glyceride. An ester of glycerol ($C_3H_8O_3$) and fatty acids.

Gymnosperm. A plant that has seeds which are not enclosed in an ovary; for example, pines, firs, and spruces.

Hemic. Organic soil material of intermediate stage of decomposition. Intermediate fiber content (U.S. system).

Hemicellulose. A degradation-resistant polymer of the sugar D-xylose, with side chains of other sugars. Hemicellulose is related structurally to *cellulose* by its similar linkage between the sugar subunits, but is somewhat less resistant to degradation than cellulose.

Heteroatom. Any element in an organic molecule other than carbon or hydrogen, for example, oxygen, nitrogen, or sulfur.

Heterocyclic. Indicating a closed ring structure in which one or more of the atoms in the ring is an element other than carbon, for example, sulfur or nitrogen.

Heterotrophic. Deriving life-sustaining energy from organic matter as distinct from autotrophic (deriving life-sustaining energy from light or inorganic substrates).

Higher plants. Vegetal organisms living on land, and mainly composed of water, carbohydrates, lipids, *lignin,* and some proteins. Some of these compounds are inherently resistant to degradation and thus are concentrated after the death of organisms; they are chiefly *lignin,* which is used in supporting tissues necessary to a better exposition to light, and cuticular waxes, which preserve the tissues from water evaporation. In contrast with higher plants, lower plants do not possess lignin and cuticular waxes.

Histosol. A soil that is greater than 50% organic matter by weight in the upper 80 cm; also a highly organic soil of low porosity.

Holocene. An epoch of the Quaternary period of the geologic time scale; from the end of the Pleistocene to the present; commenced approximately 10,000 years ago.

Horizon. In soil science, a layer of soil approximately parallel to the land surface and differing from adjacent layers in chemical, physical, and biological properties, and in characteristics such as color, structure, texture, consistency, acidity, alkalinity, and so on. The horizons result from transport and chemical processes in the soil. Some of the more common horizons are described in the following diagram taken from *Fundamentals of Soil Science*, 6th ed., by H. D. Foth. Wiley, New York, 1978.

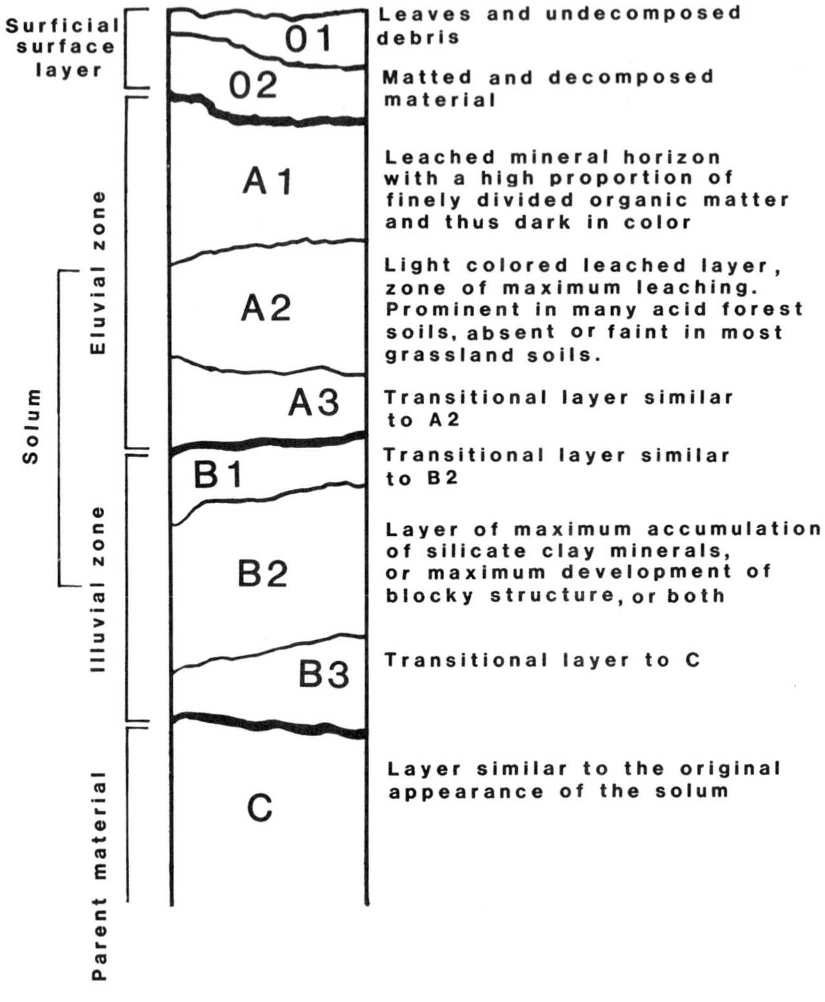

Humic acid. That fraction of *humic substances* that is not soluble in water under acid conditions (below pH 2), but becomes soluble at greater pH.

Humic substances. A general category of naturally occurring, biogenic heterogeneous organic substances that can generally be characterized as being yellow to black in color, of high molecular weight, and *refractory*.

Humification. The process of formation of *humic substances;* generally, the decomposition of organic material.

Humin. That fraction of *humic substances* that is not soluble in water at any pH value.

Humus. The organic portion of soil, brown or black in color, consisting of partially or wholly decayed plant and animal matter, that provides nutrients to plants and increases the ability of soil to retain water. This term is not entirely synonymous with *humic substances,* although it is often used as a synonym.

Hydraulic conductivity (soil). Also known as the permeability coefficient, it is the rate of flow of water (in gallons per day) through a cross-section of 1 square foot under a unit hydraulic gradient (the rate of change of pressure per unit of distance of flow).

Hydromorphic. Developed under conditions of excess moisture. Hydromorphic soils are found under conditions of poor drainage in *marshes, swamps,* seepage areas, or flats.

Hypolimnion. The lowermost layer of a lake during the summer period of lake stratification; this layer is cooler than the overlying *epilimnion.*

Illite. A group of three-layer, mica-like, and grey, light-green, or yellowish brown clay minerals, especially widely distributed in marine shales and soils derived from them; of the general formula $(H_3O\ K)_y(Al_4.Fe_4.Mg_4,Mg_6)(Si_{8-y}.Al_y)O_{20}(OH)_4$, with y less than 2.

Illuvial. Pertaining to the deposition of dissolved or particulate soil material into one area or *horizon* of the soil from another. This material is transported by the process of *eluviation.*

Imine. A class of compounds containing a nitrogen double bonded to a carbon on one side and single bonded to carbon or hydrogen on the other side.

Interflow waters. Synonymous with storm seepage waters; runoff water which infiltrates the surface soil and moves laterally toward streams. Such water is ephemeral and shallow (above the main groundwater level).

Interlamellar. Pertaining to materials between layers (commonly clay layers) such as cations, hydrated cations, hydroxides, and organic molecules.

Isoelectrophoresis. A variation of *electrophoresis* in which the medium supports a pH gradient through which a compound will migrate until it reaches its characteristic isoelectric point, that is, the pH at which the net charge on a molecule in solution is zero. Also known as isoelectric focusing.

Isoprenoids. Long-chain, branched hydrocarbons made up of isoprene subunits:

$$\left(CH_2-CH_2-CH=\overset{\overset{\displaystyle CH_3}{|}}{C} \right)$$

Isoprenoids less than 21 carbons in length are thought to be breakdown products of chlorophyll.

Isotachophoresis. A variation of *electrophoresis* set up such that ionic species move with equal velocity in an ionic band through the medium. However, within this band, the most mobile ions form the leading edge and the least mobile ions form the trailing edge, resulting in the separation of ions based on their characteristic mobility under the specified conditions.

***k'*.** Known as the capacity factor in chromatography, it is the ratio of solute amounts distributed between two phases:

$$k' = \frac{\text{total amount of solute in phase X}}{\text{total amount of solute in phase Y}}$$

Kaolinite. A common white to greyish or yellowish clay mineral of the general formula: $Al_2Si_2O_5(OH)_4$; does not appreciably expand under varying water content and does not exchange iron or magnesium.

Kerogen. Polymerized organic material of varying composition and varying molecular weight, characterized by its insolubility in nonoxidizing acids, bases, and most organic solvents; found in sedimentary environments that have been subjected to *diagenesis*.

Lactone. A cyclic ester. Hydroxy acids may exist as lactones if the hydroxyl groups are situated so that the lactone formed has a five- or six-membered ring:

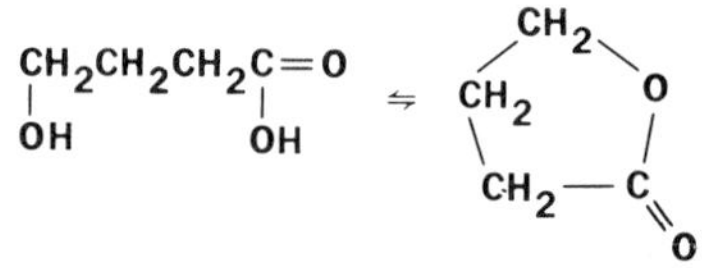

Lacustrine. Originating in, or derived from, a lake.

Ligand. A functional group, ion, or molecule bound to a central atom (e.g., metal) in a complex between chemical species called a coordination complex.

Lignin. The most abundant, natural, aromatic organic polymer that is a major structural component of wood. There is no general agreement about the structure of lignin, however, it is known to lack a regular sequence of monomers. Lignin; contains *phenolic*, hydroxyl, and *methoxyl* groups; phenols are formed when lignin decomposes.

Lignite. A low rank of coal between peat and subbituminous coal; also called brown coal (see *coalification*).

Lithology. The character of a rock or a rock formation with respect to its mineral composition, texture, color, and internal structure.

Lithosphere. The solid outer part of the earth, the crustal zone, generally considered to be about 50 miles in thickness.

Littoral. Pertaining to the shallow-water zone of lakes or seas that lies between open water and dry land, where light can penetrate to the bottom; often a habitat for rooted aquatic plants.

Low moor. *Peatland* occurring in a lower-lying topographic position equivalent to *fen* (Britain).

Macrophyte. A macroscopic plant, used in reference to macroscopic plants growing in the *littoral* zone of lakes and on river banks.

Maillard reaction. The reaction of amino groups of *amino acids,* peptides, or proteins with the "glycosidic" hydroxyl group of sugars resulting in the formation of brown pigments. Also known as the "Browning" reaction.

Marsh. An intermittently wet or continually flooded area with the surface not deeply submerged. Predominantly covered with hydrophytic plants, such as sedges, cattails, and brushes.

Meiofauna. Metazoans such as nematodes and forams whose body size is less than 1 mm, found in the top few centimeters of sediment.

Melanoidins. Brown pigments produced in *Maillard reactions.*

Mercaptan. See *thiol.*

Mesotrophic. Refers to waters intermediate between *oligotrophic* and *eutrophic.* Characterized by a moderate nutrient loading and moderate primary production of organic material by algae and/or *macrophytes,* Mesotrophic waters usually exhibit a diverse biotic community.

Metagenesis. The most advanced stage of evolution of sedimentary organic matter that follows *catagenesis* involving a high-temperature transformation of organic compounds in the sediments. Extensive "cracking" of carbon–carbon bonds occurs during metagenesis and primarily results in the production of methane (see also *diagenesis*).

Meteoric water. Water of recent atmospheric origin; waters from recent precipitation.

Methoxyl. The functional group CH_3O—.

Methylene. A carbon atom radical with two available (but unbonded) valence electrons; the other two available valence electrons are bonded to hydrogen (—H_2C—).

Micelle. A colloidal particle composed of organic molecules aggregated on the basis of solvent affinity; in polar media, micelles form with hydrophobic moieties on the inside and hydrophilic moieties directed outward.

Miocene. Referring to an epoch of the upper Tertiary period of the geologic time scale; commenced approximately 280 million years ago.

Mire. *Marsh* or *bog;* an area of wet, soggy ground. Synonymous with *peatland.*

Mollisol. A soil consisting of a relatively thick, dark-colored surface *horizon* that contains at least 0.58% organic carbon, has a base saturation of more than 50% (pH 7), and is predominantly saturated with bivalent cations.

Montmorillonite. A group of expanding-lattice clay minerals of the general formula $M_{0.33}Al_2Si_4O_{10}(OH)_2nH_2O$ where M includes one or more of the cations Na^+, Mg^{2+}, K^+, Ca^{2+}, and possibly others.

Muck. Highly decomposed organic material often with high ash content. Synonymous with *sapric* material.

Mucopolysaccharide. A highly hydrated, jelly-like substance that provides intercellular lubrication (for multicellular organism) and structural support, and acts as a flexible cement. Bacteria use mucopolysaccharides to adhere to solid surfaces. Contains *uronic acids* and *amino sugars*.

Nitrile. Any of a class of organic compounds containing carbon triple bonded to a nitrogen ($RC\equiv N$).

Nucleic acid. A polymer of nucleotide subunits linked by phosphate bridges; the nucleotides contain *purine* or *pyrimidine* bonded to *ribose* or *deoxyribose* (see chemical glossary). Nucleic acids form the basis for the genetic code in all living things.

Obligate. Referring to microorganisms that require certain environmental conditions to live, for example, obligate anaerobes require oxygen-free (anaerobic) conditions.

Olefinic. Descriptive of a class of unsaturated *aliphatic* hydrocarbons having one or more double bonds (*alkenes*).

Oligomer. A polymer containing relatively few structural subunits.

Oligocarbophilic. Growing well only on very small quantities of organic substrates.

Oligotrophic. Refers to waters low in nutrient loading with low primary production of organic material by algae and/or macrophytes. Growth in an oligotrophic water is often limited by low levels of phosphorus and nitrogen (see also *eutrophic* and *mesotrophic*).

Palynology. The study of pollen and spores.

Paraffinic. Referring to the class of compounds known as *alkanes*.

Peat. Organic material occurring in a *peatland* under wet conditions. Usually only slightly to moderately decomposed.

Peatland. An organic wetland consisting of accumulated organic matter, organic terrain (Canada), moor, or *mire* (England).

Pedo humic substances. *Humic substances* from soils.

Pedology. The study of soil in its natural position, in regard to morphology, genesis, and classification.

Pelagic. Of or relating to (or inhabiting) a zone of open, unrestricted water that is beyond the outer border of the *littoral* zone and above the *benthos*.

Peptization. To bring into colloidal suspension.

Peralkylation. The process of derivatizing all hydroxyl groups ($-OH$) in an organic molecule to alkoxyl groups ($-OC_nH_{2n+1}$).

Periphyton. Algae growing on solid surfaces such as rocks or sand grains; often important primary producers in streams.

Permethylation. The process of derivatizing all hydroxyl groups (—OH) in an organic molecule to methoxyl groups ($—OCH_3$).

Permian. Referring to the last period of the Paleozoic era on the geologic time scale; commenced approximately 280 million years ago.

Phenolase. An enzyme that promotes the oxidation of *phenolic* compounds.

Phenolic. Of, relating to, or containing a phenol group which is a hydroxyl group bonded directly to an *aromatic* ring structure; for example, naphthol is a phenolic compound:

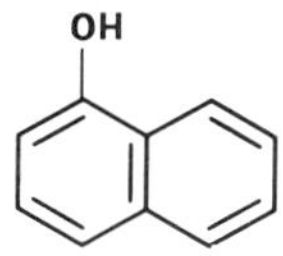

Photolysis. Chemical reaction (synthesis or degradation) induced by absorption of ultraviolet or visible radiation.

Phytoplankton. Small free-floating microorganisms dwelling in oceans, lakes, rivers, and large streams, which are capable of photosynthesis, for example, algae (sing.—phytoplankter).

Plankton. Small free-floating aquatic organisms, formed mainly of water, carbohydrates, and proteins; *phytoplankton* is generally more abundant than zooplankton (animal organisms).

Podzol. A soil consisting of a whitish-grey, highly leached A *horizon* (Podzolic), developed in cool-temperate to temperate, humid climates, under coniferous or mixed coniferous and deciduous forests. In the current soil taxonomic system, most Podzol soils (an older classification) are *Spodosols*.

Polder. A piece of previously submerged land, below the natural level of an adjacent body of water; transformed to dryland and maintained by dikes.

Polydisperse. Characterized by particles of varying size in a dispersed phase.

Polyelectrolyte. A macromolecule containing multiple ionic functional groups (either cationic or anionic).

Polypeptide. One of a group of related compounds that are polymers of *amino acids*, and as such may or may not constitute a fully functional protein:

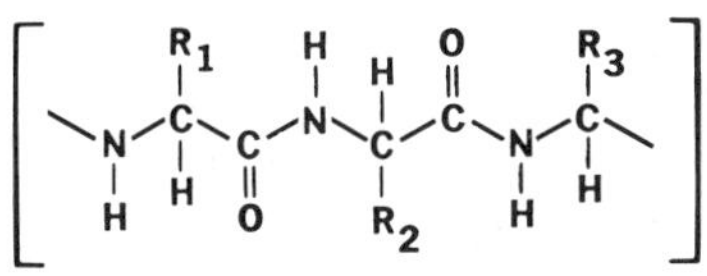

Polysaccharide. A carbohydrate polymer composed of monomeric sugar subunits, such as glucose, mannose, and fructose; commonly used as a form of energy storage in living organisms. *Cellulose* and starch are polysaccharides.

Porphyrin. A group of compounds found in all living matter which are the basis of compounds such as hemoglobin and chlorophyll. Porphyrins are derivatives of porphine, a fully conjugated cyclic structure of four pyrrole rings:

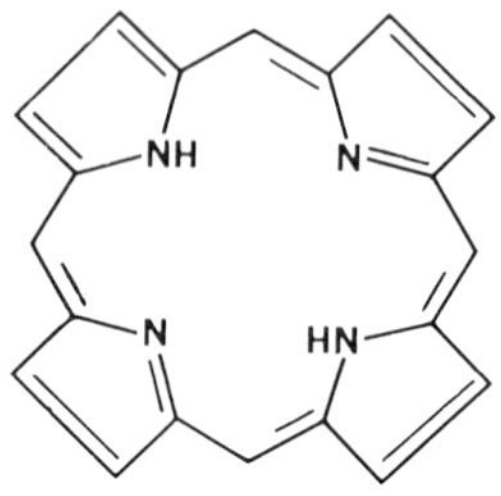

Presedimentary alteration. The first stage of degradation of organic matter, including all the biological and chemical factors of transformation or alteration from the death of organisms prior to the settling of organic remains at the surface of the sediment.

Priming effect. Stimulation of already slowly proceeding degradation processes.

Proteolysis. The hydrolysis of proteins or peptides.

Pyrite. The mineral FeS_2, commonly known as fool's gold; brass yellow or tarnished brown in color.

Pyroclastic. Pertaining to rock material formed by volcanic explosion or aerial expulsion from a volcanic vent.

Pyrophosphate Index. An index of the degree of *humification* of organic matter as measured by the determination of the absorbance of the pyrophosphate extract using a colorimeter.

Quinone. A conjugated cyclic diketone, with one to several conjugated six-membered carbon rings, for example, napthoquinone

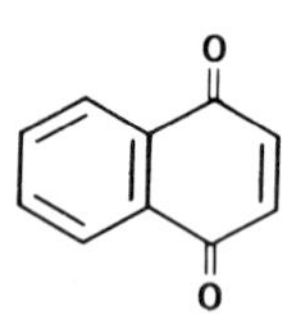

Raster. A predetermined pattern of scanning with a controlled electronic beam. The beam is directed at a screen that is rendered luminescent at each spot the beam strikes.

Recent sediment. Solid unconsolidated organic and mineral material deposited in an aquatic environment during recent geologic time and which has undergone little geothermal evolution.

Recharge water. Water from an external source which enters into the *saturated zone* of an aquifer, where all the interstitial pores are filled with water, entering either directly into the formation or indirectly by way of another formation.

Refractory. Not easily degraded.

Resin. (1) A highly cross-linked polymer. (2) Natural resins; generally high molecular weight, transparent to translucent, yellowish to brown plant secretions that are soluble in organic solvents but not in water.

Reverse-phase liquid chromatography. Describes the type of liquid chromatography that uses a nonpolar stationary phase and a polar mobile phase.

Saponification. The hydrolysis of an ester especially by alkali (e.g., NaOH) into the corresponding alcohol and the sodium salt of the corresponding acid. The process is usually carried out on fats and the sodium salt so formed is called a soap.

Sapric. Highly decomposed organic soil material with very low fiber content (U.S. system) (see also *muck*).

Sapropel. A black, unconsolidated, jelly-like ooze or sludge primarily composed of plant remains (especially algae) found slowly decomposing in an anaerobic environment in the sediments of lakes and seas.

Saprophyte. A plant that lives on decaying organic matter.

Saturated zone. The zone below the water table in an aquifer (see *unsaturated* zone).

Schiff base. An *imine,* derived by chemical condensation of aldehydes or ketones with primary amines; very weakly basic and hydrolyzed by water to form carbonyl compounds and amines.

Sedimentation environment. The environment due to different geographic, climatic, and physicochemical factors influencing the composition of mineral and organic particles which eventually settle and form the sediment.

Semiquinone. A partially reduced *quinone,* carrying an unshared electron on one of its oxygen atoms.

Sephadex. Trademark name for a chromatographic gel used in gel permeation chromatography which is composed of an extensively cross-linked gel derived from *dextran* and *epichlorohydrin.*

Sesquioxide. An oxide containing three atoms of oxygen combined with two of the other constituent element in the molecule, for example, Fe_2O_3.

Siderophore. Any one of a group of red-brown, iron-transporting biochemicals which have a characteristic absorption band at 420–440 nm and iron binding constants of about 10^{30}.

Silvicultural. Science and art of tree production.

Silyl. The radical H_3Si—.

Sod-podzolic. A *Podzol* soil in which the percent exchangeable sodium is

15% or more, which is sufficient sodium to interfere with the growth of most crop plants.

Solonetz. A soil consisting of a very thin, friable surface soil underlain by a dark, hard columnar layer usually highly alkaline; formed under subhumid to arid, cool to hot climates, and under a native vegetation of salt-tolerant plants.

Solod. One of a group of soils that has been developed from saline materials.

Soxhlet extraction. Extraction of a solid substance with a solvent (usually ether or alcohol) carried out in a distillation flask that is attached to a reflux condensor and a siphon system for drawing off the distillate.

Spin–spin coupling. The interaction between the magnetic moments of atomic nuclei within a molecule; gives rise to the splitting of otherwise singlet NMR resonance lines into multiplets.

Spodosol. A soil characterized by a whitish-grey, highly leached *A horizon,* and a B *spodic* horizon that is significantly enriched in organic matter. These soils develop in cool-temperate to temperate, humid climates under coniferous or mixed coniferous and deciduous forests. In the current soil taxonomic system, most *Podzol* soils (an older classification) are Spodosols.

Stable residue. The fraction of *humin* that is recovered after destruction of mineral phases by acid (40% HF and 6*N* HCl).

Subsidence. Loss of organic matter in organic soils due to biological oxidation, compaction, and shrinkage due to water removal.

Sulfonic acid. Any of a group of acids that contain the sulfonic group, $—SO_3H$.

Supercritical fluid. The physical state of a substance above its critical temperature (i.e., the temperature above which a gas cannot be condensed into a liquid). Supercritical fluids are unique in their properties, differing from liquids and gases but having characteristics of both.

Surfactant. A surface active agent that reduces the surface tension of water and aqueous solutions, reduces the interfacial tension between two liquids, or reduces the interfacial tension between a liquid and a solid.

Swamp. An area covered with water throughout much of the year, although the surface of the soil is usually not deeply submerged. In contrast to a *marsh,* a swamp is characterized by tree or shrub vegetation.

Synsedimentary alteration. A second stage of degradation and reworking of the organic fraction in the surficial layer of the sediment, mainly due to organisms which themselves contribute to the organic input. After this stage of degradation and reworking, the organic input is no longer modified and changes occur only by physiochemical reactions due to burial of the sediment.

Tannin. Any of a group of complex *phenolic* compounds derived from plants, characterized by their useful function of precipitating proteins. The chemistry of tannins is complex and variable.

Thiol. Any of a group of organic compounds containing an —SH moiety.

Trona. A sodium-rich mineral, $Na_2CO_3 \cdot NaHCO_3 \cdot 2H_2O$, that is commonly white, grey, or yellow in color.

Trophic level. The relative nutritional position of organisms or populations within a food web; for example, all organisms that feed on algae are at the same trophic level.

Trophogenic. Producing organic nutrients; primarily used in reference to the zone of a lake where the bulk of the photosynthesis occurs.

Unsaturated zone. The zone above the water table in an aquifer; the *vadose* zone (see also *saturated* zone).

Upwelling. The rising of cold, deep water rich in nutrients due to the earth's rotation, but strongly dependent on local factors, and causing an increase in the *planktonic* production.

Uronic acids. Any of a group of aldehyde acids that are oxidation products of sugars. The terminal carbon in a uronic acid is a carboxyl carbon rather than an alcoholic carbon as in sugar. Uronic acids occur combined in many *polysaccharides*.

Vadose zone. The zone above the water table in an aquifer; the *unsaturated zone*.

Xenobiotic. Foreign to the local biota; not an indigenous biochemical compound.

Zeeman interaction. The interaction of the magnetic moment of an unpaired electron (free radical electron) with a magnetic field resulting in two energy levels for the electron; this splitting of the electronic energy level gives rise to the transitions upon which ESR spectroscopy is based. The nuclear Zeeman interaction refers to the corresponding interaction of the magnetic moment of a nucleus with a magnetic field leading to a splitting of the (degenerate) nuclear energy states into a number of energy levels and allowing transitions to occur in NMR spectroscopy.

APPENDIX B

Glossary of Chemical Compounds

This glossary of chemical compounds is provided for the benefit of readers who may not be familiar with names of chemical compounds used in different chapters of this book. Chemical structures are presented, and in some cases other commonly used names for the compound and a brief description of the function or occurrence of the compound are included in the context of the text discussion. More detailed descriptions of most of these compounds can be found in the *Merck Index* (1983).

Acetanilide

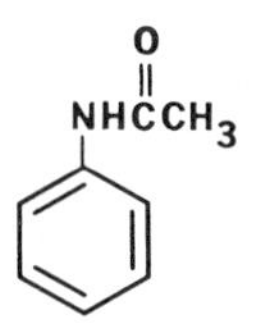

Acetic anhydride: an acetylating agent $CH_3C(=O)-O-C(=O)CH_3$

Acetol: 1-hydroxy-2-propanone $HO-CH_2C(=O)CH_3$

Acetonitrile: cyanomethane; an organic solvent $CH_3C\equiv N$

***N*-acetyl imidazole: an acetylating agent**

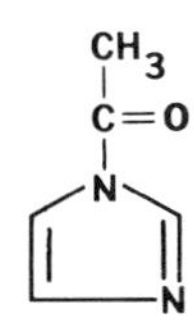

Ametryne: a triazine herbicide

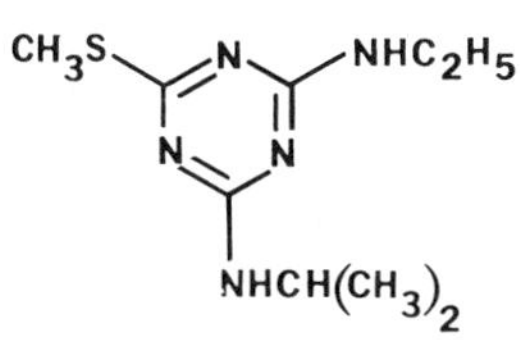

Amitrole: a herbicide

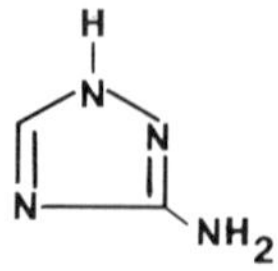

Arachidonic acid: a fatty acid essential to metabolism

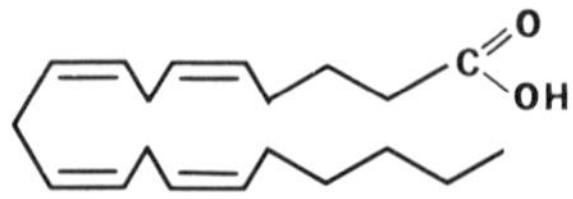

Ascorbic acid: also known as vitamin C; commonly required for growth of organisms

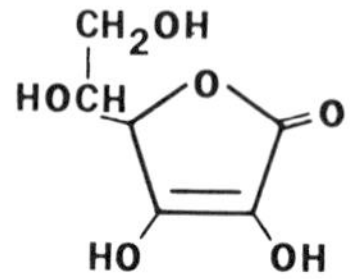

Atrazine: a triazine herbicide

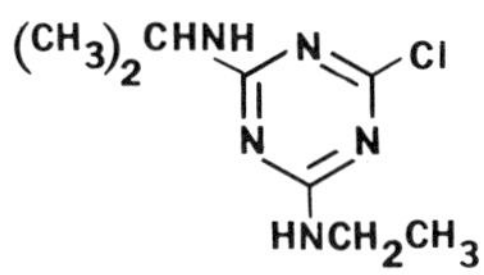

Benzenecarboxylic acids: naturally occurring compounds, such as benzoic acid

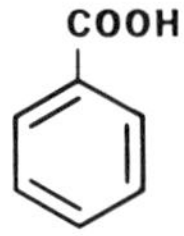

Benzene hexacarboxylic acid: also known as mellitic acid; a naturally occurring compound

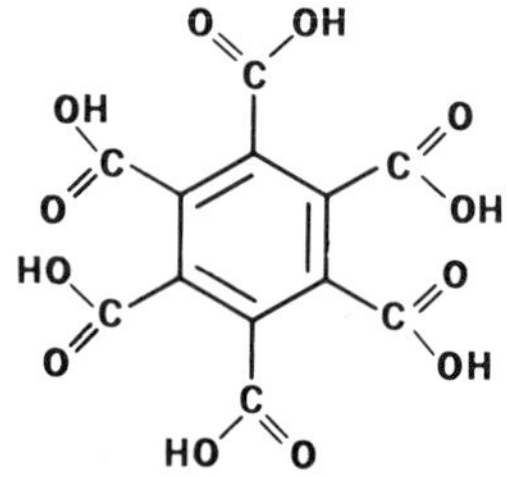

***o*-Benzoquinone: *o*-quinone; a naturally occurring compound**

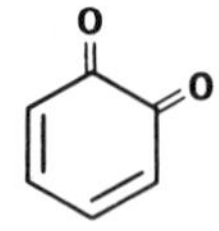

***p*-Benzoquinone: *p*-quinone; a naturally occurring compound**

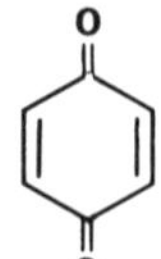

Benzylamine: α-aminotoluene

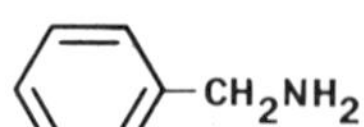

Butadiene $CH_2=CHCH=CH_2$

Butyric acid: butanoic acid; a naturally occurring compound

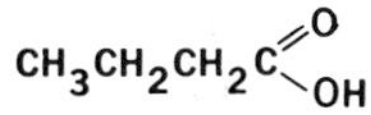

Catechol: 1,2-dihydroxybenzene; a naturally occurring compound

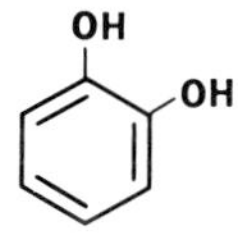

Cholesterol: the principal sterol of higher animals

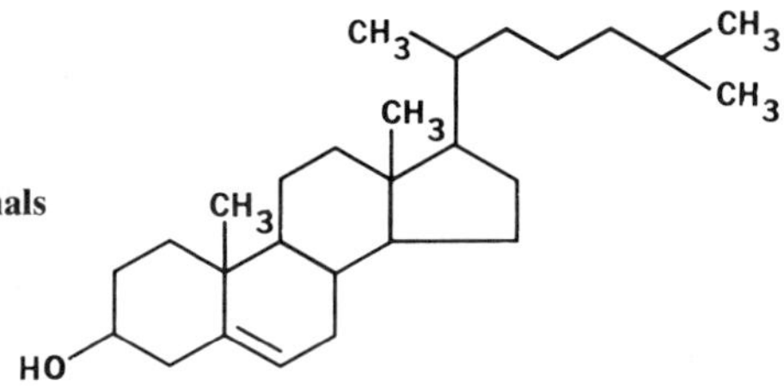

CIPC: also known as chloropropham; a carbamate herbicide

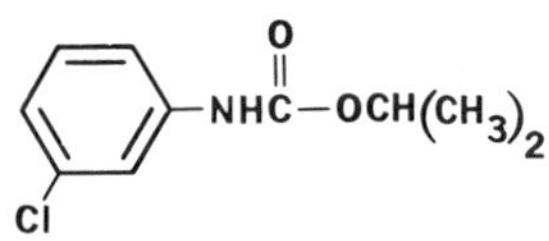

Citric acid: a naturally occurring, tribasic acid; metal-complexing agent

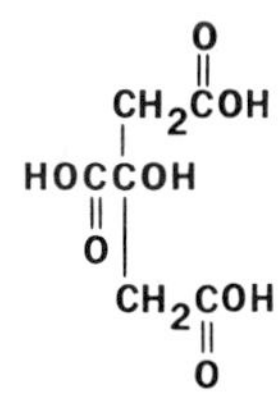

Coniferyl alcohol: a constituent of lignin

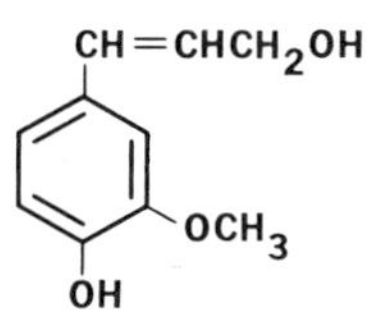

2,4-D: 2,4-dichlorophenoxyacetic acid; a herbicide

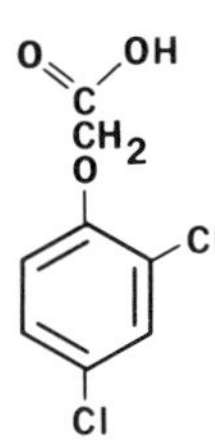

DDT: *p,p*,dichlorodiphenyltrichloroethane; a chlorinated hydrocarbon insecticide

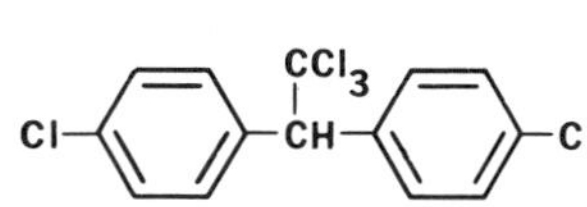

2-Deoxyribose: a pentose monosaccharide which is a constituent of the nucleotide subunits of DNA; can exist in either a ring or chain form

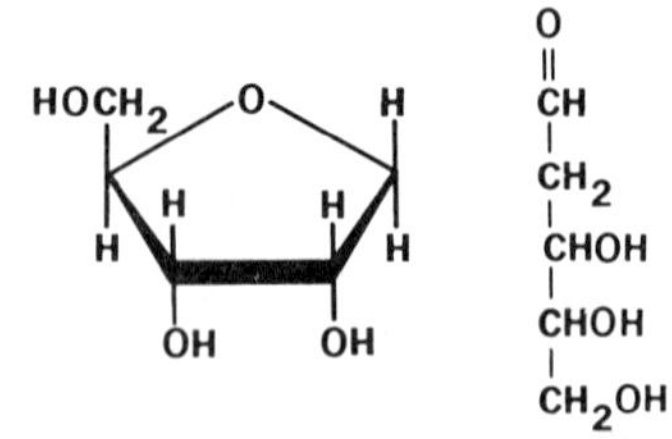

Diacyl peroxide: a group of organic peroxides $R{-}C({=}O){-}O{-}O{-}C({=}O){-}R$

Diaminoethane: ethylenediamine; an organic solvent $H_2NCH_2CH_2NH_2$

Diazomethane: azimethylene; a methylating agent $CH_2{=}N^+{=}N^-$

Dihydroxyacetone: a naturally occurring compound; the first compound in the homologous series of ketoses $HOCH_2C({=}O)CH_2OH$

3,4-Dihydroxybutanoic acid: a naturally occurring compound $HOCH_2CH(OH)CH_2COOH$

2,5-Dihydroxy-3-pentenoic acid: a naturally occurring compound $HOCH_2CH{=}CHCH(OH)COOH$

Dimethyl formamide: also known as DMF; an aprotic organic solvent $HC({=}O){-}N(CH_3)_2$

Dimethylsulfoxide: also known as DMSO; an aprotic organic solvent $(CH_3)_2S{=}O$

Dioxane: 1,4-diethylene dioxide; a water-miscible organic solvent

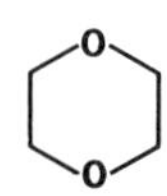

Diphenamid: an anilide herbicide

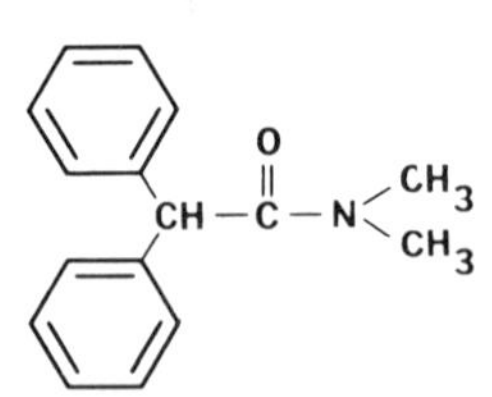

Diquat: also known as Diquat dibromide; an aquatic herbicide

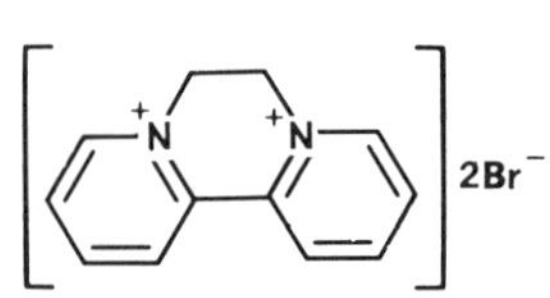

Diuron: a urea herbicide 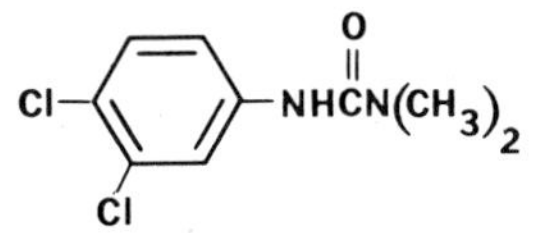

Epichlorohydrin: used as a cross-linking agent in the production of Sephadex gel 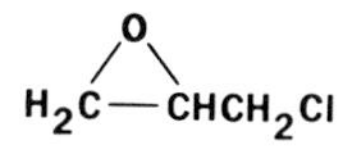

Ethylenediaminetetraacetic acid: also known as EDTA; a chelating agent for metals 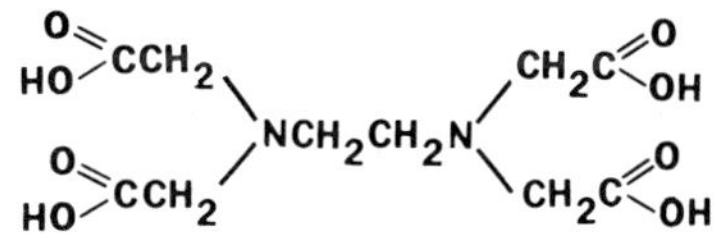

Fluoroboric acid: an inorganic reagent HBF_4

Formamide: a polar organic solvent 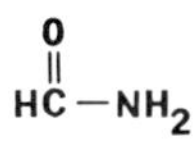

Formic acid: a naturally occurring compound 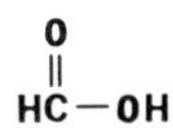

Furan: also known as furfuran; occurring in the oils of pine woods 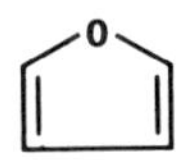

Furancarboxylic acid: 2-furoic acid; a naturally occurring compound 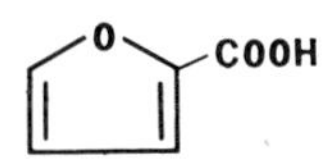

Galacturonic acid: obtained by hydrolysis of cell wall polysaccharides 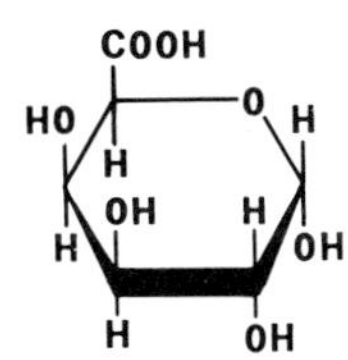

Gallic acid: obtained by the hydrolysis of tannins 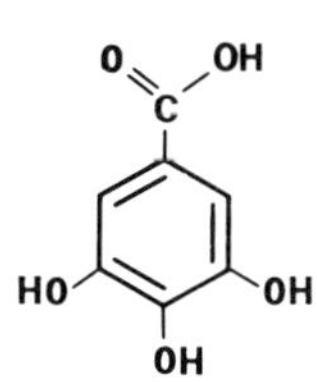

Glyceraldehyde: produced by oxidation of sugars in biological metabolism, the first compound of the homologous series of aldose monosaccharides 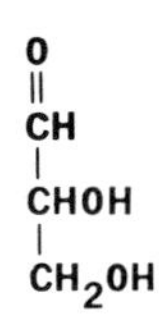

Glycolic acid: a naturally occurring compound; produced in photosynthesis 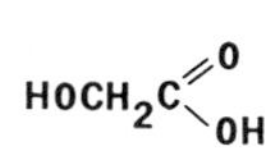

Glyoxylic acid: a naturally occurring compound, a metabolic intermediate or product of the metabolism of certain bacteria, algae, and plants

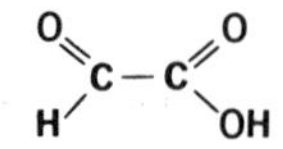

HEXAPUS: an organic metal-complexing agent

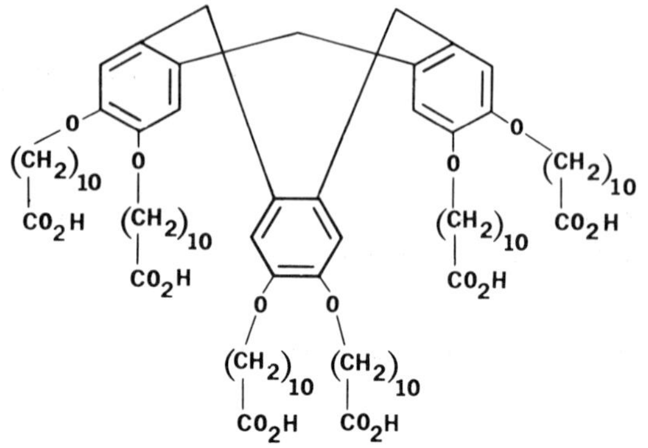

4-Hydroxybenzoic acid: a naturally occurring compound

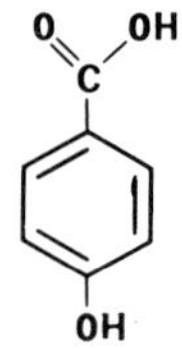

2-Hydroxybutanoic acid: a naturally occurring compound

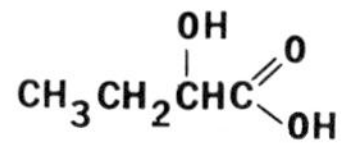

Hydroxymethyl furfural: a naturally occurring compound

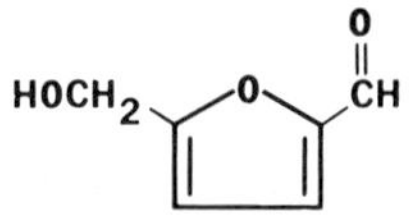

***N*-(*p*-hydroxyphenyl)glycine: a naturally occurring compound**

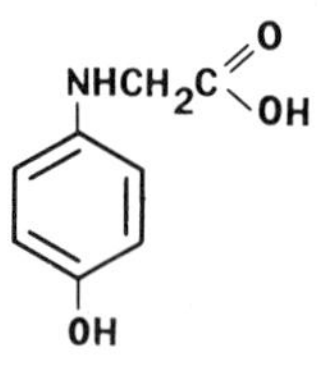

Indole: obtained in the fractionation of coal tar

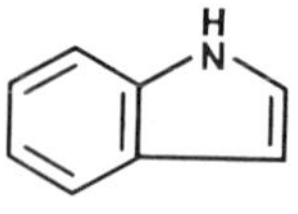

Isoamylalcohol: isopentyl alcohol; an organic solvent

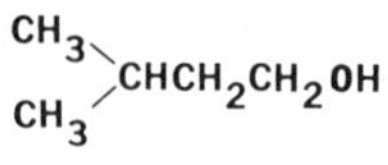

Lactic acid: a degradation product of carbohydrates, and an intermediate compound in the Krebs cycle

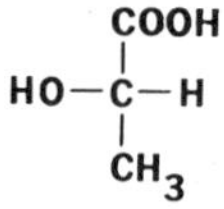

Linuron: a urea herbicide 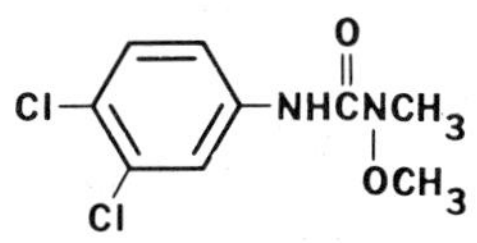

Maleic anhydride 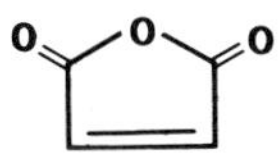

Malonic acid: 1,3-propanedioic acid; a naturally occurring compound 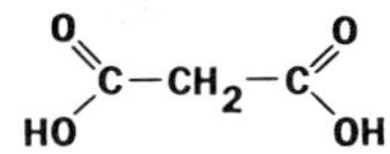

Methylethyl ketone: 2-butanone, also known as MEK; an organic solvent 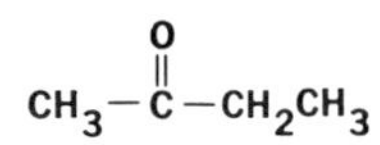

Methacrylic acid: *α*-methylacrylic acid, a naturally occurring compound 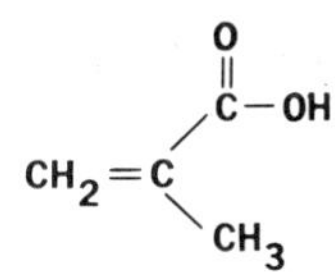

***N*-methyl-2-pyrrolidone: a water miscible organic solvent for polymeric materials** 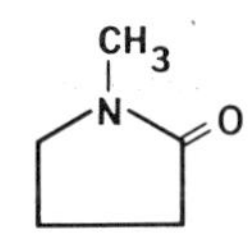

Neburon: a urea herbicide 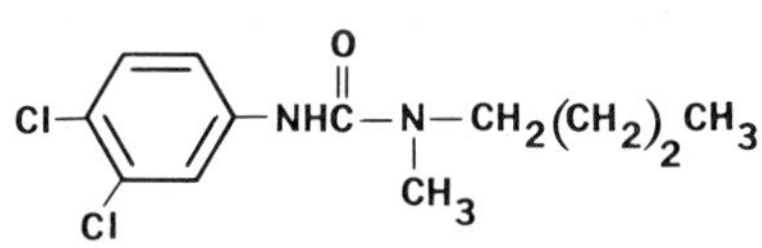

Nitrilotriacetic acid: also known as NTA, metal-complexing agent 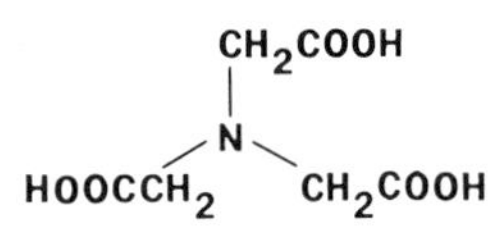

Oxalic acid: a widely occurring compound in nature 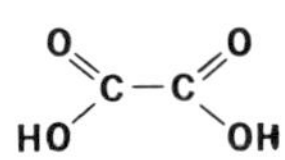

Paraquat: a bipyridinium herbicide for aquatic weeds 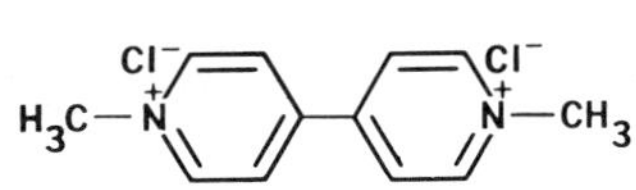

Phenazine: dibenzopyrazine; an insecticide 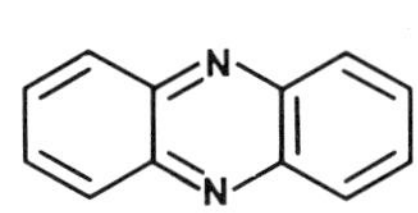

Phloroglucinol: 1,3,5-trihydroxybenzene; a structural component of many natural products 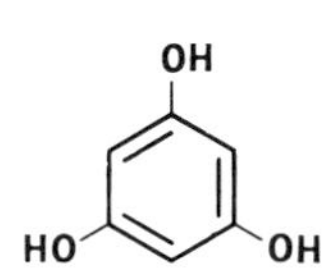

Phthalic acid: benzene-1,2-dicarboxylic acid

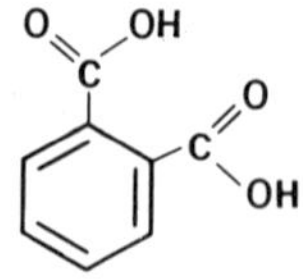

Picloram: a herbicide

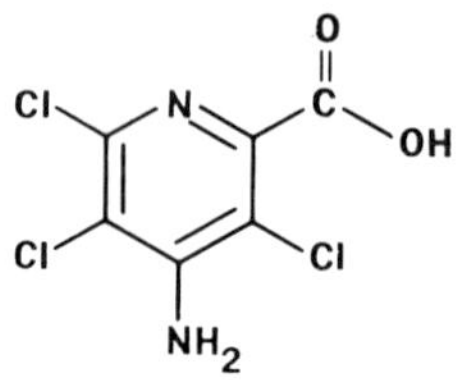

Picric acid: 2,4,6-trinitrophenol

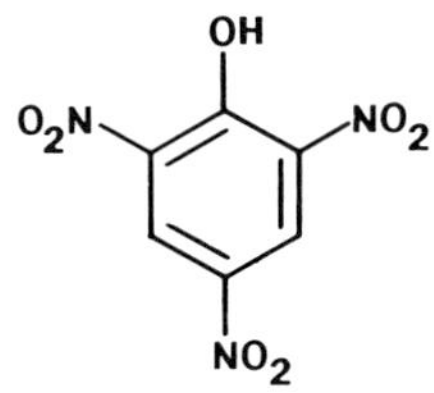

Polyacrylonitrile: orlon, a synthetic polymer

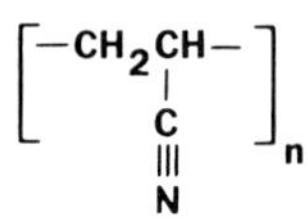

Potassium permanganate: an inorganic oxidizing agent $KMnO_4$

Prometryn: a triazine herbicide

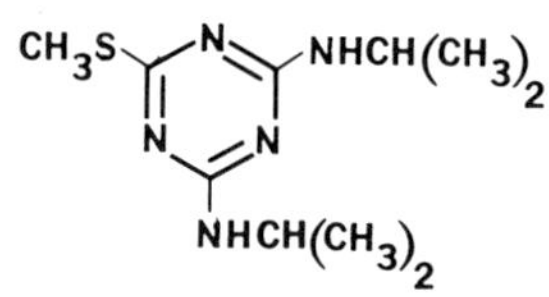

Propazine: a triazine herbicide

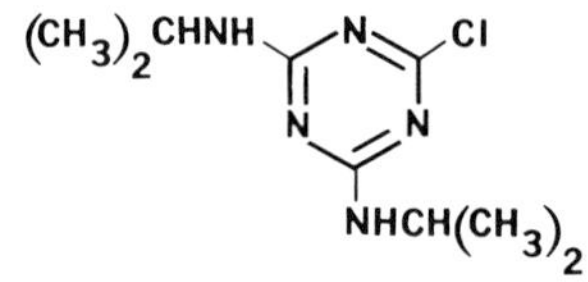

Propionic acid: propanoic acid; a naturally occurring compound

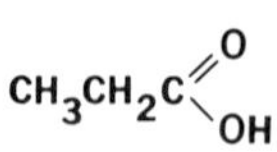

Propyl benzene

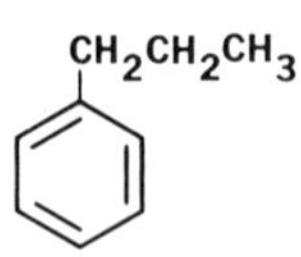

Prostaglandin: a class of physiologically active lipid acids

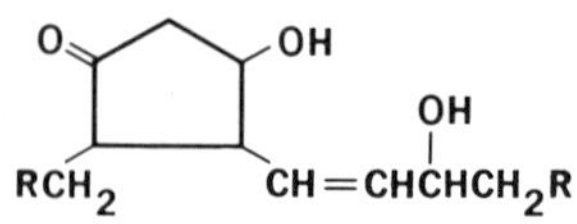

Purine: an organic base, a component of nucleic acids

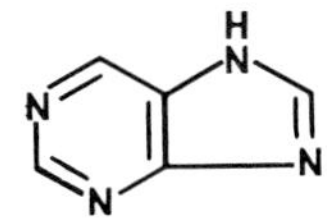

Pyridine: a basic organic solvent

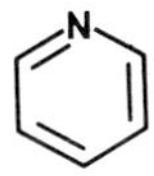

Pyrimidine: an organic base, component of nucleic acids

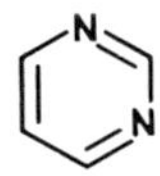

Pyrogallol: a naturally occurring compound

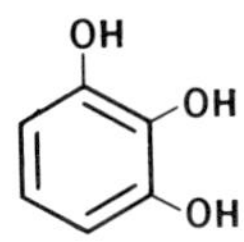

Pyrophosphoric acid: an inorganic reagent, sodium pyrophosphate is used to extract humic substances

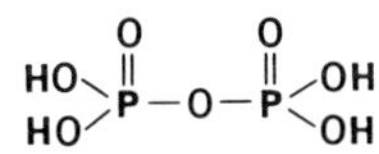

Quinhydrone: a molecular complex composed of quinone and hydroquinone

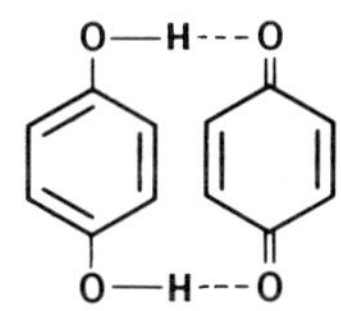

Ribose: the parent carbohydrate component in all nucleic acids; can exist in either a ring or chain form

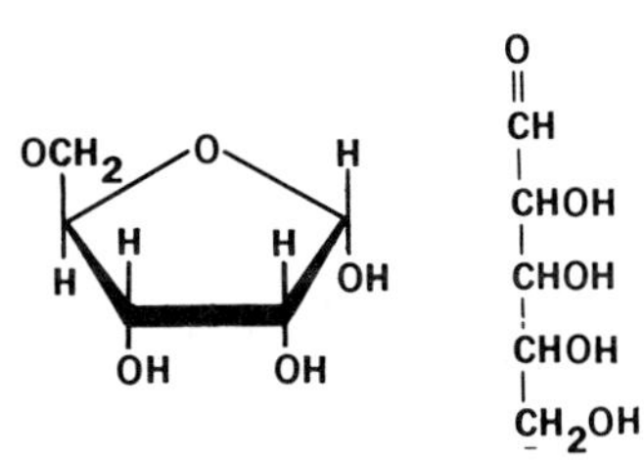

Salicylic acid: *o*-hydroxybenzoic acid; a naturally occurring compound

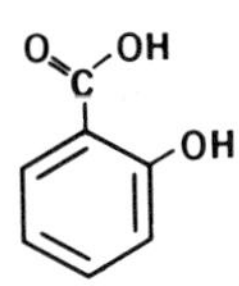

Simazine: a triazine herbicide

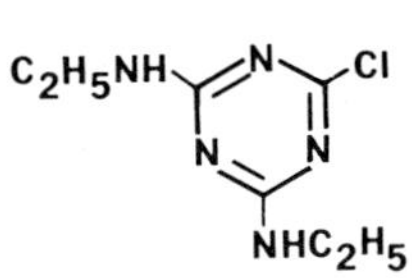

Succinic acid: an intermediate compound in respiration

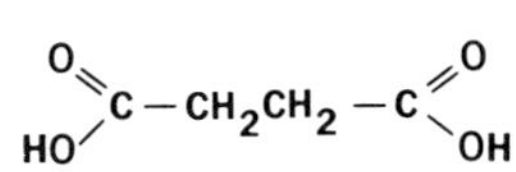

Sulfolane: a water-miscible organic solvent

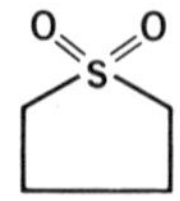

Tetrahydrofuran: also known as THF; a water-miscible organic solvent

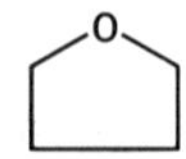

***o*-Toluidine: 2-aminotoluene**

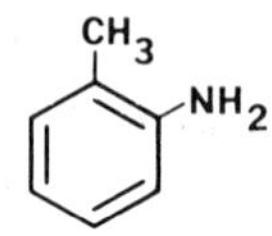

Triazole: an organic reagent

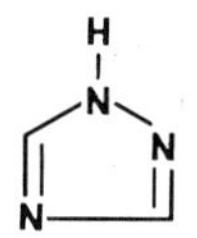

Trifluoroacetic acid: a strong organic acid

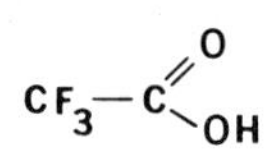

Trifluorodiazoethane: a derivatizing agent

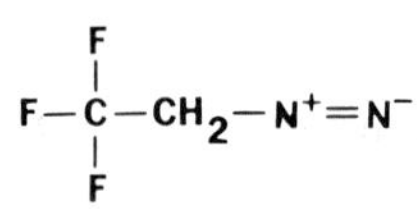

Trifluoroethylamine: an organic reagent

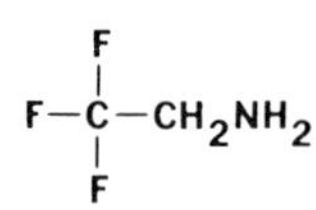

Trioctylamine: an ion-pair-forming organic solvent

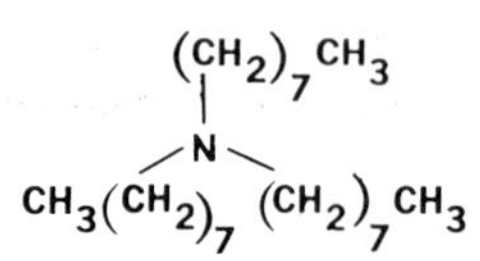

Urea: carbamide; a product of protein metabolism

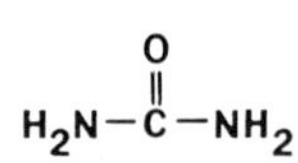

Vanillic acid: a natural product derived from lignin

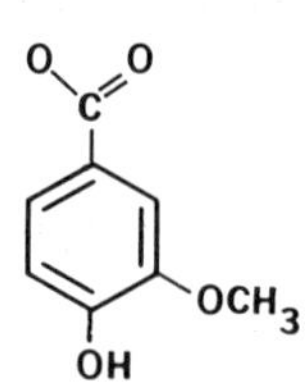

APPENDIX C

General References

The following are texts that either provide more in-depth discussion of chemical, geological, ecological, and biological concepts referred to in different chapters, or are important general references for the scientific disciplines represented in this book. These texts are recommended to readers of this book who may find themselves unfamiliar with some concepts. Many of these texts were also used in preparation of the glossaries.

CHEMISTRY

1. *Advanced Inorganic Chemistry, A Comprehensive Text,* F. Albert Cotton and Geoffrey Wilkinson, John Wiley & Sons, New York, 1980.
2. *Basic Principles of Organic Chemistry,* John D. Roberts and Marjorie C. Caserio, W. A. Benjamin, New York, 1965.
3. *Organic Chemistry,* 3rd edition, Robert T. Morrison and Robert N. Boyd, Allyn and Bacon, Boston, 1973.
4. *CRC Handbook of Chemistry and Physics,* 62nd edition, Robert C. Weast, CRC Press, Boca Raton, Florida, 1981.
5. *Fundamentals of Analytical Chemistry,* 4th edition, Douglas A. Skoog and Donald M. West, Saunders, New York, 1982.
6. *Gel Permeation Chromatography,* Klaus H. Altgelt and Leon Segal, Marcel Dekker, New York, 1971.
7. *Instrumental Methods of Analysis,* 5th edition, Hobart H. Willard, Lynne L. Merritt, and John A. Dean, D. Van Nostrand, New York, 1974.
8. *An Introduction to Separation Science,* Barry L. Karger, Lloyd R. Snyder, and Csaba Horvath, John Wiley & Sons, New York, 1973.
9. *The Merck Index,* 10th edition, Martha Windholz, Merck and Co., Rahway, N.J., 1983.
10. *Principles of Biochemistry,* Albert L. Lehninger, Worth Publishers, New York, 1982.

EARTH SCIENCE

1. *The Encyclopedia of Soil Science,* Part 1, Rhodes W. Fairbridge and Charles W. Finkel, Dowden, Hutchinson and Ross, Stroudsberg, Pennsylvania, 1979.

2. *Groundwater,* R. Allen Freeze and John A. Cherry, Prentice-Hall, Englewood Cliffs, New Jersey, 1979.
3. *Earth,* Frank Press and Raymond Siever, W. H. Freeman, San Francisco, 1982.
4. *Solutions, Minerals and Equilibria,* Robert M. Garrels and Charles L. Christ, Freeman, Cooper and Co., San Francisco, 1965.
5. *Aquatic Chemistry,* Werner Stumm and James J. Morgan, John Wiley & Sons, New York, 1981.
6. *Marine Organic Chemistry: Evolution, Composition, Interactions, and Chemistry of Organic Matter in Seawater,* E. K. Duursma and R. Dawson, Elsevier Oceanography Series No. 31, Elsevier, New York, 1981.
7. *Chemical Oceanography,* Wallace S. Broecker, Harcourt Brace Jovanovich, Inc., New York, 1974.
8. *Petroleum Geochemistry and Geology,* John M. Hunt, W. H. Freeman, San Francisco, 1979.
9. *The Chemistry of Soil Processes,* D. J. Greenland and M. H. B. Hayes, John Wiley & Sons, Chichester, 1981.
10. *The Chemistry of Soil Constituents,* D. J. Greenland and M. H. B. Hayes, John Wiley & Sons, Chichester, 1978.

ECOLOGY AND BIOLOGY

1. *Fundamentals of Ecology,* Eugene P. Odum, W. B. Saunders, Philadelphia, 1971.
2. *Ecology, The Experimental Analysis of Distribution and Abundance,* Charles J. Krebs, Harper and Row, New York, 1972.
3. *Ecosystems of the World,* Vols. 4A and 4B, *Mires: Swamp, Bog, Fen and Moor, General Studies,* editors A. D. J. Gore and David W. Goodall, Elsevier Scientific Publishing, Amsterdam, 1983.
4. *Peatlands,* P. D. Moore and D. J. Bellamy, Springer-Verlag, London, 1974.
5. *Enzyme Handbook,* Vols. 1 and 2, Thomas E. Barman, Springer-Verlag, New York, 1969.
6. *Limnology,* 2nd edition, Robert G. Wetzel, CBS College Publishing and Saunders College Publishing, Philadelphia, 1983.
7. *A Treatise on Limnology,* Vols. I, II, and III, G. Evelyn Hutchinson, John Wiley & Sons, New York, 1975.

The following texts were also used in the preparation of the glossaries for this book and may be useful to the reader unfamiliar with additional terms used in this book or with terms that are used in the literature cited in the different chapters.

1. *Chambers Dictionary of Science and Technology,* revised edition, T. C. Colocott and A. B. Dobson, Harper and Row, New York, 1974.
2. *The Condensed Chemical Dictionary,* 10th edition, revised by Gessner G. Hawley, Van Nostrand Reinhold, New York, 1981.
3. *A Dictionary of Geology,* 5th edition, John Challinor, Oxford University Press, New York, 1978.
4. *Ecology Field Glossary, A Naturalist's Vocabulary,* Walter H. Lewis, Greenwood Press, Westport, Connecticut, 1977.

5. *Geologic Time,* Don L. Eicher, Prentice-Hall, Englewood Cliffs, New Jersey, 1976.
6. *Glossary of Geology,* Margaret Gary, Robert McAfee, and Carol Wolfe, American Geologic Institute, Washington, D.C., 1972.
7. *Glossary of Soil Science Terms,* Soil Science Society of America, Madison, July 1973.
8. *International Glossary of Hydrology,* UNESCO, 1974.
9. *Radio Shack Unabridged Dictionary of Electronics,* Rudolf F. Graf, Radio Shack, a Tandy Corp., Ft. Worth, Texas, 1975.

Index

Absorbance, 98, 121
of aquatic humic substances, 94
decrease upon irradiation, 121
infrared, 196, 533
properties, prediction of, 44
spectral analysis, 183, 553
ultraviolet, 214, 218, 222, 230
visible, 214, 218
Acetylation, 540, 543
Acid Dissociation Constants, 503
in humic substances, 506
density of polar groups, 507
effect:
of charged substituents, 506
of dipolar substituents, 506
simple organic acids, 503
Acid hydrolysis, 259, 318, 321, 471
Acidic functional groups, 493
carboxyl group determination, 511
calcium acetate exchange method, 511
direct titrations, 511
in humic substances, 505
methods of analysis, 507
phenolic groups, 514
in simple organic compounds, 494
thermometric titrations, 507, 514
total acidity determination, 508
barium hydroxide method, 507, 509
direct titrations, 508
nonaqueous titrations, 510
weakly acidic groups, 514
Acidification, as precipitation method, 415
Acidity:
delocalization effects, 502
of DOM in lakes, 111, 117, 134
electrostatic effects, 498-501
in estuaries, 213, 214, 219
of soils, 71
statistical effects, 495-497
in streams, 200, 207
total, 214
see also pH
Acids, fatty, 166, 234, 237, 240
Acrylic-ester resins, 380
Adsorption:
as basis for fractionation, 405, 571
by clay, 40
of DOM, 127
of iron, 106
mechanisms for, 41
onto minerals, 105
by organic matter, 40
of phosphorus, 106
pollutants, 229
onto surfaces, 118
Adsorption chromatography:
ion exchange resins, 376
XAD resins, 91, 108, 113, 131, 379
Aggregation, 126, 218, 221, 222, 231
Alcoholic groups, 281, 331
Aldehyde groups, 129
Alfisols, 39
Algae, 123, 124, 128, 134, 275, 282, 288, 298, 301

Aliphatic content, 95, 116, 133, 199, 256
Aliphatic structures, in humin, 277, 279, 286, 296, 331
Alkaline earth ions, 217, 218, 221
Alkaline extraction, 251
Alkaline permanganate oxidation, 157
Alkyl chains, NMR evidence for, 570
Allochthonous humic material, 107
Allophanic materials, as adsorbents of humic substances, 27
Alteration:
  oxidative, 269, 370
  thermal, 177, 259, 280
Alumina, sorption by, 118, 375, 379
Alumino-silicates, 252
Aluminum, 38, 44, 201, 203, 260, 333
Americium, 244
Amide functional groups, 280, 288
Amines:
  IR spectra of, 543
  Mailard reaction, 18, 129, 169, 320
Amino acids:
  adenine, 312
  alanine, 198
  in aquatic humic substances, 214, 219, 247
  asparagine, 321
  glycine, 164
  isoleucine, 308
  leucine, 308
  neutral, 120
  serine, 307
  in soil humic substances, 580
  threonine, 307
  tyrosine, 308
Amino compounds, 70, 160, 318, 321, 323
Amino sugars, 304, 306, 324
Ammonia, 81, 197, 198, 259, 266, 304, 306, 312, 318, 321
Ammonification, 81, 305
Anion sorbent, of aquatic humics, 378
Anodic stripping voltammetry (ASV), 36, 38, 243, 244
Anthracite, 281
Apocrenic acid, 15, 20, 390
Aquatic humic substances:
  concentration methods, 370
  structure, 411
Aqueous solubility, of humic substances, 5, 8
Aromatic:
  acids, 159, 186
  carbons, 208, 279, 286
  compounds, 224
  protons, 569
  rings, 95, 118, 137, 157, 280, 572
Aromaticity, 116, 157, 160, 495, 571
  in humic substances, 160, 238, 257, 289, 331
  of humin, 275, 279, 287
Ash:
  content, 190, 252
  elemental analysis, 444
Atomic ratios, 465
  as guide to humic substance identification, 465
  in humin, 279, 295, 299
  in marine sedimentary humic substances, 252, 265, 271
Autochthonous production, of humic substances, 129
Autoxidation, of lipids, 238, 240

Bacteria, 61, 124, 264
  oligocarbophilic, 128
  saprophytic, 128
Baltic Sea, 222
Beer's law, 183, 534, 552
Bending vibration, 532, 541
Benthic filter feeding, 224
Benzenecarboxylic acids, 158, 159
Benzoic acid, 127, 160
Benzylamines, 323
Berzelius, 363
Binding:
  of colloids, 334
  by humic substances, 27
    of DDT, 101
    of metals, 36, 100, 122, 226
    of organic solutes, 204, 229
    of soil clays, 29
Biochemistry, of humus formation, 18-21
Biodegradation, of aquatic humic substances, 127
Biodeposition, in estuaries, 224
Biscayne aquifer, 87, 90, 94, 97, 98, 100, 101
Bjerrum model, electrostatic field, 498
Black Lake, 117
Black Sea, 260, 270
Black waters, particle organic carbon in, 364
Braunhuminsaure, 20
Broadkill River, 219, 221
Bromine, elemental analysis, 443
Brown humic acid, 20, 390
Browning reaction, 129, 198, 320. *See also* Maillard reaction

Cadmium, 122, 123, 203, 226, 244

Calcite precipitation, inhibition by humic substances, 125
Calcium, 100, 123, 203, 217, 260, 333
  bicarbonate, 90
  carbonate, 106, 125, 126, 216
  sulfate, 90
Carbohydrates:
  associated with humic substances:
    in estuaries, 219
    in groundwater, 96
    in lakes, 106, 130
    in lake sediments, 152, 160, 164
    in marine sediments, 269
  associated with humin, 283, 286, 301
  fucose, 166
  galactose, 166
  glucose, 166, 198, 246
  mannose, 166
  monomeric, 113
  unbound in extracts of humic substances, 465, 469, 573
  in vascular plants, 269
  xylose, 166
Carbon:
  aromatic, 236, 238, 577
  budget, 3
  composition of humic substances, 183, 252, 350
  content:
    of aquatic humic substances, 92, 99, 153, 161, 214
    of peats, 54, 58, 80
  elemental analysis, 439
  organic, in groundwater, 88
  oxygenated, 236
  paraffinic, *see* Paraffinic carbons
  relation to cation-exchange capacity, 30
  as sorbent for isolation, 375, 379
Carbonate, as competing ligand, 226
Carbon cycle, 2, 106, 148
Carbon isotopes:
  in NMR, *see* Nuclear magnetic resonance (NMR) spectroscopy
  ratios, in aquatic humic substances:
    extracted from estuaries, 115, 216
    extracted from groundwater, 97
    extracted from lake sediments, 161
    extracted from marine sediments, 262
    extracted from seawater, 233, 234, 247
  spectroscopy, 279, 285, 286
Carbonyl groups:
  content in lacustrine humic substances, 110
  in peat humic substances, 71
Carboxyl groups:
  content:
    in estuarine humic substances, 217, 222
    in groundwater humic substances, 93, 97
    in humic substances, 495
    in humin, 280, 281, 286, 288
    in lacustrine humic substances, 110
    in lake sediment humic substances, 156
    in peat humic substances, 71
    in soil humic substances, 21, 34
    in stream humic substances, 199
  determination, 511
  IR analysis, 540
  ultraviolet-visible analysis, 551
Catagenesis, 251
Cation-exchange capacity (CEC):
  of peats, 63, 81
  of soils, 29, 333
    contribution of organic matter, 30
    effect of pH, 30
    relation to pesticide adsorption, 40
Cation-exchange resins, as sorbents for isolation, 379
Cellulose:
  degradation of, 282, 290
  in peats, 63
  relation to marine sediment humic substances, 269
  relation to stream humic substances, 197
  solubility, effect of conformation, 347
Charge characteristics, fractionation on basis of, 389, 402
Chelation, 201, 203, 474
Cheluviation, 47, 60
Chemical degradation, 259
  degradative techniques, 255
  effect on sample preservation, 370
Chemical derivatives, 422
  acylation, 423
  methylation, 422
  sodium borohydride reduction, 425
  trifluoroethylalkylation, 424
Chemical fractionation methods, 415
  adsorption chromatography, 416
  precipitation, 415
  solvent extraction, 416
Chemical oxidation, 300
  of humin, 300
  to release trace metals, 37
Chemical oxidation demand (COD), 186
Chemical shift, 564
Chernozem soils, 30, 50, 306
Chesapeake Bay, 229

Chloride, as competing ligand, 226
Chlorination, drinking water, 2, 186, 207
Chlorine, elemental analysis, 443
Chloroform, as extractant, 250, 261
Cholesterol, association with riverine humic substances, 204, 229
Chromium, 203, 245
Chromophores, 95, 549, 550, 552
Classification system, of peat, 57, 85
Clathrate compounds, 83
Clay:
  and ammonia, 306
  clay-metal-humus complexes in soil, 27, 28
  contribution to cation exchange capacity of soils, 29
  and extraction of humic substances from soil, 330
  interferences in extracting humic substances from water, 366
  in marine sediments, 269
  in peats, 82
Cleavage, of humic substances by UV irradiation, 120
Climate, effect:
  on elemental composition of marine sediment humic substances, 252, 263
  on nitrogen compounds, 306, 312
Coagulation, to determine binding capacities of humic substances for metal ions, 36
Coal, 6
  anthracite, 281
  bituminous, 281
  boghead, 280, 282
  cannel, 280
  formation, 54, 280, 301
  humic acids, 49, 280, 461
  lignin in, 298
  occurrence of trace elements in, 33
  sapropelic, 280, 282, 299
Coal Creek, 446
Coalification process, 280, 298
Cobalt color comparison, as measure of DOC, 186
Cohesive energy density (CED), 341
Colligative properties, 95, 490
Colloids, in estuaries, 219, 227
  clay interferences in resin isolation, 366
  humic-clay, 28
  humic metal, 203
  iron, 221
  stability enhancement, 126
Color:
  and absorbance of groundwater humic substances, 94
  of humic substances, 1
  of lakewater, 127, 132
  in streams, 182, 190
Column chromatographic methods, for sorption, 374
Co-metabolism, of fulvic acid by bacteria, 127, 230
Competing-ligand differential spectroscopy, 203
Complexation:
  of DDT by humic substances, 229
  electron spin resonance spectroscopy studies of, 35
  by humic substances, 33, 37, 114, 548
  infrared spectroscopy studies of, 35
  of organic pollutants by humic substances, 229
  by porphyrin-type complexes, 35
  of trace metals by humic substances, 33, 37, 114, 122, 144, 200, 201, 204, 207, 226, 228, 232, 234, 243, 548
Composition, of soil humic substances, 331
Concentration methods:
  for aquatic fulvic acid, 370, 382
  ultrafiltration, 483
Condensation hypothesis:
  of coal formation, 281, 283
  reactions in formation of humic substances, 269, 330
Congress of the International Society of Soil Science (1978), 8
Contamination, of groundwater, 88
Copper:
  complexation by soil humic substances:
    in estuaries, 226, 232
    in groundwater, 99
    in lakes, 122
    in peat, 66, 78, 81, 82
    in seawater, 243
    in soil, 33, 35, 37
    in streams, 201, 203, 204
  as control of peat subsidence, 78, 81, 82
  to deactivate soil enzymes, 54, 78, 85
  toxicity and bioavailability, 122, 241
Coprecipitation, 373
  as fractionation method, 415
  of humic substances with calcite in lakes, 125
  as isolation method, 373
Copropel, 27

CP/MAS, *see* Cross-polarization magic-angle spinning (CP/MAS); Nuclear magnetic resonance (NMR) spectroscopy
Crazy Eddie Bog, 122
Crenic acid, 15, 20, 390
Cross-polarization magic-angle spinning (CP/MAS), 73, 331, 566, 578
Curie point pyrolysis/mass spectrometry, 117, 136, 260
Cuticular waxes, 262

Dating, radiocarbon, *see* Radiocarbon dating
DDT:
  in estuaries, 229
  in groundwater, 101
  in soils, 41
  in streams, 204
Decomposition:
  classes of organic materials, in peat, 57
  microbial:
    of aquatic humic substances, 121
    in peats, 76
Deforestation, effect on degrading soil humus, 3
Defunctionalization, as function of depth in sediments, 272
Degradation:
  biological, analytical methods, 259, 300, 531
  in formation of humic substances, 269
  limitations on, in marine sediments, 252
  oxidative degradation:
    in analysis of fulvic acid, 160, 186
    in peats, 75, 84
  of soil humus, 3
Depth filters, 366
Detoxification, of trace metals by humic substances, 122
Detrital food chains, in peatlands, 61
Detritus, 107, 269
Deuterated humic acid, in NMR, 545, 546
Deuterium-hydrogen exchange, 535
Deuterium oxide, infrared spectra, 544
Diagenesis:
  in lake sediments, 171, 176
  in marine sediments, 252
Dialysis, as method for ion-binding capacities, 36
Dibasic acids, as degradation products, 117
Dicarboxylic acids, titration of, 496
Diethylaminoethyl cellulose, as sorbent for aquatic humic substances, 378
Dimethylformamide, as extractant for soil humic substances, 335, 338
Dimethyl sulfoxide, as extractant for soil humic substances, 335, 338, 348, 352, 357, 360
Dipolar aprotic solvents, as extractants, 338, 340
Dipole-dipole coupling, in NMR, 565
Dipole-dipole interactions, in humic substances, 336
Dipole moment, 337
Diquat, 40, 41
Dispersion forces, 336, 340
Dissolved amino acids (DAA), patterns in lakes, 136
Dissolved organic carbon (DOC):
  analysis of, 182
  in estuaries, 214, 222, 225, 230
  in groundwater, 90
  in lakes, 110, 132, 134, 137, 140
  in natural water, 364
  in seawater, 3
  in streams, 189
Dissolved organic matter (DOM):
  in estuaries, 218
  in lakes, 106, 108, 111, 132, 143, 145
DOC, *see* Dissolved organic carbon (DOC)
DOM, *see* Dissolved organic matter (DOM)
Domain, clay, 32
Dopplerite, 27
Douala Basin, 255
Double layer, electrochemical in electrophoresis, 115
Dumas nitrogen determination techniques, 440

E_4/E_6 ratios:
  atomic ratios, in relation to, 467
  effect of HCl in, 358
  humification, in relation to, 472, 552
Electrodes, ion selective, 36, 38, 243
Electron paramagnetic resonance (EPR) spectroscopy, 203, 555
Electron spin resonance (ESR) spectroscopy, 555
  of humin, 285
  of lacustrine humic substances, 157
  of marine humic substances, 244, 247
  of peat humic substances, 71
  of soil humic substances, 35, 38
  relation to elemental composition, 472

Electrophoresis:
ampholines, 403
electrofocusing, 402
as fractionation method, 402, 407, 426
isoelectric focusing (IEP), 403, 407
isotachophoresis (ITP), 403, 407
polyocrylamide gel electrophoresis (PAGE), 402, 407
relation to elemental composition, 470
in study of estuarine humic substances, 218, 226
Electrophoretic mobilities, 217
Electrostatic forces, 335
Elemental analysis:
ash, 444
carbon, 439
effect:
of fractionation, 467
of hydrolysis, 471
halogens, 443
of humin, 258, 279, 285
hydrogen, 439
interlaboratory study, 447
of lacustrine humic substances, 108
methodology, 433
nitrogen, 440
oxygen, 441
phosphorus, 444
sample preparation and handling, 436
structural formulas from, 473
sulfur, 442
Elemental composition:
of aquatic humic and fulvic acids, 462
carbohydrate content, relation to, 463
of coal humic acids, 461
ESR properties, relation to, 472
of estuarine humic substances, 219
of groundwater humic substances, 92, 102
of humic substances from peat, 70
of humin, 70, 277
implications of, 458
of marine sediment humic substances, 251, 254, 263
of soil humic substances, 461, 463
soil types, relation to, 460
of stream humic substances, 190
Emission spectra, of lake DOM, 113
Energy flux, in lakes, 108
English Lake District, 134
Enolic hydroxyls, 333
Enzymes:
effect on humification rates, in peats, 62, 84, 85
inhibition of, 124
Epilimnetic decalcification, 125
EPR, 203, 555
Equilibrium dialysis, 203
Equivalent weight data, marine humic substances, 237
ESR, *see* Electron spin resonance (ESR) spectroscopy
Ester groups, 331
Esthwaite Water, English Lake District, 134
Estuaries, humic substances in, 211
Ether functional groups, in humin, 280
Ethylene diamine, as extractant, 348
Ethylene diamine tetra-acetic acid (EDTA), as extractant, 244
Everglades:
humin in, 289, 296
as source of organic carbon in groundwater, 90, 99, 102
Evolutionary stages, of diagenesis, 251
Exchangeable cations, in riverine humic substances, 217
Exchangeable protons, in NMR, 571
Excitation spectrum, 111, 552, 553
Exclusion chromatography, 185
Extraction:
alkali, 66, 270, 284
from marine sediment, 250
procedures, 7, 8, 27, 65, 134, 149, 154, 246
schemes, aquatic humic substances, 382
of soil, humic substances, 345
in aqueous organic acid salts, 348
in aqueous salt solutions, 347
in basic salts, 349
in DMF, 355
in DMSO, 355
in formic acid, 355
in organic solvents, 348, 352, 354
in pyrophosphate, 349
with solvent mixtures, 357
of trace metals, 37
in water, 347, 354

Fast atom bombardment, 187
Fatty acids, 166
in seawater, 234, 240
Fauna, soil, 60
Fiber content, to determine humification in peat, 76, 84
Fibric organic material, defined, 57

Fibrist, 59
Filters, 365
  characteristics, 365
    cellulose-acetate, 365
    cellulose-nitrate, 365
    chemical composition, 367
    flow characteristics, 366, 368
    glass fiber, 365
    pore size, 366
    silver membrane, 365
Filtration, 365, 382, 383
Fixation, oxygen, in sediment, 267
Flavonoids, 269
Flickering cluster model, of water structure, 346
Flocculation, of humic substances, 37, 120, 228, 230
Flora:
  littoral, 134
  microbial, 61
  as soil classification criteria, 56
Flory-Huggins solubility parameter, 343, 357
Flotation, to extract humic substances, 70
Fluorescence:
  during electrophoresis experiments, 402
  of humic substances:
    in estuaries, 222, 227, 230, 234
    in lakes, 111, 121, 127
  intensity, 111, 552
  quenching by metals, 203, 227
  spectroscopy, 234, 552
Fluorescence/absorbance ratios of lacustrine humic substances, 121
Food chains, detrital, 61
Food source, for aquatic organisms, 206
Force constant, in IR spectroscopy, 532
Forestry, peatlands used for, 75
Formamide, as extractant, 335, 338
Formation mechanism, of marine humic substances, 237
Forna, 27
Fourier transform infrared spectroscopy (FTIR), 543, 546
Fractionation:
  of aquatic humic substances, 409
  on basis:
    of adsorption, 405
    of charge characteristics, 402
    of molecular size, 394
    of precipitation, 390
    of solubility, 390
  centrifugation, 400
  chemical fractionation methods, 415
  dissolved organic carbon, 414
  effect on elemental analysis, 467
  electrofocusing, 402
  electrophoresis, 402, 426
  gel permeation chromatography, 395, 413
  in groundwater, 247
  ion-exchange, 404, 407, 418
  in lakes, 111, 132, 137
  molecular size, 413
  operational definitions, 414
  physical fractionation methods, 426
  of soil humic substances, 388
  of stable carbon isotopes, 98, 243
  ultrafiltration, 399, 413
Free radicals:
  in ESR, 555, 556
  in soil humic substances, 353
Freeze concentration, 184, 371
Freeze-drying:
  as concentration method, 185, 371
  as preservative, 371, 382, 415
FTIR, 543, 546
Fuel, peatlands mined for, 75
Fulvic acid:
  abyssal, 247
  accumulation in Spodosols, 27
  carboxyl groups, 22
  chemical properties, 21
  class structure of, 237
  coastal, 234
  defined, 5, 15, 390
  diagenetic transformations of, 48, 262
  elemental composition, of marine and terrestrial, 235
  exchange acidity, 30
  functional groups, 22
  marine, 233
  mean residence time of, 51
  pheolic hydroxyl, 22
  reactions with metal ions, 34
  riverine, 234
  role:
    in degradation of silicate minerals, 45
    in formation of Spodosols, 47
    in weathering processes, 45
  solubility characteristics of metal complexes, 36
  stability, 51
  stable free radicals of, 40
  synthetic, 240
  type structures of, 24, 238

Functional groups:
  acidic, 493
    in IR, 532
    simple organic compounds, 495
    in UV-visible, 548
  analysis, 71, 110, 389
  of fulvic acid, 270
  of humic substances, 527, 553, 558, 559
  of humin, 277
  hydrogen bonding, 534
  oxygen-containing, 160
  peat humus, 71
  of soil humic substances, 331

Galactosamine, 312
Galacturonic acid, as model for fulvic acid, 460
Gas chromatography, of degradation products, 117, 164, 260, 300
Gelbstoff, 233, 234, 375
Gel chromatography, 154, 203
Gel permeation chromatography (GPC):
  adsorption in, 395
  calibration, 399
  characterization technique for DOC:
    in estuaries, 218
    in lakes, 113, 118, 123, 131, 133
    in seawater, 240
  charge interactions, 397
  and elemental analysis, 470
  as fractionation technique, 395, 407, 413, 427
  as method for molecular weight determination, 479
  and NMR, 571
  salt-boundary effect, 397
Gley formation, 45
Global carbon cycle, 2, 106, 148
Glucosamine, 312
Glutamate, 305
Glutamine, 305, 321
Glycoside linkage, 332
Glycosylamine, 320
Gold, 78
GPC, *see* Gel permeation chromatography (GPC)
Graphite, 281
Gray humic acid (Grauhuminsaure), 20, 390
Great Lakes, 148, 152
Greenhouse substrates, peatlands mined for, 75
Green River shales, 255, 299
Groundwater, 87
  contamination, 88
  humic substance concentration in, 90, 195
  as source of stream water, 196
Guanine, 312
Gulf of Mexico, 235, 247

Haematite, 126
Halogens, 190, 246
  elemental analysis, 443
Hardwater lakes, 112
H-bonding, *see* Hydrogen, bonding
Health studies, 186
Hemicellulose, 63, 269
Hemic organic material, defined, 57
Herbicides, 42, 83, 229
  adsorption in peats, 83
  interactions with humic substances, 39, 40, 42, 229
Heterocondensates, 213
Heterocyclic nitrogen, 320
HEXAPUS, as model for humic acid, 459
Hildebrand-Scatchard equation, 344
Hildebrand solubility parameter, 341
Histosols:
  defined, 54
  and extraction, 347
  functional group analysis, 71
  humin in, 285, 288
  origin, 56
  primary production in, 61
Holocellulose, 292
Humic acids:
  abyssal, 247
  acidic nature, 15
  aquatic, 111
  carboxyl groups of, 18, 22
  cation-induced coagulation, 37
  chemical properties of, 21
  class structure of, 237
  coastal, 234
  copper-binding capacity of, 36
  defined, 5, 14, 15, 17, 390
  diagenetic transformations of, 48, 262
  distribution of, 14
  elemental composition, of marine and terrestrial, 235
  exchange acidity, 30
  formation, 271
  functional groups, 22
  marine, 233
  mean residence time of, 51

nitrogen of, 18
phenolic hydroxyl, 22
potentiometric titration of, 17
reactions with metal ions, 34
riverine, 234
role:
in degradation of silicate minerals, 45
in weathering processes, 45
solubility characteristics of metal complexes, 36
stability, 51
stable free radicals of, 40
synthetic, 239, 240
type structures of, 23, 238
Humic substances:
abundance, 149
acidic properties of, 17
ammonia fixation by, 17
associations in soil, 27
binding capacities for metal ions, 36
binding of metal ions by, 29
chemical reactions with pesticides, 44
clay-metal-humus complexes, 27
contribution to cation-exchange capacity, 29
defined, 4, 14, 17, 20, 390, 460
extraction of, 14, 27
fractionation of, 14
heterogeneous mixture, 7
humic acid/fulvic acid ratio, 26, 45
in lake sediments, 149
mean residence time of, 50
metal complexes of, 33
oxygen consumption by, 17
peat, 58
peatlands, 66
reduction properties of, 38
reference collection, 8
role:
in adsorption of pesticides, 39
in clay illuviation, 45
in degradation of silicate minerals, 45
in gley formation, 45
in soil formation, 44
in translocation of metal ions, 47
in weathering processes, 45
soil, 44, 110
sorption of nitrogen by, 15
as system of polymers, 24
translocation of sesquioxides by, 47
see also Fulvic acid; Humic acids
Humification, 110
in estuarine environments, 224, 232
in peatlands, 59
extent, 62
measurement, 62
Humin:
abundance in lake sediments, 152
in aquatic sediments, 279
aromatic structures in, 287, 289
as clay-humic acid complex, 276, 277, 288
defined, 5, 14, 15, 275, 276, 277, 284, 300, 390
formation, 271
geochemistry, 275
humic acid, relation to, 276, 300, 312
kerogen, relation to, 6, 101, 250
lignin in, 286, 290
paraffinic structures in, 286, 290
in peat, 279
polysaccharides in, 277, 279, 286, 289
proteinaceous substances in, 288
solid state NMR of, 279, 285, 286, 289, 294
stability, 51
Hydrogen:
bonding:
effect in IR spectroscopy, 534, 540, 546, 559
of lacustrine humic substances, 131
of soil humic substances, 27, 41, 42, 345
of solute-solvent interaction, 334, 336, 337, 338, 346
in water, 346
composition in humic substances, 108, 350
content:
in aquatic humic substances, 93, 161
in terrestrial organic matter, 263
effect on infrared spectra, 347, 534
elemental analysis, 439
exchangeable, in IR spectroscopy, 535
in fractionation techniques, 417
Hydrolysis:
in amino acid analysis, 247, 318, 321
effect on elemental analysis, 471
enzymatic, 124
of humin, 215, 287
of lake sediments, 148, 149, 156, 161, 166
of marine sediment, 259
in soils, 307
Hydrophilic acids:
in DOC:
in lakes, 111
in streams, 198
in water, 364
in groundwater, 100, 103, 364

Hydrophilic bases, in DOC, in water, 364
Hydrophilic functional groups, 250
Hydrophilic neutrals, in DOC, in water, 364
Hydrophilic resin matrix, 378
Hydrophobic acids, in DOC:
- in lakes, 111
- in water, 364

Hydrophobic adsorption, 42
Hydrophobic bases, in DOC, in water, 364
Hydrophobic bonding, 347
Hydrophobic effect:
- and pollutant-humic substance interactions, 229
- and sorption on XAD resins, 379

Hydrophobic/hydrophilic fractionation, 91
Hydrophobic interactions, 222
Hydrophobic material, in groundwater, 91, 364
Hydrophobic nature, of soil humic substances, 333
Hydrophobic neutrals, in DOC, in water, 364
Hydroxy-indoles, 323
Hydroxyl groups:
- in aquatic humic substances, 96, 110, 217
- in soils, 332

Hydroxyquinones, 18
Hymatomelanic acid, 20, 390
Hyperfine splitting:
- in ESR, 556
- in spectrum of fulvic acid, 557

Hypolimnion, 135

Ice structure, 346
Ijsselmeer, The Netherlands, 135
Illite, 222
- cation-exchange capacity, 30, 222

Immobilization, of nitrogen, 306
Indoles, 260, 323
Induction forces, 336
Infrared spectroscopy:
- adsorption bands, of humic substances, 542
- application to humic substances, 536
- Beer's law, 534, 552
- bending vibrations, 532, 541
- of complex mixtures, 538
- effects of metal complexation, 545
- of humin, 285
- of lake sediments, 148, 155
- of marine humic substances, 241
- of marine sediment humic substances, 251, 256, 257
- sample preservation, 535

Inhibition, steric, 124
International Humic Substances Society, 9
International Peat Society (IPS), classification, 58, 85
International Society of Soil Science, 8
Ion-dipole interactions, 222
Ion exchange:
- as bonding mechanism, 41
- capacity of riverine humic substances, 217
- as fractionation technique, 404
- as isolation method for aquatic humic substances, 376, 383
- Schubert's method, 203

Ionic strength, in estuaries, 217, 226
Ion-selective electrode (ISE) measurements, 36, 38, 243
Iron:
- colloids, 221, 223
- in estuaries, 223, 228
- hydroxide, 330, 415
- in lakes, 106, 120, 122, 123, 135
- oxide, 119, 135, 366
- in peat, 66, 81, 82, 85
- reduction, 38
- in seawater, 245
- in soils, 33, 45
- in water, 201, 203

Irradiation, ultraviolet, 118
Isobestic point, 551
Isoelectric point, 111
Isolation, of humic substances:
- from groundwater, 87, 92
- of humin, 284
- from lake sediments, 152
- from seawater, 234
- from soil, 329, 363, 382

Isoleucine, 308
Isotope ratios, 216, 234, 261. *See also* Carbon isotopes
Isotope substitution, effect on infrared spectra, 535
Isotopic analysis, 161
Isotopic composition, 261

Kaolinite, cation-exchange capacity, 30, 222
Karl Fischer water determination technique, 454
Kerguelen Islands, 253
Kerogen:
- classification, 255, 263
- formation, 1, 48, 225, 283, 298, 301

groundwater humic substances, relation to, 90, 462
humin, relation to, 6, 101, 249, 276, 279, 280, 298, 301
in marine sediments, 251, 253, 271
and NMR, 565
paraffinic structures in, 93, 298
Keto compounds, 338
Keto-indoles, 323
Ketone groups, 129, 581
Kjeldahl nitrogen determination method, 440

Labile compounds, 230
Lake Drummont, 117
Lake Haruna, 148, 149, 152, 154, 156, 158, 160, 161, 164, 166
Lake Huron, 148, 155, 166
Lake Plussee, Germany, 110, 111, 113, 114, 128, 140, 145
Lakes:
of Bermuda, 258, 289, 290, 294
of Canada, 148, 155
of Finland, 132, 133
of Germany, 118, 119, 120, 138
of Italy, 148
of Japan, 148, 149, 152, 154, 156, 158, 160, 161, 164
of Netherlands, 135
of United States, 148, 164
of Wales, 110
see also specific lakes
Lang factor, 76
Laramie-Fox Hills aquifer, 87, 93, 95, 96, 97, 102
Lawrence Lake, 121, 123, 137
Lead, 78, 122, 203
Leucine, 308
Lichen fungi, weathering by, 45
Ligands:
exchange, 42
natural organic, 233
synthetic organic, 244
Lignin:
formation of humic acids from, 19, 116, 129, 197, 224
humin, relation to, 277, 286, 290, 296, 301
lignin-protein theory, 21
in marine sediments, 262, 269
in peats, 61, 68
solvent extraction, 469
structure, 459
Lignites, 49, 56, 280
humic acids of, 14
Ligno-protein complexes, 317
Limestone aquifer, 90
Liming, of soils, 76, 77
Linuron, 229
Lipid components, in fulvic and humic acids, 148
Lipids:
fatty acids, 166, 234, 237, 240
in humin, 282
in marine sediments, 238
participation in formation of humic substances, 169
in seawater, 238
in peat, 68
triglycerides, 240
unsaturated, 233
Lithosphere, 2
Logbaba cores, 255
London forces, 336
Lyophilization, 114, 371, 383. *See also* Freeze-drying

Macrophytes, 224, 228
Madison aquifer, 87, 95, 96, 102
Magnesium, 100, 203, 217, 333, 415
Magnetic moment, 562
Magnetic resonance, 564
Magnetogyric ratio, 562
Mahakam Delta, Indonesia, 253, 260
Maillard reaction, 18, 129, 169, 320
Maleic anhydride oligomer, as model for fulvic acid, 458
Manganese:
complexation by humic materials, 33, 82, 123, 203
in estuaries, 226, 229
in filtration, hydroxides, 366
plant availability, 85
in seawater, redox, 245
Mangrove Lake, Bermuda, 258, 289, 290, 294
Mangrove swamp, humification in, 56, 59, 60, 224
Marmot Creek, 90
Melanins, *see* Maillard reaction
Melanoidins, 129, 198, 269, 283, 297
Mercury, 38, 78, 122
Mesisol, 313
Metagenesis, defined, 251
Methoxychlor, 83
Methoxyl groups, 290, 332
in aquatic humic substances, 110, 119, 281
in peat, 63
of soils, 71

Methylation, 540, 542, 543, 569
Methylene compounds, active, 494
Microaggregates, in soil, 31
Microbial interactions, 127, 130, 269, 275, 288, 301
Mineralization:
of humic substances:
in marine sediments, 264
in peatlands, 59, 82
of nitrogen, 305
Moisture determination:
effect of temperature, 452
in elemental analysis, 437
Karl Fischer reagent, 437
loss on drying, 437
Molecular size, 114, 460, 477
concentration effects in ultrafiltration, 482
effect:
of aggregation, 482, 484, 486
of interacting with membranes, 483
of pH, 486
Guinier plot, 484
humic acid fractions, 484
marine humic substances, 237, 249
measurement methods, 497
effect:
of adsorption interactions, 480
of electrostatic interactions, 480
of ionic strength, 480
electron microscopy, 485
gel permeation chromatography, 479
one light scattering, 485
scanning electron microscopy, 485
scattering of electromagnetic radiation, 483
Sephadex, 480
small-angle X-ray, 484
transmission electron microscopy, 485
ultrafiltration, 482
molecular aggregates, 481
stream humic substances, 190
Molecular size fractionation, 413, 427
gel-permeation chromatography, 413
ultrafiltration, 413
Molecular volume, 114
Molecular weight, 112, 460, 477
of aquatic humic substances, 95, 153, 213, 214, 219
distribution of fulvic and humic acids, 148
effect:
of dissociation on measurement, 490
of polydispersity on measurement, 491
fractions, 219, 571
of fulvic acid, 270
methods of measurement, 479
colligative-property measurements, 490
equilibrium ultracentrifugation, 489
sedimentation-velocity centrifugation, 489
ultracentrifugation, 487
vapor-pressure osmometry, 491
viscometry, 489
number-average, 478, 490
problems in determination, 478, 482
weight-average, 478
z-average, 479
Mollisch test, 96
Mollisol, 39, 45, 50
Molybdenum, 33, 38, 455
Monopole interactions, 335
Mucks, 66
Mucopolysaccharides, 231
Mullica River, 219
Multicomponent nature, of humic substances, 531, 553, 557, 558
Multicomponent solubility parameters, 341
Multiligand equilibria, 516
Multipole interactions, 335

Neutral polymers, solubilization, 343
New York Bight humic acids, 572
Nickel, 79, 203
Nitrate, 81
assimilation to organic nitrogen, 305
biochemistry, 304
Nitrification, 81, 306
Nitrite, 306
Nitrobacter, 306
Nitrogen:
content:
in estuaries, 214, 243
in fulvic acid, 108, 303
in groundwater humic substances, 93
in humic acid, 164, 303
in humin, 279, 295, 303
in lakes, 139
in lake sediments, 153
in marine humic substances, 235
in marine sediment humic substances, 252, 253, 254
of peats, 54, 63, 70
in soil humic substances, 224, 332, 351
of stream humic substances, 183, 190, 199, 201, 208
in terrestrial organic matter, 263
cycle, 305

fixation, 305
global distribution, 304
Nonionic macroporous sorbents, 379
Norway Sea, 572
Nuclear magnetic resonance (NMR) spectroscopy:
alkyl chains, 570
carbon-13 nuclear magnetic resonance, 573, 576
nuclear Overhauser enhancement, 573, 576
proton coupling, 574
relaxation times of carbon, 573
chemical shift, 564
cross polarization, 576
contact time, 576
cross-polarization magic-angle spinning (CP/MAS), 578
contact times, 579
free radicals, 578
incomplete transfer of polarization, 578
quantitative measure, 578
repetition rates, 579
in derivatized humic substances, 580
dipole-dipole coupling, 565
exchangeable protons, 571
Fourier transform, 566
pulse NMR, 567
sensitivity, 571
fulvic fractions, 571
gated, coupling experiment, 578
humic fractions, 571
of humic substances:
in lake sediments, 156
in marine sediments, 258
in peats, 75
in seawater, 237
in streams, 199, 207
of humin, 277, 285, 300
line broadening, 580
liquid state, 565, 573, 576
magnetic moment, 562
magnetogyric ratio, 562
marine fulvic acids, 571
marine humic acids, 571
nuclear spin, 562
polymethylene groups, 570
progressive saturation, 577
of soil humic acid, 569
solid state, 285, 566, 575, 578
CP/MAS, 566, 578
spin-lattice relaxation time, 563
spin-spin coupling, 565
spin-spin relaxation, 564
transverse relaxation time, 564
of underivatized humic substances, 567
proton NMR, 567
Nuclear Overhauser enhancement, 573, 574, 578
Nuclear spin, 562
Nucleic acids, 304, 312
Nylon, for sorption, 375, 379

Oceans:
humic substances in, 2
open, 234
Ogeechee River, Georgia, 181, 446
OH stretching, in IR, 540
Oil-shale regions, groundwater in, 90
Okefenokee Swamp, Georgia, 290
Oligocarbophilic bacteria, 128
Oligosaccharides, 332
Oligotrophic lake, 137
Oman Sea, 254, 256, 263
Optical properties, of aquatic humic substances, 154
Organic carbon, dissolved, *see* Dissolved organic carbon (DOC)
Organic matter, dissolved, *see* Dissolved organic matter (DOM)
Organic pollutants, 225, 229, 230
Organic surfactants, on filters, 367
ORGON cruises, 249, 265
Origin:
marine humic substances, 249
microbial, in lakes, 111
soil humic substances, 21, 330
stream humic substances, 193, 208
Orthophosphate, 122, 123
Oxidation:
alkaline permanganate, 157
enzyme-mediated, 129
in isolation of aquatic humic substances, 370
in seawater, 239
Oxidative cleavage, 238, 267
during transport in marine sediments, 267
Oxidative degradation, 259, 270
alkaline permanganate, 157, 206
effect on humification, in peats, 84
Oxides, transition metals, 129
Oxygen:
content:
of aquatic humic substances, 92, 99, 108, 160, 183, 235
in marine sediments, 252, 270

Oxygen, content (*Continued*)
of peat, 56
in seawater organic matter, 235, 240
elemental analysis, 441
fixation, 267
triplet, 238
Ozonation, 122

Pacific Ocean, 236
PAGE, *see* Polyacrylamide gel electrophoresis (PAGE)
Paleohumus, 48
Paraffinic carbons:
in humin, 279, 286, 288, 290, 294, 301
in marine fulvic acid, 236
Paraquat, 40, 41
Paris Basin, 255
Particulate organic carbon, 364
Peat:
autolytic degradation, 59
botanical origin of, 64
cation-exchange capacity of, 29
composition, 54, 56
dagenesis of, 280, 282
humic acid, 14, 15, 572
humin in, 276, 289, 301
humus, 66
porphyrin-type structures, 35
proton NMR, 572
uses, 58
Peatlands:
carbon dioxide effect, 58
classification, 56
cultivated, 76
defined, 53
distribution, 54
humification, 59, 75
subsidence, 54, 77, 78
retarded by copper, 78
types, 56
Peptides, 131, 138, 325, 331
Peptization, 221, 229, 230
Periodate lignin, 296
Permanganate oxidation, 186, 280
Peru, continental shelf, 261
Pesticides, 85
adsorption of, 38, 41, 54, 83
bonding mechanisms for, 41
by clay, 40
by hydrogen bonding, 42
by ligand exchange, 42
by organic matter, 38, 85
by soil, 40
by van der Waals forces, 42
chemical binding of, 44
chemical reactions of, 40
complexation by humic substances, 229
decomposition products of, 44
Petroleum, 283
hydrocarbons, 204
potential, 273
pH:
cation exchange capacity, related to, 30
effect:
on sample preservation, 370
on sorption, 379
on ultraviolet-visible absorptivity, 551
humification, related to, 76, 77
Phenol-formaldehyde weak-base resins, 378
Phenolic acidity, 258
Phenolic carbons, 286, 287
Phenolic compounds, 224
in lakewaters, 113
in marine sediments, 260, 269
in peats, 59
Phenolic content, 495
Phenolic groups, 95, 281, 331, 551
in aquatic humic substances, 110, 160
in soils, 333
Phenols:
as acidic functional groups, 494
reactions with amino acids, 317
Phenylalanine, 308
Phenyl groups, in aquatic humic substances, 110, 199, 200
Phosphorus:
elemental analysis, 444
in lakes, 106, 108, 122, 123, 144
in riverine humic substances, 190, 201, 206
Photodegradation, 118, 125
Photoreduction, of metals, 228, 245
Photosynthesis, 125
Physical fractionation methods, 426
electrophoresis, 426
gel permeation chromatography, 427
ultracentrifugation, 427
ultrafiltration, 427
Phytolysis, 121, 125, 137, 144, 205
Phytoplankton:
in estuaries, 224, 228
in lakes, 113, 122, 123, 125, 134
and marine sediment humic substances, 263, 266
in peats, 61
in seawater, 234, 241

pK_a values, 41, 503, 506
Plants, vascular, 224, 269, 280, 290, 298
Pleasant River, Maine, 190
Plon lake district, Germany, 137, 140
Plutonium, 244
Podsol humic acid, 572
Podzols, 47
Polarity, solvent, 335, 342
Polder lake, 135
Pollutants:
  cycling, 212
  degradation, 144
  in groundwater, 101
  organic, 225, 229, 230
  scavenging, 212
Polyacrylamide gel electrophoresis (PAGE), 114, 131, 140
Polyamide, for sorption, 375, 379
Polycondensation, 249, 269
Polycyclic aromatic hydrocarbons, 118, 204
Polyelectrolyte, 529, 558
  properties, 329, 343
  solubilization, 344
Polymers, solubilization, of neutral, 343
Polymer solution theory, application to soil humic substance extraction, 357
Polymethylene groups, 280, 570
Polypeptides, 312, 332
Polyphenols, 125, 197, 269
Polysaccharides:
  bonding in humic substances, 332
  in gel permeation chromatography, 399
  in humin, 277, 279, 286, 288, 289, 295, 301
  hydrolyzable sugars, 63, 260
  in lakes, 113, 131, 134, 137
Polystyrene resins, 234, 376, 379
Polyuronic acids, 283
Polyvalent cations, 29
  bridging of soil humic substances, 345
Polyvinyl pyrolidone, 108
Pore waters, estuarine sediments, 225, 228
Porphyrins, 304
Potassium, 203, 333
Precipitation:
  as basis for fractionation, 390, 415
  metal oxide, in lakes, 118
  processes, 126, 203
    metal ions, 393, 407
    organic solvents, 394, 407
    pH, 390, 407
    salting-out, 392, 407
Presedimentary alteration, 264
Preservation, of samples, 370, 382, 415
Primary productivity, 106, 212
Primary structures, of soil humic substances, 332
Proteins:
  bonding to humic substances, 332, 336
  and electrophoresis, 402
  in humin, 279, 288
  in lakes, 106, 130, 137, 138
  and nitrogen in humic substances, 304, 312, 320
  and NMR, 580
Proteolysis, 59
Protonation, as adsorption mechanism, 41
Proton binding models, 515
  aggregation, 518
  in complex mixtures, 516
  continuous models, 521
    affinity spectra, 522
    differential approach, 521
    Gaussian distribution, 524
    normal distribution, 524
  counterion condensation, 518
  Donnan potential, 521
  electrostatic effects, 519, 520
  gel phases, 521
  intrinsic pK_a models, 519
  mean fulvic acid units, 518
  simple multisite models, 518
Proton release, to determine binding capacities of humic substances for metal ions, 36
Puerto Rico, coastal waters, 227
Purification procedures, of soil humic substances, 332, 388
Purines, 312, 324
Pyridine, 340, 348, 352, 353
Pyrimidines, 312, 324
Pyrolysis, 117, 136, 161, 259, 285, 300
Pyrophosphate:
  as chelating agent, 37, 53
  index, to determine humification, 65, 70

Quinhydrone, in ESR, 556
Quinones, 59, 197, 317, 323

Radiocarbon dating, 49, 50, 98, 190, 207, 217, 247
Radionuclides, in seawater, 244
Raman spectroscopy, 548, 559
Redox, properties and reactions, 38, 245, 249

Red River aquifer, 87
  origin of humic substances, 102
Reduction, of nitrate, 306
Removal, of aquatic humic substances, from water column, 118
Resins:
  cation-exchange, 379
  methodologies, 92, 234
  nonionic resins, 379
  polyvinyl pyrolidone, 108
  sorbents, 366, 376
  strong-base ion exchange, 376
  weak base ion exchange, 378
Resuspension processes, 138
Retardation, of humification, 77, 78
Reverse osmosis, as concentration method, 373
Rivers, particulate organic carbon in, 364
Rock-Eval pyrolysis, 260
Rotary evaporation, as concentration method, 383
Rubbed fiber content, to determine humification, in peats, 65

Saanich Inlet, 225
St. Peters aquifer, 87
  molecular weight of humic substances, 95
  nitrogen content, 93
  origin of humic substances, 102
Salicylic acid, as model for humic acid, 460
Salinity, effects in estuaries, 212, 217, 227
Sandstone aquifer, 90
San Francisco Bay, 224
Sapric histosols, 357, 360
Sapric organic material, 57
Sapropel, 27, 276, 283, 289, 294, 301
Saprophytic bacteria, 128
Sargasso Sea, 235, 569
Scanning electron microscopy, 485
Scheldt estuary, 227
Schiff reaction, 19
Seawater:
  humic substances in, 233
  particulate organic carbon in, 364
Sediments:
  Amazon deep sea fan, 263
  lake, 147
  marine, 249
Selective preservation hypothesis, 290
Selectivity coefficients, for copper, 228
Semiquinone, 556, 559
Sephadex gel chromatography, 185, 312, 323, 480
Serine, 307
Shale, 276, 299
Siderophores, 123
Sierozem soils, 30
Silicates, 45, 251
Silver, 78
  as bactericide, 370
Smith Lake, Alaska, 110
Sodium, 203, 333
Sodium carbonate, 90
Sodium hydroxide, 350, 358
  hydrolysis, 148, 166, 370
Sodium pyrophosphate, as chelating agent, 28
Soil aggregation:
  Emerson's clay-domain theory, 32
  mechanisms of formation, 31
Soils, buried:
  classification of, 45
  formation of, 44
Solid state nuclear magnetic resonance spectroscopy, 7, 73, 275, 285, 566, 575
Solod soils, 306
Solonetz soils, 306
Solubility:
  in aqueous alkaline solvents, 250
  as basis for fractionation, 390
  of macromolecules, 343
  parameters, 340, 356
  of polyelectrolytes, 344
  of soil humic substances, 36
  in water, 5, 36, 336
    soil fulvic acids, 347
Solubility effect, 42
Solvation:
  in basic solvents, 349
  of soil humic substances, 331, 333, 338
Solvents:
  criteria in soil humic substances, 345
  extraction, 184, 374
  non-alkaline, 348
  properties of, 337
  solute interactions, 334
Sorption:
  on aquifer solids, 102
  onto macroporous resins, 87, 379
  of marine humic substances, 234
  methods, of concentration, 374
  of organic compounds on filters, 368
Southeastern Coastal Plain, United States, 201

Spatial distribution, DOM in lakes, 134
Spectrofluorimetry, 553
Spectroscopy, 148, 203, 234
  electron paramagnetic resonance (EPR), 203, 555
  electron spin resonance (ESR), 35, 38, 71, 285, 555
  fluorescence, 234, 552
  infrared, *see* Infrared spectroscopy
  nuclear magnetic resonance, *see* Nuclear magnetic resonance (NMR) spectroscopy
  Raman, 548, 559
  ultraviolet-visible, 549
  visible, 154
Sphagnum, 56
Spodosol, 27, 47, 313
Stable isotope analysis, 297
Straw, degradation of, 473
Stream humic substances, 181
Structural concepts:
  of marine humic substances, 237
  of soil humic substances, 332
Structural confirmation, of humic substances in freshwater, 218
  in estuaries, 226
  in solution, 333
Structural constraints, 495, 505
  on aromaticity, 495
  on carboxyl content, 495, 505
  on nitrogen acids, 505
  on phenolic content, 495, 505
  on sulfur acids, 505
Structural formulas, 473
Styrene-divinylbenzene resins, 379
Subsidence, of peatlands, 76, 77, 78
Sugars, hydrolysable, 59, 63
Sulfolane, 347, 352
Sulfonic acids, 494
Sulfur:
  content:
    of marine humic substances, 255, 505
    of peats, 54, 77
  elemental analysis, 442
  in stream humic substances, 183, 190, 201
Suspended organic carbon (SOC), 187
Suwannee River, Georgia, 98, 102, 190
Synsedimentary alteration, 264

Tamar estuary, 228
Tannins, 59, 60, 125, 269, 469
Temperature, effect on moisture determination, 452
Temporal distribution, DOM in lakes, 134
Thermal degradation studies, 259, 280
Thermodynamics, of dilute solutions, 343
Threonine, 307
Titanium, 304
Titration methods, 496, 507, 508, 510, 511
Tjeukemeer, The Netherlands, 117, 122, 127, 135
Toarcian Shales, 255
Toxicity:
  of copper, 122
  of trace metals, 122, 144, 241
Trace metals, 225
  anodic stripping voltammetry, 38
  in coal, 33
  complexation, *see* Complexation, of trace metals by humic substances
  interactions, 33, 34, 122, 201
  ion-selective electrode measurements of, 38
  in sediments, 37
  in soils, 37
  speciation in soil solution, 37
  toxicity, 122, 144, 241
Transition metal oxides, 129
Translocation, of mineral matter, 47, 60
Transport:
  of humic substances, 6, 60, 66, 267
  of metal ions, 52
  of planktonic organic matter, 270
Trihalomethanes, 186, 207
Trona water, 90

Ulmic acid, 20, 390
Ultracentrifugation, 400, 427, 487
Ultrafiltration, 95, 185, 399, 407, 413, 427, 482
Ultraviolet:
  absorbance, 214, 218, 222, 230
  irradiation, 112, 118, 234
  spectroscopy, 234, 549
Unsaturation, 495, 505
Upper Klamath Lake, 131
Urease, 539
Uronic acids, 269, 573

Vacuum distillation, as concentration method, 371
Vanadium, reduction, 38
van der Waals forces, 27, 41, 42, 221, 325, 336
Viscosity:
  measurements of, 218
  of solvents, 340

Voltammetry, anodic stripping, *see* Anodic stripping voltammetry (ASV)

Water:
- in histosols, 56
- repellancy, of soil, 32
- retention, in peats, 63, 77
- solution properties, 346
- structure, 346

Water table, maintenance, in peatlands, 76, 77
Wax, leaf, 60
Weathering, of rocks and minerals, 33
- role:
  - of humic substances, 33, 45
  - of microorganisms, 46

Williamson River system, Oregon, 113
Williams River, 131
Woody tissues, 290, 298

XAD resins, 186, 207, 379, 382
Xenobiotic compounds, 197, 205

Yampa River, Colorado, 446

Zeeman effect, 555
Zinc, 33, 79, 81, 82, 85, 122, 203, 227, 244
Zinc dust distillation, 332